Remote Sensing and Image Interpretation

Seventh Edition

Remote Sensing and Image Interpretation

Seventh Edition

Thomas M. Lillesand, **Emeritus**
University of Wisconsin—Madison

Ralph W. Kiefer, **Emeritus**
University of Wisconsin—Madison

Jonathan W. Chipman
Dartmouth College

WILEY

Vice President and Publisher	*Petra Recter*
Executive Editor	*Ryan Flahive*
Sponsoring Editor	*Marian Provenzano*
Editorial Assistant	*Kathryn Hancox*
Associate Editor	*Christina Volpe*
Assistant Editor	*Julia Nollen*
Senior Production Manager	*Janis Soo*
Production Editor	*Bharathy Surya Prakash*
Marketing Manager	*Suzanne Bochet*
Photo Editor	*James Russiello*
Cover Design	*Kenji Ngieng*
Cover Photo	*Quantum Spatial and Washington State DOT*

This book was set in 10/12 New Aster by Laserwords and printed and bound by Courier Westford.

Founded in 1807, John Wiley & Sons, Inc. has been a valued source of knowledge and understanding for more than 200 years, helping people around the world meet their needs and fulfill their aspirations. Our company is built on a foundation of principles that include responsibility to the communities we serve and where we live and work.

In 2008, we launched a Corporate Citizenship Initiative, a global effort to address the environmental, social, economic, and ethical challenges we face in our business. Among the issues we are addressing are carbon impact, paper specifications and procurement, ethical conduct within our business and among our vendors, and community and charitable support. For more information, please visit our website: www.wiley.com/go/citizenship.

Library of Congress Cataloging-in-Publication Data

Lillesand, Thomas M.
Remote sensing and image interpretation / Thomas M. Lillesand, Ralph W. Kiefer, Jonathan W. Chipman. — Seventh edition.
pages cm
Includes bibliographical references and index.
ISBN 978-1-118-34328-9 (paperback)
1. Remote sensing. I. Kiefer, Ralph W. II. Chipman, Jonathan W. III. Title.
G70.4.L54 2015
621.36'78—dc23

2014046641

Printed in the United States of America
10 9 8 7 6 5 4 3 2 1

PREFACE

This book is designed to be primarily used in two ways: as a textbook in introductory courses in remote sensing and image interpretation and as a reference for the burgeoning number of practitioners who use geospatial information and analysis in their work. Rapid advances in computational power and sensor design are allowing remote sensing and its kindred technologies, such as geographic information systems (GIS) and the Global Positioning System (GPS), to play an increasingly important role in science, engineering, resource management, commerce, and other fields of human endeavor. Because of the wide range of academic and professional settings in which this book might be used, we have made this discussion "discipline neutral." That is, rather than writing a book heavily oriented toward a single field such as business, ecology, engineering, forestry, geography, geology, urban and regional planning, or water resource management, we approach the subject in such a manner that students and practitioners in any discipline should gain a clear understanding of remote sensing systems and their virtually unlimited applications. In short, anyone involved in geospatial data acquisition and analysis should find this book to be a valuable text and reference.

The world has changed dramatically since the first edition of this book was published, nearly four decades ago. Students may read this new edition in an ebook format on a tablet or laptop computer whose processing power and user interface are beyond the dreams of the scientists and engineers who pioneered the use of computers in remote sensing and image interpretation in the 1960s and early 1970s. The book's readers have diversified as the field of remote sensing has become a truly international activity, with countries in Asia, Africa, and Latin America contributing at all levels from training new remote sensing analysts, to using geospatial technology in managing their natural resources, to launching and operating new earth observation satellites. At the same time, the proliferation of high-resolution image-based visualization platforms—from Google Earth to Microsoft's Bing Maps—is in a sense turning everyone with access to the Internet into an "armchair remote-sensing aficionado." Acquiring the expertise to produce informed, reliable *interpretations* of all this newly available imagery, however, takes time and effort. To paraphrase the words attributed to Euclid, there is no royal road to image analysis—developing these skills still requires a solid grounding in the principles of electromagnetic radiation, sensor design, digital image processing, and applications.

This edition of the book strongly emphasizes digital image acquisition and analysis, while retaining basic information about earlier analog sensors and methods (from which a vast amount of archival data exist, increasingly valuable as a source for studies of long-term change). We have expanded our coverage of lidar systems and of 3D remote sensing more generally, including digital photogrammetric methods such as structure-from-motion (SFM). In keeping with the changes sweeping the field today, images acquired from uninhabited aerial system (UAS) platforms are now included among the figures and color plates, along with images from many of the new optical and radar satellites that have been launched since the previous edition was published. On the image analysis side, the continuing improvement in computational power has led to an increased emphasis on techniques that take advantage of high-volume data sets, such as those dealing with neural network classification, object-based image analysis, change detection, and image time-series analysis.

While adding in new material (including many new images and color plates) and updating our coverage of topics from previous editions, we have also made some improvements to the organization of the book. Most notably, what was formerly Chapter 4—on visual image interpretation—has been split. The first sections, dealing with methods for visual image interpretation, have been brought into Chapter 1, in recognition of the importance of visual interpretation throughout the book (and the field). The remainder of the former Chapter 4 has been moved to the end of the book and expanded into a new, broader review of applications of remote sensing not limited to visual methods alone. In addition, our coverage of radar and lidar systems has been moved ahead of the chapters on digital image analysis methods and applications of remote sensing.

Despite these changes, we have also endeavored to retain the traditional strengths of this book, which date back to the very first edition. As noted above, the book is deliberately "discipline neutral" and can serve as an introduction to the principles, methods, and applications of remote sensing across many different subject areas. There is enough material in this book for it to be used in many different ways. Some courses may omit certain chapters and use the book in a one-semester or one-quarter course; the book may also be used in a two-course sequence. Others may use this discussion in a series of modular courses, or in a shortcourse/workshop format. Beyond the classroom, the remote sensing practitioner will find this book an enduring reference guide—technology changes constantly, but the fundamental principles of remote sensing remain the same. We have designed the book with these different potential uses in mind.

As always, this edition stands upon the shoulders of those that preceded it. Many individuals contributed to the first six editions of this book, and we thank them again, collectively, for their generosity in sharing their time and expertise. In addition, we would like to acknowledge the efforts of all the expert reviewers who have helped guide changes in this edition and previous editions. We thank the reviewers for their comments and suggestions.

Illustration materials for this edition were provided by: Dr. Sam Batzli, USGS WisconsinView program, University of Wisconsin—Madison Space Science and Engineering Center; Ruediger Wagner, Vice President of Imaging, Geospatial Solutions Division and Jennifer Bumford, Marketing and Communications, Leica Geosystems; Philipp Grimm, Marketing and Sales Manager, ILI GmbH; Jan Schoderer, Sales Director UltraCam Business Unit and Alexander Wiechert, Business Director, Microsoft Photogrammetry; Roz Brown, Media Relations Manager, Ball Aerospace; Rick Holasek, NovaSol; Stephen Lich and Jason Howse, ITRES, Inc.; Qinghua Guo and Jacob Flanagan, UC-Merced; Dr. Thomas Morrison, Wake Forest University; Dr. Andrea Laliberte, Earthmetrics, Inc.; Dr. Christoph Borel-Donohue, Research Associate Professor of Engineering Physics, U.S. Air Force Institute of Technology; Elsevier Limited, the German Aerospace Center (DLR), Airbus Defence & Space, the Canadian Space Agency, Leica Geosystems, and the U.S. Library of Congress. Dr. Douglas Bolger, Dartmouth College, and Dr. Julian Fennessy, Giraffe Conservation Foundation, generously contributed to the discussion of wildlife monitoring in Chapter 8, including the giraffe telemetry data used in Figure 8.24. Our particular thanks go to those who kindly shared imagery and information about the Oso landslide in Washington State, including images that ultimately appeared in a figure, a color plate, and the front and back covers of this book; these sources include Rochelle Higgins and Susan Jackson at Quantum Spatial, Scott Campbell at the Washington State Department of Transportation, and Dr. Ralph Haugerud of the U.S. Geological Survey.

Numerous suggestions relative to the photogrammetric material contained in this edition were provided by Thomas Asbeck, CP, PE, PLS; Dr. Terry Keating, CP, PE, PLS; and Michael Renslow, CP, RPP.

We also thank the many faculty, academic staff, and graduate and undergraduate students at Dartmouth College and the University of Wisconsin—Madison who made valuable contributions to this edition, both directly and indirectly.

Special recognition is due our families for their patient understanding and encouragement while this edition was in preparation.

Finally, we want to encourage you, the reader, to use the knowledge of remote sensing that you might gain from this book to literally make the world a better place. Remote sensing technology has proven to provide numerous scientific, commercial, and social benefits. Among these is not only the efficiency it brings to the day-to-day decision-making process in an ever-increasing range of applications, but also the potential this field holds for improving the stewardship of earth's resources and the global environment. This book is intended to provide a technical foundation for you to aid in making this tremendous potential a reality.

Thomas M. Lillesand
Ralph W. Kiefer
Jonathan W. Chipman

This book is dedicated to the peaceful application of remote sensing in order to maximize the scientific, social, and commercial benefits of this technology for all humankind.

Contents

1 Concepts and Foundations of Remote Sensing 1

2 Elements of Photographic Systems 85

6 Microwave and Lidar Sensing 385

7 Digital Image Analysis 485

1 Concepts and Foundations of Remote Sensing

1.1 INTRODUCTION

Remote sensing is the science and art of obtaining information about an object, area, or phenomenon through the analysis of data acquired by a device that is not in contact with the object, area, or phenomenon under investigation. As you read these words, you are employing remote sensing. Your eyes are acting as sensors that respond to the light reflected from this page. The "data" your eyes acquire are impulses corresponding to the amount of light reflected from the dark and light areas on the page. These data are analyzed, or interpreted, in your mental computer to enable you to explain the dark areas on the page as a collection of letters forming words. Beyond this, you recognize that the words form sentences, and you interpret the information that the sentences convey.

In many respects, remote sensing can be thought of as a reading process. Using various sensors, we remotely collect *data* that may be analyzed to obtain *information* about the objects, areas, or phenomena being investigated. The remotely collected data can be of many forms, including variations in force distributions, acoustic wave distributions, or electromagnetic energy distributions. For example, a gravity meter acquires data on variations in the distribution of the force of gravity. Sonar, like a bat's navigation system, obtains data on variations in acoustic wave distributions. Our eyes acquire data on variations in electromagnetic energy distributions.

Overview of the Electromagnetic Remote Sensing Process

This book is about *electromagnetic* energy sensors that are operated from airborne and spaceborne platforms to assist in inventorying, mapping, and monitoring earth resources. These sensors acquire data on the way various earth surface features emit and reflect electromagnetic energy, and these data are analyzed to provide information about the resources under investigation.

Figure 1.1 schematically illustrates the generalized processes and elements involved in electromagnetic remote sensing of earth resources. The two basic processes involved are *data acquisition* and *data analysis*. The elements of the data acquisition process are energy sources (*a*), propagation of energy through the atmosphere (*b*), energy interactions with earth surface features (*c*), retransmission of energy through the atmosphere (*d*), airborne and/or spaceborne sensors (*e*), resulting in the generation of sensor data in pictorial and/or digital form (*f*). In short, we use sensors to record variations in the way earth surface features reflect and emit electromagnetic energy. The data analysis process (*g*) involves examining the data using various viewing and interpretation devices to analyze pictorial data and/or a computer to analyze digital sensor data. Reference data about the resources being studied (such as soil maps, crop statistics, or field-check data) are used

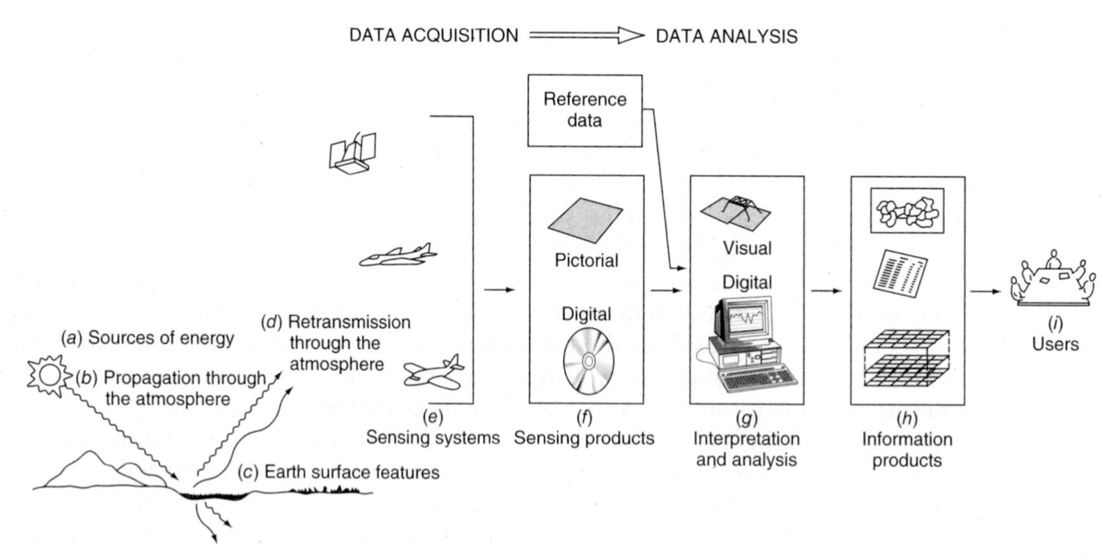

Figure 1.1 Electromagnetic remote sensing of earth resources.

when and where available to assist in the data analysis. With the aid of the reference data, the analyst extracts information about the type, extent, location, and condition of the various resources over which the sensor data were collected. This information is then compiled (*h*), generally in the form of maps, tables, or digital spatial data that can be merged with other "layers" of information in a *geographic information system* (*GIS*). Finally, the information is presented to users (*i*), who apply it to their decision-making process.

Organization of the Book

In the remainder of this chapter, we discuss the basic principles underlying the remote sensing process. We begin with the fundamentals of electromagnetic energy and then consider how the energy interacts with the atmosphere and with earth surface features. Next, we summarize the process of acquiring remotely sensed data and introduce the concepts underlying digital imagery formats. We also discuss the role that reference data play in the data analysis procedure and describe how the spatial location of reference data observed in the field is often determined using *Global Positioning System* (*GPS*) methods. These basics will permit us to conceptualize the strengths and limitations of "real" remote sensing systems and to examine the ways in which they depart from an "ideal" remote sensing system. We then discuss briefly the rudiments of GIS technology and the spatial frameworks (coordinate systems and datums) used to represent the positions of geographic features in space. Because visual examination of imagery will play an important role in every subsequent chapter of this book, this first chapter concludes with an overview of the concepts and processes involved in visual interpretation of remotely sensed images. By the end of this chapter, the reader should have a grasp of the foundations of remote sensing and an appreciation for the close relationship among remote sensing, GPS methods, and GIS operations.

Chapters 2 and 3 deal primarily with photographic remote sensing. Chapter 2 describes the basic tools used in acquiring aerial photographs, including both analog and digital camera systems. Digital videography is also treated in Chapter 2. Chapter 3 describes the photogrammetric procedures by which precise spatial measurements, maps, digital elevation models (DEMs), orthophotos, and other derived products are made from airphotos.

Discussion of nonphotographic systems begins in Chapter 4, which describes the acquisition of airborne multispectral, thermal, and hyperspectral data. In Chapter 5 we discuss the characteristics of spaceborne remote sensing systems and examine the principal satellite systems used to collect imagery from reflected and emitted radiance on a global basis. These satellite systems range from the Landsat and SPOT series of moderate-resolution instruments, to the latest generation of high-resolution commercially operated systems, to various meteorological and global monitoring systems.

Chapter 6 is concerned with the collection and analysis of radar and lidar data. Both airborne and spaceborne systems are discussed. Included in this latter category are such systems as the ALOS, Envisat, ERS, JERS, Radarsat, and ICESat satellite systems.

In essence, from Chapter 2 through Chapter 6, this book progresses from the simplest sensing systems to the more complex. There is also a progression from short to long wavelengths along the electromagnetic spectrum (see Section 1.2). That is, discussion centers on photography in the ultraviolet, visible, and near-infrared regions, multispectral sensing (including thermal sensing using emitted long-wavelength infrared radiation), and radar sensing in the microwave region.

The final two chapters of the book deal with the manipulation, interpretation, and analysis of images. Chapter 7 treats the subject of digital image processing and describes the most commonly employed procedures through which computer-assisted image interpretation is accomplished. Chapter 8 presents a broad range of applications of remote sensing, including both visual interpretation and computer-aided analysis of image data.

Throughout this book, the International System of Units (SI) is used. Tables are included to assist the reader in converting between SI and units of other measurement systems.

Finally, a Works Cited section provides a list of references cited in the text. It is not intended to be a compendium of general sources of additional information. Three appendices provided on the publisher's website (http://www.wiley.com/college/lillesand) offer further information about particular topics at a level of detail beyond what could be included in the text itself. Appendix A summarizes the various concepts, terms, and units commonly used in radiation measurement in remote sensing. Appendix B includes sample coordinate transformation and resampling procedures used in digital image processing. Appendix C discusses some of the concepts, terminology, and units used to describe radar signals.

1.2 ENERGY SOURCES AND RADIATION PRINCIPLES

Visible light is only one of many forms of electromagnetic energy. Radio waves, ultraviolet rays, radiant heat, and X-rays are other familiar forms. All this energy is inherently similar and propagates in accordance with basic wave theory. As shown in Figure 1.2, this theory describes electromagnetic energy as traveling in a harmonic, sinusoidal fashion at the "velocity of light" c. The distance from one wave peak to the next is the *wavelength* λ, and the number of peaks passing a fixed point in space per unit time is the wave *frequency* v.

From basic physics, waves obey the general equation

$$c = v\lambda \tag{1.1}$$

Because c is essentially a constant $(3 \times 10^8\,\text{m/sec})$, frequency v and wavelength λ for any given wave are related inversely, and either term can be used to

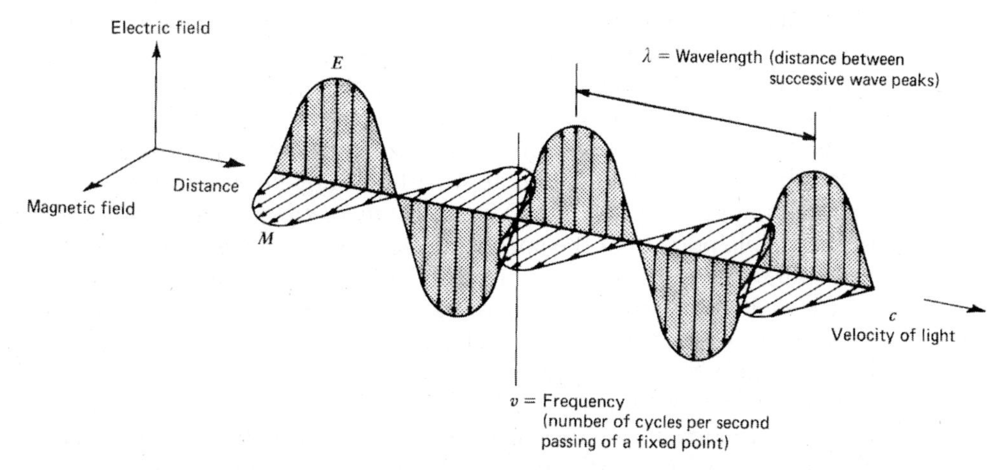

Figure 1.2 Electromagnetic wave. Components include a sinusoidal electric wave (E) and a similar magnetic wave (M) at right angles, both being perpendicular to the direction of propagation.

characterize a wave. In remote sensing, it is most common to categorize electromagnetic waves by their wavelength location within the *electromagnetic spectrum* (Figure 1.3). The most prevalent unit used to measure wavelength along the spectrum is the *micrometer* (μm). A micrometer equals 1×10^{-6} m.

Although names (such as "ultraviolet" and "microwave") are generally assigned to regions of the electromagnetic spectrum for convenience, there is no clear-cut dividing line between one nominal spectral region and the next. Divisions of the spectrum have grown from the various methods for sensing each type of radiation more so than from inherent differences in the energy characteristics of various wavelengths. Also, it should be noted that the portions of the

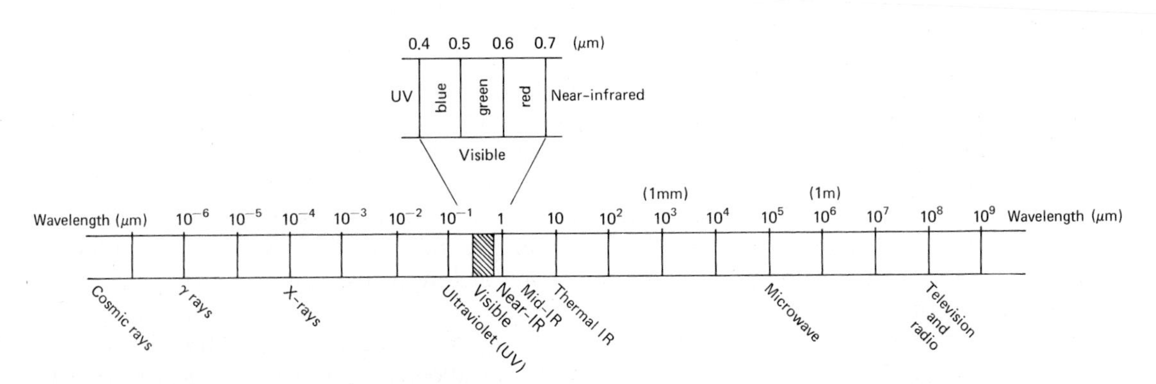

Figure 1.3 Electromagnetic spectrum.

electromagnetic spectrum used in remote sensing lie along a continuum characterized by magnitude changes of many powers of 10. Hence, the use of logarithmic plots to depict the electromagnetic spectrum is quite common. The "visible" portion of such a plot is an extremely small one, because the spectral sensitivity of the human eye extends only from about 0.4 μm to approximately 0.7 μm. The color "blue" is ascribed to the approximate range of 0.4 to 0.5 μm, "green" to 0.5 to 0.6 μm, and "red" to 0.6 to 0.7 μm. *Ultraviolet* (*UV*) energy adjoins the blue end of the visible portion of the spectrum. Beyond the red end of the visible region are three different categories of *infrared* (*IR*) waves: *near IR* (from 0.7 to 1.3 μm), *mid IR* (from 1.3 to 3 μm; also referred to as *shortwave IR* or *SWIR*), and *thermal IR* (beyond 3 to 14 μm, sometimes referred to as *longwave IR*). At much longer wavelengths (1 mm to 1 m) is the *microwave* portion of the spectrum.

Most common sensing systems operate in one or several of the visible, IR, or microwave portions of the spectrum. *Within the IR portion of the spectrum, it should be noted that only thermal-IR energy is directly related to the sensation of heat; near- and mid-IR energy are not.*

Although many characteristics of electromagnetic radiation are most easily described by wave theory, another theory offers useful insights into how electromagnetic energy interacts with matter. This theory—the particle theory—suggests that electromagnetic radiation is composed of many discrete units called *photons* or *quanta*. The energy of a quantum is given as

$$Q = hv \tag{1.2}$$

where

Q = energy of a quantum, joules (J)
h = Planck's constant, 6.626×10^{-34} J sec
v = frequency

We can relate the wave and quantum models of electromagnetic radiation behavior by solving Eq. 1.1 for v and substituting into Eq. 1.2 to obtain

$$Q = \frac{hc}{\lambda} \tag{1.3}$$

Thus, we see that the energy of a quantum is inversely proportional to its wavelength. *The longer the wavelength involved, the lower its energy content.* This has important implications in remote sensing from the standpoint that naturally emitted long wavelength radiation, such as microwave emission from terrain features, is more difficult to sense than radiation of shorter wavelengths, such as emitted thermal IR energy. The low energy content of long wavelength radiation means that, in general, systems operating at long wavelengths must "view" large areas of the earth at any given time in order to obtain a detectable energy signal.

The sun is the most obvious source of electromagnetic radiation for remote sensing. However, *all* matter at temperatures above absolute zero (0 K, or −273°C) continuously emits electromagnetic radiation. Thus, terrestrial objects are also sources of radiation, although it is of considerably different magnitude and spectral composition than that of the sun. How much energy any object radiates is, among other things, a function of the surface temperature of the object. This property is expressed by the *Stefan–Boltzmann law*, which states that

$$M = \sigma T^4 \tag{1.4}$$

where

M = total radiant exitance from the surface of a material, watts (W) m^{-2}
σ = *Stefan–Boltzmann constant*, 5.6697×10^{-8} W m^{-2} K^{-4}
T = absolute temperature (K) of the emitting material

The particular units and the value of the constant are not critical for the student to remember, yet it is important to note that the total energy emitted from an object varies as T^4 and therefore increases very rapidly with increases in temperature. Also, it should be noted that this law is expressed for an energy source that behaves as a *blackbody*. A blackbody is a hypothetical, ideal radiator that totally absorbs and reemits all energy incident upon it. Actual objects only approach this ideal. We further explore the implications of this fact in Chapter 4; suffice it to say for now that the energy emitted from an object is primarily a function of its temperature, as given by Eq. 1.4.

Just as the total energy emitted by an object varies with temperature, the spectral distribution of the emitted energy also varies. Figure 1.4 shows energy distribution curves for blackbodies at temperatures ranging from 200 to 6000 K. The units on the ordinate scale ($\text{W}\,\text{m}^{-2}\mu\text{m}^{-1}$) express the radiant power coming from a blackbody per 1-μm spectral interval. Hence, the *area* under these curves equals the total radiant exitance, M, and the curves illustrate graphically what the Stefan–Boltzmann law expresses mathematically: The higher the temperature of the radiator, the greater the total amount of radiation it emits. The curves also show that there is a shift toward shorter wavelengths in the peak of a blackbody radiation distribution as temperature increases. The *dominant wavelength*, or wavelength at which a blackbody radiation curve reaches a maximum, is related to its temperature by *Wien's displacement law*,

$$\lambda_m = \frac{A}{T} \tag{1.5}$$

where

λ_m = wavelength of maximum spectral radiant exitance, μm
A = 2898 μm K
T = temperature, K

Thus, for a blackbody, the wavelength at which the maximum spectral radiant exitance occurs varies inversely with the blackbody's absolute temperature. We observe this phenomenon when a metal body such as a piece of iron is heated. As the object becomes progressively hotter, it begins to glow and its color

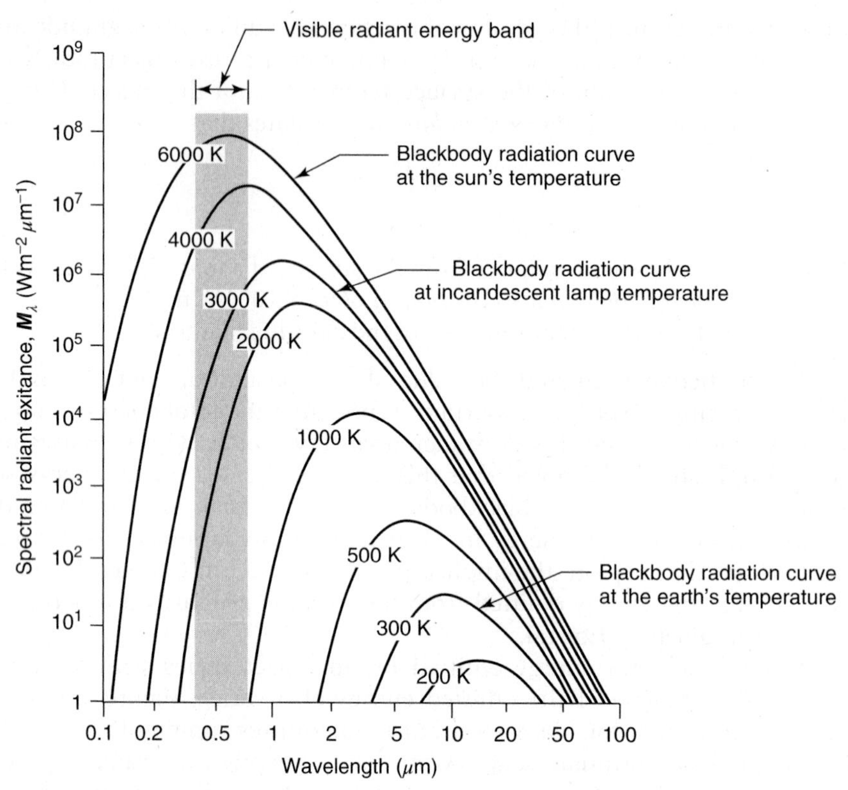

Figure 1.4 Spectral distribution of energy radiated from blackbodies of various temperatures. (Note that spectral radiant exitance M_λ is the energy emitted per unit wavelength interval. Total radiant exitance M is given by the area under the spectral radiant exitance curves.)

changes successively to shorter wavelengths—from dull red to orange to yellow and eventually to white.

The sun emits radiation in the same manner as a blackbody radiator whose temperature is about 6000 K (Figure 1.4). Many incandescent lamps emit radiation typified by a 3000 K blackbody radiation curve. Consequently, incandescent lamps have a relatively low output of blue energy, and they do not have the same spectral constituency as sunlight.

The earth's ambient temperature (i.e., the temperature of surface materials such as soil, water, and vegetation) is about 300 K (27°C). From Wien's displacement law, this means the maximum spectral radiant exitance from earth features occurs at a wavelength of about 9.7 μm. Because this radiation correlates with terrestrial heat, it is termed "thermal infrared" energy. This energy can neither be seen nor photographed, but it can be sensed with such thermal devices as radiometers and scanners (described in Chapter 4). By comparison, the sun has a much higher energy peak that occurs at about 0.5 μm, as indicated in Figure 1.4.

Our eyes—and photographic sensors—are sensitive to energy of this magnitude and wavelength. Thus, when the sun is present, we can observe earth features by virtue of *reflected* solar energy. Once again, the longer wavelength energy *emitted* by ambient earth features can be observed only with a nonphotographic sensing system. The general dividing line between reflected and emitted IR wavelengths is approximately 3 μm. Below this wavelength, reflected energy predominates; above it, emitted energy prevails.

Certain sensors, such as radar systems, supply their own source of energy to illuminate features of interest. These systems are termed "active" systems, in contrast to "passive" systems that sense naturally available energy. A very common example of an active system is a camera utilizing a flash. The same camera used in sunlight becomes a passive sensor.

1.3 ENERGY INTERACTIONS IN THE ATMOSPHERE

Irrespective of its source, all radiation detected by remote sensors passes through some distance, or *path length*, of atmosphere. The path length involved can vary widely. For example, space photography results from sunlight that passes through the full thickness of the earth's atmosphere twice on its journey from source to sensor. On the other hand, an airborne thermal sensor detects energy emitted directly from objects on the earth, so a single, relatively short atmospheric path length is involved. The net effect of the atmosphere varies with these differences in path length and also varies with the magnitude of the energy signal being sensed, the atmospheric conditions present, and the wavelengths involved.

Because of the varied nature of atmospheric effects, we treat this subject on a sensor-by-sensor basis in other chapters. Here, we merely wish to introduce the notion that the atmosphere can have a profound effect on, among other things, the intensity and spectral composition of radiation available to any sensing system. These effects are caused principally through the mechanisms of atmospheric *scattering* and *absorption*.

Scattering

Atmospheric scattering is the unpredictable diffusion of radiation by particles in the atmosphere. *Rayleigh scatter* is common when radiation interacts with atmospheric molecules and other tiny particles that are much smaller in diameter than the wavelength of the interacting radiation. The effect of Rayleigh scatter is inversely proportional to the fourth power of wavelength. Hence, there is a much stronger tendency for short wavelengths to be scattered by this mechanism than long wavelengths.

A "blue" sky is a manifestation of Rayleigh scatter. In the absence of scatter, the sky would appear black. But, as sunlight interacts with the earth's atmosphere, it scatters the shorter (blue) wavelengths more dominantly than the other visible wavelengths. Consequently, we see a blue sky. At sunrise and sunset, however, the sun's rays travel through a longer atmospheric path length than during midday. With the longer path, the scatter (and absorption) of short wavelengths is so complete that we see only the less scattered, longer wavelengths of orange and red.

Rayleigh scatter is one of the primary causes of "haze" in imagery. Visually, haze diminishes the "crispness," or "contrast," of an image. In color photography, it results in a bluish-gray cast to an image, particularly when taken from high altitude. As we see in Chapter 2, haze can often be eliminated or at least minimized by introducing, in front of the camera lens, a filter that does not transmit short wavelengths.

Another type of scatter is *Mie scatter*, which exists when atmospheric particle diameters essentially equal the wavelengths of the energy being sensed. Water vapor and dust are major causes of Mie scatter. This type of scatter tends to influence longer wavelengths compared to Rayleigh scatter. Although Rayleigh scatter tends to dominate under most atmospheric conditions, Mie scatter is significant in slightly overcast ones.

A more bothersome phenomenon is *nonselective scatter*, which comes about when the diameters of the particles causing scatter are much larger than the wavelengths of the energy being sensed. Water droplets, for example, cause such scatter. They commonly have a diameter in the range 5 to 100 μm and scatter all visible and near- to mid-IR wavelengths about equally. Consequently, this scattering is "nonselective" with respect to wavelength. In the visible wavelengths, equal quantities of blue, green, and red light are scattered; hence fog and clouds appear white.

Absorption

In contrast to scatter, atmospheric absorption results in the effective loss of energy to atmospheric constituents. This normally involves absorption of energy at a given wavelength. The most efficient absorbers of solar radiation in this regard are water vapor, carbon dioxide, and ozone. Because these gases tend to absorb electromagnetic energy in specific wavelength bands, they strongly influence the design of any remote sensing system. The wavelength ranges in which the atmosphere is particularly transmissive of energy are referred to as *atmospheric windows*.

Figure 1.5 shows the interrelationship between energy sources and atmospheric absorption characteristics. Figure 1.5*a* shows the spectral distribution of the energy emitted by the sun and by earth features. These two curves represent the most common sources of energy used in remote sensing. In Figure 1.5*b*, spectral regions in which the atmosphere blocks energy are shaded. Remote sensing data acquisition is limited to the nonblocked spectral regions, the atmospheric windows. Note in Figure 1.5*c* that the spectral sensitivity range of the eye (the

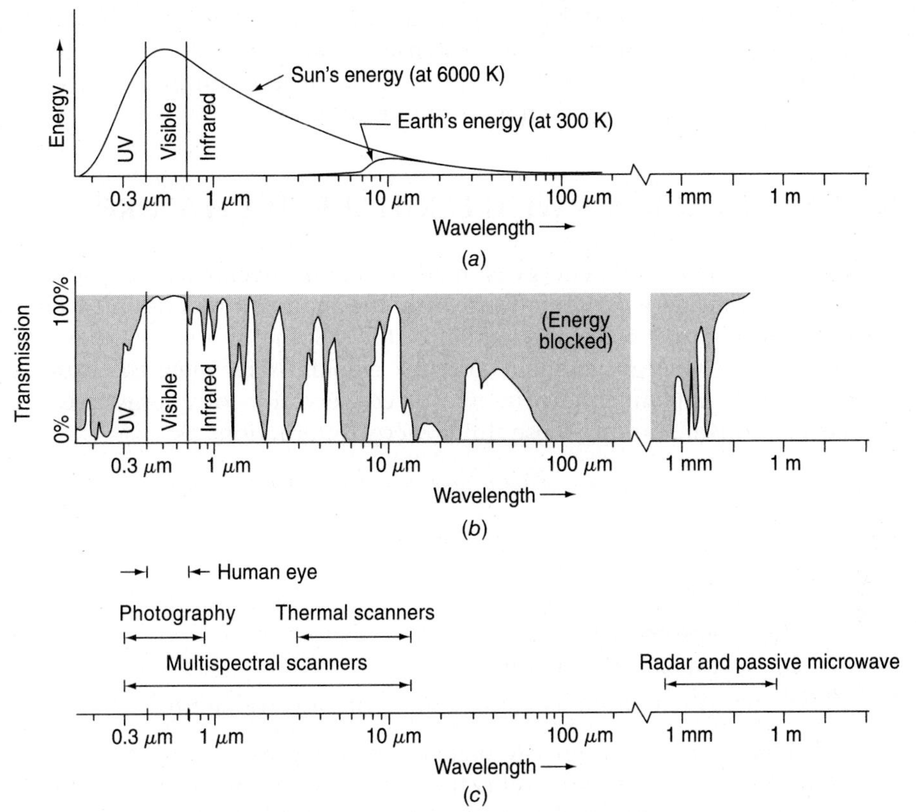

Figure 1.5 Spectral characteristics of (*a*) energy sources, (*b*) atmospheric transmittance, and (*c*) common remote sensing systems. (Note that wavelength scale is logarithmic.)

"visible" range) coincides with both an atmospheric window and the peak level of energy from the sun. Emitted "heat" energy from the earth, shown by the small curve in (*a*), is sensed through the windows at 3 to 5 μm and 8 to 14 μm using such devices as *thermal sensors*. *Multispectral sensors* observe simultaneously through multiple, narrow wavelength ranges that can be located at various points in the visible through the thermal spectral region. *Radar* and *passive microwave systems* operate through a window in the region 1 mm to 1 m.

The important point to note from Figure 1.5 is the interaction and the interdependence between the primary sources of electromagnetic energy, the atmospheric windows through which source energy may be transmitted to and from earth surface features, and the spectral sensitivity of the sensors available to detect and record the energy. One cannot select the sensor to be used in any given remote sensing task arbitrarily; one must instead consider (1) the spectral sensitivity of the sensors available, (2) the presence or absence of atmospheric windows in the spectral range(s) in which one wishes to sense, and (3) the source, magnitude, and spectral composition of the energy available in these ranges. Ultimately, however, the choice of spectral range of the sensor must be based on the manner in which the energy interacts with the features under investigation. It is to this last, very important, element that we now turn our attention.

1.4 ENERGY INTERACTIONS WITH EARTH SURFACE FEATURES

When electromagnetic energy is incident on any given earth surface feature, three fundamental energy interactions with the feature are possible. These are illustrated in Figure 1.6 for an element of the volume of a water body. Various fractions of the energy incident on the element are *reflected, absorbed,* and/or *transmitted*. Applying the principle of conservation of energy, we can state the interrelationship among these three energy interactions as

$$E_I(\lambda) = E_R(\lambda) + E_A(\lambda) + E_T(\lambda) \tag{1.6}$$

where

E_I = incident energy
E_R = reflected energy
E_A = absorbed energy
E_T = transmitted energy

with all energy components being a function of wavelength λ.

Equation 1.6 is an energy balance equation expressing the interrelationship among the mechanisms of reflection, absorption, and transmission. Two points concerning this relationship should be noted. First, the proportions of energy reflected, absorbed, and transmitted will vary for different earth features, depending on their material type and condition. These differences permit us to distinguish different features on an image. Second, the wavelength dependency means that, even within a given feature type, the proportion of reflected, absorbed, and

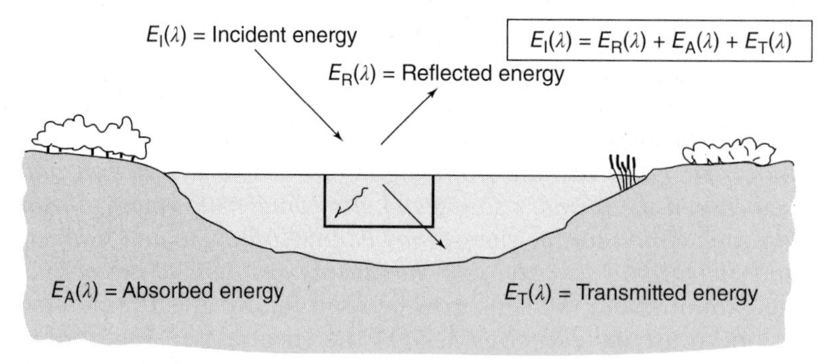

Figure 1.6 Basic interactions between electromagnetic energy and an earth surface feature.

transmitted energy will vary at different wavelengths. Thus, two features may be indistinguishable in one spectral range and be very different in another wavelength band. Within the visible portion of the spectrum, these spectral variations result in the visual effect called *color*. For example, we call objects "blue" when they reflect more highly in the blue portion of the spectrum, "green" when they reflect more highly in the green spectral region, and so on. Thus, the eye utilizes spectral variations in the magnitude of reflected energy to discriminate between various objects. Color terminology and color mixing principles are discussed further in Section 1.12.

Because many remote sensing systems operate in the wavelength regions in which reflected energy predominates, the reflectance properties of earth features are very important. Hence, it is often useful to think of the energy balance relationship expressed by Eq. 1.6 in the form

$$E_R(\lambda) = E_I(\lambda) - [E_A(\lambda) + E_T(\lambda)] \tag{1.7}$$

That is, the reflected energy is equal to the energy incident on a given feature reduced by the energy that is either absorbed or transmitted by that feature.

The reflectance characteristics of earth surface features may be quantified by measuring the portion of incident energy that is reflected. This is measured as a function of wavelength and is called *spectral reflectance*, ρ_λ. It is mathematically defined as

$$\begin{aligned} \rho_\lambda &= \frac{E_R(\lambda)}{E_I(\lambda)} \\ &= \frac{\text{energy of wavelength } \lambda \text{ reflected from the object}}{\text{energy of wavelength } \lambda \text{ incident upon the object}} \times 100 \end{aligned} \tag{1.8}$$

where ρ_λ is expressed as a percentage.

A graph of the spectral reflectance of an object as a function of wavelength is termed a *spectral reflectance curve*. The configuration of spectral reflectance curves gives us insight into the spectral characteristics of an object and has a strong influence on the choice of wavelength region(s) in which remote sensing data are acquired for a particular application. This is illustrated in Figure 1.7, which shows highly generalized spectral reflectance curves for deciduous versus coniferous trees. Note that the curve for each of these object types is plotted as a "ribbon" (or "envelope") of values, not as a single line. This is because spectral reflectances vary somewhat within a given material class. That is, the spectral reflectance of one deciduous tree species and another will never be identical, nor will the spectral reflectance of trees of the same species be exactly equal. We elaborate upon the variability of spectral reflectance curves later in this section.

In Figure 1.7, assume that you are given the task of selecting an airborne sensor system to assist in preparing a map of a forested area differentiating deciduous versus coniferous trees. One choice of sensor might be the human eye. However, there is a potential problem with this choice. The spectral reflectance curves for each tree

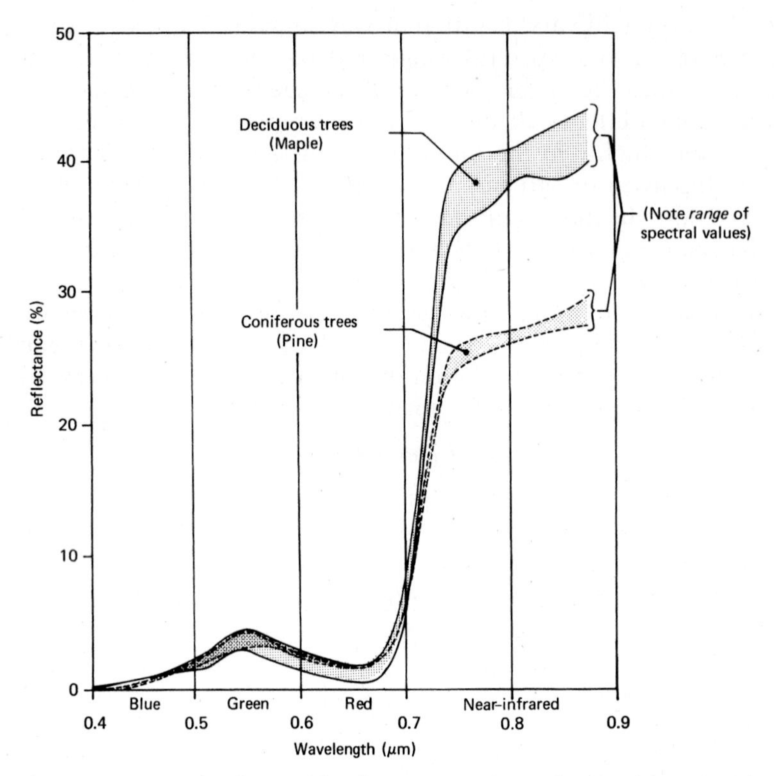

Figure 1.7 Generalized spectral reflectance envelopes for deciduous (broad-leaved) and coniferous (needle-bearing) trees. (Each tree type has a range of spectral reflectance values at any wavelength.) (Adapted from Kalensky and Wilson, 1975.)

type overlap in most of the visible portion of the spectrum and are very close where they do not overlap. Hence, the eye might see both tree types as being essentially the same shade of "green" and might confuse the identity of the deciduous and coniferous trees. Certainly one could improve things somewhat by using spatial clues to each tree type's identity, such as size, shape, site, and so forth. However, this is often difficult to do from the air, particularly when tree types are intermixed. How might we discriminate the two types on the basis of their spectral characteristics alone? We could do this by using a sensor that records near-IR energy. A specialized digital camera whose detectors are sensitive to near-IR wavelengths is just such a system, as is an analog camera loaded with black and white IR film. On near-IR images, deciduous trees (having higher IR reflectance than conifers) generally appear much lighter in tone than do conifers. This is illustrated in Figure 1.8, which shows stands of coniferous trees surrounded by deciduous trees. In Figure 1.8*a* (visible spectrum), it is virtually impossible to distinguish between tree types, even though the conifers have a distinctive conical shape whereas the deciduous trees have rounded crowns. In Figure 1.8*b* (near IR), the coniferous trees have a

Figure 1.8 Low altitude oblique aerial photographs illustrating deciduous versus coniferous trees. (*a*) Panchromatic photograph recording reflected sunlight over the wavelength band 0.4 to 0.7 μm. (*b*) Black-and-white infrared photograph recording reflected sunlight over 0.7 to 0.9 μm wavelength band. (Author-prepared figure.)

distinctly darker tone. On such an image, the task of delineating deciduous versus coniferous trees becomes almost trivial. In fact, if we were to use a computer to analyze digital data collected from this type of sensor, we might "automate" our entire mapping task. Many remote sensing data analysis schemes attempt to do just that. For these schemes to be successful, the materials to be differentiated must be spectrally separable.

Experience has shown that many earth surface features of interest can be identified, mapped, and studied on the basis of their spectral characteristics. Experience has also shown that some features of interest cannot be spectrally separated. Thus, to utilize remote sensing data effectively, one must know and understand the spectral characteristics of the particular features under investigation in any given application. Likewise, one must know what factors influence these characteristics.

Spectral Reflectance of Earth Surface Feature Types

Figure 1.9 shows typical spectral reflectance curves for many different types of features: healthy green grass, dry (non-photosynthetically active) grass, bare soil (brown to dark-brown sandy loam), pure gypsum dune sand, asphalt, construction concrete (Portland cement concrete), fine-grained snow, clouds, and clear lake water. The lines in this figure represent *average* reflectance curves compiled by measuring a large sample of features, or in some cases *representative* reflectance measurements from a single typical example of the feature class. Note how distinctive the curves are for each feature. In general, the configuration of these curves is an indicator of the type and condition of the features to which they apply. Although the reflectance of individual features can vary considerably above and below the lines shown here, these curves demonstrate some fundamental points concerning spectral reflectance.

For example, spectral reflectance curves for healthy green vegetation almost always manifest the "peak-and-valley" configuration illustrated by green grass in Figure 1.9. The valleys in the visible portion of the spectrum are dictated by the pigments in plant leaves. Chlorophyll, for example, strongly absorbs energy in the wavelength bands centered at about 0.45 and 0.67 μm (often called the "chlorophyll absorption bands"). Hence, our eyes perceive healthy vegetation as green in color because of the very high absorption of blue and red energy by plant leaves and the relatively high reflection of green energy. If a plant is subject to some form of stress that interrupts its normal growth and productivity, it may decrease or cease chlorophyll production. The result is less chlorophyll absorption in the blue and red bands. Often, the red reflectance increases to the point that we see the plant turn yellow (combination of green and red). This can be seen in the spectral curve for dried grass in Figure 1.9.

As we go from the visible to the near-IR portion of the spectrum, the reflectance of healthy vegetation increases dramatically. This spectral feature, known as the *red edge*, typically occurs between 0.68 and 0.75 μm, with the exact position depending on the species and condition. Beyond this edge, from about

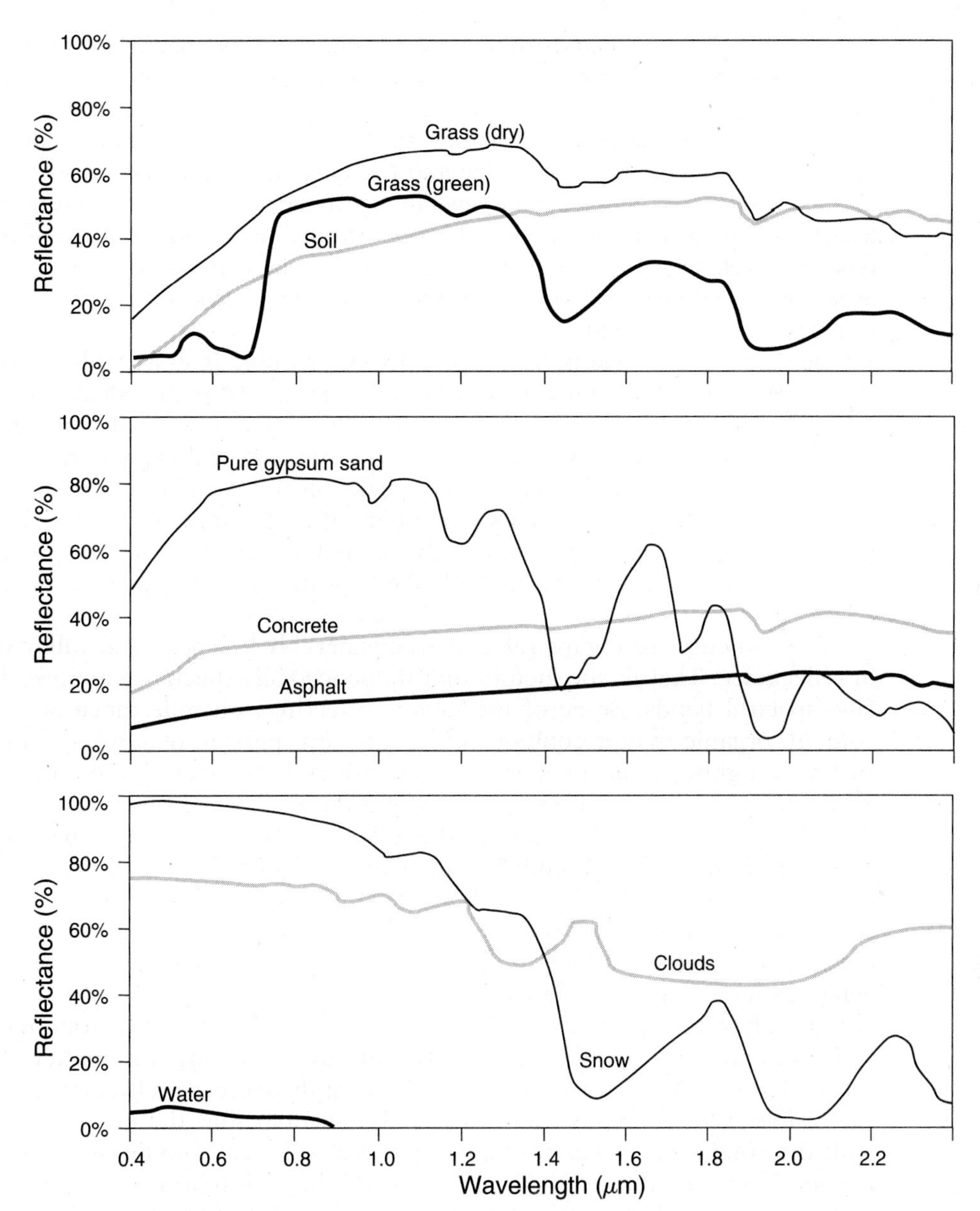

Figure 1.9 Spectral reflectance curves for various features types. (Original data courtesy USGS Spectroscopy Lab, Johns Hopkins University Spectral Library, and Jet Propulsion Laboratory [JPL]; cloud spectrum from Bowker et al., after Avery and Berlin, 1992. JPL spectra © 1999, California Institute of Technology.)

0.75 to 1.3 μm (representing most of the near-IR range), a plant leaf typically reflects 40 to 50% of the energy incident upon it. Most of the remaining energy is transmitted, because absorption in this spectral region is minimal (less than 5%). Plant reflectance from 0.75 to 1.3 μm results primarily from the internal structure of plant leaves. Because the position of the red edge and the magnitude of the near-IR reflectance beyond the red edge are highly variable among plant species, reflectance measurements in these ranges often permit us to discriminate between species, even if they look the same in visible wavelengths. Likewise, many plant stresses alter the reflectance in the red edge and the near-IR region, and sensors operating in these ranges are often used for vegetation stress detection. Also, multiple layers of leaves in a plant canopy provide the opportunity for multiple transmissions and reflections. Hence, the near-IR reflectance increases with the number of layers of leaves in a canopy, with the maximum reflectance achieved at about eight leaf layers (Bauer et al., 1986).

Beyond 1.3 μm, energy incident upon vegetation is essentially absorbed or reflected, with little to no transmittance of energy. Dips in reflectance occur at 1.4, 1.9, and 2.7 μm because water in the leaf absorbs strongly at these wavelengths. Accordingly, wavelengths in these spectral regions are referred to as *water absorption bands*. Reflectance peaks occur at about 1.6 and 2.2 μm, between the absorption bands. Throughout the range beyond 1.3 μm, leaf reflectance is approximately inversely related to the total water present in a leaf. This total is a function of both the moisture content and the thickness of a leaf.

The soil curve in Figure 1.9 shows considerably less peak-and-valley variation in reflectance. That is, the factors that influence soil reflectance act over less specific spectral bands. Some of the factors affecting soil reflectance are moisture content, organic matter content, soil texture (proportion of sand, silt, and clay), surface roughness, and presence of iron oxide. These factors are complex, variable, and interrelated. For example, the presence of moisture in soil will decrease its reflectance. As with vegetation, this effect is greatest in the water absorption bands at about 1.4, 1.9, and 2.7 μm (clay soils also have hydroxyl absorption bands at about 1.4 and 2.2 μm). Soil moisture content is strongly related to the soil texture: Coarse, sandy soils are usually well drained, resulting in low moisture content and relatively high reflectance; poorly drained fine-textured soils will generally have lower reflectance. Thus, the reflectance properties of a soil are consistent only within particular ranges of conditions. Two other factors that reduce soil reflectance are surface roughness and content of organic matter. The presence of iron oxide in a soil will also significantly decrease reflectance, at least in the visible wavelengths. In any case, it is essential that the analyst be familiar with the conditions at hand. Finally, because soils are essentially opaque to visible and infrared radiation, it should be noted that soil reflectance comes from the uppermost layer of the soil and may not be indicative of the properties of the bulk of the soil.

Sand can have wide variation in its spectral reflectance pattern. The curve shown in Figure 1.9 is from a dune in New Mexico and consists of roughly 99% gypsum with trace amounts of quartz (Jet Propulsion Laboratory, 1999). Its absorption and reflectance features are essentially identical to those of its parent material, gypsum. Sand derived from other sources, with differing mineral compositions, would have a spectral reflectance curve indicative of its parent material. Other factors affecting the spectral response from sand include the presence or absence of water and of organic matter. Sandy soil is subject to the same considerations listed in the discussion of soil reflectance.

As shown in Figure 1.9, the spectral reflectance curves for asphalt and Portland cement concrete are much flatter than those of the materials discussed thus far. Overall, Portland cement concrete tends to be relatively brighter than asphalt, both in the visible spectrum and at longer wavelengths. It is important to note that the reflectance of these materials may be modified by the presence of paint, soot, water, or other substances. Also, as materials age, their spectral reflectance patterns may change. For example, the reflectance of many types of asphaltic concrete may increase, particularly in the visible spectrum, as their surface ages.

In general, snow reflects strongly in the visible and near infrared, and absorbs more energy at mid-infrared wavelengths. However, the reflectance of snow is affected by its grain size, liquid water content, and presence or absence of other materials in or on the snow surface (Dozier and Painter, 2004). Larger grains of snow absorb more energy, particularly at wavelengths longer than 0.8 μm. At temperatures near 0°C, liquid water within the snowpack can cause grains to stick together in clusters, thus increasing the effective grain size and decreasing the reflectance at near-infrared and longer wavelengths. When particles of contaminants such as dust or soot are deposited on snow, they can significantly reduce the surface's reflectance in the visible spectrum.

The aforementioned absorption of mid-infrared wavelengths by snow can permit the differentiation between snow and clouds. While both feature types appear bright in the visible and near infrared, clouds have significantly higher reflectance than snow at wavelengths longer than 1.4 μm. Meteorologists can also use both spectral and bidirectional reflectance patterns (discussed later in this section) to identify a variety of cloud properties, including ice/water composition and particle size.

Considering the spectral reflectance of water, probably the most distinctive characteristic is the energy absorption at near-IR wavelengths and beyond. In short, water absorbs energy in these wavelengths whether we are talking about water features per se (such as lakes and streams) or water contained in vegetation or soil. Locating and delineating water bodies with remote sensing data are done most easily in near-IR wavelengths because of this absorption property. However, various conditions of water bodies manifest themselves primarily in visible wavelengths. The energy–matter interactions at these wavelengths are very complex and depend on a number of interrelated factors. For example, the reflectance from a water body can stem from an interaction with the water's surface (specular reflection), with material suspended in the water, or with the bottom of the depression containing the water body. Even with deep water where bottom effects are negligible, the reflectance properties of a water body are a function of not only the water per se but also the material in the water.

Clear water absorbs relatively little energy having wavelengths less than about 0.6 μm. High transmittance typifies these wavelengths with a maximum in the blue-green portion of the spectrum. However, as the turbidity of water changes (because of the presence of organic or inorganic materials), transmittance—and therefore reflectance—changes dramatically. For example, waters containing large quantities of suspended sediments resulting from soil erosion normally have much higher visible reflectance than other "clear" waters in the same geographic area. Likewise, the reflectance of water changes with the chlorophyll concentration involved. Increases in chlorophyll concentration tend to decrease water reflectance in blue wavelengths and increase it in green wavelengths. These changes have been used to monitor the presence and estimate the concentration of algae via remote sensing data. Reflectance data have also been used to determine the presence or absence of tannin dyes from bog vegetation in lowland areas and to detect a number of pollutants, such as oil and certain industrial wastes.

Figure 1.10 illustrates some of these effects, using spectra from three lakes with different bio-optical properties. The first spectrum is from a clear, oligotrophic lake with a chlorophyll level of 1.2 μg/l and only 2.4 mg/l of dissolved organic carbon (DOC). Its spectral reflectance is relatively high in the blue-green portion of the spectrum and decreases in the red and near infrared. In contrast,

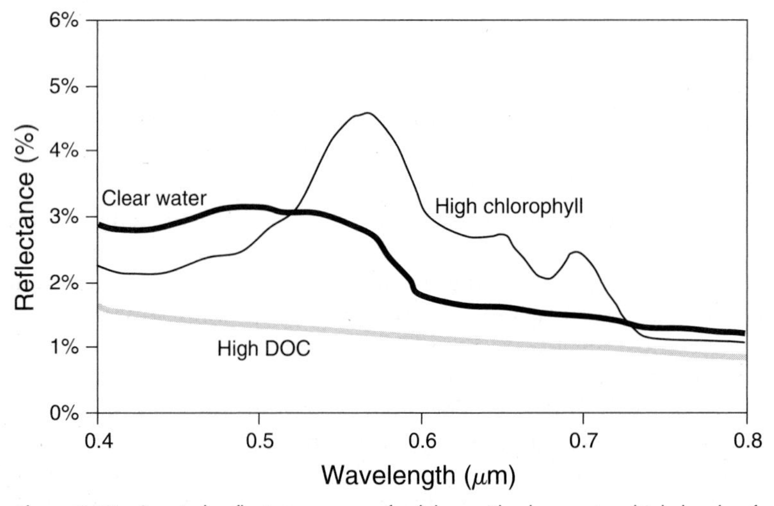

Figure 1.10 Spectral reflectance curves for lakes with clear water, high levels of chlorophyll, and high levels of dissolved organic carbon (DOC).

the spectrum from a lake experiencing an algae bloom, with much higher chlorophyll concentration (12.3 μg/l), shows a reflectance peak in the green spectrum and absorption in the blue and red regions. These reflectance and absorption features are associated with several pigments present in algae. Finally, the third spectrum in Figure 1.10 was acquired on an ombrotrophic bog lake, with very high levels of DOC (20.7 mg/l). These naturally occurring tannins and other complex organic molecules give the lake a very dark appearance, with its reflectance curve nearly flat across the visible spectrum.

Many important water characteristics, such as dissolved oxygen concentration, pH, and salt concentration, cannot be observed directly through changes in water reflectance. However, such parameters sometimes correlate with observed reflectance. In short, there are many complex interrelationships between the spectral reflectance of water and particular characteristics. One must use appropriate reference data to correctly interpret reflectance measurements made over water.

Our discussion of the spectral characteristics of vegetation, soil, and water has been very general. The student interested in pursuing details on this subject, as well as factors influencing these characteristics, is encouraged to consult the various references contained in the Works Cited section located at the end of this book.

Spectral Response Patterns

Having looked at the spectral reflectance characteristics of vegetation, soil, sand, concrete, asphalt, snow, clouds, and water, we should recognize that these broad feature types are often spectrally separable. However, the degree of separation between types varies among and within spectral regions. For example, water and vegetation might reflect nearly equally in visible wavelengths, yet these features are almost always separable in near-IR wavelengths.

Because spectral responses measured by remote sensors over various features often permit an assessment of the type and/or condition of the features, these responses have often been referred to as *spectral signatures*. Spectral reflectance and spectral emittance curves (for wavelengths greater than 3.0 μm) are often referred to in this manner. The physical radiation measurements acquired over specific terrain features at various wavelengths are also referred to as the spectral signatures for those features.

Although it is true that many earth surface features manifest very distinctive spectral reflectance and/or emittance characteristics, these characteristics result in spectral "response patterns" rather than in spectral "signatures." The reason for this is that the term *signature* tends to imply a pattern that is absolute and unique. This is not the case with the spectral patterns observed in the natural world. As we have seen, spectral response patterns measured by remote sensors may be quantitative, but they are not absolute. They may be distinctive, but they are not necessarily unique.

We have already looked at some characteristics of objects that influence their spectral response patterns. *Temporal effects* and *spatial effects* can also enter into any given analysis. Temporal effects are any factors that change the spectral characteristics of a feature over time. For example, the spectral characteristics of many species of vegetation are in a nearly continual state of change throughout a growing season. These changes often influence when we might collect sensor data for a particular application.

Spatial effects refer to factors that cause the same types of features (e.g., corn plants) at a given point in *time* to have different characteristics at different geographic *locations*. In small-area analysis the geographic locations may be meters apart and spatial effects may be negligible. When analyzing satellite data, the locations may be hundreds of kilometers apart where entirely different soils, climates, and cultivation practices might exist.

Temporal and spatial effects influence virtually all remote sensing operations. These effects normally complicate the issue of analyzing spectral reflectance properties of earth resources. Again, however, temporal and spatial effects might be the keys to gleaning the information sought in an analysis. For example, the process of *change detection* is premised on the ability to measure temporal effects. An example of this process is detecting the change in suburban development near a metropolitan area by using data obtained on two different dates.

An example of a useful spatial effect is the change in the leaf morphology of trees when they are subjected to some form of stress. For example, when a tree becomes infected with Dutch elm disease, its leaves might begin to cup and curl, changing the reflectance of the tree relative to healthy trees that surround it. So, even though a spatial effect might cause differences in the spectral reflectances of the same type of feature, this effect may be just what is important in a particular application.

Finally, it should be noted that the apparent spectral response from surface features can be influenced by shadows. While an object's spectral reflectance (a ratio of reflected to incident energy, see Eq. 1.8) is not affected by changes in illumination, the absolute amount of energy reflected does depend on illumination conditions. Within a shadow, the total reflected energy is reduced, and the spectral response is shifted toward shorter wavelengths. This occurs because the incident energy within a shadow comes primarily from Rayleigh atmospheric scattering, and as discussed in Section 1.3, such scattering primarily affects short wavelengths. Thus, in visible-wavelength imagery, objects inside shadows will tend to appear both darker and bluer than if they were fully illuminated. This effect can cause problems for automated image classification algorithms; for example, dark shadows of trees on pavement may be misclassified as water. The effects of illumination geometry on reflectance are discussed in more detail later in this section, while the impacts of shadows on the image interpretation process are discussed in Section 1.12.

Atmospheric Influences on Spectral Response Patterns

In addition to being influenced by temporal and spatial effects, spectral response patterns are influenced by the atmosphere. Regrettably, the energy recorded by a sensor is always modified to some extent by the atmosphere between the sensor and the ground. We will indicate the significance of this effect on a sensor-by-sensor basis throughout this book. For now, Figure 1.11 provides an initial frame of reference for understanding the nature of atmospheric effects. Shown in this figure is the typical situation encountered when a sensor records reflected solar energy. The atmosphere affects the "brightness," or *radiance*, recorded over any given point on the ground in two almost contradictory ways. First, it attenuates (reduces) the energy illuminating a ground object (and being reflected from the object). Second, the atmosphere acts as a reflector itself, adding a scattered, extraneous *path radiance* to the signal detected by the sensor. By expressing these two atmospheric effects mathematically, the total radiance recorded by the sensor may be related to the reflectance of the ground object and the incoming radiation or *irradiance* using the equation

$$L_{\text{tot}} = \frac{\rho ET}{\pi} + L_p \tag{1.9}$$

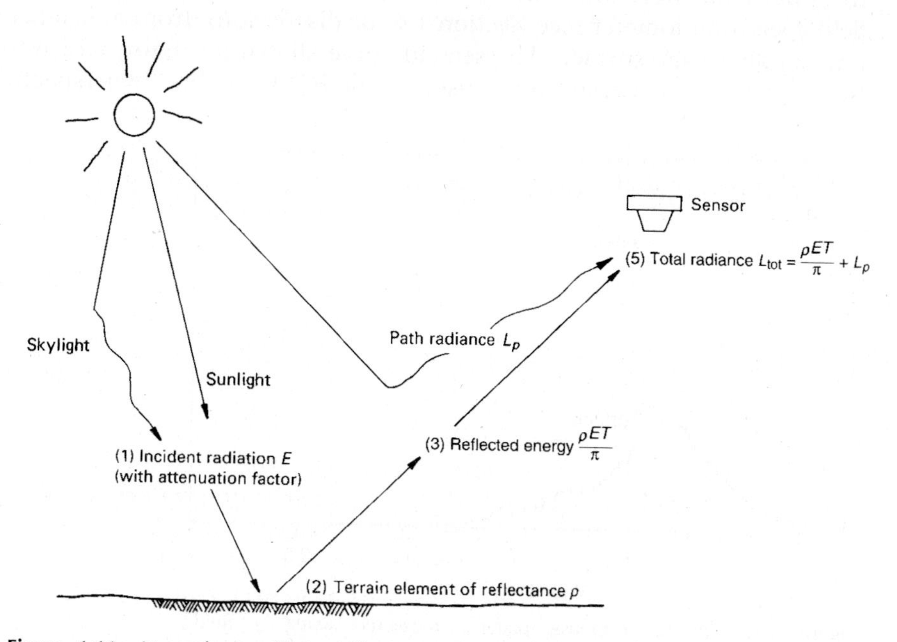

Figure 1.11 Atmospheric effects influencing the measurement of reflected solar energy. Attenuated sunlight and skylight (E) is reflected from a terrain element having reflectance ρ. The attenuated radiance reflected from the terrain element ($\rho ET/\pi$) combines with the path radiance (L_p) to form the total radiance (L_{tot}) recorded by the sensor.

where

L_{tot} = total spectral radiance measured by sensor
ρ = reflectance of object
E = irradiance on object, incoming energy
T = transmission of atmosphere
L_p = path radiance, from the atmosphere and not from the object

It should be noted that all of the above factors depend on wavelength. Also, as shown in Figure 1.11, the irradiance (E) stems from two sources: (1) directly reflected "sunlight" and (2) diffuse "skylight," which is sunlight that has been previously scattered by the atmosphere. The relative dominance of sunlight versus skylight in any given image is strongly dependent on weather conditions (e.g., sunny vs. hazy vs. cloudy). Likewise, irradiance varies with the seasonal changes in solar elevation angle (Figure 7.4) and the changing distance between the earth and sun.

For a sensor positioned close to the earth's surface, the path radiance L_p will generally be small or negligible, because the atmospheric path length from the surface to the sensor is too short for much scattering to occur. In contrast, imagery from satellite systems will be more strongly affected by path radiance, due to the longer atmospheric path between the earth's surface and the spacecraft. This can be seen in Figure 1.12, which compares two spectral response patterns from the same area. One "signature" in this figure was collected using a handheld field spectroradiometer (see Section 1.6 for discussion), from a distance of only a few cm above the surface. The second curve shown in Figure 1.12 was collected by the Hyperion hyperspectral sensor on the EO-1 satellite (hyperspectral systems

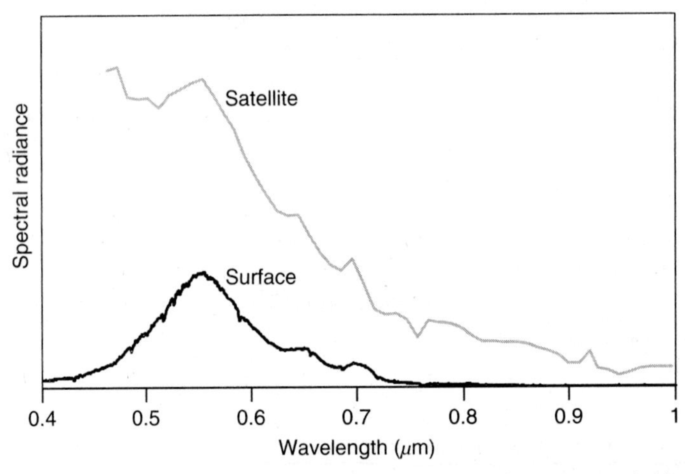

Figure 1.12 Spectral response patterns measured using a field spectroradiometer in close proximity to the earth's surface, and from above the top of the atmosphere (via the Hyperion instrument on EO-1). The difference between the two "signatures" is caused by atmospheric scattering and absorption in the Hyperion image.

are discussed in Chapter 4, and the Hyperion instrument is covered in Chapter 5). Due to the thickness of the atmosphere between the earth's surface and the satellite's position above the atmosphere, this second spectral response pattern shows an elevated signal at short wavelengths, due to the extraneous path radiance.

In its raw form, this near-surface measurement from the field spectroradiometer could not be directly compared to the measurement from the satellite, because one is observing *surface reflectance* while the other is observing the so-called *top of atmosphere* (*TOA*) reflectance. Before such a comparison could be performed, the satellite image would need to go through a process of *atmospheric correction*, in which the raw spectral data are modified to compensate for the expected effects of atmospheric scattering and absorption. This process, discussed in Chapter 7, generally does not produce a perfect representation of the spectral response curve that would actually be observed at the surface itself, but it can produce a sufficiently close approximation to be suitable for many types of analysis.

Readers who might be interested in obtaining additional details about the concepts, terminology, and units used in radiation measurement may wish to consult Appendix A.

Geometric Influences on Spectral Response Patterns

The geometric manner in which an object reflects energy is an important consideration. This factor is primarily a function of the surface roughness of the object. *Specular* reflectors are flat surfaces that manifest mirror-like reflections, where the angle of reflection equals the angle of incidence. *Diffuse* (or *Lambertian*) reflectors are rough surfaces that reflect uniformly in all directions. Most earth surfaces are neither perfectly specular nor perfectly diffuse reflectors. Their characteristics are somewhat between the two extremes.

Figure 1.13 illustrates the geometric character of specular, near-specular, near-diffuse, and diffuse reflectors. The category that describes any given surface is dictated by the surface's roughness *in comparison to the wavelength of the energy being sensed*. For example, in the relatively long wavelength radio range, a sandy beach can appear smooth to incident energy, whereas in the visible portion of the spectrum, it appears rough. In short, when the wavelength of incident energy is much smaller than the surface height variations or the particle sizes that make up a surface, the reflection from the surface is diffuse.

Diffuse reflections contain spectral information on the "color" of the reflecting surface, whereas specular reflections generally do not. *Hence, in remote sensing, we are most often interested in measuring the diffuse reflectance properties of terrain features*.

Because most features are not perfect diffuse reflectors, however, it becomes necessary to consider the viewing and illumination geometry. Figure 1.14 illustrates the relationships that exist among *solar elevation*, *azimuth angle*, and *viewing angle*. Figure 1.15 shows some typical geometric effects that can influence the

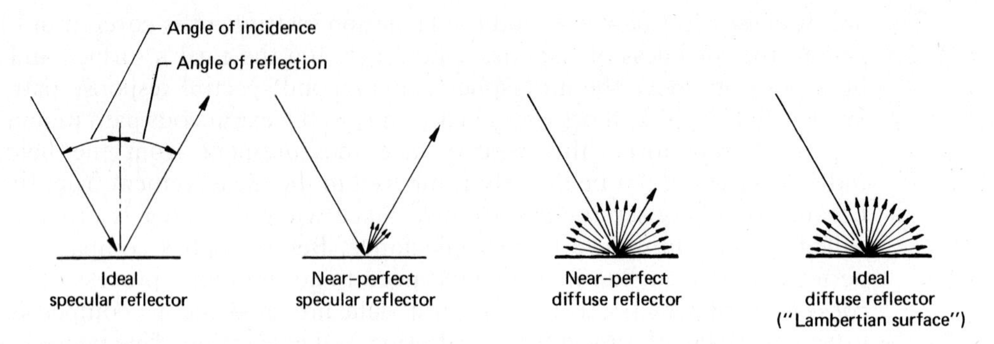

Figure 1.13 Specular versus diffuse reflectance. (We are most often interested in measuring the diffuse reflectance of objects.)

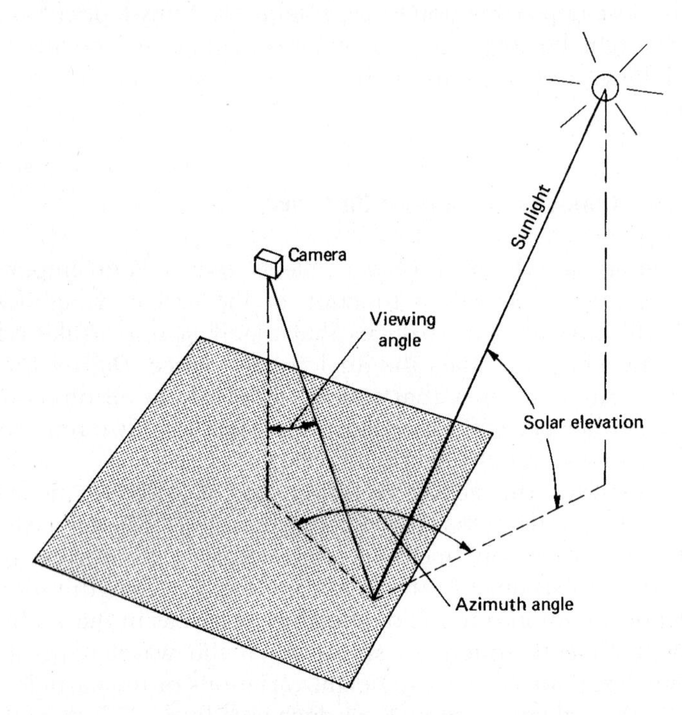

Figure 1.14 Sun-object-image angular relationship.

apparent reflectance in an image. In (*a*), the effect of *differential shading* is illustrated in profile view. Because the sides of features may be either sunlit or shaded, variations in brightness can result from identical ground objects at different locations in the image. The sensor receives more energy from the sunlit side of the tree at *B* than from the shaded side of the tree at *A*. Differential shading is clearly a function of solar elevation and object height, with a stronger effect at

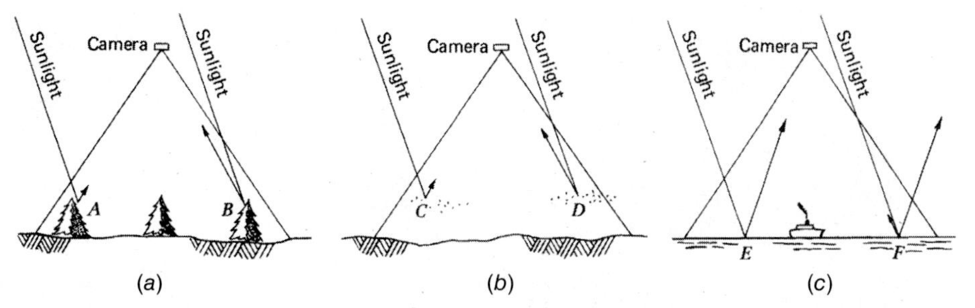

Figure 1.15 Geometric effects that cause variations in focal plane irradiance: (*a*) differential shading, (*b*) differential scattering, and (*c*) specular reflection.

low solar angles. The effect is also compounded by differences in slope and aspect (slope orientation) over terrain of varied relief.

Figure 1.15*b* illustrates the effect of *differential atmospheric scattering*. As discussed earlier, backscatter from atmospheric molecules and particles adds light (path radiance) to that reflected from ground features. The sensor records more atmospheric backscatter from area *D* than from area *C* due to this geometric effect. In some analyses, the variation in this path radiance component is small and can be ignored, particularly at long wavelengths. However, under hazy conditions, differential quantities of path radiance often result in varied illumination across an image.

As mentioned earlier, specular reflections represent the extreme in directional reflectance. When such reflections appear, they can hinder analysis of the imagery. This can often be seen in imagery taken over water bodies. Figure 1.15*c* illustrates the geometric nature of this problem. Immediately surrounding point *E* on the image, a considerable increase in brightness would result from specular reflection. A photographic example of this is shown in Figure 1.16, which includes areas of specular reflection from the right half of the large lake in the center of the image. These mirrorlike reflections normally contribute little information about the true character of the objects involved. For example, the small water bodies just below the larger lake have a tone similar to that of some of the fields in the area. Because of the low information content of specular reflections, they are avoided in most analyses.

The most complete representation of an object's geometric reflectance properties is the *bidirectional reflectance distribution function* (*BRDF*). This is a mathematical description of how reflectance varies for all combinations of illumination and viewing angles at a given wavelength (Schott, 2007). The BRDF for any given feature can approximate that of a Lambertian surface at some angles and be non-Lambertian at other angles. Similarly, the BRDF can vary considerably with wavelength. A variety of mathematical models (including a provision for wavelength dependence) have been proposed to represent the BRDF (Jupp and Strahler, 1991).

Figure 1.16 Aerial photograph containing areas of specular reflection from water bodies. This image is a portion of a summertime photograph taken over Green Lake, Green Lake County, WI. Scale 1:95,000. Cloud shadows indicate direction of sunlight at time of exposure. Reproduced from color IR original. (NASA image.)

Figure 1.17*a* shows graphic representations of the BRDF for three objects, each of which can be visualized as being located at the point directly beneath the center of one of the hemispheres. In each case, illumination is from the south (located at the back right, in these perspective views). The brightness at any point on the hemisphere indicates the relative reflectance of the object at a given viewing angle. A perfectly diffuse reflector (Figure 1.17*a*, top) has uniform reflectance in all directions. At the other extreme, a specular reflector (bottom) has very high reflectance in the direction directly opposite the source of illumination, and very low reflectance in all other directions. An intermediate surface (middle) has somewhat elevated reflectance at the specular angle but also shows some reflectance in other directions.

Figure 1.17*b* shows a geometric reflectance pattern dominated by backscattering, in which reflectance is highest when viewed from the same direction as the source of illumination. (This is in contrast to the intermediate and specular examples from Figure 1.17*a*, in which forward scattering predominates.) Many natural surfaces display this pattern of backscattering as a result of the differential shading (Figure 1.16*a*). In an image of a relatively uniform surface, there may be a localized area of increased brightness (known as a "hotspot"), located where the azimuth and zenith angles of the sensor are the same as those of the

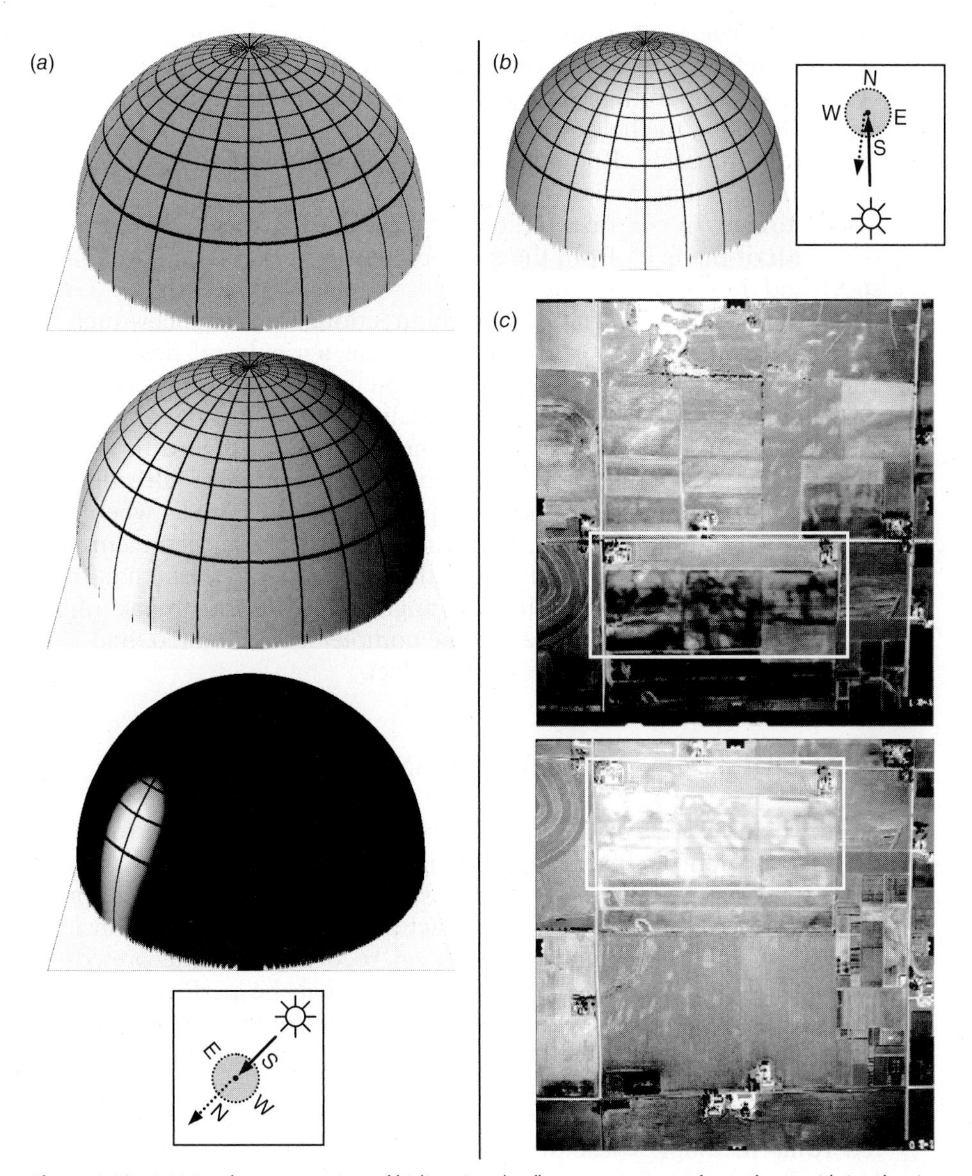

Figure 1.17 (*a*) Visual representation of bidirectional reflectance patterns, for surfaces with Lambertian (top), intermediate (middle), and specular (bottom) characteristics (after Campbell, 2002). (*b*) Simulated bidirectional reflectance from an agricultural field, showing a "hotspot" when viewed from the direction of solar illumination. (*c*) Differences in apparent reflectance in a field, when photographed from the north (top) and the south (bottom). (Author-prepared figure.)

sun. The existence of the hotspot is due to the fact that the sensor is then viewing only the sunlit portion of all objects in the area, without any shadowing.

An example of this type of hotspot is shown in Figure 1.17*c*. The two aerial photographs shown were taken mere seconds apart, along a single north–south flight line. The field delineated by the white box has a great difference in its apparent reflectance in the two images, despite the fact that no actual changes occurred on the ground during the short interval between the exposures. In the top photograph, the field was being viewed from the north, opposite the direction of solar illumination. Roughness of the field's surface results in differential shading, with the camera viewing the shadowed side of each small variation in the field's surface. In contrast, the bottom photograph was acquired from a point to the south of the field, from the same direction as the solar illumination (the hotspot), and thus appears quite bright.

To summarize, variations in bidirectional reflectance—such as specular reflection from a lake, or the hotspot in an agricultural field—can significantly affect the appearance of objects in remotely sensed images. These effects cause objects to appear brighter or darker solely as a result of the angular relationships among the sun, the object, and the sensor, without regard to any actual reflectance differences on the ground. Often, the impact of directional reflectance effects can be minimized by advance planning. For example, when photographing a lake when the sun is to the south and the lake's surface is calm, it may be preferable to take the photographs from the east or west, rather than from the north, to avoid the sun's specular reflection angle. However, the impact of varying bidirectional reflectance usually cannot be completely eliminated, and it is important for image analysts to be aware of this effect.

1.5 DATA ACQUISITION AND DIGITAL IMAGE CONCEPTS

To this point, we have discussed the principal sources of electromagnetic energy, the propagation of this energy through the atmosphere, and the interaction of this energy with earth surface features. These factors combine to produce energy "signals" from which we wish to extract information. We now consider the procedures by which these signals are detected, recorded, and interpreted.

The *detection* of electromagnetic energy can be performed in several ways. Before the development and adoption of electronic sensors, analog film-based cameras used chemical reactions on the surface of a light-sensitive film to detect energy variations within a scene. By developing a photographic film, we obtained a *record* of its detected signals. Thus, the film acted as both the detecting and the recording medium. These pre-digital photographic systems offered many advantages: They were relatively simple and inexpensive and provided a high degree of spatial detail and geometric integrity.

Electronic sensors generate an electrical signal that corresponds to the energy variations in the original scene. A familiar example of an electronic sensor is a handheld digital camera. Different types of electronic sensors have different designs of detectors, ranging from charge-coupled devices (CCDs, discussed in Chapter 2) to the antennas used to detect microwave signals (Chapter 6). Regardless of the type of detector, the resulting data are generally recorded onto some magnetic or optical computer storage medium, such as a hard drive, memory card, solid-state storage unit or optical disk. Although sometimes more complex and expensive than film-based systems, electronic sensors offer the advantages of a broader spectral range of sensitivity, improved calibration potential, and the ability to electronically store and transmit data.

In remote sensing, the term *photograph* historically was reserved exclusively for images that were *detected* as well as recorded on film. The more generic term *image* was adopted for any pictorial representation of image data. Thus, a pictorial record from a thermal scanner (an electronic sensor) would be called a "thermal image," *not* a "thermal photograph," because film would not be the original detection mechanism for the image. Because the term *image* relates to any pictorial product, all photographs are images. Not all images, however, are photographs.

A common exception to the above terminology is use of the term *digital photography*. As we describe in Section 2.5, digital cameras use electronic detectors rather than film for image detection. While this process is not "photography" in the traditional sense, "digital photography" is now the common way to refer to this technique of digital data collection.

We can see that the data interpretation aspects of remote sensing can involve analysis of pictorial (image) and/or digital data. *Visual interpretation* of pictorial image data has long been the most common form of remote sensing. Visual techniques make use of the excellent ability of the human mind to qualitatively evaluate spatial patterns in an image. The ability to make subjective judgments based on selected image elements is essential in many interpretation efforts. Later in this chapter, in Section 1.12, we discuss the process of visual image interpretation in detail.

Visual interpretation techniques have certain disadvantages, however, in that they may require extensive training and are labor intensive. In addition, *spectral characteristics* are not always fully evaluated in visual interpretation efforts. This is partly because of the limited ability of the eye to discern tonal values on an image and the difficulty of simultaneously analyzing numerous spectral images. In applications where spectral patterns are highly informative, it is therefore preferable to analyze *digital*, rather than pictorial, image data.

The basic character of digital image data is illustrated in Figure 1.18. Although the image shown in (*a*) appears to be a continuous-tone photograph, it is actually composed of a two-dimensional array of discrete *picture elements*, or *pixels*. The intensity of each pixel corresponds to the average brightness, or radiance, measured electronically over the ground area corresponding to each pixel. A total of 500 rows and 400 columns of pixels are shown in Figure 1.18*a*. Whereas the individual pixels are virtually impossible to discern in (*a*), they are readily observable in the enlargements shown in (*b*) and (*c*). These enlargements correspond to sub-areas located near the center of (*a*). A 100 row × 80 column enlargement is shown in (*b*) and a 10 row × 8 column enlargement is included in (*c*). Part (*d*) shows the individual *digital number (DN)*—also referred to as the

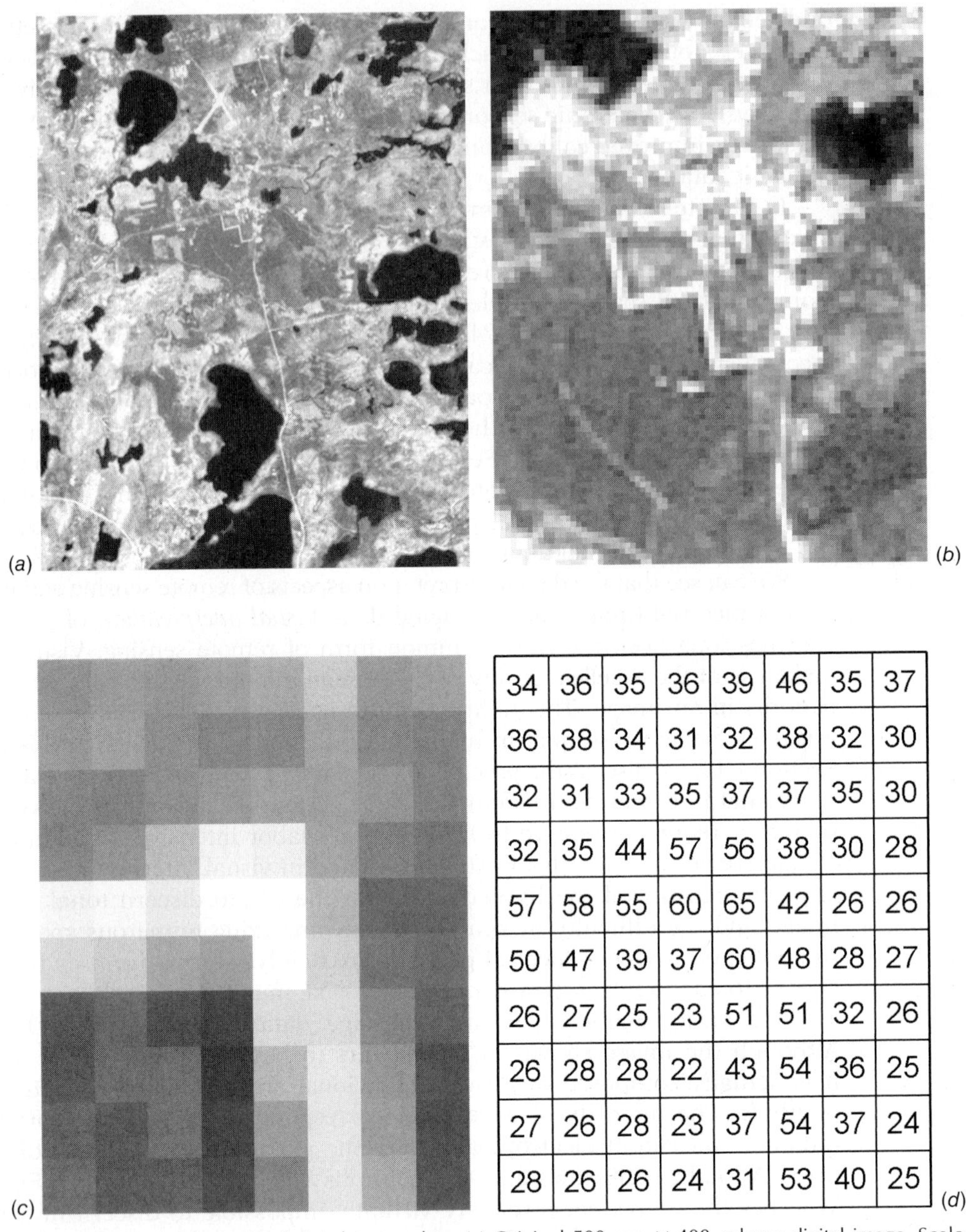

34	36	35	36	39	46	35	37
36	38	34	31	32	38	32	30
32	31	33	35	37	37	35	30
32	35	44	57	56	38	30	28
57	58	55	60	65	42	26	26
50	47	39	37	60	48	28	27
26	27	25	23	51	51	32	26
26	28	28	22	43	54	36	25
27	26	28	23	37	54	37	24
28	26	26	24	31	53	40	25

Figure 1.18 Basic character of digital image data. (*a*) Original 500 row × 400 column digital image. Scale 1:200,000. (*b*) Enlargement showing 100 row × 80 column area of pixels near center of (*a*). Scale 1:40,000. (*c*) 10 row × 8 column enlargement. Scale 1:4,000. (*d*) Digital numbers corresponding to the radiance of each pixel shown in (*c*). (Author-prepared figure.)

"brightness value" or "pixel value"—corresponding to the average radiance measured in each pixel shown in (*c*). These values result from quantizing the original electrical signal from the sensor into positive integer values using a process called *analog-to-digital (A-to-D) signal conversion*. (The A-to-D conversion process is discussed further in Chapter 4.)

Whether an image is acquired electronically or photographically, it may contain data from a single spectral band or from multiple spectral bands. The image shown in Figure 1.18 was acquired using a single broad spectral band, by integrating all energy measured across a range of wavelengths (a process analogous to photography using "black-and-white" film). Thus, in the digital image, there is a single DN for each pixel. It is also possible to collect "color" or *multispectral* imagery, whereby data are collected simultaneously in several spectral bands. In the case of a color photograph, three separate sets of detectors (or, for analog cameras, three layers within the film) each record radiance in a different range of wavelengths.

In the case of a digital multispectral image, each pixel includes multiple DNs, one for each spectral band. For example, as shown in Figure 1.19, one pixel in a digital image might have values of 88 in the first spectral band, perhaps representing blue wavelengths, 54 in the second band (green), 27 in the third (red), and so on, all associated with a single ground area.

When viewing this multi-band image, it is possible to view a single band at a time, treating it as if it were a discrete image, with brightness values proportional to DN as in Figure 1.18. Alternatively, and more commonly, three bands from the image can be selected and displayed simultaneously in shades of red, green, and blue, to create a *color composite* image, whether on a computer monitor or in a hard copy print. If the three bands being displayed were originally detected by the sensor in the red, green, and blue wavelength ranges of the visible spectrum, then this composite will be referred to as a *true-color* image, because it will approximate the natural combination of colors that would be seen by the human eye. Any other combination of bands—perhaps involving bands acquired in wavelengths outside the visible spectrum—will be referred to as a *false-color* image. One common false-color combination of spectral bands involves displaying near-IR, red, and green bands (from the sensor) in red, green, and blue, respectively, on the display device. Note that, in all cases, these three-band composite images involve displaying some combination of bands from the sensor in red, green, and blue on the display device because the human eye perceives color as a mixture of these three primary colors. (The principles of color perception and color mixing are described in more detail in Section 1.12.)

With multi-band digital data, the question arises of how to organize the data. In many cases, each band of data is stored as a separate file or as a separate block

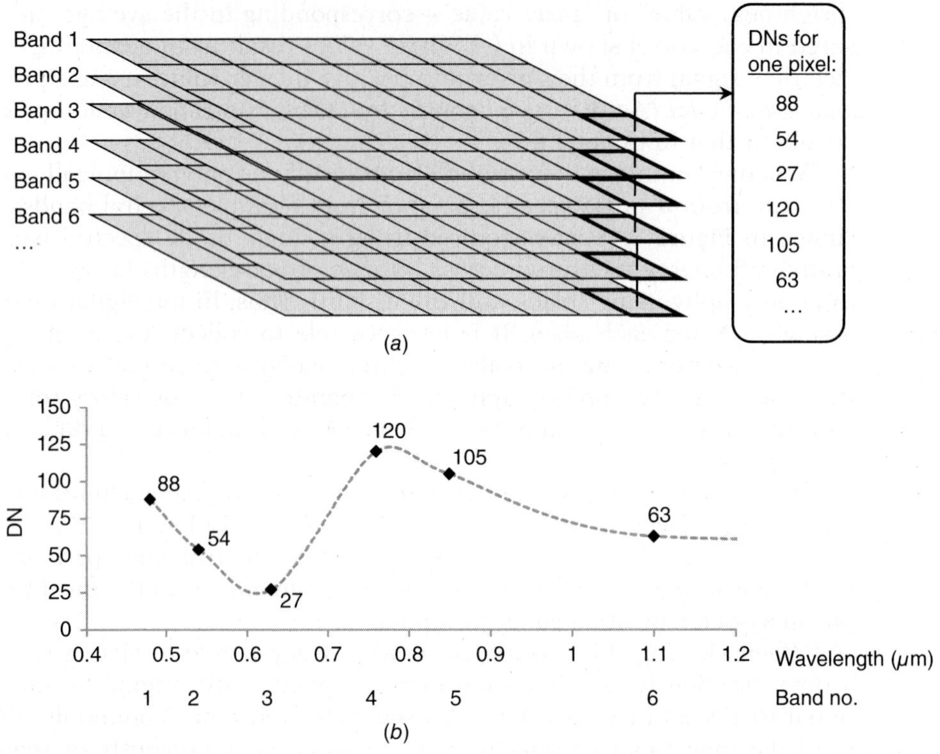

Figure 1.19 Basic character of multi-band digital image data. (*a*) Each band is represented by a grid of cells or pixels; any given pixel has a set of DNs representing its value in each band. (*b*) The spectral signature for the pixel highlighted in (*a*), showing band number and wavelength on the X axis and pixel DN on the Y axis. Values between the wavelengths of each spectral band, indicated by the dashed line in (*b*), are not measured by this sensor and would thus be unknown.

of data within a single file. This format is referred to as *band sequential* (*BSQ*) format. It has the advantage of simplicity, but it is often not the optimal choice for efficient display and visualization of data, because viewing even a small portion of the image requires reading multiple blocks of data from different "places" on the computer disk. For example, to view a true-color digital image in BSQ format, with separate files used to store the red, green, and blue spectral bands, it would be necessary for the computer to read blocks of data from three locations on the storage medium.

An alternate method for storing multi-band data utilizes the *band interleaved by line* (*BIL*) format. In this case, the image data file contains first a line of data from band 1, then the same line of data from band 2, and each subsequent band. This block of data consisting of the first line from each band is then followed by the second line of data from bands 1, 2, 3, and so forth.

The third common data storage format is *band interleaved by pixel* (*BIP*). This is perhaps the most widely used format for three-band images, such as those from most consumer-grade digital cameras. In this format, the file contains each band's measurement for the first pixel, then each band's measurement for the next pixel, and so on. The advantage of both BIL and BIP formats is that a computer can read and process the data for small portions of the image much more rapidly, because the data from all spectral bands are stored in closer proximity than in the BSQ format.

Typically, the DNs constituting a digital image are recorded over such numerical ranges as 0 to 255, 0 to 511, 0 to 1023, 0 to 2047, 0 to 4095 or higher. These ranges represent the set of integers that can be recorded using 8-, 9-, 10-, 11-, and 12-bit binary computer coding scales, respectively. (That is, $2^8 = 256$, $2^9 = 512$, $2^{10} = 1024$, $2^{11} = 2048$, and $2^{12} = 4096$.) The technical term for the number of bits used to store digital image data is *quantization level* (or *color depth*, when used to describe the number of bits used to display a color image). As discussed in Chapter 7, with the appropriate calibration coefficients these integer DNs can be converted to more meaningful physical units such as spectral reflectance, radiance, or normalized radar cross section.

Elevation Data

Increasingly, remote sensing instruments are used to collect three-dimensional spatial data, in which each observation has a *Z* coordinate representing elevation, along with the *X* and *Y* coordinates used to represent the horizontal position of the pixel's column and row. Particularly when collected over broad areas, these elevation data may represent the *topography*, the three-dimensional shape of the land surface. In other cases (usually, at finer spatial scales), these elevation data may represent the three-dimensional shapes of objects on or above the ground surface, such as tree crowns in a forest, or buildings in a city. Elevation data may be derived from the analysis of raw measurements from many types of remote sensing instruments, including photographic systems, multispectral sensors, radar systems, and lidar systems.

Elevation data may be represented in many different formats. Figure 1.20*a* shows a small portion of a traditional contour map, from the U.S. Geological Survey's 7.5-minute (1:24,000-scale) quadrangle map series. In this map, topographic elevations are indicated by contour lines. Closely spaced lines indicate steep terrain, while flat areas like river floodplains have more widely spaced contours.

Figure 1.20*b* shows a *digital elevation model* (*DEM*). Note that the white rectangle in (*b*) represents the much smaller area shown in (*a*). The DEM is similar to a digital image, with the DN at each pixel representing a surface elevation rather than a radiance value. In (*b*), the brightness of each pixel is represented as being proportional to its elevation, so light-toned areas are topographically higher and dark-toned areas are lower. The region shown in this map consists of highly dissected terrain, with a complex network of river valleys; one major valley runs

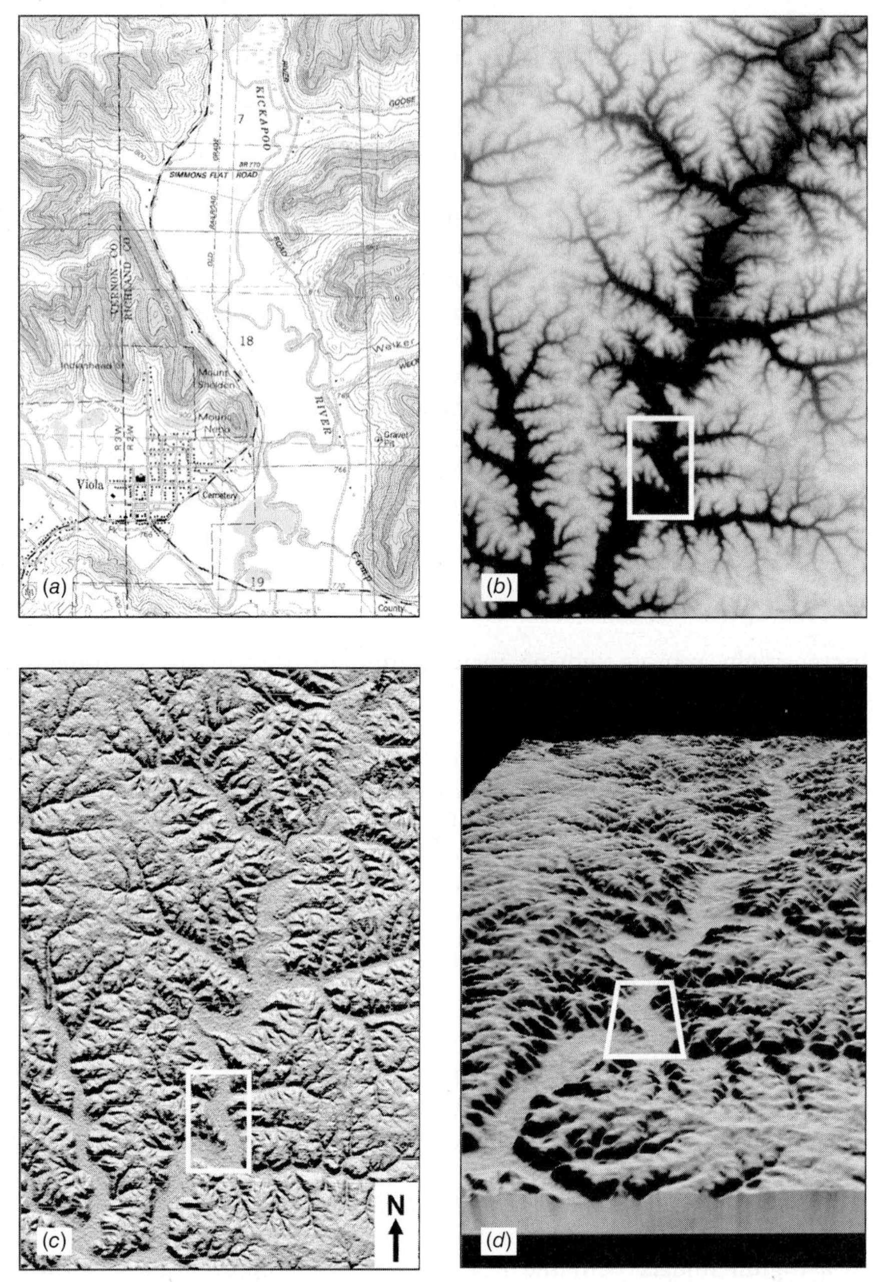

Figure 1.20 Representations of topographic data. (*a*) Portion of USGS 7.5-minute quadrangle map, showing elevation contours. Scale 1:45,000. (*b*) Digital elevation model, with brightness proportional to elevation. Scale 1:280,000. (*c*) Shaded-relief map derived from (*b*), with simulated illumination from the north. Scale 1:280,000. (*d*) Three-dimensional perspective view, with shading derived from (*c*). Scale varies in this projection. White rectangles in (*b*), (*c*), and (*d*) indicate area enlarged in (*a*). (Author-prepared figure.)

from the upper right portion of (*b*) to the lower center, with many tributary valleys branching off from each side.

Figure 1.20*c* shows another way of visualizing topographic data using *shaded relief*. This is a simulation of the pattern of shading that would be expected from a three-dimensional surface under a given set of illumination conditions. In this case, the simulation includes a primary source of illumination located to the north, with a moderate degree of diffuse illumination from other directions to soften the intensity of the shadows. Flat areas will have uniform tone in a shaded relief map. Slopes facing toward the simulated light source will appear bright, while slopes facing away from the light will appear darker.

To aid in visual interpretation, it is often preferable to create shaded relief maps with illumination from the top of the image, regardless of whether that is a direction from which solar illumination could actually come in the real world. When the illumination is from other directions, particularly from the bottom of the image, an untrained analyst may have difficulty correctly perceiving the landscape; in fact, the topography may appear inverted. (This effect is illustrated in Figure 1.29.)

Figure 1.20*d* shows yet another method for visualizing elevation data, a *three-dimensional perspective view*. In this example, the shaded relief map shown in (*c*) has been "draped" over the DEM, and a simulated view has been created based on a viewpoint located at a specified position in space (in this case, above and to the south of the area shown). This technique can be used to visualize the appearance of a landscape as seen from some point of interest. It is possible to "drape" other types of imagery over a DEM; perspective views created using an aerial photograph or high-resolution satellite image may appear quite realistic. Animation of successive perspective views created along a user-defined flight line permits the development of simulated "fly-throughs" over an area.

The term "digital elevation model" or DEM can be used to describe any image where the pixel values represent elevation (Z) coordinates. Two common subcategories of DEMs are a *digital terrain model* (*DTM*) and a *digital surface model* (*DSM*). A DTM (sometimes referred to as a "bald-earth DEM") records the elevation of the bare land surface, without any vegetation, buildings, or other features above the ground. In contrast, a DSM records the elevation of whatever the uppermost surface is at every location; this could be a tree crown, the roof of a building, or the ground surface (where no vegetation or structures are present). Each of these models has its appropriate uses. For example, a DTM would be useful for predicting runoff in a watershed after a rainstorm, because streams will flow over the ground surface rather than across the top of the forest canopy. In contrast, a DSM could be used to measure the size and shape of objects on the terrain, and to calculate intervisibility (whether a given point B can be seen from a reference point A).

Figure 1.21 compares a DSM and DTM for the same site, using airborne lidar data from the Capitol Forest area in Washington State (Andersen, McGaughey, and Reutebuch, 2005). In Figure 1.21*a*, the uppermost lidar points have been used

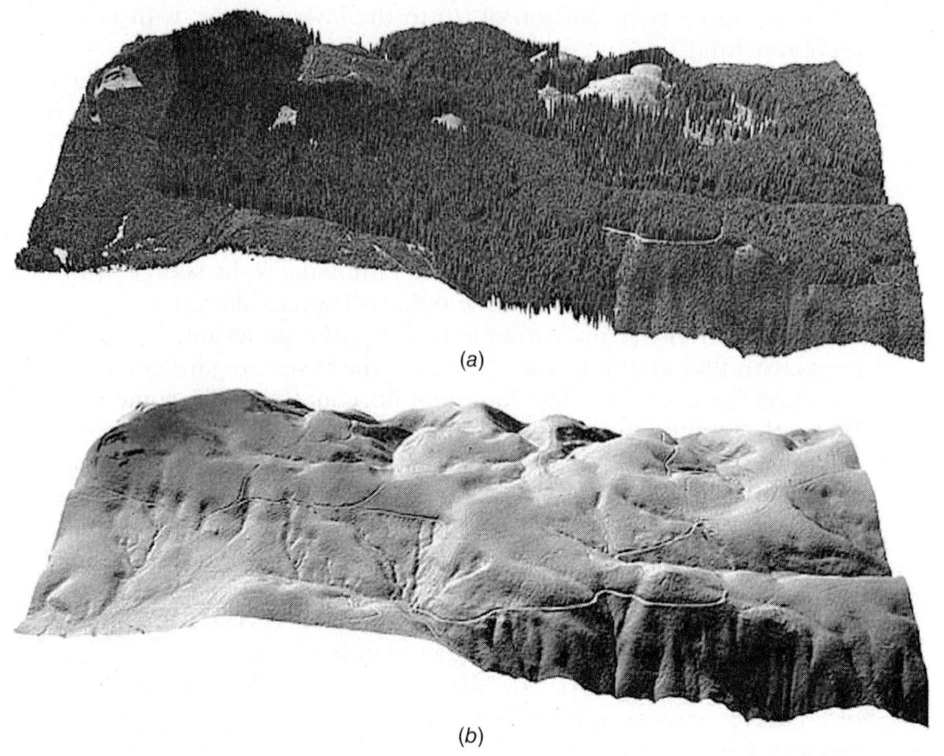

Figure 1.21 Airborne lidar data of the Capitol Forest site, Washington State. (a) Digital surface model (DSM) showing tops of tree crowns and canopy gaps. (*b*) Digital terrain model (DTM) showing hypothetical bare earth surface. (From Andersen et al., 2006; courtesy Ward Carlson, USDA Forest Service PNW Research Station.)

to create a DSM showing the elevation of the upper surface of the forest canopy, the presence of canopy gaps, and, in many cases, the shape of individual tree crowns. In Figure 1.21*b*, the lowermost points have been used to create a DTM, showing the underlying ground surface if all vegetation and structures were removed. Note the ability to detect fine-scale topographic features, such as small gullies and roadcuts, even underneath a dense forest canopy (Andersen et al., 2006).

Plate 1 shows a comparison of a DSM (*a*) and DTM (*b*) for a wooded area in New Hampshire. The models were derived from airborne lidar data acquired in early December. This site is dominated by a mix of evergreen and deciduous tree species, with the tallest (pines and hemlocks) exceeding 40 m in height. Scattered clearings in the center and right side are athletic fields, parkland, and former ski slopes now being taken over by shrubs and small trees. With obscuring vegetation removed, the DTM in (*b*) shows a variety of glacial and post-glacial landforms, as well as small roads, trails, and other constructed features. Also, by subtracting the elevations in (*b*) from those in (*a*), it is possible to calculate the height of the forest canopy above ground level at each point. The result, shown in (*c*), is referred to as a *canopy height model* (*CHM*). In this model, the ground surface has been flattened, so that all remaining variation represents differences in height of the trees relative to the ground. Lidar and other high-resolution 3D data are widely used for this type of canopy height analysis (Clark et al., 2004). (See Sections 6.23 and 6.24 for more discussion.)

Increasingly, elevation data are being used for analysis not just in the form of highly processed DEM, but in the more basic form of a *point cloud*. A point cloud is simply a data set containing many three-dimensional point locations, each representing a single measurement of the (X, Y, Z) coordinates of an object or surface. The positions, spacing, intensity, and other characteristics of the points in this cloud can be analyzed using sophisticated 3D processing algorithms to extract information about features (Rutzinger et al., 2008).

Further discussion of the acquisition, visualization, and analysis of elevation data, including DEMs and point clouds, can be found in Chapters 3 and 6, under the discussion of photogrammetry, interferometric radar, and lidar systems.

1.6 REFERENCE DATA

As we have indicated in the previous discussion, rarely, if ever, is remote sensing employed without the use of some form of *reference data*. The acquisition of reference data involves collecting measurements or observations about the objects, areas, or phenomena that are being sensed remotely. These data can take on any of a number of different forms and may be derived from a number of sources. For example, the data needed for a particular analysis might be derived from a soil survey map, a water quality laboratory report, or an aerial photograph. They may also stem from a "field check" on the identity, extent, and condition of agricultural crops, land uses, tree species, or water pollution problems. Reference data may also involve field measurements of temperature and other physical and/or chemical properties of various features. The geographic positions at which such field measurements are made are often noted on a map base to facilitate their location in a corresponding remote sensing image. Usually, GPS receivers are used to determine the precise geographic position of field observations and measurements (as described in Section 1.7).

Reference data are often referred to by the term *ground truth*. This term is not meant literally, because many forms of reference data are not collected on the ground and can only approximate the truth of actual ground conditions. For example, "ground" truth may be collected in the air, in the form of detailed aerial photographs used as reference data when analyzing less detailed high altitude or satellite imagery. Similarly, the "ground" truth will actually be "water" truth if we are studying water features. In spite of these inaccuracies, ground truth is a widely used term for reference data.

Reference data might be used to serve any or all of the following purposes:

1. To aid in the analysis and interpretation of remotely sensed data.
2. To calibrate a sensor.
3. To verify information extracted from remote sensing data.

Hence, reference data must be collected in accordance with the principles of statistical sampling design appropriate to the particular application.

Reference data can be very expensive and time consuming to collect properly. They can consist of either *time-critical* and/or *time-stable* measurements. Time-critical measurements are those made in cases where ground conditions change rapidly with time, such as in the analysis of vegetation condition or water pollution events. Time-stable measurements are involved when the materials under observation do not change appreciably with time. For example, geologic applications often entail field observations that can be conducted at any time and that would not change appreciably from mission to mission.

One form of reference data collection is the ground-based measurement of the reflectance and/or emittance of surface materials to determine their spectral response patterns. This might be done in the laboratory or in the field using the principles of *spectroscopy*. Spectroscopic measurement procedures can involve the use of a variety of instruments. Often, a *spectroradiometer* is used in such measurement procedures. This device measures, as a function of wavelength, the energy coming from an object within its view. It is used primarily to prepare spectral reflectance curves for various objects.

In laboratory spectroscopy, artificial sources of energy might be used to illuminate objects under study. In the laboratory, other field parameters such as viewing geometry between object and sensor are also simulated. More often, therefore, in situ field measurements are preferred because of the many variables of the natural environment that influence remote sensor data that are difficult, if not impossible, to duplicate in the laboratory.

In the acquisition of field measurements, spectroradiometers may be operated in a number of modes, ranging from handheld to helicopter or aircraft mounted. Figures 1.10 and 1.12, in Section 1.4 of this chapter, both contain examples of measurements acquired using a handheld field spectroradiometer. Figure 1.22 illustrates a highly portable instrument that is well suited for handheld operation. Through a fiber-optic input, this particular system acquires a continuous spectrum by recording data in over 1000 narrow bands simultaneously (over the range 0.35 to 2.5 μm). The unit is typically transported in a backpack carrier with provision for integrating the spectrometer with a notebook computer. The computer provides for flexibility in data acquisition, display, and storage. For example, reflectance spectra can be displayed in real

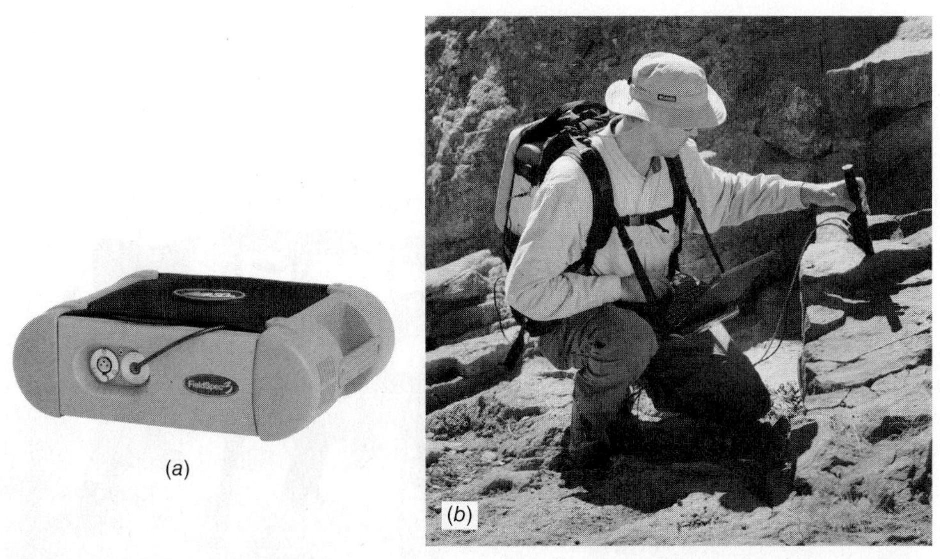

Figure 1.22 ASD, Inc. FieldSpec Spectroradiometer: (*a*) the instrument; (*b*) instrument shown in field operation. (Courtesy ASD, Inc.)

time, as can computed reflectance values within the wavelength bands of various satellite systems. In-field calculation of band ratios and other computed values is also possible. One such calculation might be the normalized difference vegetation index (NDVI), which relates the near-IR and visible reflectance of earth surface features (Chapter 7). Another option is matching measured spectra to a library of previously measured samples. The overall system is compatible with a number of post-processing software packages and also affords Ethernet, wireless, and GPS compatibility as well.

Figure 1.23 shows a versatile all-terrain instrument platform designed primarily for collecting spectral measurements in agricultural cropland environments. The system provides the high clearance necessary for making measurements over mature row crops, and the tracked wheels allow access to difficult landscape positions. Several measurement instruments can be suspended from the system's telescopic boom. Typically, these include a spectroradiometer, a remotely operated digital camera system, and a GPS receiver (Section 1.7). While designed primarily for data collection in agricultural fields, the long reach of the boom makes this device a useful tool for collecting spectral data over such targets as emergent vegetation found in wetlands as well as small trees and shrubs.

Using a spectroradiometer to obtain spectral reflectance measurements is normally a three-step process. First, the instrument is aimed at a *calibration panel* of known, stable reflectance. The purpose of this step is to quantify the

Figure 1.23 All-terrain instrument platform designed for collecting spectral measurements in agricultural cropland environments. (Courtesy of the University of Nebraska-Lincoln Center for Advanced Land Management Information Technologies.)

incoming radiation, or irradiance, incident upon the measurement site. Next, the instrument is suspended over the target of interest and the radiation reflected by the object is measured. Finally, the spectral reflectance of the object is computed by ratioing the reflected energy measurement in each band of observation to the incoming radiation measured in each band. Normally, the term *reflectance factor* is used to refer to the result of such computations. A reflectance factor is defined formally as the ratio of the radiant flux actually reflected by a sample surface to that which would be reflected into the same sensor geometry by an ideal, perfectly diffuse (Lambertian) surface irradiated in exactly the same way as the sample.

Another term frequently used to describe the above type of measurement is *bidirectional reflectance factor*: one direction being associated with the sample viewing angle (usually 0° from normal) and the other direction being that of the sun's illumination (defined by the solar zenith and azimuth angles; see Section 1.4). In the bidirectional reflectance measurement procedure described above, the sample and the reflectance standard are measured sequentially. Other approaches exist in which the incident spectral irradiance and reflected spectral radiance are measured simultaneously.

1.7 THE GLOBAL POSITIONING SYSTEM AND OTHER GLOBAL NAVIGATION SATELLITE SYSTEMS

As mentioned previously, the location of field-observed reference data is usually determined using a *global navigation satellite system* (*GNSS*). GNSS technology is also used extensively in such other remote sensing activities as navigating aircraft during sensor data acquisition and geometrically correcting and referencing raw image data. The first such system, the U.S. *Global Positioning System* (*GPS*) was originally developed for military purposes, but has subsequently become ubiquitous in many civil applications worldwide, from vehicle navigation to surveying, and location-based services on cellular phones and other personal electronic devices. Other GNSS "constellations" have been or are being developed as well, a trend that will greatly increase the accuracy and reliability of GNSS for end-users over the next decade.

The U.S. Global Positioning System includes at least 24 satellites rotating around the earth in precisely known orbits, with subgroups of four or more satellites operating in each of six different orbit planes. Typically, these satellites revolve around the earth approximately once every 12 hours, at an altitude of approximately 20,200 km. With their positions in space precisely known at all times, the satellites transmit time-encoded radio signals that are recorded by ground-based receivers and can be used to aid in positioning and navigation. The nearly circular orbital planes of the satellites are inclined about 60° from the equator and are spaced every 60° in longitude. This means that, in the absence of obstructions from the terrain or nearby buildings, an observer at any point on the earth's surface can receive the signal from at least four GPS satellites at any given time (day or night).

International Status of GNSS Development

Currently, the U.S. Global Positioning System has only one operational counterpart, the Russian *GLONASS* system. The full GLONASS constellation consists of 24 operational satellites, a number that was reached in October 2011. In addition, a fully comprehensive European GNSS constellation, *Galileo*, is scheduled for completion by 2020 and will include 30 satellites. The data signals provided by Galileo will be compatible with those from the U.S. GPS satellites, resulting in a greatly increased range of options for GNSS receivers and significantly improved accuracy. Finally, China has announced plans for the development of its own *Compass* GNSS constellation, to include 30 satellites in operational use by 2020. The future for these and similar systems is an extremely bright and rapidly progressing one.

GNSS Data Processing and Corrections

The means by which GNSS signals are used to determine ground positions is called *satellite ranging*. Conceptually, the process simply involves measuring the time required for signals transmitted by at least four satellites to reach the ground receiver. Knowing that the signals travel at the speed of light (3×10^8 m/sec in a vacuum), the distance from each satellite to the receiver can be computed using a form of three-dimensional triangulation. In principle, the signals from only four satellites are needed to identify the receiver's location, but in practice it is usually desirable to obtain measurements from as many satellites as practical.

GNSS measurements are potentially subject to numerous sources of error. These include *clock bias* (caused by imperfect synchronization between the high-precision atomic clocks present on the satellites and the lower-precision clocks used in GNSS receivers), uncertainties in the satellite orbits (known as *satellite ephemeris errors*), errors due to atmospheric conditions (signal velocity depends on time of day, season, and angular direction through the atmosphere), receiver errors (due to such influences as electrical noise and signal-matching errors), and multipath errors (reflection of a portion of the transmitted signal from objects not in the straight-line path between the satellite and receiver).

Such errors can be compensated for (in great part) using *differential* GNSS measurement methods. In this approach, simultaneous measurements are made by a stationary base station receiver (located over a point of precisely known position) and one (or more) roving receivers moving from point to point. The positional errors measured at the base station are used to refine the position measured by the rover(s) at the same instant in time. This can be done either by bringing the data from the base and rover together in a post-processing mode after the field observations are completed or by instantaneously broadcasting the base station corrections to the rovers. The latter approach is termed *real-time differential* GNSS positioning.

In recent years, there have been efforts to improve the accuracy of GNSS positioning through the development of regional networks of high-precision base stations, generally referred to as *satellite-based augmentation systems* (*SBAS*). The data from these stations are used to derive spatially explicit correction factors that are then broadcast in real time, allowing advanced receiver units to determine their positions with a higher degree of accuracy. One such SBAS network, the *Wide Area Augmentation System* (*WAAS*), consists of approximately 25 ground reference stations distributed across the United States that continuously monitor GPS satellite transmissions. Two main stations, located on the U.S. east and west coasts, collect the data from the reference stations and create a composited correction message that is location specific. This message is then broadcast through one of two *geostationary* satellites, satellites occupying a fixed position over the equator. Any WAAS-enabled GPS unit can receive these correction signals. The GPS receiver then determines which correction data are appropriate at the current location.

The WAAS signal reception is ideal for open land, aircraft, and marine applications, but the position of the relay satellites over the equator makes it difficult to receive the signals at high latitudes or when features such as trees and mountains obstruct the view of the horizon. In such situations, GPS positions can sometimes actually contain more error with WAAS correction than without. However, in unobstructed operating conditions where a strong WAAS signal is available, positions are normally accurate to within 3 m or better.

Paralleling the deployment of the WAAS system in North America are the Japanese *Multi-functional Satellite Augmentation System* (*MSAS*) in Asia, the *European Geostationary Navigation Overlay Service* (*EGNOS*) in Europe, and proposed future SBAS networks such as India's *GPS Aided Geo-Augmented Navigation* (*GAGAN*) system. Like WAAS, these SBAS systems use geostationary satellites to transmit data for real-time differential correction.

In addition to the regional SBAS real-time correction systems such as WAAS, some nations have developed additional networks of base stations that can be used for post-processing GNSS data for differential correction (i.e., high-accuracy corrections made after data collection, rather than in real time). One such system is the U.S. National Geodetic Survey's *Continuously Operating Reference Stations* (*CORS*) network. More than 1800 sites in the cooperative CORS network provide GNSS reference data that can be accessed via the Internet and used in post-processing for differential correction.

With the development of new satellite constellations, and new resources for real-time and post-processed differential correction, GNSS-based location services are expected to become even more widespread in industry, resource management, and consumer technology applications in the coming years.

1.8 CHARACTERISTICS OF REMOTE SENSING SYSTEMS

Having introduced some basic concepts, we now have the elements necessary to characterize a remote sensing system. In so doing, we can begin to appreciate some of the problems encountered in the design and application of the various sensing systems examined in subsequent chapters. In particular, the design and operation of every real-world sensing system represents a series of compromises, often in response to the limitations imposed by physics and by the current state of technological development. When we consider the process from start to finish, users of remote sensing systems need to keep in mind the following factors:

1. **The energy source.** All passive remote sensing systems rely on energy that originates from sources other than the sensor itself, typically in the form of either reflected radiation from the sun or emitted radiation from earth surface features. As already discussed, the spectral distribution of reflected sunlight and self-emitted energy is far from uniform. Solar energy levels obviously vary with respect to time and location, and different earth surface materials emit energy with varying degrees of efficiency. While we have some control over the sources of energy for active systems such as radar and lidar, those sources have their own particular

characteristics and limitations, as discussed in Chapter 6. Whether employing a passive or active system, the remote sensing analyst needs to keep in mind the nonuniformity and other characteristics of the energy source that provides illumination for the sensor.

2. **The atmosphere.** The atmosphere normally compounds the problems introduced by energy source variation. To some extent, the atmosphere always modifies the strength and spectral distribution of the energy received by a sensor. It restricts where we can look spectrally, and its effects vary with wavelength, time, and place. The importance of these effects, like source variation effects, is a function of the wavelengths involved, the sensor used, and the sensing application at hand. Elimination of, or compensation for, atmospheric effects via some form of calibration is particularly important in those applications where repetitive observations of the same geographic area are involved.
3. **The energy–matter interactions at the earth's surface.** Remote sensing would be simple if every material reflected and/or emitted energy in a unique, known way. Although spectral response patterns such as those in Figure 1.9 play a central role in detecting, identifying, and analyzing earth surface materials, the spectral world is full of ambiguity. Radically different material types can have great spectral similarity, making identification difficult. Furthermore, the general understanding of the energy–matter interactions for earth surface features is at an elementary level for some materials and virtually nonexistent for others.
4. **The sensor.** An ideal sensor would be highly sensitive to all wavelengths, yielding spatially detailed data on the absolute brightness (or radiance) from a scene as a function of wavelength, throughout the spectrum, across wide areas on the ground. This "supersensor" would be simple and reliable, require virtually no power or space, be available whenever and wherever needed, and be accurate and economical to operate. At this point, it should come as no surprise that an ideal "supersensor" does not exist. No single sensor is sensitive to all wavelengths or energy levels. All real sensors have fixed limits of spatial, spectral, radiometric and temporal resolution.

   The choice of a sensor for any given task always involves trade-offs. For example, photographic systems generally have very fine spatial resolution, providing a detailed view of the landscape, but they lack the broad spectral sensitivity obtainable with nonphotographic systems. Similarly, many nonphotographic systems are quite complex optically, mechanically, and/or electronically. They may have restrictive power, space, and stability requirements. These requirements often dictate the type of *platform*, or vehicle, from which a sensor can be operated. Platforms can range from stepladders to aircraft (fixed-wing or helicopters) to satellites.

   In recent years, *uninhabited aerial vehicles* (*UAVs*) have become an increasingly important platform for remote sensing data acquisition.

While the development of UAV technology for military applications has received a great deal of attention from the media, such systems are also ideally suited for many civilian applications, particularly in environmental monitoring, resource management, and infrastructure management (Laliberte et al., 2010). UAVs can range from palm-size radio-controlled airplanes and helicopters to large-size aircraft weighing tens of tons and controlled from thousands of km away. They also can be completely controlled through human intervention, or they can be partially or fully autonomous in their operation. Figure 1.24 shows two different types of UAVs used for environmental applications of remote sensing. In 1.24*(a)*, the Ikhana UAV is a fixed-wing aircraft based on the design of a military UAV but operated by NASA for civilian scientific research purposes. (The use of Ikhana for monitoring wildfires is discussed in Section 4.10, and imagery from this system is illustrated in Figure 4.34 and Plate 9.) In contrast, the UAV shown in 1.24*(b)* is a vertical takeoff UAV, designed in the form of a helicopter. In this photo the UAV is carrying a lightweight hyperspectral sensor used to map marine environments such as seagrass beds and coral reefs.

Depending on the sensor–platform combination needed in a particular application, the acquisition of remote sensing data can be a very expensive endeavor, and there may be limitations on the times and places that data can be collected. Airborne systems require detailed flight planning in advance, while data collection from satellites is limited by the platform's orbit characteristics.

5. **The data processing and supply system.** The capability of current remote sensors to generate data far exceeds the capacity to handle these data. This is generally true whether we consider "manual" image interpretation procedures or digital analyses. Processing sensor data into an interpretable format can be—and often is—an effort entailing considerable thought, hardware, time, and experience. Also, many data users would like to receive their data immediately after acquisition by the sensor in order to make the timely decisions required in certain applications (e.g., agricultural crop management, disaster assessment). Fortunately, the distribution of remote sensing imagery has improved dramatically over the past two decades. Some sources now provide in-flight data processing immediately following image acquisition, with near real-time data downloaded over the Internet. In some cases, users may work with imagery and other spatial data in a *cloud computing environment*, where the data and/or software are stored remotely, perhaps even distributed widely across the Internet. At the opposite extreme—particularly for highly specialized types of imagery or for experimental or newly developed remote sensing systems—it may take weeks or months before data are made available, and the user may need to acquire not just the data but highly specialized or custom software for data processing. Finally, as discussed

Figure 1.24 Uninhabited aerial vehicles (UAVs) used for environmental applications of remote sensing. (*a*) NASA's Ikhana UAV, with imaging sensor in pod under left wing. (Photo courtesy NASA Dryden Flight Research Center and Jim Ross.) (*b*) A vertical takeoff UAV mapping seagrass and coral reef environments in Florida. (Photo courtesy Rick Holasek and NovaSol.)

in Section 1.6, most remote sensing applications require the collection and analysis of additional reference data, an operation that may be complex, expensive, and time consuming.

6. **The users of remotely sensed data.** Central to the successful application of any remote sensing system is the person (or persons) using the remote sensor data from that system. The "data" generated by remote sensing procedures become "information" only if and when someone understands their generation, knows how to interpret them, and knows how best to use them. *A thorough understanding of the problem at hand is paramount to the productive application of any remote sensing methodology. Also, no single combination of data acquisition and analysis procedures will satisfy the needs of all data users.*

Whereas the interpretation of aerial photography has been used as a practical resource management tool for nearly a century, other forms of remote sensing are relatively new, technically complex, and unconventional means of acquiring information. In earlier years, these newer forms of remote sensing had relatively few satisfied users. Since the late 1990s, however, as new applications continue to be developed and implemented, increasing numbers of users are becoming aware of the potentials, as well as the limitations, of remote sensing techniques. As a result, remote sensing has become an essential tool in many aspects of science, government, and business alike.

One factor in the increasing acceptance of remote sensing imagery by end-users has been the development and widespread adoption of easy-to-use geovisualization systems such as Google Maps and Earth, NASA's WorldWinds, and other web-based image services. By allowing more potential users to become comfortable and familiar with the day-to-day use of aerial and satellite imagery, these and other software tools have facilitated the expansion of remote sensing into new application areas.

1.9 SUCCESSFUL APPLICATION OF REMOTE SENSING

The student should now begin to appreciate that successful use of remote sensing is premised on the *integration* of multiple, interrelated data sources and analysis procedures. No single combination of sensor and interpretation procedure is appropriate to all applications. The key to designing a successful remote sensing effort involves, at a minimum, (1) clear definition of the problem at hand, (2) evaluation of the potential for addressing the problem with remote sensing techniques, (3) identification of the remote sensing data acquisition procedures appropriate to the task, (4) determination of the data interpretation procedures to be employed and the reference data needed, and (5) identification of the criteria by which the quality of information collected can be judged.

All too often, one (or more) of the above components of a remote sensing application is overlooked. The result may be disastrous. Many programs exist with little or no means of evaluating the performance of remote sensing systems in terms of information quality. Many people have acquired burgeoning quantities of remote sensing data with inadequate capability to interpret them. In some cases an inappropriate decision to use (or *not* to use) remote sensing has been made, because the problem was not clearly defined and the constraints or opportunities associated with remote sensing methods were not clearly understood. A clear articulation of the information requirements of a particular problem and the extent to which remote sensing might meet these requirements in a timely manner is paramount to any successful application.

The success of many applications of remote sensing is improved considerably by taking a *multiple-view* approach to data collection. This may involve *multistage* sensing, wherein data about a site are collected from multiple altitudes. It may involve *multispectral* sensing, whereby data are acquired simultaneously in several spectral bands. Or, it may entail *multitemporal* sensing, where data about a site are collected on more than one occasion.

In the multistage approach, satellite data may be analyzed in conjunction with high altitude data, low altitude data, and ground observations (Figure 1.25). Each successive data source might provide more detailed information over smaller geographic areas. Information extracted at any lower level of observation may then be extrapolated to higher levels of observation.

A commonplace example of the application of multistage sensing techniques is the detection, identification, and analysis of forest disease and insect problems. From space images, the image analyst could obtain an overall view of the major vegetation categories involved in a study area. Using this information, the areal extent and position of a particular species of interest could be determined and representative subareas could be studied more closely at a more refined stage of imaging. Areas exhibiting stress on the second-stage imagery could be delineated. Representative samples of these areas could then be field checked to document the presence and particular cause of the stress.

After analyzing the problem in detail by ground observation, the analyst would use the remotely sensed data to extrapolate assessments beyond the small study areas. By analyzing the large-area remotely sensed data, the analyst can determine the severity and geographic extent of the disease problem. Thus, while the question of specifically *what* the problem is can generally be evaluated only by detailed ground observation, the equally important questions of *where*, *how much*, and *how severe* can often be best handled by remote sensing analysis.

In short, more information is obtained by analyzing multiple views of the terrain than by analysis of any single view. In a similar vein, multispectral imagery provides more information than data collected in any single spectral band. When the signals recorded in the multiple bands are analyzed in conjunction with each other, more information becomes available than if only a single band were used or if the multiple bands were analyzed independently. The multispectral approach

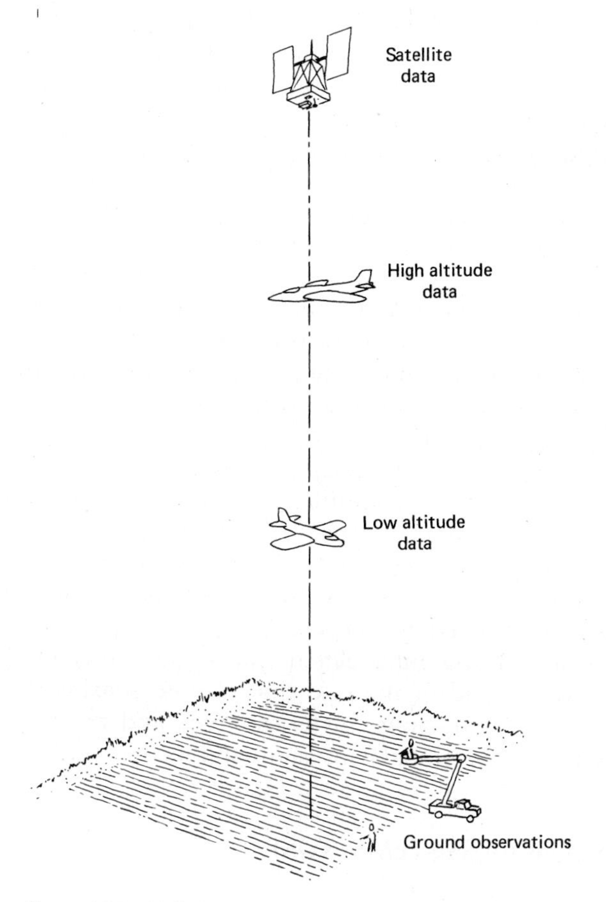

Figure 1.25 Multistage remote sensing concept.

forms the heart of numerous remote sensing applications involving discrimination of earth resource types, cultural features, and their condition.

Again, multitemporal sensing involves sensing the same area at multiple times and using changes occurring with time as discriminants of ground conditions. This approach is frequently taken to monitor land use change, such as suburban development in urban fringe areas. In fact, regional land use surveys might call for the acquisition of multisensor, multispectral, multistage, multitemporal data to be used for multiple purposes!

In any approach to applying remote sensing, not only must the right mix of data acquisition and data interpretation techniques be chosen, but the right mix of remote sensing and "conventional" techniques must also be identified. The student must recognize that remote sensing is a tool best applied in concert with others; it is not an end in itself. In this regard, remote sensing data are currently being used extensively in computer-based GISs (Section 1.10). The GIS environment permits the synthesis, analysis, and communication of virtually unlimited sources and types of biophysical and socioeconomic data—as long as they can be geographically referenced. Remote sensing can be thought of as the "eyes" of such systems providing repeated, synoptic (even global) visions of earth resources from an aerial or space vantage point.

Remote sensing affords us the capability to literally see the invisible. We can begin to see components of the environment on an ecosystem basis, in that remote sensing data can transcend the cultural boundaries within which much of our current resource data are collected. Remote sensing also transcends disciplinary boundaries. It is so broad in its application that nobody "owns" the field. Important contributions are made to—and benefits derived from—remote sensing by both the "hard" scientist interested in basic research and the "soft" scientist interested in its operational application.

There is little question that remote sensing will continue to play an increasingly broad and important role in the scientific, governmental, and commercial sectors alike. The technical capabilities of sensors, space platforms, data communication and distribution systems, GPSs, digital image processing systems, and GISs are improving on almost a daily basis. At the same time, we are witnessing the evolution of a spatially enabled world society. Most importantly, we are becoming increasingly aware of how interrelated and fragile the elements of our global resource base really are and of the role that remote sensing can play in inventorying, monitoring, and managing earth resources and in modeling and helping us to better understand the global ecosystem and its dynamics.

1.10 GEOGRAPHIC INFORMATION SYSTEMS (GIS)

We anticipate that the majority of individuals using this book will at some point in their educational backgrounds and/or professional careers have experience with geographic information systems. The discussion below is provided as a brief introduction to such systems primarily for those readers who might lack such background.

Geographic information systems are computer-based systems that can deal with virtually any type of information about features that can be referenced by geographical location. These systems are capable of handling both *locational data* and *attribute data* about such features. That is, not only do GISs permit the automated mapping or display of the locations of features, but also these systems provide a capability for recording and analyzing descriptive characteristics ("attributes") of the features. For example, a GIS might contain not only a map of the locations of roads but also a database of descriptors about each road. These attributes might include information such as road width, pavement type, speed limit, number of traffic lanes, date of construction, and so on. Table 1.1 lists other examples of attributes that might be associated with a given point, line, or area feature.

TABLE 1.1 Example Point, Line, and Area Features and Typical Attributes Contained in a GIS[a]

Point feature	Well (depth, chemical constituency)
Line feature	Power line (service capacity, age, insulator type)
Area feature	Soil mapping unit (soil type, texture, color, permeability)

[a]Attributes shown in parentheses.

The data in a GIS may be kept in individual standalone files (e.g., "shapefiles"), but increasingly a *geodatabase* is used to store and manage spatial data. This is a type of *relational database*, consisting of tables with attributes in columns and data records in rows (Table 1.2), and explicitly including locational information for each record. While database implementations vary, there are certain desirable characteristics that will improve the utility of a database in a GIS. These characteristics include flexibility, to allow a wide range of database queries and operations; reliability, to avoid accidental loss of data; security, to limit access to authorized users; and ease of use, to insulate the end user from the details of the database implementation.

One of the most important benefits of a GIS is the ability to spatially interrelate multiple types of information stemming from a range of sources. This concept is illustrated in Figure 1.26, where we have assumed that a hydrologist wishes to use a GIS to study soil erosion in a watershed. As shown, the system contains data from a range of source maps (*a*) that have been geocoded on a cell-by-cell basis to form a series of *layers* (*b*), all in geographic registration. The analyst can then manipulate and overlay the information contained in, or derived from, the various data layers. In this example, we assume that assessing the potential for soil erosion throughout the watershed involves the simultaneous cell-by-cell consideration of three types of data derived from the original data layers: slope, soil erodibility, and surface runoff potential. The slope information can be computed from the elevations in the topography layer. The erodibility, which is an attribute associated with each soil type, can be extracted from a relational database management system incorporated in the GIS. Similarly, the runoff potential is an attribute associated with each land cover type (land cover data can

TABLE 1.2 Relational Database Table Format

ID Number[a]	Street Name	Lanes	Parking	Repair Date	. . .
143897834	"Maple Ct"	2	Yes	2012/06/10	. . .
637292842	"North St"	2	Seasonal	2006/08/22	. . .
347348279	"Main St"	4	Yes	2015/05/15	. . .
234538020	"Madison Ave"	4	No	2014/04/20	. . .

[a]Each data record, or "tuple," has a unique identification, or ID, number.

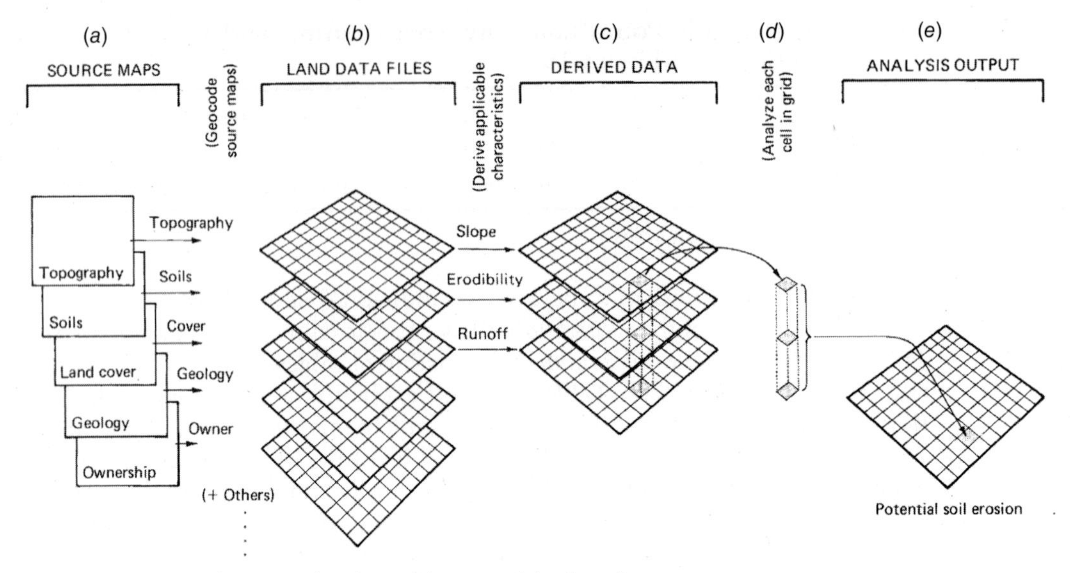

Figure 1.26 GIS analysis procedure for studying potential soil erosion.

be obtained through interpretation of aerial photographs or satellite images). The analyst can use the system to interrelate these three sources of derived data (*c*) in each grid cell and use the result to locate, display, and/or record areas where combinations of site characteristics indicate high soil erosion potential (i.e., steep slopes and highly erodible soil–land cover combinations).

The above example illustrates the GIS analysis function commonly referred to as *overlay analysis*. The number, form, and complexity of other data analyses possible with a GIS are virtually limitless. Such procedures can operate on the system's spatial data, the attribute data, or both. For example, *aggregation* is an operation that permits combining detailed map categories to create new, less detailed categories (e.g., combining "jack pine" and "red pine" categories into a single "pine" category). *Buffering* creates a zone of specified width around one or more features (e.g., the area within 50 m of a stream). *Network analysis* permits such determinations as finding the shortest path through a street network, determining the stream flows in a drainage basin, or finding the optimum location for a fire station. *Intervisibility* operations use elevation data to permit *viewshed mapping* of what terrain features can be "seen" from a specified location. Similarly, many GISs permit the generation of *perspective views* portraying terrain surfaces from a viewing position other than vertical.

One constraint on the use of multiple layers in a GIS is that the spatial scale at which each of the original source maps was compiled must be compatible with the others. For example, in the analysis shown in Figure 1.26 it would be inappropriate to incorporate both soil data from very high resolution aerial photographs of a single township and land cover digitized from a highly generalized map of the entire nation. Another common constraint is that the compilation dates of different source maps must be reasonably close in time. For example, a GIS analysis of wildlife habitat might yield incorrect conclusions if it were based on land cover data that are many years out of date. On the other hand, since other types of spatial data are less changeable over time, the map compilation date might not be as important for a layer such as topography or bedrock geology.

Most GISs use two primary approaches to represent the locational component of geographic information: a *raster* (grid-based) or *vector* (point-based) format. The raster data model that was used in our soil erosion example is illustrated in Figure 1.27*b*. In this approach, the location of geographic objects or conditions is defined by the row and column position of the cells they occupy. The value stored for each cell indicates the type of object or condition that is found at that location over the entire cell. Note that the finer the grid cell size used, the more geographic specificity there is in the data file. A coarse grid requires less data storage space but will provide a less precise geographic description of the original data. Also, when using a very coarse grid, several data types and/or attributes may occur in each cell, but the cell is still usually treated as a single homogeneous unit during analysis.

The vector data model is illustrated in Figure 1.27*c*. Using this format, feature boundaries are converted to straight-sided polygons that approximate the original regions. These polygons are encoded by determining the coordinates of their vertices, called *points* or *nodes*, which can be connected to form *lines* or *arcs*. *Topological coding* includes "intelligence" in the data structure relative to the spatial relationship (connectivity and adjacency) among features. For example, topological coding keeps track of which arcs share common nodes and what polygons are to the left and right of a given arc. This information facilitates such spatial operations as overlay analysis, buffering, and network analysis.

Raster and vector data models each have their advantages and disadvantages. Raster systems tend to have simpler data structures; they afford greater computational efficiency in such operations as overlay analysis; and they represent

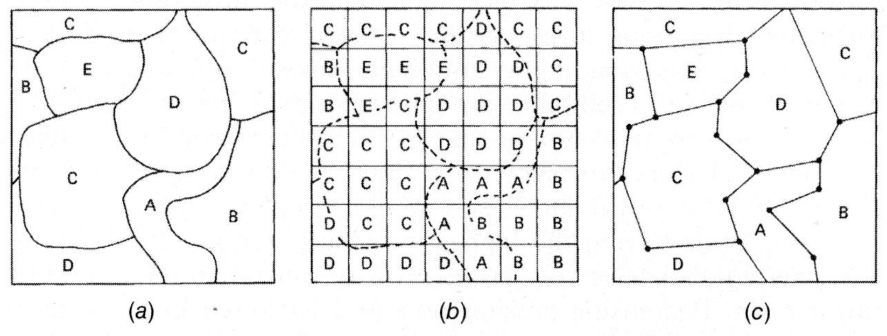

Figure 1.27 Raster versus vector data formats: (*a*) landscape patches, (*b*) landscape represented in raster format, and (*c*) landscape represented in vector format.

features having high spatial variability and/or "blurred boundaries" (e.g., between pure and mixed vegetation zones) more effectively. On the other hand, raster data volumes are relatively greater; the spatial resolution of the data is limited to the size of the cells comprising the raster; and the topological relationships among spatial features are more difficult to represent. Vector data formats have the advantages of relatively lower data volumes, better spatial resolution, and the preservation of topological data relationships (making such operations as network analysis more efficient). However, certain operations (e.g., overlay analysis) are more complex computationally in a vector data format than in a raster format.

As we discuss frequently throughout this book, digital remote sensing images are collected in a raster format. Accordingly, digital images are inherently compatible spatially with other sources of information in a raster domain. Because of this, "raw" images can be easily included directly as layers in a raster-based GIS. Likewise, such image processing procedures as automated land cover classification (Chapter 7) result in the creation of interpreted or derived data files in a raster format. These derived data are again inherently compatible with the other sources of data represented in a raster format. This concept is illustrated in Plate 2, in which we return to our earlier example of using overlay analysis to assist in soil erosion potential mapping for an area in western Dane County, Wisconsin. Shown in (*a*) is an automated land cover classification that was produced by processing Landsat Thematic Mapper (TM) data of the area. (See Chapter 7 for additional information on computer-based land cover classification.) To assess the soil erosion potential in this area, the land cover data were merged with information on the intrinsic erodibility of the soil present (*b*) and with land surface slope information (*c*). These latter forms of information were already resident in a GIS covering the area. Hence, all data could be combined for analysis in a mathematical model, producing the soil erosion potential map shown in (*d*). To assist the viewer in interpreting the landscape patterns shown in Plate 2, the GIS was also used to visually enhance the four data sets with topographic shading based on a DEM, providing a three-dimensional appearance.

For the land cover classification in Plate 2*a*, water is shown as dark blue, nonforested wetlands as light blue, forested wetlands as pink, corn as orange, other row crops as pale yellow, forage crops as olive, meadows and grasslands as yellow-green, deciduous forest as green, evergreen forest as dark green, low-intensity urban areas as light gray, and high-intensity urban areas as dark gray. In (*b*), areas of low soil erodibility are shown in dark brown, with increasing soil erodibility indicated by colors ranging from orange to tan. In (*c*), areas of increasing steepness of slope are shown as green, yellow, orange, and red. The soil erosion potential map (*d*) shows seven colors depicting seven levels of potential soil erosion. Areas having the highest erosion potential are shown in dark red. These areas tend to have row crops growing on inherently erodible soils with sloping terrain. Decreasing erosion potential is shown in a spectrum of colors from orange through yellow to green. Areas with the lowest erosion potential are indicated in dark green. These include forested regions, continuous-cover crops, and grasslands growing on soils with low inherent erodibility, and flat terrain.

Remote sensing images (and information extracted from such images), along with GPS data, have become primary data sources for modern GISs. Indeed, the boundaries between remote sensing, GIS, and GPS technology have become blurred, and these combined fields will continue to revolutionize how we inventory, monitor, and manage natural resources on a day-to-day basis. Likewise, these technologies are assisting us in modeling and understanding biophysical processes at all scales of inquiry. They are also permitting us to develop and communicate cause-and-effect "what-if" scenarios in a spatial context in ways never before possible.

1.11 SPATIAL DATA FRAMEWORKS FOR GIS AND REMOTE SENSING

If one is examining an image purely on its own, with no reference to any outside source of spatial information, there may be no need to consider the type of coordinate system used to represent locations within the image. In many cases, however, analysts will be comparing points in the image to GPS-located reference data, looking for differences between two images of the same area, or importing an image into a GIS for quantitative analysis. In all these cases, it is necessary to know how the column and row coordinates of the image relate to some real-world map coordinate system.

Because the shape of the earth is approximately spherical, locations on the earth's surface are often described in an angular coordinate or *geographical* system, with latitude and longitude specified in degrees (°), minutes (′), and seconds (″). This system originated in ancient Greece, and it is familiar to many people today. Unfortunately, the calculation of distances and areas in an angular coordinate system is complex. More significantly, it is impossible to accurately represent the three-dimensional surface of the earth on the two-dimensional planar surface of a map or image without introducing distortion in one or more of the following elements: shape, size, distance, and direction. Thus, for many purposes we wish to mathematically transform angular geographical coordinates into a planar, or Cartesian ($X-Y$) coordinate system. The result of this transformation process is referred to as a *map projection*.

While many types of map projections have been defined, they can be grouped into several broad categories based either on the geometric models used or on the spatial properties that are preserved or distorted by the transformation. Geometric models for map projection include cylindrical, conic, and azimuthal or planar surfaces. From a map user's perspective, the spatial properties of map projections may be more important than the geometric model used. A *conformal* map projection preserves angular relationships, or shapes, within local areas; over large areas, angles and shapes become distorted. An *azimuthal* (or *zenithal*) projection preserves absolute directions relative to the central point of projection. An *equidistant* projection preserves equal distances, for some but not all points—scale is constant either for all distances along meridians or for all distances from one or two points. An *equal-area* (or equivalent) projection preserves equal areas. Because a detailed explanation of the relationships among these properties is beyond the scope of this discussion, suffice it to say that no two-dimensional map projection can accurately preserve all of these properties, but certain subsets of these characteristics can be preserved in a single projection. For example, the azimuthal equidistant projection preserves both direction and distance—but only relative to the central point of the projection; directions and distances between other points are not preserved.

In addition to the map projection associated with a given image, GIS data layer, or other spatial data set, it is also often necessary to consider the *datum* used with that map projection. A datum is a mathematical definition of the three-dimensional solid (generally a slightly flattened ellipsoid) used to represent the surface of the earth. The actual planet itself has an irregular shape that does not correspond perfectly to any ellipsoid. As a result, a variety of different datums have been described; some designed to fit the surface well in one particular region (such as the North American Datum of 1983, or NAD83) and others designed to best approximate the planet as a whole. Most GISs require that both a map projection and a datum be specified before performing any coordinate transformations.

To apply these concepts to the process of collecting and working with remotely sensed images, most such images are initially acquired with rows and columns of pixels aligned with the flight path (or orbit track) of the imaging platform, be it a satellite, an aircraft, or a UAV. Before the images can be mapped, or used in combination with other spatial data, they need to be *georeferenced*. Historically, this process typically involved identification of visible control points whose true geographic coordinates were known. A mathematical model would then be used to transform the row and column coordinates of the raw image into a defined map coordinate system. In recent years, remote sensing platforms have been outfitted with sophisticated systems to record their exact position and angular orientation. These systems, incorporating an *inertial measurement unit* (*IMU*) and/or multiple onboard GPS units, enable highly precise modeling of the viewing geometry of the sensor, which in turn is used for *direct georeferencing* of the sensor data—relating them to a defined map projection without the necessity of additional ground control points.

Once an image has been georeferenced, it may be ready for use with other spatial information. On the other hand, some images may have further geometric distortions, perhaps caused by varying terrain, or other factors. To remove these distortions, it may be necessary to orthorectify the imagery, a process discussed in Chapters 3 and 7.

1.12 VISUAL IMAGE INTERPRETATION

When we look at aerial and space images, we see various objects of different sizes, shapes, and colors. Some of these objects may be readily identifiable while others may not, depending on our own individual perceptions and experience. When we can identify what we see on the images and communicate this information to others, we are practicing *visual image interpretation*. The images contain raw image *data*. These data, when processed by a human interpreter's brain, become usable *information*.

Image interpretation is best learned through the experience of viewing hundreds of remotely sensed images, supplemented by a close familiarity with the environment and processes being observed. Given this fact, no textbook alone can fully train its readers in image interpretation. Nonetheless, Chapters 2 through 8 of this book contain many examples of remote sensing images, examples that we hope our readers will peruse and interpret. To aid in that process, the remainder of this chapter presents an overview of the principles and methods typically employed in image interpretation.

Aerial and space images contain a detailed record of features on the ground at the time of data acquisition. An image interpreter systematically examines the images and, frequently, other supporting materials such as maps and reports of field observations. Based on this study, an interpretation is made as to the physical nature of objects and phenomena appearing in the images. Interpretations may take place at a number of levels of complexity, from the simple recognition of objects on the earth's surface to the derivation of detailed information regarding the complex interactions among earth surface and subsurface features. Success in image interpretation varies with the training and experience of the interpreter, the nature of the objects or phenomena being interpreted, and the quality of the images being utilized. Generally, the most capable image interpreters have keen powers of observation coupled with imagination and a great deal of patience. In addition, it is important that the interpreter have a thorough understanding of the phenomenon being studied as well as knowledge of the geographic region under study.

Elements of Image Interpretation

Although most individuals have had substantial experience in interpreting "conventional" photographs in their daily lives (e.g., newspaper photographs), the interpretation of aerial and space images often departs from everyday image interpretation in three important respects: (1) the portrayal of features from an overhead, often unfamiliar, vertical perspective; (2) the frequent use of wavelengths outside of the visible portion of the spectrum; and (3) the depiction of the earth's surface at unfamiliar scales and resolutions (Campbell and Wynne, 2011). While these factors may be insignificant to the experienced image interpreter, they can represent a substantial challenge to the novice image analyst! However, even this challenge continues to be mitigated by the extensive use of aerial and space imagery in such day-to-day activities as navigation, GIS applications, and weather forecasting.

A systematic study of aerial and space images usually involves several basic characteristics of features shown on an image. The exact characteristics useful for any specific task and the manner in which they are considered depend on the field of application. However, most applications consider the following basic characteristics, or variations of them: shape, size, pattern, tone (or hue), texture, shadows, site, association, and spatial resolution (Olson, 1960).

Shape refers to the general form, configuration, or outline of individual objects. In the case of stereoscopic images, the object's *height* also defines its shape. The shape of some objects is so distinctive that their images may be identified solely from this criterion. The Pentagon building near Washington, DC, is a classic example. All shapes are obviously not this diagnostic, but every shape is of some significance to the image interpreter.

Size of objects on images must be considered in the context of the image scale. A small storage shed, for example, might be misinterpreted as a barn if size were not considered. Relative sizes among objects on images of the same scale must also be considered.

Pattern relates to the spatial arrangement of objects. The repetition of certain general forms or relationships is characteristic of many objects, both natural and constructed, and gives objects a pattern that aids the image interpreter in recognizing them. For example, the ordered spatial arrangement of trees in an orchard is in distinct contrast to that of natural forest tree stands.

Tone (or *hue*) refers to the relative brightness or color of objects on an image. Figure 1.8 showed how relative photo tones could be used to distinguish between deciduous and coniferous trees on black and white infrared photographs. Without differences in tone or hue, the shapes, patterns, and textures of objects could not be discerned.

Texture is the frequency of tonal change on an image. Texture is produced by an aggregation of unit features that may be too small to be discerned individually on the image, such as tree leaves and leaf shadows. It is a product of their individual shape, size, pattern, shadow, and tone. It determines the overall visual "smoothness" or "coarseness" of image features. As the scale of the image is reduced, the texture of any given object or area becomes progressively finer and ultimately disappears. An interpreter can often distinguish between features with similar reflectances based on their texture differences. An example would be the smooth texture of green grass as contrasted with the rough texture of green tree crowns on medium-scale airphotos.

Shadows are important to interpreters in two opposing respects: (1) The shape or outline of a shadow affords an impression of the profile view of objects (which aids interpretation) and (2) objects within shadows reflect little light and are difficult to discern on an image (which hinders interpretation). For example, the shadows cast by various tree species or cultural features (bridges, silos, towers, poles, etc.) can definitely aid in their identification on airphotos. In some cases, the shadows cast by large animals can aid in their identification. Figure 1.28 is a large-scale aerial photograph taken under low sun angle conditions that shows camels and their shadows in Saudi Arabia. Note that the camels themselves can be seen at the "base" of their shadows. Without the shadows, the animals could be counted, but identifying them specifically as camels could be difficult. Also, the shadows resulting from subtle variations in terrain elevations, especially in the case of low sun angle images, can aid in assessing natural topographic variations that may be diagnostic of various geologic landforms.

As a general rule, the shape of the terrain is more easily interpreted when shadows fall toward the observer. This is especially true when images are examined monoscopically, where relief cannot be seen directly, as it can be in stereoscopic images. In Figure 1.29*a*, a large ridge with numerous side valleys can be seen in the center of the image. When this image is inverted (i.e., turned such that the shadows fall away from the observer), as in (*b*), the result is a confusing image that almost seems to have a valley of sorts running through the center of the image (from bottom to top). This arises because one "expects" light sources to generally be above objects (ASPRS, 1997, p. 73). The orientation of shadows with respect to the observer is less important for interpreting images of buildings, trees, or animals (as in Figure 1.28) than for interpreting the terrain.

Site refers to topographic or geographic location and is a particularly important aid in the identification of vegetation types. For example, certain tree species would be expected to occur on well-drained upland sites, whereas other tree species would be expected to occur on poorly drained lowland sites. Also, various tree species occur only in certain geographic areas (e.g., redwoods occur in California, but not in Indiana).

Figure 1.28 Vertical aerial photograph showing camels that cast long shadows under a low sun angle in Saudi Arabia. Black-and-white rendition of color original. (© George Steinmetz/Corbis.)

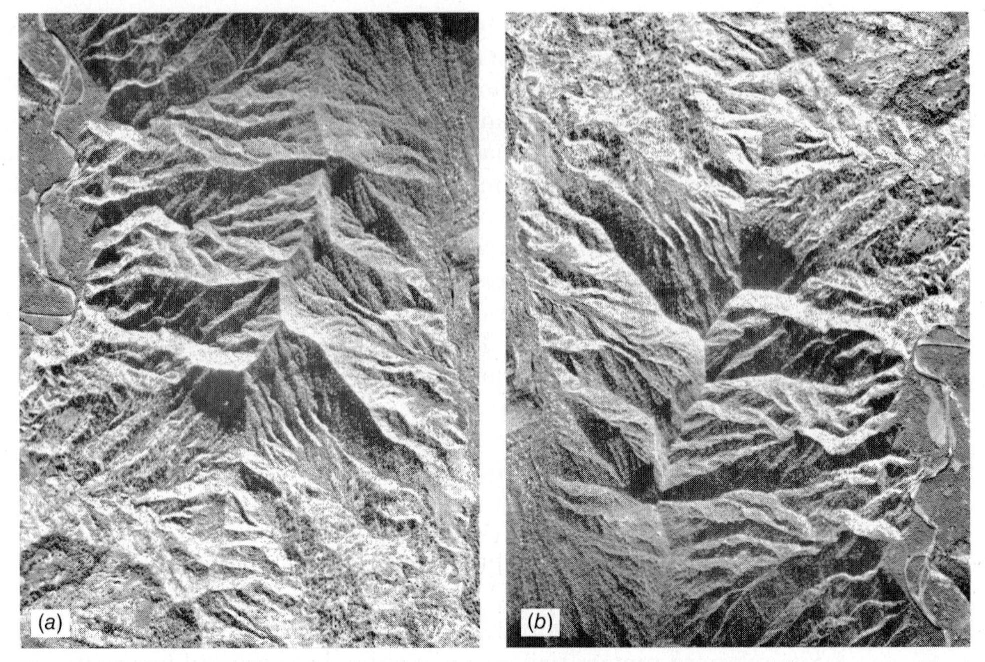

Figure 1.29 Photograph illustrating the effect of shadow direction on the interpretability of terrain. Island of Kauai, Hawaii, mid-January. Scale 1:48,000. (*a*) Shadows falling toward observer, (*b*) same image turned such that shadows are falling away from observer. (Courtesy USDA–ASCS panchromatic photograph.)

Association refers to the occurrence of certain features in relation to others. For example, a Ferris wheel might be difficult to identify if standing in a field near a barn but would be easy to identify if in an area recognized as an amusement park.

Spatial resolution depends on many factors, but it always places a practical limit on interpretation because some objects are too small or have too little contrast with their surroundings to be clearly seen on the image.

Other factors, such as image scale, spectral resolution, radiometric resolution, date of acquisition, and even the condition of images (e.g., torn or faded historical photographic prints) also affect the success of image interpretation activities.

Many of the above elements of image interpretation are illustrated in Figure 1.30. Figure 1.30*a* is a nearly full-scale copy of a 230 × 230-mm airphoto that was produced at an original scale of 1:28,000 (or 1 cm = 280 m). Parts (*b*) through (*e*) of Figure 1.30 show four scenes extracted and enlarged from this airphoto. Among the land cover types in Figure 1.30*b* are water, trees, suburban houses, grass, a divided highway, and a drive-in theater. Most of the land cover types are easily identified in this figure. The drive-in theater could be difficult for inexperienced interpreters to identify, but a careful study of the elements of image interpretation leads to its identification. It has a unique *shape* and *pattern*. Its *size* is consistent with a drive-in theater (note the relative size of the cars on

Figure 1.30 Aerial photographic subscenes illustrating the elements of image interpretation, Detroit Lakes area, Minnesota, mid-October. (*a*) Portion of original photograph of scale 1:32,000; (*b*) and (*c*) enlarged to a scale of 1:4,600; (*d*) enlarged to a scale of 1:16,500; (*e*) enlarged to a scale of 1:25,500. North is to the bottom of the page. (Courtesy KBM, Inc.)

Figure 1.30 *(Continued)*

the highway and the parking spaces of the theater). In addition to the curved rows of the parking area, the *pattern* also shows the projection building and the screen. The identification of the screen is aided by its *shadow*. It is located in *association* with a divided highway, which is accessed by a short roadway.

Many different land cover types can be seen in Figure 1.30*c*. Immediately noticeable in this photograph, at upper left, is a feature with a superficially similar appearance to the drive-in theater. Careful examination of this feature, and the surrounding grassy area, leads to the conclusion that this is a baseball diamond. The trees that can be seen in numerous places in the photograph are casting shadows of their trunks and branches because the mid-October date of this photograph is a time when deciduous trees are in a leaf-off condition. Seen in the right one-third of the photograph is a residential area. Running top to bottom through the center of the

Figure 1.30 (*Continued*)

image is a commercial area with buildings that have a larger size than the houses in the residential area and large parking areas surrounding these larger buildings.

Figure 1.30*d* shows two major linear features. Near the bottom of the photograph is a divided highway. Running diagonally from upper left to lower right is an airport runway 1390 m long (the scale of this figure is 1:16,500, and the length of this linear feature is 8.42 cm at this scale). The terminal area for this airport is near the bottom center of Figure 1.30*d*.

Figure 1.30*e* illustrates natural versus constructed features. The water body at *a* is a natural feature, with an irregular shoreline and some surrounding wetland areas (especially visible at the narrow end of the lake). The water body at *b* is part of a sewage treatment plant; the "shoreline" of this feature has unnaturally straight sections in comparison with the water body shown at *a*.

Image Interpretation Strategies

As previously mentioned, the image interpretation process can involve various levels of complexity, from a simple direct recognition of objects in the scene to the inference of site conditions. An example of direct recognition would be the identification of a highway interchange. Assuming the interpreter has some experience with the vertical perspective of aerial and space images, recognition of a highway interchange should be a straightforward process. On the other hand, it may often be necessary to infer, rather than directly observe, the characteristics of features based on their appearance on images. In the case of a buried gas pipeline, for example, the actual pipeline cannot be seen, but there are often changes at the ground surface caused by the buried pipeline that are visible on aerial and space images. Soils are typically better drained over the pipeline because of the sand and gravel used for backfill, and the presence of a buried pipeline can often be inferred by the appearance of a light-toned linear streak across the image. Also, the interpreter can take into account the probability of certain ground cover types occurring on certain dates at certain places. Knowledge of the crop development stages (*crop calendar*) for an area would determine if a particular crop is likely to be visible on a particular date. For example, corn, peas, and winter wheat would each have a significant vegetative ground cover on different dates. Likewise, in a particular growing region, one crop type may be present over a geographic area many times larger than that of another crop type; therefore, the probability of occurrence of one crop type would be much greater than another.

In a sense, the image interpretation process is like the work of a detective trying to put all the pieces of evidence together to solve a mystery. For the interpreter, the mystery might be presented in terms of trying to understand why certain areas in an agricultural field look different from the rest of that field. At the most general level, the interpreter must recognize the area under study as an agricultural field. Beyond this, consideration might be made as to whether the crop present in the field is a row crop (e.g., corn) or a continuous cover crop (e.g., alfalfa). Based on the crop calendar and regional growing conditions, a decision might be made that the crop is indeed corn, rather than another row crop, such as soybeans. Furthermore, it might be noted that the anomalously appearing areas in the field are associated with areas of slightly higher topographic relief relative to the rest of the field. With knowledge of the recent local weather conditions, the interpreter might infer that the anomalously appearing areas are associated with drier soil conditions and the corn in these areas is likely drought stressed. Hence, the interpreter uses the process of *convergence of evidence* to successively increase the accuracy and detail of the interpretation.

Image Interpretation Keys

The image interpretation process can often be facilitated through the use of *image interpretation keys*. Keys can be valuable training aids for novice interpreters and provide useful reference or refresher materials for more experienced interpreters. An image interpretation key helps the interpreter evaluate the information presented on aerial and space images in an organized and consistent manner. It provides guidance about the correct identification of features or conditions on the images. Ideally, a key consists of two basic parts: (1) a collection of annotated or captioned images (preferably stereopairs) illustrative of the features or conditions to be identified and (2) a graphic or word description that sets forth in some systematic fashion the image recognition characteristics of those features or conditions. Two general types of image interpretation keys exist, differentiated by the method of presentation of diagnostic features. A *selective key* contains numerous example images with supporting text. The interpreter selects the example that most nearly resembles the feature or condition found on the image under study.

An *elimination key* is arranged so that the interpretation proceeds step by step from the general to the specific and leads to the elimination of all features or conditions except the one being identified. Elimination keys often take the form of *dichotomous keys* where the interpreter makes a series of choices between two alternatives and progressively eliminates all but one possible answer. Figure 1.31 shows a dichotomous key prepared for the identification of fruit and nut crops in the Sacramento Valley, California. The use of elimination keys can lead to more positive

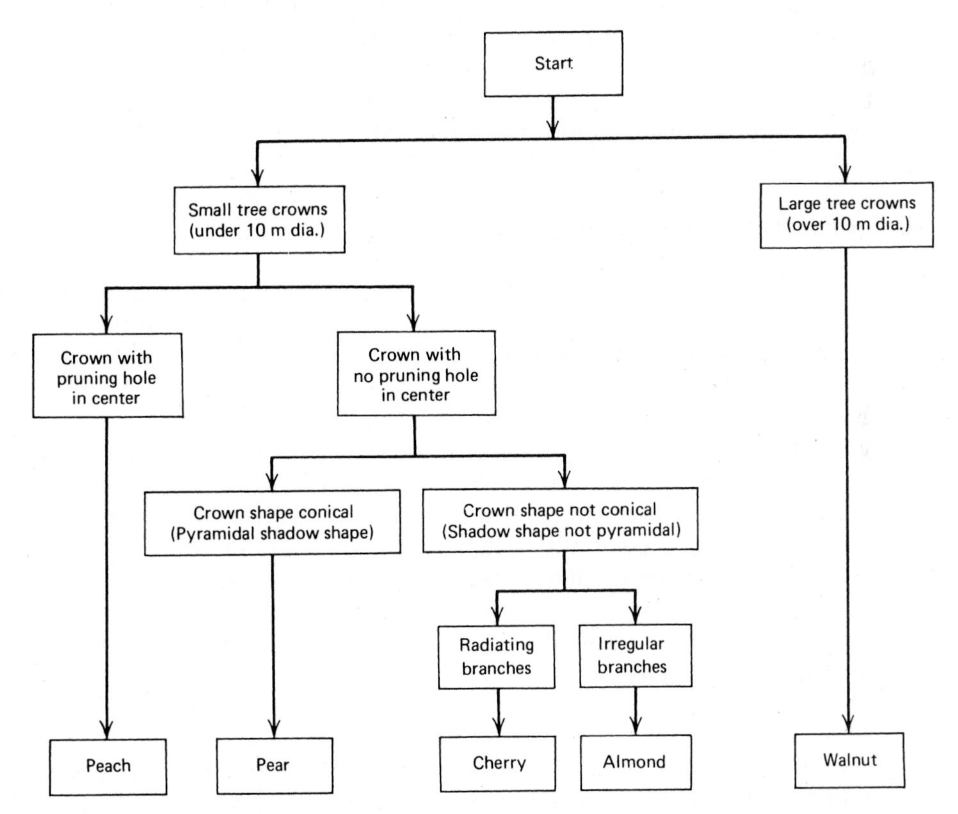

Figure 1.31 Dichotomous airphoto interpretation key to fruit and nut crops in the Sacramento Valley, CA, designed for use with 1:6,000 scale panchromatic aerial photographs. (Adapted from American Society of Photogrammetry, 1983. Copyright © 1975, American Society of Photogrammetry. Reproduced with permission.)

answers than selective keys but may result in erroneous answers if the interpreter is forced to make an uncertain choice between two unfamiliar image characteristics.

As a generalization, keys are more easily constructed and more reliably utilized for cultural feature identification (e.g., houses, bridges, roads, water towers) than for vegetation or landform identification. However, a number of keys have been successfully employed for agricultural crop identification and tree species identification. *Such keys are normally developed and used on a region-by-region and season-by-season basis because the appearance of vegetation can vary widely with location and season.*

Wavelengths of Sensing

The band(s) of the electromagnetic energy spectrum selected for aerial and space imaging affects the amount of information that can be interpreted from the images. Numerous examples of this are scattered throughout this book. The general concepts of multiband imagery were discussed in Section 1.5. To explain how the combinations of colors shown in an image relate to the various bands of data recorded by a sensor, we next turn our attention to the principles of color, and how combinations of colors are perceived.

Color Perception and Color Mixing

Color is among the most important elements of image interpretation. Many features and phenomena in an image can best be identified and interpreted through examination of subtle differences in color. As discussed in Section 1.5, multi-wavelength remote sensing images may be displayed in either true- or false-color combinations. Particularly for interpreting false-color imagery, an understanding of the principles of color perception and color mixing is essential.

Light falling on the retina of the human eye is sensed by rod and cone cells. There are about 130 million rod cells, and they are 1000 times more light sensitive than the cone cells. When light levels are low, human vision relies on the rod cells to form images. All rod cells have the same wavelength sensitivity, which peaks at about 0.55 μm. Therefore, human vision at low light levels is monochromatic. It is the cone cells that determine the colors the eye sees. There are about 7 million cone cells; some sensitive to blue energy, some to green energy, and some to red energy. The *trichromatic theory of color vision* explains that when the blue-sensitive, green-sensitive, and red-sensitive cone cells are stimulated by different amounts of light, we perceive color. When all three types of cone cells are stimulated equally, we perceive white light. Other theories of color vision have been proposed. The *opponent process of color vision* hypothesizes that color vision involves three mechanisms, each responding to a pair of so-called opposites: white–black, red–green, and blue–yellow. This theory is based on many psychophysical observations and states that colors are formed by a *hue cancellation method*. The hue cancellation method is based on the observation that when certain colors are mixed together, the resulting colors are not what would be intuitively expected. For example, when red and green are mixed together, they produce yellow, not reddish green. (For further information, see Robinson et al., 1995.)

In the remainder of this discussion, we focus on the trichromatic theory of color vision. Again, this theory is based on the concept that we perceive all colors by synthesizing various amounts of just three (blue, green, and red).

Blue, green, and red are termed *additive primaries*. Plate 3*a* shows the effect of projecting blue, green, and red light in partial superimposition. Where all three beams overlap, the visual effect is white because all three of the eyes' receptor systems are stimulated equally. Hence, white light can be thought of as the mixture of blue, green, and red light. Various combinations of the three additive primaries can be used to produce other colors. As illustrated, when red light and green light are mixed, yellow light is produced. Mixture of blue and red light results in the production of magenta light (bluish-red). Mixing blue and green results in cyan light (bluish-green).

Yellow, magenta, and cyan are known as the *complementary colors*, or *complements*, of blue, green, and red light. Note that the complementary color for any given primary color results from mixing the remaining two primaries.

Like the eye, color television and computer monitors operate on the principle of additive color mixing through use of blue, green, and red elements on the screen. When viewed at a distance, the light from the closely spaced screen elements forms a continuous color image.

Whereas color television and computer monitors simulate different colors through *additive* mixture of blue, green, and red *lights*, color film photography is based on the principle of *subtractive* color mixture using superimposed yellow, magenta, and cyan *dyes*. These three dye colors are termed the *subtractive primaries*, and each results from subtracting one of the additive primaries from white light. That is, yellow dye absorbs the blue component of white light. Magenta dye absorbs the green component of white light. Cyan dye absorbs the red component of white light.

The subtractive color-mixing process is illustrated in Plate 3*b*. This plate shows three circular filters being held in front of a source of white light. The filters contain yellow, magenta, and cyan dye. The yellow dye absorbs blue light from the white background and transmits green and red. The magenta dye absorbs green light and transmits blue and red. The cyan dye absorbs red light and transmits blue and green. The superimposition of magenta and cyan dyes results in the passage of only blue light from the background. This comes about because the magenta dye absorbs the green component of the white background, and the cyan dye absorbs the red component. Superimposition of the yellow and cyan dyes results in the perception of green. Likewise, superimposition of yellow and magenta dyes results in the perception of red. Where all three dyes overlap, all light from the white background is absorbed and black results.

In color film photography, and in color printing, various proportions of yellow, magenta, and cyan dye are superimposed to control the proportionate amount of blue, green, and red light that reaches the eye. Hence, the subtractive mixture of yellow, magenta, and cyan dyes on a photograph is used to control the additive mixture of blue, green, and red light reaching the eye of the observer. To accomplish this, color film is manufactured with three emulsion layers that are sensitive to blue, green, and red light but contain yellow, magenta, and cyan dye after processing (see Section 2.4).

In digital color photography, the photosites in the detector array are typically covered with a blue, green, or red filter, resulting in independent recording of the three additive primary colors (see Section 2.5).

When interpreting color images, the analyst should keep in mind the relationship between the color of a feature in the imagery, the color mixing process that would produce that color, and the sensor's wavelength ranges that are assigned to the three primary colors used in that mixing process. It is then possible to work backwards to infer the spectral properties of the feature on the landscape. For example, if a feature in a false-color image has a yellow hue when displayed on a computer monitor, that feature can be assumed to have a relatively high reflectance in the wavelengths that are being displayed in the monitor's red and green color planes, and a relatively low reflectance in the wavelength displayed in blue (because yellow results from the additive combination of red plus green light). Knowing the spectral sensitivity of each of the sensor's spectral bands, the analyst can interpret the color as a spectral response pattern (see Section 1.4) that is characterized by high reflectance at two wavelength ranges and low reflectance at a third. The analyst can then use this information to draw inferences about the nature and condition of the feature shown in the false-color image.

Temporal Aspects of Image Interpretation

The temporal aspects of natural phenomena are important for image interpretation because such factors as vegetative growth and soil moisture vary during the year. For crop identification, more positive results can be achieved by obtaining images at several times during the annual growing cycle. Observations of local vegetation emergence and recession can aid in the timing of image acquisition for natural vegetation mapping. In addition to seasonal variations, weather can cause significant short-term changes. Because soil moisture conditions may change dramatically during the day or two immediately following a rainstorm, the timing of image acquisition for soil studies is very critical.

Another temporal aspect of importance is the comparison of *leaf-off* photography with *leaf-on* photography. Leaf-off conditions are preferable for applications in which it is important to be able to see as much detail as possible underneath trees. Such applications include activities such as topographic mapping and urban feature identification. Leaf-on conditions are preferred for vegetation mapping. Figure 1.32*a* illustrates leaf-on photography. Here considerable detail of the ground surface underneath the tree crowns is obscured. Figure 1.32*b*

Figure 1.32 Comparison of leaf-on photography with leaf-off photography. Gladstone, OR. (*a*) Leaf-on photograph exposed in summer. (*b*) Leaf-off photograph exposed in spring. Scale 1:1,500. (Courtesy Oregon Metro.)

illustrates leaf-off photography. Here, the ground surface underneath the tree crowns is much more visible than in the leaf-on photograph. Because leaf-off photographs are typically exposed in spring or fall, there are longer shadows in the image than with leaf-on photography, which is typically exposed in summer. (Shadow length also varies with time of day.) Also, these photographs illustrate leaf-on and leaf-off conditions in an urban area where most of the trees are deciduous and drop their leaves in fall (leaf-off conditions). The evergreen trees in the images (e.g., lower right) maintain their needles throughout the year and cast dark shadows. Hence, there would not be leaf-off conditions for such trees.

Image Spatial Resolution and Ground Sample Distance

Every remote sensing system has a limit on how small an object on the earth's surface can be and still be "seen" by a sensor as being separate from its surroundings. This limit, called the *spatial resolution* of a sensor, is an indication of how well a sensor can record spatial detail. In some cases, the *ground sample distance* (*GSD*), or ground area represented by a single pixel in a digital image, may correspond closely to the spatial resolution of that sensor. In other cases, the ground sample distance may be larger or smaller than the sensor's spatial resolution, perhaps as a result of the A-to-D conversion process or of digital image manipulation such as resampling (see Chapter 7 and Appendix B). This distinction between spatial resolution and ground sample distance is subtle but important. For the sake of simplicity, the following discussion treats the GSD in a digital image as being equivalent to the spatial resolution of the sensor that produced the image, but note that in actual images the sampling distance may be larger or smaller than the spatial resolution.

Figure 1.33 illustrates, in the context of a digital image, the interplay between the spatial resolution of a sensor and the spatial variability present in a ground scene. In (*a*), a single pixel covers only a small area of the ground (on the order of

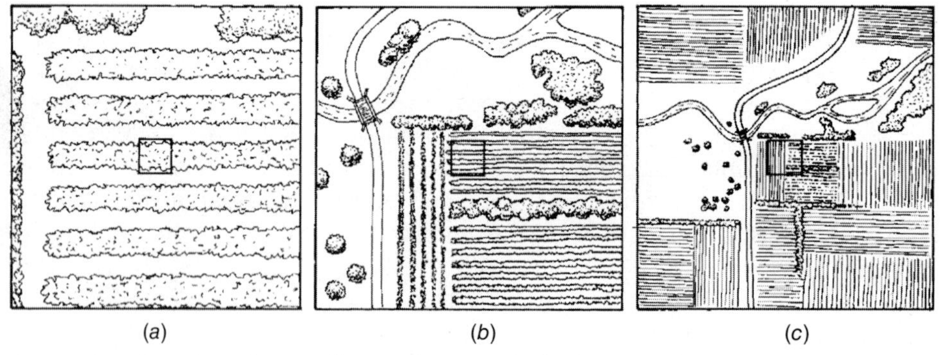

Figure 1.33 Ground resolution cell size effect: (*a*) small, (*b*) intermediate, and (*c*) large ground resolution cell size.

the width of the rows of the crop shown). In (*b*), a coarser ground resolution is depicted and a single pixel integrates the radiance from both the crop rows and the soil between them. In (*c*), an even coarser resolution results in a pixel measuring the average radiance over portions of two fields. Thus, depending on the spatial resolution of the sensor and the spatial structure of the ground area being sensed, digital images comprise a range of "pure" and "mixed" pixels. In general, the larger the percentage of mixed pixels, the more limited is the ability to record and extract spatial detail in an image. This is illustrated in Figure 1.34, in which the same area has been imaged over a range of different ground resolution cell sizes.

Further discussion of the spatial resolution of remote sensing systems—including the factors that determine spatial resolution and the methods used for measuring or calculating a system's resolution—can be found in Chapter 2 (for camera systems), Chapters 4 and 5 (for airborne and spaceborne multispectral and thermal sensors), and Chapter 6 (for radar systems).

Other Forms of Resolution Important in Image Interpretation

It should be noted that there are other forms of resolution that are important characteristics of remote sensing images. These include the following:

Spectral resolution, referring to a sensor's ability to distinguish among different ground features based on their spectral properties. Spectral resolution depends upon the number, wavelength location, and narrowness of the spectral bands in which a sensor collects image data. The bands in which any sensor collects data can range from a single broad band (for *panchromatic* images), a few broad bands (for *multispectral* images), or many very narrow bands (for *hyperspectral* images).

Radiometric resolution, referring to the sensor's ability to differentiate among subtle variations in brightness. Does the sensor divide the range from the "brightest" pixel to "darkest" pixel that can be recorded in an image (the *dynamic range*) into 256, or 512, or 1024 gray level values? The finer the radiometric resolution is, the greater the quality and interpretability of an image. (See also the discussion of quantization and digital numbers in Section 1.5.)

Temporal resolution, referring to the ability to detect changes over shorter or longer periods of time. Most often, this term is used in reference to a sensor that produces a time-series of multiple images. This could be a satellite system with a defined 16-day or 26-day orbital repeat cycle, or a tripod-mounted camera with a timer that collects one image every hour to serve as reference data. The importance of rapid and/or repeated coverage of an area varies dramatically with the application at hand. For example, in disaster response applications, temporal resolution might outweigh the importance of some, or all, of the other types of resolution we have summarized above.

In subsequent chapters, as new remote sensing systems are introduced and discussed, the reader should keep in mind these multiple resolutions that determine whether a given system would be suitable for a particular application.

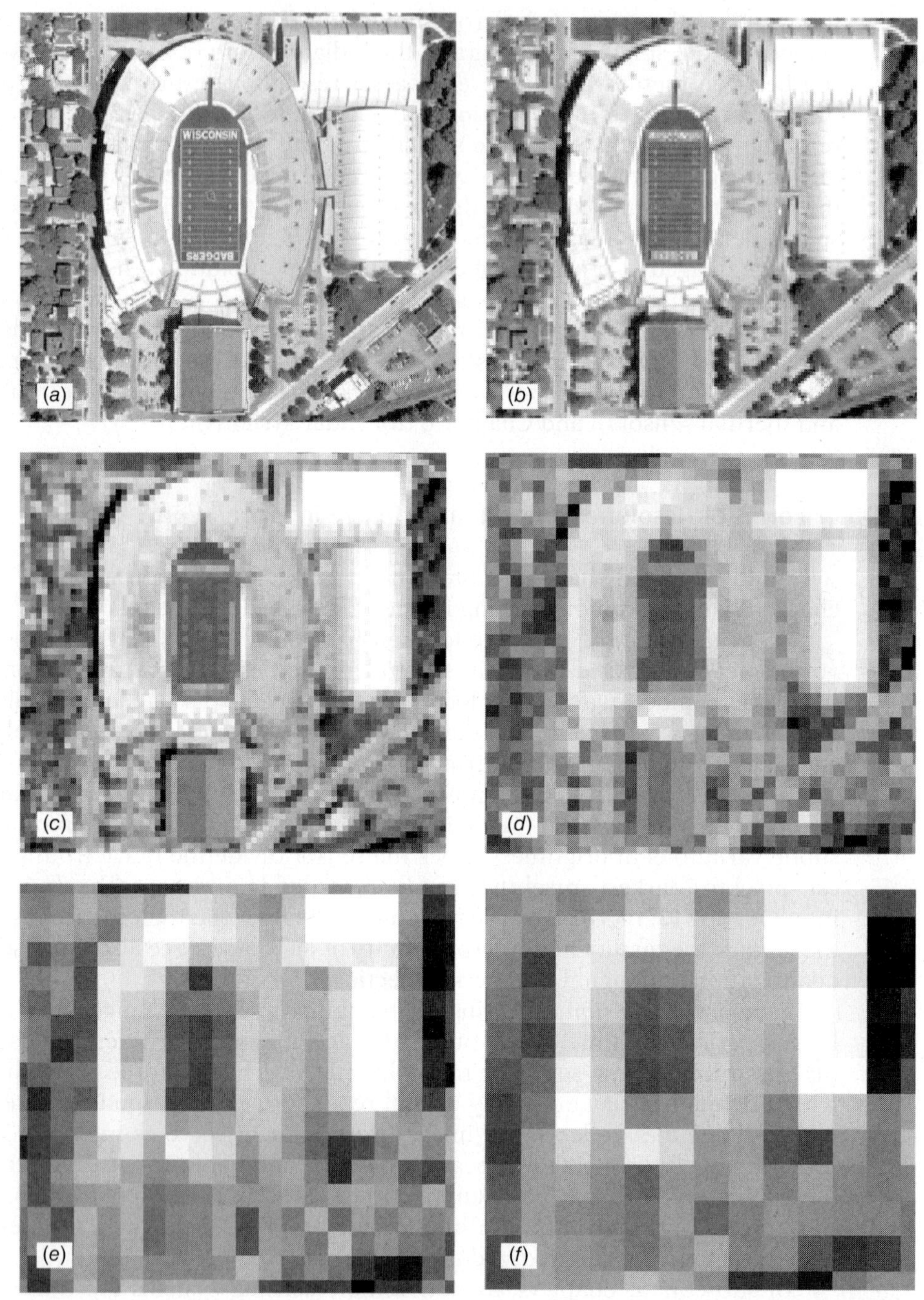

Figure 1.34 Ground resolution cell size effect on ability to extract detail from a digital image. Shown is a portion of the University of Wisconsin-Madison campus, including Camp Randall Stadium and vicinity, at a ground resolution cell size (per pixel) of: (*a*) 1 m, (*b*) 2.5 m, (*c*) 5 m, (*d*) 10 m, (*e*) 20 m, and (*f*) 30 m, and an enlarged portion of the image at (*g*) 0.5 m, (*h*) 1 m, and (*i*) 2.5 m. (Courtesy University of Wisconsin-Madison, Environmental Remote Sensing Center, and NASA Affiliated Research Center Program.)

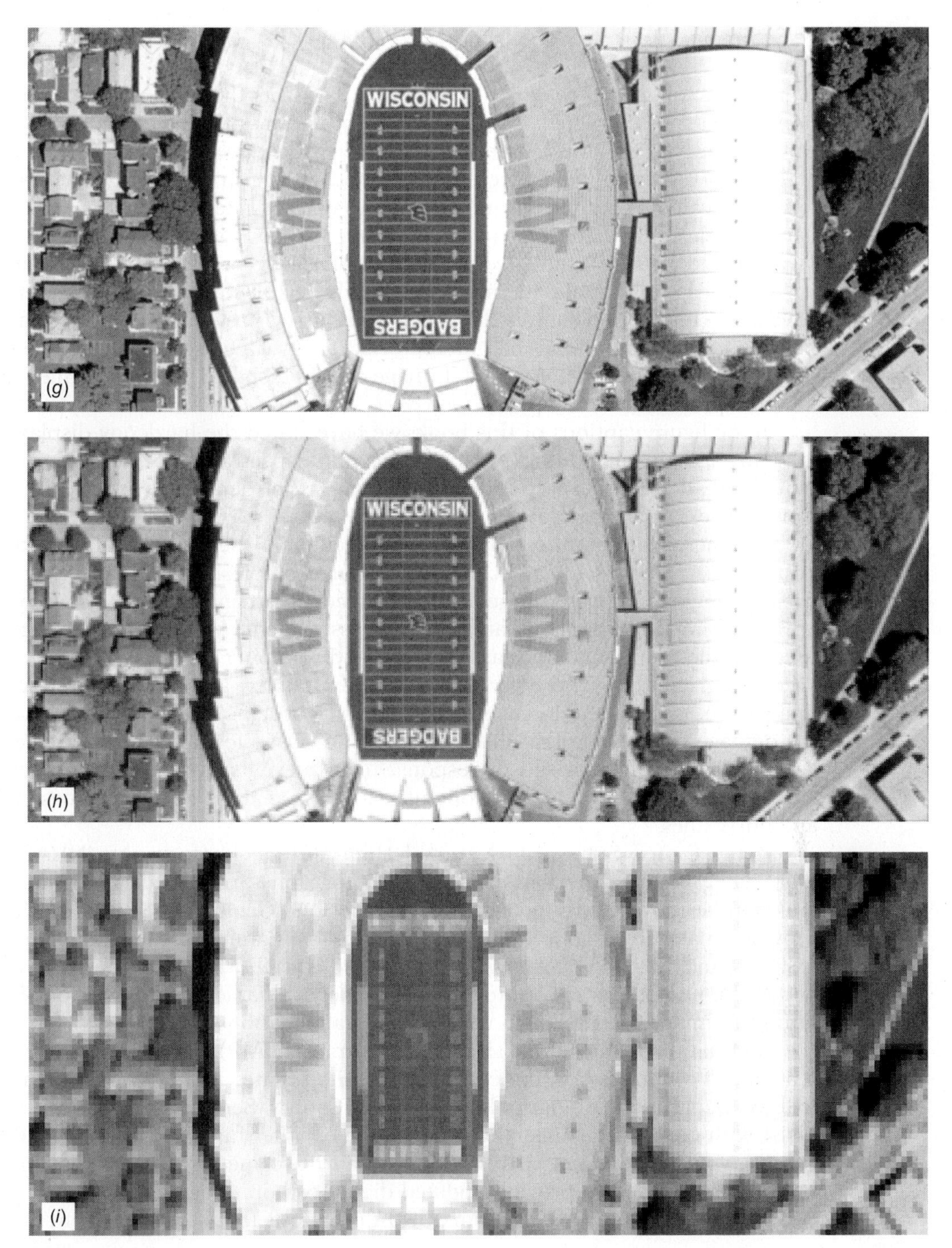

Figure 1.34 *(Continued)*

Image Scale

Image scale, discussed in detail in Section 3.3, affects the level of useful information that can be extracted from aerial and space images. The scale of an image can be thought of as the relationship between a distance measured on the image and the corresponding distance on the ground. Although terminology with regard to image scale has not been standardized, we can consider that *small-scale* images have a scale of 1:50,000 or smaller, *medium-scale* images have a scale between 1:12,000 and 1:50,000, and *large-scale* airphotos have a scale of 1:12,000 or larger.

In the case of digital data, images do not have a fixed scale per se; rather, they have a specific ground sample distance, as discussed previously (and illustrated in Figures 1.33 and 1.34), and can be reproduced at various scales. Thus, one could refer to the *display scale* of a digital image as it is displayed on a computer monitor or as printed in hardcopy.

In the figure captions of this book, we have stated the hardcopy display scale of many images—including photographic, multispectral, and radar images—so that the reader can develop a feel for the degree of detail that can be extracted from images of varying scales.

As generalizations, the following statements can be made about the appropriateness of various image scales for resource studies. Small-scale images are used for regional surveys, large-area resource assessment, general resource management planning, and large-area disaster assessment. Medium-scale images are used for the identification, classification, and mapping of such features as tree species, agricultural crop type, vegetation community, and soil type. Large-scale images are used for the intensive monitoring of specific items such as surveys of the damage caused by plant disease, insects, or tree blowdown. Large-scale images are also used for emergency response to such events as hazardous waste spills and planning search and rescue operations in association with tornadoes, floods, and hurricanes.

In the United States, the National High-Altitude Photography (NHAP) program, later renamed the National Aerial Photography Program (NAPP), was a federal multiagency activity coordinated by the U.S. Geological Survey (USGS). It provided nationwide photographic coverage at nominal scales ranging from 1:80,000 and 1:58,000 (for NHAP) to 1:40,000 (for NAPP). The archive of NHAP and NAPP photos has proven to be an extremely valuable ongoing source of medium-scale images supporting a wide range of applications.

The National Agriculture Imagery Program (NAIP) acquires peak growing season leaf-on imagery in the continental United States and delivers this imagery to U.S. Department of Agriculture (USDA) County Service Centers in order to assist with crop compliance and a multitude of other farm programs. NAIP imagery is typically acquired with GSDs of one to two meters. The one-meter GSD imagery is intended to provide updated digital orthophotography. The two-meter GSD imagery is intended to support USDA programs that require current imagery acquired during the agricultural growing season but do not require high horizontal accuracy. NAIP photographs are also useful in many non-USDA applications, including real estate, recreation, and land use planning.

Approaching the Image Interpretation Process

There is no single "right" way to approach the image interpretation process. The specific image products and interpretation equipment available will, in part, influence how a particular interpretation task is undertaken. Beyond these factors, the specific goals of the task will determine the image interpretation process employed. Many applications simply require the image analyst to identify and count various discrete objects occurring in a study area. For example, counts may be made of such items as motor vehicles, residential dwellings, recreational watercraft, or animals. Other applications of the interpretation process often involve the identification of anomalous conditions. For example, the image analyst might survey large areas looking for such features as failing septic systems, sources of water pollution entering a stream, areas of a forest stressed by an insect or disease problem, or evidence of sites having potential archaeological significance.

Many applications of image interpretation involve the delineation of discrete areal units throughout images. For example, the mapping of land use, soil types, or forest types requires the interpreter to outline the boundaries between areas of one type versus another. Such tasks can be problematic when the boundary is not a discrete edge, but rather a "fuzzy edge" or gradation from one type of area to another, as is common with natural phenomena such as soils and natural vegetation.

Two extremely important issues must be addressed before an interpreter undertakes the task of delineating separate areal units on remotely sensed images. The first is the definition of the *classification system* or criteria to be used to separate the various categories of features occurring in the images. For example, in mapping land use, the interpreter must fix firmly in mind what specific characteristics determine if an area is "residential," "commercial," or "industrial." Similarly, the forest type mapping process must involve clear definition of what constitutes an area to be delineated in a particular species, height, or crown density class.

The second important issue in delineation of discrete areal units on images is the selection of the *minimum mapping unit* (MMU) to be employed in the process. This refers to the smallest size areal entity to be mapped as a discrete area. Selection of the MMU will determine the extent of detail conveyed by an interpretation. This is illustrated in Figure 1.35. In (*a*), a small MMU results in a much more detailed interpretation than does the use of a large MMU, as illustrated in (*b*).

Once the classification system and MMU have been determined, the interpreter can begin the process of delineating boundaries between feature types. Experience suggests that it is advisable to delineate the most highly contrasting feature types first and to work from the general to the specific. For example, in a

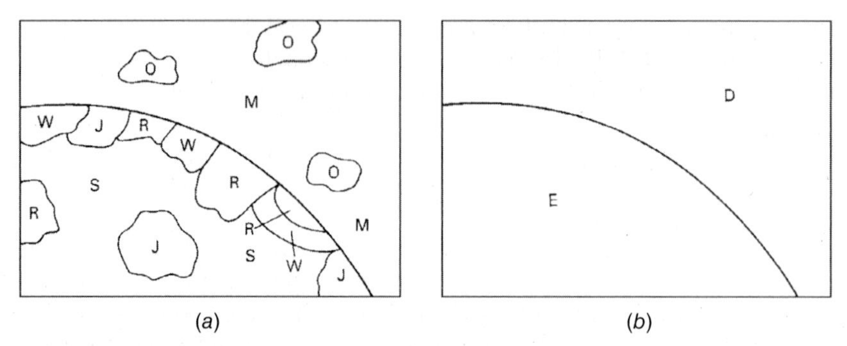

Figure 1.35 Influence of minimum mapping unit size on interpretation detail. (*a*) Forest types mapped using a small MMU: O, oak; M, maple; W, white pine; J, jack pine; R, red pine; S, spruce. (*b*) Forest types mapped using a large MMU: D, deciduous; E, evergreen.

land use mapping effort, it would be better to separate "urban" from "water" and "agriculture" before separating more detailed categories of each of these feature types based on subtle differences.

In certain applications, the interpreter might choose to delineate *photomorphic regions* as part of the delineation process. These are regions of reasonably uniform tone, texture, and other image characteristics. When initially delineated, the feature type identity of these regions may not be known. Field observations or other ground truth can then be used to verify the identity of each region. Regrettably, there is not always a one-to-one correspondence between the appearance of a photomorphic region and a mapping category of interest. However, the delineation of such regions often serves as a stratification tool in the interpretation process and can be valuable in applications such as vegetation mapping (where photomorphic regions often correspond directly to vegetation classes of interest).

Basic Equipment for Visual Interpretation

Visual image interpretation equipment generally serves one of several fundamental purposes: viewing images, making measurements on images, performing image interpretation tasks, and transferring interpreted information to base maps or digital databases. Basic equipment for viewing images and transferring interpreted information is described here. Equipment involved in performing measuring and mapping tasks will be described in Chapter 3.

The airphoto interpretation process typically involves the utilization of stereoscopic viewing to provide a three-dimensional view of the terrain. Some space images are also analyzed stereoscopically. The stereo effect is possible because we have binocular vision. That is, since we have two eyes that are slightly separated, we continually view the world from two slightly different perspectives. Whenever objects lie at different distances in a scene, each eye sees a slightly different view of the objects. The differences between the two views are synthesized by the mind to provide depth perception. Thus, the two views provided by our separated eyes enable us to see in three dimensions.

Vertical aerial photographs are often taken along flight lines such that successive images overlap by at least 50% (see Figure 3.2). This overlap also provides two views taken from separated positions. By viewing the left image of a pair with the left eye and the right image with the right eye, we obtain a three-dimensional view of the terrain surface. The process of stereoscopic viewing can be done using a traditional *stereoscope*, or using various methods for stereoscopic viewing on computer monitors. This book contains many *stereopairs*, or *stereograms*, which can be viewed in three dimensions using a lens stereoscope such as shown in Figure 1.36. An average separation of about 58 mm between common points has been used in the stereograms in this book. The exact spacing varies somewhat because of the different elevations of the points.

Typically, images viewed in stereo manifest *vertical exaggeration*. This is caused by an apparent difference between the vertical scale and the horizontal scale of the stereomodel. Because the vertical scale appears to be larger than the horizontal scale, objects in the stereomodel appear to be too tall. A related consideration is that slopes in the stereomodel will appear to be steeper than they actually are. (The geometric terms and concepts used in this discussion of vertical exaggeration are explained in more detail in Chapter 3.)

Figure 1.36 Simple lens stereoscope. (Author-prepared figure.)

Many factors contribute to vertical exaggeration, but the primary cause is the lack of equivalence between the original, in-flight, *photographic base–height ratio,* B/H' (Figure 3.24*a*), and the corresponding, in-office, *stereoviewing base–height ratio,* b_e/h_e (Figure 1.37). The perceived vertical exaggeration in the stereomodel is approximately the ratio of these two ratios. The photographic base–height ratio is the ratio of the *air base* distance between the two exposure stations to the flying height above the average terrain elevation. The stereoviewing base–height ratio is the ratio between the viewer's *eye base* (b_e) to the distance from the eyes at which the stereomodel is perceived (h_e). The perceived vertical exaggeration, *VE,* is approximately the ratio of the photographic base–height ratio to the stereoviewing base–height ratio,

$$VE \cong \frac{B/H'}{b_e/h_e} \tag{1.10}$$

where $b_e/h_e = 0.15$ for most observers.

In short, vertical exaggeration varies directly with the photographic base–height ratio.

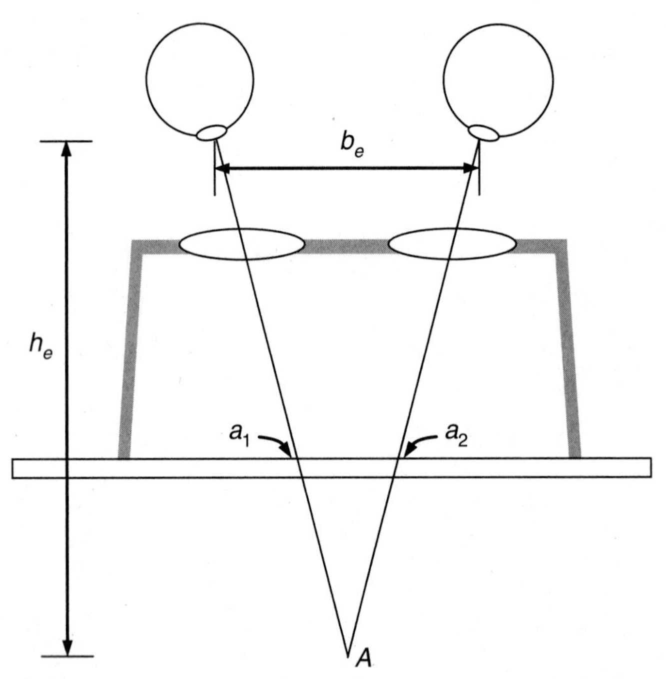

Figure 1.37 Major cause of vertical exaggeration perceived when viewing a stereomodel. There is a lack of equivalence between the photographic base–height ratio when the photographs are taken and the stereoviewing base–height ratio when the photographs are viewed with a stereoscope.

While vertical exaggeration is often misleading to novice image interpreters, it is often very useful in the interpretation process. This is due to the fact that subtle variation in the heights of image features is more readily discriminated when exaggerated. As we note in Chapter 3, large base–height ratios also improve the quality of many photogrammetric measurements made from vertical aerial photographs.

Figure 1.38 can be used to test stereoscopic vision. When this diagram is viewed through a stereoscope, the rings and other objects should appear to be at varying distances from the observer. Your stereovision ability can be evaluated by filling in Table 1.3 (answers are in the second part of the table). People whose eyesight is very weak in one eye may not have the ability to see in stereo. This will preclude three-dimensional viewing of the stereograms in this book. However, many people with essentially monocular vision have become proficient photo interpreters. In fact, many forms of interpretation involve monocular viewing with such basic equipment as handheld magnifying glasses or tube magnifiers (2× to 10× lenses mounted in a transparent stand).

Some people will be able to view the stereograms in this book without a stereoscope. This can be accomplished by holding the book about 25 cm from your eyes and allowing the view of each eye to drift into a straight-ahead viewing position (as when looking at objects at an infinite distance) while still maintaining focus on the stereogram. When the two images have fused into one, the stereogram will be seen in three dimensions. Most persons will find stereoviewing without proper stereoscopes to be a tiring procedure, producing eyestrain. It is, however, a useful technique to employ when stereoscopes are not available.

Several types of analog stereoscopes are available, utilizing lenses or a combination of lenses, mirrors, and prisms. *Lens stereoscopes*, such as the one shown in Figure 1.36, are portable and comparatively inexpensive. Most are small instruments with folding legs. The lens spacing can usually be adjusted from about

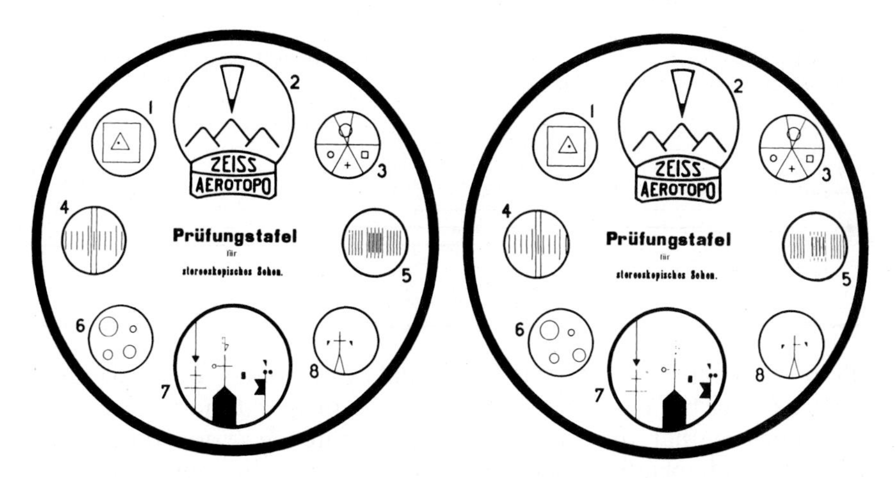

Figure 1.38 Stereoscopic vision test. (Courtesy Carl Zeiss, Inc.)

TABLE 1.3 Stereovision Test for Use with Figure 1.38

PART I

Within the rings marked 1 through 8 are designs that appear to be at different elevations. Using "1" to designate the highest elevation, write down the depth order of the designs. It is possible that two or more designs may be at the same elevation. In this case, use the same number for all designs at the same elevation.

Ring 1		**Ring 6**	
Square	(2)	Lower left circle	( )
Marginal ring	(1)	Lower right circle	( )
Triangle	(3)	Upper right circle	( )
Point	(4)	Upper left circle	( )
		Marginal ring	( )
Ring 7		**Ring 3**	
Black flag with ball	( )	Square	( )
Marginal ring	( )	Marginal ring	( )
Black circle	( )	Cross	( )
Arrow	( )	Lower left circle	( )
Tower with cross	( )	Upper left circle	( )
Double cross	( )		
Black triangle	( )		
Black rectangle	( )		

PART II

Indicate the relative elevations of the rings 1 through 8.

( ) ( ) ( ) ( ) ( ) ( ) ( ) ( )
Highest Lowest

PART III

Draw profiles to indicate the relative elevations of the letters in the words "prufungstafel" and "stereoskopisches sehen."

(Answers to Stereovision Test on next page)

45 to 75 mm to accommodate individual eye spacings. Lens magnification is typically 2 power but may be adjustable. The principal disadvantage of small lens stereoscopes is that the images must be quite close together to be positioned properly underneath the lenses. Because of this, the interpreter cannot view the entire stereoscopic area contained by the overlapping aerial photographs without

TABLE 1.3 *(Continued)*

PART I

Ring 1		**Ring 6**	
Square	(2)	Lower left circle	(4)
Marginal ring	(1)	Lower right circle	(5)
Triangle	(3)	Upper right circle	(1)
Point	(4)	Upper left circle	(3)
		Marginal ring	(2)
Ring 7		**Ring 3**	
Black flag with ball	(5)	Square	(4)
Marginal ring	(1)	Marginal ring	(2)
Black circle	(4)	Cross	(3)
Arrow	(2)	Lower left circle	(1)
Tower with cross	(7)	Upper left circle	(5)
Double cross	(2)		
Black triangle	(3)		
Black rectangle	(6)		

PART II

(7) (6) (5) (1) (4) (2)[a] (3)[a] (8)

Highest Lowest

PART III

PRUFUNGSTAFEL STEREOSKOPISCHES SEHEN

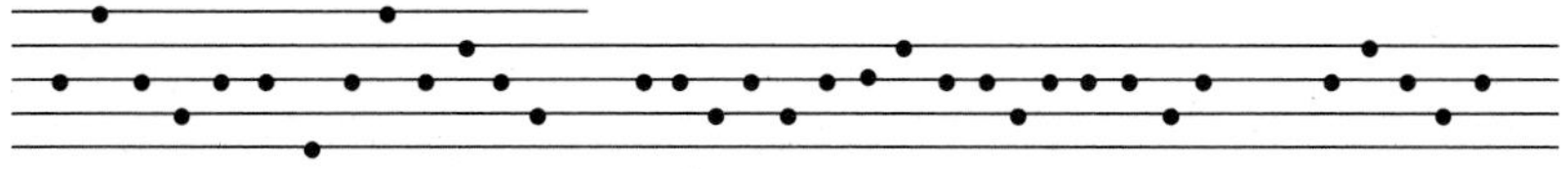

[a]Rings 2 and 3 are at the same elevation.

raising the edge of one of the photographs. More advanced devices for viewing hardcopy stereoscopic image pairs include *mirror stereoscopes* (larger stereoscopes that use a combination of prisms and mirrors to separate the lines of sight from each of the viewer's eyes) and *zoom stereoscopes,* expensive precision instruments used for viewing stereopairs under variable magnification.

With the proliferation of digital imagery and software for viewing and analyzing digital images, analog stereoscopes have been replaced in the laboratory (if not in the field) by various computer hardware configurations for stereoviewing. These devices are discussed in Section 3.10.

Relationship between Visual Image Interpretation and Computer Image Processing

In recent years, there has been increasing emphasis on the development of quantitative, computer-based processing methods for analyzing remotely sensed data. As will be discussed in Chapter 7, those methods have become increasingly sophisticated and powerful. Despite these advances, computers are still somewhat limited in their ability to evaluate many of the visual "clues" that are readily apparent to the human interpreter, particularly those referred to as image *texture*. Therefore, visual and numerical techniques should be seen as complementary in nature, and consideration must be given to which approach (or combination of approaches) best fits a particular application.

The discussion of visual image interpretation in this section has of necessity been brief. As mentioned at the start of this section, the skill of image interpretation is best learned interactively, through the experience of interpreting many images. In the ensuing chapters of this book, we provide many examples of remotely sensed images, from aerial photographs to synthetic aperture radar images. We hope that the reader will apply the principles and concepts discussed in this section to help interpret the features and phenomena illustrated in those images.

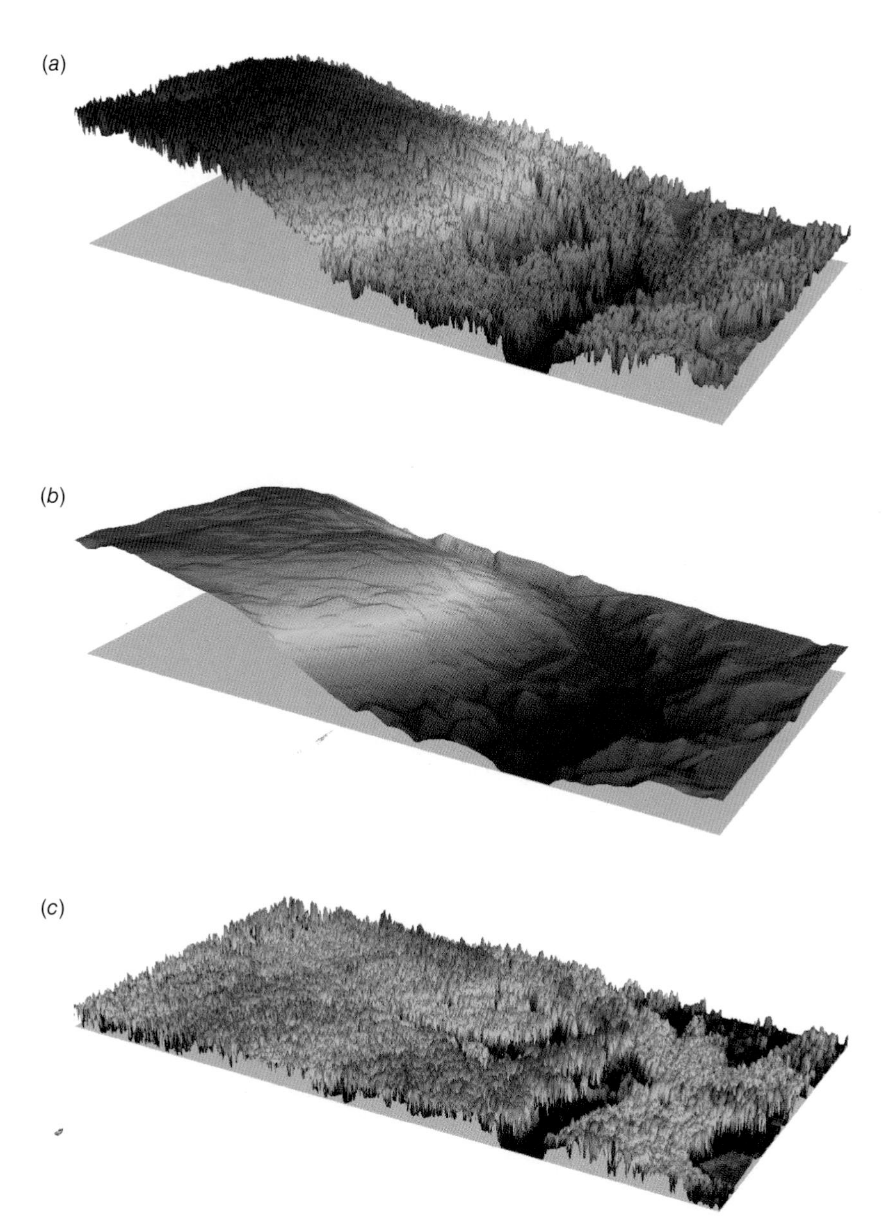

Plate 1 Digital elevation models of a forested hillside in New Hampshire. (*a*) Digital surface model (DSM). (*b*) Digital terrain model (DTM). (*c*) Canopy height model, derived by subtracting elevations in (*b*) from corresponding elevations in (*a*). (Courtesy Dartmouth College, Laboratory for Geographic Information Science and Applied Spatial Analysis.) (For major discussion, see Sections 1.5 and 6.24.)

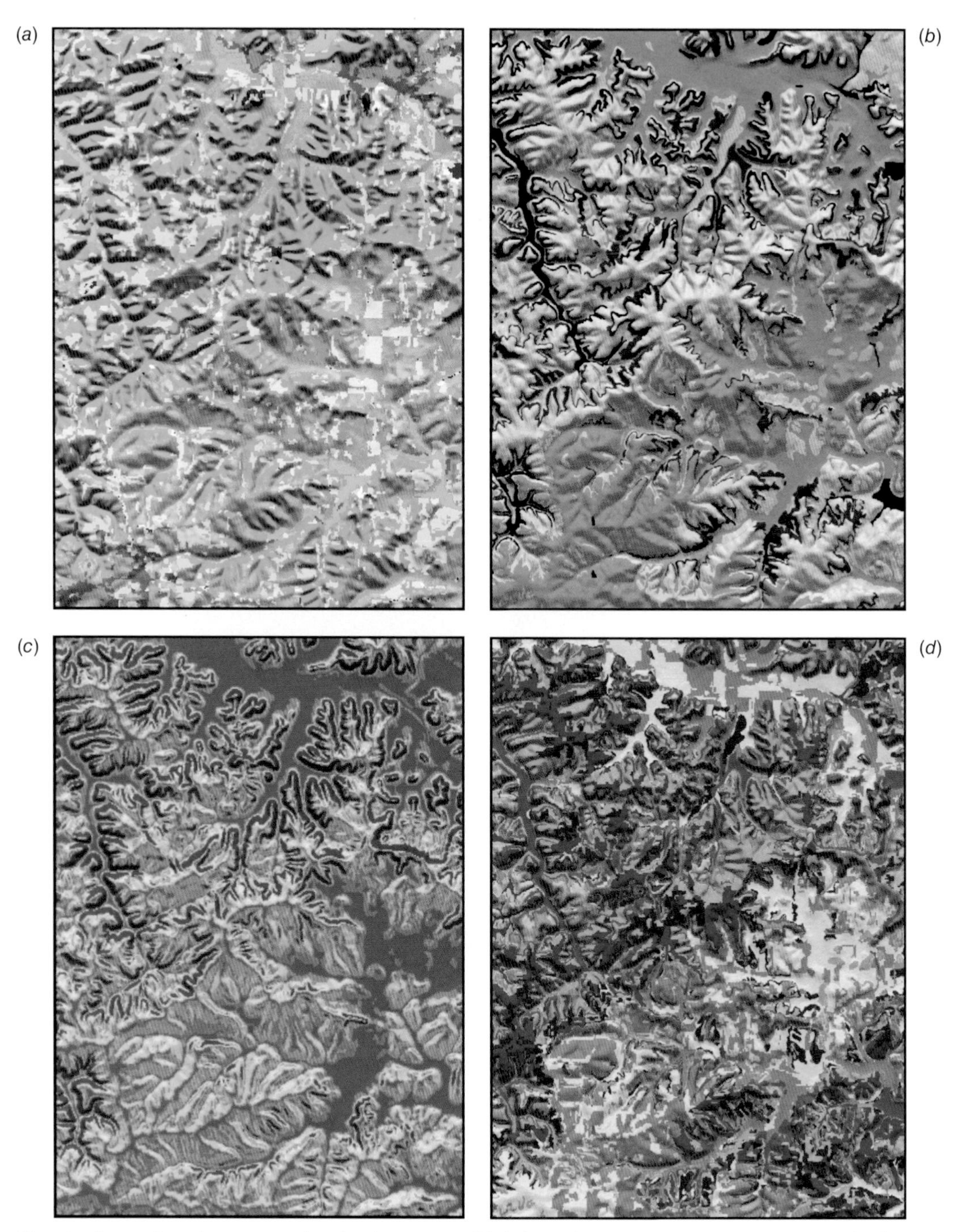

Plate 2 Integration of remote sensing data in a geographic information system, Dane County, WI: (*a*) land cover; (*b*) soil erodibility; (*c*) slope; (*d*) soil erosion potential. (Relief shading added in order to enhance interpretability of figure.) Scale 1:140,000. (Author-prepared figure.) (For major discussion, see Section 1.10.)

(a)

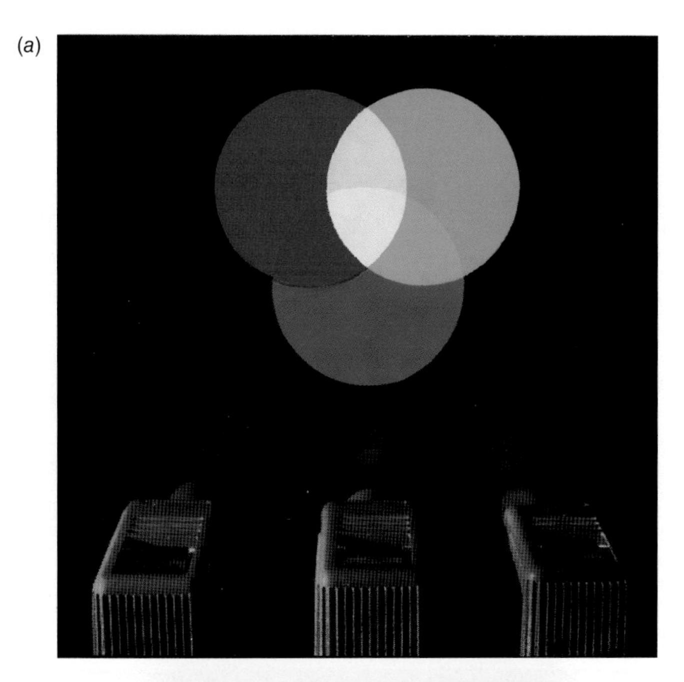

(b)

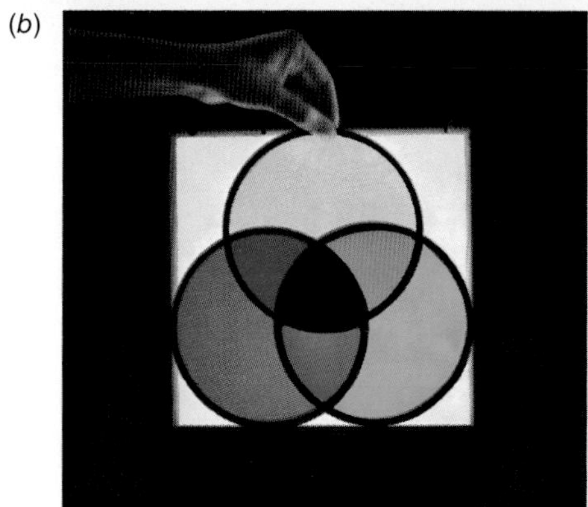

Plate 3 Color-mixing processes. (*a*) Color additive process—operative when lights of different colors are superimposed. (*b*) Color subtractive process—operative when dyes of different colors are superimposed. (Courtesy Eastman Kodak Company.) (For major discussion, see Section 1.12.)

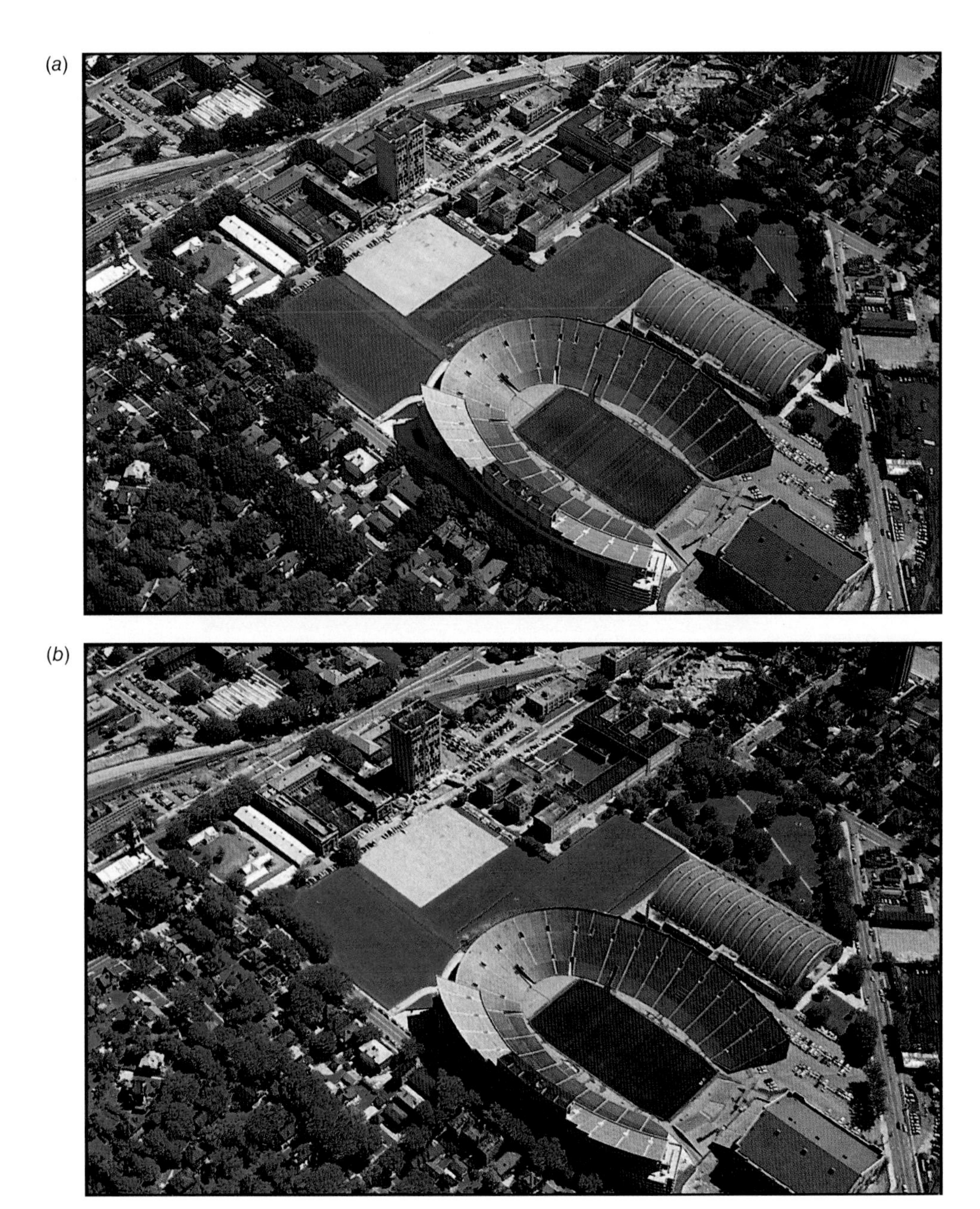

Plate 4 Oblique normal color (*a*) and color infrared (*b*) aerial photographs showing a portion of the University of Wisconsin-Madison campus. The football field has artificial turf with low near-IR reflectance. (Author-prepared figure.) (For major discussion, see Section 2.4.)

Plate 5 Oblique normal color (*a*) and color infrared (*b*) aerial photographs showing flowing lava on the flank of Kilauea Volcano, Hawaii. The orange tones on the color infrared photograph represent infrared energy emitted from the flowing lava. The pink tones represent sunlight reflected from the living vegetation. (Author-prepared figure.) (For major discussion, see Section 2.4.)

(a)

(b)

Plate 6 Color (*a*) and color infrared (*b*) aerial photographs taken with a Leica RCD30 medium-format digital camera system in support of a power line corridor mapping project. (Courtesy Leica Geosystems, Geospatial Solutions Division.) (For major discussion, see Section 2.6.)

(a)

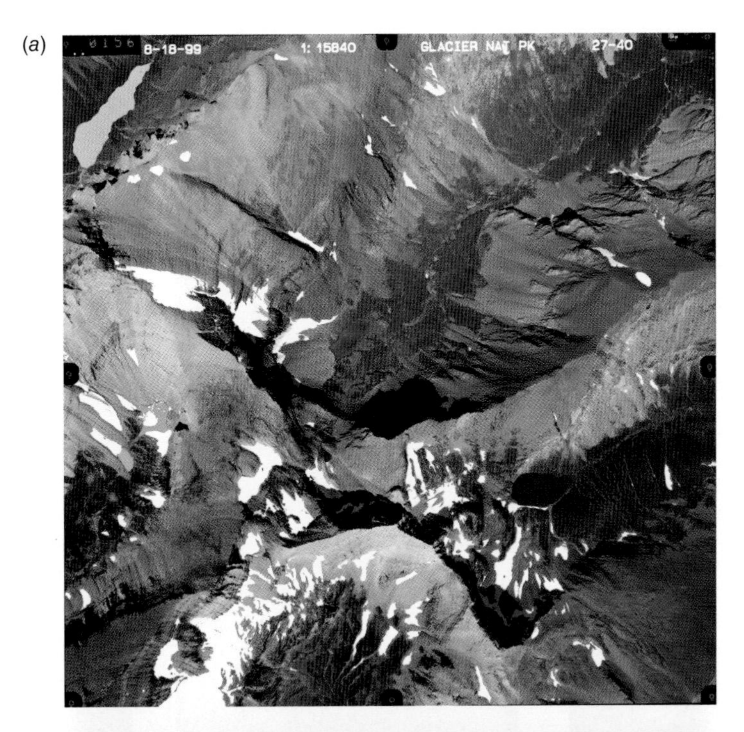

(b)

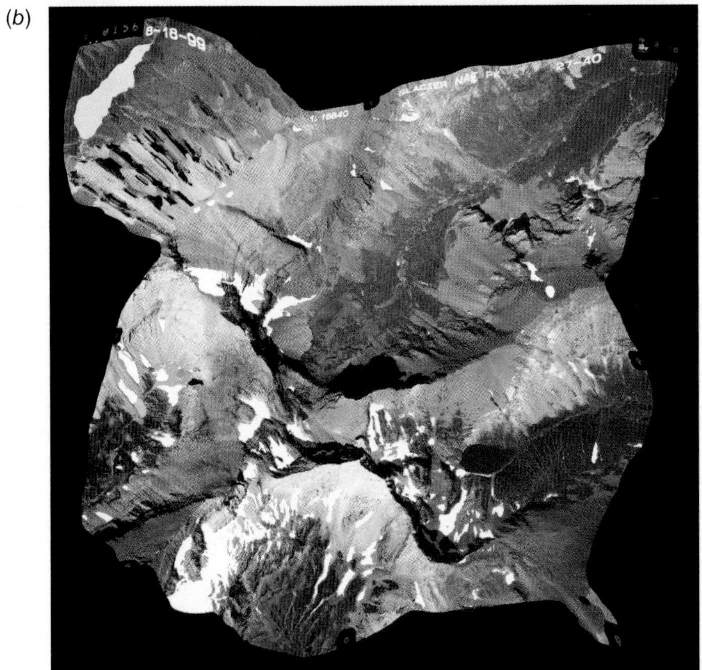

Plate 7 Uncorrected color photograph (*a*) and digital orthophoto (*b*) showing an area of high relief in Glacier National Park. Scale of (*b*) 1:64,000. (Courtesy Dr. Frank L. Scarpace and the USGS Upper Midwest Environmental Sciences Center.) (For major discussion, see Section 3.10.)

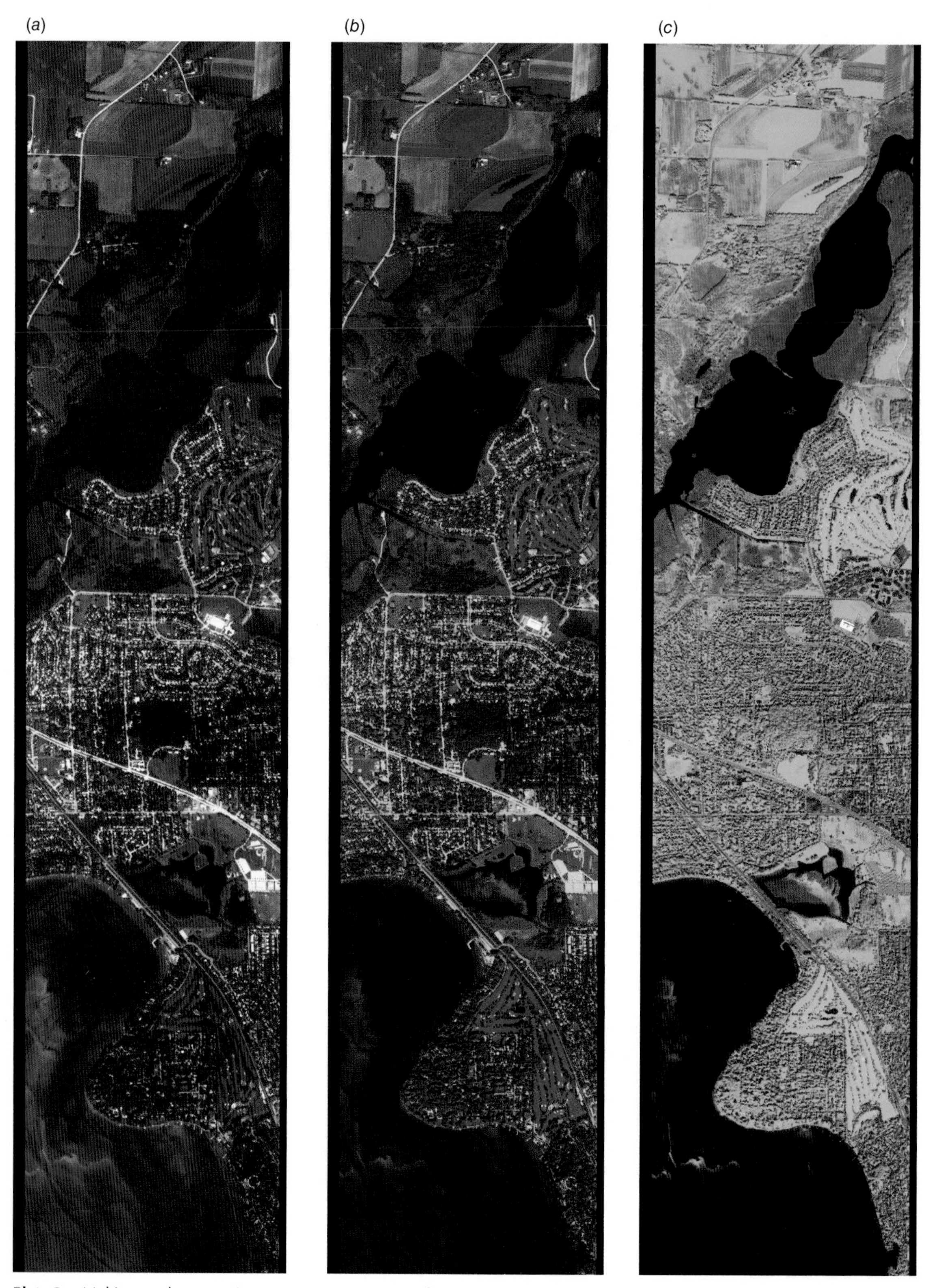

Plate 8 Multispectral scanner images, Dane County, WI, late September: (*a*) normal color composite, ATLAS bands 4, 2, and 1 (red, green, and blue) shown in Figure 4.7; (*b*) color IR composite, bands 6, 4, and 2 (near IR, red, and green); (*c*) color mid-IR composite, bands 8, 6, and 4 (mid-IR, near IR, and red). Scale 1:50,000. (Courtesy NASA Stennis Space Center.) (For major discussion, see Section 4.4).

(a)

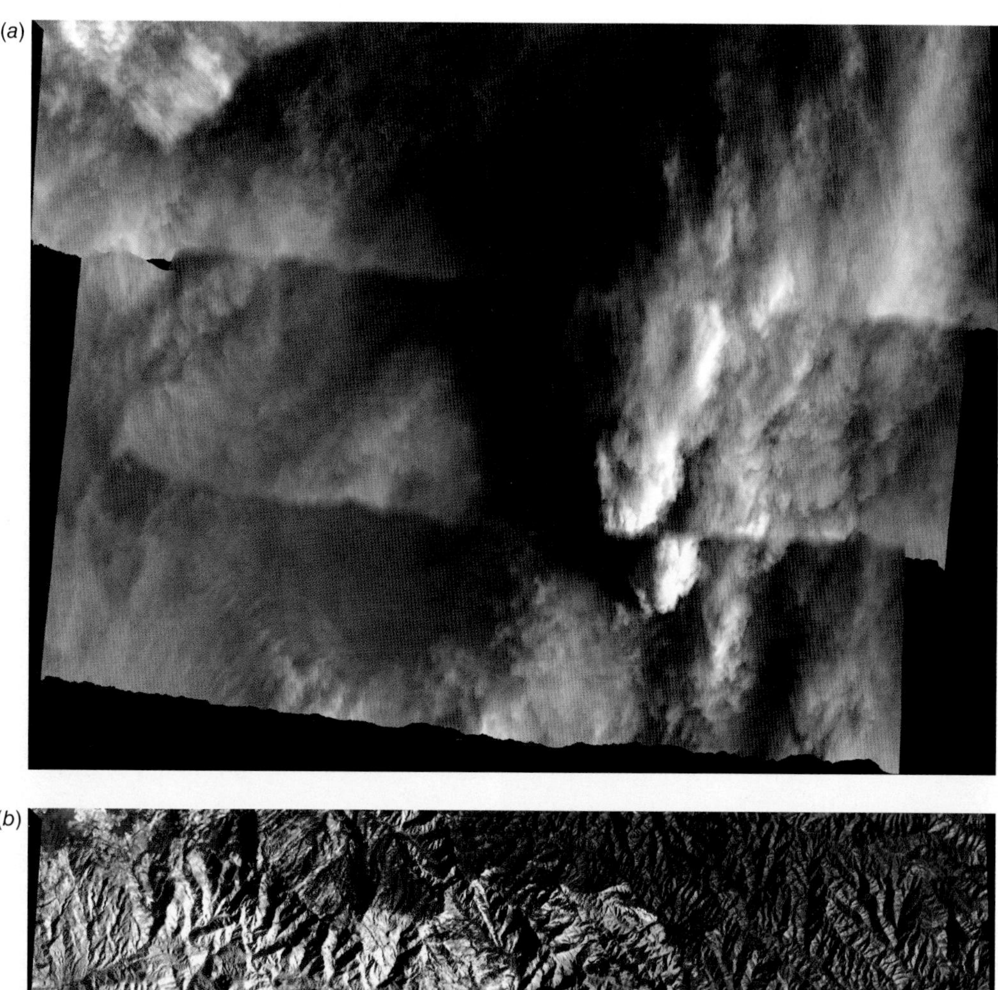

(b)

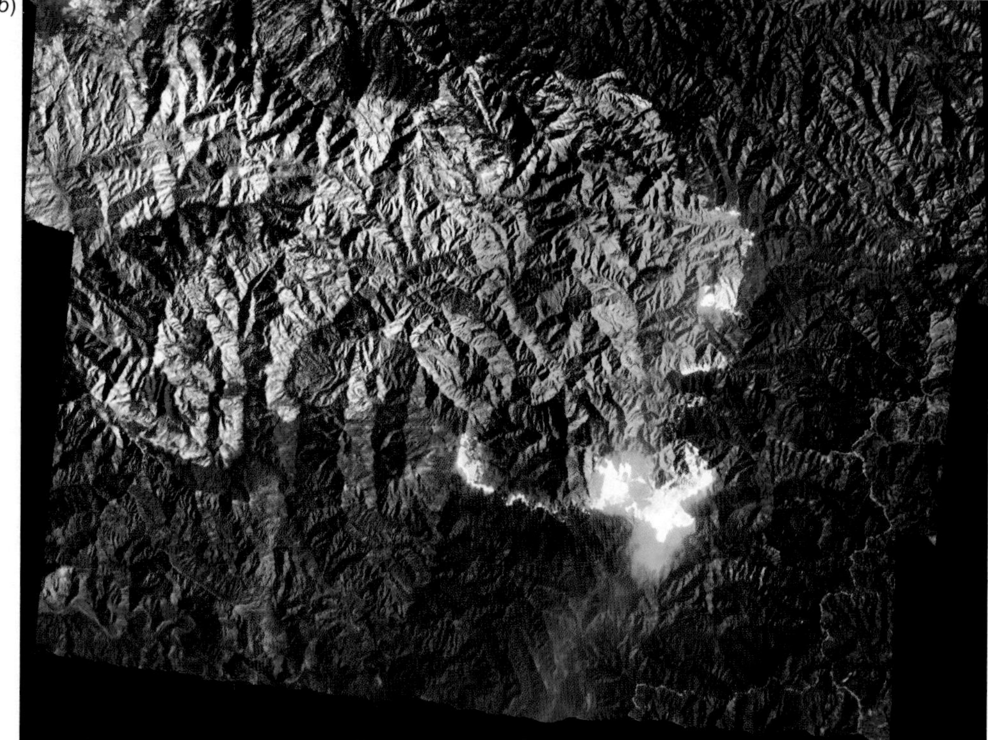

Plate 9 Images of the Zaca Fire near Santa Barbara, CA, acquired by the Autonomous Modular Sensor (AMS) on NASA's Ikhana UAV. (*a*) True-color composite of bands 5, 3, and 2 (red, green, and blue). (*b*) False-color composite of bands 12, 9, and 10 (thermal IR, mid-IR, and mid-IR). (Courtesy NASA Airborne Science Program.) (For major discussion see Section 4.8.)

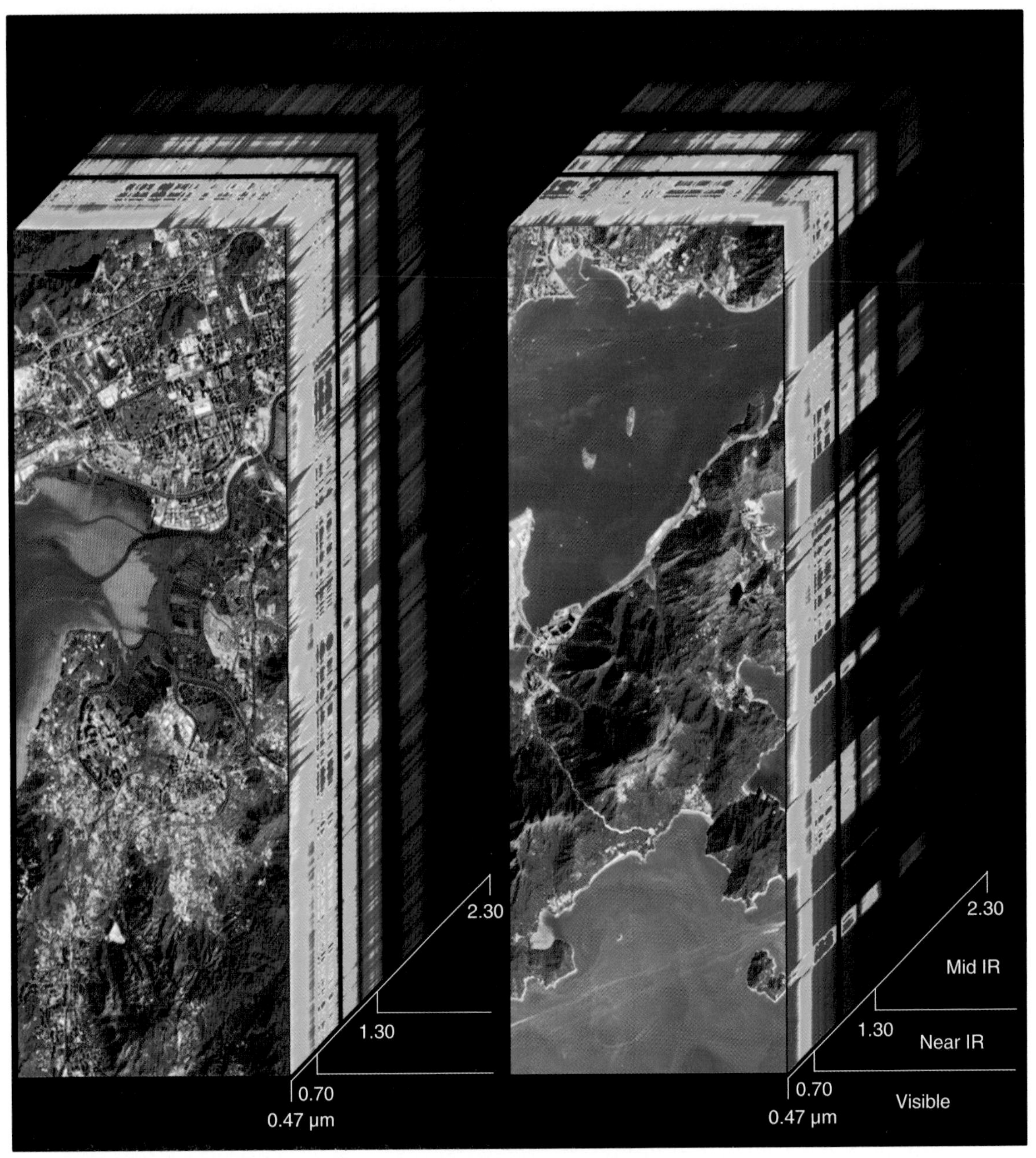

Plate 10 Isometric view of hyperspectral image cubes from two consecutive segments of an EO-1 Hyperion image, Lantau and Shenzhen, China. (Author-prepared figure.) (For major discussion, see Section 4.13.)

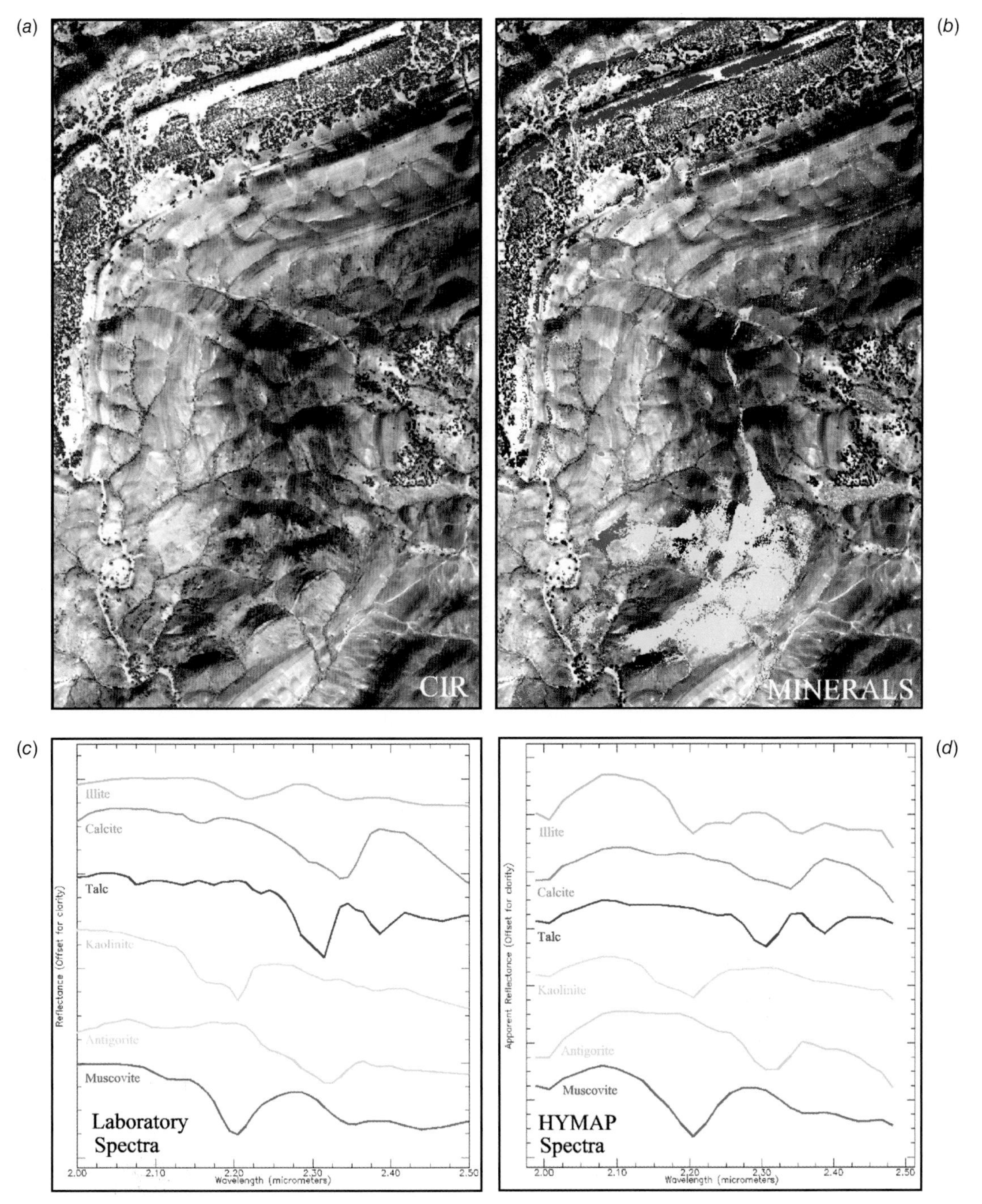

Plate 11 HyMap hyperspectral scanner data and spectral reflectance curves: (*a*) color IR composite of three hyperspectral bands; (*b*) location of six minerals displayed on a grayscale image; (*c*) laboratory spectral reflectance curves of selected minerals; (*d*) spectral reflectance curves of selected minerals as determined from hyperspectral data. Scale 1:40,000. (Reproduced with permission from the American Society for Photogrammetry and Remote Sensing, Bethesda, MD; www.asprs.org.) (For major discussion, see Section 4.13.)

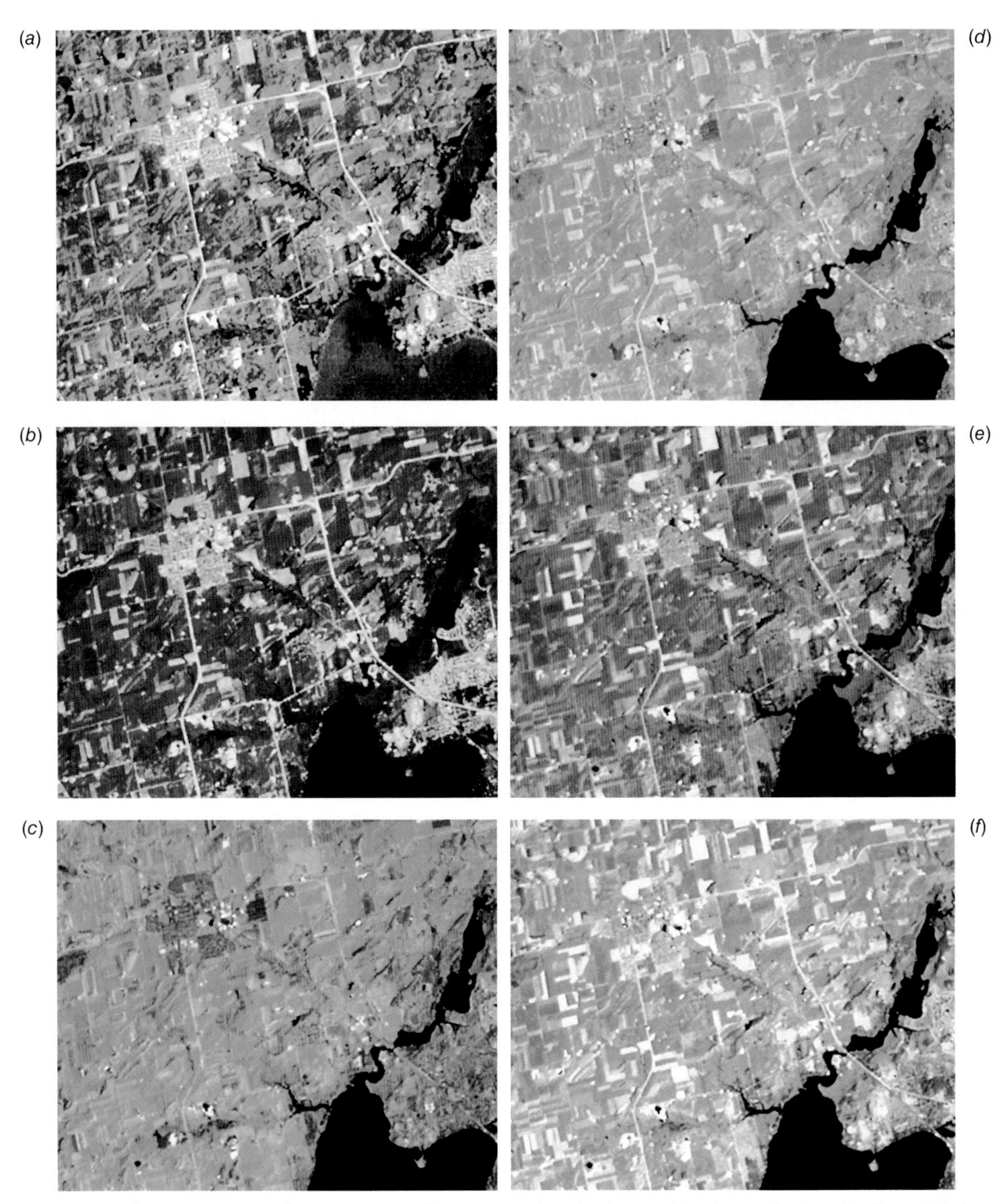

Plate 12 Color composite Landsat TM images illustrating six band-color combinations, suburban Madison, WI, late August. Scale 1:200,000. (Author-prepared figure.) (See Section 5.4, Table 5.4, for the specific band-color combinations shown here.)

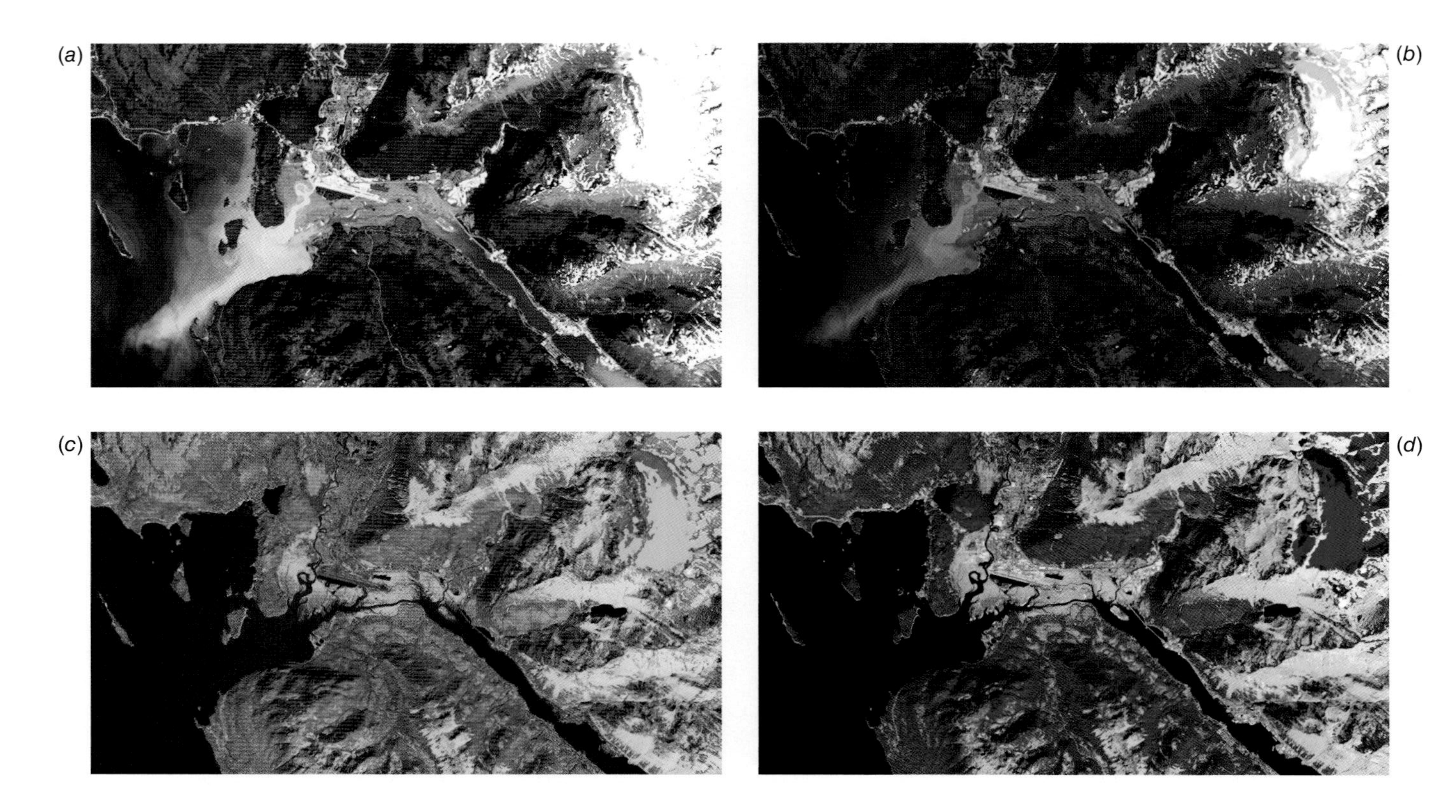

Plate 13 Selected color composites produced from Landsat-8 data acquired August 1, 2013, Juneau, AK, and vicinity. Scale 1:223,000. (*a*) Bands 4 (red), 3 (green), and 2 (blue) as RGB, normal color composite; (*b*) bands 5 (NIR), 4 (red), and 3 (green) as RGB, color-infrared composite; (*c*) bands 6 (SWIR 1), 5 (NIR), and 4 (red) as RGB, false color composite; (*d*) bands 7 (SWIR 2), 6 (SWIR 1), and 5 (NIR), false color composite. (Author-prepared figure.) (For major discussion, see Section 5.5.)

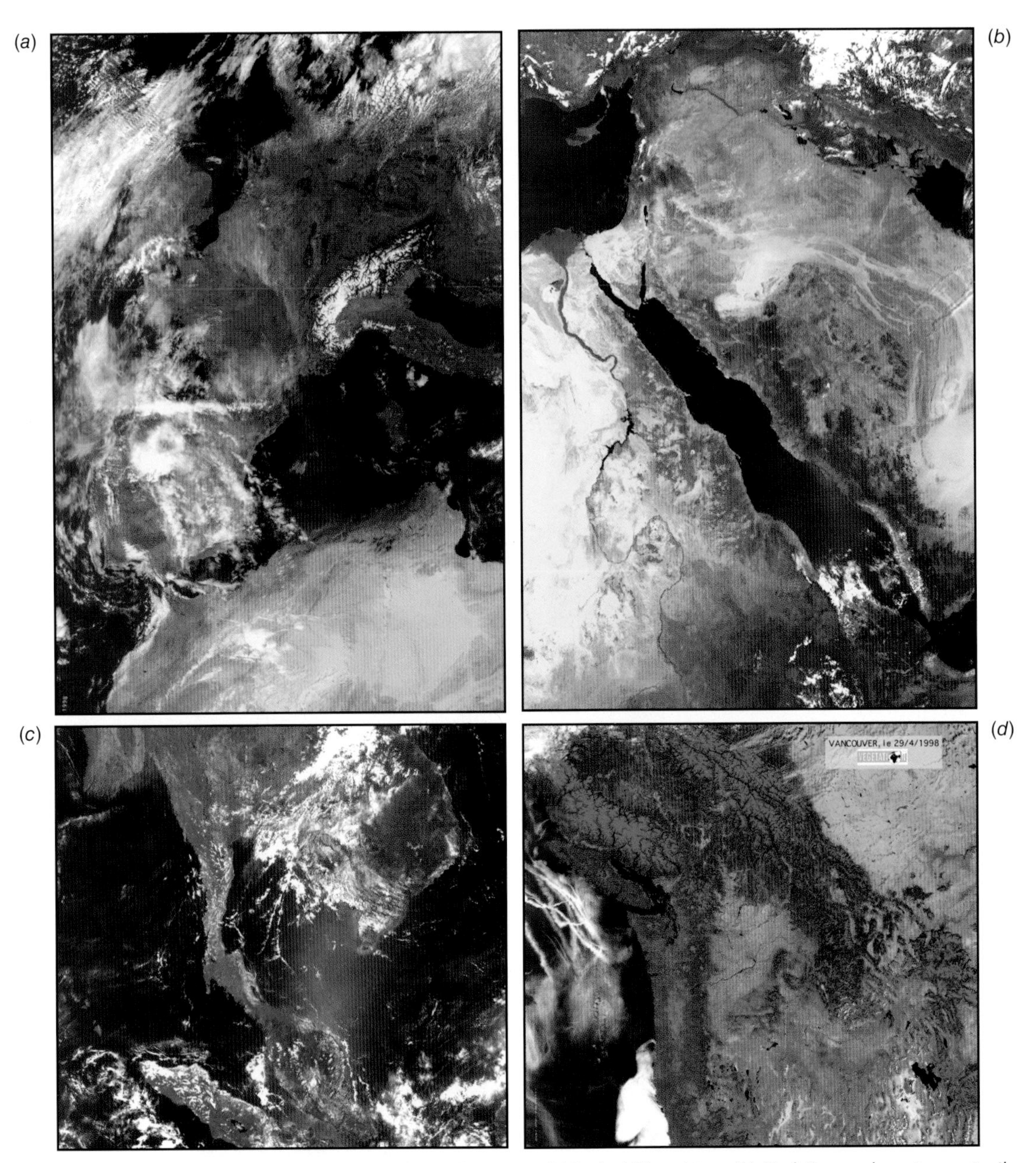

Plate 14 SPOT-4 Vegetation images: (*a*) Western Europe and North Africa, May; (*b*) Red Sea and environs, April; (*c*) southeast Asia, April; (*d*) northwest United States and western Canada, April. (*a*), (*b*), and (*c*) are composites of the blue, red, and near-IR bands, displayed as blue, green, and red, respectively. (*d*) is a composite of the red, near-IR, and mid-IR bands, displayed as blue, green, and red, respectively. (Distribution Astrium Services/Spot Image S.A. France.) (For major discussion, see Section 5.7.)

(a)

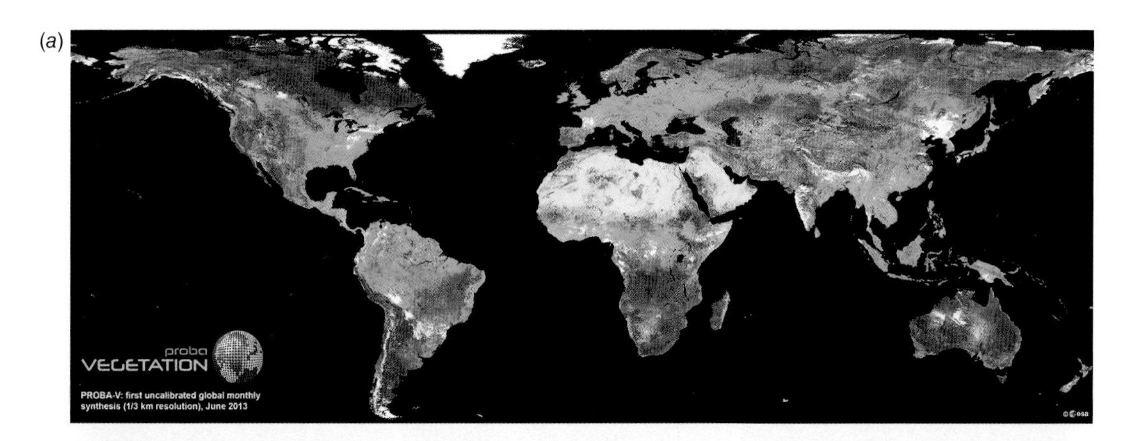

(b)

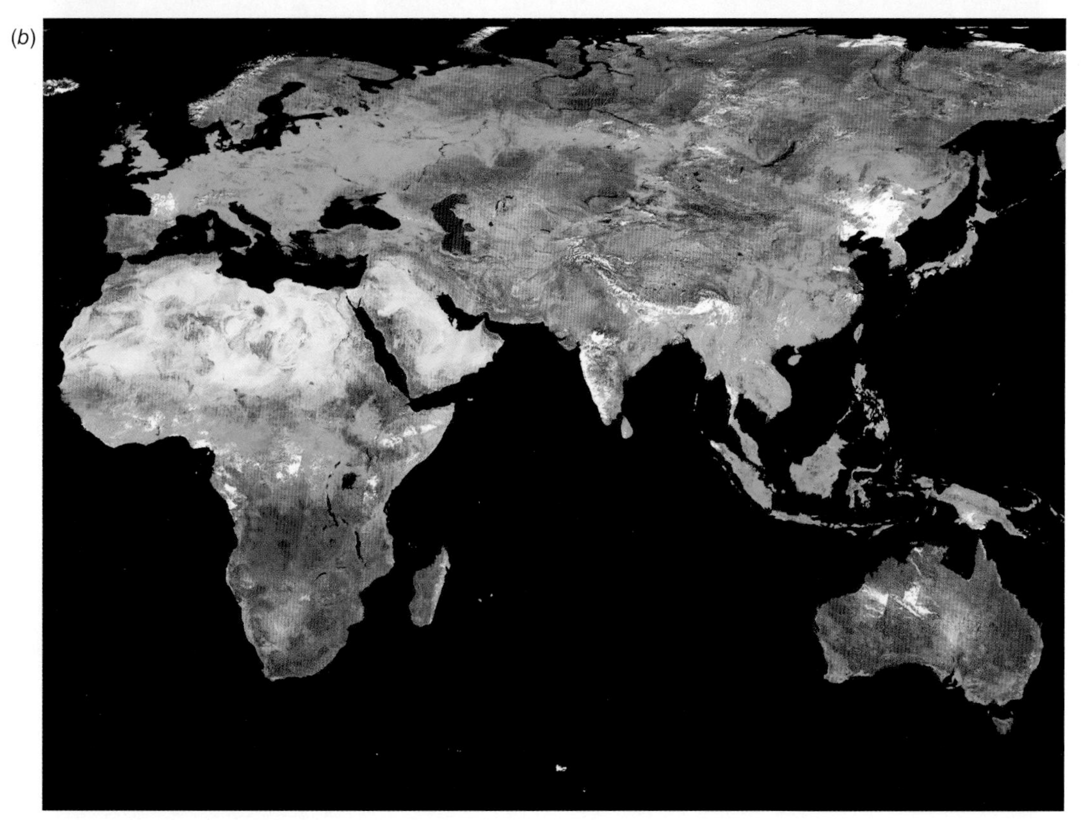

Plate 15 First uncalibrated global mosaic of vegetation derived from Proba-V data, June 2013. (*a*) Complete mosaic. (*b*) Regional subscene at a larger scale. (Courtesy ESA.) (For major discussion, see Section 5.11.)

Plate 16 Merger of IKONOS multispectral and panchromatic images, for an agricultural area in eastern Wisconsin, mid-June. (*a*) IKONOS 4-m-resolution multispectral image. (*b*) IKONOS 1-m-resolution panchromatic image. (*c*) Merged multispectral and panchromatic image, using spectral merging techniques, with 1-m effective resolution. Scale 1:3,000. (Courtesy DigitalGlobe.) (For major discussion, see Sections 5.12 and 7.6.)

Plate 17 High resolution satellite imagery from QuickBird showing a portion of the Port of Hamburg, Germany. The image is the result of applying pan-sharpening techniques to a true-color composite of multispectral bands 1, 2, and 3 shown as blue, green, and red, respectively, and has an effective resolution of 61 cm. (Courtesy DigitalGlobe.) (For major discussion, see Sections 5.12 and 7.6.)

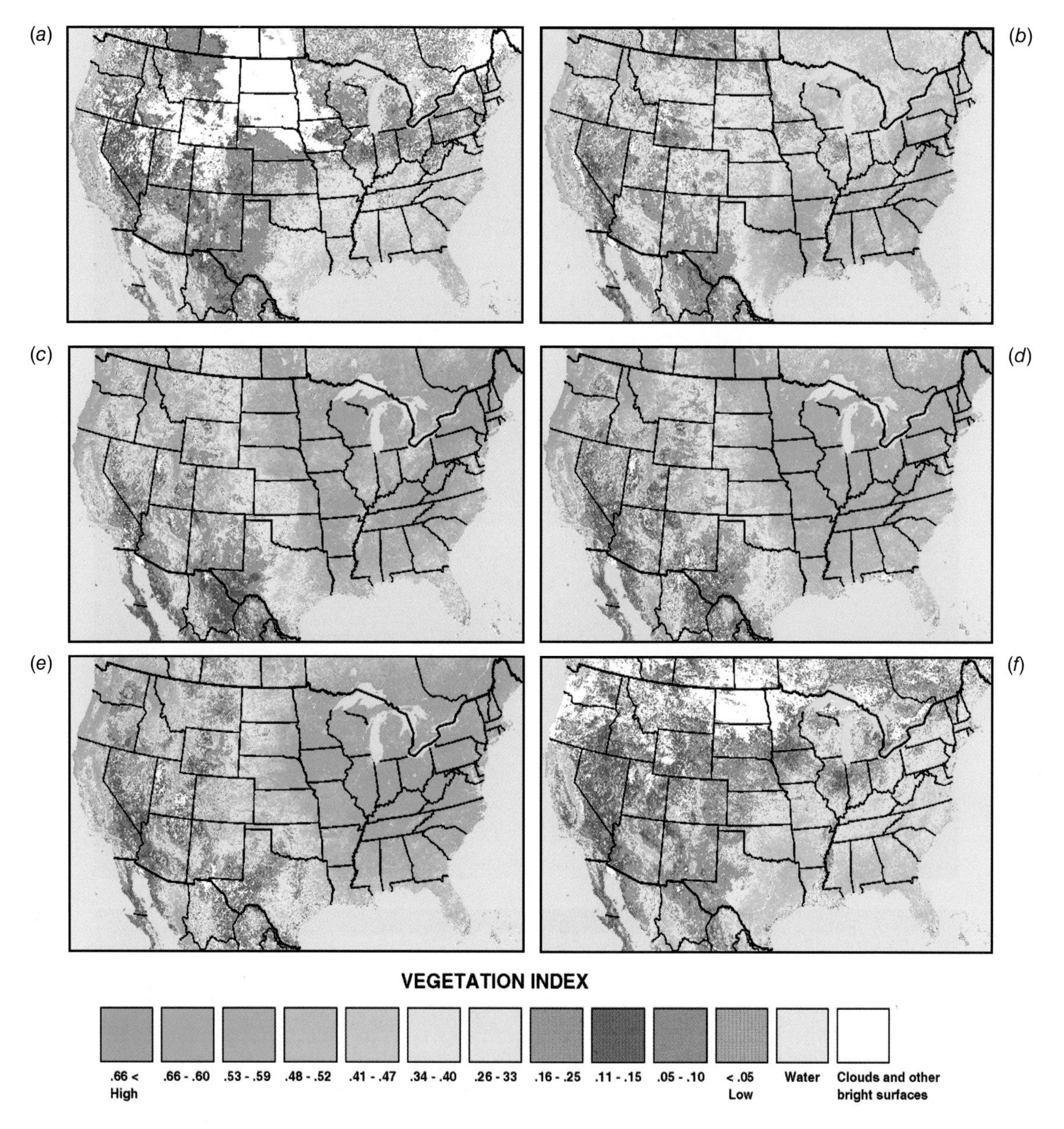

Plate 18 Color-coded normalized difference vegetation index (NDVI) images of the conterminous United States for six 2-week periods: (*a*) February 27–March 12; (*b*) April 24–May 7; (*c*) June 19–July 2; (*d*) July 17–30; (*e*) August 14–27; (*f*) November 6–19. (Courtesy USGS EROS Data Center.) (For major discussion, see Sections 5.15 and 7.19.)

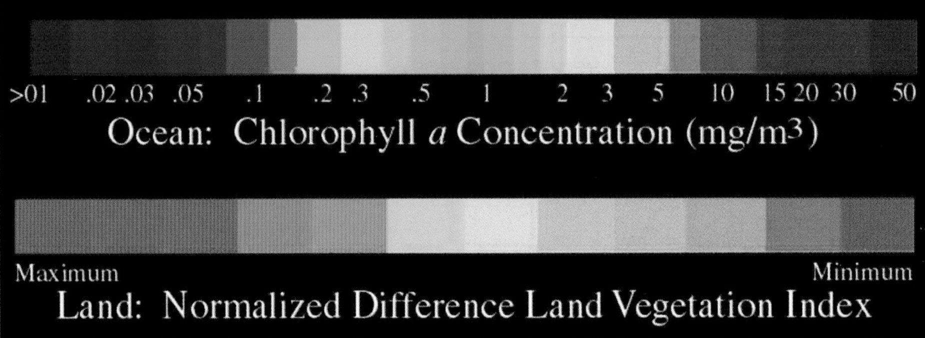

Plate 19 SeaWIFS image showing ocean biogeochemistry and land vegetation. Composite image of data from September through July. Orthographic projection. (Courtesy SeaWIFS Project, NASA/Goddard Space Flight Center, and DigitalGlobe.) (For major discussion, see Section 5.18.)

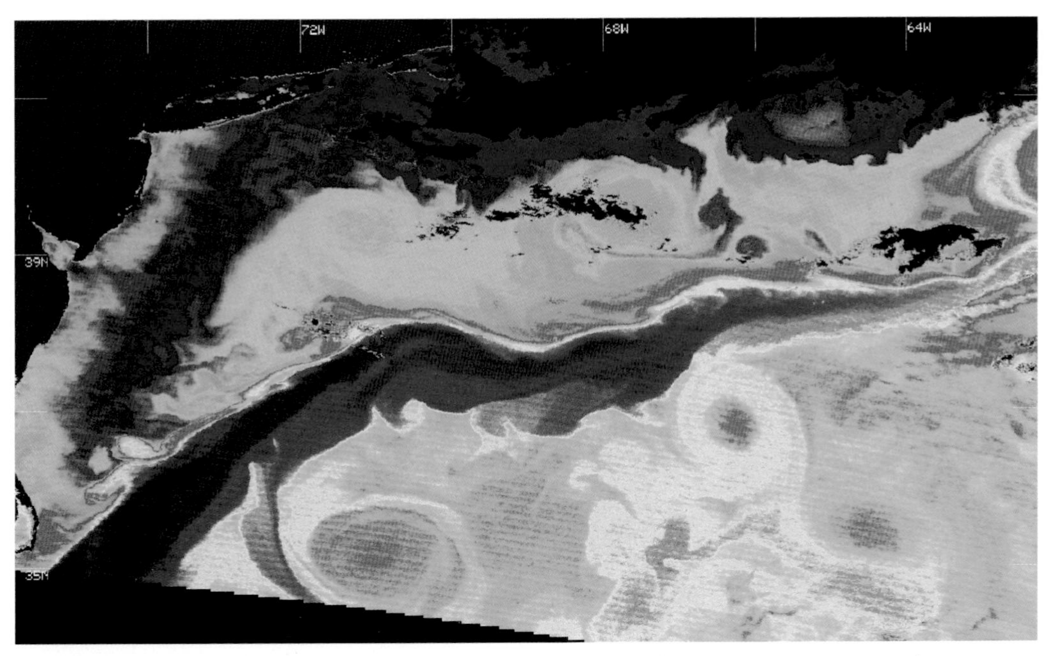

Plate 20 Sea surface temperature map of the western Atlantic, derived from Terra MODIS imagery, showing warm temperatures associated with the Gulf Stream. (Courtesy RSMAS, University of Miami.) (For major discussion, see Section 5.19.)

Plate 21 Lake surface temperature map of the Great Lakes, April 18, 2006, from Aqua MODIS imagery. Shorter wavelength colors (e.g., blue) portray cooler temperatures and longer wavelength colors (e.g., red) show warmer temperatures. (Author-prepared figure.) (For major discussion, see Section 5.19.)

Plate 22 ASTER data, Death Valley, CA, early March. (*a*) VNIR (visible and near-IR) bands 1, 2, and 3 displayed as blue, green, and red, respectively. (*b*) SWIR (mid-IR) bands 5, 7, and 9 displayed as blue, green, and red, respectively. (*c*) TIR (thermal IR) bands 10, 12, and 13 displayed as blue, green, and red, respectively. Scale 1:350,000. (Author-prepared figure.) (For major discussion, see Section 5.19.)

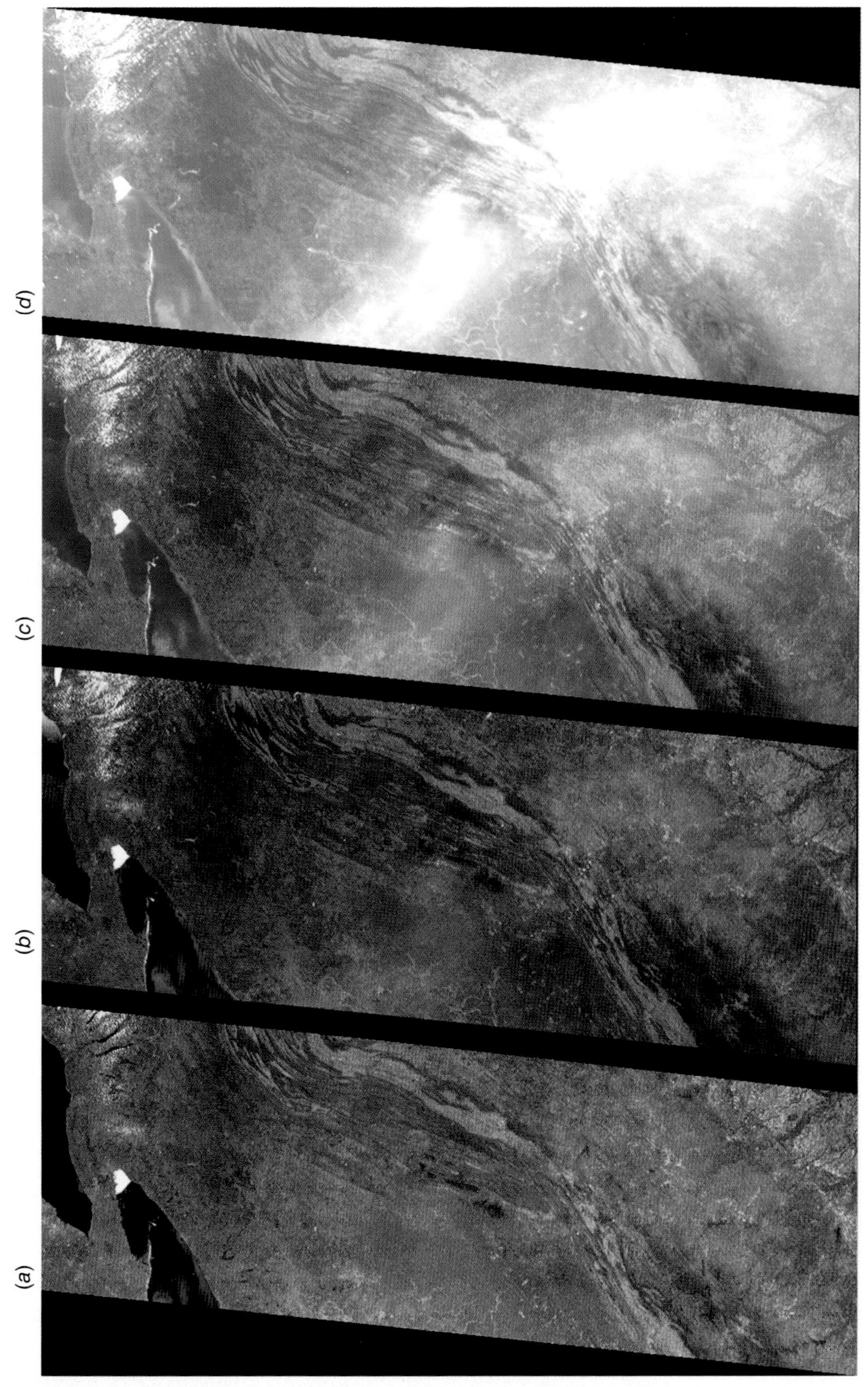

Plate 23 Multiangle images acquired by MISR over the eastern United States: (*a*) 0° (nadir-viewing), (*b*) 45.6° forward, (*c*) 60.0° forward, (*d*) 70.5° forward. (Courtesy NASA.) (For major discussion, see Section 5.19.)

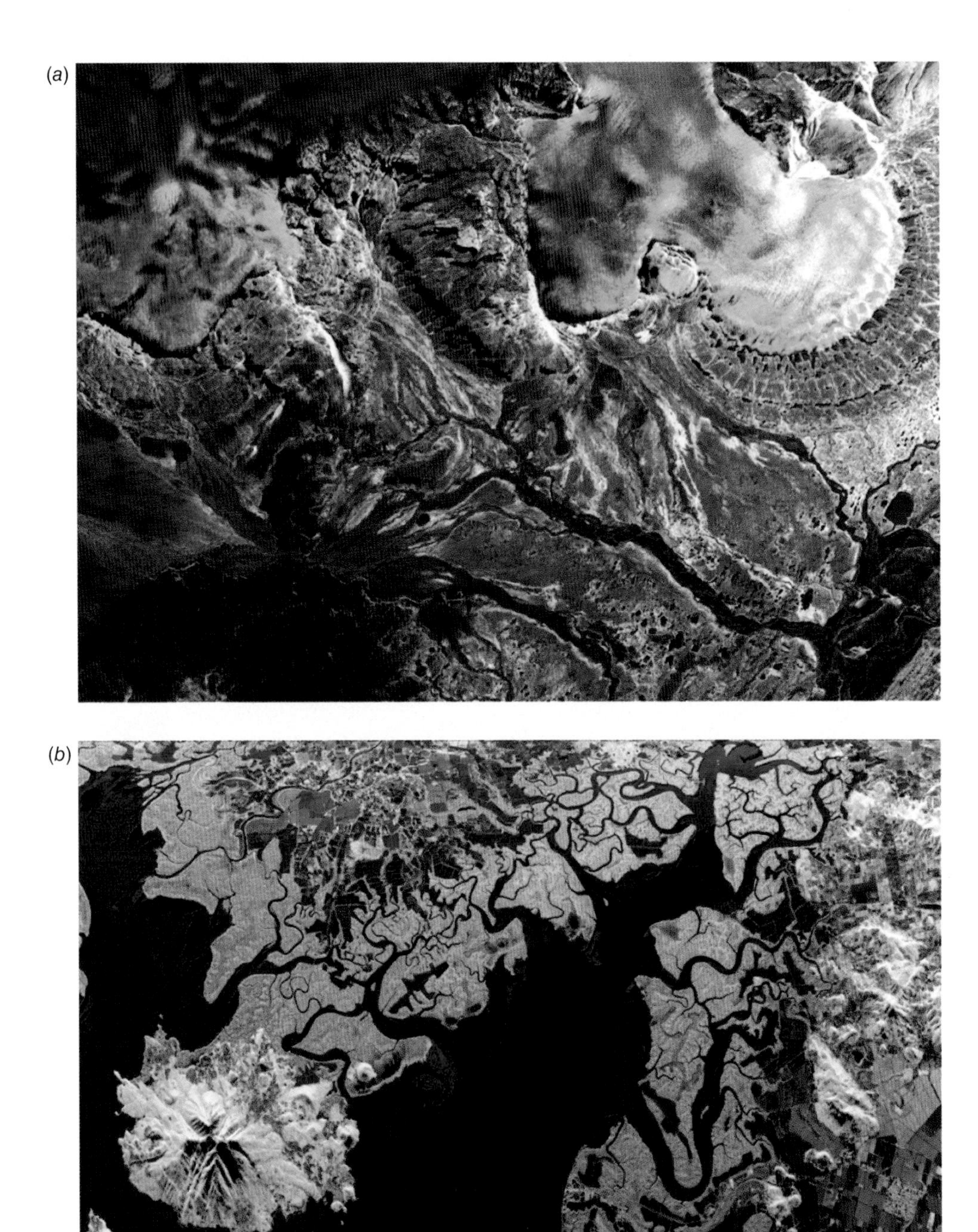

Plate 24 Polarimetric L-band radar images acquired from an uninhabited aerial vehicle (UAV). (*a*) Hofjokull (glacier), Iceland, June. Nominal scale 1:125,000. (*b*) Bahia de San Loranzo, Honduras, February. Nominal scale 1:300,000. (Courtesy NASA/JPL/Caltech.) (For major discussion, see Section 6.8.)

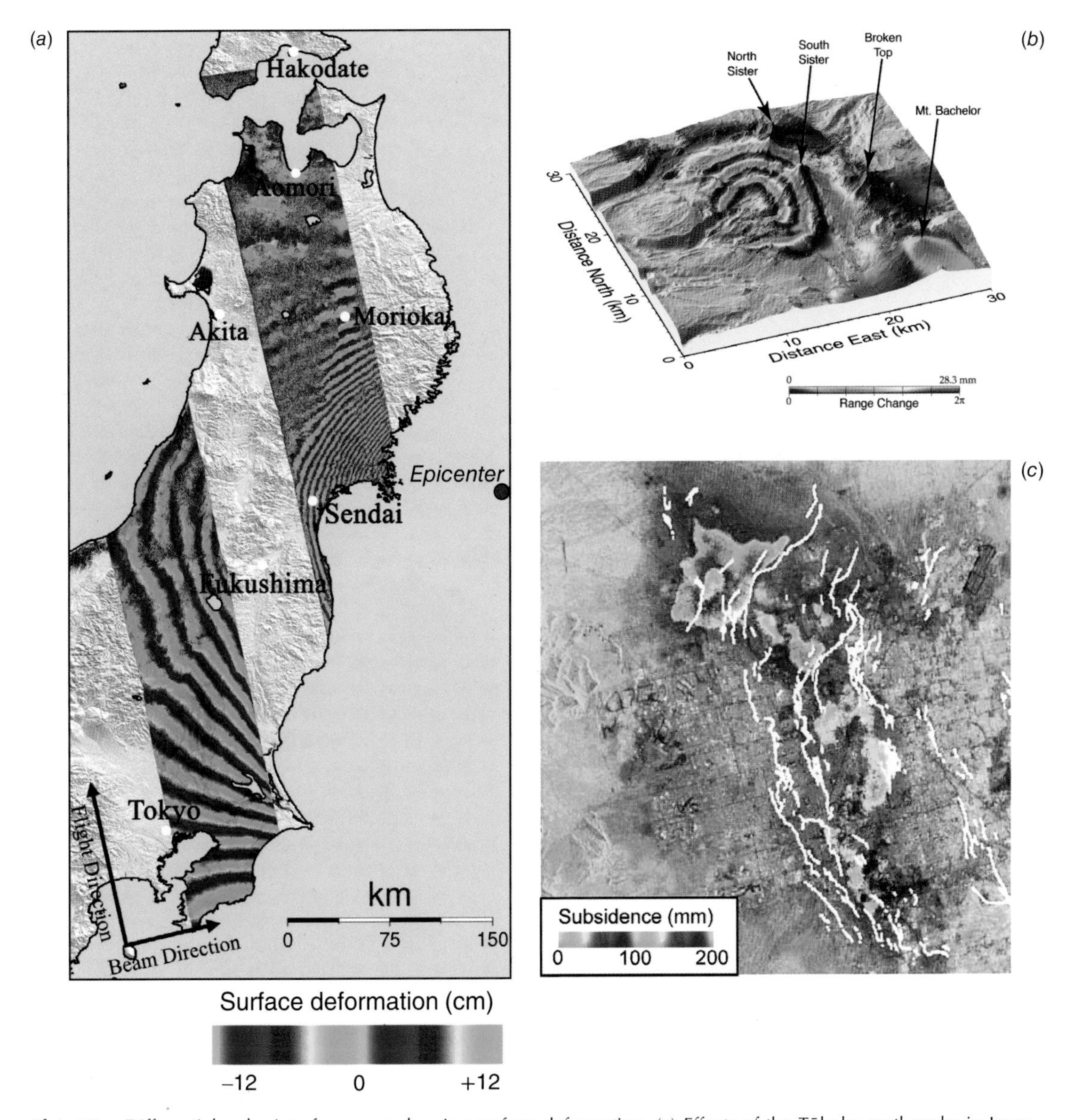

Plate 25 Differential radar interferograms showing surface deformation. (*a*) Effects of the Tōhoku earthquake in Japan, March 11, 2011. (Courtesy JAXA.) (*b*) Ground uplift from magma accumulation below South Sister volcano, OR. (Courtesy C. Wicks, USGS.) (*c*) Subsidence in Las Vegas, NV. White lines indicate surface faults. (Courtesy Stanford University Radar Interferometry Group.) (For major discussion, see Section 6.9.)

Plate 26 SIR-C color composite SAR image of a volcano-dominated landscape in central Africa, October 1994. Scale 1:360,000. (Courtesy NASA/JPL/Caltech.) (For major discussion, see Section 6.11.)

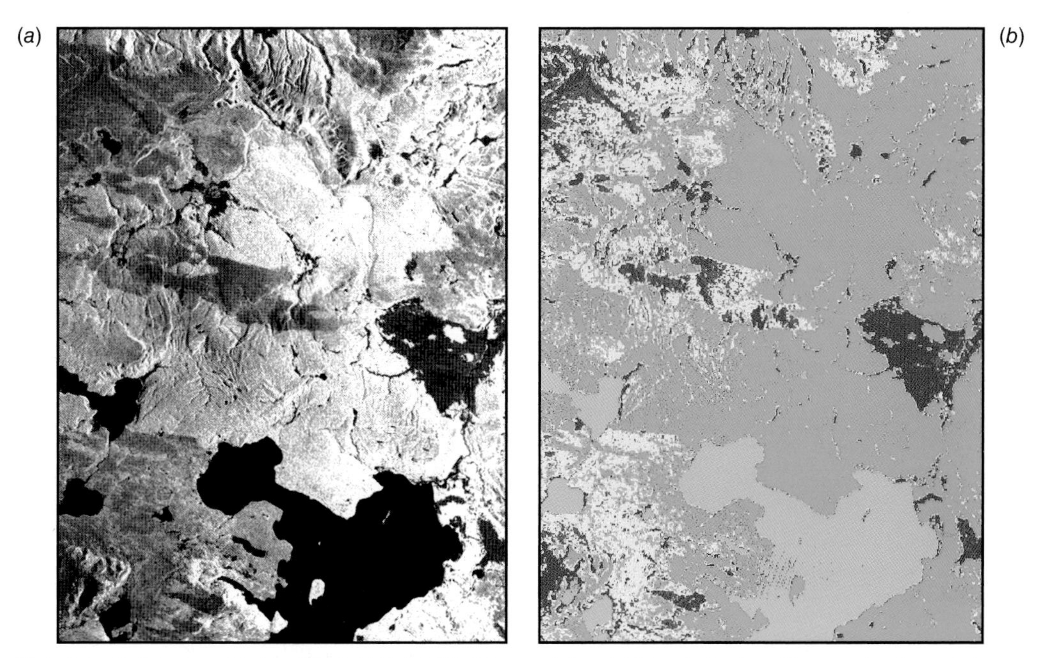

Plate 27 Yellowstone National Park, WY: (*a*) SIR-C image, L-band, VH-polarization; (*b*) map of estimated above-ground biomass derived from (*a*). (Courtesy NASA/JPL/Caltech.) (For major discussion, see Section 6.11.)

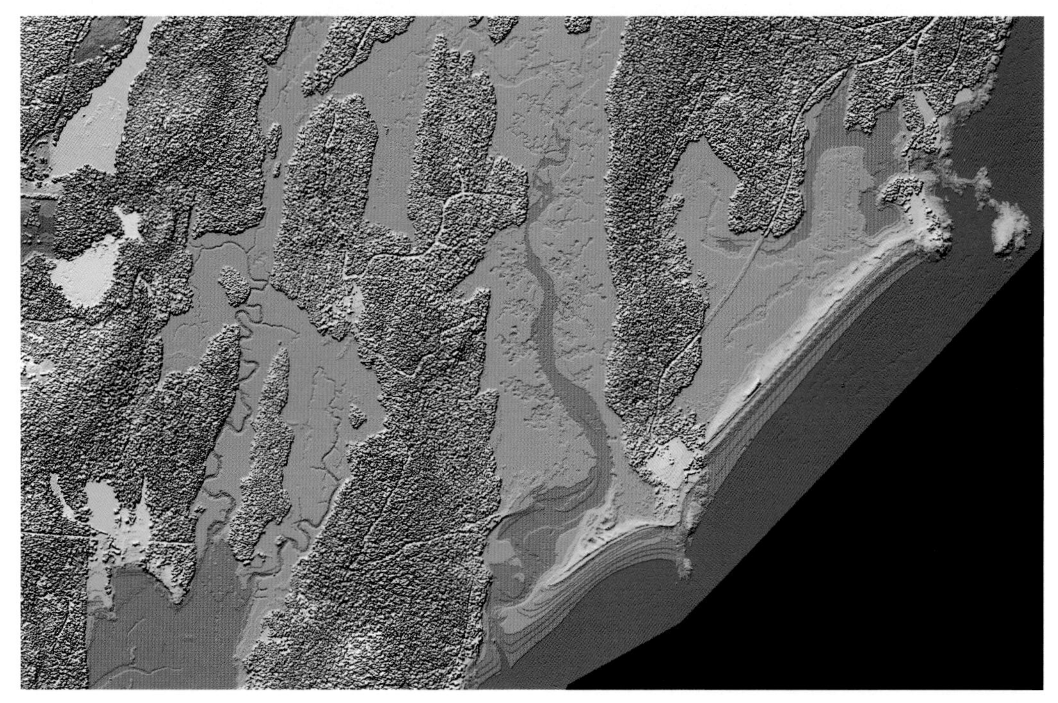

Plate 28 Shaded-relief DEM from airborne lidar data, Sagadahoc County, ME. Nearshore waters of the Gulf of Maine appear in blue, low-lying areas (primarily tidal salt marshes) appear in green, and higher elevations appear in colors ranging from light green to white. (Author-prepared figure.) (For major discussion, see Section 6.24.)

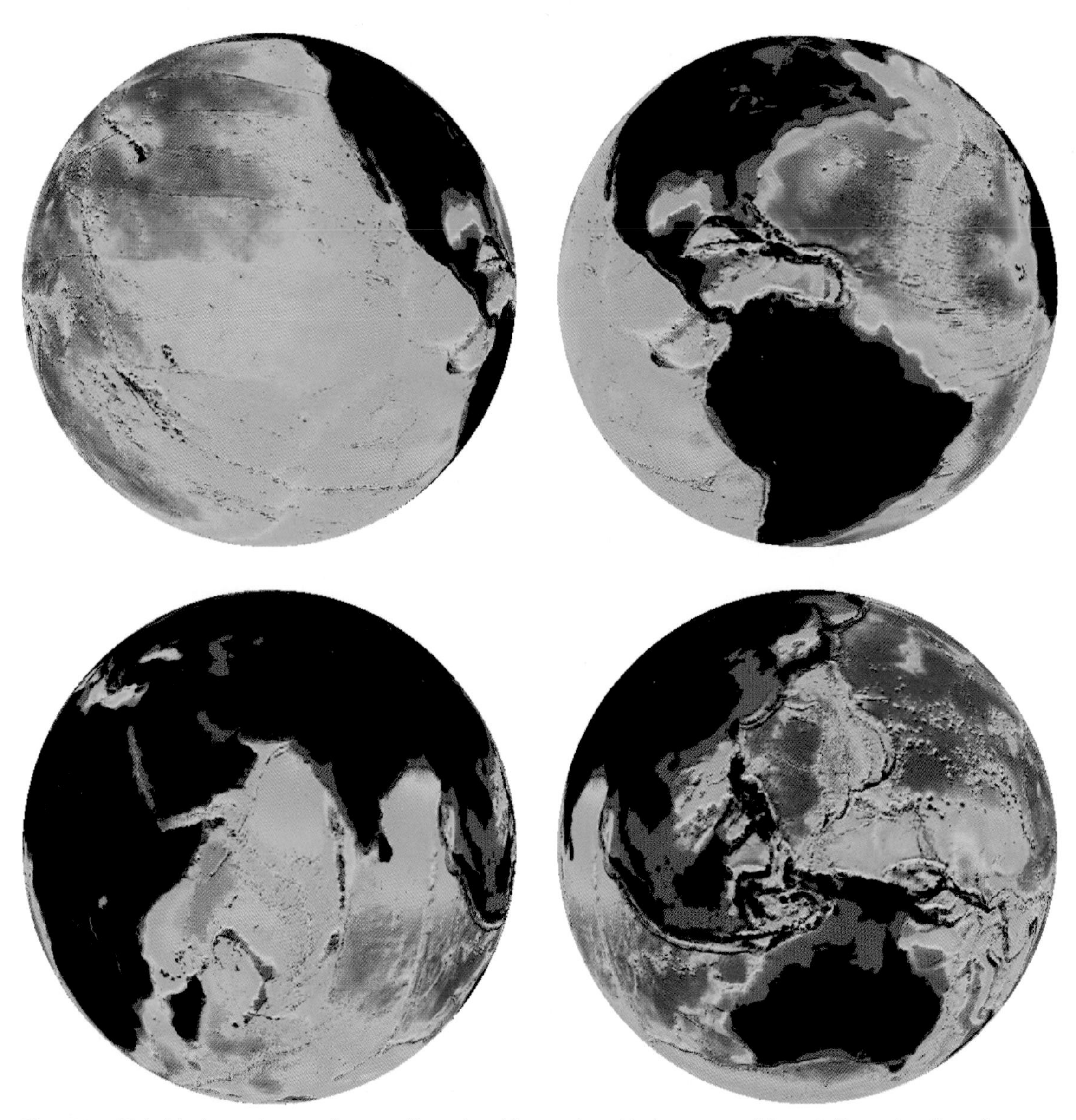

Plate 29 Global bathymetric maps from satellite radar altimetry data. (Author-prepared figure.) (For major discussion, see Section 6.21.)

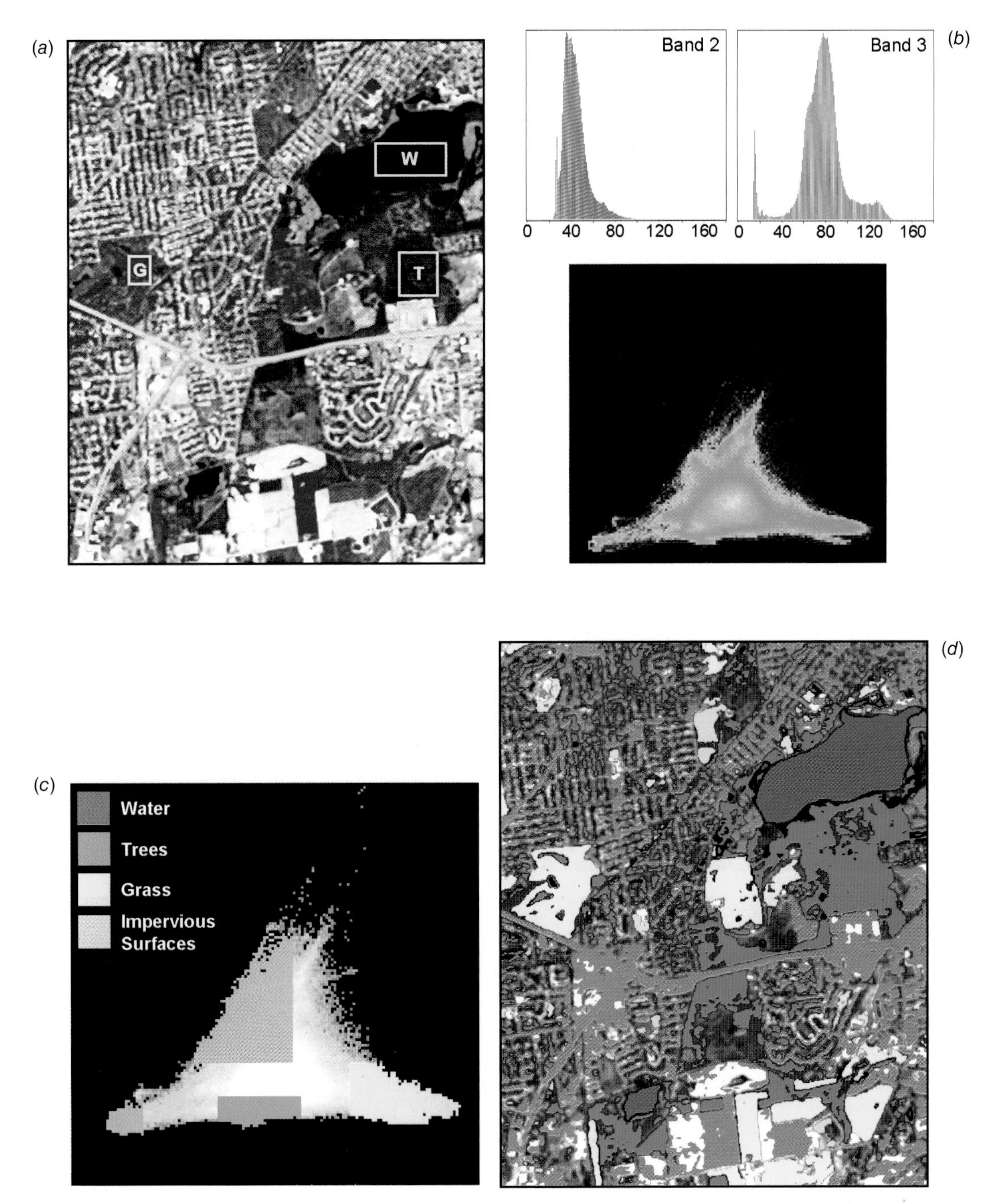

Plate 30 Sample interactive preliminary classification procedure used in the supervised training set refinement process: (*a*) original SPOT HRV color infrared composite image including selected training areas; (*b*) histograms and scatter diagram for band 2 (red) and band 3 (near-IR); (*c*) parallelepipeds associated with the initial training areas showing their locations in the band 2/band 3 scatter diagram; (*d*) partially completed classification superimposed on band 2 of the original images. (Author-prepared figure.) (For major discussion, see Section 7.10.)

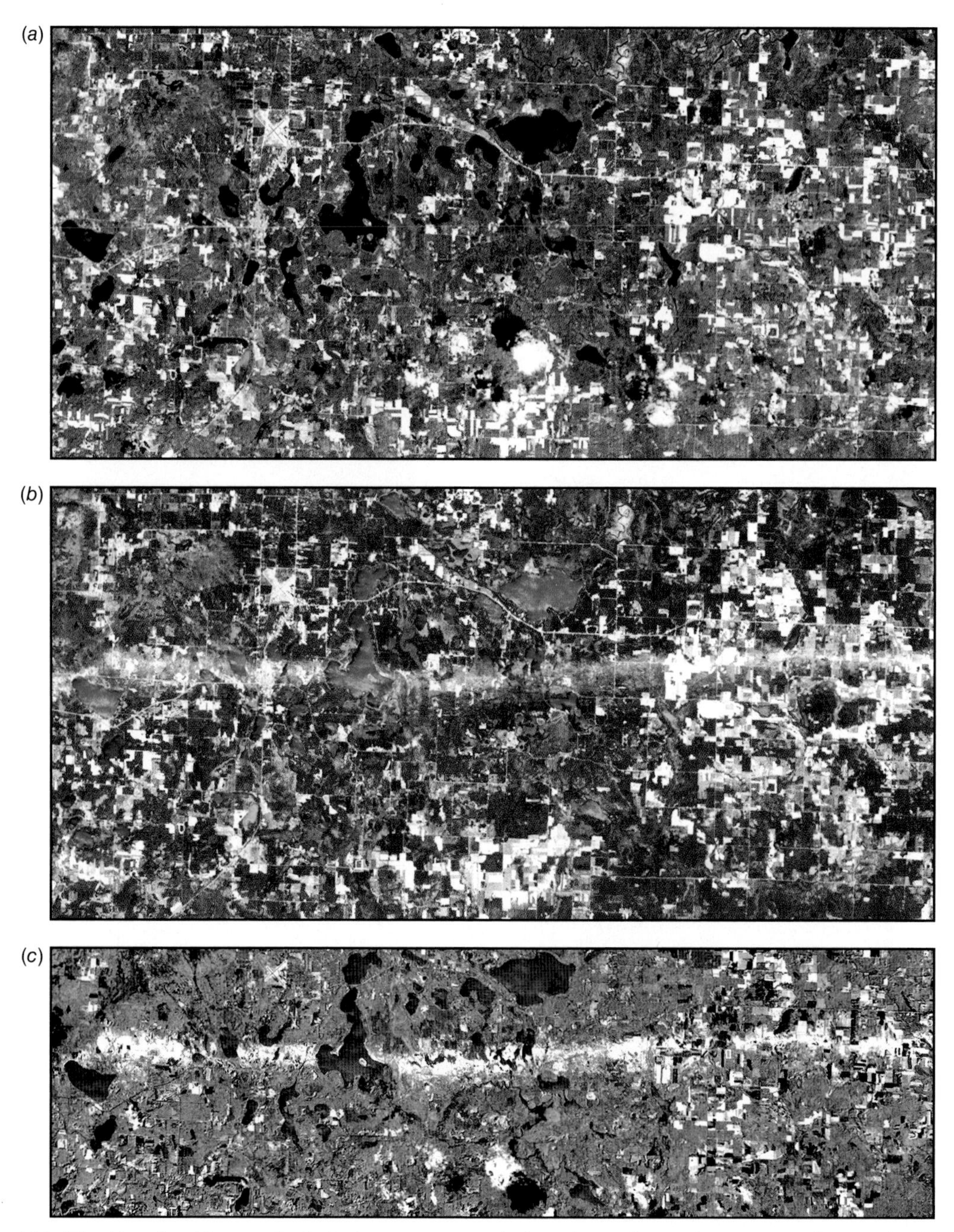

Plate 31 Use of multitemporal principal components analysis to detect change due to a tornado occurring in Burnett County in northwestern Wisconsin. (*a*) "Before" image acquired approximately one month prior to the tornado. (*b*) "After" image acquired one day after the tornado. (*c*) "Change" image depicting the damage path as portrayed in the second principal component derived from a composite of the "before" and "after" images. Scale 1:270,000. (Courtesy UW-Madison Environmental Remote Sensing Center.) (For major discussion, see Section 7.18.)

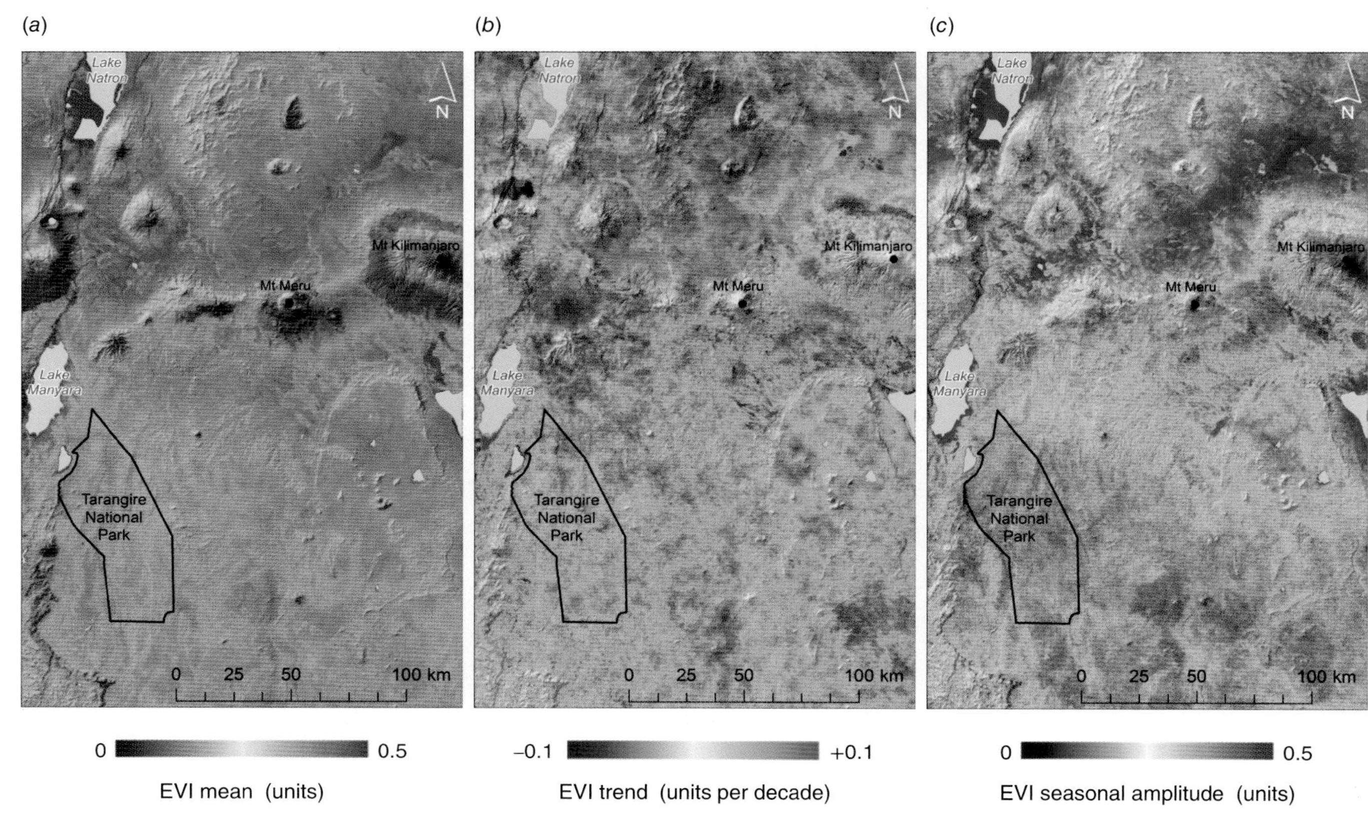

Plate 32 Landscape phenology from time series of MODIS enhanced vegetation index (EVI) data, for Tarangire region in northern Tanzania (2000–2011). (*a*) Long-term mean of EVI. (*b*) Decadal trend in EVI. (*c*) Amplitude of seasonal cycle. (Author-prepared figure.) (For major discussion, see Section 7.19.)

(a)

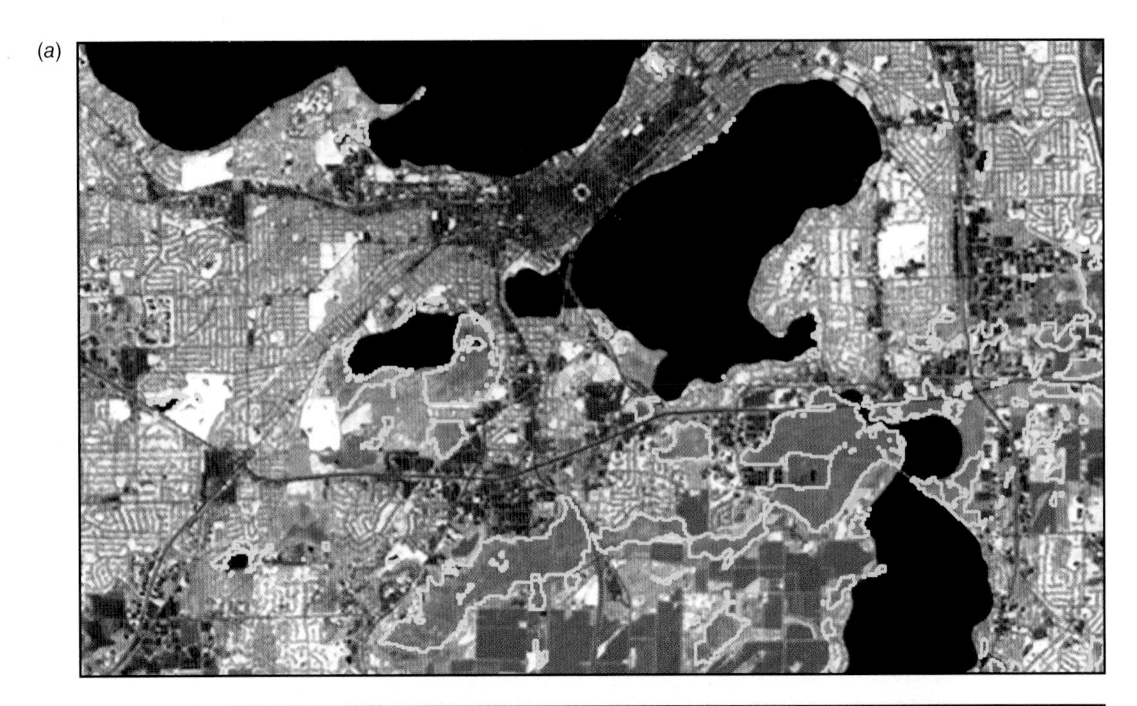

(b)

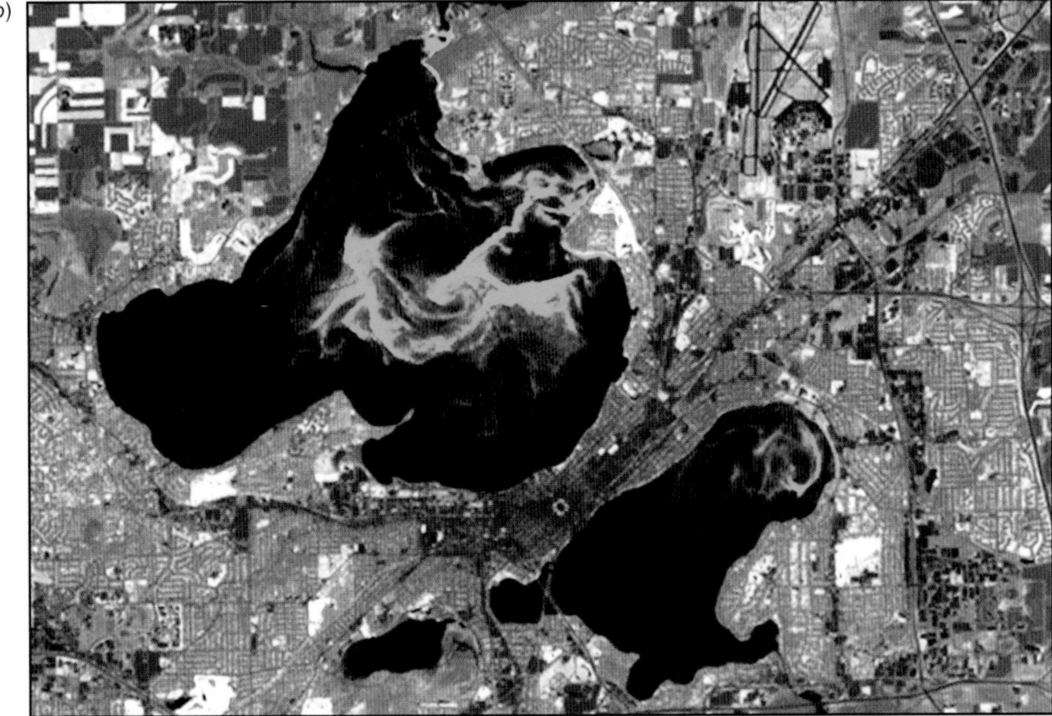

Plate 33 Multitemporal NDVI data merging. (*a*) Using multitemporal data as an aid in mapping invasive plant species, in this case reed canary grass (shown in red to pink tones) in wetlands. This color composite results from the merger of NDVI values derived from Landsat-7 ETM+ images of southern Wisconsin acquired on March 7 (blue), April 24 (green), and October 15 (red), respectively. Scale 1:120,000. (*b*) Monitoring algal blooms in portions of the two northernmost lakes shown. NDVI values derived from Landsat-7 ETM+ images acquired on April 24 (blue), October 31 (green), and October 15 (red), respectively. Scale 1:130,000. (Author-prepared figure.) (For major discussion, see Section 7.20.)

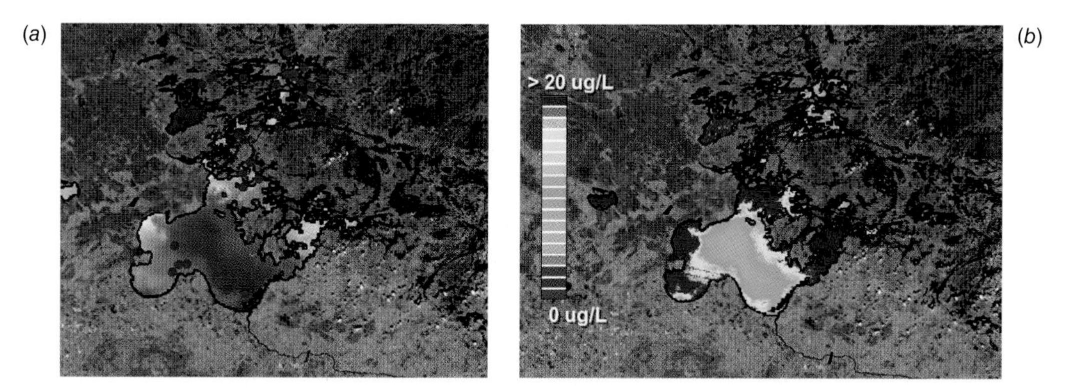

Plate 34 Biophysical modeling of chlorophyll concentration in lakes in Minnesota and Canada. (*a*) MODIS true-color composite (bands 1, 4, and 3) for lake pixels only, superimposed on grayscale background, with sampling locations identified (red dots). (*b*) Resulting pixel-level map of chlorophyll concentration. (Author-prepared figure.) (For major discussion, see Section 7.22.)

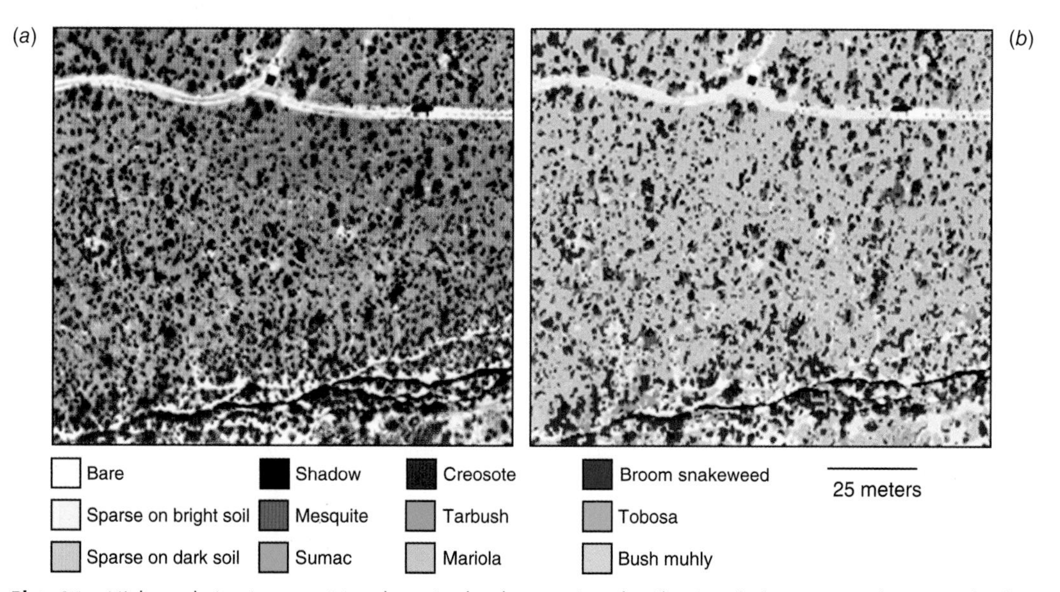

Plate 35 High resolution imagery (*a*) and species-level vegetation classification (*b*) from a UAV, for a rangeland in New Mexico, USA. (From Laliberte et al., 2011 [doi:10.3390/rs3112529]; Courtesy of the journal *Remote Sensing*.) (For major discussion, see Section 8.6.)

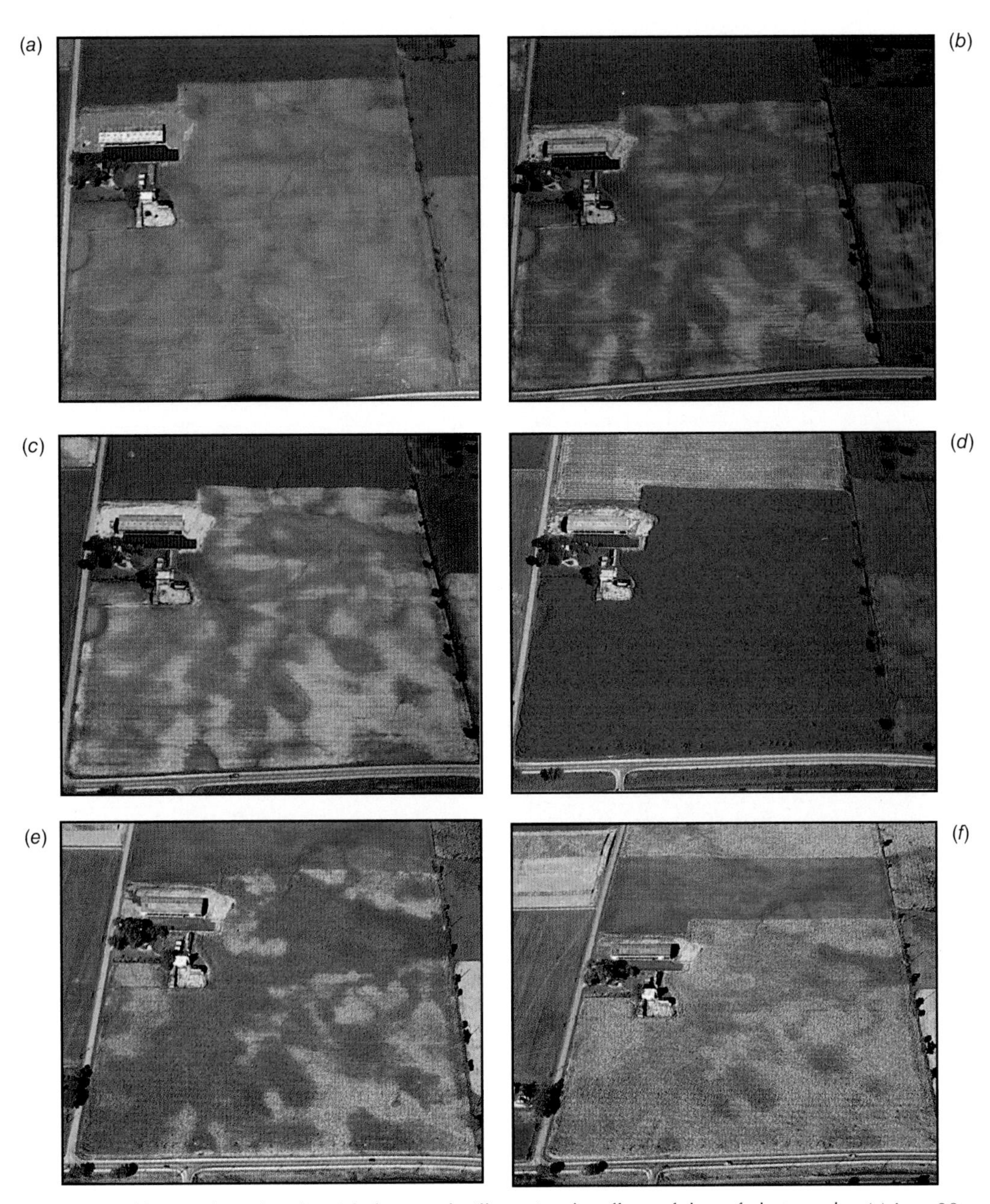

Plate 36 Oblique color infrared aerial photographs illustrating the effects of date of photography: (*a*) June 30; (*b*) July 1; (*c*) July 2; (*d*) August 11; (*e*) September 17; (*f*) October 8. Dane County, WI. Approximate horizontal scale at photo center is 1:7600. (Author-prepared figure.) (For major discussion, see Section 8.3.)

Plate 37 Digital camera image showing the aftermath of an F4 tornado striking Haysville, KS, on May 3, 1999. Scale 1:2300. (Courtesy Emerge.) (For major discussion, see Section 8.14.)

Plate 38 Volcanic eruption at Bárðarbunga, Iceland, seen in a Landsat-8 OLI image acquired on September 6, 2014. Landsat OLI bands 7, 5, and 3 displayed as red, green, and blue, respectively. (*a*) Scale 1:275,000. (*b*) Scale 1:1,050,000. (Author-prepared figure.) (For major discussion, see Section 8.14.)

(a)

(b)

Plate 39 QuickBird images of tsunami damage near Gleebruk, Indonesia. (*a*) Pre-tsunami image, April 12, 2004. (*b*) Post-tsunami image, January 2, 2005. (Courtesy DigitalGlobe.) (For major discussion, see Section 8.14.)

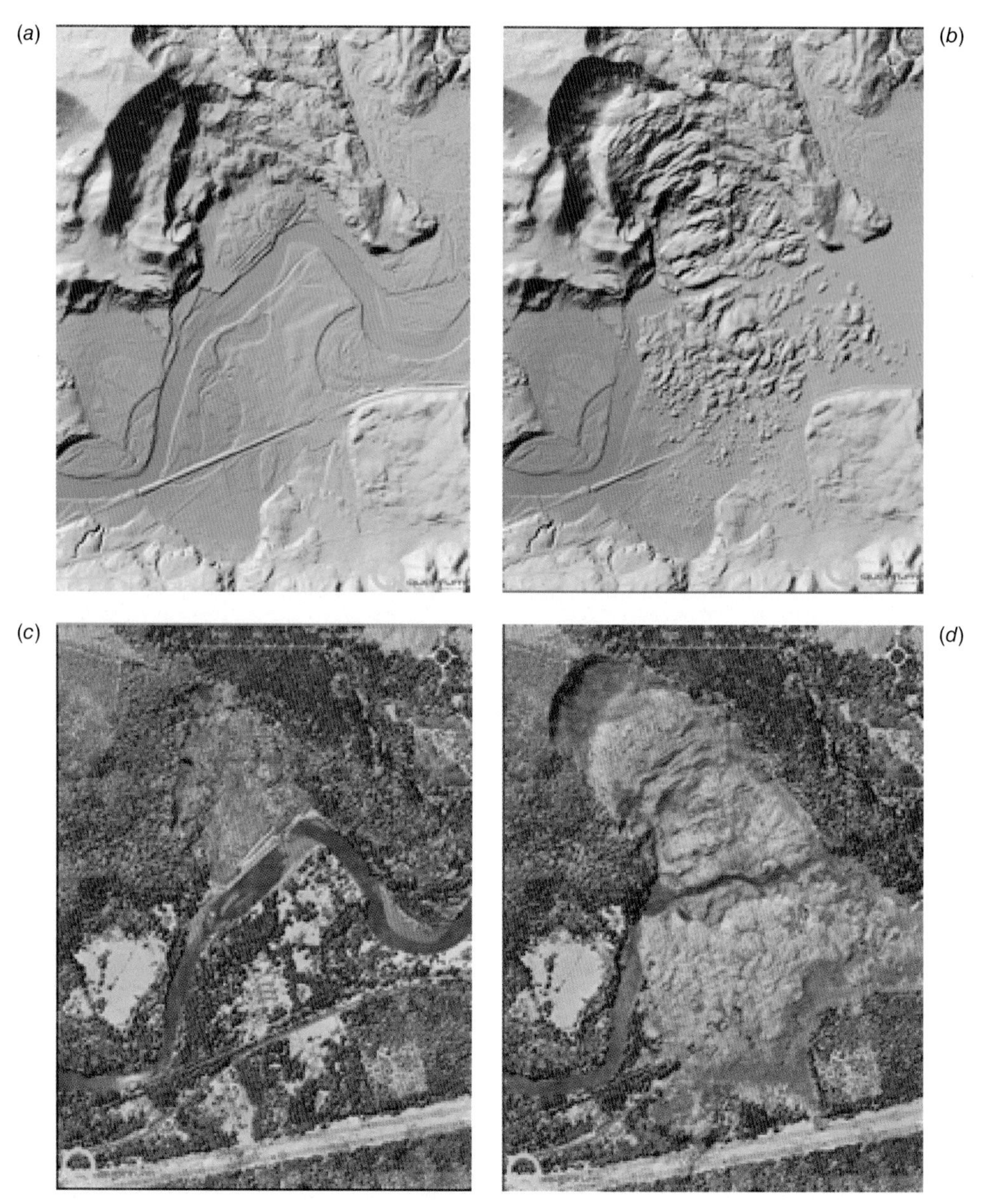

Plate 40 Oso landslide, Washington, March 2014. The landslide covered about 120 ha, with a volume of about 4,000,000 cubic meters of material moved downslope. (*a*) April 2013 lidar image; (*b*) lidar image two days after the slide; (*c*) April 2013 normal color aerial photograph; (*d*) normal color aerial photograph two days after the slide. Scale 1:23,000. (Image processing by Quantum Spatial. Courtesy Washington State Department of Transportation.) (For major discussion, see Section 8.14.)

2 ELEMENTS OF PHOTOGRAPHIC SYSTEMS

2.1 INTRODUCTION

One of the most common, versatile, and economical forms of remote sensing is aerial photography. Historically, most aerial photography has been film-based. In recent years, however, digital photography has become the most dominant form of newly collected photographic imagery. In this chapter, unless otherwise specified, when we use the term "aerial photography," we are referring to both film and digital aerial photography. The basic advantages that aerial photography affords over on-the-ground observation include:

1. **Improved vantage point.** Aerial photography gives a bird's-eye view of large areas, enabling us to see earth surface features in their spatial context. In short, aerial photography permits us to look at the "big picture" in which objects of interest reside. It is often difficult, if not impossible, to obtain this view of the environment through on-the-ground observation. With aerial photography, we also see the "whole picture" in that *all*

observable earth surface features are recorded simultaneously. Completely different information might be extracted by different people looking at a photograph. The hydrologist might concentrate on surface water bodies, the geologist on bedrock structure, the agriculturalist on soil or crop type, and so on.

2. **Capability to stop action.** Unlike the human eye, photographs can give us a "stop action" view of dynamic conditions. For example, aerial photographs are very useful in studying dynamic phenomena such as floods, moving wildlife populations, traffic, oil spills, and forest fires.
3. **Permanent recording.** Aerial photographs are virtually permanent records of existing conditions. As such, these records can be studied at leisure, under office rather than field conditions. A single image can be studied by a large number of users. Airphotos can also be conveniently compared against similar data acquired at previous times, so that changes over time can be monitored easily.
4. **Broadened spectral sensitivity.** With photography, invisible UV and near-IR energy can be detected and subsequently recorded in the form of a visible image; hence with photography we can see certain phenomena the eye cannot.
5. **Increased spatial resolution and geometric fidelity.** With the proper selection of camera, sensor, and flight parameters, we are able to record more spatial detail on a photograph than we can see with the unaided eye. With proper ground reference data, we can also obtain accurate measurements of positions, distances, directions, areas, heights, volumes, and slopes from airphotos.

This and the following chapter detail and illustrate the above characteristics of aerial photographs. In this chapter, we describe the various materials and methods used to *acquire* aerial photographs. In Chapter 3 we examine various aspects of *measuring* and *mapping* with airphotos (photogrammetry).

2.2 EARLY HISTORY OF AERIAL PHOTOGRAPHY

Photography was born in 1839 with the public disclosure of the pioneering photographic processes of Nicephore Niepce, William Henry Fox Talbot, and Louis Jacques Mande Daguerre. As early as 1840, Argo, Director of the Paris Observatory, advocated the use of photography for topographic surveying. The first known aerial photograph was taken in 1858 by a Parisian photographer named Gaspard-félix Tournachon. Known as "Nadar," he used a tethered balloon to obtain the photograph over Val de Bievre, near Paris.

As an outgrowth of their use in obtaining meteorological data, kites were used to obtain aerial photographs beginning in about 1882. The first aerial photograph taken from a kite is credited to an English meteorologist, E. D. Archibald. By

1890, A. Batut of Paris had published a textbook on the latest state of the art. In the early 1900s the kite photography of an American, G. R. Lawrence, brought him worldwide attention. On May 28, 1906, he photographed San Francisco approximately six weeks after the great earthquake and fire (Figure 2.1*a*). He

Figure 2.1 "San Francisco in Ruins," photographed by G. R. Lawrence on May 8, 1906, approximately six weeks after the great earthquake and fire. (*a*) Note the backlighting provided by the low sun angle when the image was taken (sun in upper right of photograph). (*b*) This enlargement includes a small area located in the center-left of (*a*) approximately 4.5 cm from the left edge and 1.9 cm from the bottom of the photograph. (A bright smoke plume is included in this area.) The damage to the various buildings present in this area is readily observable in this enlargement. (Courtesy Library of Congress.)

hoisted his personally constructed panoramic camera, which had a field of view of approximately 130°, some 600 m above San Francisco Bay using a train of 17 kites flown from a naval ship. He also designed a stabilizing mechanism for his large camera, which reportedly weighed 22 kg and employed celluloid-film plate negatives about 0.5 by 1.2 m in size (Baker, 1989). The impressive amount of detail recorded by his camera can be seen in Figure 2.1*b*.

The airplane, which had been invented in 1903, was not used as a camera platform until 1908, when a photographer accompanied Wilbur Wright and took the first aerial motion pictures (over Le Mans, France). Obtaining aerial photographs became a much more practical matter with the airplane than it had been with kites and balloons. Photography from aircraft received heightened attention in the interest of military reconnaissance during World War I, when more than one million aerial reconnaissance photographs were taken. After World War I, former military photographers founded aerial survey companies, and widespread aerial photography of the United States began. In 1934, the American Society of Photogrammetry (now the American Society for Photogrammetry and Remote Sensing) was founded as a scientific and professional organization dedicated to advancing this field.

In 1937, the U.S. Department of Agriculture's Agricultural Stabilization and Conservation Service (USDA-ASCS) began photographing selected counties of the United States on a repetitive basis.

During World War II, in a facility located 56 km from London, a team of some 2000 trained interpreters analyzed tens of millions of stereoscopic aerial photographs to discover German rocket launch sites and analyze the damage of Allied bombing raids. They used a stereoplotter (Section 3.10) to make accurate measurements from the aerial photographs.

The film-based archives of aerial photographs begun in the United States (and elsewhere) in the 1930s are now recognized as an invaluable record for monitoring landscape change. This record continues to build, but increasingly, digital cameras are being used to accomplish this purpose. The transition from the use of film-based to digital aerial cameras took approximately a decade. Development of digital cameras in the civil sector began during the 1990s. The first widely available digital aerial cameras were unveiled at the 2000 International Society for Photogrammetry and Remote Sensing (ISPRS) Congress held in Amsterdam, The Netherlands. Currently, nearly a dozen manufacturers of digital aerial cameras exist, and digital acquisition procedures overwhelmingly dominate those that are film-based. However, we treat both film-based and digital photographic systems in the following discussion for a number of reasons, including the facts that many film-based aerial photographic systems are still in use and many historical film photographs are now being converted to a digital format by means of scanning the film using a scanning densitometer to enable further digital analysis. Hence, it is important to understand the basic characteristics of such imagery in many applications.

2.3 PHOTOGRAPHIC BASICS

The basics of photography using both film and digital cameras are similar in many ways. In this section, we describe the simple camera, focus, exposure, geometric factors influencing exposure, and filters.

The Simple Camera

The cameras used in the early days of photography were often no more than a light-tight box with a pinhole at one end and the light-sensitive material to be exposed positioned against the opposite end (Figure 2.2*a*). The amount of exposure of the film was controlled by varying the time the pinhole was allowed to pass light. Often, exposure times were in hours because of the low sensitivity of the photographic materials available and the limited light-gathering capability of the pinhole design. In time, the pinhole camera was replaced by the simple lens camera, shown in Figure 2.2*b*. By replacing the pinhole with a lens, it became possible to enlarge the hole through which light rays from an object were collected to form an image, thereby allowing more light to reach the film in a given amount of time. In addition to the lens, an adjustable *diaphragm* and an adjustable *shutter* were introduced. The diaphragm controls the diameter of the lens opening during film exposure, and the shutter controls the duration of exposure.

The design and function of modern adjustable cameras is conceptually identical to that of the early simple lens camera. To obtain sharp, properly exposed photographs with such systems, they must be focused and the proper exposure settings must be made. We describe each of these operations separately.

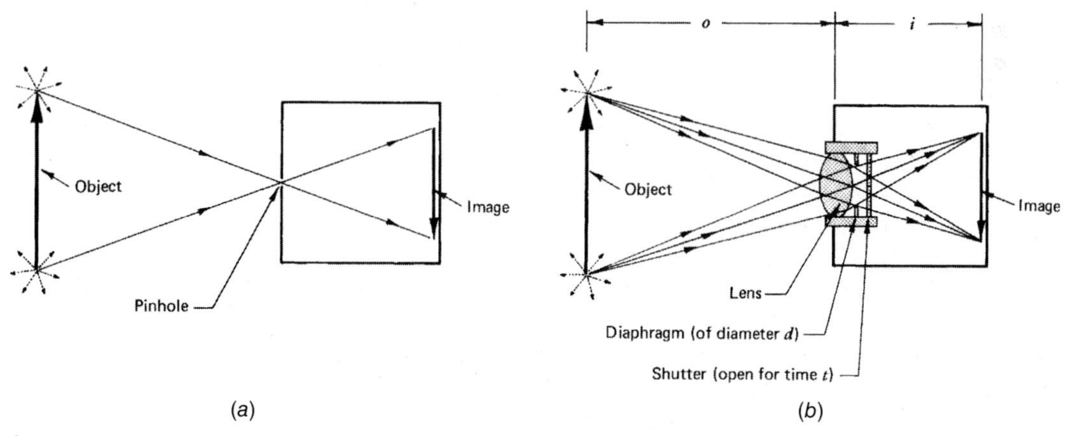

Figure 2.2 Comparison between (*a*) pinhole and (*b*) simple lens cameras.

Focus

Three parameters are involved in focusing a camera: the focal length of the camera lens, f, the distance between the lens and the object to be photographed, o, and the distance between the lens and the image plane, i. The focal length of a lens is the distance from the lens at which parallel light rays are focused to a point. (Light rays coming from an object at an infinite distance are parallel.) Object distance o and image distance i are shown in Figure 2.2*b*. When a camera is properly focused, the relationship among the focal length, object distance, and image distance is

$$\frac{1}{f} = \frac{1}{o} + \frac{1}{i} \tag{2.1}$$

Because f is a constant for any given lens, as object distance o for a scene changes, image distance i must change. This is done by moving the camera lens with respect to the image plane (film or electronic sensor). When focused on an object at a discrete distance, a camera can image over a range just beyond and in front of this distance with acceptable focus. This range is commonly referred to as the *depth of field*.

In aerial photography the object distances involved are effectively infinite. Hence the $1/o$ term in Eq. 2.1 goes to zero and i must equal f. Thus, most aerial cameras are manufactured with their image plane precisely located at a *fixed* distance f from their lens.

Exposure

The exposure[1] at any point in the image plane of a camera is determined by the irradiance at that point multiplied by the exposure time, expressed by

$$E = \frac{sd^2t}{4f^2} \tag{2.2}$$

where

E = exposure, J mm^{-2}
s = scene brightness, J mm^{-2} sec^{-1}
d = diameter of lens opening, mm
t = exposure time, sec
f = lens focal length, mm

It can be seen from Eq. 2.2 that, for a given camera and scene, the exposure can be varied by changing the camera shutter speed t and/or the diameter of the lens opening d. Various combinations of d and t will yield equivalent exposures.

---

[1]The internationally accepted symbol for exposure is H. To avoid confusion with the use of this symbol for flying height, we use E to represent "exposure" in our discussion of photographic systems. Elsewhere, E is used as the internationally accepted symbol for "irradiance."

EXAMPLE 2.1

A camera with a 40-mm-focal-length lens is producing properly exposed images with a lens opening diameter of 5 mm and an exposure time of $\frac{1}{125}$ sec (condition 1). If the lens opening is increased to 10 mm and the scene brightness does not change, what exposure time should be used to maintain proper exposure (condition 2)?

Solution

We wish to maintain the same exposure for conditions 1 and 2. Hence,

$$E_1 = \frac{s_1(d_1)^2 t_1}{4(f_1)^2} = \frac{s_2(d_2)^2 t_2}{4(f_2)^2} = E_2$$

Canceling constants, we obtain

$$(d_1)^2 t_1 = (d_2)^2 t_2$$

or

$$t_2 = \frac{(d_1)^2 t_1}{(d_2)^2} = \frac{5^2}{10^2} \cdot \frac{1}{125} = \frac{1}{500} \text{ sec}$$

The diameter of the lens opening of a camera is determined by adjusting the diaphragm to a particular *aperture setting*, or *f-stop*. This is defined by

$$F = \text{f-stop} = \frac{\text{lens focal length}}{\text{lens opening diameter}} = \frac{f}{d} \tag{2.3}$$

As can be seen in Eq. 2.3, as the f-stop number increases, the diameter of the lens opening decreases and, accordingly, the image exposure decreases. Because the *area* of the lens opening varies as the square of the diameter, the change in exposure with f-stop is proportional to the square root of the f-stop. Shutter speeds are normally established in sequential multiples of 2 ($\frac{1}{125}$ sec, $\frac{1}{250}$ sec, $\frac{1}{500}$ sec, $\frac{1}{1000}$ sec, ...). Thus, f-stops vary as the square root of 2 (f/1.4, f/2, f/2.8, f/4,...). Note, for example, that when the value of the f-stop is 2, it is written as f/2.

The interplay between f-stops and shutter speeds is well known to photographers. For constant exposure, an incremental change in shutter speed setting must be accompanied by an incremental change in f-stop setting. For example, the exposure obtained at $\frac{1}{500}$ sec and f/1.4 could also be obtained at $\frac{1}{250}$ sec and f/2. Short exposure times allow one to "stop action" and prevent blurring when photographing moving objects (or when the camera is moving, as in the case of aerial photography). Large lens-opening diameters (small f-stop numbers) allow more light to reach the image plane and are useful under low-light conditions. Small lens-opening diameters (large f-stop numbers) yield greater depth of field. The f-stop corresponding to the largest lens opening diameter is called the *lens speed*. The larger the lens-opening diameter (smaller f-stop number), the "faster" the lens is.

Using f-stops, Eq. 2.2 can be simplified to

$$E = \frac{st}{4F^2} \tag{2.4}$$

where $F = \text{f-stop} = f/d$.

Equation 2.4 is a convenient means of summarizing the interrelationship among exposure, scene brightness, exposure time, and f-stop. This relationship may be used in lieu of Eq. 2.2 to determine various f-stop and shutter speed settings that result in identical exposures.

EXAMPLE 2.2

A camera is producing properly exposed images when the lens aperture setting is f/8 and the exposure time is $\frac{1}{125}$ sec (condition 1). If the lens aperture setting is changed to f/4, and the scene brightness does not change, what exposure time should be used to yield a proper exposure (condition 2)? (Note that this is simply a restatement of the condition of Example 2.1.)

Solution

We wish to maintain the *same exposure* for conditions 1 and 2. With the scene brightness the same in each case,

$$E_1 = \frac{s_1 t_1}{4(F_1)^2} = \frac{s_2 t_2}{4(F_2)^2} = E_2$$

Canceling constants,

$$\frac{t_1}{(F_1)^2} = \frac{t_2}{(F_2)^2}$$

and

$$t_2 = \frac{t_1 (F_2)^2}{(F_1)^2} = \frac{1}{125} \cdot \frac{4^2}{8^2} = \frac{1}{500} \text{ sec}$$

Geometric Factors Influencing Exposure

Images are formed because of variations in scene brightness values over the area photographed. Ideally, in aerial photography, such variations would be related solely to variations in ground object type and/or condition. This assumption is a great oversimplification because many factors that have nothing to do with the type or condition of a ground feature can and do influence exposure. Because these factors influence exposure measurements but have nothing to do with true changes in ground cover type or condition, we term them *extraneous effects*. Extraneous effects are of two general types: geometric and atmospheric. Atmospheric effects were introduced in Sections 1.3 and 1.4; here we discuss the major geometric effects that influence film exposure.

Probably the most important geometric effect influencing exposure is *exposure falloff*. This extraneous effect is a variation in focal plane exposure purely associated with the distance an image point is from the image center. Because of falloff, *a ground scene of spatially uniform reflectance does not produce spatially uniform exposure in the focal plane*. Instead, for a uniform ground scene, exposure at the focal plane is at a maximum at the center of the image and decreases with radial distance from the center.

The factors causing falloff are depicted in Figure 2.3, which shows an image being acquired over a ground area assumed to be of uniform brightness. For a beam of light coming from a point directly on the optical axis, exposure E_0 is directly proportional to the area, A, of the lens aperture and inversely proportional to the square of the focal length of the lens, f^2. However, for a beam exposing a point at an angle θ off the optical axis, exposure E_θ is reduced from E_0 for three reasons:

1. The effective light-collecting area of the lens aperture, A, decreases in proportion to $\cos\theta$ when imaging off-axis areas ($A_\theta = A \cos\theta$).
2. The distance from the camera lens to the focal plane, f_θ, increases as $1/\cos\theta$ for off-axis points, $f_\theta = f/\cos\theta$. Because exposure varies inversely as the square of this distance, there is an exposure reduction of $\cos^2\theta$.
3. The effective size of an image area element, dA, projected perpendicular to the beam decreases in proportion to $\cos\theta$ when the element is located off-axis, $dA_\theta = dA \cos\theta$.

Combining the above effects, the overall theoretical reduction in exposure for an off-axis point is

$$E_\theta = E_0 \cos^4\theta \qquad (2.5)$$

where

θ = angle between the optical axis and the ray to the off-axis point
E_θ = exposure at the off-axis point
E_0 = exposure that would have resulted if the point had been located at the optical axis

The systematic effect expressed by the above equation is compounded by differential transmittance of the lens and by *vignetting effects* in the camera optics. Vignetting refers to internal shadowing resulting from the lens mounts and other

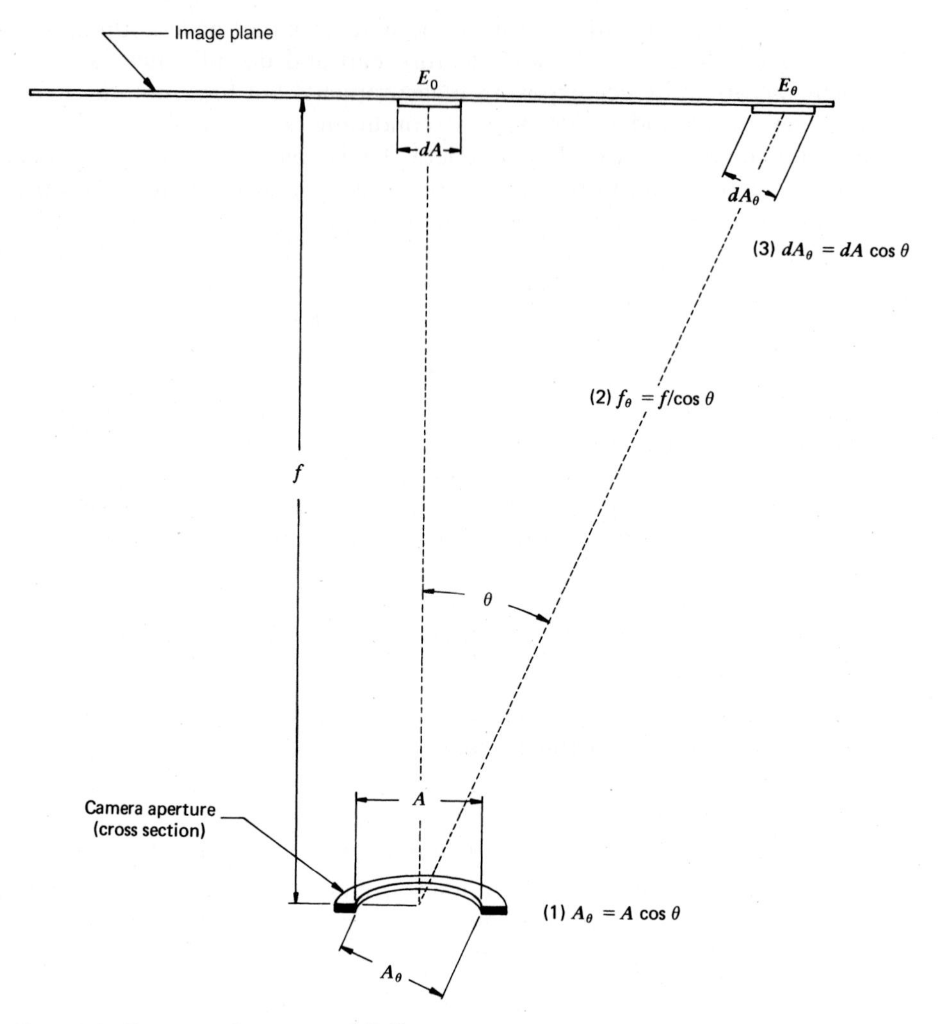

Figure 2.3 Factors causing exposure falloff.

aperture surfaces within the camera. The effect of vignetting varies from camera to camera and varies with aperture setting for any given camera.

Falloff and vignetting are normally mitigated at the time of exposure by using antivignetting filters. When such filters are not used, or when they fail to negate the exposure variations completely, it is appropriate to correct off-axis exposure values by normalizing them to the value they would possess had they been at the center of the photograph. This is done through the application of a correction model that is determined (for a given f-stop) by a radiometric calibration of the camera. This calibration essentially involves photographing a scene of uniform brightness, measuring exposure at various θ locations, and identifying the relationship that best describes the falloff. For most cameras this relationship takes on the form

$$E_\theta = E_0 \cos^n \theta \tag{2.6}$$

Because modern cameras are normally constructed in such a way that their actual falloff characteristics are much less severe than the theoretical $\cos^4$ falloff, n in the above equation is normally in the range 1.5 to 4. All exposure values measured off-axis are then corrected in accordance with the falloff characteristics of the particular camera in use.

Unique to digital cameras is yet another source of focal plane exposure variation, *pixel vignetting*. This results from the fact that most digital camera sensors are angle-dependent at the individual pixel level. Light striking such sensors at a right angle results in a signal that is greater than the same illumination striking the sensor at an oblique angle. Most digital cameras have an internal image–processing capability to apply corrections for this and the other natural, optical, and mechanical vignetting effects. These corrections are normally applied during the process of converting the raw sensor data to a standard format such as TIFF or JPEG.

The location of an object within a scene can also affect the resulting exposure, as was discussed in Section 1.4.

Filters

Through the use of filters, we can be selective about which wavelengths of energy reflected from a scene we allow to reach the image plane (film or electronic). Filters are transparent (glass or gelatin) materials that, by absorption or reflection, eliminate or reduce the energy reaching the image plane in selected portions of the photographic spectrum. They are placed in the optical path of a camera in front of the lens.

Aerial camera filters consist mainly of organic dyes suspended in glass or in a dried gelatin film. Filters are most commonly designated by *Kodak Wratten* filter numbers. They come in a variety of forms having a variety of spectral transmittance properties. The most commonly used spectral filters are *absorption filters*. As their name indicates, these filters absorb and transmit energy of selected wavelengths. A "yellow" filter, for example, absorbs blue energy incident upon it and transmits green and red energy. The green and red energy combine to form yellow—the color we would see when looking through the filter if it is illuminated by white light (see Plate 3*b*).

Absorption filters are often used to permit differentiation of objects with nearly identical spectral response patterns in major portions of the photographic spectrum. For example, two objects may appear to reflect the same color when viewed only in the visible portion of the spectrum but may have different reflection characteristics in the UV or near-IR region.

Figure 2.4 illustrates generalized spectral reflectance curves for natural grass and artificial turf. Because the artificial turf is manufactured with a green color to

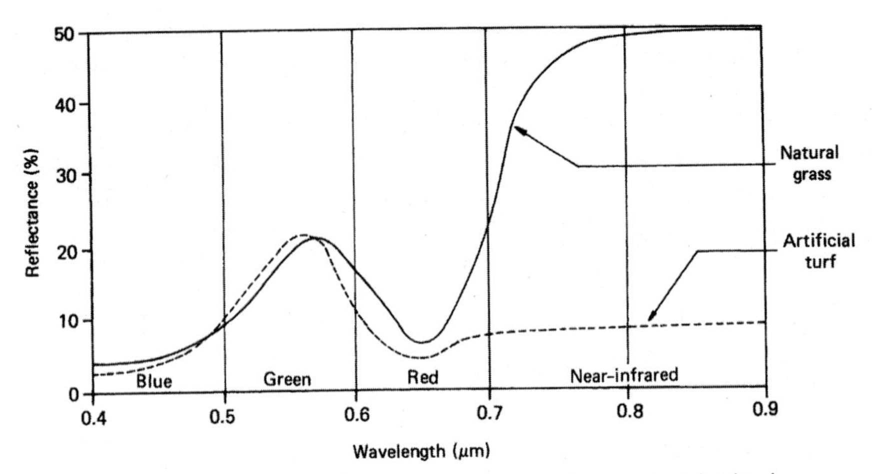

Figure 2.4 Generalized spectral reflectance curves for natural grass and artificial turf.

visually resemble natural grass, the reflectance in blue, green, and red is similar for both surfaces. However, the natural grass reflects very highly in the near IR whereas the artificial turf does not. In Figure 2.5, we illustrate distinguishing between natural grass and artificial turf using photography in selected bands. Illustrated in (*a*) is an image in the visible part of the spectrum (0.4 to 0.7 μm) that records natural grass (outside the stadium) and artificial turf (inside the stadium) with a similar tone. Illustrated in (*b*) is an image where the scene reflectance is filtered such that only wavelengths longer than 0.7 μm are recorded, with the result that the natural grass has a light tone (high IR reflectance) and the artificial turf a very dark tone (low IR reflectance). The filter used in such photography, which selectively absorbs energy below a certain wavelength, is referred to as a *short wavelength blocking* filter or a *high pass* filter.

When one is interested in sensing the energy in only an isolated narrow portion of the spectrum, a *bandpass* filter may be used. Wavelengths above and below a specific range are blocked by such a filter. The spectral transmittance curves for a typical high pass filter and a bandpass filter are illustrated in Figure 2.6. Several high pass and bandpass filters may be used simultaneously to selectively photograph various wavelength bands as separate images.

There is a large selection of filters from which to choose for any given application. Manufacturers' literature describes the spectral transmittance properties of each available type (e.g., Eastman Kodak Company, 1990). It should be noted that *low pass* absorption filters are not available. *Interference* filters must be used when short wavelength transmittance is desired. These filters reflect rather than absorb unwanted energy. They are also used when extremely narrow bandpass characteristics are desired.

Antivignetting filters are often used to improve the uniformity of exposure throughout an image. As described earlier in this section, there is a geometrically

Figure 2.5 Simultaneous oblique aerial photographs showing the effect of filtration on discrimination of ground objects. Illustrated in (*a*) is an image in the visible part of the spectrum (0.4 to 0.7 μm) that records natural grass (outside the stadium) and artificial turf (inside the stadium) with a similar tone. Illustrated in (*b*) is an image in which the scene reflectance is filtered such that only wavelengths longer than 0.7 μm are recorded, with the result that the natural grass has a light tone and the artificial turf a very dark tone. (Author-prepared figure.)

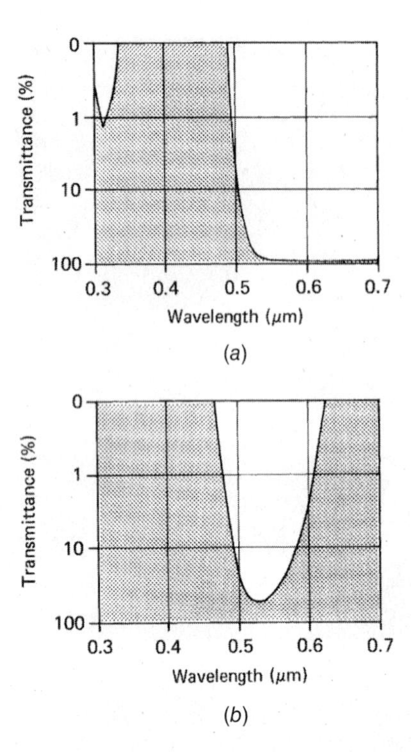

Figure 2.6 Typical transmission curves for filter types commonly used in aerial photography: (*a*) typical high pass filter (Kodak Wratten No. 12); (*b*) typical bandpass filter (Kodak Wratten No. 58). (Adapted from Eastman Kodak Company, 1992.)

based decrease in illumination with increasing distance from the center of a photograph. To negate the effect of this illumination *falloff*, antivignetting filters are designed to be strongly absorbing in their central area and progressively transparent in their circumferential area. To reduce the number of filters used, and thereby the number of between-filter reflections possible, antivignetting features are sometimes built into haze and other absorption filters.

Polarizing filters are sometimes used in photography. Unpolarized light vibrates in all directions perpendicular to the direction of propagation. As shown in Figure 2.7, a linear polarizing filter transmits only light vibrating in one plane. Unpolarized light striking various natural surfaces can be reflected with at least

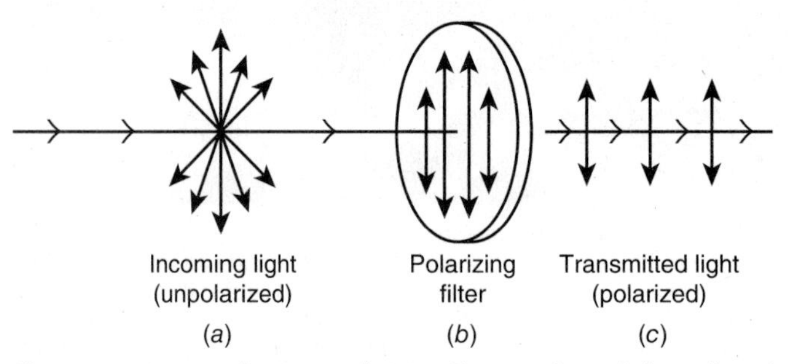

Figure 2.7 Process of using a polarizing filter to polarize light. (Adapted from American Society of Photogrammetry, 1968.)

partial polarization. The amount of polarization depends on the angle of incidence of the incoming light (see Figure 1.13), the nature of the surface, and the viewing angle. It also depends on the azimuth angle (see Figure 1.14). Polarizing filters can be used to reduce reflections in photographs of smooth surfaces (e.g., glass windows and water bodies). The effect of polarization is the greatest when the angle of reflection is approximately 35%. This often limits the use of polarizing filters in vertical aerial photography.

2.4 FILM PHOTOGRAPHY

In film photography, the resulting images can be either black and white or color, depending on the film used. Black and white photographs are typically exposed on negative film from which positive paper prints are made. Color images are typically recorded directly on film, which is processed to a positive transparency (although color negative films exist as well).

In black and white photography using negative film and positive prints, each of these materials consists of a light-sensitive photographic *emulsion* coated onto a *base*. The generalized cross sections of black-and-white film and print paper are shown in Figure 2.8*a* and *b*. In both cases, the emulsion consists of a thin layer of light-sensitive silver halide crystals, or grains, held in place by a solidified gelatin.

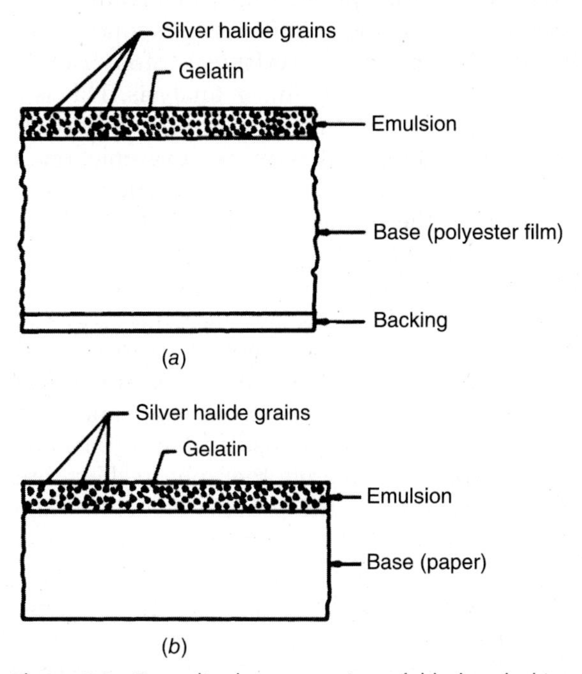

Figure 2.8 Generalized cross section of black-and-white photographic materials: (*a*) film and (*b*) print paper. (Adapted from Eastman Kodak Company, 1992.)

Paper is the base material for paper prints. Various plastics are used for film bases. When exposed to light, the silver halide crystals within an emulsion undergo a photochemical reaction, forming an invisible *latent image*. Upon treatment with suitable agents in the *development process*, these exposed silver salts are reduced to silver grains that appear black, forming a visible image. The number of black crystals at any point on the film is proportional to the exposure at that point. Those areas on the negative that were not exposed are clear after processing because crystals in these areas are dissolved as part of the development process. Those areas of the film that were exposed become various shades of gray, depending on the amount of exposure. Hence a "negative" image of reversed tonal rendition is produced.

Most aerial photographic paper prints are produced using the negative-to-positive sequence and a *contact printing procedure*. Here, the film is exposed and processed as usual, resulting in a negative of reversed scene geometry and brightness. The negative is then placed in emulsion-to-emulsion contact with print paper. Light is passed through the negative, thereby exposing the print paper. When processed, the image on the print is a positive representation of the original ground scene at the size of the negative.

Film Density and Characteristic Curves

The *radiometric* characteristics of aerial photographs determine how a specific film—exposed and processed under specific conditions—responds to scene energy of varying intensity. Knowledge of these characteristics is often useful, and sometimes essential, to the process of photographic image analysis. This is particularly true when one attempts to establish a quantitative relationship between the tonal values on an image and some ground phenomenon. For example, one might wish to measure the darkness, or optical density, of a transparency at various image points in a cornfield and correlate these measurements with a ground-observed parameter such as crop yield. If a correlation exists, the relationship could be used to predict crop yield based on photographic density measurements at other points in the scene. Such an effort can be successful only if the radiometric properties of the particular film under analysis are known. Even then, the analysis must be undertaken with due regard for such extraneous sensing effects as differing levels of illumination across a scene, atmospheric haze, and so on. If these factors can be sufficiently accounted for, considerable information can be extracted from the tonal levels expressed on a photograph. In short, image density measurements may sometimes be used in the process of determining the type, extent, and condition of ground objects. In this section, we discuss the interrelationships between film exposure and film density, and we explain how *characteristic curves* (plots of film density versus log exposure) are analyzed.

A photograph can be thought of as a visual record of the response of many small detectors to energy incident upon them. The energy detectors in a photographic record on film are the silver halide grains in the film emulsion, and the energy causing the response of these detectors is referred to as a film's *exposure*. During the instant that a photograph is exposed, the different reflected energy levels in the scene irradiate the film for the same length of time. A scene is visible on a processed film only because of the irradiance differences that are caused by the reflectance differences among scene elements. Thus, *film exposure at a point in a photograph is directly related to the reflectance of the object imaged at that point. Theoretically, film exposure varies linearly with object reflectance, with both being a function of wavelength.*

There are many ways of quantifying and expressing film exposure. Most photographic literature uses units of the form *meter-candle-second* (*MCS*) or *ergs/cm*^{2}. The student first "exposed" to this subject might feel hopelessly lost in understanding unit equivalents in photographic radiometry. This comes about because many exposure calibrations are referenced to the sensitivity response of the human eye, through definition of a "standard observer." Such observations are termed *photometric* and result in photometric, rather than radiometric, units. To avoid unnecessary confusion over how exposure is measured and expressed in absolute terms, we will deal with *relative* exposures and not be directly concerned about specifying any absolute units.

The result of exposure at a point on a film, after development, is a silver deposit whose darkening, or light-stopping, qualities are systematically related to the amount of exposure at that point. One measure of the "darkness" or "lightness" at a given point on a film is *opacity* O. Because most quantitative remote sensing image analyses involve the use of negatives or diapositives, opacity is determined through measurement of film *transmittance* T. As shown in Figure 2.9, transmittance T is the ability of a film to pass light. At any given point p, the transmittance is

$$T_p = \frac{\text{light passing through the film at point } p}{\text{total light incident upon the film at point } p} \tag{2.7}$$

Opacity O at point p is

$$O_p = \frac{1}{T_p} \tag{2.8}$$

Although transmittance and opacity adequately describe the "darkness" of a film emulsion, it is often convenient to work with a logarithmic expression, *density*. This is an appropriate expression, because the human eye responds to light levels nearly logarithmically. Hence, there is a nearly linear relationship between image density and its visual tone. Density D at a point p is defined as the common logarithm of film opacity at that point:

$$D_p = \log(O_p) = \log\left(\frac{1}{T_p}\right) \tag{2.9}$$

Instruments designed to measure density by shining light through film transparencies are called *transmission densitometers*. Density measurements may also be

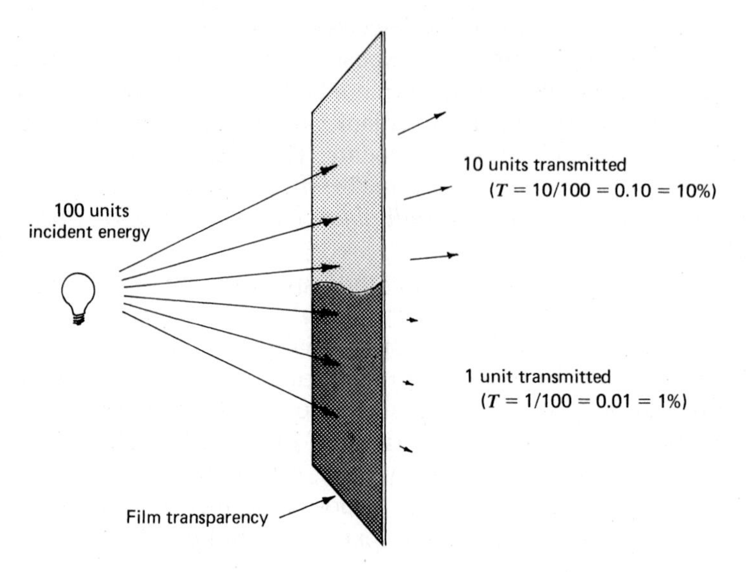

Figure 2.9 Film transmittance. To measure transmittance, a negative or positive transparency is illuminated from one side and the light transmitted through the image is measured on the other. Shown is a section of an image having a transmittance of 0.10 (or 10%) at one image point and 0.01 (or 1%) at another.

made from paper prints with a *reflectance densitometer*, but more precise measurements can be made on the original film material. When analyzing density on a transparency, the process normally involves placing the film in a beam of light that passes through it. The darker an image is, the smaller the amount of light that is allowed to pass, the lower the transmittance, the higher the opacity, and the higher the density. Some sample values of transmittance, opacity, and density are indicated in Table 2.1.

TABLE 2.1 Sample Transmittance, Opacity, and Density Values

Percent Transmittance	T	O	O
100	1.0	1	0.00
50	0.50	2	0.30
25	0.25	4	0.60
10	0.1	10	1.00
1	0.01	100	2.00
0.1	0.001	1000	3.00

There are some basic differences between the nature of light absorptance in black-and-white versus color films. Densities measured on black-and-white film are controlled by the amount of developed silver in the image areas of measurement. In color photography, the processed image contains no silver, and densities are caused by the absorption characteristics of three dye layers in the film: yellow, magenta, and cyan. The image analyst is normally interested in investigating the image density of each of these dye layers separately. Hence, color film densities are normally measured through each of three filters chosen to isolate the spectral regions of maximum absorption of the three film dyes.

An essential task in quantitative film analysis is to relate image density values measured on a photograph to the exposure levels that produced them. This is done to establish the cause (exposure) and effect (density) relationship that characterizes a given photograph.

Because density is a logarithmic parameter, it is convenient to also deal with exposure E in logarithmic form (log E). If one plots density values as a function of the log E values that produced them, curves similar to that shown in Figure 2.10 will be obtained.

The curve shown in Figure 2.10 is for a typical black-and-white negative film. Every film has a unique *D*–log *E curve*, from which many of the characteristics of the film may be determined. Because of this, these curves are known as *characteristic curves*. Characteristic curves are different for different film types, for different manufacturing batches within a film type, and even for films of the same batch. Manufacturing, handling, storage, and processing conditions all affect the response

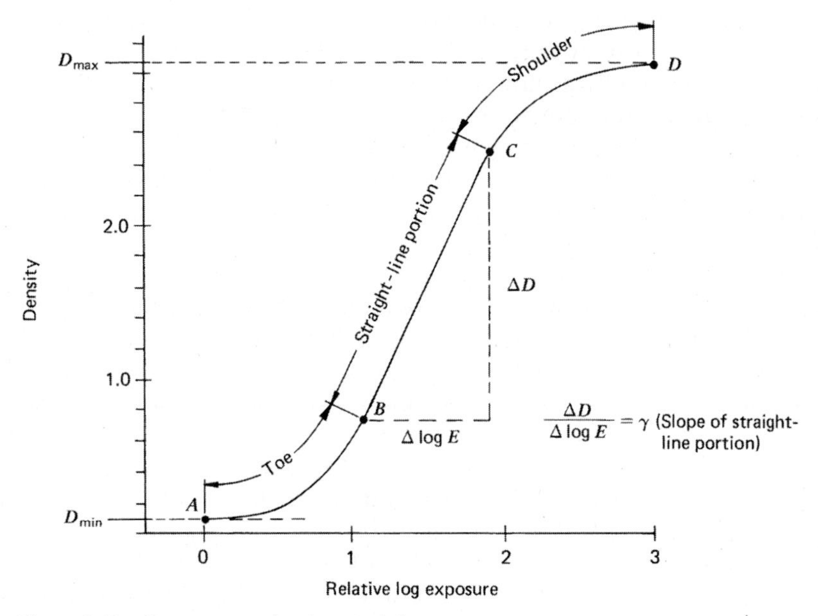

Figure 2.10 Components of a characteristic curve.

of a film (indicated by its D–log E curve). In the case of color film, characteristic curves also differ between one emulsion layer and another.

Figure 2.10 illustrates the various film response characteristics extractable from a D–log E curve. The curve shown is typical of a black-and-white negative film (similar characteristics are found for each layer of a color film). There are three general divisions to the curve. First, as the exposure increases from that of point A to that of point B, the density increases from a minimum, D_{min}, at an increasing rate. This portion of the curve is called the *toe*. As exposure increases from point B to point C, changes in density are nearly linearly proportional to changes in log exposure. This region is called the *straight-line portion* of the curve. Finally, as log exposure increases from point C to point D, the density increases at a decreasing rate. This portion is known as the *shoulder* of the curve. The shoulder terminates at a maximum density, D_{max}. Remember that this curve applies to a negative film. For a positive film, the relationship is reversed. That is, density decreases with increasing exposure.

It should be noted that even in areas of a film where there is no exposure, a minimum density D_{min} results from two causes: (1) The plastic base of the film has some density D_{base} and (2) some density develops even when an unexposed emulsion is processed. The range of densities a film provides is simply the difference between D_{max} and D_{min}.

Another important characteristic of D–log E curves is the slope of the linear portion of the curve. This slope is called *gamma* (γ) and is expressed as

$$\gamma = \frac{\Delta D}{\Delta \log E} \tag{2.10}$$

Gamma is an important determinant of the *contrast* of a film. While the term *contrast* has no rigid definition, in general, the higher the gamma, the higher the contrast of a film. With high contrast film, a given scene exposure range is distributed over a large density range; the reverse is true of low contrast film. For example, consider a photograph taken of a light gray and a dark gray object. On high contrast film, the two gray levels may lie at the extremes of the density scale, resulting in nearly white and nearly black images on the processed photograph. On low contrast film, both gray values would lie at nearly the same point on the density scale, showing the two objects in about the same shade of gray.

An important basic characteristic of a film is its *speed*, which expresses the sensitivity of the film to light. This parameter is graphically represented by the horizontal position of the characteristic curve along the log E axis. A "fast" film (more light-sensitive film) is one that will accommodate low exposure levels (i.e., it lies farther to the left on the log E axis). For a given level of scene energy, a fast film will require a shorter exposure time than will a slow film. This is advantageous in aerial photography, because it reduces image blur due to flight motion.

The term *exposure latitude* expresses the range of log E values that will yield an acceptable image on a given film. For most films, good results are obtained when scenes are recorded over the linear portion of the D–log E curves and a fraction of the toe of the curve (Figure 2.10). Features recorded on the extremes of the toe or shoulder of the curve will be underexposed or overexposed. In these areas, different exposure levels will be recorded at essentially the same density, making discrimination difficult. (As we discuss in Section 2.5, digital camera sensors are not limited in this manner. They generally have much greater exposure latitude than film, and they produce linear responses across their entire *dynamic range*—the range between the minimum and maximum scene brightness levels that can be detected simultaneously in the same image.)

Image density is measured with an instrument called a *densitometer* (or *microdensitometer* when small film areas are measured). With *spot* densitometers, different reading positions on the image are located by manually translating the image under analysis with respect to the measurement optics. These devices are convenient in applications where conventional visual interpretation is supported by taking a small number of density readings at discrete points of interest in an image. Applications calling for density measurements throughout an entire image dictate the use of *scanning densitometers*.

Most scanning densitometers are *flatbed* systems. The most common form uses a linear array of charge-coupled devices (CCDs; Section 2.5). Optics focus the CCD array on one horizontal line of the subject at a time and step vertically to scan the next line.

The output from a scanning film densitometer is a digital image composed of pixels whose size is determined by the aperture of the source/receiving optics used during the scanning process. The output is converted to a series of digital numbers. This A-to-D conversion process normally results in recording the density data on any of several forms of computer storage media.

Scanners with the highest radiometric resolution can record a film density range of 0 to 4, which means that the difference between the lightest and darkest detectable intensities is 10,000:1. Typically, monochromatic scanning is done with a 12-bit resolution, which results in 4096 levels of gray. Color scanning is typically 36- to 48-bit scanning (12 to 16 bits per color), resulting in many billions of colors that can be resolved. Output resolutions can be varied, depending on user needs.

Flatbed scanners are often employed in softcopy photogrammetry operations (Section 3.9). Such scanners typically permit scanning at continuously variable resolutions (e.g., 7 to 224 μm) over formats as large as 250–275 mm, with a geometric accuracy of better than 2 μm in each axis of scanning. Many are designed to scan individual transparencies and prints of various sizes as well as uncut rolls of film. Roll-film-capable scanners can support multi-day unattended scanning operations.

Black-and-White Films

Black-and-white aerial photographs are normally made with either *panchromatic* film or *infrared-sensitive* film. The generalized spectral sensitivities for each of these film types are shown in Figure 2.11. As can be seen from Figure 2.11, the

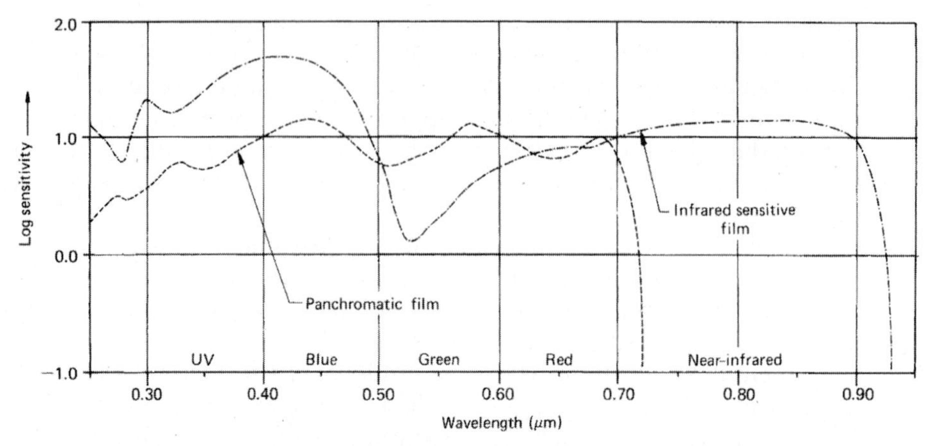

Figure 2.11 Generalized spectral sensitivities for panchromatic and black-and-white IR-sensitive films. (Adapted from Eastman Kodak Company, 1992.)

spectral sensitivity of panchromatic film extends over the UV and the visible portions of the spectrum. Infrared-sensitive film is sensitive not only to UV and visible energy but also to near-IR energy.

The use of black-and-white IR photography to distinguish between deciduous and coniferous trees was illustrated in Figure 1.8. Other applications of black-and-white aerial photography are described in Chapter 8. Here, we simply want the reader to become familiar with the spectral sensitivities of these materials.

It is of interest to note what determines the "boundaries" of the spectral sensitivity of black-and-white film materials. We can photograph over a range of about 0.3 to 0.9 μm. The 0.9-μm limit stems from the photochemical instability of emulsion materials that are sensitive beyond this wavelength. (Certain films used for scientific experimentation are sensitive out to about 1.2 μm and form the only exception to this rule. These films are not commonly available and typically require long exposure times, making them unsuitable for aerial photography.) Black-and-white films, both panchromatic and black-and-white IR, are usually exposed through a yellow (blue-absorbing) filter to reduce the effects of atmospheric haze.

Figure 2.12 shows a comparison between panchromatic and black-and-white IR aerial photographs. The image tones shown in this figure are very typical. Healthy green vegetation reflects much more sunlight in the near IR than in the visible part of the spectrum; therefore, it appears lighter in tone on black-and-white infrared photographs than on panchromatic photographs. Note, for example, that in Figure 2.12 the trees are much lighter toned in (*b*) than in (*a*). Note also that the limits of the stream water and the presence of water and wet soils in the fields can be seen more distinctly in the black-and-white IR photograph (*b*). Water and wet soils typically have a much darker tone in black-and-white IR photographs than in panchromatic photographs because sunlight reflection from water and wet soils in the near IR is considerably less than in the visible part of the electromagnetic spectrum.

Figure 2.12 Comparison of panchromatic and black-and-white IR aerial photographs. Flooding of Bear Creek, northwest Alabama (scale 1:9000): (*a*) panchromatic film with a Wratten No. 12 (yellow) filter; (*b*) black-and-white IR film with a Wratten No. 12 filter. (Courtesy Mapping Services Branch, Tennessee Valley Authority.)

As might be suspected from Figure 2.11, the 0.3-μm limit to photography is determined by something other than film sensitivity. In fact, virtually all photographic emulsions are sensitive in this UV portion of the spectrum. The problem with photographing at wavelengths shorter than about 0.4 μm is twofold: (1) The atmosphere absorbs or scatters much of this energy and (2) glass camera lenses absorb such energy. But photographs can be acquired in the 0.3- to 0.4-μm range if extremes of altitude and unfavorable atmospheric conditions are avoided. Furthermore, some improvement in image quality is realized if quartz camera lenses are used.

To date, the applications of aerial UV photography have been limited in number, due primarily to strong atmospheric scattering of UV energy. A notable exception is the use of UV photography in monitoring oil films on water. Minute traces of floating oil, often invisible on other types of photography, can be detected in UV photography.

Color Film

Although black-and-white panchromatic film was, for a long time, the standard film type for aerial photography, remote sensing applications increasingly involve the use of color film. The major advantage to the use of color is the fact that the human eye can discriminate many more shades of color than it can tones of gray. As we illustrate in subsequent chapters, this capability is essential in many applications of airphoto interpretation.

The basic cross-sectional structure and spectral sensitivity of color film are shown in Figure 2.13. As shown in Figure 2.13*a*, the top film layer is sensitive to blue light, the second layer to green and blue light, and the third to red and blue light. Because these bottom two layers have blue sensitivity as well as the desired green and red sensitivities, a blue-absorbing filter layer is introduced between the first and second photosensitive layers. This filter layer blocks the passage of blue light beyond the blue-sensitive layer. This effectively results in selective sensitization of each of the film layers to the blue, green, and red primary colors. The yellow (blue-absorbing) filter layer has no permanent effect on the appearance of the film because it is dissolved during processing.

In order to eliminate atmospheric scatter caused by UV energy, color film is normally exposed through a UV-absorbing (haze) filter.

From the standpoint of spectral sensitivity, the three layers of color film can be thought of as three black-and-white silver halide emulsions (Figure 2.13*b*). Again, the colors physically present in each of these layers after the film is processed are *not* blue, green, and red. Rather, after processing, the blue-sensitive layer contains yellow dye, the green-sensitive layer contains magenta dye, and the red-sensitive layer contains cyan dye (see Figure 2.13*a*). The amount of dye introduced in each layer is inversely related to the intensity of the corresponding primary light present in the scene photographed. When viewed in composite, the dye layers produce the visual sensation of the original scene.

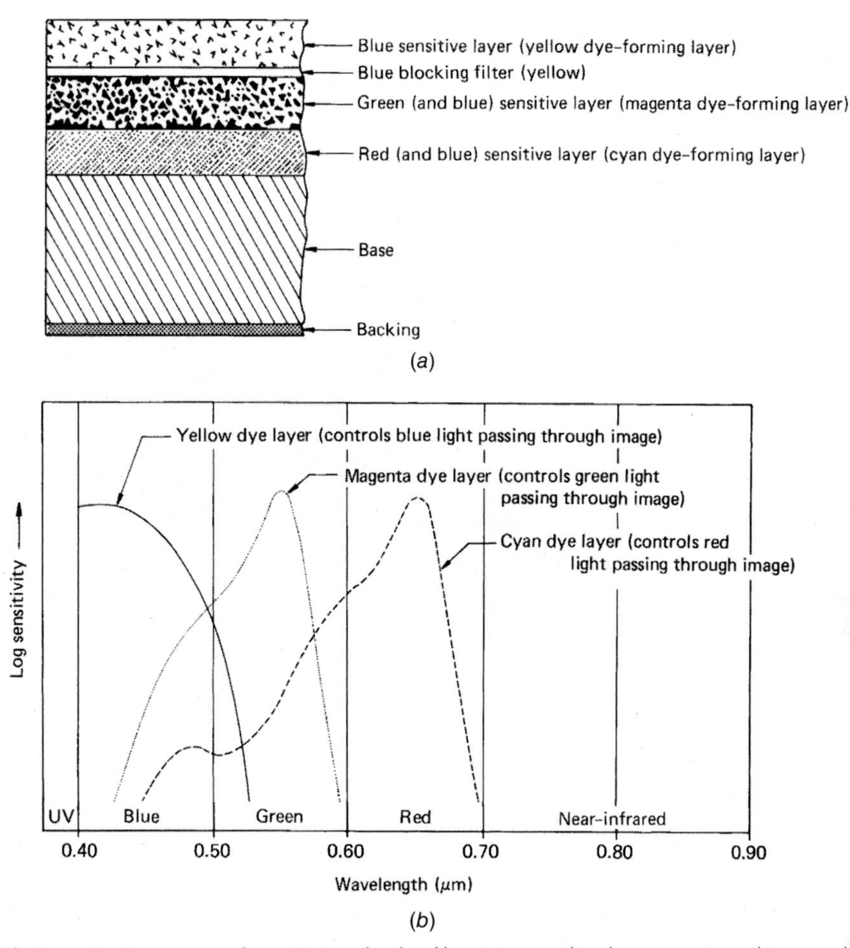

Figure 2.13 Structure and sensitivity of color film: (*a*) generalized cross section; (*b*) spectral sensitivities of the three dye layers. (Adapted from Eastman Kodak Company, 1992.)

The manner in which the three dye layers of color film operate is shown in Figure 2.14. For purposes of illustration, the original scene is represented schematically in (*a*) by a row of boxes that correspond to scene reflectance in four spectral bands: blue, green, red, and near IR. During exposure, the blue-sensitive layer is activated by the blue light, the green-sensitive layer is activated by the green light, and the red-sensitive layer is activated by the red light, as shown in (*b*). No layer is activated by the near-IR energy because the film is not sensitive to near-IR energy. During processing, dyes are introduced into each sensitivity layer in *inverse* proportion to the intensity of light recorded in each layer. Hence the more intense the exposure of the blue layer to blue light, the less yellow dye is introduced in the image and the more magenta and cyan dyes are introduced. This is shown in (*c*),

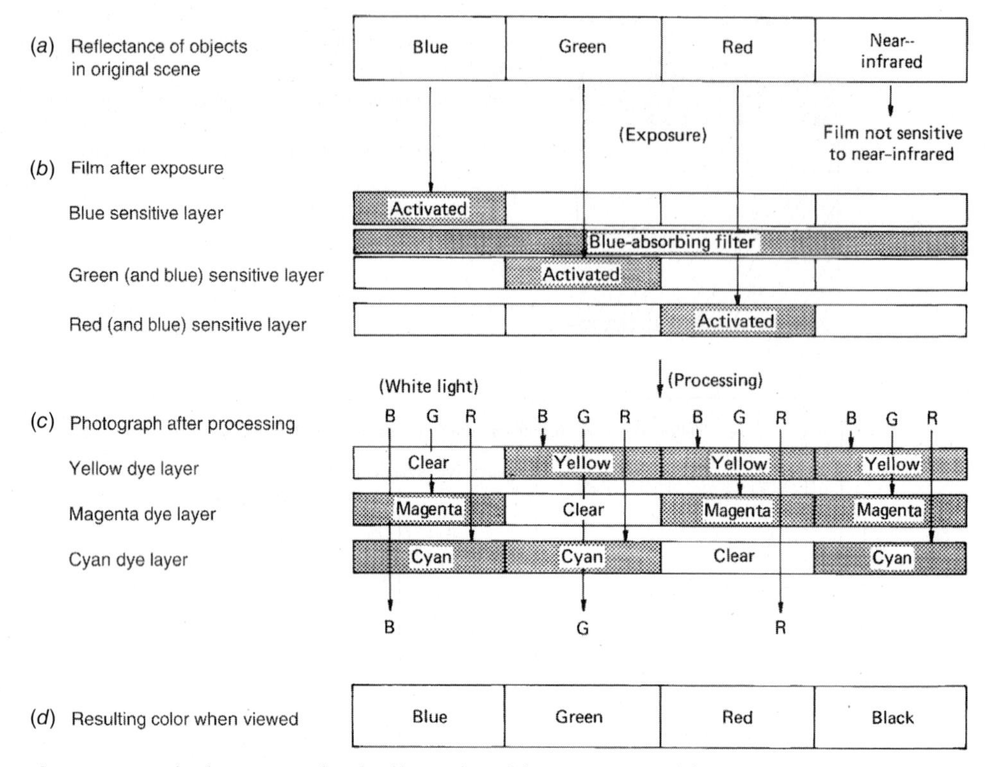

Figure 2.14 Color formation with color film. (Adapted from Eastman Kodak Company, 1992.)

where, for blue light, the yellow dye layer is clear and the other two layers contain magenta and cyan dyes. Likewise, green exposure results in the introduction of yellow and cyan dyes, and red exposure results in the introduction of yellow and magenta dyes. When the developed image is viewed with a white light source (*d*), we perceive the colors in the original scene through the subtractive process. Where a blue object was present in the scene, the magenta dye subtracts the green component of the white light, the cyan dye subtracts the red component, and the image appears blue. Green and red are produced in an analogous fashion. Other colors are produced in accordance with the proportions of blue, green, and red present in the original scene.

Color Infrared Film

The assignment of a given dye color to a given spectral sensitivity range is a film manufacturing parameter that can be varied arbitrarily. The color of the dye developed in any given emulsion layer need not bear any relationship to the color of light to which the layer is sensitive. Any desired portions of the photographic spectrum, including the near IR, can be recorded on color film with any color assignment.

In contrast to "normal" color film, *color IR* film is manufactured to record green, red, and the photographic portion (0.7 to 0.9 μm) of the near-IR scene energy in its three emulsion layers. The dyes developed in each of these layers are again yellow, magenta, and cyan. The result is a "false color" film in which blue images result from objects reflecting primarily green energy, green images result from objects reflecting primarily red energy, and red images result from objects reflecting primarily in the near-IR portion of the spectrum.

The basic structure and spectral sensitivity of color IR film are shown in Figure 2.15. (Note that there are some overlaps in the sensitivities of the layers.) The

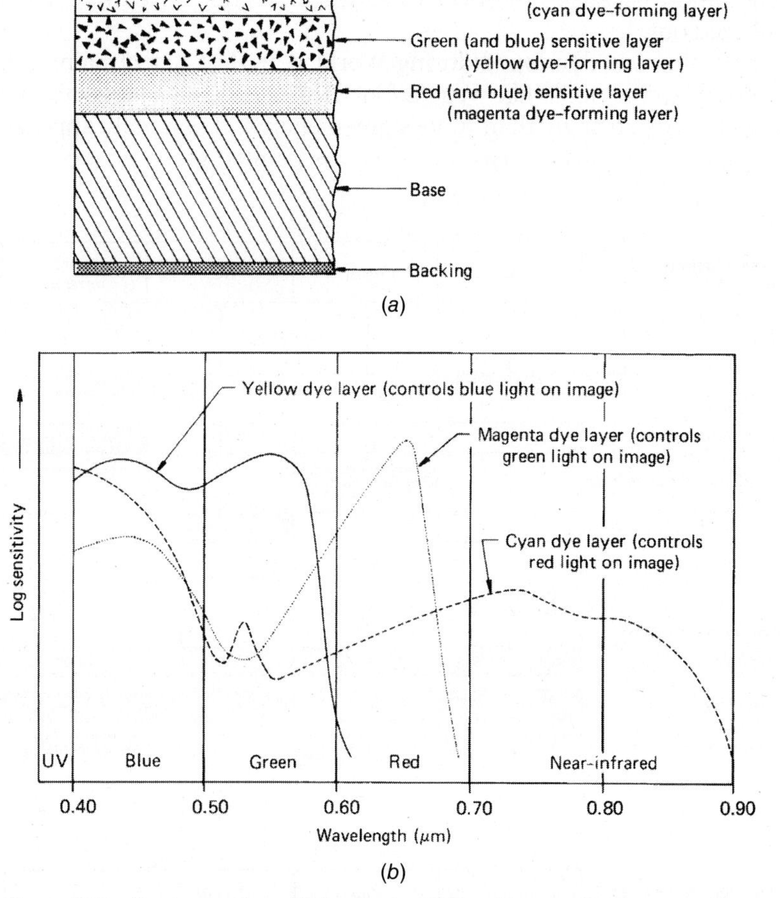

Figure 2.15 Structure and sensitivity of color IR film: (*a*) generalized cross section; (*b*) spectral sensitivities of the three dye layers. (Adapted from Eastman Kodak Company, 1992.)

process by which the three primary colors are reproduced with such films is shown in Figure 2.16. Various combinations of the primary colors and complementary colors, as well as black and white, can also be reproduced on the film, depending on scene reflectance. For example, an object with a high reflectance in both green and near IR would produce a magenta image (blue plus red). It should be noted that most color IR films are designed to be used with a yellow (blue-absorbing) filter over the camera lens. As shown in Figure 2.17, the yellow filter blocks the passage of any light having a wavelength below about 0.5 μm. This means that the blue (and UV) scene energy is not permitted to reach the film, a fact that aids in the interpretation of color IR imagery. If a yellow (blue-absorbing) filter were not used, it would be very difficult to ascribe any given image color to a particular ground reflectance because of the nearly equal sensitivity of all layers of the film to blue energy. The use of a blue-absorbing filter has the further advantage of improving haze penetration because the effect of Rayleigh scatter is reduced when the blue light is filtered out.

Color IR film was developed during World War II to detect painted targets that were camouflaged to look like vegetation. Because healthy vegetation reflects IR energy much more strongly than it does green energy, it generally appears in various tones of red on color IR film. However, objects painted green generally have low IR

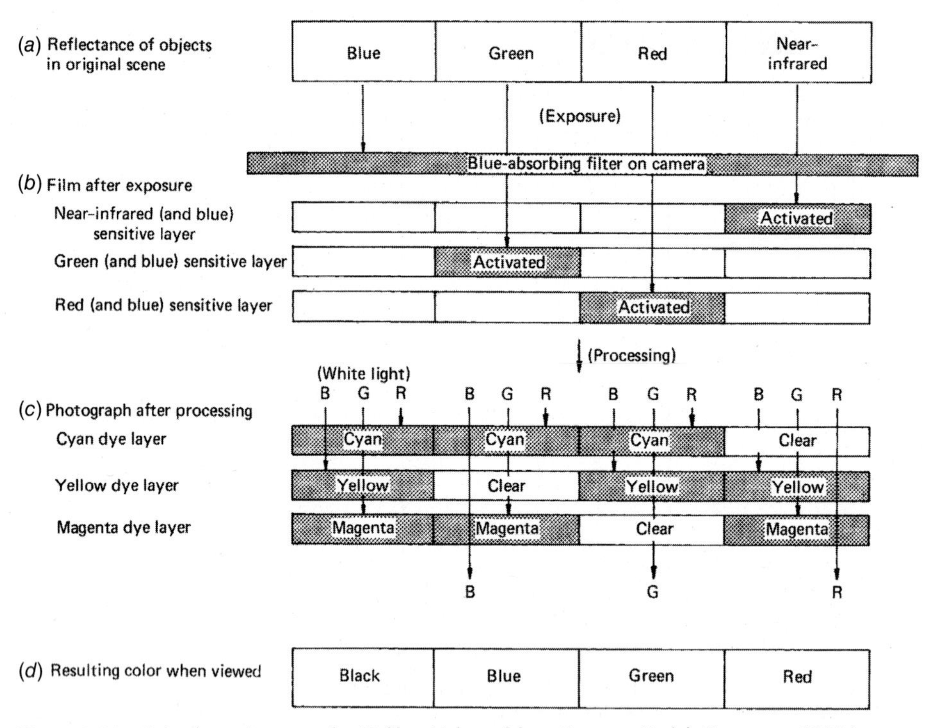

Figure 2.16 Color formation on color IR film. (Adapted from Eastman Kodak Company, 1992.)

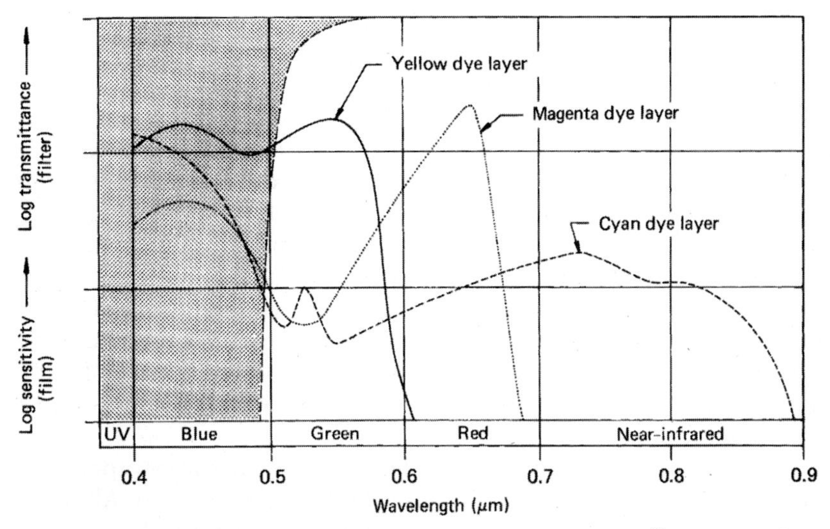

Figure 2.17 Spectral sensitivity of color IR film with Kodak Wratten® No. 12 (yellow) filter. (Adapted from Eastman Kodak Company, 1990 and 1992.)

reflectance. Thus, they appear blue on the film and can be readily discriminated from healthy green vegetation. Because of its genesis, color IR film has often been referred to as "camouflage detection film." With its vivid color portrayal of near-IR energy, color IR film has become an extremely useful film for resource analyses.

Plate 4 illustrates normal color (*a*) and color IR (*b*) aerial photographs of a portion of the University of Wisconsin–Madison campus. The grass, tree leaves, and football field reflect more strongly in the green than in the blue or red and thus appear green in the natural color photograph. The healthy grass and tree leaves reflect much more strongly in the near-IR than in the green or red and thus appear red in the color IR photograph. The football field has artificial turf that does not reflect well in the near-IR and thus does not appear red. The large rectangular gravel parking area adjacent to the natural grass practice fields appears a light brown in the normal color photograph and nearly white in the color IR photograph. This means it has a high reflectance in green, red, and near-IR. The red-roofed buildings appear a greenish-yellow on the color IR film, which means that they reflect highly in the red and also have some near-IR reflectance. The fact that the near-IR-sensitive layer of the film also has some sensitivity to red (Figure 2.17) also contributes to the greenish-yellow color of the red roofs when photographed on the color IR film.

Almost every aerial application of color IR photography deals with photographing *reflected sunlight*. The amount of energy *emitted* from the earth at ambient temperature (around 300 K) is insignificant, in the range of 0.4 to 0.9 μm, and hence cannot be photographed. This means that color IR film cannot, for example, be used to detect the temperature difference between two water bodies or between wet and dry soils. As explained in Chapter 4, electronic sensors (such as radiometers or thermal scanners) operating in the wavelength range 3 to 5 or 8 to 14 μm can be used to distinguish between temperatures of such objects.

The energy *emitted* from extremely hot objects such as flames from burning wood (forest fires or burning buildings) or flowing lava *can* be photographed on color and color IR film. Figure 2.18 shows blackbody radiation curves for earth features at an ambient temperature of 27°C (300 K) and flowing lava at 1100°C (1373 K). As calculated from Wien's displacement law (Eq. 1.5), the peak wavelength of the emitted energy is 9.7 μm for the earth features at 27°C and 2.1 μm for lava at 1100°C. When the spectral distribution of emitted energy is calculated, it is found that the energy emitted from the features at 27°C is essentially zero over the range of photographic wavelengths. In the case of flowing lava at 1100°C, the emitted energy in the range of IR photography (0.5 to 0.9 μm) is sufficient to be recorded on photographic films.

Plate 5 shows normal color (*a*) and color IR (*b*) aerial photographs of flowing lava on the flank of Kilauea Volcano on the Island of Hawaii. Although the emitted energy can be seen as a faint orange glow on the normal color photograph, it is more clearly evident on the color IR film. The orange tones on the color IR photograph represent IR energy *emitted* from the flowing lava. The pink tones represent sunlight *reflected* from the living vegetation (principally tree ferns). Keep in mind that it is *only* when the temperature of a feature is extremely high that IR film will record energy emitted by an object. At all other times, the film is

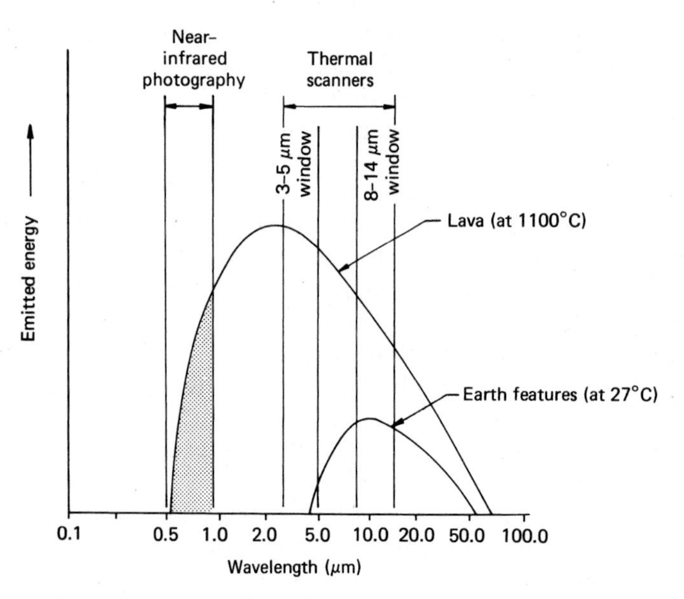

Figure 2.18 Blackbody radiation curves for earth surface features (at 27°C) and flowing lava (at 1100°C).

responding to *reflected* IR energy that is not directly related to the temperature of the feature.

2.5 DIGITAL PHOTOGRAPHY

Since approximately 2010, the acquisition and analysis of digital aerial photography has virtually replaced analog film-based methods on an operational basis. Also, digital methods are continuing to evolve at a very rapid pace. Here we describe the technology of digital photography as of the year 2014. In Section 2.6 we describe using digital cameras for aerial photography.

Digital versus Analog Photography

As shown in Figure 2.19, the fundamental differences between digital and film-based aerial photography begin with the use of photosensitive solid state sensors rather than the silver halide crystals present in film to capture images. Typically, digital cameras incorporate a two-dimensional array of silicon semiconductors consisting of either charge-coupled device (CCD) or complementary metal-oxide semiconductor (CMOS) detectors. Each detector (or *photosite location*) in the array senses the energy radiating from one pixel in the image field. When this energy strikes the surface of the detector, a small electrical charge is produced,

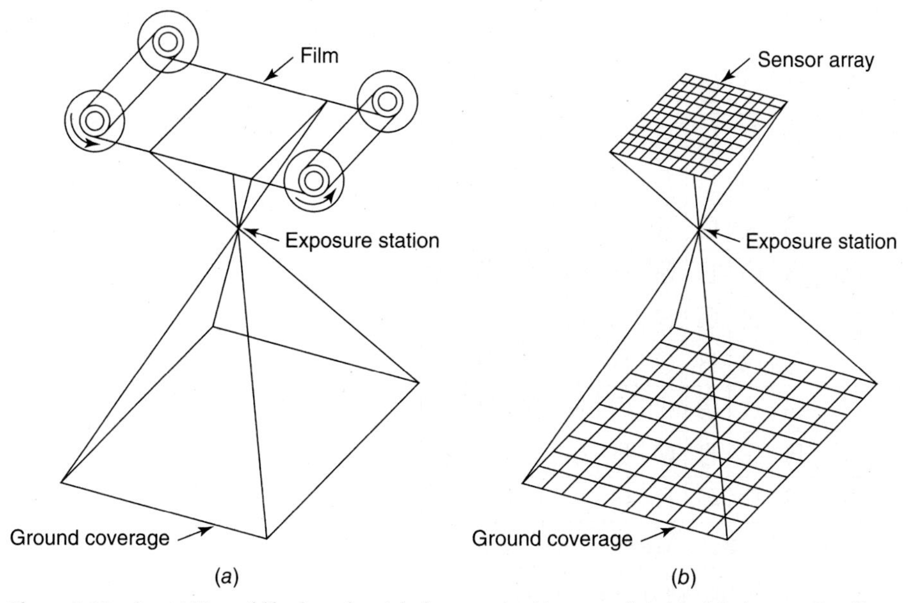

Figure 2.19 Acquisition of film-based aerial photography (*a*) versus digital aerial photography (*b*).

with the magnitude of the charge being proportional to the scene brightness within the pixel. The charge is converted to a voltage value, which is in turn converted to a digital brightness value. This process results in the creation of digital brightness value for each photosite pixel in the array.

It should be noted that the CCD or CMOS semiconductors designed for use in remote sensing applications are much more sensitive to brightness variations coming from the image field than are the silver halide crystals present in film. At the same time, their response is linear (not S-shaped like a film's *D*-Log *E* curve). Also, the electronic sensors can measure over a much wider dynamic range of scene brightness values than can film. Thus, the many advantages of digital photography over analog methods begin with improved sensitivity, linearity, and dynamic range of image data capture.

CCD and CMOS image sensors are monochromatic. To obtain full-color data, each photosite of the sensor array is typically covered with a blue, green, or red filter. Usually, photosites are square, with alternating blue-, green-, and red-sensitive sites arranged in a *Bayer pattern* (Figure 2.20). Half of the filters in this pattern are green; the remainder are blue or red, in equal numbers. (This over-weighted allocation of green-sensitive pixels takes advantage of the green peak of solar radiation and the increased sensitivity of the human eye to green.) To assign blue, green, and red values to each photosite, the two missing colors at each photosite are interpolated from surrounding photosites of the same color (see Section 7.2 and Appendix B for a discussion of resampling schemes for such interpolation). The resulting data set is a two-dimensional array of discrete pixels, with each pixel having three DNs representing the scene brightness in each spectral

G	B	G	B	G	B	G	B	G	B
R	G	R	G	R	G	R	G	R	G
G	B	G	B	G	B	G	B	G	B
R	G	R	G	R	G	R	G	R	G
G	B	G	B	G	B	G	B	G	B
R	G	R	G	R	G	R	G	R	G
G	B	G	B	G	B	G	B	G	B
R	G	R	G	R	G	R	G	R	G
G	B	G	B	G	B	G	B	G	B
R	G	R	G	R	G	R	G	R	G

Figure 2.20 Bayer pattern of blue, green, and red CCD filters. Note that one-half of these filters are green, one-quarter are blue, and one-quarter are red.

band of sensing. The *color depth*, or *quantization level*, of CCD or CMOS sensors is typically 8 to 14 bits (256 to 16,384 gray levels per band).

An alternative to the use of Bayer pattern sensor arrays is a three-layer *Foveon X3 CMOS sensor* that has three photodetectors (blue, green, and red) at every pixel location. This is based on the natural property of silicon to absorb different wavelengths of light at different depths. By placing photodetectors at different depths in the CMOS sensor, blue, green, and red energy can be sensed independently, similar to the three layers of color film. Conceptually, this should result in sharper images, better color, and freedom from color artifacts resulting from interpolation (resampling) that are possible with Bayer-array sensors. However, recent improvements in Bayer pattern interpolation algorithms have dampened demand for three-layer sensors.

There are several other means through which multiple multispectral bands (e.g., blue, green, red, and near-IR) can be acquired simultaneously with a digital camera. For example, many consumer cameras have been designed to rapidly record three black-and-white versions of the same scene through three different internal filters. These individual images are subsequently color-composited using the color additive principles (see Section 1.12). Another approach is to use multiple single-band camera heads, each equipped with its own lens, filter, and CCD array. Yet another approach is to use one camera head, but to use some form of beamsplitter to separate the incoming energy into the desired discrete spectral bands of sensing and record each of these bands using a separate CCD. We illustrate various combinations of these approaches in Section 2.6. We should also point out that we limit our discussion in this chapter to area array, or frame, camera designs. A description of the use of linear array sensors to obtain digital photographic data is included in Section 4.3.

Advantages of Digital Photography

Among the major advantages that digital photography provides relative to analog or film-based photography are the following:

1. **Enhanced image capture capabilities.** We earlier noted the superior sensitivity, linearity, and dynamic range of digital cameras. These cameras have no moving parts and can collect images very rapidly, using very short inter-image intervals (2 seconds or less). This enables the collection of photography at higher flight speeds and provides a means to obtain stereoscopic coverage with very high overlap (up to 90%). Overlap of up to 90% and sidelap of up to 60% is often acquired in support of multi-ray photogrammetric applications (Section 3.10). The marginal cost of obtaining the additional photos to provide such high amounts of overlap and sidelap is much lower using digital rather than film-based systems.

Digital images can also be collected over a very broad range of ambient lighting conditions. This is particularly helpful under low-light conditions, even shortly after the sun has set. Consequently, digital photographs can be collected over extended times of day and times of the year (depending on latitude). Also, they can often be collected under overcast conditions that might thwart the collection of properly exposed analog photography. Thus, digital methods open a broader temporal window of opportunity to collect useful photography.

2. **Reduced time and complexity of creating primary data products.** Digital photography circumvents the chemical processing required with film-based photography. It also eliminates the need to scan hardcopy images to create digital products. Digital photographs are available virtually in real time. Derivative data products such as digital orthophotos (Section 3.10) can also be generated very quickly, with some or all of the data processing performed in-flight. Such timeliness is extremely valuable in such applications as disaster response.
3. **Intrinsic compatibility with complementary digital technologies.** There is a myriad of advantages to digital photographic data that stem simply from the fact that such data are intrinsically compatible with many other complementary digital technologies. These complementary technologies include, but are not limited to: digital photogrammetry, multisensor data integration, digital image processing and analysis, GIS integration and analysis, data compression, distributed processing and mass storage (including cloud computing architectures), mobile applications, Internet communication and distribution, hard and softcopy display and visualization, and spatially enabled decision support systems. The synergy among these technologies continues to greatly increase the efficiency, automation, and applicability of digital photography workflows that are supporting a rapidly increasing number of applications.

2.6 AERIAL CAMERAS

Aerial photographs can be made with virtually any type of camera. Many successful applications have employed aerial photographs made from light aircraft with handheld cameras. For example, the photographs in Plates 4 and 5 were made in this manner. Most aerial photographic remote sensing endeavors, however, entail the use of aerial photography made with precision-built aerial cameras. These cameras are specifically designed to expose a large number of photographs in rapid succession with the ultimate in geometric fidelity.

There are many different models of aerial cameras currently in use. Here, we discuss *single-lens frame* film cameras, *panoramic* film cameras, and small-, medium-, and large-format digital cameras.

Single-Lens Frame Film Cameras

Single-lens frame cameras are by far the most common film cameras in use today for aerial photography. They are used almost exclusively in obtaining aerial photographs for remote sensing in general and photogrammetric mapping purposes in particular. *Mapping* cameras (often referred to as *metric* or *cartographic* cameras) are single-lens frame cameras designed to provide extremely high geometric image quality. They employ a low distortion lens system held in a fixed position relative to the plane of the film. The film format size (the nominal size of each image) is commonly a square 230 mm on a side. The total width of the film used is 240 mm, and the film magazine capacity ranges up to film lengths of 120 m. A frame of imagery is acquired with each opening of the camera shutter, which is generally tripped automatically at a set frequency by an electronic device called an *intervalometer*. Figure 2.21 illustrates a typical aerial mapping camera and its associated gyrostabilized suspension mount.

Although mapping cameras typically use film with a 230×230 mm image format, other cameras with different image sizes have been built. For example, a special-purpose camera built for NASA called the *Large Format Camera* (LFC) had a 230×460 mm image format. It was used on an experimental basis to photograph

Figure 2.21 Intergraph RMK TOP Aerial Survey Camera System. Camera operation is computer driven and can be integrated with a GPS unit. (Courtesy Intergraph Corporation.)

the earth from the space shuttle (Doyle, 1985). (See Figures 3.4 and 8.16 for examples of LFC photography.)

Recall that for an aerial camera the distance between the center of the lens system and the film plane is equal to the focal length of the lens. It is at this fixed distance that light rays coming from an effectively infinite distance away from the camera come to focus on the film. (Most mapping cameras cannot be focused for use at close range.) For mapping purposes, 152-mm-focal-length lenses are most widely used. Lenses with 90 and 210 mm focal lengths are also used for mapping. Longer focal lengths, such as 300 mm, are used for very high altitude applications. Frame camera lenses are somewhat loosely termed as being either (1) *normal angle* (when the angular field of view of the lens system is up to 75°), (2) *wide angle* (when the field of view is 75° to 100°), and (3) *superwide angle* (when the field of view is greater than 100°) (angle measured along image diagonal).

Figure 2.22 illustrates how the angular field of view for a typical film-mapping camera is determined. It is defined as the angle (θ_d) subtended at the rear nodal point of the camera lens by a diagonal (d) of the image format. The size of one-half of this angle is related to one-half of the diagonal dimension of the format as follows

$$\frac{\theta_d}{2} = \tan^{-1}\left(\frac{d}{2f}\right)$$

which can be restated as

$$\theta_d = 2\tan^{-1}\left(\frac{d}{2f}\right) \tag{2.11}$$

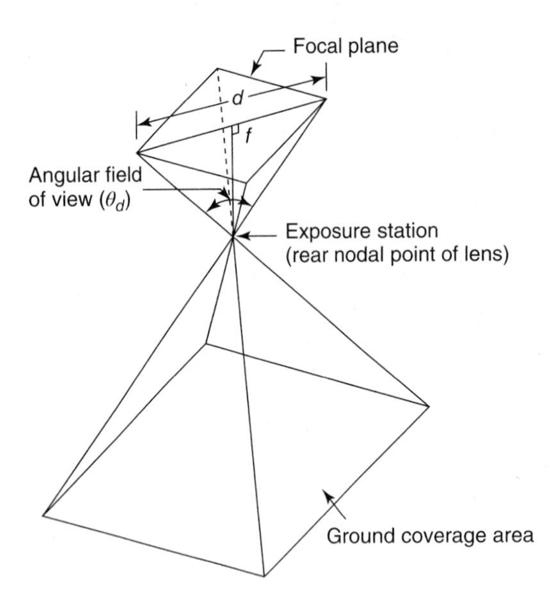

Figure 2.22 Angular field of view of a traditional square-format aerial mapping camera.

EXAMPLE 2.3

Mapping cameras that have a format size of 230 mm (9 in) square are often fitted with 152-mm-focal-length (6 in) lenses. Determine the angular field of view for such a combination of format size and focal length.

Solution

From Eq. 2.11

$$\theta_d = 2\tan^{-1}\left(\frac{\sqrt{230^2 + 230^2}}{2(152)}\right) = 94^\circ$$

Accordingly, this lens would be referred to as a wide angle lens.

Figure 2.23 illustrates the principal components of a single-lens frame mapping camera. The *lens cone assembly* includes the *lens*, *filter*, *shutter*, and *diaphragm*. The lens is generally composed of multiple lens elements that gather the light rays from a scene and bring them to focus in the *focal plane*. The filter serves any of the various functions enumerated in the previous section. The shutter and diaphragm (typically located between lens elements) control film exposure. The shutter controls the duration of exposure (from $\frac{1}{100}$ to $\frac{1}{1000}$ sec) while the diaphragm forms an aperture that can be varied in size. The camera *body* typically houses an electrical film drive mechanism for advancing the film, flattening the film during exposure, cocking the shutter, and tripping the shutter. The *camera magazine* holds the film supply and takeup reels, the film-advancing mechanism, and the film-flattening mechanism. Film flattening during exposure is often accomplished by drawing the film against a vacuum plate lying behind the focal plane. The focal plane is the plane in which the film is exposed. The *optical axis* of the camera is perpendicular to the film plane and extends through the center of the lens system.

During the time a frame camera shutter is opened for exposure of a photograph, aircraft motion causes the image to blur. To negate this effect, most frame cameras have built-in *image motion compensation*. This works by moving the film across the focal plane at a rate just equal to the rate of image movement. The camera system illustrated in Figure 2.21 incorporates this capability.

Shown in Figure 2.24 is a *vertical photograph* made with a mapping camera whose optical axis was directed as nearly vertical as possible at the instant of exposure. Note the appearance of the four *fiducial marks* at the middle of the image sides. (As illustrated in Figure 3.12, some mapping cameras incorporate corner fiducials.) These marks define the frame of reference for spatial measurements made from such aerial photos (explained in Chapter 3). Lines connecting

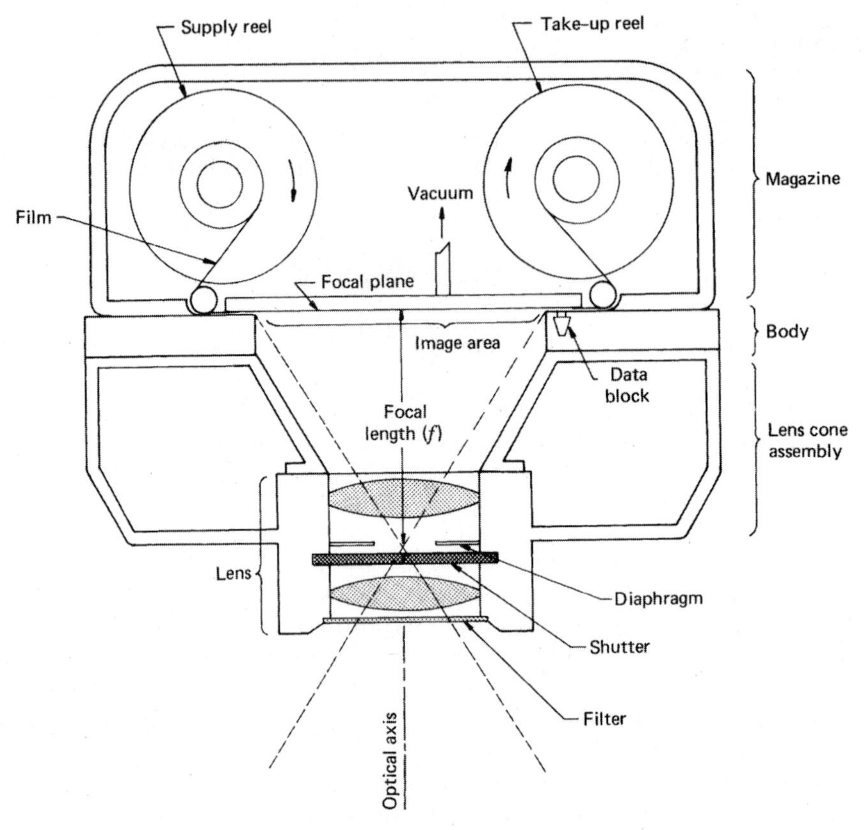

Figure 2.23 Principal components of a single-lens frame mapping camera.

opposite fiducial marks intersect at a photograph's *principal point*. As part of the manufacturer's calibration of a mapping camera, the camera focal length, the distances between fiducial marks, and the exact location of the principal point are precisely determined.

Panoramic Film Cameras

Of primarily historical interest is the panoramic film camera. This camera views only a comparatively narrow angular field at any given instant through a narrow slit. Ground areas are covered by either rotating the camera lens or rotating a prism in front of the lens. Figure 2.25 illustrates the design using lens rotation.

In Figure 2.25, the terrain is scanned from side to side, transverse to the direction of flight. The film is exposed along a curved surface located at the focal distance from the rotating lens assembly, and the angular coverage of the camera can

Figure 2.24 Vertical aerial photograph taken with a 230 × 230-mm precision mapping film camera showing Langenburg, Germany. Note the camera fiducial marks on each side of image. Data blocks (on left of image) record image identification, clock, level bubble, and altimeter. Frame number is recorded in lower left corner of image. Scale 1:13,200. (Courtesy Carl Zeiss.)

extend from horizon to horizon. The exposure slit moves along the film as the lens rotates, and the film is held fixed during a given exposure. After one scan is completed, the film is advanced for the next exposure.

Panoramic cameras incorporating the rotating prism design contain a fixed lens and a flat film plane. Scanning is accomplished by rotating the prism in front of the lens, yielding imagery geometrically equivalent to that of the rotating lens camera.

Figure 2.26 illustrates the large area of coverage characteristic of panoramic photography. The distortions inherent in panoramic imaging are also apparent in the figure. Areas near the two ends of the photograph are compressed. This scale

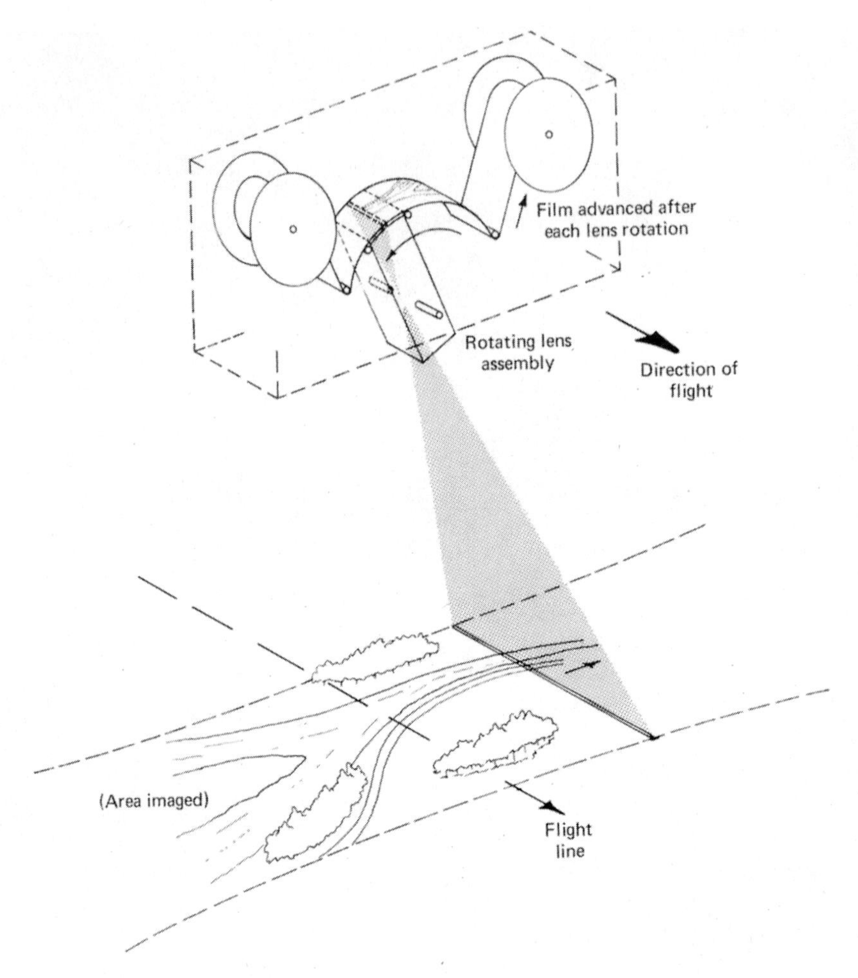

Figure 2.25 Operating principle of a panoramic camera.

variation, called *panoramic distortion*, is a result of the cylindrical shape of the focal plane and the nature of scanning. Also, *scan positional distortion* is introduced in panoramic imaging due to forward motion of the aircraft during the time a scan is made.

Digital Camera Format Sizes

While the available format sizes of film-based aerial mapping cameras have been standardized for many decades, there is no such standardization in the case of

Figure 2.26 Panoramic photograph with 180° scan angle. Note large area of coverage and geometric distortion. (Courtesy USAF Rome Air Development Center.)

digital camera format sizes. That is, film has been available in a limited number of widths (e.g., 35 mm, 70 mm, and 240 mm). Large format analog cameras typically employ an image format size of 230 mm × 230 mm recorded on 240-mm film. There is no such standardization in the case of digital camera formats. They take on many shapes and sizes. The across-track and along-track format dimensions of a digital camera sensor are determined by the number and physical size of the pixels present in each direction of the sensor array. The across-track dimension can be found from

$$d_{xt} = n_{xt} \times p_d \tag{2.12}$$

where

d_{xt} = across-track dimension of the sensor array
n_{xt} = number of pixels present in the across-track direction
p_d = physical dimension of each pixel

The along-track dimension of the sensor can be determined in a similar fashion from

$$d_{at} = n_{at} \times p_d \tag{2.13}$$

where

d_{at} = along-track dimension of the sensor array
n_{at} = number of pixels present in the along-track direction
p_d = physical dimension of each pixel

EXAMPLE 2.4

A digital camera's sensor consists of pixels that have a physical size of 5.6 μm (or 0.0056 mm). The number of pixels in the across-track direction is 16,768, and the number of pixels in the along-track direction is 14,016. Determine the across-track and along-track dimensions of the camera's sensor.

Solution

From Eq. 2.12

$$d_{xt} = n_{xt} \times p_d = 16{,}768 \text{ pixels} \times 0.0056 \text{ mm/pixel} = 93.90 \text{ mm}$$

From Eq. 2.13

$$d_{at} = n_{at} \times p_d = 14{,}016 \text{ pixels} \times 0.0056 \text{ mm/pixel} = 78.49 \text{ mm}$$

Early in their development, digital cameras were classified as being small-, medium-, or large-format based on the size of their detector arrays. Originally, small-format cameras featured array sizes of up to 24 mm × 36 mm (equivalent to 35 mm film cameras). Medium-format cameras employed array sizes greater than 24 mm × 36 mm up to 60 mm × 90 mm. Large-format cameras included array sizes greater than 60 mm × 90 mm. Over time, the dividing lines between the various formats have shifted such that what defines a "small-," "medium-," or "large-format" digital camera defies exact definition, particularly when multiple lenses and detector arrays can be used to stitch together large "virtual images."

Currently, there is a trend toward using the term "large-format" to refer loosely to any camera designed specifically for wide-area mapping applications. "Medium-format" digital cameras are those that are generally used to collect images of moderate areas or to augment the data collected by another sensor, such as a lidar system. "Small-format" cameras are those used for small-area mapping and might be operated from small aerial platforms.

Space limits exhaustive discussion of the myriad of digital aerial cameras currently available. Here we briefly describe only representative examples of each format type, with emphasis on large-format systems used for wide-area mapping.

Small-Format Digital Cameras

The digital camera options for acquiring small-format aerial photography (SFAP) are virtually limitless. These range from miniature cameras that fit in a model rocket to the numerous consumer-grade compact digital cameras, professional-grade digital single-lens reflex (DSLR) cameras, and industrial cameras. Industrial cameras, often used in quality assurance of manufactured goods, must interface with a separate CPU to operate. Most other systems have built-in CPUs. While many of these camera types produce images having spatial resolution that meets or exceeds that of the traditional 35-mm or 70-mm film cameras they have replaced, they often lack the geometric design and stability needed for precise photogrammetric mapping and analysis. They are nonetheless useful in a broad range of applications wherein precise positioning and measurement is not required.

Historically, the dominant platform options for collecting SFAP were small fixed-wing aircraft and helicopters. Current platforms include, but are not limited to, ultralight aircraft, gliders, motorized paragliders, UAVs, blimps, tethered balloons, and kites. Stationary towers and poles are also used to collect small-format aerial photography.

Those readers wishing to obtain additional information about the acquisition and analysis of SFAP may wish to consult such references on the subject as Aber et al. (2010) and Warner et al. (1996).

Medium-Format Digital Cameras

One example of a medium-format digital camera used for aerial photography is the Trimble Applanix Digital Sensor System (DSS), which can be configured with two cameras to acquire both color and color-IR images. In addition to the camera system(s), the DSS incorporates a GPS receiver to record the position from which each image is acquired at the instant of exposure and an *Inertial Measurement Unit* (*IMU*), which determines the angular orientation, or attitude, of each exposure. The GPS and IMU observations permit *direct georeferencing* (Section 3.9) of the image data in order to produce geometrically corrected orthophotos and orthomosaics (also described in Chapter 3). The DSS was among the first medium format camera systems to employ such technology. Most mapping-quality aerial camera systems now afford direct georeferencing capabilities through integration with a GPS receiver and an IMU. (Plate 37 illustrates a sample DSS photograph.)

Figure 2.27 is an example of another medium-format digital camera system. It shows the Leica RCD30 system. This is a 60-megapixel camera that produces co-registered four-band imagery in the blue, green, red, and near-IR spectral regions using a beam splitter and a single camera head. The beam splitter separates the incoming RGB energy from the near-IR portion of the incoming energy. A high precision Bayer array is used to record the RGB portion of the spectrum, and a second CCD of identical size records the near-IR image. Plate 6 illustrates simultaneous color and color-IR images collected by this system in support of power line corridor mapping.

Figure 2.27 Leica RCD30 Camera System. Shown from left to right are the operator controller, camera head, and camera controller. (Courtesy Leica Geosystems, Geospatial Solutions Division.)

Shown in Figure 2.28 is an IGI medium-format DigiCAM system. The camera is shown in the upper-right of this figure. Also shown (in a counter-clockwise direction from the camera) are the sensor management unit (SMU) that controls the camera, a multipurpose touch screen display, filters that control the acquisition of color or color-IR image data, and an exchangeable hot swap solid state drive (SSD) storage unit. The camera can be equipped with a 40-, 50-, or 60-megapixel sensor and a broad selection of different lenses. The system is designed to be modular in character with multi-camera configurations available in groups

Figure 2.28 IGI DigiCAM system. (Courtesy IGI GmbH.)

of two to five cameras and SMUs, all mounted in the same pod and operated simultaneously from the same touch-screen display. The cameras can also be configured to acquire nadir and/or oblique images. In Section 3.1 we illustrate the Penta DigiCAM, a five-camera system designed to acquire four oblique images and one nadir image simultaneously. DigiCAMS can also be integrated with available IGI thermal and lidar sensors.

Large-Format Digital Cameras

Figure 2.29 illustrates an example of a large-format digital aerial mapping camera, the Z/I DMCII_{250}. This system incorporates five synchronously operated CCD array cameras. One is a 250 megapixel panchromatic camera incorporating a $17{,}216 \times 14{,}656$ array of 5.6 μm-sized CCDs and a 112 mm-focal-length lens. The system uses 14-bit data recording and has a GSD of 2.5 cm when operated from 500 m above ground.

Four multispectral cameras are also included in the system. All have 42-megapixel arrays (6846×6096 pixels) of 7.2 μm-sized CCDs. Each has a dedicated filter for imaging in the blue, green, red, and near-IR, respectively. These cameras

Figure 2.29 Z/I DMCII_{250} Camera. (Courtesy Leica Geosystems, Geospatial Solutions Division.)

have 45-mm-focal-length lenses. The ground footprint for the multispectral and panchromatic cameras is approximately equivalent because of the different focal length lenses used in each system. Automated forward image motion compensation and rapid frame rates enable the system to be operated at low altitudes for detailed surveys as well as high altitudes for regional projects.

Figure 2.30 is a grayscale reproduction of a color image obtained with a $DMCII_{250}$ camera over a portion of Aalen, Germany. Note the swimming pool that appears near the top-center of Figure 2.30*a*. This pool is 50 m in length. The area including the pool is enlarged in Figure 2.30*b*. This enlargement demonstrates the spatial resolution present in Figure Figure 2.30*a*. Clearly observable in the enlargement are such features as ripples on the pool's surface caused by the movement of swimmers, individuals sitting on the edge of the pool, wet footprints near some of the pool's ladders, and sunbathers present on the lawn areas near the bottom of the enlargement.

Microsoft/Vexel has developed a family of large-format digital cameras for aerial photography. The largest of these to date is the UltraCam Eagle, depicted in Figure 2.31. Shown in this figure is the sensor head as well as the operator touch screen interface panel and keyboard, an on-board solid state storage unit (left) that accepts exchangeable storage modules, and a portable power supply (right). Not visible in this figure are the various components of the system that are integrated into the sensor head. These include the camera heads, a GPS, IMU, and flight management system.

This camera system includes a total of eight camera heads. Four of these are used to produce a single large panchromatic image, and the other four camera heads are used to obtain multispectral blue, green, red, and near-IR images. The panchromatic image results from data collected by nine CCD arrays that produce overlapping sub-images that are "monolithically stitched" together during data post processing to form a single composite photograph. The panchromatic images are $20{,}010 \times 13{,}080$ pixels in size. The multispectral images are $6{,}670 \times 4{,}360$ pixels in size. The physical pixel size of the CCD sensors used to produce both the panchromatic and multispectral images is 5.2 μm and they produce data with 14-bit (or better) radiometric resolution.

The UltraCam Eagle features a very wide-image footprint in the across-track flight direction. It also provides the flexibility of exchangeable lens systems having three different focal lengths to facilitate data collection at various flying heights. The first lens system incorporates an 80-mm-focal-length panchromatic and a 27-mm-focal-length multispectral lens. The second includes a 100-mm-focal-length panchromatic lens and a 33-mm-focal-length multispectral lens. The third lens system includes a 210-mm-focal-length panchromatic lens and a 70-mm-focal-length multispectral lens.

Note that the above focal length selections directly impact the GSD and total angular field of view of the various images collected at a given altitude. For example, if the system is equipped with the first choice of lens systems and panchromatic images are acquired from a flying height above ground of 1000 m, the GSD

Figure 2.30 Z/I DMCII$_{250}$ aerial photographs depicting a portion of Aalen, Germany. Note in (*a*) the appearance of a 50-m-long swimming pool near the top/center of this image. The area containing the swimming pool is enlarged in (*b*). (Courtesy Leica Geosystems, Geospatial Solutions Division.)

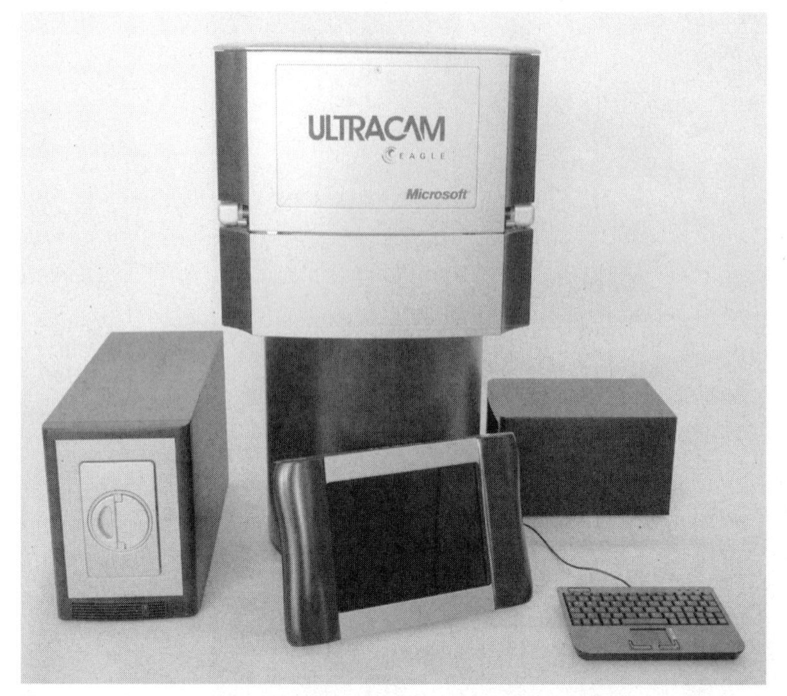

Figure 2.31 Microsoft/Vexel UltraCam Eagle Camera System. (Courtesy Microsoft Photogrammetry Division.)

of the images is 6.5 cm, and the total angular field of view across-track is 66°. In comparison, if the longer focal length lens system is employed at the same flying height, the panchromatic images would have a GSD of approximately 2.5 cm and a total angular field of view across-track of 28°. These differences highlight the interplay among focal length, flying height, physical pixel size, physical sensor array size, GSD, and angular field of view in the collection of digital aerial camera photography for various applications. The basic geometric character of this interplay is described below.

Geometric Elements of Area Array Digital Cameras

Figure 2.32 depicts a side view of a square pixel contained in a two-dimensional digital camera sensor array and the ground area within which this pixel collects incoming energy. From similar triangles,

$$\frac{GSD}{H'} = \frac{p_d}{f}$$

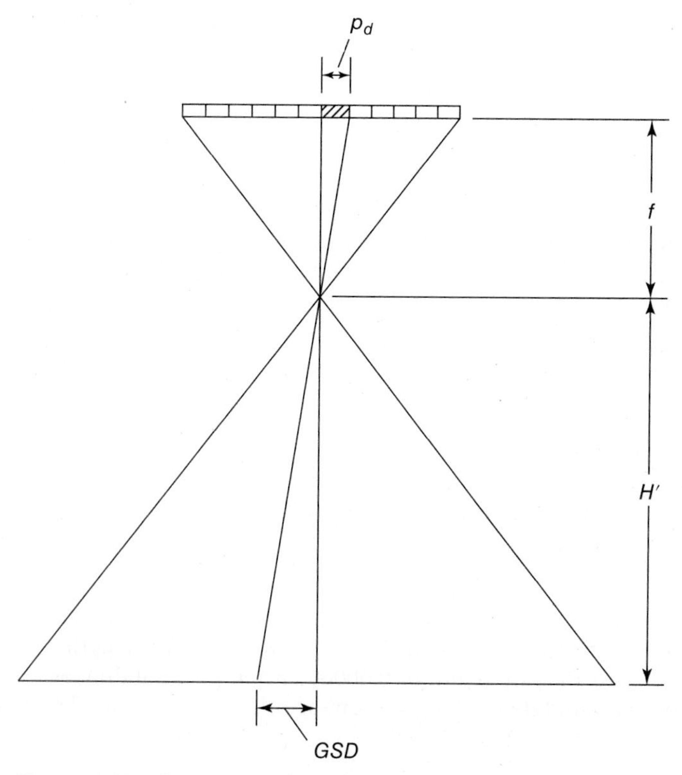

Figure 2.32 Geometric relationship among the physical pixel dimension (p_d), focal length (f), flying height above ground (H'), and ground sampling distance (GSD) for a digital array camera.

Rearranging yields,

$$GSD = \frac{H' p_d}{f} \tag{2.14}$$

where

GSD = ground sampling distance
H' = flying height above ground
p_d = physical size of the pixels comprising the sensor array

EXAMPLE 2.5

Assume that a digital camera equipped with a 70-mm-focal-length lens is being operated from a flying height above ground of 1000 m. If the physical size of the pixels comprising the camera's sensor is 6.9 μm (or 0.00069 cm), what is the GSD of the resulting digital photographs?

Solution

From Eq. 2.14

$$GSD = \frac{H' p_d}{f} = \frac{1000 \text{ m} \times 0.00069 \text{ cm}}{0.7 \text{ m}} = 9.98 \text{ cm}$$

For flight planning purposes, it is often necessary to determine what flying height above ground will yield photography of a desired GSD for a given camera system. This can be done by simply rearranging Eq. 2.14 such that

$$H' = \frac{(GSD) \times f}{p_d} \tag{2.15}$$

EXAMPLE 2.6

A GSD of 7.5 cm is required for a particular application. If the digital camera to be used for the aerial photographic mission is equipped with an 80-mm-focal-length lens and has a sensor consisting of pixels that are 6 μm (or 0.00060 cm) in physical dimension, determine the flying height above ground that will result in images having the desired GSD.

Solution

From Eq. 2.15

$$H' = \frac{(GSD) \times f}{p_d} = \frac{7.5 \text{ cm} \times 0.080 \text{ m}}{0.00060 \text{ cm}} = 1000 \text{ m}$$

Another important geometric characteristic of a digital area array camera is its angular field of view. Because area array sensors are often rectangular rather than square, consideration must be given to the angular field of view in both the across-track and along-track directions. With reference to Figure 2.33, the angle θ_{xt} defines the total across-track angular field of view of a digital camera sensor. The size of one-half of this angle is related to one-half of the dimension of the sensor in the across-track direction as follows

$$\frac{\theta_{xt}}{2} = \tan^{-1}\left(\frac{d_{xt}}{2f}\right)$$

which can be restated as

$$\theta_{xt} = 2\tan^{-1}\left(\frac{d_{xt}}{2f}\right) \tag{2.16}$$

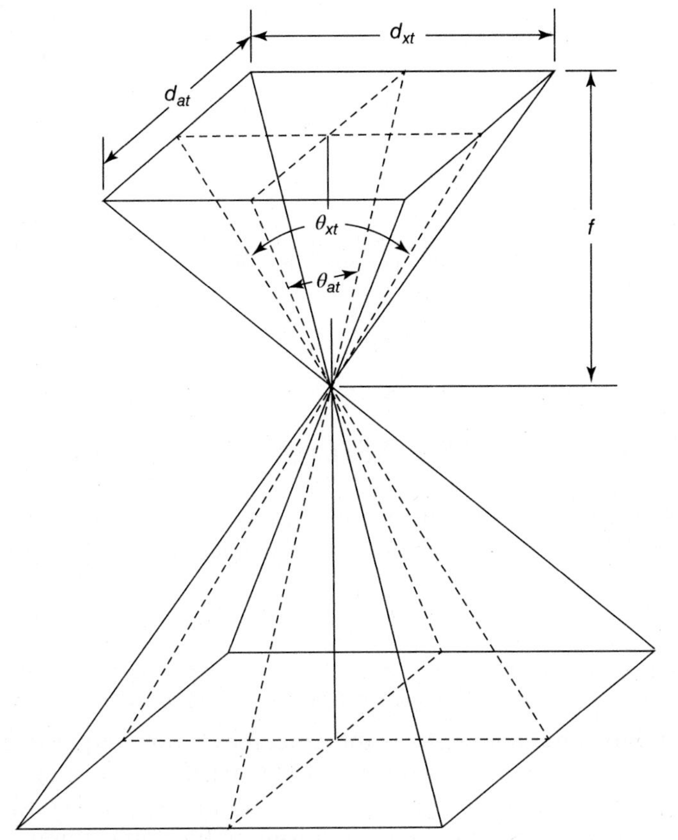

Figure 2.33 Across-track (θ_{xt}) and along-track (θ_{at}) angular fields of view for an area-array digital camera.

A similar derivation for the total along-track angular field of view, θ_{at}, yields

$$\theta_{at} = 2\tan^{-1}\left(\frac{d_{at}}{2f}\right) \tag{2.17}$$

EXAMPLE 2.7

If the dimensions of a digital camera sensor are 10.40 cm in the across-track direction and 6.81 cm in the along-track direction, and the camera has a 80-mm-focal-length lens, what are the across-track and along-track total angular fields of view for the camera?

Solution

From Eq. 2.16

$$\theta_{xt} = 2\tan^{-1}\left(\frac{10.40\text{ cm}}{2(8)\text{ cm}}\right) = 66^\circ$$

From Eq. 2.17

$$\theta_{at} = 2\tan^{-1}\left(\frac{6.81\text{ cm}}{2(8)\text{ cm}}\right) = 46^\circ$$

2.7 SPATIAL RESOLUTION OF CAMERA SYSTEMS

Spatial resolution is an expression of the optical quality of an image produced by a particular camera system. Here we describe the spatial resolution of both film and digital camera systems.

Film Camera Systems

The resolution of film camera systems is influenced by a host of parameters, such as the resolving power of the film and camera lens used to obtain an image, any uncompensated image motion during exposure, the atmospheric conditions present at the time of image exposure, the conditions of film processing, and so on. Some of these elements are quantifiable. For example, we can measure the resolving power of a film by photographing a standard test chart. Such a chart is shown in Figure 2.34. It consists of groups of three parallel lines separated by spaces equal to the width of the lines. Successive groups systematically decrease in size within the chart. The resolving power of a film is the reciprocal of the center-to-center distance (in millimeters) of the lines that are just "distinguishable" in the test chart image when viewed under a microscope. Hence, film resolving power is expressed in units of lines per millimeter. Film resolving power is sometimes referred to in units of "line pairs" per millimeter. In this case, the term *line pair* refers to a line (white) and a space (black) of equal width, as shown in Figure 2.34. The terms *lines per millimeter* and *line pairs per millimeter* refer to the same line spacing and can be used interchangeably. Film resolving power is specified at a particular contrast ratio between the lines and their background. This is done because resolution is very strongly influenced by contrast. For example, a black-and-white negative film might have a resolution of 125 line pairs/mm at a contrast ratio of 1000:1, but have a resolution of only 50 line pairs/mm at a contrast ratio of 1.6:1. Historically, a broad array of film emulsion types (negative,

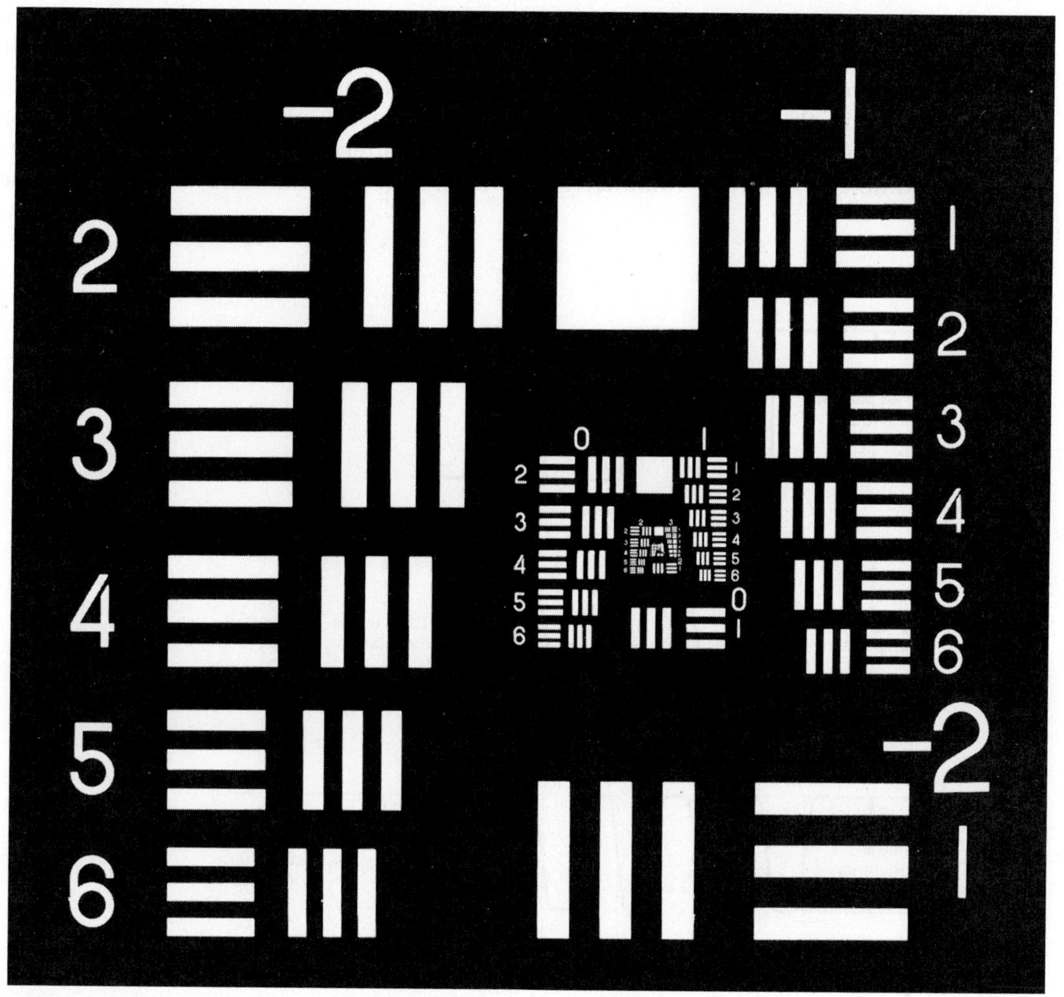

Figure 2.34 USAF resolving power test chart. (Courtesy Teledyne-Gurley Co.)

positive, black-and-white, black-and-white-infrared, color, and color-infrared) having various resolutions were readily available commercially. With the trend toward increased use of digital cameras to obtain aerial photography, the availability of various film types and resolutions has decreased. (Readers who wish to investigate the current availability of various aerial films are encouraged to consult such websites as www.kodak.com and www.agfa.com.)

An alternative method of determining film resolution that eliminates the subjectivity of deciding when lines are just "distinguishable" is the construction of a film's *modulation transfer function* (*MTF*). In this method, a scanning densitometer

(Section 2.4) is used to scan across images of a series of "square-wave" test patterns similar to the one shown in Figure 2.35*a*. An ideal film would exactly record not only the brightness variation (modulation) of the test pattern but also the distinct edges in the pattern. For actual films, the fidelity of the film recording process depends upon the spatial frequency of the pattern. For test patterns with a small number of lines per millimeter, the maximum and minimum brightness values as

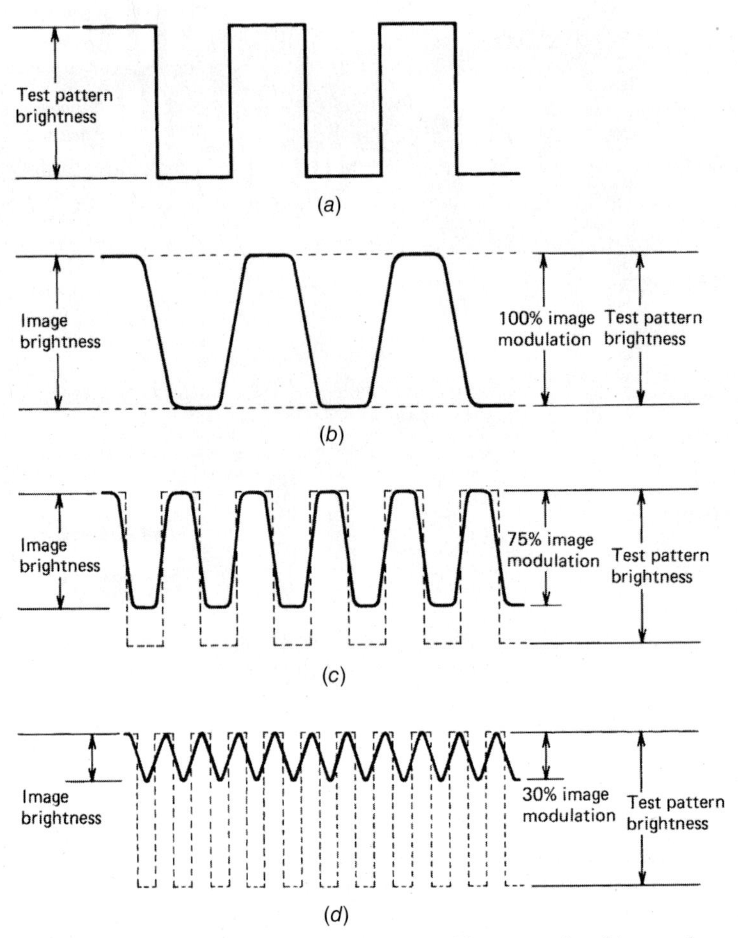

Figure 2.35 (*a*) Square-wave test pattern. (*b*) Modulation transfer of image of test pattern shown in (*a*). (*c, d*) Modulation transfer of images of test patterns having higher spatial frequency. [Note in (*b*) that 100% modulation occurs, but the image shows a reduction in edge sharpness as compared with the test pattern.] In (*c*), edge sharpness is further reduced, and, in addition, modulation transfer is reduced to 75% of that of the test pattern. In (*d*), further sharpness is lost and modulation transfer is reduced to 30%. (Adapted from Wolf et al., 2013.)

measured from the film image (Figure 2.35*b*) might correspond exactly with those of the test pattern. At the spatial frequency of this test pattern, the film's modulation transfer is said to be 100%. However, note that the test pattern edges are somewhat rounded on the film image. As the line width and spacing of the test pattern are reduced, density scans across the film image of the test pattern will produce both reduced modulations and increased edge rounding. This is illustrated in Figures 2.35*c* and *d* (showing 75 and 30% modulation transfer, respectively). By measuring film densities across many such patterns of progressively higher spatial frequency, a complete curve for the modulation transfer function can be constructed (Figure 2.36). Again, this curve expresses the fidelity with which images of features over a range of different sizes or spatial frequencies can be recorded by a given film.

The resolution, or modulation transfer, of any given film is primarily a function of the size distribution of the silver halide grains in the emulsion. In general, the higher the granularity of a film, the lower its resolving power. However, films of higher granularity are generally more sensitive to light, or *faster*, than those having lower granularity. Hence, there is often a trade-off between film "speed" and resolution.

The resolving power of any particular camera–film *system* can be measured by flying over and photographing a large bar target array located on the ground. The imagery thus obtained incorporates the image degradation realized in flight resulting from such factors as atmospheric effects and residual

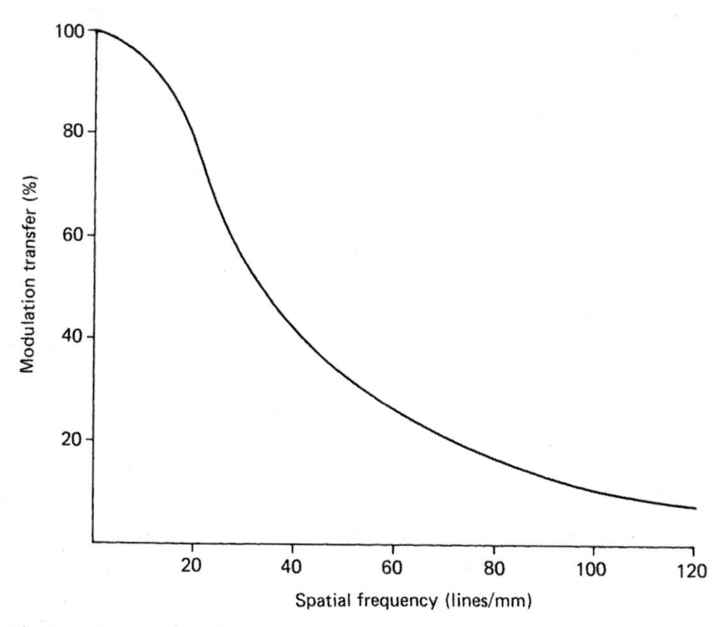

Figure 2.36 Curve of modulation transfer function (MTF). (Adapted from Wolf et al., 2013.)

image motion during exposure (including that due to camera vibrations). The advantage to this is that we can begin to judge the *dynamic* spatial resolution of the total photographic system instead of the *static* resolution of any one of its components.

The effects of scale and resolution can be combined to express image quality in terms of a *ground resolution distance* (*GRD*). This distance extrapolates the dynamic system resolution on a film to a ground distance. We can express this as

$$\text{GRD} = \frac{\text{reciprocal of image scale}}{\text{system resolution}} \tag{2.18}$$

For example, a photograph at a scale of 1:50,000 taken with a system having a dynamic resolution of 40 lines/mm would have a ground resolution distance of

$$\text{GRD} = \frac{50{,}000}{40} = 1250 \text{ mm} = 1.25 \text{ m}$$

This result assumes that we are dealing with an original film at the scale at which it was exposed. Enlargements would show some loss of image definition in the printing and enlarging process.

In summary, the ground resolution distance provides a framework for comparing the expected capabilities of various film-based images to record spatial detail. However, this and any other measure of spatial resolution must be used with caution because many unpredictable variables enter into what can and cannot be detected, recognized, or identified on an aerial photograph.

Digital Camera Systems

As with film camera systems, the resolution of digital camera systems can be measured by photographing a test chart. Figure 2.37 illustrates a resolution test chart developed specifically for digital cameras. However, a modulation transfer function approach is the most common and accepted means used to assess a digital camera system's spatial resolution. Without going into the mathematical details of how this is accomplished, suffice it to say that the MTF process we described earlier is refined by using a smooth sine-wave variation from white to black as opposed to the abrupt alternate black-and-white lines of the square wave pattern. (The units of resolution measurement become cycles/mm.) This results in a much more accurate portrayal of resolution at high spatial frequencies (the square wave test chart results in an overly optimistic estimation of resolution). In addition, use of the sine-wave variation permits the mathematical ability to cascade (multiply in series) the MTF curves of the various components of the camera system along with the environmental conditions of imaging to arrive at a composite measure of resolution performance of the total imaging system. For example, the MTFs for a lens set at a certain f/stop, a given sensor array, a

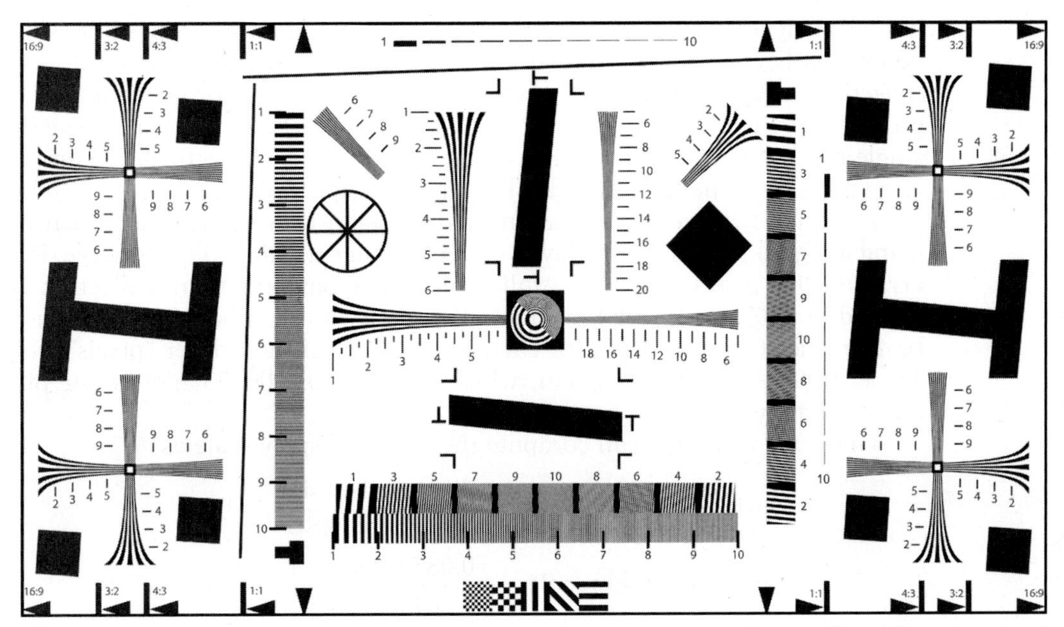

Figure 2.37 ISO 12233 resolution test chart for testing digital camera system resolution. (Adapted from International Standards Organization.)

particular atmospheric condition, and so forth, can be multiplied together. Then for any assumed ground scene frequency and modulation, one is able to estimate the modulation of the resulting image. If the MTF for any camera system component or imaging variable changes, the system MTF can be recalculated, and a new estimate of the modulation of the image can be made. Thus the MTFs become a valuable tool for comparing the relative performance of different digital camera systems.

Recall that in the case of film-based images, the key factor involved in extrapolating a system's dynamic image resolution to a ground resolution distance (GRD) is the scale of the original image. However, digital image files do not have a single scale per se. They can be displayed and printed at many different scales. In general, digital images are not displayed or printed at the original scale of the camera because it is typically too small to be useful. Also, the ground area covered by each pixel in the image does not change as the scale of a digital display or printed image changes. *Because of this, the ground sample distance (GSD), rather than the image scale, is the appropriate metric for determining the interpretability of a digital image.*

We have to qualify the above statement from the perspective that the term "GSD" is often used very loosely. As mentioned earlier (Section 1.12), the GSD of an image file may or may not contain pixels whose ground dimension is equal to the native spatial resolution of a given sensor (the ground resolution cell size). It has been suggested that to avoid this type of ambiguity, three types of GSD can be defined: collection GSD, product GSD, and display GSD (Comer et al., 1998). A *collection GSD* is simply the linear dimension of a single pixel's footprint on the ground (see Figure 2.32 and Eq. 2.14). A *product GSD* is the linear dimension of a single pixel's footprint in a digital image product after all rectification and resampling procedures have occurred (Chapter 7 and Appendix B).

A *display GSD* refers to the linear dimension of a single pixel when printed using a digital printer or displayed on a computer monitor. For example, if a CCD array is 3000 pixels wide and a full-scene image having a 0.4m collection GSD is printed at 300 dots (pixels) per inch, the total width of the printed image would be 10 inches (printing 1 dot for each pixel). Each of the printer "pixels" would be 1" per 300 pixels in dimension, which equates to 0.00333 inches per pixel, or approximately 0.085 mm per pixel.

Given the above we can compute the printer display scale as

$$\text{printer display scale} = \frac{\text{display pixel size}}{\text{collection GSD}} \tag{2.19}$$

$$\text{printer display scale} = \frac{0.085\text{ mm}}{0.4\text{ m}} = \frac{1}{4706} \quad \text{or} \quad 1{:}4706$$

When displaying an image file on a computer monitor, the *dot pitch* (the distance between display dots of the monitor) controls the image size and scale. A typical dot pitch is on the order of 0.3 mm. For the above example, such a dot pitch would yield the following monitor display scale (when displaying 1 dot for each pixel),

$$\text{monitor display scale} = \frac{\text{display pixel size}}{\text{collection GSD}} \tag{2.20}$$

$$\text{monitor display scale} = \frac{0.3\text{ mm}}{0.4\text{ m}} = \frac{1}{1333} \quad \text{or} \quad 1{:}1333$$

In short, the scale of a digital image is variable, and if two image or map products are produced at the same scale but a different GSD, the product having the smaller GSD will portray more interpretive spatial detail. It will also consist of a greater number of pixels, resulting in a trade-off between spatial detail and data handling challenges.

Detection, Recognition, and Identification of Objects

With both film and digital photography, spatial resolution values are difficult to interpret in a practical sense. Our interest in measuring a system's resolution goes beyond determining the ability of a system to record distinct images of small, nearly contiguous objects of a given shape on a test chart. We are interested not only in object *detection* but also in object *recognition* and *identification*. Hence, "spatial resolution" defies precise definition. At the detection level, the objective is to discern separate objects discretely. At the recognition level, we attempt to determine what objects are—for example, trees versus row crops. At the identification level, we more specifically identify objects—for example, oak trees versus corn.

In the above context, it is important to note that a GSD of 1 m does not mean that one can identify or even recognize an object on the ground that is 1 m in size on the ground. One pixel might suffice to only indicate that something is present. For visual identification of a particular object, 5 to 10 or many more pixels may be required (depending upon the GSD used and the nature of the object). Also, while small GSDs generally aid in visual image interpretation, they may or may not aid in various digital image interpretation activities. For example, the accuracy of digital image classification through the application of spectral pattern recognition (Chapter 7) is influenced by the choice of GSD. Classification procedures that work well at a given GSD may not work as well at either a larger or smaller GSD.

2.8 AERIAL VIDEOGRAPHY

Aerial videography is a form of electronic imaging whereby analog or digital video signals are recorded on various storage media. Cameras used for aerial videography can have many configurations, including single-band cameras, multiband cameras, or multiple single-band cameras, and can sense in the visible, near-IR, and mid-IR wavelengths (not all cameras have this full range).

Analog video systems were commonplace prior to the mid-1980s, when digital video was first introduced. Today, digital systems are the norm, and digital video recording is afforded by many devices, ranging from still cameras that include an optional video capture mode to camera phones, laptops, tablets, surveillance cameras, personal media players, and many other devices. Video systems flown in light aircraft for recording large areas are typically either consumer-grade or professional-grade digital video cameras.

Like digital still-frame cameras, digital video cameras employ either CCD or CMOS semiconductor two-dimensional detector arrays that convert the incoming energy at each photosite into a digital brightness value. The primary difference between a still camera and a video camera is simply the rapid rate at which individual image frames are captured by a video camera in order to record scene and subject motion in a smooth and continuous fashion. There are two frame rate collection standards for video cameras. The PAL standard specifies 25 frames per second, and the NTSC standard specifies 30 frames per 1.001 second (about 29.97 frames per second). In comparison, standard film-based movies are produced at 24 frames per second. Because of the high bit rate associated with collecting and recording video data, the data are typically subject to some for of image compression (Chapter 7) according to various industry compression standards.

Depending on camera design, digital video data are captured using one of two approaches: interlaced or progressive. *Interlaced data capture* involves recording the data from alternating sets of photosite lines. Each frame is captured by successively scanning the odd-numbered lines, and then the even-numbered lines. *Progressive data capture* involves recording the data from all photosite lines in the array at the same instant. Digital video data are typically stored using hard drives or solid state storage devices.

Digital video recording is typically done following Standard Definition (SD) or High Definition (HD) standards, using a variety of image compression standards. Standard Definition digital video cameras typically record with an image format of 4:3 (horizontal to vertical size) or 16:9. High Definition digital video cameras use the 16:9 format. Standard Definition video with a 4:3 image format typically consists of 640 (horizontal) $\times$ 480 pixels (vertical). High Definition video consists of up to 1920 $\times$ 1080 pixels.

One of the more recent trends in videography is the use of *4K resolution* systems for digital video capture and display. The terminology "4K" derives from the fact that these systems typically have an approximate resolution of 4,000 pixels in the horizontal direction. (Note that the reference to the horizontal resolution of these systems is different from the traditional categorization of HDTV systems in accordance with their vertical resolution—1080 pixels, for example.) Various current (2014) 4K video cameras, projectors, and monitors incorporate resolutions on the order of 4096 pixels $\times$ 2160 pixels, or approximately 8.8 megapixels per frame.

Aerial videography has been used for applications similar to those for which small- and medium-format film and digital cameras are used. Often, video imagery is acquired over small sample areas to obtain "ground truth" to aid in the recognition of ground conditions in coarser resolution data sets (e.g., moderate resolution satellite data). It is also very useful in applications where successive coverage of rapidly changing conditions is required. These include such activities as monitoring oil spills, fires, floods, storm damage, and other natural disasters. Videography is also well-suited to monitoring linear features such as roads, power lines, and rivers. In such applications, two cameras are often used in tandem. One might employ wide-angle viewing, while the second might provide a more detailed view of smaller areas. Alternatively, one camera might obtain a nadir view while the other provides forward-looking oblique imagery. Video surveys of coastal areas are often accomplished by flying offshore parallel to the shoreline and acquiring high oblique imagery perpendicular to the shoreline.

During the course of video data collection, ground objects are viewed from multiple vantage points as they pass under or alongside the aircraft. Thus, there are multiple view angle-sun angle combinations that can be analyzed. Often, the continuous change in perspective afforded by video imagery results in a pseudo 3D effect. While either multiple cameras or a specialized dual lens video camera can be used to collect true stereoscopic coverage, such imagery is more the exception than the rule.

Among the advantages of aerial videography are its relatively low cost and its flexibility in terms of scheduling data acquisition with short lead times. Video data collection supported by GPS and IMU system integration provides an effective means to georeference the resulting imagery for subsequent display, analysis, and GIS operations. Video imagery can also be transmitted and disseminated very rapidly, often in real time. Video images are also ideal for capturing and "freezing" objects and events that are in motion. Finally, the inclusion of an audio recording capability in a video system provides for in-flight narration and commentary by the pilot and trained observers as an aid to documenting and interpreting the image data.

2.9 CONCLUSION

As we indicated earlier in this chapter, aerial film photography has historically been the most widely used form of remote sensing due to its general availability, geometric integrity, versatility, and economy. However, the use of film-based photography has decreased dramatically since the introduction of the first commercial large-format digital camera in 2000. Digital photography has replaced film-based photography almost completely. At the same time, the availability and resolution of satellite imagery continue to improve and remote sensing from earth orbit is, in many cases, supplementing or replacing the use of aerial photography for a variety of remote sensing applications. Thus, as a remote sensing data user, it is important to understand the full range of sensor systems that are available and appropriate for any given application. (High spatial resolution satellite remote sensing systems are discussed in Chapter 5.)

3 BASIC PRINCIPLES OF PHOTOGRAMMETRY

3.1 INTRODUCTION

Photogrammetry is the science and technology of obtaining spatial measurements and other geometrically reliable derived products from photographs. Historically, photogrammetric analyses involved the use of hardcopy photographic products such as paper prints or film transparencies. Today, most photogrammetric procedures involve the use of digital, or *softcopy*, photographic data products. In certain cases, these digital products might result from high resolution scanning of hardcopy photographs (e.g., historical photographs). However, the overwhelming majority of digital photographs used in modern photogrammetric applications come directly from digital cameras. In fact, photogrammetric processing of digital camera data is often accomplished in-flight such that digital orthophotos, digital elevation models (DEMs), and other GIS data products are available immediately after, or even during, an aerial photographic mission.

While the physical character of hardcopy and digital photographs is quite different, the basic geometric principles used to analyze them photogrammetrically are identical. In fact, it is often easier to visualize and understand these principles in a hardcopy context and then extend them to the softcopy environment. This is the approach we adopt in this chapter. Hence, our objective in this discussion is to not only prepare the reader to be able to make basic measurements from hardcopy photographic images, but also to understand the underlying principles of modern digital (softcopy) photogrammetry. We stress *aerial* photogrammetric techniques and procedures in this discussion, but the same general principles hold for *terrestrial* (ground-based) and *space*-based operations as well.

In this chapter, we introduce only the most basic aspects of the broad subject of photogrammetry. (More comprehensive and detailed treatment of the subject of photogrammetry is available in such references as ASPRS, 2004; Mikhail et al., 2001; and Wolf et al., 2013.) We limit our discussion to the following photogrammetric activities.

1. **Determining the scale of a vertical photograph and estimating horizontal ground distances from measurements made on a vertical photograph.** The scale of a photograph expresses the mathematical relationship between a distance measured on the photo and the corresponding horizontal distance measured in a ground coordinate system. Unlike maps, which have a single constant scale, aerial photographs have a range of scales that vary in proportion to the elevation of the terrain involved. Once the scale of a photograph is known at any particular elevation, ground distances at that elevation can be readily estimated from corresponding photo distance measurements.
2. **Using area measurements made on a vertical photograph to determine the equivalent areas in a ground coordinate system.** Computing ground areas from corresponding photo area measurement is simply an extension of the above concept of scale. The only difference is that whereas ground distances and photo distances vary linearly, ground areas and photo areas vary as the square of the scale.
3. **Quantifying the effects of relief displacement on vertical aerial photographs.** Again unlike maps, aerial photographs in general do not show the true plan or top view of objects. The images of the tops of objects appearing in a photograph are displaced from the images of their bases. This is known as *relief displacement* and causes any object standing above the terrain to "lean away" from the principal point of a photograph radially. Relief displacement, like scale variation, precludes the use of aerial photographs directly as maps. However, reliable ground measurements and maps can be obtained from vertical photographs if photo measurements are analyzed with due regard for scale variations and relief displacement.
4. **Determining object heights from relief displacement measurements.** While relief displacement is usually thought of as an image distortion that must be dealt with, it can also be used to estimate the heights of objects appearing on a photograph. As we later illustrate, the magnitude of relief displacement depends on the flying height, the distance from the photo principal point to the feature, and the height of the feature. Because

these factors are geometrically related, we can measure an object's relief displacement and radial position on a photograph and thereby determine the height of the object. This technique provides limited accuracy but is useful in applications where only approximate object heights are needed.

5. **Determining object heights and terrain elevations by measuring image parallax.** The previous operations are performed using vertical photos individually. Many photogrammetric operations involve analyzing images in the area of overlap of a stereopair. Within this area, we have two views of the same terrain, taken from different vantage points. Between these two views, the relative positions of features lying closer to the camera (at higher elevation) will change more from photo to photo than the positions of features farther from the camera (at lower elevation). This change in relative position is called *parallax*. It can be measured on overlapping photographs and used to determine object heights and terrain elevations.

6. **Determining the elements of exterior orientation of aerial photographs.** In order to use aerial photographs for photogrammetric mapping purposes, it is necessary to determine six independent parameters that describe the position and angular orientation of each photograph at the instant of its exposure relative to the origin and orientation of the ground coordinate system used for mapping. These six variables are called the *elements of exterior orientation*. Three of these are the 3D position (X, Y, Z) of the center of the photocoordinate axis system at the instant of exposure. The remaining three are the 3D rotation angles (ω, ϕ, κ) related to the amount and direction of tilt in each photo at the instant of exposure. These rotations are a function of the orientation of the platform and camera mount when the photograph is taken. For example, the wings of a fixed-wing aircraft might be tilted up or down, relative to horizontal. Simultaneously, the camera might be tilted up or down toward the front or rear of the aircraft. At the same time, the aircraft might be rotated into a headwind in order to maintain a constant heading.

   We will see that there are two major approaches that can be taken to determine the elements of exterior orientation. The first involves the use of *ground control* (photo-identifiable points of known ground coordinates) together with a mathematical procedure called *aerotriangulation*. The second approach entails *direct georeferencing*, which involves the integration of GPS and IMU observations to determine the position and angular orientation of each photograph (Sections 1.11 and 2.6). We treat both of these approaches at a conceptual level, with a minimum of mathematical detail.

7. **Production of maps, DEMs, and orthophotos.** "Mapping" from aerial photographs can take on many forms. Historically, topographic maps were produced using hardcopy stereopairs placed in a device called a *stereoplotter*. With this type of instrument the photographs are mounted in special projectors that can be mutually oriented to precisely correspond

to the angular tilts (ω, ϕ, κ) present when the photographs were taken. Each of the projectors can also be translated in x, y, and z such that a reduced-size model is created that exactly replicates the exterior orientation of each of the photographs comprising the stereopair. (The scale of the resulting stereomodel is determined by the "air base" distance between the projectors chosen by the instrument operator.) When viewed stereoscopically, the model can be used to prepare an analog or digital planimetric map having no tilt or relief distortions. In addition, topographic contours can be integrated with the planimetric data and the height of individual features appearing in the model can be determined.

Whereas a stereoplotter is designed to transfer *map* information, without distortions, from stereo photographs, a similar device can be used to transfer *image* information, with distortions removed. The resulting undistorted image is called an *orthophotograph* (or *orthophoto*). Orthophotos combine the geometric utility of a map with the extra "real-world image" information provided by a photograph. The process of creating an orthophoto depends on the existence of a reliable DEM for the area being mapped. The DEMs are usually prepared photogrammetrically as well. In fact, *photogrammetric workstations* generally provide the integrated functionality for such tasks as generating DEMs, digital orthophotos, topographic maps, perspective views, and "fly-throughs," as well as the extraction of spatially referenced GIS data in two or three dimensions.

8. **Preparing a flight plan to acquire vertical aerial photography.** Whenever new photographic coverage of a project area is to be obtained, a photographic flight mission must be planned. As we will discuss, mission planning software highly automates this process. However, most readers of this book are, or will become, consumers of image data rather than providers. Such individuals likely will not have direct access to flight planning software and will appropriately rely on professional data suppliers to jointly design a mission to meet their needs. There are also cases where cost and logistics might dictate that the data consumer and the data provider are one in the same!

   Given the above, it is important for image analysts to understand at least the basic rudiments of mission planning to in order to facilitate such activities as preliminary estimation of data volume (as it influences both data collection and analysis), choosing among alternative mission parameters, and ensuring a reasonable fit between the information needs of a given project and the data collected to meet those needs. Decisions have to be made relative to such mission elements as image scale or ground sample distance (GSD), camera format size and focal length, and desired image overlap. The analyst can then determine such geometric factors as the appropriate flying height, the distance between image centers, the direction and spacing of flight lines, and the total number of images required to cover the project area.

Each of these photogrammetric operations is covered in separate sections in this chapter. We first discuss some general geometric concepts that are basic to these techniques.

3.2 BASIC GEOMETRIC CHARACTERISTICS OF AERIAL PHOTOGRAPHS

Geometric Types of Aerial Photographs

Aerial photographs are generally classified as either vertical or oblique. *Vertical photographs* are those made with the camera axis directed as vertically as possible. However, a "truly" vertical aerial photograph is rarely obtainable because of the previously described unavoidable angular rotations, or tilts, caused by the angular attitude of the aircraft at the instant of exposure. These unavoidable tilts cause slight (1° to 3°) unintentional inclination of the camera optical axis, resulting in the acquisition of *tilted photographs*.

Virtually all photographs are tilted. When tilted unintentionally and slightly, tilted photographs can be analyzed using the simplified models and methods appropriate to the analysis of truly vertical photographs without the introduction of serious error. This is done in many practical applications where approximate measurements suffice (e.g., determining the ground dimensions of a flat agricultural field based on photo measurements made with a scale or digitizing tablet). However, as we will discuss, precise digital photogrammetric procedures employ methods and models that rigorously account for even very small angles of tilt with no loss of accuracy.

When aerial photographs are taken with an intentional inclination of the camera axis, *oblique photographs* result. *High oblique photographs* include an image of the horizon, and *low oblique photographs* do not. In this chapter, we emphasize the geometric aspects of acquiring and analyzing vertical aerial photographs given their extensive historical and continuing use for large-area photogrammetric mapping. However, the use of oblique aerial photographs has increased rapidly in such applications as urban mapping and disaster assessment. With their "side-view" character, oblique photographs afford a more natural perspective in comparison to the top view perspective of vertical aerial photographs. This can greatly facilitate the image interpretation process (particularly for those individuals having limited training or experience in image interpretation). Various examples of oblique aerial photographs are interspersed throughout this book.

Photogrammetric measurements can also be made from oblique aerial photographs if they are acquired with this purpose in mind. This is often accomplished using multiple cameras that are shuttered simultaneously. Figure 3.1 illustrates such a configuration, the IGI Penta-DigiCAM system. This five-camera system employs one nadir-pointing camera, two cameras viewing obliquely in opposite

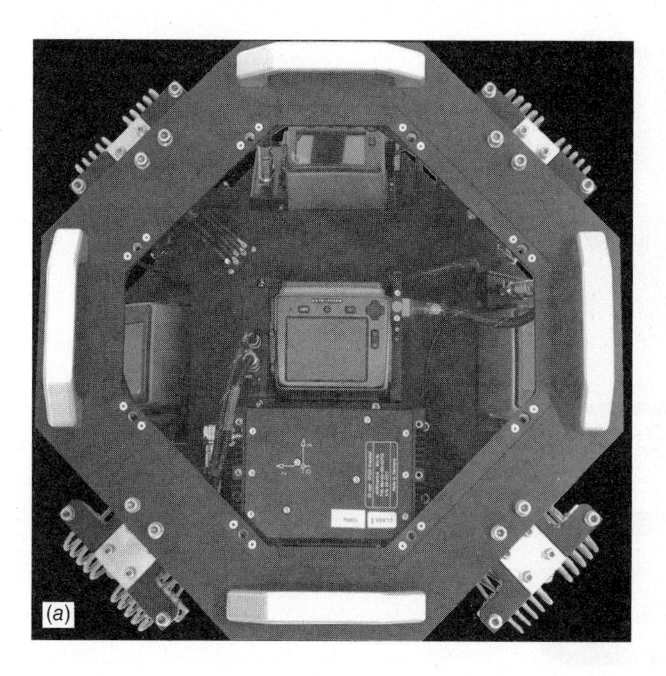

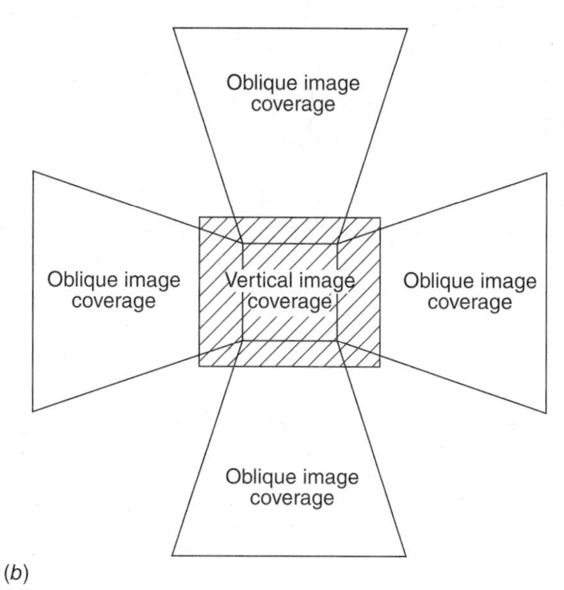

Figure 3.1 IGI Penta-DigiCAM system: (*a*) early version of system with camera back removed to illustrate the angular orientation of the nadir and oblique cameras (lower oblique camera partially obscured by IMU mounted to camera at center-bottom; (*b*) Maltese Cross ground coverage pattern resulting from the system; (*c*) later version of system installed in a gyro-stabilized mount with data storage units and GNSS/IMU system mounted on top. (Courtesy of IGI GmbH.)

(c)

Figure 3.1 *(Continued)*

directions in the cross-track direction, and two viewing obliquely in opposite directions along-track. (We earlier illustrated a single DigiCam system in Section 2.6.) Figure 3.1 also depicts the "Maltese Cross" ground coverage pattern that results with each simultaneous shutter release of the Penta-DigiCAM. Successive, overlapping composites of this nature can be acquired to depict the top, front, back, and sides of all ground features in a study area. In addition, the oblique orientation angles from vertical can be varied to meet the need of various applications of the photography.

Taking Vertical Aerial Photographs

Most vertical aerial photographs are taken with frame cameras along *flight lines*, or *flight strips*. The line traced on the ground directly beneath the aircraft during acquisition of photography is called the *nadir line*. This line connects the image centers of the vertical photographs. Figure 3.2 illustrates the typical character of the photographic coverage along a flight line. Successive photographs are generally taken with some degree of *endlap*. Not only does this lapping ensure total coverage along a flight line, but an endlap of at least 50% is essential for total *stereoscopic coverage* of a project area. Stereoscopic coverage consists of adjacent pairs of

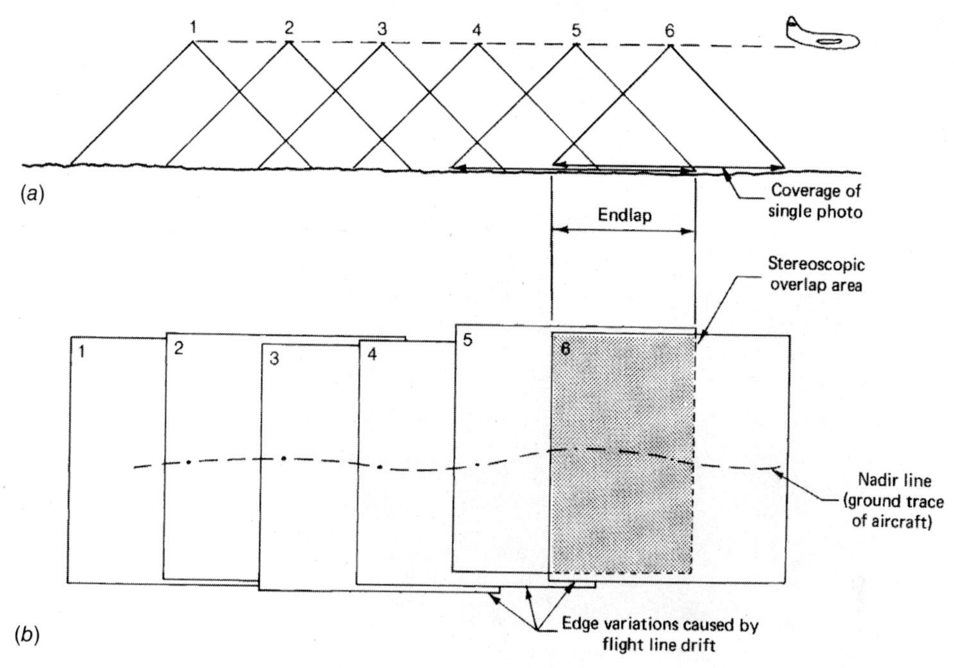

Figure 3.2 Photographic coverage along a flight strip: (*a*) conditions during exposure; (*b*) resulting photography.

overlapping vertical photographs called *stereopairs*. Stereopairs provide two different perspectives of the ground area in their region of endlap. When images forming a stereopair are viewed through a stereoscope, each eye psychologically occupies the vantage point from which the respective image of the stereopair was taken in flight. The result is the perception of a three-dimensional *stereomodel*. As pointed out in Chapter 1, most applications of aerial photographic interpretation entail the use of stereoscopic coverage and stereoviewing.

Successive photographs along a flight strip are taken at intervals that are controlled by the camera *intervalometer* or software-based sensor control system. The area included in the overlap of successive photographs is called the *stereoscopic overlap area*. Typically, successive photographs contain 55 to 65% overlap to ensure at least 50% endlap over varying terrain, in spite of unintentional tilt. Figure 3.3 illustrates the ground coverage relationship of successive photographs forming a stereopair having approximately a 60% stereoscopic overlap area.

The ground distance between the photo centers at the times of exposure is called the *air base*. The ratio between the air base and the flying height above ground determines the *vertical exaggeration* perceived by photo interpreters. The larger the *base–height ratio*, the greater the vertical exaggeration.

Figure 3.3 Acquisition of successive photographs yielding a stereopair. (Courtesy Leica Geosystems.)

Figure 3.4 shows Large Format Camera photographs of Mt. Washington and vicinity, New Hampshire. These stereopairs illustrate the effect of varying the percentage of photo overlap and thus the base–height ratio of the photographs. These photographs were taken from the Space Shuttle at an orbital altitude of 364 km. The stereopair in (*a*) has a base–height ratio of 0.30. The stereopair in (*b*) has a base–height ratio of 1.2 and shows much greater apparent relief (greater vertical exaggeration) than (*a*).

This greater apparent relief often aids in visual image interpretation. Also, as we will discuss later, many photogrammetric mapping operations depend upon accurate determination of the position at which rays from two or more photographs intersect in space. Rays associated with larger base–height ratios intersect at larger (closer to being perpendicular) angles than do those associated with the smaller (closer to being parallel) angles associated with smaller base–height ratios. Thus larger base–height ratios result in more accurate determination of ray intersection positions than do smaller base–height ratios.

Most project sites are large enough for multiple-flight-line passes to be made over the area to obtain complete stereoscopic coverage. Figure 3.5 illustrates how adjacent strips are photographed. On successive flights over the area, adjacent strips have a *sidelap* of approximately 30%. Multiple strips comprise what is called a *block* of photographs.

Figure 3.4 Large Format Camera stereopairs, Mt. Washington and vicinity, New Hampshire; scale 1:800,000 (1.5 times enlargement from original image scale): (*a*) 0.30 base–height ratio; (*b*) 1.2 base–height ratio. (Courtesy NASA and ITEK Optical Systems.)

As discussed in Section 3.1, planning an aerial photographic mission is usually accomplished with the aid of flight-planning software. Such software is guided by user input of such factors as the area to be covered by the mission; the required photo scale or GSD, overlap, and endlap; camera-specific parameters such as format size and focal length; and the ground speed of the aircraft to be used. A DEM and other background maps of the area to be covered are often integrated into such systems as well, providing a 2D or 3D planning environment. Flight-planning software is also usually closely coupled to, or integrated with, the flight navigation and guidance software, display systems, and sensor control and monitoring systems used during image acquisition. In this manner, in-flight displays visualize the approach to the mission area, the individual flight lines and photo centers, and the turns at the ends of flight lines. If a portion of a flight line

Figure 3.4 *(Continued)*

is missed (e.g., due to cloud cover), guidance is provided back to the area to be reflown. Such robust flight management and control systems provide a high degree of automation to the navigation and guidance of the aircraft and the operation of the camera during a typical mission.

Geometric Elements of a Vertical Photograph

The basic geometric elements of a hardcopy vertical aerial photograph taken with a single-lens frame camera are depicted in Figure 3.6. Light rays from terrain objects are imaged in the plane of the film negative after intersecting at the camera lens exposure station, L. The negative is located behind the lens at a distance equal to the lens focal length, f. Assuming the size of a paper print positive (or film positive) is equal to that of the negative, positive image positions can be depicted diagrammatically in front of the lens in a plane located at a distance f. This rendition is

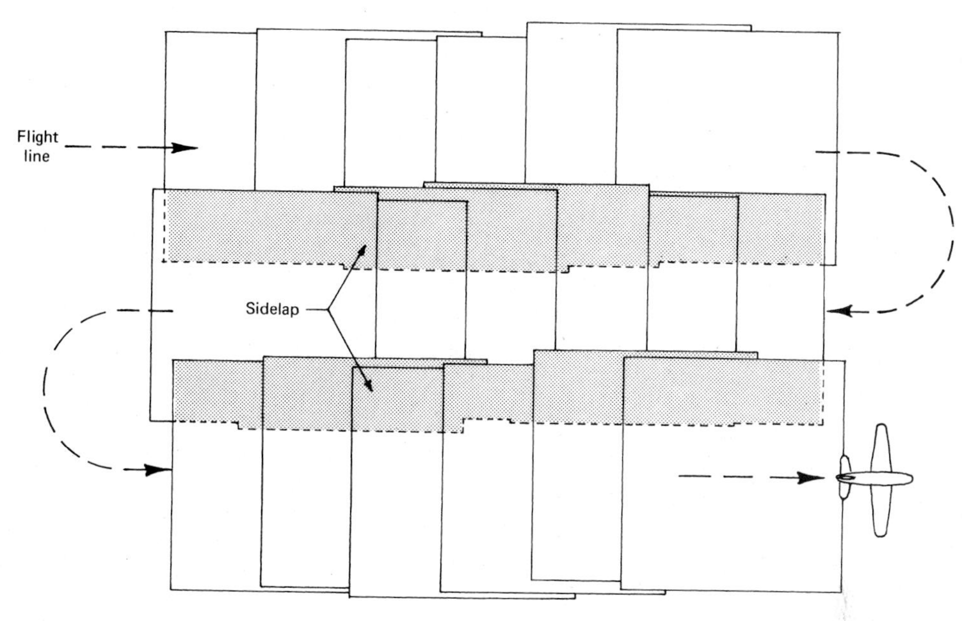

Figure 3.5 Adjacent flight lines over a project area.

appropriate in that most photo positives used for measurement purposes are contact printed, resulting in the geometric relationships shown.

The x and y coordinate positions of image points are referenced with respect to axes formed by straight lines joining the opposite fiducial marks (see Figure 2.24) recorded on the positive. The x axis is arbitrarily assigned to the fiducial axis most nearly coincident with the line of flight and is taken as positive in the forward direction of flight. The positive y axis is located 90° counterclockwise from the positive x axis. Because of the precision with which the fiducial marks and the lens are placed in a metric camera, the photocoordinate origin, o, can be assumed to coincide exactly with the *principal point*, the intersection of the lens optical axis and the film plane. The point where the prolongation of the optical axis of the camera intersects the terrain is referred to as the *ground principal point*, O. Images for terrain points A, B, C, D, and E appear geometrically reversed on the negative at a', b', c', d', and e' and in proper geometric relationship on the positive at a, b, c, d, and e. (Throughout this chapter we refer to points on the image with lowercase letters and corresponding points on the terrain with uppercase letters.)

The xy photocoordinates of a point are the perpendicular distances from the xy coordinate axes. Points to the right of the y axis have positive x coordinates and points to the left have negative x coordinates. Similarly, points above the x axis have positive y coordinates and those below have negative y coordinates.

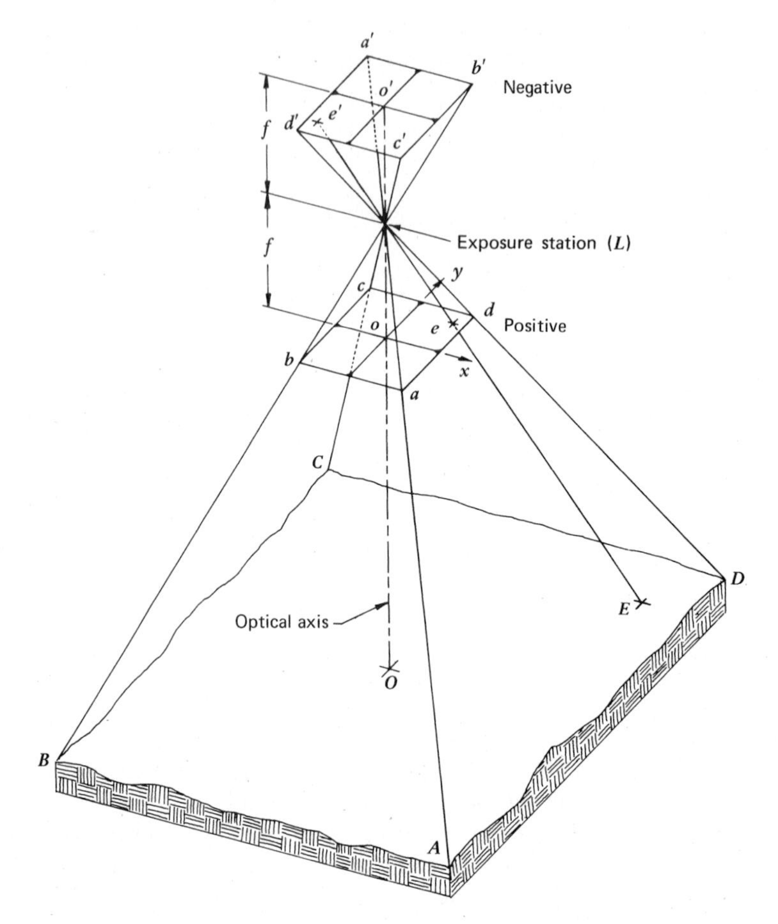

Figure 3.6 Basic geometric elements of a vertical photograph.

Photocoordinate Measurement

Measurements of photocoordinates may be obtained using any one of many measurement devices. These devices vary in their accuracy, cost, and availability. For rudimentary photogrammetric problems—where low orders of measurement accuracy are acceptable—a triangular *engineer's scale* or *metric scale* may be used. When using these scales, measurement accuracy is generally improved by taking the average of several repeated measurements. Measurements are also generally more accurate when made with the aid of a magnifying lens.

In a softcopy environment, photocoordinates are measured using a raster image display with a cursor to collect "raw" image coordinates in terms of row and column values within the image file. The relationship between the row and column coordinate system and the camera's fiducial axis coordinate system is determined through the development of a mathematical *coordinate transformation* between the two systems. This process requires that some points have their coordinates known in both systems. The fiducial marks are used for this purpose in that their positions in the focal plane are determined during the calibration of the camera, and they can be readily measured in the row and column coordinate system. (Appendix B contains a description of the mathematical form of the *affine coordinate transformation*, which is often used to interrelate the fiducial and row and column coordinate systems.)

Irrespective of what approach is used to measure photocoordinates, these measurements contain errors of varying sources and magnitudes. These errors stem from factors such as camera lens distortions, atmospheric refraction, earth curvature, failure of the fiducial axes to intersect at the principal point, and shrinkage or expansion of the photographic material on which measurements are made. Sophisticated photogrammetric analyses include corrections for all these errors. For simple measurements made on paper prints, such corrections are usually not employed because errors introduced by slight tilt in the photography will outweigh the effect of the other distortions.

3.3 PHOTOGRAPHIC SCALE

One of the most fundamental and frequently used geometric characteristics of hardcopy aerial photographs is that of *photographic scale*. A photograph "scale," like a map scale, is an expression that states that one unit (any unit) of distance on a photograph represents a specific number of units of actual ground distance. Scales may be expressed as *unit equivalents*, *representative fractions*, or *ratios*. For example, if 1 mm on a photograph represents 25 m on the ground, the scale of the photograph can be expressed as 1 mm=25 m (unit equivalents), or $\frac{1}{25,000}$ (representative fraction), or 1:25,000 (ratio).

Quite often the terms "large scale" and "small scale" are confused by those not working with expressions of scale on a routine basis. For example, which photograph would have the "larger" scale—a 1:10,000 scale photo covering several city blocks or a 1:50,000 photo that covers an entire city? The intuitive answer is often that the photo covering the larger "area" (the entire city) is the larger scale product. This is not the case. The larger scale product is the 1:10,000 image because it shows ground features at a larger, more detailed, size. The 1:50,000 scale photo of the entire city would render ground features at a much smaller, less detailed size. Hence, in spite of its larger ground coverage, the 1:50,000 photo would be termed the smaller scale product.

A convenient way to make scale comparisons is to remember that the same objects are smaller on a "smaller" scale photograph than on a "larger" scale photo. Scale comparisons can also be made by comparing the magnitudes of the representative fractions involved. (That is, $\frac{1}{50,000}$ is smaller than $\frac{1}{10,000}$.)

The most straightforward method for determining photo scale is to measure the corresponding photo and ground distances between any two points. This requires that the points be mutually identifiable on both the photo and a map. The scale S is then computed as the ratio of the photo distance d to the ground distance D,

$$S = \text{photo scale} = \frac{\text{photo distance}}{\text{ground distance}} = \frac{d}{D} \tag{3.1}$$

EXAMPLE 3.1

Assume that two road intersections shown on a photograph can be located on a 1:25,000 scale topographic map. The measured distance between the intersections is 47.2 mm on the map and 94.3 mm on the photograph. (*a*) What is the scale of the photograph? (*b*) At that scale, what is the length of a fence line that measures 42.9 mm on the photograph?

Solution

(*a*) The ground distance between the intersections is determined from the map scale as

$$0.0472 \text{ m} \times \frac{25{,}000}{1} = 1180 \text{ m}$$

By direct ratio, the photo scale is

$$S = \frac{0.0943 \text{ m}}{1180 \text{ m}} = \frac{1}{12{,}513} \quad or \ 1\!:\!12{,}500$$

(Note that because only three significant, or meaningful, figures were present in the original measurements, only three significant figures are indicated in the final result.)

(*b*) The ground length of the 42.9-mm fence line is

$$D = \frac{d}{S} = 0.0429 \text{ m} \div \frac{1}{12{,}500} = 536.25 \text{ m} \quad \text{or} \quad 536 \text{ m}$$

For a vertical photograph taken over flat terrain, scale is a function of the focal length f of the camera used to acquire the image and the flying height above the ground, H', from which the image was taken. In general,

$$\text{Scale} = \frac{\text{camera focal length}}{\text{flying height above terrain}} = \frac{f}{H'} \tag{3.2}$$

Figure 3.7 illustrates how we arrive at Eq. 3.2. Shown in this figure is the side view of a vertical photograph taken over flat terrain. Exposure station L is at an aircraft

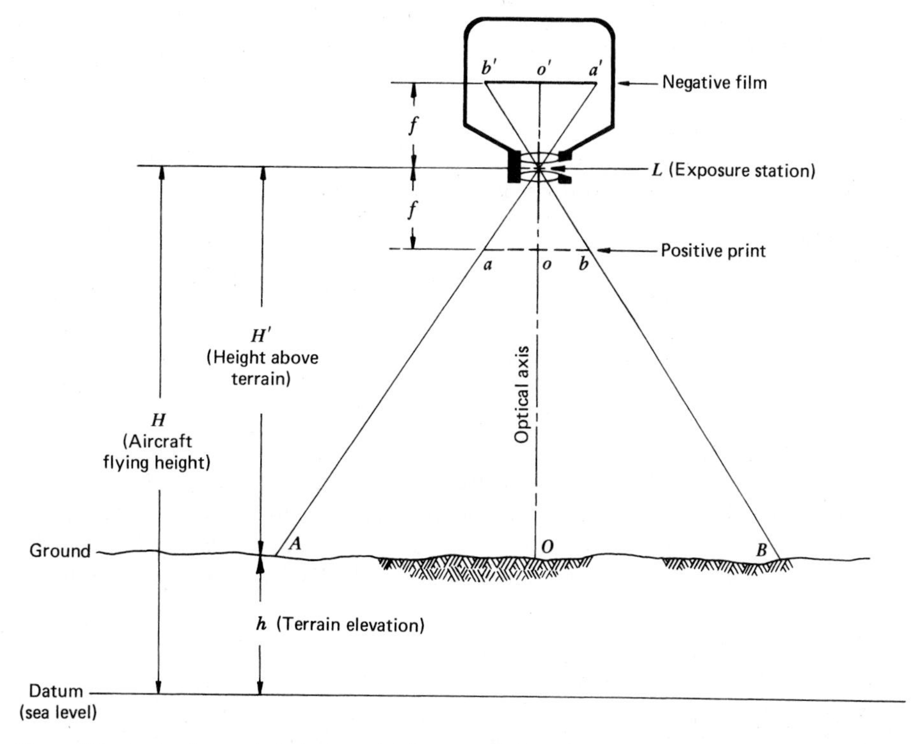

Figure 3.7 Scale of a vertical photograph taken over flat terrain.

flying height H above some *datum*, or arbitrary base elevation. The datum most frequently used is mean sea level. If flying height H and the elevation of the terrain h are known, we can determine H' by subtraction ($H' = H - h$). If we now consider terrain points A, O, and B, they are imaged at points a', o', and b' on the negative film and at a, o, and b on the positive print. We can derive an expression for photo scale by observing similar triangles *Lao* and *LAO*, and the corresponding photo ($\overline{ao}$) and ground ($\overline{AO}$) distances. That is,

$$S = \frac{\overline{ao}}{\overline{AO}} = \frac{f}{H'} \tag{3.3}$$

Equation 3.3 is identical to our scale expression of Eq. 3.2. Yet another way of expressing these equations is

$$S = \frac{f}{H - h} \tag{3.4}$$

Equation 3.4 is the most commonly used form of the scale equation.

EXAMPLE 3.2

A camera equipped with a 152-mm-focal-length lens is used to take a vertical photograph from a flying height of 2780 m above mean sea level. If the terrain is flat and located at an elevation of 500 m, what is the scale of the photograph?

Solution

$$\text{Scale} = \frac{f}{H-h} = \frac{0.152 \text{ m}}{2780 \text{ m} - 500 \text{ m}} = \frac{1}{15{,}000} \quad \text{or} \quad 1{:}15{,}000$$

The most important principle expressed by Eq. 3.4 is that photo scale is a function of terrain elevation h. Because of the level terrain, the photograph depicted in Figure 3.7 has a constant scale. However, *photographs taken over terrain of varying elevation will exhibit a continuous range of scales associated with the variations in terrain elevation*. Likewise, tilted and oblique photographs have nonuniform scales.

EXAMPLE 3.3

Assume that a vertical photograph was taken at a flying height of 5000 m above sea level using a camera with a 152-mm-focal-length lens. (*a*) Determine the photo scale at points *A* and *B*, which lie at elevations of 1200 and 1960 m. (*b*) What ground distance corresponds to a 20.1-mm photo distance measured at each of these elevations?

Solution

(*a*) By Eq. 3.4

$$S_A = \frac{f}{H-h_A} = \frac{0.152 \text{ m}}{5000 \text{ m} - 1200 \text{ m}} = \frac{1}{25{,}000} \quad \text{or } 1{:}25{,}000$$

$$S_B = \frac{f}{H-h_B} = \frac{0.152 \text{ m}}{5000 \text{ m} - 1960 \text{ m}} = \frac{1}{20{,}000} \quad \text{or } 1{:}20{,}000$$

(*b*) The ground distance corresponding to a 20.1-mm photo distance is

$$D_A = \frac{d}{S_A} = 0.0201 \text{ m} \div \frac{1}{25{,}000} = 502.5 \text{ m} \quad \text{or} \quad 502 \text{ m}$$

$$D_B = \frac{d}{S_B} = 0.0201 \text{ m} \div \frac{1}{20{,}000} = 402 \text{ m}$$

Often it is convenient to compute an *average scale* for an entire photograph. This scale is calculated using the average terrain elevation for the area imaged. Consequently, it is exact for distances occurring at the average elevation and is approximate at all other elevations. Average scale may be expressed as

$$S_{\text{avg}} = \frac{f}{H - h_{\text{avg}}} \tag{3.5}$$

where h_{avg} is the average elevation of the terrain shown in the photograph.

The result of photo scale variation is geometric distortion. All points on a *map* are depicted in their true relative horizontal (planimetric) positions, but points on a *photo* taken over varying terrain are displaced from their true "map positions." This difference results because a map is a scaled *orthographic* projection of the ground surface, whereas a vertical photograph yields a *perspective* projection. The differing nature of these two forms of projection is illustrated in Figure 3.8. As shown, a

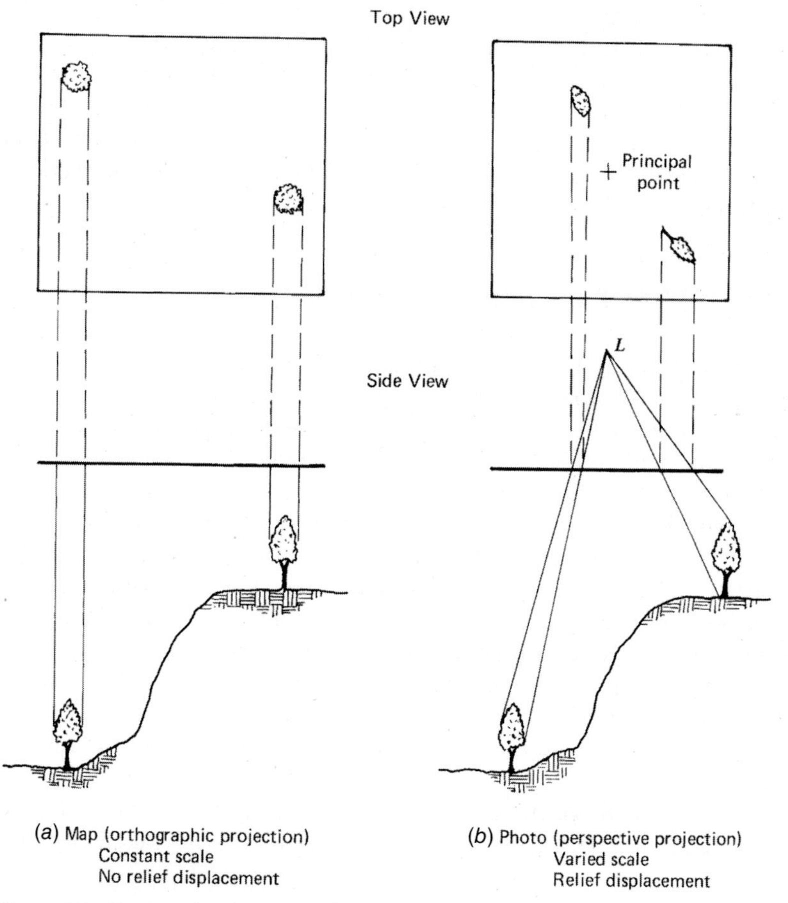

Figure 3.8 Comparative geometry of (*a*) a map and (*b*) a vertical aerial photograph. Note differences in size, shape, and location of the two trees.

map results from projecting vertical rays from ground points to the map sheet (at a particular scale). A photograph results from projecting converging rays through a common point within the camera lens. Because of the nature of this projection, any variations in terrain elevation will result in scale variation *and* displaced image positions.

On a map we see a top view of objects in their true relative horizontal positions. On a photograph, areas of terrain at the higher elevations lie closer to the camera at the time of exposure and therefore appear larger than corresponding areas lying at lower elevations. Furthermore, the tops of objects are always displaced from their bases (Figure 3.8). This distortion is called *relief displacement* and causes any object standing above the terrain to "lean" away from the principal point of a photograph radially. We treat the subject of relief displacement in Section 3.6.

By now the reader should see that the only circumstance wherein an aerial photograph can be treated as if it were a map directly is in the case of a vertical photograph imaging uniformly flat terrain. This is rarely the case in practice, and the image analyst must always be aware of the potential geometric distortions introduced by such influences as tilt, scale variation, and relief displacement. Failure to deal with these distortions will often lead, among other things, to a lack of geometric "fit" among image-derived and nonimage data sources in a GIS. However, if these factors are properly addressed photogrammetrically, extremely reliable measurements, maps, and GIS products can be derived from aerial photography.

3.4 GROUND COVERAGE OF AERIAL PHOTOGRAPHS

The ground coverage of a photograph is, among other things, a function of camera format size. For example, an image taken with a camera having a 230 × 230-mm format (on 240-mm film) has about 17.5 times the ground area coverage of an image of equal scale taken with a camera having a 55 × 55-mm format (on 70-mm film) and about 61 times the ground area coverage of an image of equal scale taken with a camera having a 24 × 36-mm format (on 35-mm film). As with photo scale, the ground coverage of photography obtained with any given format is a function of focal length and flying height above ground, H'. For a constant flying height, the width of the ground area covered by a photo varies inversely with focal length. Consequently, photos taken with shorter focal length lenses have larger areas of coverage (and smaller scales) than do those taken with longer focal length lenses. For any given focal length lens, the width of the ground area covered by a photo varies directly with flying height above terrain, with image scale varying inversely with flying height.

The effect that flying height has on ground coverage and image scale is illustrated in Figures 3.9*a*, *b*, and *c*. These images were all taken over Chattanooga,

Figure 3.9 (*a*) Scale 1:210,000 vertical aerial photograph showing Chattanooga, TN. This figure is a 1.75× reduction of an original photograph taken with f = 152.4 mm from 18,300 m flying height. (NASA photograph.) (*b*) Scale 1:35,000 vertical aerial photograph providing coverage of area outlined in (*a*). This figure is a 1.75× reduction of an original photograph taken with f = 152.4 mm from 3050 m flying height. (*c*) Scale 1:10,500 vertical aerial photograph providing coverage of area outlined in (*b*). This figure is a 1.75× reduction of an original photograph taken with f = 152.4 mm from 915 m flying height. (Courtesy Mapping Services Branch, Tennessee Valley Authority.)

Figure 3.9 *(Continued)*

Tennessee, with the same camera type equipped with the same focal length lens but from three different altitudes. Figure 3.9*a* is a high-altitude, small-scale image showing virtually the entire Chattanooga metropolitan area. Figure 3.9*b* is a lower altitude, larger scale image showing the ground area outlined in Figure 3.9*a*. Figure 3.9*c* is a yet lower altitude, larger scale image of the area outlined in Figure 3.9*b*. Note the trade-offs between the ground area covered by an image and the object detail available in each of the photographs.

Figure 3.9 (*Continued*)

3.5 AREA MEASUREMENT

The process of measuring areas using aerial photographs can take on many forms. The accuracy of area measurement is a function of not only the measuring device used, but also the degree of image scale variation due to relief in the terrain and tilt in the photography. Although large errors in area determinations can result even with vertical photographs in regions of moderate to high relief, accurate measurements may be made on vertical photos of areas of low relief.

Simple scales may be used to measure the area of simply shaped features. For example, the area of a rectangular field can be determined by simply measuring its length and width. Similarly, the area of a circular feature can be computed after measuring its radius or diameter.

EXAMPLE 3.4

A rectangular agricultural field measures 8.65 cm long and 5.13 cm wide on a vertical photograph having a scale of 1:20,000. Find the area of the field at ground level.

Solution

$$\text{Ground length} = \text{photo length} \times \frac{1}{S} = 0.0865 \text{ m} \times 20{,}000 = 1730 \text{ m}$$

$$\text{Ground width} = \text{photo width} \times \frac{1}{S} = 0.0513 \text{ m} \times 20{,}000 = 1026 \text{ m}$$

$$\text{Ground area} = 1730 \text{ m} \times 1026 \text{ m} = 1{,}774{,}980 \text{ m}^2 = 177 \text{ ha}$$

The ground area of an irregularly shaped feature is usually determined by measuring the area of the feature on the photograph. The photo area is then converted to a ground area from the following relationship:

$$\text{Ground area} = \text{photo area} \times \frac{1}{S^2}$$

EXAMPLE 3.5

The area of a lake is 52.2 cm^2 on a 1:7500 vertical photograph. Find the ground area of the lake.

Solution

$$\text{Ground area} = \text{photo area} \times \frac{1}{S^2} = 0.00522 \text{ m}^2 \times 7500^2 = 293{,}625 \text{ m}^2 = 29.4 \text{ ha}$$

Numerous methods can be used to measure the area of irregularly shaped features on a photograph. One of the simplest techniques employs a transparent grid overlay consisting of lines forming rectangles or squares of known area. The grid is placed over the photograph and the area of a ground unit is estimated by counting grid units that fall within the unit to be measured. Perhaps the most widely used grid overlay is a *dot grid* (Figure 3.10). This grid, composed of uniformly spaced dots, is superimposed over the photo, and the dots falling within the region to be measured are counted. From knowledge of the dot density of the grid, the photo area of the region can be computed.

Figure 3.10 Transparent dot grid overlay. (Author-prepared figure.)

EXAMPLE 3.6

A flooded area is covered by 129 dots on a 25-dot/cm^2 grid on a 1:20,000 vertical aerial photograph. Find the ground area flooded.

Solution

$$\text{Dot density} = \frac{1\ \text{cm}^2}{25\ \text{dots}} \times 20{,}000^2 = 16{,}000{,}000\ \text{cm}^2/\text{dot} = 0.16\ \text{ha/dot}$$

$$\text{Ground area} = 129\ \text{dots} \times 0.16\ \text{ha/dot} = 20.6\ \text{ha}$$

The dot grid is an inexpensive tool and its use requires little training. When numerous regions are to be measured, however, the counting procedure becomes quite tedious. An alternative technique is to use a digitizing tablet. These devices are interfaced with a computer such that area determination simply involves tracing around the boundary of the region of interest and the area can be read out directly. When photographs are available in softcopy format, area measurement often involves digitizing from a computer monitor using a mouse or other form of cursor control. The process of digitizing directly from a computer screen is called *heads-up digitizing* because the image analyst can view the original image and the digitized features being compiled simultaneously in one place. The heads-up, or *on-screen,* approach is not only more comfortable, it also affords the ability to digitally zoom in on features to be digitized, and it is much easier to detect mistakes made while pointing at the digitized features and to perform any necessary remeasurement.

3.6 RELIEF DISPLACEMENT OF VERTICAL FEATURES

Characteristics of Relief Displacement

In Figure 3.8, we illustrated the effect of relief displacement on a photograph taken over varied terrain. In essence, an increase in the elevation of a feature causes its position on the photograph to be displaced radially outward from the principal point. Hence, when a vertical feature is photographed, relief displacement causes the top of the feature to lie farther from the photo center than its base. As a result, vertical features appear to lean away from the center of the photograph.

The pictorial effect of relief displacement is illustrated by the aerial photographs shown in Figure 3.11. These photographs depict the construction site of the Watts Bar Nuclear Plant adjacent to the Tennessee River. An operating coal-fired steam plant with its fan-shaped coal stockyard is shown in the upper right of Figure 3.11*a*; the nuclear plant is shown in the center. Note particularly the two

Figure 3.11 Vertical photographs of the Watts Bar Nuclear Power Plant Site, near Kingston, TN. In (*a*) the two plant cooling towers appear near the principal point and exhibit only slight relief displacement. The towers manifest severe relief displacement in (*b*). (Courtesy Mapping Services Branch, Tennessee Valley Authority.)

large cooling towers adjacent to the plant. In (*a*) these towers appear nearly in top view because they are located very close to the principal point of this photograph. However, the towers manifest some relief displacement because the top tower appears to lean somewhat toward the upper right and the bottom tower toward the lower right. In (*b*) the towers are shown at a greater distance from the principal point. Note the increased relief displacement of the towers. We now see more of a

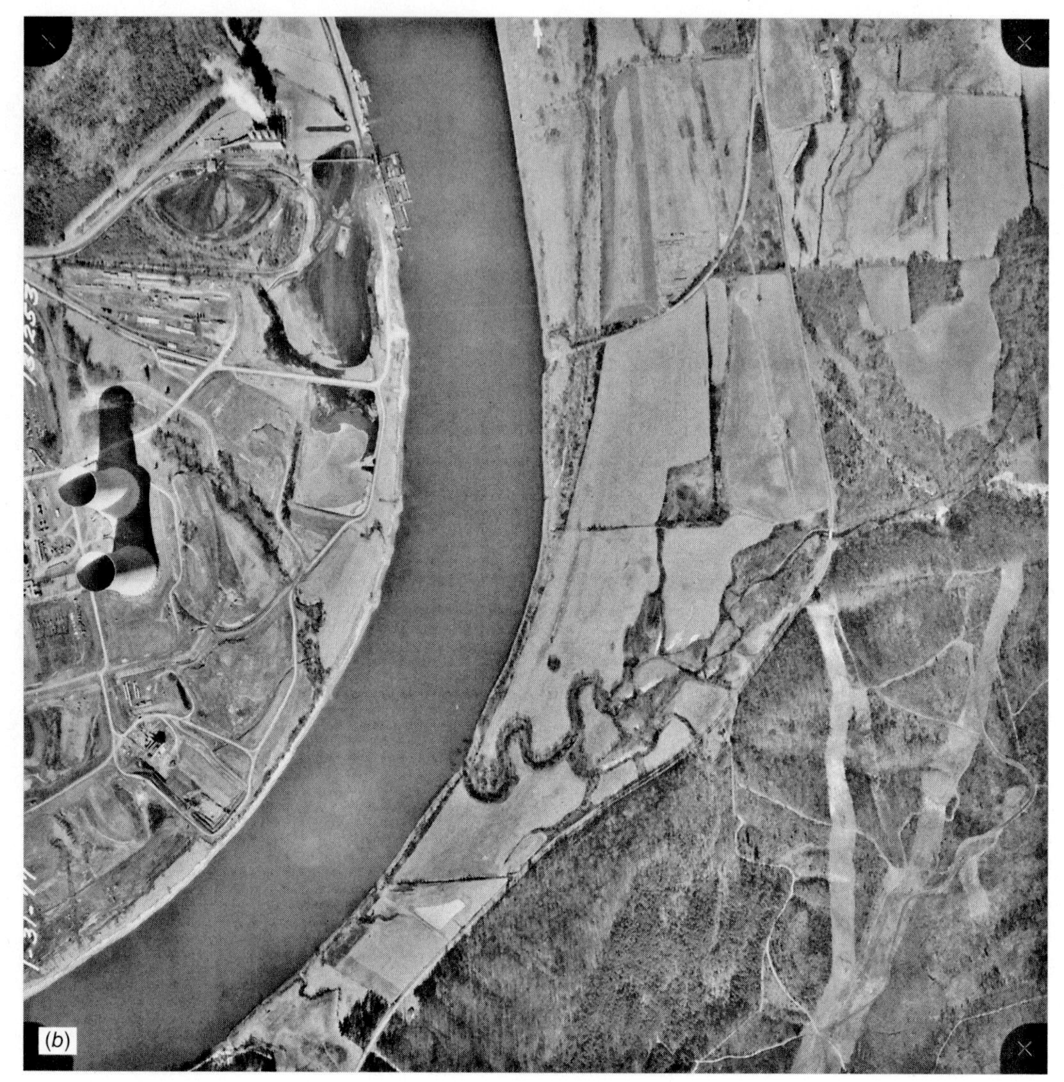

Figure 3.11 (*Continued*)

"side view" of the objects because the images of their tops are displaced farther than the images of their bases. These photographs illustrate the radial nature of relief displacement and the increase in relief displacement with an increase in the radial distance from the principal point of a photograph.

The geometric components of relief displacement are illustrated in Figure 3.12, which shows a vertical photograph imaging a tower. The photograph is taken from flying height H above datum. When considering the relief displacement of a vertical feature, it is convenient to arbitrarily assume a datum plane placed at the base of the feature. If this is done, the flying height H must be correctly referenced to this same datum, *not* mean sea level. Thus, in Figure 3.12 the height of the tower (whose base is at datum) is h. Note that the top of the tower, A, is imaged at a in the photograph whereas the base of the tower, A', is imaged at a'. That is, the image of the top of the tower is radially displaced by the distance d from that of the bottom. The distance d is the relief displacement of the tower. The equivalent distance projected to datum is D. The distance from the photo principal point to the top of the tower is r. The equivalent distance projected to datum is R.

We can express d as a function of the dimensions shown in Figure 3.12. From similar triangles $AA'A''$ and LOA'',

$$\frac{D}{h} = \frac{R}{H}$$

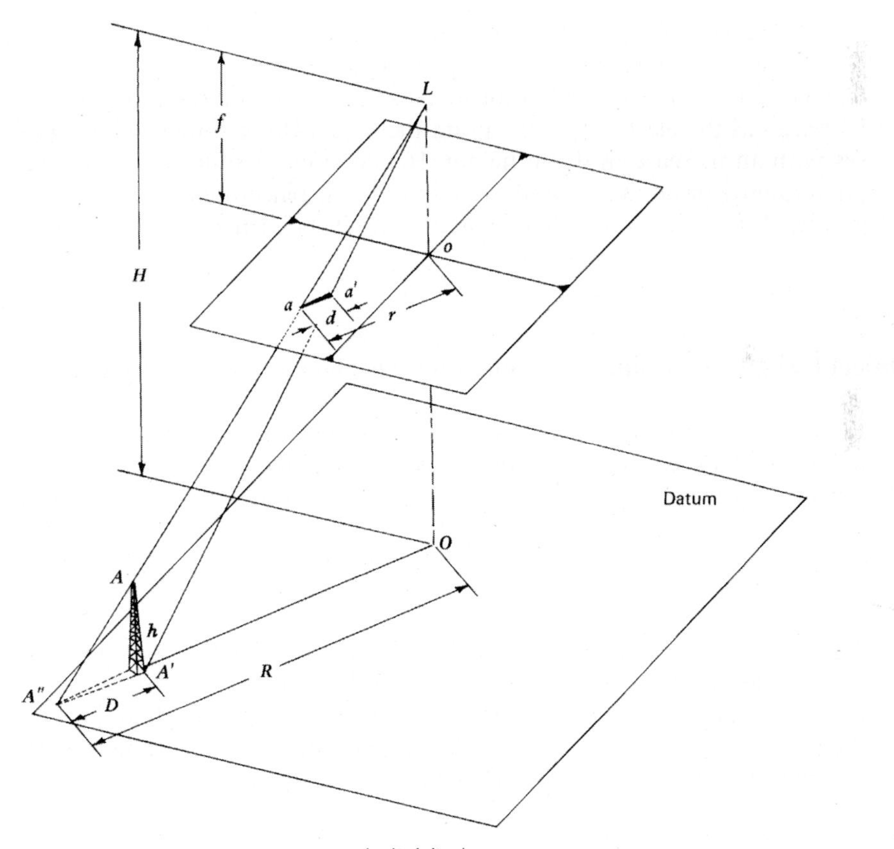

Figure 3.12 Geometric components of relief displacement.

Expressing distances D and R at the scale of the photograph, we obtain

$$\frac{d}{h} = \frac{r}{H}$$

Rearranging the above equation yields

$$d = \frac{rh}{H} \tag{3.6}$$

where

d = relief displacement
r = radial distance on the photograph from the principal point to the displaced image point
h = height above datum of the object point
H = flying height above the same datum chosen to reference h

An analysis of Eq. 3.6 indicates mathematically the nature of relief displacement seen pictorially. That is, relief displacement of any given point increases as the distance from the principal point increases (this can be seen in Figure 3.11), and it increases as the elevation of the point increases. Other things being equal, it decreases with an increase in flying height. Hence, under similar conditions high altitude photography of an area manifests less relief displacement than low altitude photography. Also, there is no relief displacement at the principal point (since $r = 0$).

Object Height Determination from Relief Displacement Measurement

Equation 3.6 also indicates that relief displacement increases with the feature height h. This relationship makes it possible to indirectly measure heights of objects appearing on aerial photographs. By rearranging Eq. 3.6, we obtain

$$h = \frac{dH}{r} \tag{3.7}$$

To use Eq. 3.7, both the top and base of the object to be measured must be clearly identifiable on the photograph and the flying height H must be known. If this is the case, d and r can be measured on the photograph and used to calculate the object height h. (When using Eq. 3.7, it is important to remember that H must be referenced to the elevation of the base of the feature, not to mean sea level.)

EXAMPLE 3.7

For the photo shown in Figure 3.12, assume that the relief displacement for the tower at A is 2.01 mm, and the radial distance from the center of the photo to the top of the tower is 56.43 mm. If the flying height is 1220 m above the base of the tower, find the height of the tower.

Solution

By Eq. 3.7

$$h = \frac{dH}{r} = \frac{2.01 \text{ mm } (1220 \text{ m})}{56.43 \text{ mm}} = 43.4 \text{ m}$$

While measuring relief displacement is a very convenient means of calculating heights of objects from aerial photographs, the reader is reminded of the assumptions implicit in the use of the method. We have assumed use of truly vertical photography, accurate knowledge of the flying height, clearly visible objects, precise location of the principal point, and a measurement technique whose accuracy is consistent with the degree of relief displacement involved. If these assumptions are reasonably met, quite reliable height determinations may be made using single prints and relatively unsophisticated measuring equipment.

Correcting for Relief Displacement

In addition to calculating object heights, quantification of relief displacement can be used to correct the image positions of terrain points appearing in a photograph. Keep in mind that terrain points in areas of varied relief exhibit relief displacements as do vertical objects. This is illustrated in Figure 3.13. In this figure, the datum plane has been set at the average terrain elevation (not at mean sea level). If all terrain points were to lie at this common elevation, terrain points A and B would be located at A' and B' and would be imaged at points a' and b' on the photograph. Due to the varied relief, however, the position of point A is shifted radially outward on the photograph (to a), and the position of point B is shifted radially inward (to b). These changes in image position are the relief displacements of points A and B. Figure 3.13*b* illustrates the effect they have on the geometry of the photo. Because A' and B' lie at the same terrain elevation, the image line $a'b'$ accurately represents the scaled horizontal length and directional orientation of the ground line AB.

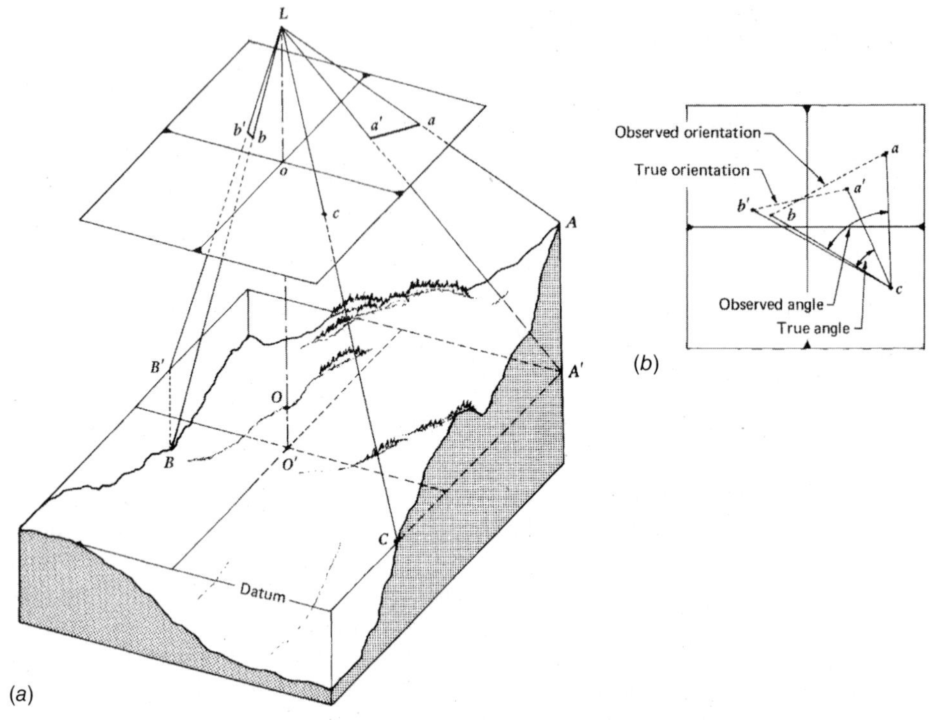

Figure 3.13 Relief displacement on a photograph taken over varied terrain: (*a*) displacement of terrain points; (*b*) distortion of horizontal angles measured on photograph.

When the relief displacements are introduced, the resulting line *ab* has a considerably altered length and orientation.

Angles are also distorted by relief displacements. In Figure 3.13*b*, the horizontal ground angle *ACB* is accurately expressed by *a*′*cb*′ on the photo. Due to the displacements, the distorted angle *acb* will appear on the photograph. Note that, because of the radial nature of relief displacements, angles about the origin of the photo (such as *aob*) will not be distorted.

Relief displacement can be corrected for by using Eq. 3.6 to compute its magnitude on a point-by-point basis and then laying off the computed displacement distances radially (in reverse) on the photograph. This procedure establishes the datum-level image positions of the points and removes the relief distortions, resulting in planimetrically correct image positions at datum scale. This scale can be determined from the flying height above datum ($S=f/H$). Ground lengths, directions, angles, and areas may then be directly determined from these corrected image positions.

EXAMPLE 3.8

Referring to the vertical photograph depicted in Figure 3.13, assume that the radial distance r_a to point A is 63.84 mm and the radial distance r_b to point B is 62.65 mm. Flying height H is 1220 m above datum, point A is 152 m above datum, and point B is 168 m below datum. Find the radial distance and direction one must lay off from points a and b to plot a' and b'.

Solution

By Eq. 3.6

$$d_a = \frac{r_a h_a}{H} = \frac{63.84 \text{ mm} \times 152 \text{ m}}{1220 \text{ m}} = 7.95 \text{ mm} \qquad \text{(plot inward)}$$

$$d_b = \frac{r_b h_b}{H} = \frac{62.65 \text{ mm} \times (-168 \text{ m})}{1220 \text{ m}} = -8.63 \text{ mm} \qquad \text{(plot outward)}$$

3.7 IMAGE PARALLAX

Characteristics of Image Parallax

Thus far we have limited our discussion to photogrammetric operations involving only single vertical photographs. Numerous applications of photogrammetry incorporate the analysis of stereopairs and use of the principle of *parallax*. The term *parallax* refers to the apparent change in relative positions of stationary objects caused by a change in viewing position. This phenomenon is observable when one looks at objects through a side window of a moving vehicle. With the moving window as a frame of reference, objects such as mountains at a relatively great distance from the window appear to move very little within the frame of reference. In contrast, objects close to the window, such as roadside trees, appear to move through a much greater distance.

In the same way that the close trees move relative to the distant mountains, terrain features close to an aircraft (i.e., at higher elevation) will appear to move relative to the lower elevation features when the point of view changes between successive exposures. These relative displacements form the basis for three-dimensional viewing of overlapping photographs. In addition, they can be measured and used to compute the elevations of terrain points.

Figure 3.14 illustrates the nature of parallax on overlapping vertical photographs taken over varied terrain. Note that the relative positions of points A and B change with the change in viewing position (in this case, the exposure station). Note also that the *parallax displacements occur only parallel to the line of flight*. In theory,

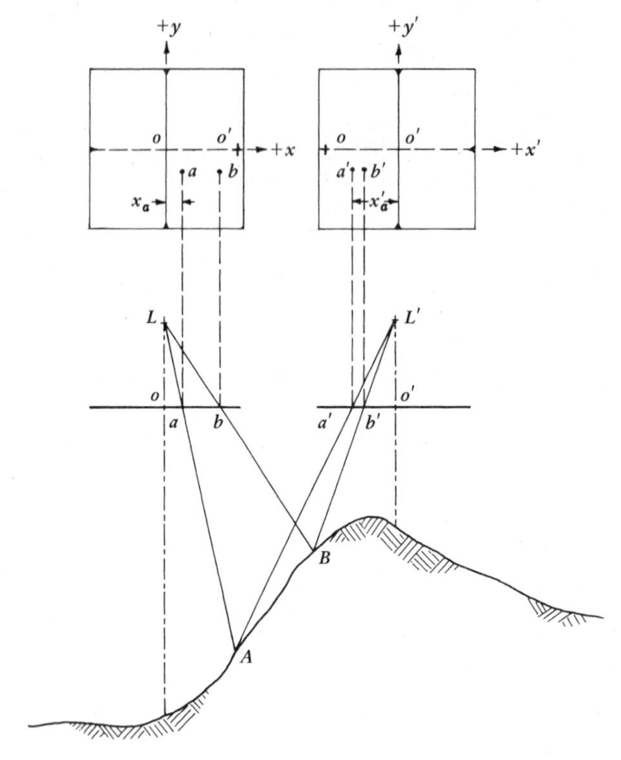

Figure 3.14 Parallax displacements on overlapping vertical photographs.

the direction of flight should correspond precisely to the fiducial x axis. In reality, however, unavoidable changes in the aircraft orientation will usually slightly offset the fiducial axis from the flight axis. The true flight line axis may be found by first locating on a photograph the points that correspond to the image centers of the preceding and succeeding photographs. These points are called the *conjugate principal points*. A line drawn through the principal points and the conjugate principal points defines the flight axis. As shown in Figure 3.15, all photographs except those on the ends of a flight strip normally have two sets of flight axes. This happens because the aircraft's path between exposures is usually slightly curved. In Figure 3.15, the flight axis for the stereopair formed by photos 1 and 2 is flight axis 12. The flight axis for the stereopair formed by photos 2 and 3 is flight axis 23.

The line of flight for any given stereopair defines a photocoordinate x axis for use in parallax measurement. Lines drawn perpendicular to the flight line and passing through the principal point of each photo form the photographic y axes for parallax measurement. The parallax of any point, such as A in Figure 3.15, is expressed

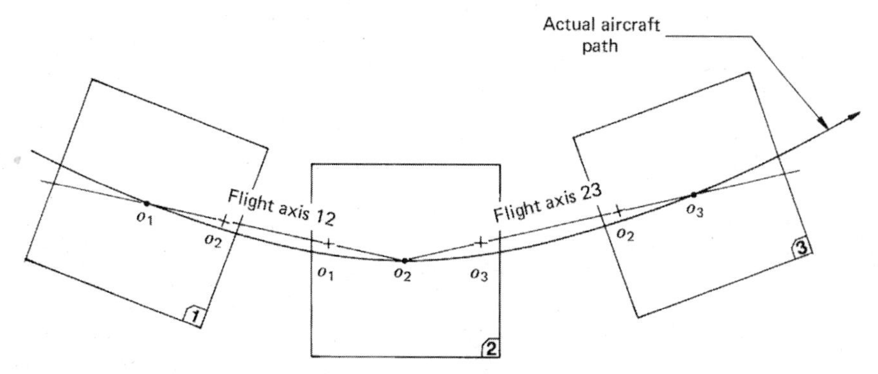

Figure 3.15 Flight line axes for successive stereopairs along a flight strip. (Curvature of aircraft path is exaggerated.)

in terms of the flight line coordinate system as

$$p_a = x_a - x'_a \tag{3.8}$$

where

p_a = parallax of point A
x_a = measured x coordinate of image a on the left photograph of the stereopair
x'_a = x coordinate of image a' on the right photograph

The x axis for each photo is considered positive to the right of each photo principal point. This makes x'_a a negative quantity in Figure 3.14.

Object Height and Ground Coordinate Location from Parallax Measurement

Figure 3.16 shows overlapping vertical photographs of a terrain point, A. Using parallax measurements, we may determine the elevation at A and its ground coordinate location. Referring to Figure 3.16*a*, the horizontal distance between exposure stations L and L' is called B, the *air base*. The triangle in Figure 3.16*b* results from superimposition of the triangles at L and L' in order to graphically depict the nature of parallax p_a as computed from Eq. 3.8 algebraically. From similar triangles $La'_x a_x$ (Figure 3.16*b*) and $LA_x L'$ (Figure 3.16*a*)

$$\frac{p_a}{f} = \frac{B}{H - h_A}$$

from which

$$H - h_A = \frac{Bf}{p_a} \tag{3.9}$$

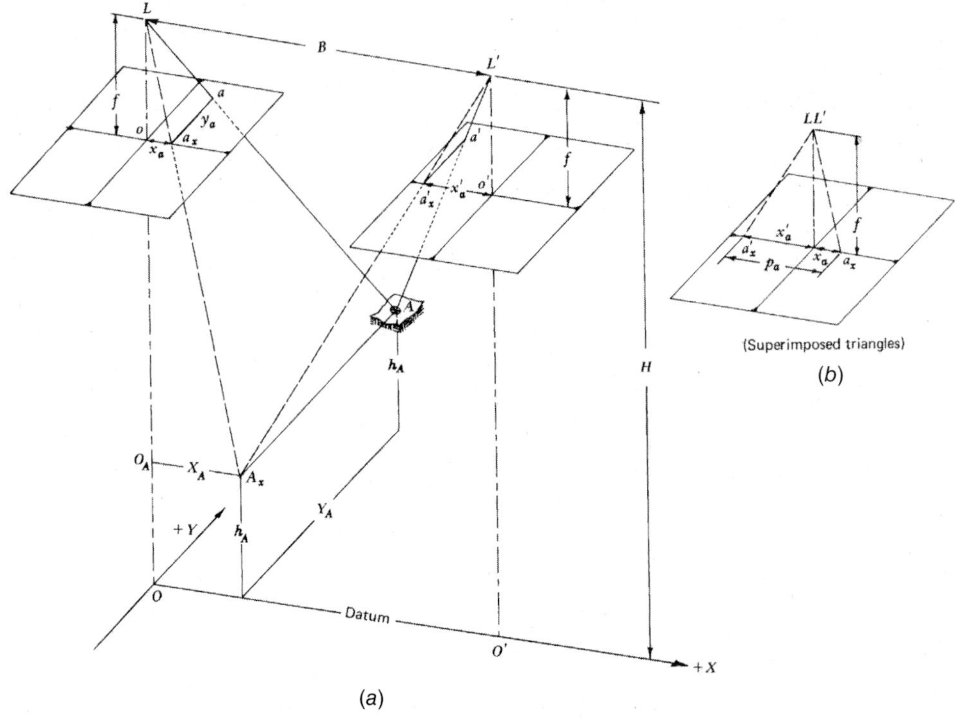

Figure 3.16 Parallax relationships on overlapping vertical photographs: (*a*) adjacent photographs forming a stereopair; (*b*) superimposition of right photograph onto left.

Rearranging yields

$$h_A = H - \frac{Bf}{p_a} \tag{3.10}$$

Also, from similar triangles LO_AA_x and Loa_x,

$$\frac{X_A}{H - h_A} = \frac{x_a}{f}$$

from which

$$X_A = \frac{x_a(H - h_A)}{f}$$

and substituting Eq. 3.9 into the above equation yields

$$X_A = B\frac{x_a}{p_a} \tag{3.11}$$

A similar derivation using y coordinates yields

$$Y_A = B\frac{y_a}{p_a} \tag{3.12}$$

Equations 3.10 to 3.12 are commonly known as the *parallax equations*. In these equations, X and Y are ground coordinates of a point with respect to an arbitrary coordinate system whose origin is vertically below the left exposure station and with positive X in the direction of flight; p is the parallax of the point in question; and x and y are the photocoordinates of the point on the left-hand photo. The major assumptions made in the derivation of these equations are that the photos are truly vertical and that they are taken from the same flying height. If these assumptions are sufficiently met, a complete survey of the ground region contained in the photo overlap area of a stereopair can be made.

EXAMPLE 3.9

The length of line AB and the elevation of its endpoints, A and B, are to be determined from a stereopair containing images a and b. The camera used to take the photographs has a 152.4-mm lens. The flying height was 1200 m (average for the two photos) and the air base was 600 m. The measured photographic coordinates of points A and B in the "flight line" coordinate system are $x_a = 54.61$ mm, $x_b = 98.67$ mm, $y_a = 50.80$ mm, $y_b = -25.40$ mm, $x'_a = -59.45$ mm, and $x'_b = -27.39$ mm. Find the length of line AB and the elevations of A and B.

Solution

From Eq. 3.8

$$p_a = x_a - x'_a = 54.61 - (-59.45) = 114.06 \text{ mm}$$

$$p_b = x_b - x'_b = 98.67 - (-27.39) = 126.06 \text{ mm}$$

From Eqs. 3.11 and 3.12

$$X_A = B\frac{x_a}{p_a} = \frac{600 \times 54.61}{114.06} = 287.27 \text{ m}$$

$$X_B = B\frac{x_b}{p_b} = \frac{600 \times 98.67}{126.06} = 469.63 \text{ m}$$

$$Y_A = B\frac{y_a}{p_a} = \frac{600 \times 50.80}{114.06} = 267.23 \text{ m}$$

$$Y_B = B\frac{y_b}{p_b} = \frac{600 \times (-25.40)}{126.06} = -120.89 \text{ m}$$

Applying the Pythagorean theorem yields

$$AB = [(469.63 - 287.27)^2 + (-120.89 - 267.23)^2]^{1/2} = 428.8 \text{ m}$$

From Eq. 3.10, the elevations of A and B are

$$h_A = H - \frac{Bf}{p_a} = 1200 - \frac{600 \times 152.4}{114.06} = 398 \text{ m}$$

$$h_B = H - \frac{Bf}{p_b} = 1200 - \frac{600 \times 152.4}{126.06} = 475 \text{ m}$$

In many applications, the *difference* in elevation between two points is of more immediate interest than is the actual value of the elevation of either point. In such cases, the change in elevation between two points can be found from

$$\Delta h = \frac{\Delta p H'}{p_a} \tag{3.13}$$

where

Δh = difference in elevation between two points whose parallax *difference* is Δp
H' = flying height above the lower point
p_a = parallax of the higher point

Using this approach in our previous example yields

$$\Delta h = \frac{12.00 \times 802}{126.06} = 77 \text{ m}$$

Note this answer agrees with the value computed above.

Parallax Measurement

To this point in our discussion, we have said little about how parallax measurements are made. In Example 3.9 we assumed that x and x' for points of interest were measured directly on the left and right photos, respectively. Parallaxes were then calculated from the algebraic differences of x and x', in accordance with Eq. 3.8. This procedure becomes cumbersome when many points are analyzed, because two measurements are required for each point.

Figure 3.17 illustrates the principle behind methods of parallax measurement that require only a single measurement for each point of interest. If the two photographs constituting a stereopair are fastened to a base with their flight lines aligned, the distance D remains constant for the setup, and the parallax of a point can be

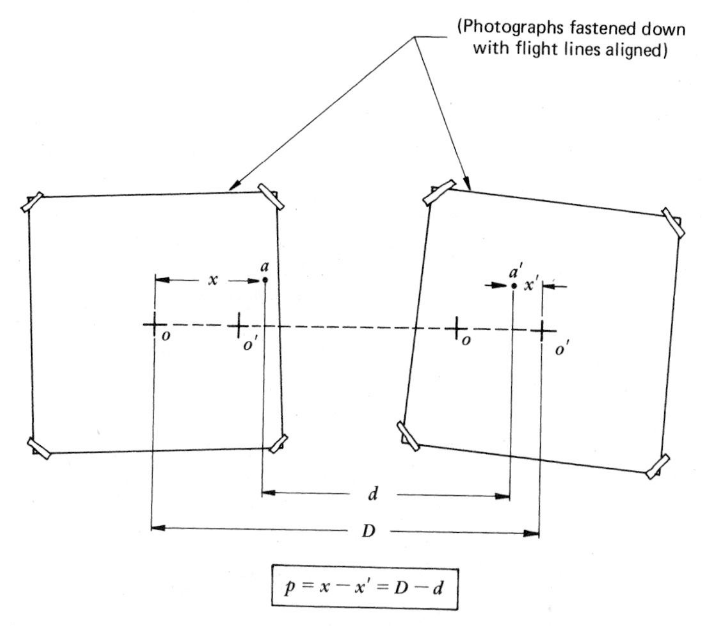

Figure 3.17 Alignment of a stereopair for parallax measurement.

derived from measurement of the single distance d. That is, $p = D - d$. Distance d can be measured with a simple scale, *assuming a and a' are identifiable*. In areas of uniform photo tone, individual features may not be identifiable, making the measurement of d very difficult.

Employing the principle illustrated in Figure 3.17, a number of devices have been developed to increase the speed and accuracy of parallax measurement. These devices also permit parallax to be easily measured in areas of uniform photo tone. All employ stereoscopic viewing and the principle of the *floating mark*. This principle is illustrated in Figure 3.18. While viewing through a stereoscope, the image analyst uses a device that places small identical marks over each photograph. These marks are normally dots or crosses etched on transparent material. The marks—called *half marks*—are positioned over similar areas on the left-hand photo and the right-hand photo. The left mark is seen only by the left eye of the analyst and the right mark is seen only by the right eye. The relative positions of the half marks can be shifted along the direction of flight until they visually fuse together, forming a single mark that appears to "float" at a specific level in the stereomodel. The apparent elevation of the floating mark varies with the spacing between the half marks. Figure 3.18 illustrates how the fused marks can be made to float and can actually be set on the terrain at particular points in the stereomodel. Half-mark positions (a, b), (a, c), and (a, d) result in floating-mark positions in the model at B, C, and D.

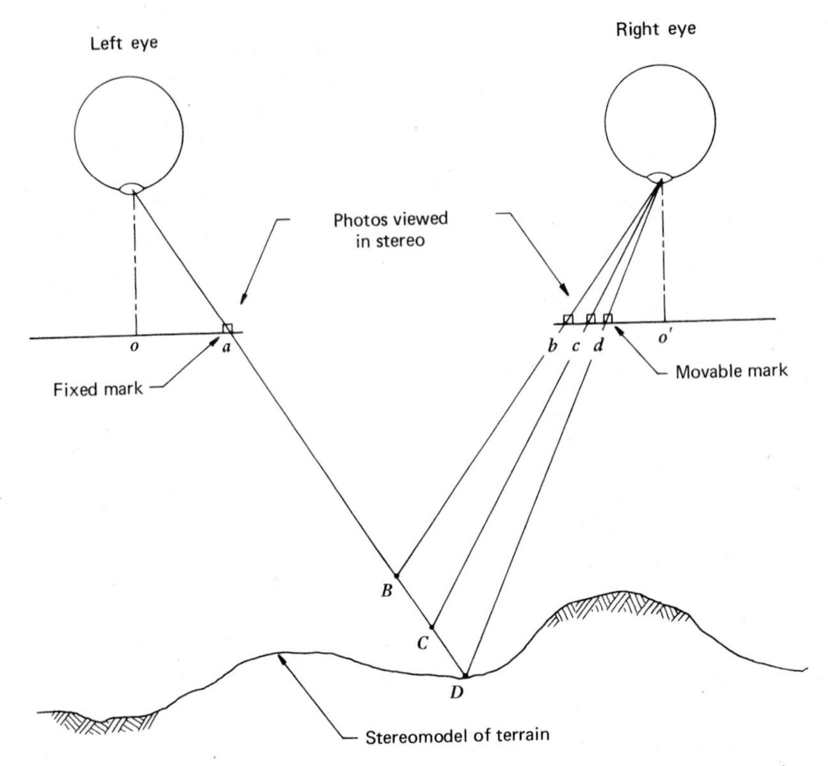

Figure 3.18 Floating-mark principle. (Note that only the right half mark is moved to change the apparent height of the floating mark in the stereomodel.)

A very simple device for measuring parallax is the *parallax wedge*. It consists of a transparent sheet of plastic on which are printed two converging lines or rows of dots (or graduated lines). Next to one of the converging lines is a scale that shows the horizontal distance between the two lines at each point. Consequently, these graduations can be thought of as a series of distance d measurements as shown in Figure 3.17.

Figure 3.19 shows a parallax wedge set up for use. The wedge is positioned so that one of the converging lines lies over the left photo in a stereopair and one over the right photo. When viewed in stereo, the two lines fuse together over a portion of their length, forming a single line that appears to float in the stereomodel. Because the lines on the wedge converge, the floating line appears to slope through the stereoscopic image.

Figure 3.20 illustrates how a parallax wedge might be used to determine the height of a tree. In Figure 3.20*a*, the position of the wedge has been adjusted until the sloping line appears to intersect the top of the tree. A reading is taken from the scale at this point (58.55 mm). The wedge is then positioned such that the line

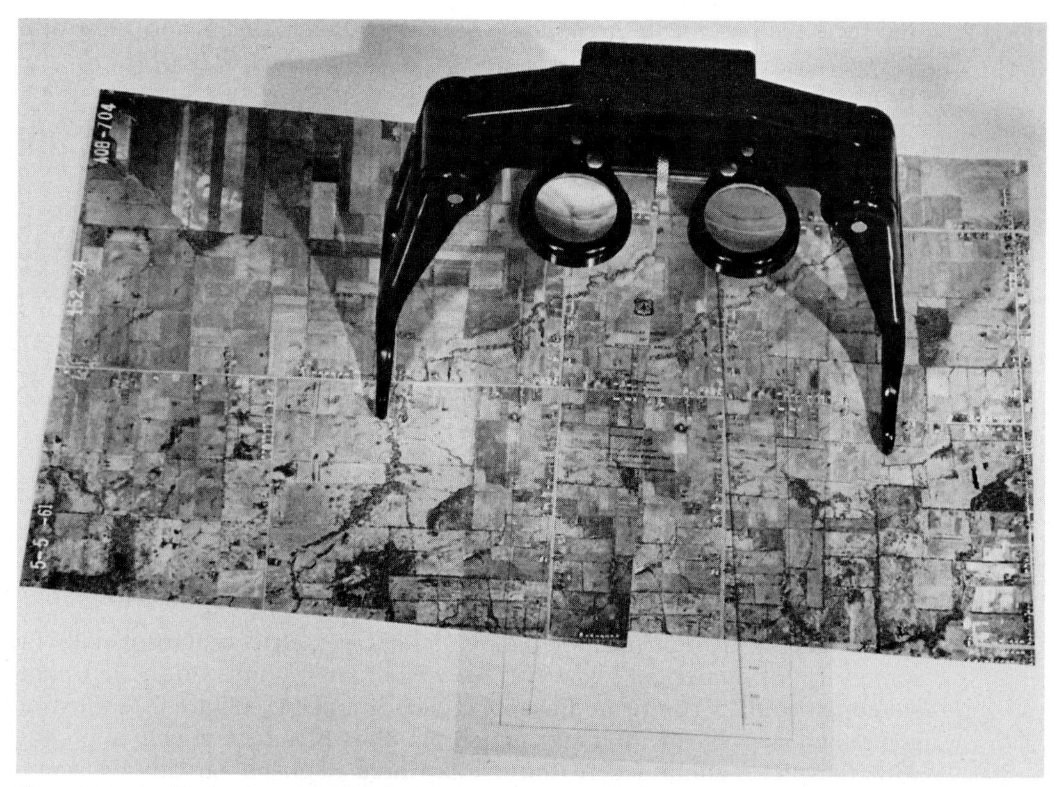

Figure 3.19 Parallax wedge oriented under lens stereoscope. (Author-prepared figure.)

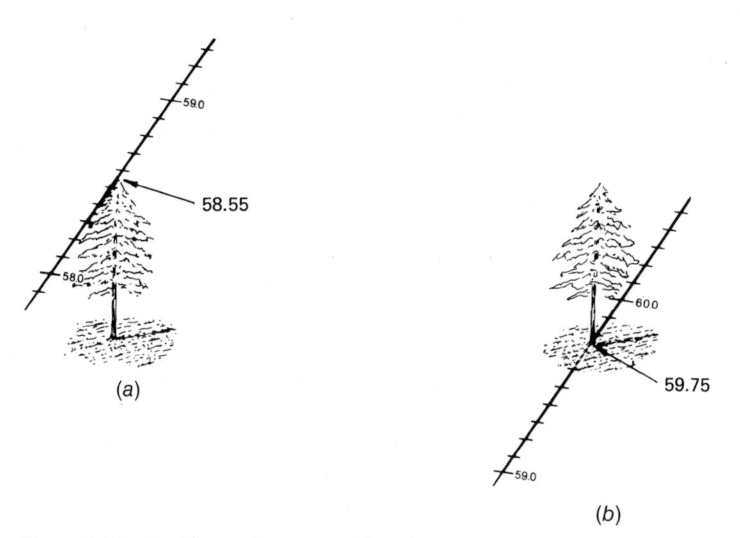

Figure 3.20 Parallax wedge oriented for taking a reading on (*a*) the top and (*b*) the base of a tree.

intersects the base of the tree, and a reading is taken (59.75 mm). The difference between the readings (1.20 mm) is used to determine the tree height.

EXAMPLE 3.10

The flying height for an overlapping pair of photos is 1600 m above the ground and p_a is 75.60 mm. Find the height of the tree illustrated in Figure 3.20.

Solution

From Eq. 3.13

$$\Delta h = \frac{\Delta p H'}{p_a}$$

$$= \frac{1.20 \times 1600}{75.60} = 25 \text{ m}$$

Parallax measurement in softcopy photogrammetric systems usually involves some form of numerical *image correlation* to match points on the left photo of a stereopair to their conjugate images on the right photo. Figure 3.21 illustrates the general concept of digital image matching. Shown is a stereopair of photographs with the pixels contained in the overlap area depicted (greatly exaggerated in size). A *reference window* on the left photograph comprises a local neighborhood of pixels around a fixed location. In this case the reference window is square and 5×5 pixels in size. (Windows vary in size and shape based on the particular matching technique.)

A *search window* is established on the right-hand photo of sufficient size and general location to encompass the conjugate image of the central pixel of the reference window. The initial location of the search window can be determined based on the location of the reference window, the camera focal length, and the size of the area of overlap. A subsearch "moving window" of pixels (Chapter 7) is then systematically moved pixel by pixel about the rows and columns of the search window, and the numerical correlation between the digital numbers within the reference and subsearch windows at each location of the moving subsearch window is computed. The conjugate image is assumed to be at the location where the correlation is a maximum.

There are various types of algorithms that can be used to perform image matching. (One approach employs the simple principle of *epipolar* geometry to minimize the number of unnecessary computations made in the search process. This involves using a small search window that is moved only along the straight line direction in which the parallax of any point occurs.) The details of these procedures are not as

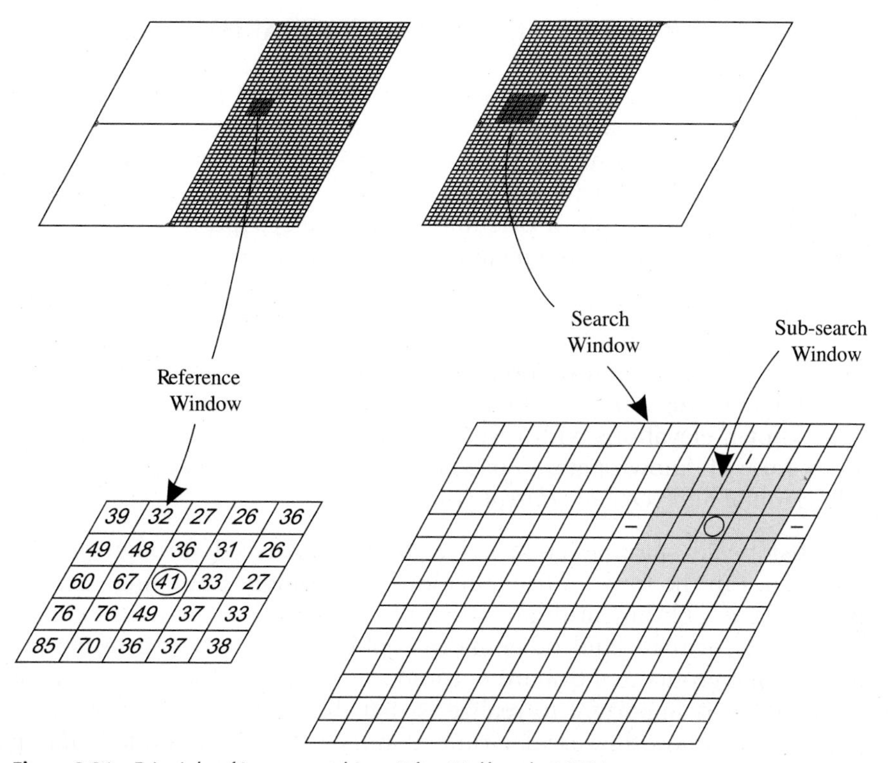

Figure 3.21 Principle of image matching. (After Wolf et al., 2013.)

important as the general concept of locating the conjugate image point for all points on a reference image. The resulting photocoordinates can then be used in the various parallax equations earlier described (Eqs. 3.10 to 3.13). However, the parallax equations assume perfectly vertical photography and equal flying heights for all images. This simplifies the geometry and hence the mathematics of computing ground positions from the photo measurements. However, softcopy systems are not constrained to the above assumptions. Such systems employ mathematical models of the imaging process that readily handle variations in the flying height and attitude of each photograph. As we discuss in Section 3.9, the relationship among image coordinates, ground coordinates, the exposure station position, and angular orientation of each photograph is normally described by a series of *collinearity equations*. They are used in the process of analytical *aerotriangulation*, which involves determining the X, Y, Z, ground coordinates of individual points based on photocoordinate measurements.

3.8 GROUND CONTROL FOR AERIAL PHOTOGRAPHY

Many photogrammetric activities involve the use of a form of ground reference data known as *ground control*. Ground control consists of physical points on the ground whose positions are known in terms of mapping system coordinates. As we discuss in the following section, one important role of ground control is to aid in determining the exterior orientation parameters (position and angular orientation) of a photograph at the instant of exposure.

Ground control points must be mutually identifiable on the ground and on a photograph. Such points may be *horizontal control points*, *vertical control points*, or both. Horizontal control point positions are known planimetrically in some XY coordinate systems (e.g., a state plane coordinate system). Vertical control points have known elevations with respect to a level datum (e.g., mean sea level). A single point with known planimetric position and known elevation can serve as both a horizontal and vertical control point.

Historically, ground control has been established through ground-surveying techniques in the form of triangulation, trilateration, traversing, and leveling. Currently, the establishment of ground control is aided by the use of GPS procedures. The details of these and other more sophisticated surveying techniques used for establishing ground control are not important for the reader of this book to understand. What is important to realize is that the positions of ground control points, irrespective of how they are determined, must be highly accurate in that photogrammetric measurements can only be as a reliable as the ground control on which they are based. Depending on the number and measurement accuracy of the control points required in a particular project, the costs of obtaining ground control measurements can be substantial.

As mentioned, ground control points must be clearly identifiable both on the ground and on the photography being used. Ideally, they should be located in locally flat areas and free of obstructions such as buildings or overhanging trees. When locating a control point in the field, potential obstructions can usually be identified simply by occupying the prospective control point location, looking up 45° from horizontal and rotating one's view horizontally through 360°. This entire field of view should be free of obstructions. Often, control points are selected and surveyed after photography has been taken, thereby ensuring that the points are identifiable on the image. Cultural features, such as road intersections, are often used as control points in such cases. If a ground survey is made prior to a photo mission, control points may be premarked with artificial targets to aid in their identification on the photography. Crosses that contrast with the background land cover make ideal control point markers. Their size is selected in accordance with the scale of the photography to be flown and their material form can be quite variable. In many cases, markers are made by simply painting white crosses on roadways.

Alternatively, markers can be painted on contrasting sheets of Masonite, plywood, or heavy cloth.

3.9 DETERMINING THE ELEMENTS OF EXTERIOR ORIENTATION OF AERIAL PHOTOGRAPHS

As previously stated (in Section 3.1), in order to use aerial photographs for any precise photogrammetric mapping purposes, it is first necessary to determine the six independent parameters that describe the position and angular orientation of the photocoordinate axis system of each photograph (at the instant the photograph was taken) relative to the origin and angular orientation of the ground coordinate system used for mapping. The process of determining the exterior orientation parameters for an aerial photograph is called *georeferencing*. Georeferenced images are those for which 2D photo coordinates can be projected to the 3D ground coordinate reference system used for mapping and vice versa.

For a frame sensor, such as a frame camera, a single exterior orientation applies to an entire image. With line scanning or other dynamic imaging systems, the exposure station position and orientation change with each image line. The process of georeferencing is equally important in establishing the geometric relationship between image and ground coordinates for such sensors as lidar, hyperspectral scanners, and radar as it is in aerial photography.

In the remainder of this section, we discuss the two basic approaches taken to georeference frame camera images. The first is *indirect georeferencing*, which makes use of ground control and a procedure called *aerotriangulation* to "back out" computed values for the six exterior orientation parameters of all photographs in a flight strip or block. The second approach is *direct georeferencing*, wherein these parameters are measured directly through the integration of airborne GPS and inertial measurement unit (IMU) observations.

Indirect Georeferencing

Figure 3.22 illustrates the relationship between the 2D (x, y) photocoordinate system and the 3D (X, Y, Z) ground coordinate system for a typical photograph. This figure also shows the six elements of exterior orientation: the 3D ground coordinates of the exposure station (L) and the 3D rotations of the tilted photo plane (ω, ϕ, and κ) relative to an equivalent perfectly vertical photograph. Figure 3.22 also shows what is termed the *collinearity condition*: the fact that the exposure station of any photograph, any object point in the ground coordinate system, and its photographic image all lie on a straight line. This condition holds irrespective of the angular tilt of a photograph. The condition can also be expressed mathematically in terms of *collinearity equations*. These equations describe the relationships among image coordinates, ground coordinates, the exposure station position, and

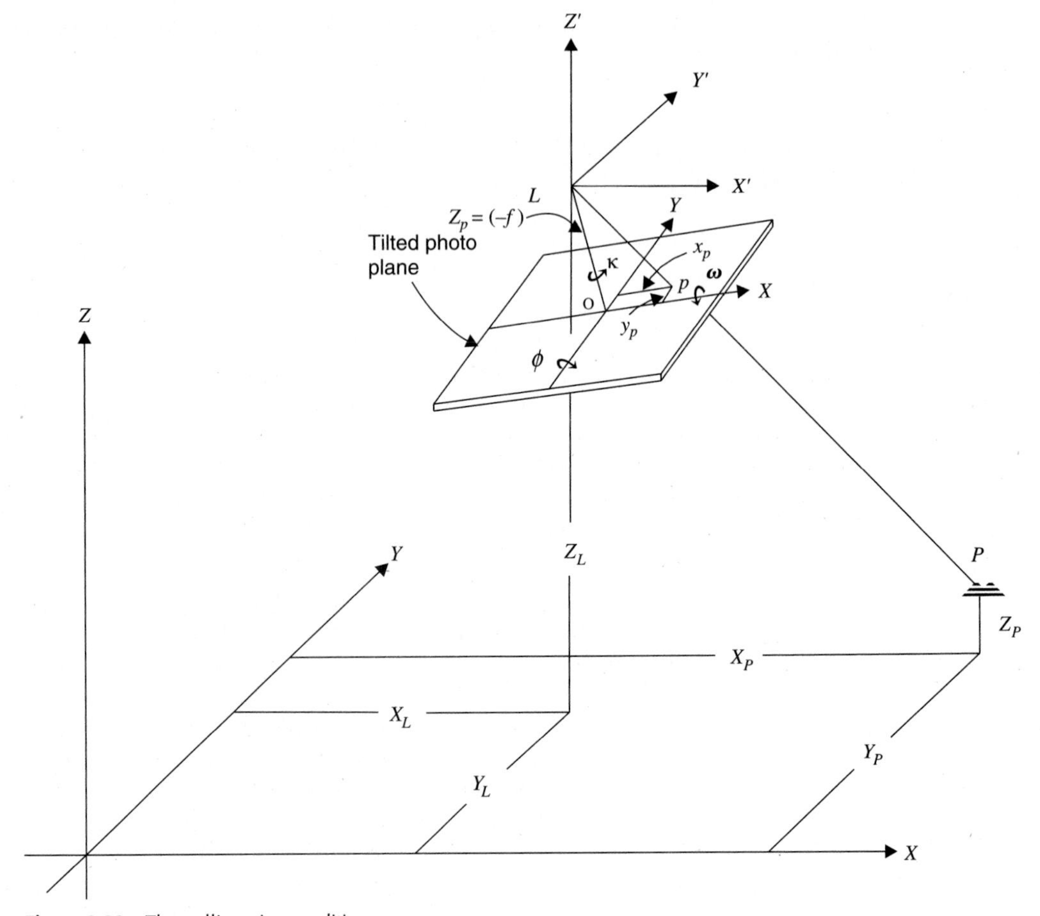

Figure 3.22 The collinearity condition.

the angular orientation of a photograph as follows:

$$x_p = -f\left[\frac{m_{11}(X_p - X_L) + m_{12}(Y_p - Y_L) + m_{13}(Z_p - Z_L)}{m_{31}(X_p - X_L) + m_{32}(Y_p - Y_L) + m_{33}(Z_p - Z_L)}\right] \tag{3.14}$$

$$y_p = -f\left[\frac{m_{21}(X_p - X_L) + m_{22}(Y_p - Y_L) + m_{23}(Z_p - Z_L)}{m_{31}(X_p - X_L) + m_{32}(Y_p - Y_L) + m_{33}(Z_p - Z_L)}\right] \tag{3.15}$$

where

x_p, y_p = image coordinates of any point p
f = focal length
X_P, Y_P, Z_P = ground coordinates of point P

X_L, Y_L, Z_L = ground coordinates of exposure station L
$m_{11}, \ldots, m_{33}$ = coefficients of a 3×3 rotation matrix defined by the angles ω, ϕ, and κ that transforms the ground coordinate system to the image coordinate system

The above equations are nonlinear and contain nine unknowns: the exposure station position (X_L, Y_L, Z_L), the three rotation angles (ω, ϕ, and κ, which are imbedded in the m coefficients), and the three object point coordinates (X_P, Y_P, Z_P) for points other than control points.

A detailed description of how the collinearity equations are solved for the above unknowns is beyond the scope of this discussion. Suffice it to say that the process involves linearizing the equations using Taylor's theorem and then using ground control points to solve for the six unknown elements of exterior orientation. In this process, photocoordinates for at least three *XYZ* ground control points are measured. In that Eqs. 3.14 and 3.15 yield two equations for each control point, three such points yield six equations that can be solved simultaneously for the six unknowns. If more than three control points are available, then more than six equations can be formed and a least squares solution for the unknowns is performed. The aim of using more than three control points is to increase the accuracy of the solution through redundancy and to prevent incorrect data from going undetected.

At this juncture, we have described how ground control is applied to georeferencing of only a single photograph. However, most projects involve long strips or blocks of aerial photos; thus, it would be cost prohibitive to use field measurements to obtain the ground coordinates of a minimum of three ground control points in each image. Dealing with this more general situation (in the absence of using a GPS and IMU to accomplish data acquisition) typically involves the process of *analytical aerotriangulation* (Graham and Koh, 2002). This procedure permits "bridging" between overlapping photographs along a strip and between sidelapping photographs within a block. This is done by establishing *pass points* between photographs along a strip and *tie points* between images in overlapping strips. Pass points and tie points are simply well-defined points in overlapping image areas that can be readily and reliably found in multiple images through automated image matching. Their photocoordinate values are measured in each image in which they appear.

Figure 3.23 illustrates a small block of photographs consisting of two flight lines with five photographs included in each flight line. This block contains images of 20 pass points, five tie points, and six ground control points, for a total of 31 object points. As shown, pass points are typically established near the center, left-center edge, and right-center edge of each image along the direction of flight. If the forward overlap between images is at least 60%, pass points selected in these locations will appear in two or three images. Tie points are located near the top and bottom of sidelapping images and will appear on four to six images. The table included in the lower portion of Figure 3.23 indicates the number of photographs in which images of each object point appear. As can be seen from this table, the grand total of object point images to be measured in this block is 94, each yielding an x and y photocoordinate value, for a total of 188 observations. There are also 18 direct observations for the 3D ground coordinates of the six

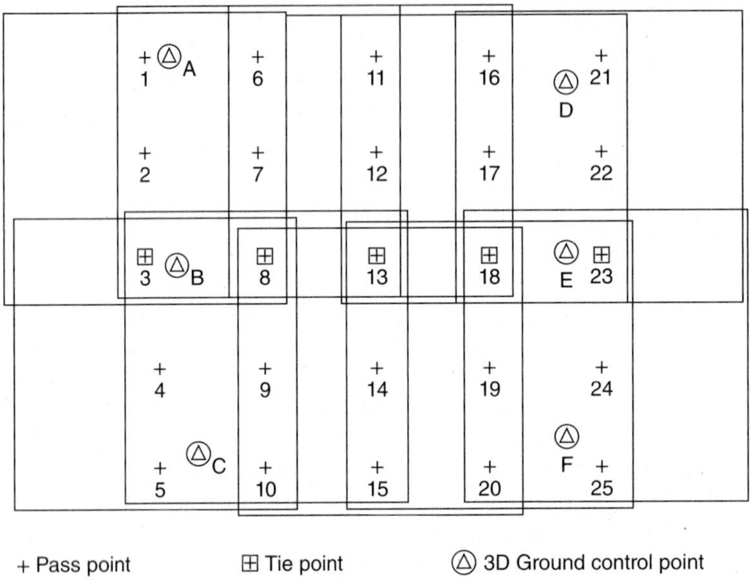

+ Pass point ⊞ Tie point ⓐ 3D Ground control point

Point ID	No. of photos containing point	Point ID	No. of photos containing point
1	2	14	3
A	2	15	3
2	2	16	3
3	4	17	3
B	4	18	6
4	2	19	3
5	2	20	3
C	2	21	2
6	3	D	2
7	3	22	2
8	6	E	4
9	3	23	4
10	3	24	2
11	3	F	2
12	3	25	2
13	6	Total no. of image points = 94	

Figure 3.23 A small (10-photo) block of photographs illustrating the typical location of pass points, tie points, and ground control points whose ground coordinates are computed through the aerotriangulation process.

control points. However, many systems allow for errors in the ground control values, so they are adjusted along with the photocoordinate measurements.

The nature of analytical aerotriangulation has evolved significantly over time, and many variations of how it is accomplished exist. However, all methods involve writing equations (typically collinearity equations) that express the elements of exterior orientation of each photo in a block in terms of camera constants (e.g., focal length, principal point location, lens distortion), measured photo coordinates, and ground control coordinates. The equations are then solved simultaneously to compute the unknown exterior orientation parameters for all photos in a block and the ground coordinates of the pass points and tie points (thus increasing the spatial density of the control available to accomplish subsequent mapping tasks). For the small photo block considered here, the number of unknowns in the solution consists of the X, Y, and Z object space coordinates of all object points ($3\times31=93$) and the six exterior orientation parameters for each of the photographs in the block ($6\times10=60$). Thus, the total number of unknowns in this relatively small block is $93 + 60=153$.

The above process by which all photogrammetric measurements in a block are related to ground control values in one massive solution is often referred to as *bundle adjustment*. The "bundle" inferred by this terminology is the conical bundle of light rays that pass through the camera lens at each exposure station. In essence, the bundles from all photographs are adjusted simultaneously to best fit all the ground control points, pass points, and tie points in a block of photographs of virtually any size.

With the advent of airborne GPS over a decade ago, the aerotriangulation process was greatly streamlined and improved. By including the GPS coordinates of the exposure station for each photo in a block in a bundle adjustment, the need for ground control was greatly reduced. Each exposure station becomes an additional control point. It is not unusual to employ only 10 to 12 control points to control a block of hundreds of images when airborne GPS is employed (D.F. Maune, 2007).

Direct Georeferencing

The currently preferred method for determining the elements of exterior orientation is direct georeferencing. As stated earlier, this approach involves processing the raw measurements made by an airborne GPS together with IMU data to calculate both the position and angular orientation of each image. The GPS data afford high *absolute* accuracy information on position and velocity. At the same time, IMU data provide very high *relative* accuracy information on position, velocity, and angular orientation. However, the absolute accuracy of IMUs tends to degrade with the time when operated in a stand-alone mode. This is where the integration of the GPS and IMU data takes on importance. The high accuracy GPS position information is used to control the IMU position error, which in turn controls the IMU's orientation error.

Another advantage to GPS/IMU integration is that GPS readings are collected at discrete intervals, whereas an IMU provides continuous positioning. Hence the IMU data can be used to smooth out random errors in the positioning data. Finally, the IMU can serve as a backup system for the GPS if it loses its "lock" on satellite signals for a short period of time as the aircraft is maneuvered during the photographic mission.

To say that direct positioning and orientation systems have had a major influence on how modern photogrammetric operations are performed is a severe understatement. The direct georeferencing approach in many circumstances eliminates the need for ground control surveys (except to establish a base station for the airborne GPS operations) and the related process of aerotriangulation. The need to manually select pass and tie points is eliminated by using autocorrelation methods to select hundreds of points per stereomodel. This translates to improving the efficiency, cost, and timeliness of delivery of geospatial data products. Not having to perform extensive ground surveys can also improve personnel safety in circumstances such as working in treacherous terrain or the aftermath of a natural disaster.

Finally, direct georeferencing still has its own limitations and considerations. The spatial relationships among the GPS antenna, IMU, and the mapping camera have to be well calibrated and controlled. Also, a small mapping project could actually cost more to accomplish using this approach due to the fixed costs of acquiring and operating such a system. Current systems are very accurate at small mapping scales. Depending on photo acquisition parameters and accuracy requirements, the procedure might not be appropriate for mapping at very large scales. Some limited ground control is also recommended for quality control purposes. Likewise, flight timing should be optimized to ensure strong GPS signals from a number of satellites, which may narrow the acquisition time window for flight planning purposes.

3.10 PRODUCTION OF MAPPING PRODUCTS FROM AERIAL PHOTOGRAPHS

Elementary Planimetric Mapping

Photogrammetric mapping can take on many forms, depending upon the nature of the photographic data available, the instrumentation and/or software used, and the form and accuracy required in any particular mapping application. Many applications only require the production of *planimetric maps*. Such maps portray the plan view (X and Y locations) of natural and cultural features of interest. They do not represent the contour or relief (Z elevations) of the terrain, as do *topographic maps*.

Planimetric mapping with hardcopy images can often be accomplished with relatively simple and inexpensive methods and equipment, particularly when relief effects are minimal and the ultimate in positional accuracy is not required. In such cases, an analyst might use such equipment as an optical transfer scope to transfer the locations of image features to a map base. This is done by pre-plotting the position of several photo control points on a map sheet at the desired scale. Then the image is scaled, stretched, rotated, and translated to optically fit (as closely as possible) the plotted positions of the control points on the map base. Once the orientation of the image to the map base is accomplished, the locations of other features of interest in the image are transferred to the map.

Planimetric features can also be mapped from hardcopy images with the aid of a table digitizer. In this approach, control points are again identified whose XY coordinates are known in the ground coordinate system and whose xy coordinates are then measured in the digitizer axis system. This permits the formulation of a two-dimensional coordinate transformation (Section 3.2 and Appendix B) to relate the digitizer xy coordinates to the ground XY coordinate system. This transformation is then used to relate the digitizer coordinates of features other than the ground control points to the ground coordinate mapping system.

The above control point measurement and coordinate transformation approach can also be applied when digital, or softcopy, image data are used in the mapping process. In this case, the row and column xy coordinate of a pixel in the image file is related to the XY ground coordinate system via control point measurement. Heads-up digitizing is then used to obtain the xy coordinates of the planimetric features to be mapped from the image, and these are transformed into the ground coordinate mapping system.

We stress that the accuracy of the ground coordinates resulting from either tablet or heads-up digitizing can be highly variable. Among the many factors that can influence this accuracy are the number and spatial distribution of the control points, the accuracy of the ground control, the accuracy of the digitizer (in tablet digitizing), the accuracy of the digitizing process, and the mathematical form of coordinate transformation used. Compounding all of these factors are the potential effects of terrain relief and image tilt. For many applications, the accuracy of these approaches may suffice and the cost of implementing more sophisticated photogrammetric procedures can be avoided. However, when higher-order accuracy is required, it may only be achievable through softcopy mapping procedures employing stereopairs (or larger strips or blocks) of georeferenced photographs.

Evolution from Hardcopy to Softcopy Viewing and Mapping Systems

Stereopairs represent a fundamental unit from which mapping products are derived from aerial photographs. Historically, planimetric and topographic maps were generated from hardcopy stereopairs using a device called a *stereoplotter*. While very few hardcopy stereoplotters are in use today, they serve as a convenient graphical means to illustrate the functionality of current softcopy-based systems.

Figure 3.24 illustrates the principle underlying the design and operation of a *direct optical projection stereoplotter*. Shown in (*a*) are the conditions under which

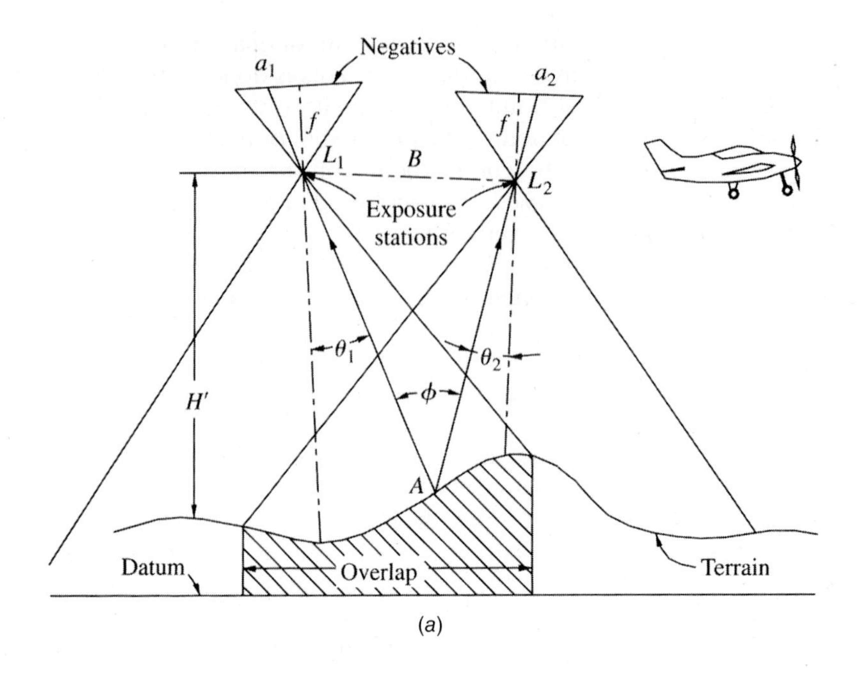

(a)

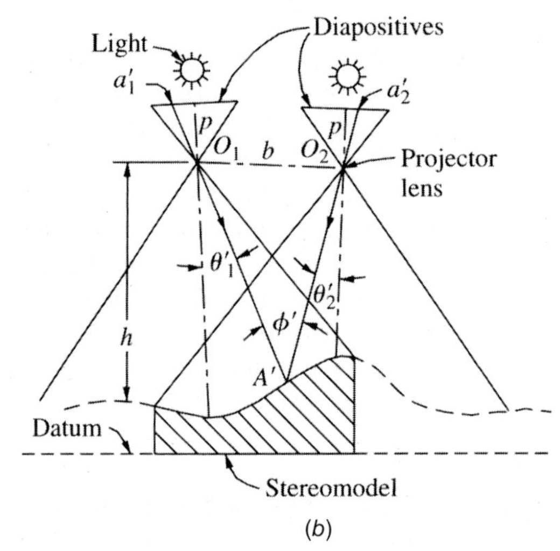

(b)

Figure 3.24 Fundamental concept of stereoplotter instrument design: (*a*) exposure of stereopair in flight; (*b*) projection in stereoplotter. (From P. R. Wolf, B. Dewitt, and B. Wilkinson, 2013, *Elements of Photogrammetry with Applications in GIS*, 4th ed., McGraw-Hill. Reproduced with the permission of The McGraw-Hill Companies.)

a stereopair is exposed in-flight. Note that the flying height for each exposure station is slightly different and that the camera's optical axis is not perfectly vertical when the photos are exposed. Also note the angular relationships between the light rays coming from point *A* on the terrain surface and recorded on each of the two negatives.

As shown in (*b*), the negatives are used to produce diapositives (transparencies printed on glass or film transparencies "sandwiched" between glass plates), which are placed in two stereoplotter projectors. Light rays are then projected through both the left and right diapositives. When the rays from the left and right images intersect below the projectors, they form a *stereomodel*, which can be viewed and measured in stereo. To aid in creating the stereomodel, the projectors can be rotated about, and translated along, their x, y, and z axes. In this way, the diapositives can be positioned and rotated such that they bear the exact relative angular orientation to each other in the projectors as the negatives did when they were exposed in the camera at the two exposure stations. The process of establishing this angular relationship in the stereoplotter is called *relative orientation*, and it results in the creation of a miniature 3D stereomodel of the overlap area.

Relative orientation of a stereomodel is followed by *absolute orientation*, which involves scaling and leveling the model. The desired scale of the model is produced by varying the base distance, b, between the projectors. The scale of the resulting model is equal to the ratio b/B. Leveling of the model can be accomplished by rotating both projectors together about the X direction and Y direction of the mapping coordinate system.

Once the model is oriented, the X, Y, and Z ground coordinates of any point in the overlap area can be obtained by bringing a reference floating mark in contact with the model at that point. This reference mark can be translated in the X and Y directions throughout the model, and it can be raised and lowered in the Z direction. In preparing a topographic map, natural or cultural features are mapped planimetrically by tracing them with the floating mark, while continuously raising and lowering the mark to maintain contact with the terrain. Contours are compiled by setting the floating mark at the desired elevation of a contour and moving the floating mark along the terrain so that it just maintains contact with the surface of the model. Typically, the three-dimensional coordinates of all points involved in the map compilation process are recorded digitally to facilitate subsequent automated mapping, GIS data extraction, and analysis.

It should be noted that stereoplotters recreate the elements of exterior orientation in the original images forming the stereomodel, and the stereoplotting operation focuses on the *intersections* of rays from conjugate points (rather than the distorted positions of these points themselves on the individual photos). In this manner, the effects of tilt, relief displacement, and scale variation inherent in the original photographs are all negated in the stereoplotter map compilation process.

Direct optical projection stereoplotters employed various techniques to project and view the stereomodel. In order to see stereo, the operator's eyes had to view each image of the stereopair separately. *Anaglyphic systems* involved projecting one photo through a cyan filter and the other through a red filter. By viewing the model through eyeglasses having corresponding color filters, the operator's left eye would see only the left photo and the right eye would see only the right photo. Other approaches to stereo viewing included the use of polarizing filters in place of colored filters, or placing shutters over the projectors to alternate display of the left and right images as the operator viewed the stereomodel through a synchronized shutter system.

Again, direct optical projection plotters represented the first generation of such systems. Performing the relative and absolute orientation of these instruments was an iterative and sometimes trial-and-error process, and mapping planimetric features and contours with such systems was very tedious. As time went by, stereoplotter designs evolved from being direct optical devices to optical-mechanical, analytical, and now softcopy systems.

Softcopy-based systems entered the commercial marketplace in the early 1990s. Early in their development, the dominant data source used by these systems was aerial photography that had been scanned by precision photogrammetric scanners. Today, these systems primarily process digital camera data. They incorporate high quality displays affording 3D viewing. Like their predecessors, softcopy systems employ various means to enforce the stereoscopic viewing condition that the left eye of the image analyst only sees the left image of a stereopair and the right eye only sees the right image. These include, but are not limited to, anaglyphic and polarization systems, as well as split-screen and rapid flicker approaches. The split screen technique involves displaying the left image on the left side of a monitor and the right image on the right side. The analyst then views the images through a stereoscope. The rapid flicker approach entails high frequency (120 Hz) alternation between displaying the left image alone and then the right alone on the monitor. The analyst views the display with a pair of electronically shuttered glasses that are synchronized to be alternately clear or opaque on the left or right side as the corresponding images are displayed.

In addition to affording a 3D viewing capability, softcopy photogrammetric workstations must have very robust computational power and large data storage capabilities. However, these hardware requirements are not unique to photogrammetric workstations. What is unique about photogrammetric workstations is the diversity, modularity, and integration of the suite of software these systems typically incorporate to generate photogrammetric mapping products.

The collinearity condition is frequently the basis for many softcopy analysis procedures. For example, in the previous section of this discussion we illustrated the use of the collinearity equations to georeference individual photographs. Collinearity is also frequently used to accomplish relative and absolute orientation of stereopairs. Another very important application of the collinearity equations is their incorporation in the process of *space intersection*.

Space Intersection and DEM Production

Space intersection is a procedure by which the *X, Y,* and *Z* coordinates of any point in the overlap of a stereopair of tilted photographs can be determined. As shown in Figure 3.25, space intersection is premised on the fact that corresponding rays from overlapping photos intersect at a unique point. Image correlation (Section 3.7) can be used to match any given point with its conjugate image to determine the point of intersection.

Space intersection makes use of the fact that a total of four collinearity equations can be written for each point in the overlap area. Two of these equations relate to the point's x and y coordinates on the left-hand photo; two result from the x' and y' coordinates measured on the right-hand photo. If the exterior

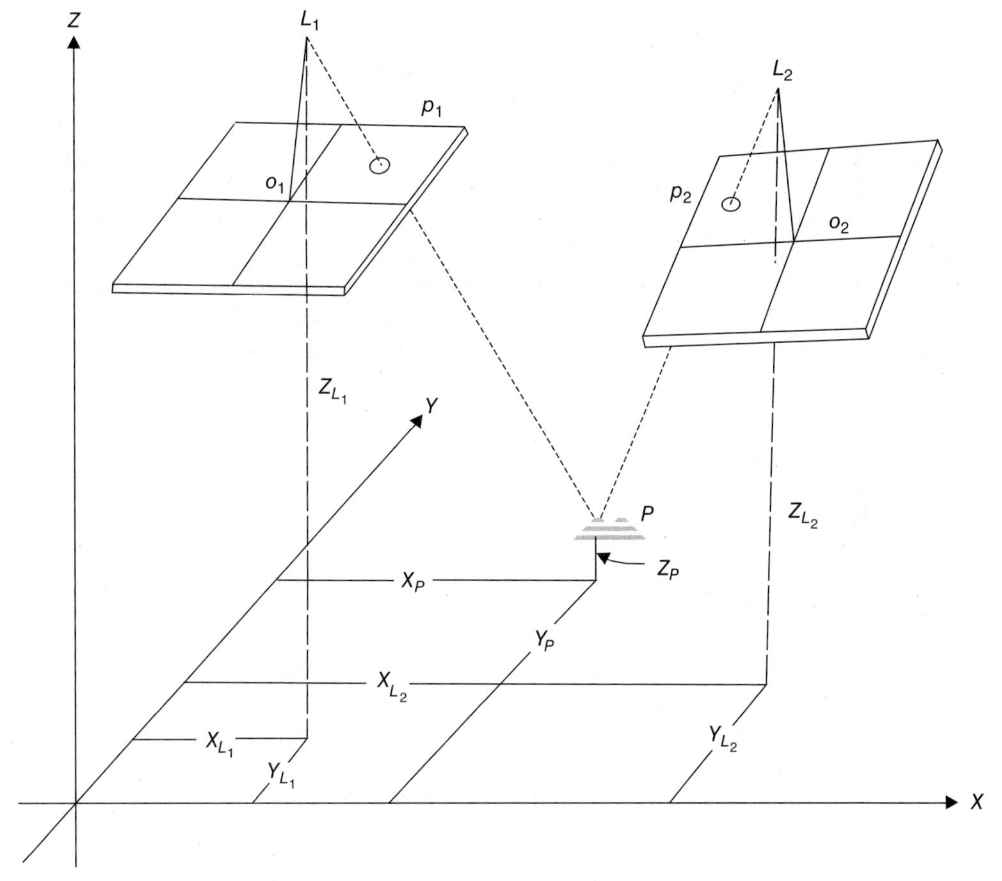

Figure 3.25 Space intersection.

orientation of both photos is known, then the only unknowns in each equation are X, Y, and Z for the point under analysis. Given four equations for three unknowns, a least squares solution for the ground coordinates of each point can be performed. This means that an image analyst can stereoscopically view, and extract the planimetric positions and elevations of, any point in the stereomodel. These data can serve as direct input to a GIS or CAD system.

Systematic sampling of elevation values throughout the overlap area can form the basis for DEM production with softcopy systems. While this process is highly automated, the image correlation involved is often far from perfect. This normally leads to the need to edit the resulting DEM, but this hybrid DEM compilation process is still a very useful one. The process of editing and reviewing a DEM is greatly facilitated in a softcopy-based system in that all elevation points (and even contours if they are produced) can be superimposed on the 3D view of the original stereomodel to aid in inspection of the quality of elevation data.

Digital/Orthophoto Production

As implied by their name, orthophotos are orthographic photographs. They do not contain the scale, tilt, and relief distortions characterizing normal aerial photographs. In essence, orthophotos are "photomaps." Like maps, they have one scale (even in varying terrain), and like photographs, they show the terrain in actual detail (not by lines and symbols). Hence, orthophotos give the resource analyst the "best of both worlds"—a product that can be readily interpreted like a photograph but one on which true distances, angles, and areas may be measured directly. Because of these characteristics, orthophotos make excellent base maps for compiling data to be input to a GIS or overlaying and updating data already incorporated in a GIS. Orthophotos also enhance the communication of spatial data, since data users can often relate better to an orthophoto than a conventional line and symbol map or display.

The primary inputs to the production of a digital orthophoto are a conventional, perspective digital photograph and a DEM. The objective of the production process is to compensate for the effects of tilt, relief displacement, and scale variation by reprojecting the original image orthographically. One way to think about this reprojection process is envisioning that the original image is projected over the DEM, resulting in a "draped" image in which elevation effects are minimized.

There are various means that can be used to create orthophotos. Here we illustrate the process of *backward projection*. In this process, we start with point locations in the DEM and find their corresponding locations on the original photograph. Figure 3.26 illustrates the process. At each ground position in the DEM (X_P, Y_P, Z_P) the associated image point can be computed through the collinearity equations as x_p, y_p. The brightness value of the image at that point is then inserted into an output array, and the process is repeated for every line and column

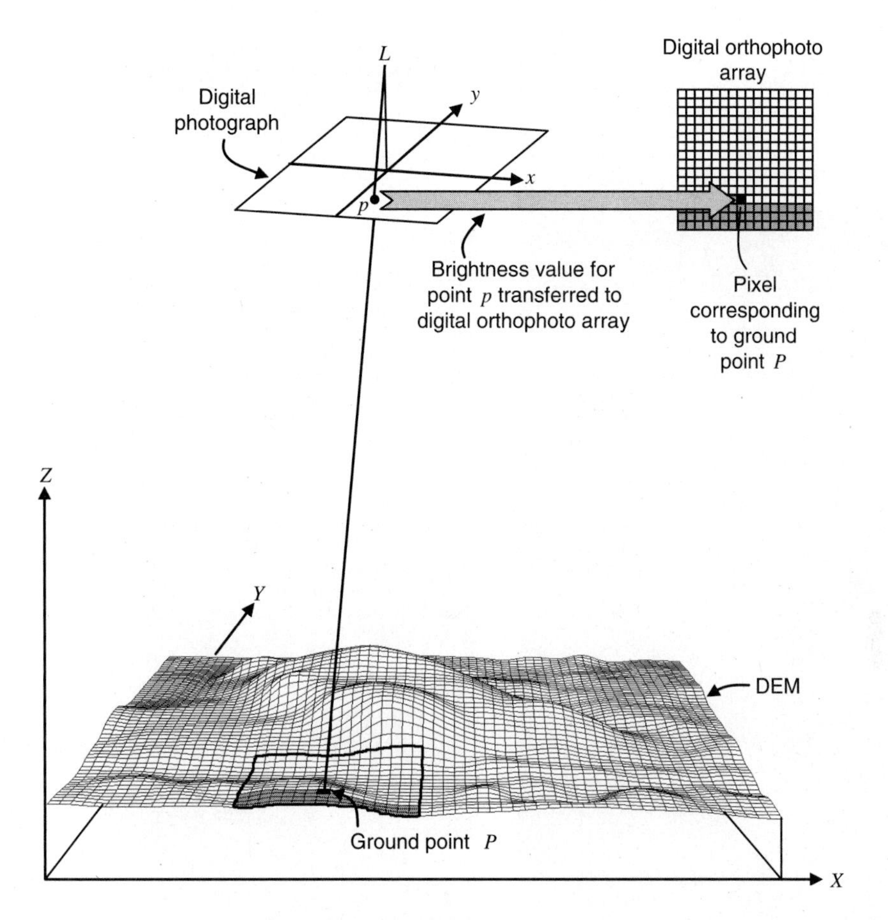

Figure 3.26 Digital orthophoto production process.

position in the DEM to form the entire digital orthophoto. A minor complication in this whole process is the fact that rarely will the photocoordinate value (x_p, y_p) computed for a given DEM cell be exactly centered over a pixel in the original digital input image. Accordingly, the process of *resampling* (Chapter 7 and Appendix B) is employed to determine the best brightness value to assign to each pixel in the orthophoto based on a consideration of the brightness values of a neighborhood of pixels surrounding each computed photocoordinate position (x_p, y_p).

Figure 3.27 illustrates the influence of the above reprojection process. Figure 3.27*a* is a conventional (perspective) photograph of a power line clearing traversing a hilly forested area. The excessively crooked appearance of the linear clearing is due to relief displacement. Figure 3.27*b* is a portion of an orthophoto covering the same area. The relief effects have been removed and the true path of the power line is shown.

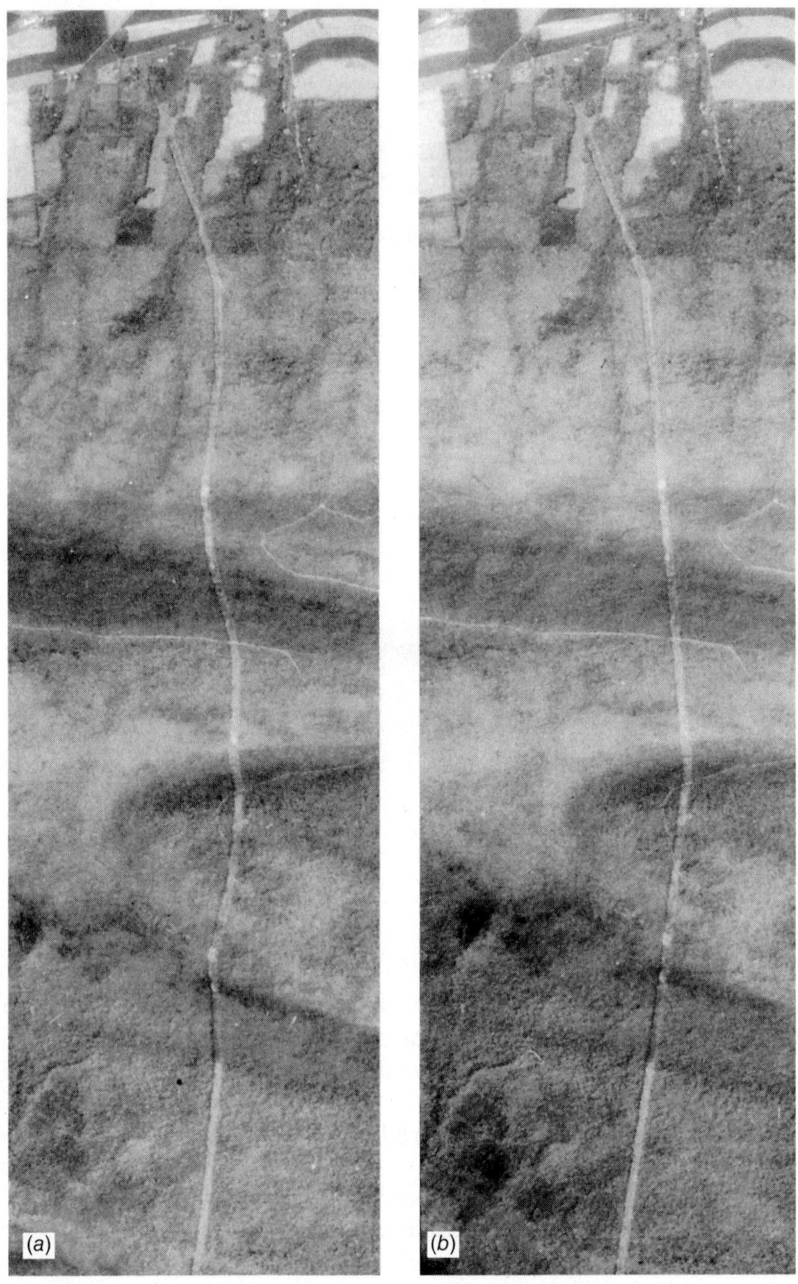

Figure 3.27 Portion of (*a*) a perspective photograph and (*b*) an orthophoto showing a power line clearing traversing hilly terrain. (Note the excessive crookedness of the power line clearing in the perspective photo that is eliminated in the orthophoto.) (Courtesy USGS.)

Orthophotos alone do not convey topographic information. However, they can be used as base maps for contour line overlays prepared in a separate stereoplotting operation. The result of overprinting contour information on an orthophoto is a *topographic orthophotomap*. Much time is saved in the preparation of such maps because the instrument operator need not map the planimetric data in the map compilation process. Figure 3.28 illustrates a portion of a topographic orthophotomap.

Orthophotos may be viewed stereoscopically when they are paired with *stereomates*. These products are photographs made by *introducing* image parallax as a function of known terrain elevations obtained during the production of their corresponding orthophoto. Figure 3.29 illustrates an orthophoto and a corresponding stereomate that may be viewed stereoscopically. These products were generated as a part of a stereoscopic orthophotomapping program undertaken by the Canadian Forest Management Institute. The advantage of such products is that they combine the attributes of an orthophoto with the benefits of stereo observation. (Note that Figure 3.27 can also be viewed in stereo. This figure consists of an orthophoto and

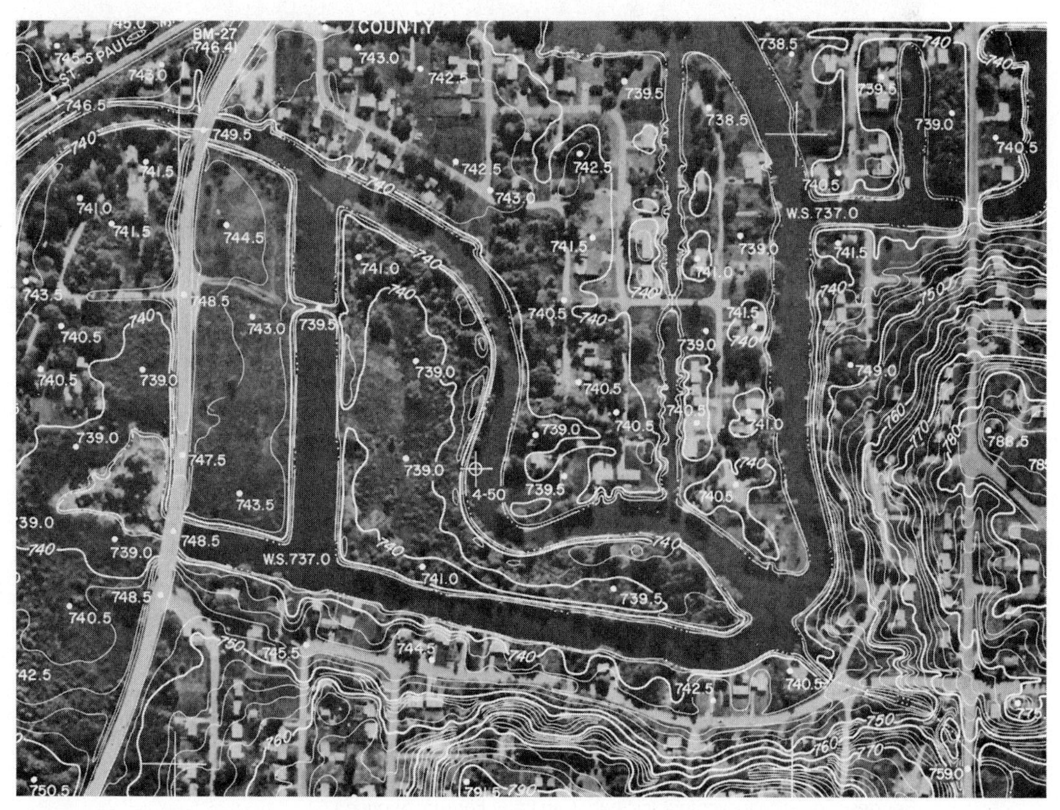

Figure 3.28 Portion of a 1:4800 topographic orthophotomap. Photography taken over the Fox Chain of Lakes, IL. (Courtesy Alster and Associates, Inc.)

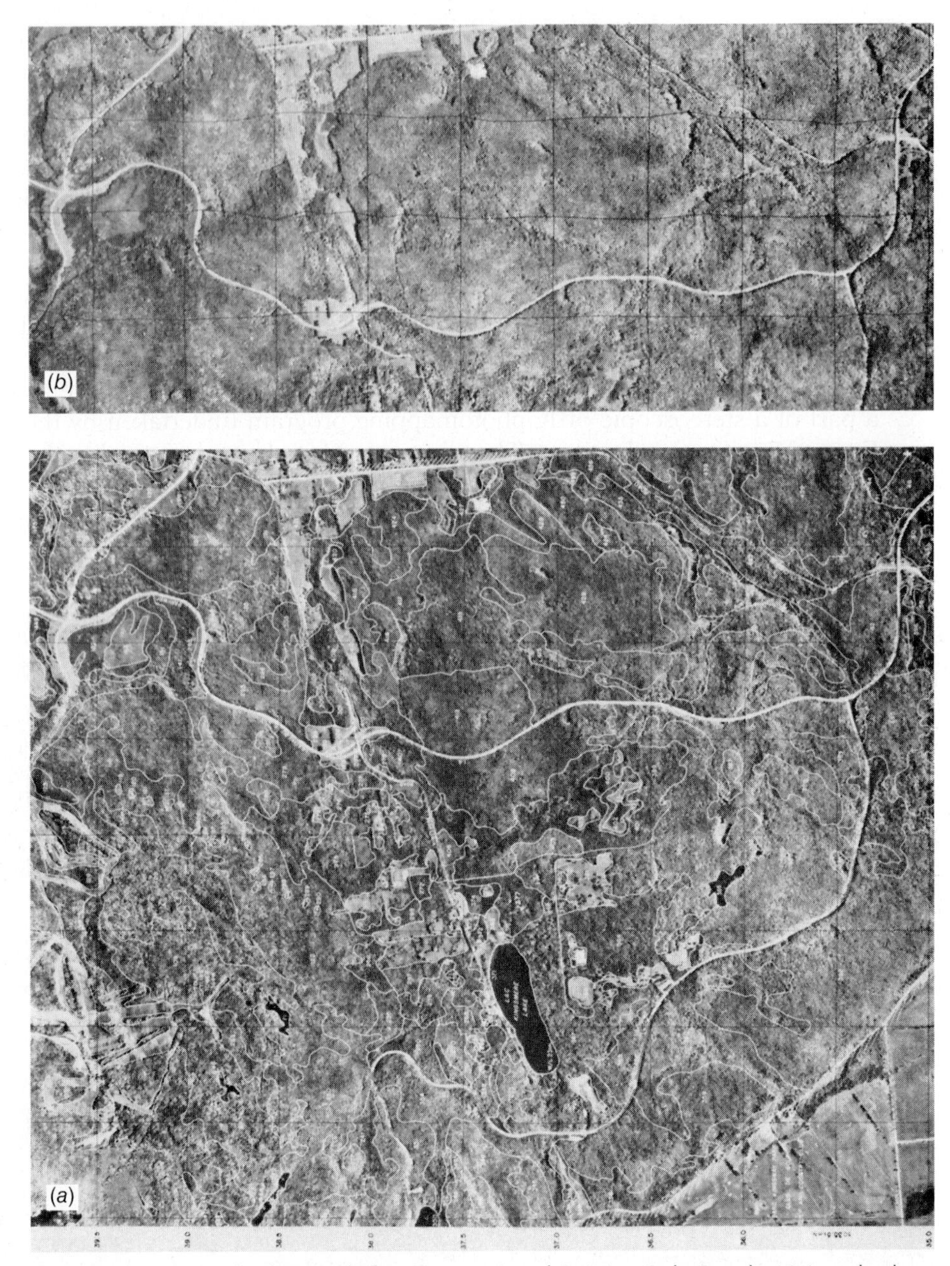

Figure 3.29 Stereo orthophotograph showing a portion of Gatineau Park, Canada: (*a*) An orthophoto and (*b*) a stereomate provide for three-dimensional viewing of the terrain. Measurements made from, or plots made on, the orthophoto have map accuracy. Forest-type information is overprinted on this scene along with a Universal Transverse Mercator (UTM) grid. Note that the UTM grid is square on the orthophoto but is distorted by the introduction of parallax on the stereomate. Scale 1:38,000. (Courtesy Forest Management Institute, Canadian Forestry Service.)

one of the two photos comprising the stereopair from which the orthophoto was produced.)

One caveat we wish to note here is that tall objects such as buildings will still appear to lean in an orthophoto if these features are not included in the DEM used in the orthophoto production process. This effect can be particularly troublesome in urban areas. The effect can be overcome by including building outline elevations in the DEM, or minimized by using only the central portion of a photograph, where relief displacement of vertical features is at a minimum.

Plate 7 is yet another example of the need for, and influence of, the distortion correction provided through the orthophoto production process. Shown in (*a*) is an original uncorrected color photograph taken over an area of high relief in Glacier National Park. The digital orthophoto corresponding to the uncorrected photograph in (*a*) is shown in (*b*). Note the locational errors that would be introduced if GIS data were developed from the uncorrected image. GIS analysts are encouraged to use digital orthophotos in their work whenever possible. Two major federal sources of such data in the United States are the U.S. Geological Survey (USGS) National Digital Orthophoto Program (NDOP) and the USDA National Agriculture Imagery Program (NAIP).

Figures 3.30 and 3.31 illustrate the visualization capability afforded by merging digital orthophoto data with DEM data. Figure 3.30 shows a perspective

Figure 3.30 Perspective view of a rural area generated digitally by draping orthophoto image data over a digital elevation model of the same area. (Courtesy University of Wisconsin-Madison, Environmental Remote Sensing Center, and NASA Affiliated Research Center Program.)

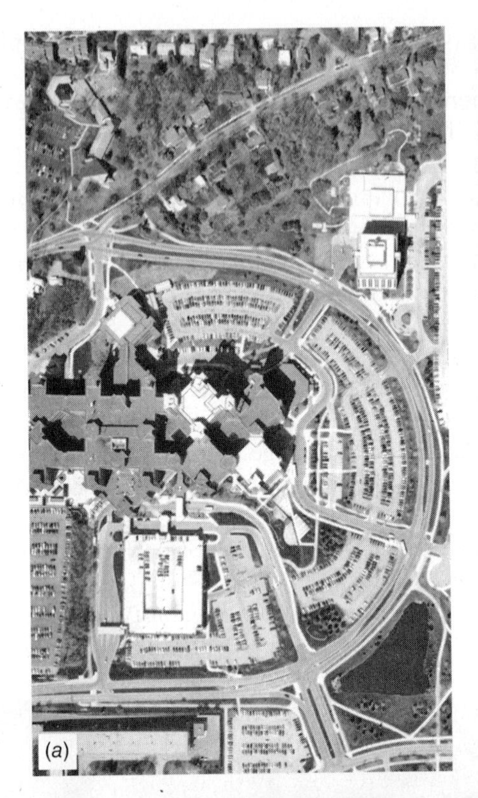

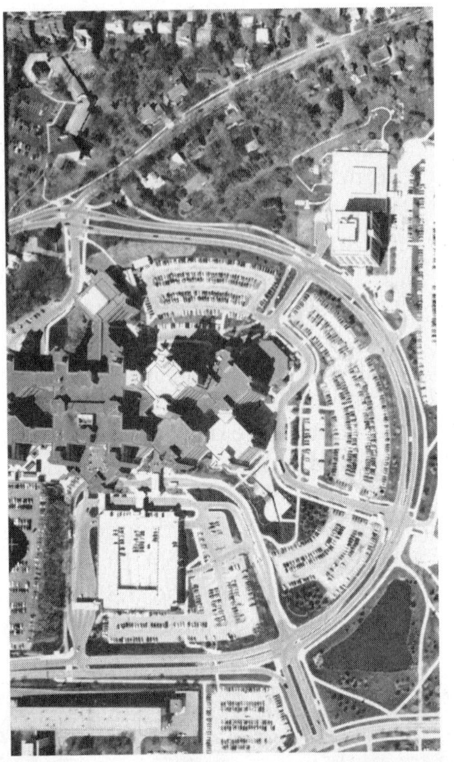

Figure 3.31 Vertical stereopair (*a*) covering the ground area depicted in the perspective view shown in (*b*). The image of each building face shown in (*b*) was extracted automatically from the photograph in which that face was shown with the maximum relief displacement in the original block of aerial photographs covering the area. (Courtesy University of Wisconsin-Madison, Campus Mapping Project.)

view of a rural area located near Madison, Wisconsin. This image was created by draping digital orthophoto data over a DEM of the same area. Figure 3.31 shows a stereopair (*a*) and a perspective view (*b*) of the Clinical Science Center, located on the University of Wisconsin-Madison campus. The images of the various faces of the buildings shown in (*b*) were extracted from the original digitized aerial photograph in which that face was displayed with the greatest relief displacement in accordance with the direction of the perspective view. This process is done automatically from among all of the relevant photographs of the original block of aerial photographs covering the area of interest.

Multi-Ray Photogrammetry

The term *multi-ray photogrammetry* refers to any photogrammetric mapping operation that exploits the redundancy (increased robustness and accuracy) afforded by analyzing a large number of mutually overlapping photographs simultaneously. This terminology is relatively new, but multi-ray photogrammetry is not inherently a new technology. Rather, it is an extension of existing principles and techniques that has come about with the widespread adoption of digital cameras and fully digital mapping work flows.

Most multi-ray operations are based on aerial photographs that are acquired with a very high overlap (80–90%) and sidelap (as much as 60%). Multi-ray flight patterns result in providing from 12 to 15 rays to image each point on the ground. The frame interval rates and storage capacity of modern digital cameras facilitate the collection of such coverage with relatively little marginal cost in comparison to the traditional approach of obtaining stereo coverage with 60% overlap and 20–30% sidelap. The benefit afforded by the multi-ray approach is not only increased accuracy, but also more complete (near total) automation of many procedures.

One of the most important applications of multi-ray imagery is in the process of *dense image matching*. This involves using multiple rays from multiple images to find conjugate matching locations in a stereomodel. Usually, on the order of 50% of all pixels in a model can be accurately matched completely automatically in this manner. In relatively invariant terrain this accuracy drops below 50%, but in highly structured areas, such as cities, accuracies well above 50% are often realized. This leads to very high point densities in the production of DEMs and point clouds. For example, if a GSD of 10 cm is used to acquire the multi-ray photography, this equates to 100 pixels per square meter on the ground. Thus, 50% accuracy in the image matching process would result in a point density of 50 points per square meter on the ground.

Another important application of multi-ray photogrammetry is the production of *true orthophotos*. Earlier in this section we discussed the use of the backward projection process to create a digital orthophoto using a single photograph and a DEM. This process works very well for ground features that are at the ground elevation of the DEM used. However, we also noted that elevated features (e.g., buildings, trees, overpasses) that are not included in the DEM still manifest relief displacement as a function of their elevation above ground and their image distance from the principal point of the original photograph. Tall features located at a distance from the principal point can also completely block or occlude the appearance of ground areas that are in the "shadow" of these features during ray projection. Figure 3.32 illustrates the nature of this problem and its solution

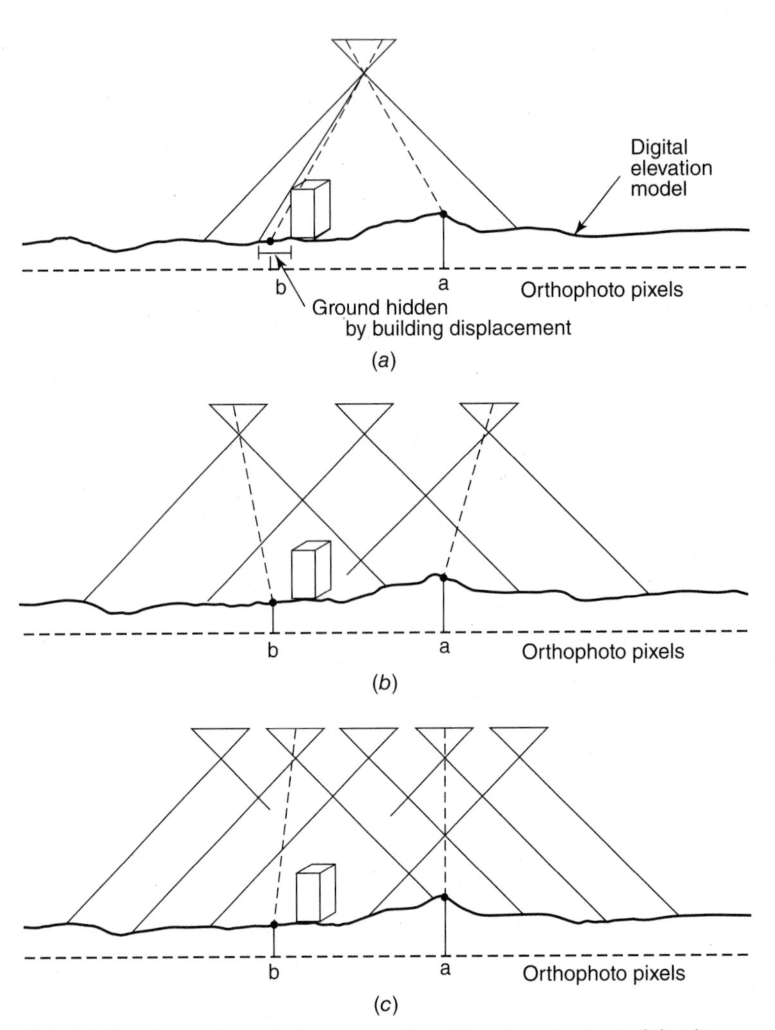

Figure 3.32 Comparison among approaches taken to extract pixel brightness numbers for a conventional orthophoto produced using a single photograph (*a*), a true orthophoto using multiple photographs acquired with 60% overlap (*b*), and a true orthophoto using multiple photographs acquired with 80% overlap (*c*). (After Jensen, 2007).

through the use of more than one original photograph to produce a true digital orthophoto composite image.

Shown in Figure 3.32*a* is the use of a single photograph to produce a conventional orthophoto. The brightness value to be used to portray pixel *a* in the orthophoto is obtained by starting with the ground (X, Y, Z) position of point *a* known from the DEM, then using the collinearity equations to project up to the photograph to determine the x, y photocoordinates of point *a*, and then using resampling to interpolate an appropriate brightness value to use in the orthophoto to depict point *a*. This procedure works well for point *a* because there is no vertical feature obstructing the ray between point *a* on the ground and the photo exposure station. This is not the case for point *b*, which is in a ground area that is obscured by the relief displacement of the nearby building. In this situation, the brightness value that would be placed in the orthophoto for the ground position of point *b* would be that of the roof of the building, and the side of the building would be shown where the roof should be. The ground area near point *b* would not be shown at all in the orthophoto. Clearly, the severity of such relief displacement effects increases both with the height of the vertical feature involved and the feature's distance from the ground principal point.

Figure 3.32*b* illustrates the use of three successive aerial photographs obtained along a flight line to mitigate the effects of relief displacement in the digital orthophoto production process. In this case, the nominal overlap between the successive photos is the traditional value of 60%. The outlines of ground-obstructing features are identified and recorded using traditional stereoscopic feature extraction tools. The brightness values used to portray all other pixel positions in the orthophoto are automatically interpolated from the photograph having the best view of each position. For point *a* the best view is from the closest exposure station to that point, Exposure Station #3. For point *b*, the best view is obtained from Exposure Station #1. The software used for the orthophoto compilation process analyzes the DEM and feature data available for the model to determine that the view of the ground for pixel *b* is obscured from Exposure Station #2. In this manner, each pixel in the orthophoto composite is assigned the brightness value from the corresponding position in the photograph acquired from the closest exposure station affording an unobstructed view of the ground.

Figure 3.32*c* illustrates a multi-ray solution to the true orthophoto production process. In this case, the overlap along the flight strip of images is increased to 80% (or more). This results in several more closely spaced exposure stations being available to cover a given study area. In this way the ray projections to each pixel in the orthophoto become much more vertical and parallel, as if all rays are projected nearly orthographically (directly straight downward) at every point in the orthophoto. An automated DSM is used to create the true orthophoto. In such images, building rooftops are shown in their correct planimetric location, directly above the associated building foundation (with no lean). None of the sides of buildings are shown, and the ground areas around all sides of buildings are shown in their correct location.

3.11 FLIGHT PLANNING

Frequently, the objectives of a photographic remote sensing project can only be met through procurement of new photography of a study area. These occasions can arise for many reasons. For example, photography available for a particular area could be outdated for applications such as land use mapping. In addition, available photography may have been taken in the wrong season. For example, photography acquired for topographic mapping is usually flown in the fall or spring to minimize vegetative cover. This photography will likely be inappropriate for applications involving vegetation analysis.

In planning the acquisition of new photography, there is always a trade-off between cost and accuracy. At the same time, the availability, accuracy, and cost of alternative data sources are continually changing as remote sensing technology advances. This leads to such decisions as whether analog or digital photography is appropriate. For many applications, high resolution satellite data may be an acceptable and cost-effective alternative to aerial photography. Similarly, lidar data might be used in lieu of, or in addition to, aerial photography. Key to making such decisions is specifying the nature and accuracy of the end product(s) required for the application at hand. For example, the required end products might range from hardcopy prints to DEMs, planimetric and topographic maps, thematic digital GIS datasets, and orthophotos, among many others.

The remainder of this discussion assumes that aerial photography has been judged to best serve the needs of a given project, and the task at hand is to develop a flight plan for acquiring the photography over the project's study area.

As previously mentioned, flight planning software is generally used for this purpose. Here we illustrate the basic computational considerations and procedures embedded in such software by presenting two "manual" example solutions to the flight planning process. We highlight the geometric aspects of preparing a flight plan for both a film-based camera mission and a digital camera mission of the same study area. Although we present two solutions using the same study area, we do not mean to imply that the two mission designs yield photography of identical quality and utility. They are simply presented as two representative examples of the flight planning process.

Before we address the geometric aspects of photographic mission planning, we stress that one of the most important parameters in an aerial mission is beyond the control of even the best planner—the weather. In most areas, only a few days of the year are ideal for aerial photography. In order to take advantage of clear weather, commercial aerial photography firms will fly many jobs in a single day, often at widely separated locations. Flights are usually scheduled between 10 a.m. and 2 p.m. for maximum illumination and minimum shadow, although digital cameras that provide high sensitivity under low light conditions can be used for missions conducted as late as sunset, or shortly thereafter, and under heavily overcast conditions. However, as previously mentioned, mission timing is often optimized to ensure strong GPS signals from a number of satellites, which may narrow the acquisition time window. In addition, the mission planner may need to accommodate such mission-specific constraints as maximum allowable building lean in orthophotos produced from the photography, occlusions in urban areas, specular reflections over areas covered by water, vehicular traffic volumes at the time of imaging, and civil and military air traffic control restrictions. Overall, a great deal of time, effort, and expense go into the planning and execution of a photographic mission. In many respects, it is an art as well as a science.

The parameters needed for the geometric design of a film-based photographic mission are (1) the focal length of the camera to be used, (2) the film format size, (3) the photo scale desired, (4) the size of the area to be photographed, (5) the average elevation of the area to be photographed, (6) the overlap desired, (7) the sidelap desired, and (8) the ground speed of the aircraft to be used. When designing a digital camera photographic mission, the required parameters are the same, except the number and physical dimension of the pixels in the sensor array are needed in lieu of the film format size, and the GSD for the mission is required instead of a mission scale.

Based on the above parameters, the mission planner prepares computations and a flight map that indicate to the flight crew (1) the flying height above datum from which the photos are to be taken; (2) the location, direction, and number of flight lines to be made over the area to be photographed; (3) the time interval between exposures; (4) the number of exposures on each flight line; and (5) the total number of exposures necessary for the mission.

When flight plans are computed manually, they are normally portrayed on a map for the flight crew. However, old photography or even a satellite image may be used for this purpose. The computations prerequisite to preparing flight plans for a film-based and a digital camera mission are given in the following two examples, respectively.

EXAMPLE 3.11

A study area is 10 km wide in the east–west direction and 16 km long in the north–south direction (see Figure 3.33). A camera having a 152.4-mm-focal-length lens and a 230-mm format is to be used. The desired photo scale is 1:25,000 and the nominal endlap and sidelap are to be 60 and 30%. Beginning and ending flight lines are to be positioned along the boundaries of the study area. The only map available for the area is at a scale of 1:62,500. This map indicates that the average terrain elevation is 300 m above datum. Perform the computations necessary to develop a flight plan and draw a flight map.

Solution

(a) Use north–south flight lines. Note that using north–south flight lines minimizes the number of lines required and consequently the number of aircraft turns and realignments necessary. (Also, flying in a cardinal direction often facilitates the identification of roads, section lines, and other features that can be used for aligning the flight lines.)

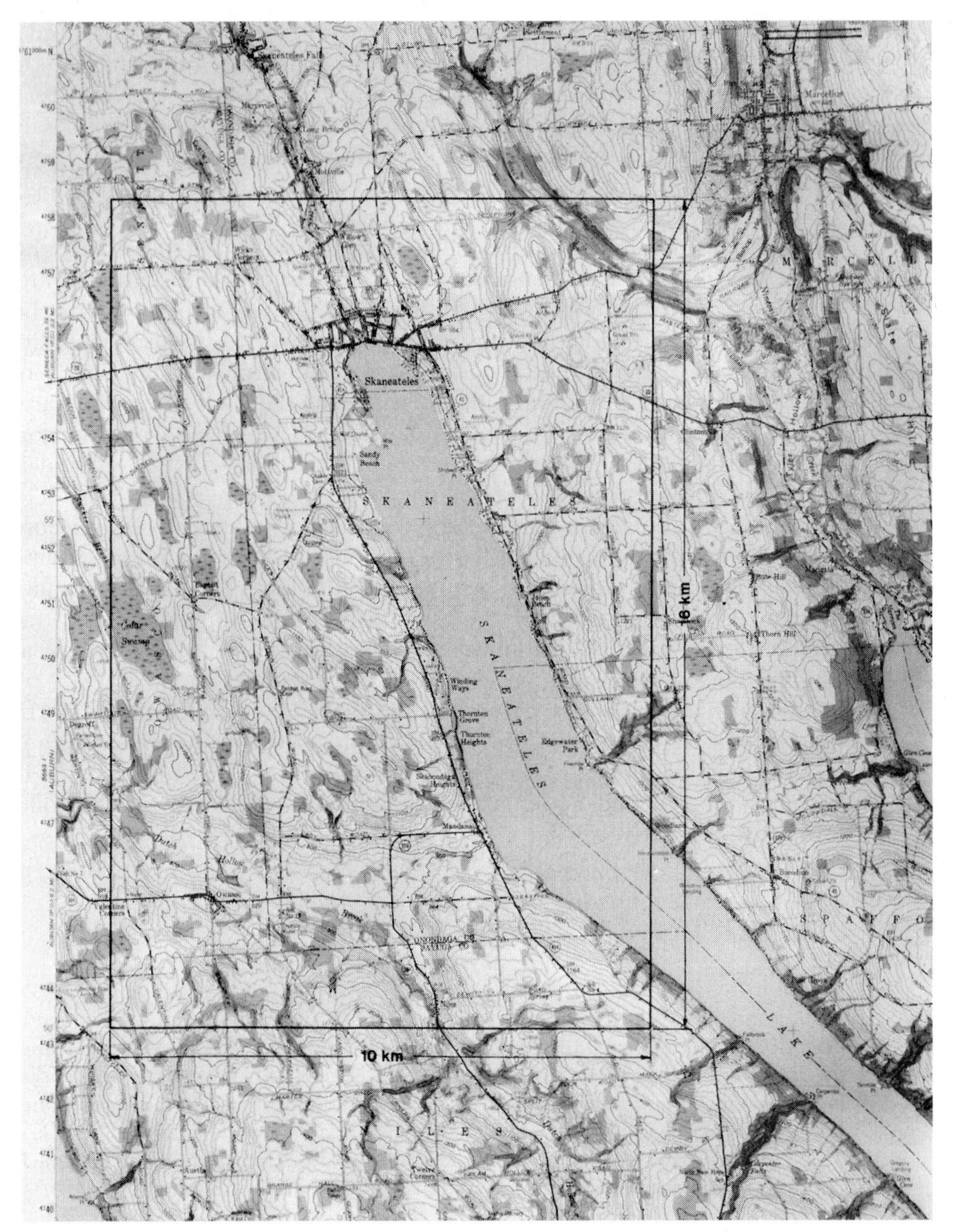

Figure 3.33 A 10 × 16-km study area over which photographic coverage is to be obtained. (Author-prepared figure.)

(b) Find the flying height above terrain ($H' = f/S$) and add the mean site elevation to find flying height above mean sea level:

$$H = \frac{f}{S} + h_{\text{avg}} = \frac{0.1524 \text{ m}}{1/25{,}000} + 300 \text{ m} = 4110 \text{ m}$$

(c) Determine ground coverage per image from film format size and photo scale:

$$\text{Coverage per photo} = \frac{0.23 \text{ m}}{1/25{,}000} = 5750 \text{ m on a side}$$

(d) Determine ground separation between photos on a line for 40% advance per photo (i.e., 60% endlap):

$$0.40 \times 5750 \text{ m} = 2300 \text{ m between photo centers}$$

(e) Assuming an aircraft speed of 160 km/hr, the time between exposures is

$$\frac{2300 \text{ m/photo}}{160 \text{ km/hr}} \times \frac{3600 \text{ sec/hr}}{1000 \text{ m/km}} = 51.75 \text{ sec} \quad \text{(use 51 sec)}$$

(f) Because the intervalometer can only be set in even seconds (this varies between models), the number is rounded off. By rounding down, at least 60% coverage is ensured. Recalculate the distance between photo centers using the reverse of the above equation:

$$51 \text{ sec/photo} \times 160 \text{ km/hr} \times \frac{1000 \text{ m/km}}{3600 \text{ sec/hr}} = 2267 \text{ m}$$

(g) Compute the number of photos per 16-km line by dividing this length by the photo advance. Add one photo to each end and round the number up to ensure coverage:

$$\frac{16{,}000 \text{ m/line}}{2267 \text{ m/photo}} + 1 + 1 = 9.1 \text{ photos/line} \quad \text{(use 10)}$$

(h) If the flight lines are to have a sidelap of 30% of the coverage, they must be separated by 70% of the coverage:

$$0.70 \times 5750 \text{ m coverage} = 4025 \text{ m between flight lines}$$

(i) Find the number of flight lines required to cover the 10-km study area width by dividing this width by distance between flight lines (note: this division gives number of spaces between flight lines; add 1 to arrive at the number of lines):

$$\frac{10{,}000 \text{ m width}}{4025 \text{ m/flight line}} + 1 = 3.48 \quad \text{(use 4)}$$

The adjusted spacing between lines for using four lines is

$$\frac{10{,}000 \text{ m}}{4 - 1 \text{ spaces}} = 3333 \text{ m/space}$$

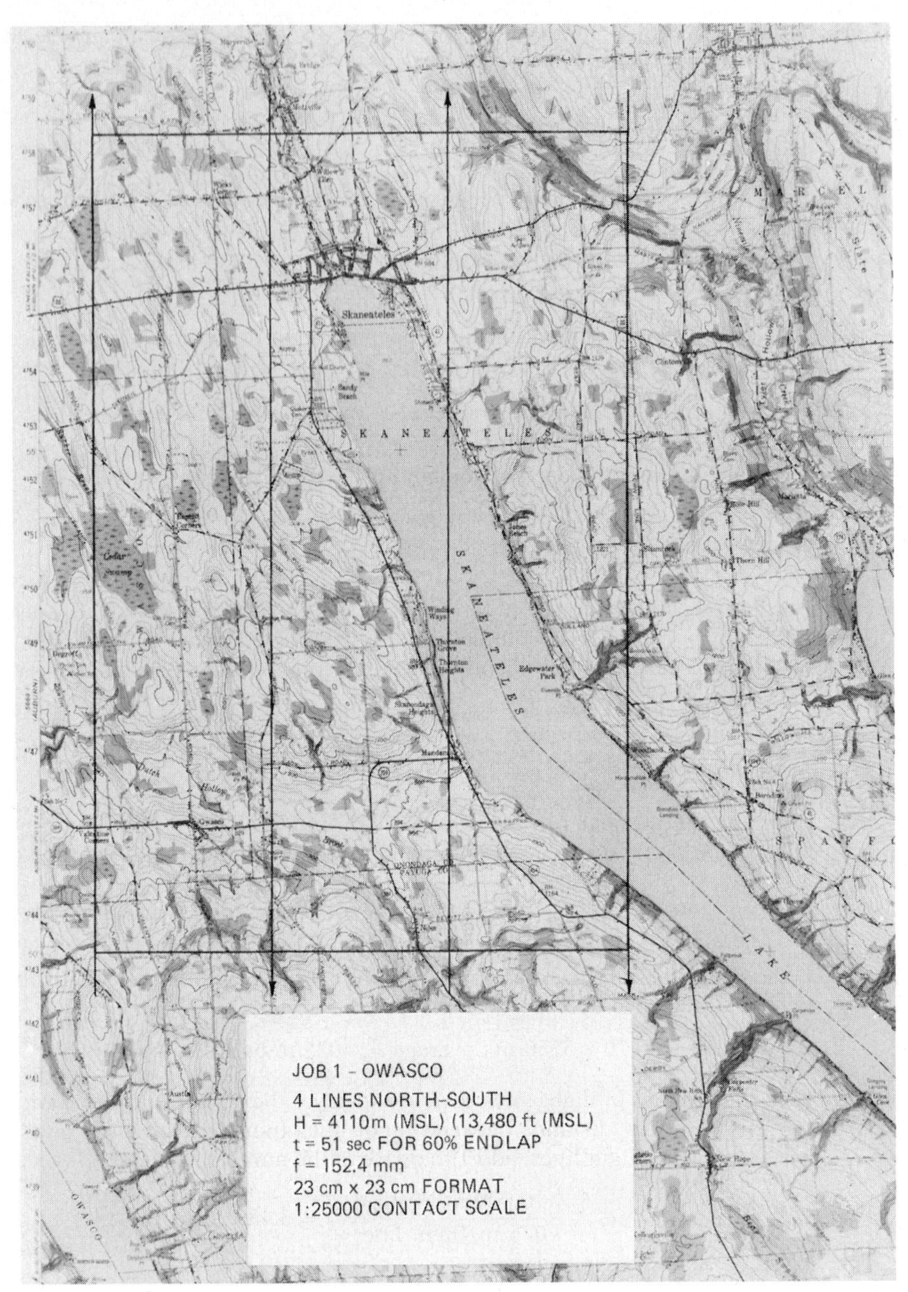

Figure 3.34 Flight map for Example 3.11. (Lines indicate centers of each flight line to be followed.) (Author-prepared figure.)

(j) Find the spacing of flight lines on the map (1:62,500 scale):

$$3333 \text{ m} \times \frac{1}{62{,}500} = 53.3 \text{ mm}$$

(k) Find the total number of photos needed:

$$10 \text{ photos/line} \times 4 \text{ lines} = 40 \text{ photos}$$

(Note: The first and last flight lines in this example were positioned coincident with the boundaries of the study area. This provision ensures complete coverage of the area under the "better safe than sorry" philosophy. Often, a savings in film, flight time, and money is realized by experienced flight crews by moving the first and last lines in toward the middle of the study area.)

The above computations would be summarized on a flight map as shown in Figure 3.34. In addition, a set of detailed specifications outlining the material, equipment, and procedures to be used for the mission would be agreed upon prior to the mission. These specifications typically spell out the requirements and tolerances for flying the mission, the form and quality of the products to be delivered, and the ownership rights to the original images. Among other things, mission specifications normally include such details as mission timing, ground control and GPS/IMU requirements, camera calibration characteristics, film and filter type, exposure conditions, scale tolerance, endlap, sidelap, tilt and image-to-image flight line orientation (crab), photographic quality, product indexing, and product delivery schedules.

EXAMPLE 3.12

Assume that it is desired to obtain panchromatic digital camera coverage of the same study area described in the previous example. Also assume that a GSD of 25 cm is required given the mapping accuracy requirements of the mission. The digital camera to be used for the mission has a panchromatic CCD that includes 20,010 pixels in the across-track direction, and 13,080 pixels in the along-track direction. The physical size of each pixel is 5.2 μm (0.0052 mm). The camera is fitted with an 80-mm-focal-length lens. As in the previous example, stereoscopic coverage is required that has 60% overlap and 30% sidelap. The aircraft to be used to conduct the mission will be operated at a nominal speed of 260 km/hr. Perform the computations necessary to develop a preliminary flight plan for this mission in order to estimate flight parameters.

Solution

(a) As in the previous example, use north–south flight lines.

(b) Find the flying height above terrain $\left(H' = \frac{(GSD)\times f}{p_d}\right)$ and add the mean elevation to find the flying height above mean sea level:

$$H = \frac{(GSD)\times f}{p_d} + h_{avg} = \frac{0.25 \text{ m} \times 80 \text{ mm}}{0.0052 \text{ mm}} + 300 \text{ m} = 4146 \text{ m}$$

(c) Determine the across-track ground coverage of each image:
From Eq. 2.12, the across-track sensor dimension is

$$d_{xt} = n_{xt} \times p_d = 20{,}010 \times 0.0052 \text{ mm} = 104.05 \text{ mm}$$

Dividing by the image scale, the across-track ground coverage distance is

$$\frac{d_{xt}H'}{f} = \frac{104.05 \text{ mm} \times 3846 \text{ m}}{80 \text{ mm}} = 5002 \text{ m}$$

(d) Determine the along-track ground coverage of each image:
From Eq. 2.13, the along-track sensor dimension is

$$d_{at} = n_{at} \times p_d = 13{,}080 \times 0.0052 \text{ mm} = 68.02 \text{ mm}$$

Dividing by the image scale, the along-track ground coverage distance is

$$\frac{d_{at}H'}{f} = \frac{68.02 \text{ mm} \times 3846 \text{ m}}{80 \text{ mm}} = 3270 \text{ m}$$

(e) Determine the ground separation between photos along-track for 40% advance (i.e., 60% endlap):

$$0.40 \times 3270 \text{ m} = 1308 \text{ m}$$

(f) Determine the interval between exposures for a flight speed of 260 km/hr:

$$\frac{1308 \text{ m/photo}}{260 \text{ km/hr}} \times \frac{3600 \text{ sec/hr}}{1000 \text{ m/km}} = 18.11 \text{ sec (use 18)}$$

(g) Compute the number of photos per 16 km line by dividing the line length by the photo advance. Add one photo to each end and round up to ensure coverage:

$$\frac{16{,}000 \text{ m/line}}{1308 \text{ m/photo}} + 1 + 1 = 14.2 \text{ photos/line (use 15)}$$

(h) If the flight lines are to have sidelap of 30% of the across-track coverage, they must be separated by 70% of the coverage:

$$0.70 \times 5002 \text{ m} = 3501 \text{ m}$$

(i) Find the number of flight lines required to cover the 10-km study area width by dividing this width by the distance between flight lines (Note: This division gives number

of spaces between flight lines; add 1 to arrive at the number of lines):

$$\frac{10{,}000 \text{ m}}{3501 \text{ m/line}} + 1 = 3.86 \text{ flight lines}$$

Use 4 flight lines.

(j) Find the total number of photos needed:

$$15 \text{ photos/line} \times 4 \text{ lines} = 60 \text{ photos}$$

As is the case with the acquisition of analog aerial photography, a flight plan for acquiring digital photography is accompanied by a set of detailed specifications stating the requirements and tolerances for flying the mission, preparing image products, ownership rights, and other considerations. These specifications would generally parallel those used for film-based missions. However, they would also address those specific considerations that are related to digital data capture. These include, but are not limited to, use of single versus multiple camera heads, GSD tolerance, radiometric resolution of the imagery, geometric and radiometric image pre-processing requirements, and image compression and storage formats. Overall, the goal of such specifications is to not only ensure that the digital data resulting from the mission are not only of high quality, but also that they are compatible with the hardware and software to be used to store, process, and supply derivative products from the mission imagery.

3.12 CONCLUSION

As we have seen in this chapter, photogrammetry is a very large and rapidly changing subject. Historically, most photogrammetric operations were analog in nature, involving the physical projection and measurement of hardcopy images with the aid of precise optical or mechanical equipment. Today, all-digital photogrammetric workflows are the norm. In addition, most softcopy photogrammetric systems also readily provide some functionality for handling various forms of non-photographic imagery (e.g., lidar, line scanner data, satellite imagery). With links to GIS and image processing software, modern softcopy photogrammetric systems represent highly integrated systems for spatial data capture, manipulation, analysis, storage, display, and output.

4 MULTISPECTRAL, THERMAL, AND HYPERSPECTRAL SENSING

4.1 INTRODUCTION

In Section 1.5 we briefly described multispectral imagery, which consists of image data selectively acquired in multiple spectral bands. The film-based, digital, and video camera systems described in Chapter 2 might be considered as simple multispectral sensors, because they can be used to acquire imagery in three or four wavelength bands ranging from 0.3 to 0.9 μm. In this chapter, we look at a class of instruments known as *multispectral scanners*. These instruments can be designed to collect data in many more spectral bands and over a wider range of the electromagnetic spectrum. Utilizing different types of electronic detectors, multispectral scanners can extend the range of sensing from 0.3 to approximately 14 μm. (This includes the UV, visible, near-IR, mid-IR, and thermal-IR spectral regions.) Furthermore, multispectral scanner systems can sense in very narrow bands.

We begin our discussion in this chapter with a treatment of how multispectral scanner images are acquired physically, covering both across-track and along-track scanning systems. We examine the basic processes, operating principles, and geometric characteristics of these systems. After our treatment of multispectral scanning, we discuss thermal images. A thermal image can be thought of as merely a particular kind of multispectral image consisting of data acquired only in the thermal portion of the spectrum (in one or more bands). However, thermal images must be interpreted with due regard for the basic thermal radiation principles involved. We discuss these principles as we describe how thermal images can be interpreted visually, calibrated radiometrically, and processed digitally. We conclude the chapter with an introduction to *hyperspectral sensing*, the acquisition of images in many (often hundreds) very narrow, contiguous spectral bands throughout the visible, near-IR, and mid-IR portions of the spectrum.

In this chapter we stress *airborne* scanning systems. However, as we see in Chapter 5, the operating principles of multispectral, thermal, and hyperspectral sensors operated from space platforms are essentially identical to the airborne systems described in this chapter.

4.2 ACROSS-TRACK SCANNING

Airborne multispectral scanner systems build up two-dimensional images of the terrain for a swath beneath the aircraft. There are two different ways in which this can be done—using *across-track* (*whiskbroom*) scanning or *along-track* (*pushbroom*) scanning.

Figure 4.1 illustrates the operation of an across-track, or whiskbroom, scanner. Using a rotating or oscillating mirror, such systems scan the terrain along *scan lines* that are at right angles to the flight line. This allows the scanner to repeatedly measure the energy from one side of the aircraft to the other. Data are collected within an arc below the aircraft typically of 90° to 120°. Successive scan lines are covered as the aircraft moves forward, yielding a series of contiguous, or just touching, narrow strips of observation comprising a two-dimensional image of rows (scan lines) and columns. After being reflected off the rotating or oscillating mirror, the incoming energy is separated into several spectral components that are sensed independently. Figure 4.1*b* illustrates an example of a system that records both thermal and nonthermal wavelengths. A *dichroic grating* is used to separate these two forms of energy. The nonthermal wavelength component is directed from the grating through a prism (or diffraction grating) that splits the energy into a continuum of UV, visible, and near-IR wavelengths. At the same time, the dichroic grating disperses the thermal component of the incoming signal into its constituent wavelengths. By placing an array of electro-optical detectors at the proper geometric positions behind the grating and the prism, the incoming beam is essentially "pulled apart" into multiple spectral bands, each of which is measured independently. Each detector is designed to have its peak spectral sensitivity in a specific wavelength band. Figure 4.1 illustrates a five-band scanner, but as we will see later (Section 4.13), scanners with hundreds of bands are available.

At any instant in time, the scanner "sees" the energy within the system's *instantaneous field of view (IFOV)*. The IFOV is normally expressed as the cone angle within which incident energy is focused on the detector. (See β in Figure 4.1.)

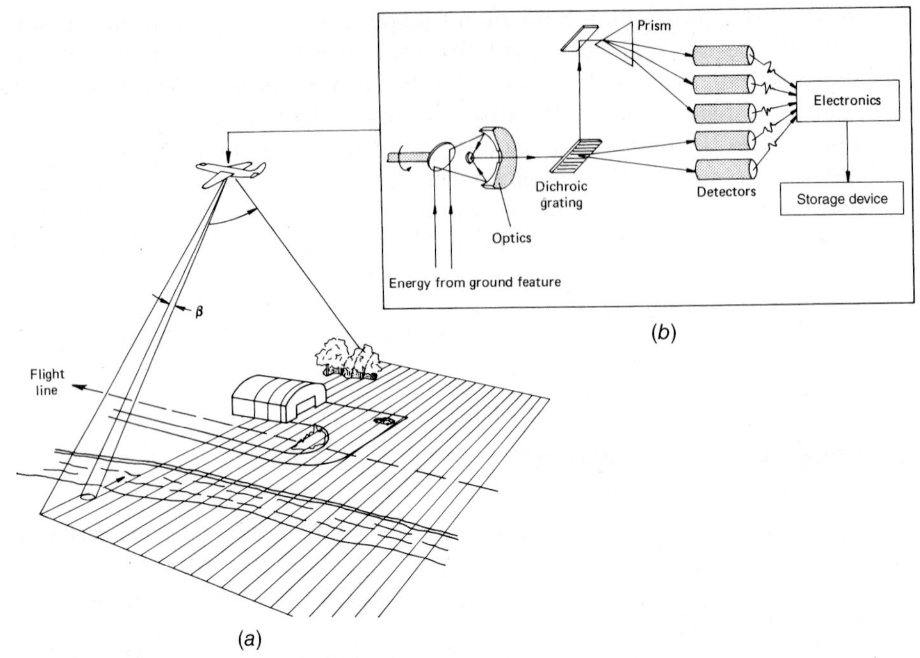

Figure 4.1 Across-track, or whiskbroom, multispectral scanner system operation: (*a*) scanning procedure during flight; (*b*) scanner schematic.

The angle β is determined by the instrument's optical system and size of its detectors. All energy propagating toward the instrument within the IFOV contributes to the detector response at any instant. Hence, more than one land cover type or feature may be included in the IFOV at any given instant, and the composite signal response will be recorded. Thus, an image typically contains a combination of "pure" and "mixed" pixels, depending upon the IFOV and the spatial complexity of the ground features.

Figure 4.2 illustrates the segment of the ground surface observed when the IFOV of a scanner is oriented directly beneath the aircraft. This area can be expressed as a circle of diameter D given by

$$D = H'\beta \tag{4.1}$$

where

D = diameter of circular ground area viewed
H' = flying height above terrain
β = IFOV of system (expressed in radians)

The ground segment sensed at any instant is called the *ground resolution element* or *ground resolution cell*. The diameter D of the ground area sensed at any instant in time is loosely referred to as the system's *spatial resolution*. For example, the spatial resolution of a scanner having a 2.5-milliradian (mrad) IFOV and

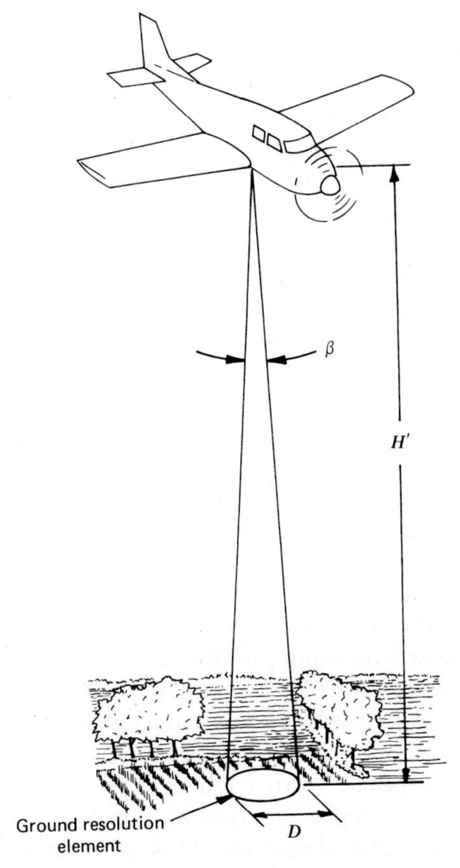

Figure 4.2 Instantaneous field of view and resulting ground area sensed directly beneath an aircraft by a multispectral scanner.

being operated from 1000 m above the terrain can be found from Eq. 4.1 as $D = 1000\,\text{m} \times (2.5 \times 10^{-3}\,\text{rad}) = 2.5\,\text{m}$. That is, the ground resolution cell would be 2.5 m in diameter directly under the aircraft. (Depending on the optical properties of the system used, ground resolution cells directly beneath the aircraft can be either circular or square.) The size of the ground resolution cell increases symmetrically on each side of the nadir as the distance between the scanner and the ground resolution cell increases. Hence, the ground resolution cells are larger toward the edge of the image than near the middle. This causes a scale distortion that often must be accounted for in image interpretation or mathematically compensated for in image generation. (We discuss such distortions in Section 4.6.)

The IFOV for airborne multispectral scanner systems typically ranges from about 0.5 to 5 mrad. A small IFOV is desirable to record fine spatial detail. On the other hand, a larger IFOV means a greater quantity of total energy is focused on a detector as the scanner's mirror sweeps across a ground resolution cell. This permits more sensitive scene radiance measurements due to higher signal levels. The result is an improvement in the *radiometric resolution*, or the ability to discriminate very slight energy differences. Thus, there is a trade-off between high spatial resolution and high radiometric resolution in the design of multispectral scanner systems. A large IFOV yields a signal that is much greater than the background electronic *noise* (extraneous, unwanted responses) associated with any given system. Thus, other things being equal, a system with a large IFOV will have a higher *signal-to-noise ratio* than will one with a small IFOV. Again, a large IFOV results in a longer *dwell time*, or residence time of measurement, over any given ground area. What is sacrificed for these higher signal levels is spatial resolution. In a similar vein, the signal-to-noise ratio can be increased by broadening the wavelength band over which a given detector operates. What is sacrificed in this case is *spectral resolution*, that is, the ability to discriminate fine spectral differences.

Again, what we have described above as a scanner's ground resolution cell (the scanner's IFOV projected onto the ground) is often simply called a system's "resolution" (we use this term often in this chapter and in Chapter 5, especially in tables describing system characteristics).

As described in Section 1.5, digital images are created by quantizing an analog electrical signal, a process known as analog-to-digital (A-to-D) conversion. Consider a single scan line from an across-track scanner, such as one of the scan lines shown in Figure 4.1. The electromagnetic energy entering the sensor is split into multiple spectral bands, with each wavelength range being directed toward a specific detector. The electrical response from each detector is a continuous analog signal. Figure 4.3 is a graphical representation of the A-to-D conversion process for the signal from one of these detectors. This continuous signal is sampled at a set time interval (ΔT) and recorded numerically at each sample

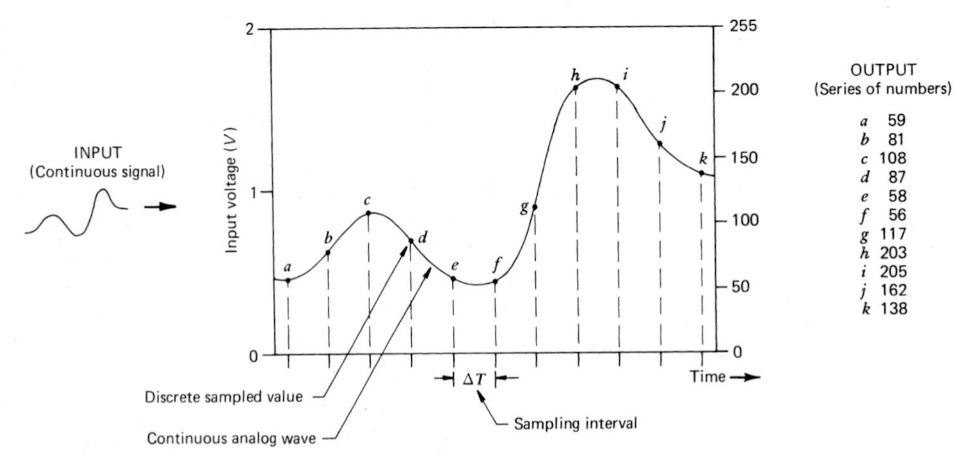

Figure 4.3 Analog-to-digital conversion process.

point $(a, b, \ldots, j, k)$. The sampling rate for a particular signal is determined by the highest frequency of change in the signal. The sampling rate must be at least twice as high as the highest frequency present in the original signal in order to adequately represent the variation in the signal.

In Figure 4.3 we illustrate the incoming sensor signal in terms of an electrical voltage value ranging between 0 and 2 V. The digital number (DN) output values are integers ranging from 0 to 255. Accordingly, a sampled voltage of 0.46 recorded by the sensor [at (a) in Figure 4.3] would be recorded as a DN of 59. (The DNs measured at the other sampling points along the signal are shown along the right side of the figure.)

Most scanners use square detectors, and across-track scan lines are sampled such that they are represented by a series of just-touching pixels projected to ground level. The ground track of the aircraft ideally advances a distance just equal to the size of the resolution cell between rotations of the scanning mirror. This results in sampling the ground without gaps or overlaps.

The ground distance between adjacent sampling points in a digital scanner image need not necessarily exactly equal the dimensions of the IFOV projected onto the ground. The ground pixel size, or *ground sample distance* (GSD), is determined by the sampling time interval ΔT. Figure 4.4 illustrates a case where the sampling time interval results in a ground pixel width that is smaller than the width of the scanner's IFOV projected onto the ground. One real-world system that exemplifies this case is the Landsat Multispectral Scanner (MSS), discussed in Chapter 5. The Landsat MSS has a projected IFOV approximately 79 m wide, but its signal is sampled to yield a GSD of 57 m. This illustrates that there can often be a difference between the projection of a scanner's IFOV onto the ground and the GSD. Often the term *resolution* is loosely used to refer to either of these dimensions. In general, if the projected IFOV and GSD differ, the ability to resolve features in the image will be limited by the coarser of these two resolutions.

While on the topic of resolution, we must also point out that the meaning of the term with respect to images acquired by film cameras is not the same as that with reference to electro-optical systems.

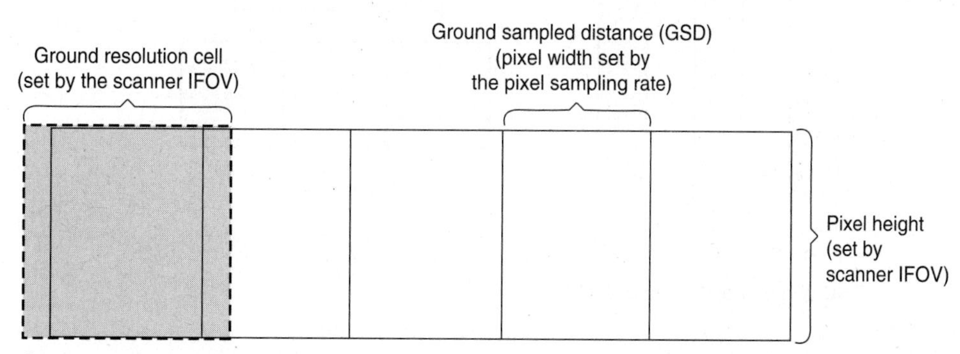

Figure 4.4 Ground sampled distance concept.

The concept of photographic film resolution (described in Section 2.7) is based on being able to distinguish two objects from each other. When lines and spaces of equal width are wide enough that they can just be distinguished on the film, the center-to-center line spacing is called the film resolution. When these lines are projected onto the ground, the resulting distance is called the ground resolution distance (GRD), as expressed by Eq. 2.18 and shown in Figure 4.5*a*. Applying this concept to a scanner's ground IFOV, we can see that it gives a misleading value of the ability of the scanner to resolve ground objects in that such objects must be separated from each other by at least the GSD to be distinguished individually (Figure 4.5*b*). Consequently, the nominal resolution of an electro-optical system must be approximately doubled to compare it with the GRD of a film-based camera system. For example, a 1-m-resolution scanner system would have the approximate detection capability of a 2-m-resolution film camera system. Looking at this concept another way, the optimal GRD that can be achieved with a digital imaging system is approximately twice its GSD.

As explained in Section 2.7, the image analyst is interested not only in object detection but also in object recognition and identification. In the case of scanner images, the larger the number of pixels that make up the image of an object on the ground, the more information that can be determined about that object. For example, one pixel on a vehicle might just be identified as an object on the ground (and this assumes adequate contrast between the vehicle and its background). Two pixels on the same object might be adequate to determine the object's orientation. Three pixels might permit identification of the object as a vehicle. And it might be possible to identify the type of vehicle when it includes five or more pixels. Thus, the effective spatial resolution of a digital imaging system depends on a number of factors in addition to its GSD (e.g., the nature of the ground scene, optical distortions, image motion, illumination and viewing geometry, and atmospheric effects).

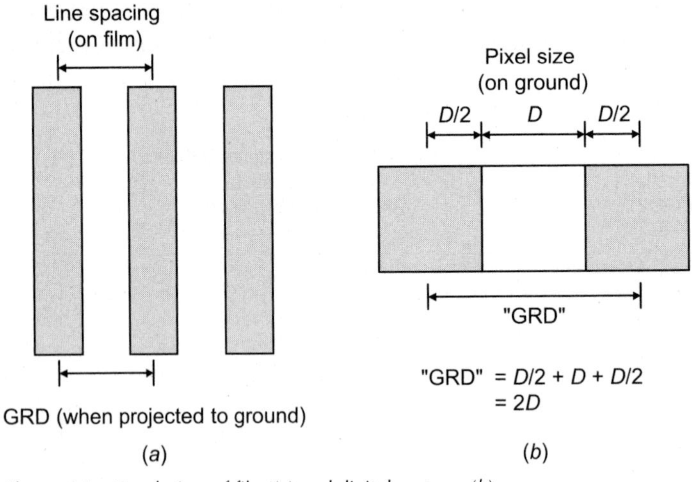

Figure 4.5 Resolution of film (*a*) and digital systems (*b*).

4.3 ALONG-TRACK SCANNING

As with across-track systems, along-track, or pushbroom, scanners record multispectral image data along a swath beneath an aircraft. Also similar is the use of the forward motion of the aircraft to build up a two-dimensional image by recording successive scan lines that are oriented at right angles to the flight direction. However, there is a distinct difference between along-track and across-track systems in the manner in which each scan line is recorded. In an along-track system there is no scanning mirror. Instead, a *linear array* of detectors is used (Figure 4.6). Linear arrays typically consist of numerous *charge-coupled devices (CCDs)* positioned end to end. As illustrated in Figure 4.6, each detector element is dedicated to sensing the energy in a single column of data. The size of the ground resolution cell is determined by the IFOV of a single detector projected onto the ground. The GSD in the across-track direction is set by the detector IFOV. The GSD in the along-track direction is again set by the sampling interval (ΔT) used for the A-to-D signal conversion. Normally, the sampling results in just-touching square pixels comprising the image.

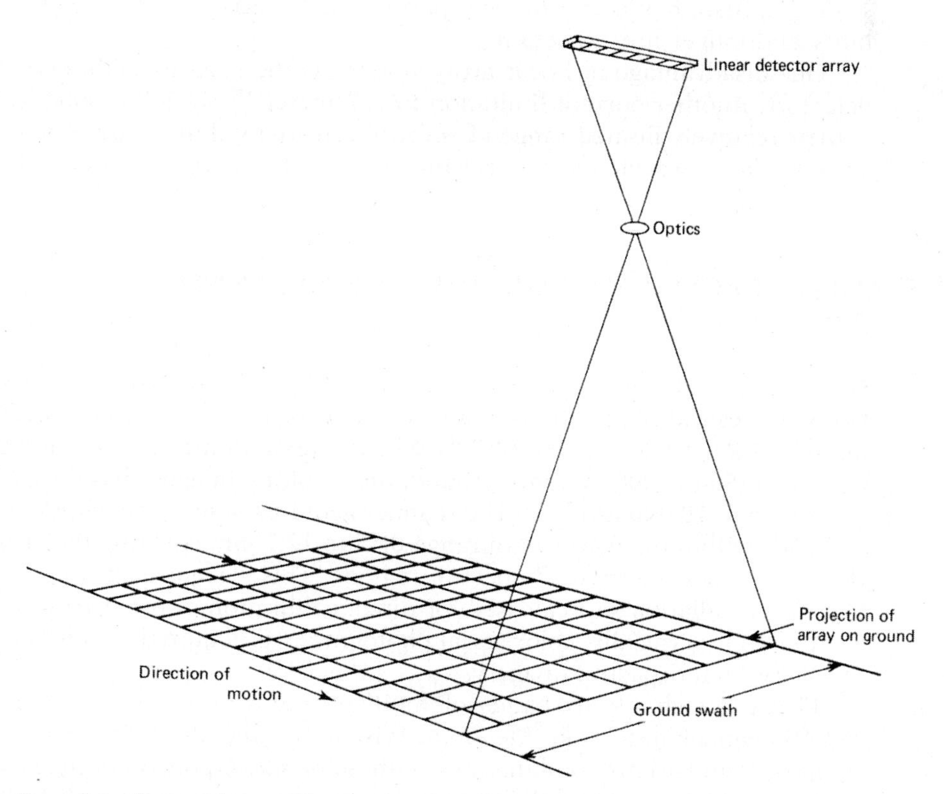

Figure 4.6 Along-track, or pushbroom, scanner system operation.

Linear array CCDs are designed to be very small, and a single array may contain over 10,000 individual detectors. Each spectral band of sensing requires its own linear array. Normally, the arrays are located in the focal plane of the scanner such that each scan line is viewed by all arrays simultaneously.

Linear array systems afford a number of advantages over across-track mirror scanning systems. First, linear arrays provide the opportunity for each detector to have a longer dwell time over which to measure the energy from each ground resolution cell. This enables a stronger signal to be recorded (and, thus, a higher signal-to-noise ratio) and a greater range in the signal levels that can be sensed, which leads to better radiometric resolution. In addition, the geometric integrity of linear array systems is greater because of the fixed relationship among detector elements recording each scan line. The geometry along each row of data (scan line) is similar to an individual photo taken by an aerial mapping camera. The geometric errors introduced into the sensing process by variations in the scan mirror velocity of across-track scanners are not present in along-track scanners. Because linear arrays are solid-state microelectronic devices, along-track scanners are generally smaller in size and weight and require less power for their operation than across-track scanners. Also, having no moving parts, a linear array system has higher reliability and longer life expectancy.

One disadvantage of linear array systems is the need to calibrate many more detectors. Another current limitation to commercially available solid-state arrays is their relatively limited range of spectral sensitivity. Linear array detectors that are sensitive to wavelengths longer than the mid IR are not readily available.

4.4 EXAMPLE ACROSS-TRACK MULTISPECTRAL SCANNER AND IMAGERY

One example of an airborne across-track multispectral scanner is NASA's Airborne Terrestrial Applications Scanner (ATLAS). This system has an IFOV of 2.0 mrad and a total scan angle of 73°. When operated on a LearJet aircraft at altitudes of 1250 to 5000 m above ground, the resulting imagery has a spatial resolution of 2.5 to 10.0 m at nadir. The scanner acquires digital data simultaneously in 15 bands within the wavelength range 0.45 to 12.2 μm, covering the visible, near-IR, mid-IR, and thermal-IR spectral regions. This system includes two onboard blackbody radiation sources used for direct calibration of the thermal bands (see Section 4.11). Typically, color aerial photography is acquired simultaneously with the ATLAS multispectral imagery.

Figure 4.7 shows six bands of multispectral scanner data acquired by the ATLAS scanner near Lake Mendota, Wisconsin, and Plate 8 shows three color composites of various combinations of these bands. A portion of the lake appears in the lower left corner of the imagery; on the late September date of this image, a large algal bloom was occurring in the lake. Residential neighborhoods occupy the center of the image area, while the upper third shows agricultural fields,

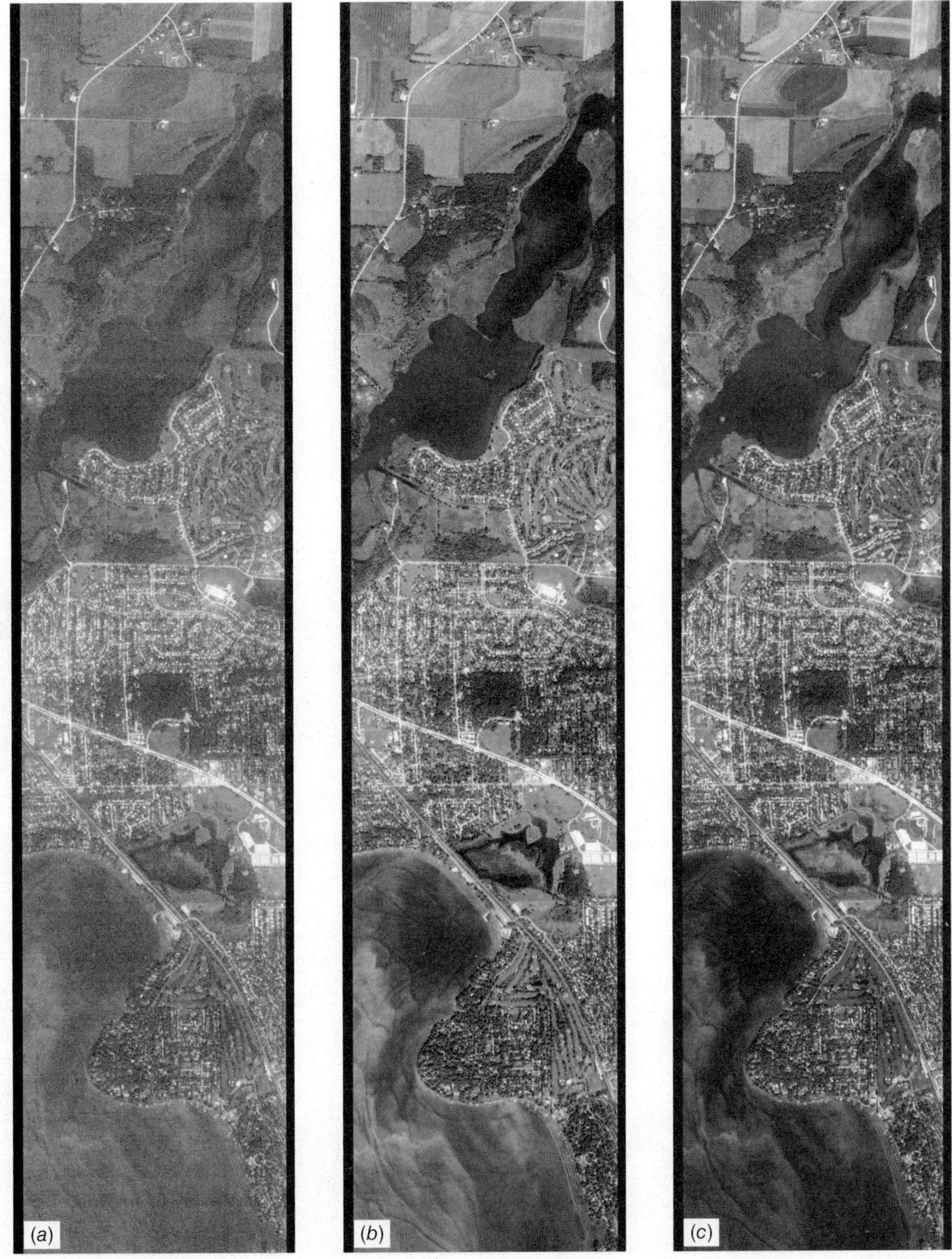

Figure 4.7 Six bands from a 15-band multispectral scanner image, Dane County, Wisconsin, late September: (*a*) band 1, 0.45 to 0.52 μm (blue); (*b*) band 2, 0.52 to 0.60 μm (green); (*c*) band 4, 0.63 to 0.69 μm (red); (*d*) band 6, 0.76 to 0.90 μm (near IR); (*e*) band 8, 2.08 to 2.35 μm (mid IR); (*f*) band 12, 9.0 to 9.4 μm (thermal IR). Scale 1:50,000. (Courtesy NASA Stennis Space Center.)

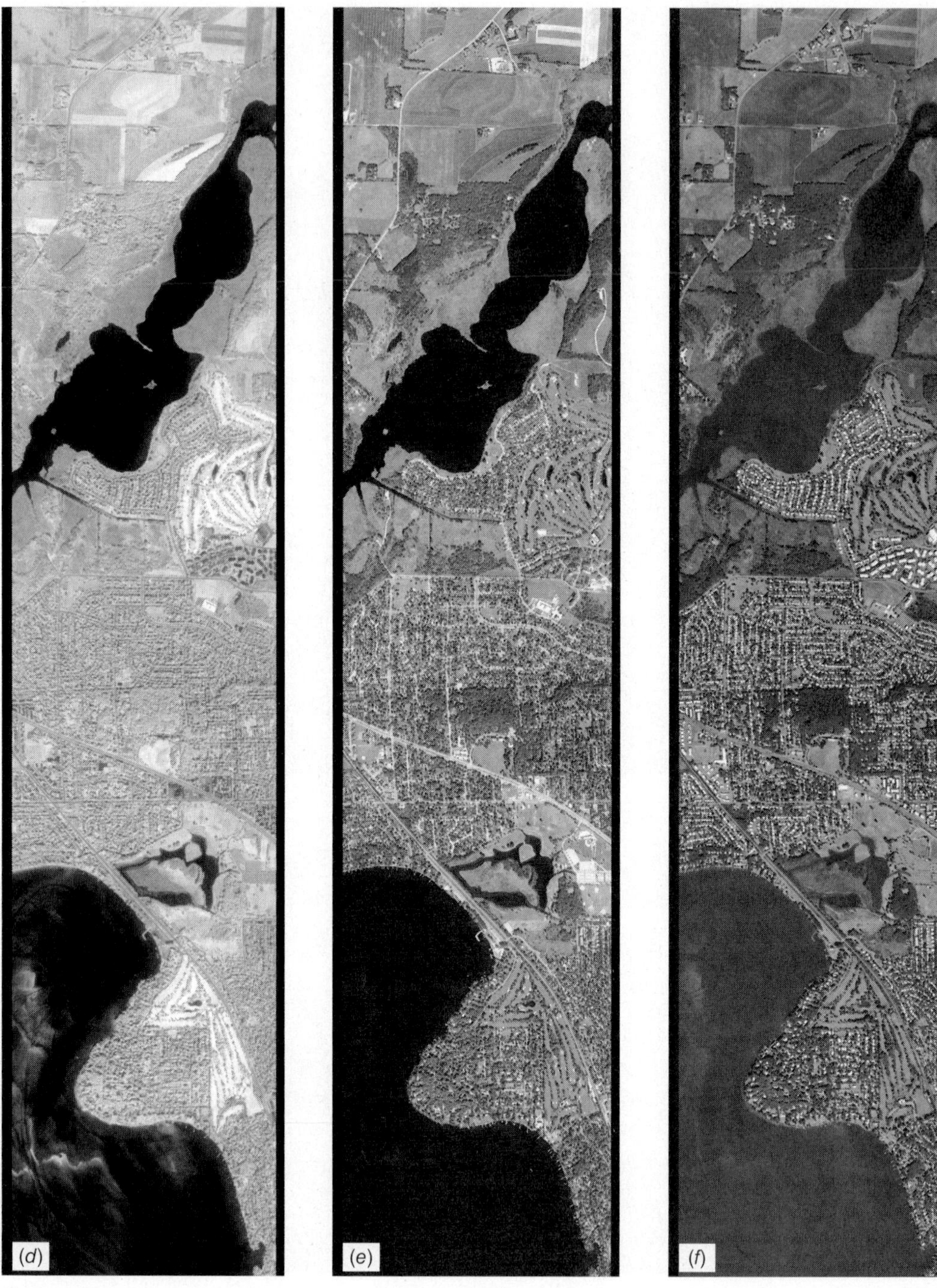

Figure 4.7 *(Continued)*

a portion of the Yahara River, and adjacent wetlands (marshes). As is typical with multispectral data, the three bands in the visible part of the spectrum (bands 1, 2, and 4) show a great deal of correlation; that is, objects or areas that are light toned in one visible band are light toned in the other two visible bands. There are, of course, some variations in reflectance, which is why different colors are seen in the color composite (Plate 8*a*). Overall, the blue band has a hazy appearance with less contrast, due to the influence of atmospheric scattering and path radiance (Section 1.4), while the green and red bands show progressively less of these effects.

In the near IR, healthy vegetation is much lighter toned than in the visible bands, resulting in very light tones in Figure 4.7*d* where such vegetation is present (such as certain agricultural fields at the top of the image and golf courses and parks in the central and lower portions of the image). As was seen in the photographs in Figure 2.12, water is generally extremely dark in the near IR—compare the image tone of the Yahara River in Figure 4.7*a* with its tone in 4.7*d*—and this is true of the mid-IR band (*e*) as well. The anomalous appearance of the lake in the visible and near-IR, with lighter tones than would normally be expected from water, is due to relatively dense concentrations of algae in and on the surface of the lake, while the darker linear features cutting through these patches represent wakes from the passage of boats that have stirred up the water column and temporarily dissipated the otherwise widespread algae at or near the surface. Note that there are many distinct differences between the near-IR and mid-IR bands, for example, in the agricultural fields at upper right. A recently harvested field appears very bright in the mid IR (*e*), but less bright in the near IR (*d*) due to its bare soil surface. In the thermal band (*f*), the river and lake appear darker than land areas, indicating that they are cooler than the surrounding land at the mid-day hour this image was acquired (Section 4.10). Trees likewise appear darker (cooler) than grassy areas, while many rooftops appear very bright due to their relatively high temperature. (The interpretation of thermal imagery is discussed in more detail later in this chapter.)

In the normal color and color IR combinations of multispectral bands shown in Plate 8, different feature types are seen as different colors. In these late-September images, healthy vegetation has a generally greenish hue in the normal color composite (8*a*), a red hue in the first false color composite (*b*), and once again a green hue in the second false color composite (*c*). This difference between (*b*) and (*c*) results from the fact that the near-IR band, which is highly indicative of vegetation, is shown in red in (*b*) and in green in (*c*). Likewise, in (*b*) and (*c*), algae floating on the surface of the lake have a bright red or green hue respectively, due to the reflectance of near-IR light, while algae deeper within the water column lack this color (due to absorption of near IR by the intervening water). Cattails and other marsh vegetation along the river can be differentiated from other cover types on the basis of their hue in the three color composites, and differences in soil and vegetation conditions among the agricultural fields can also be seen. Again, it is this difference in spectral reflectance that aids in the discrimination of cover types visually and in the process of automated image classification (discussed in Chapter 7).

4.5 EXAMPLE ALONG-TRACK MULTISPECTRAL SCANNER AND IMAGERY

An example of an airborne along-track scanner is the *Leica ADS100 Airborne Digital Sensor,* manufactured by Leica Geosystems (Figure 4.8). The Leica ADS100 incorporates a total of 13 linear arrays in three groups, with 20,000 pixels per line. One group of four linear arrays is inclined forward at an angle of 25.6° off nadir, another group of five lines is nadir-viewing, and the remaining group of four linear arrays is inclined backward at 17.7° off nadir. Data from these arrays can be used for stereoscopic viewing and for photogrammetric analyses. There are three possible stereoscopic combinations of the forward-, nadir-, and backward-viewing arrays, with angles of 17.7° (using the nadir and rear-viewing arrays), 25.6° (using the nadir and forward-viewing arrays), and 43.3° (using the rear-viewing and forward-viewing arrays). These three combinations provide stereoscopic base-height ratios of 0.3, 0.5, and 0.8, respectively.

Each group of linear arrays includes blue, green, red, and near-infrared bands, while the nadir-viewing group also includes a second line of green-sensitive CCDs

Figure 4.8 The Leica ADS100 along-track scanner system. (Courtesy Leica Geosystems.)

Figure 4.9 Leica ADS40 panchromatic images of the Texas state capitol, Austin, TX. Flight direction is left to right. (*a*) Forward-viewing array, 28.4° off nadir; (*b*) nadir-viewing array; (*c*) backward-viewing array, 14.2° off nadir. (Courtesy Leica Geosystems.)

offset one-half pixel from the first line. This permits a higher-resolution image to be acquired at nadir by treating the two green arrays as if they were a single line of 40,000 CCDs. The visible-wavelength bands have sensitivities in the ranges of 0.435 to 0.495 μm, 0.525 to 0.585 μm, and 0.619 to 0.651 μm, while the near-IR band covers the range from 0.808 to 0.882 μm.

The Leica ADS100 sensor has a wide dynamic range and records data with 12-bit radiometric resolution (4096 gray-level values). All 13 linear arrays share a single lens system with a focal length of 62.5 mm and a total field of view ranging from 65° across-track (for the forward-viewing arrays) to 77° (for the nadir-viewing arrays). At a flying height of 1250 meters above ground level, the spatial resolution at nadir is 10 cm.

Figure 4.9 shows an example of along-track panchromatic imagery acquired over the Texas state capitol in Austin. These images were acquired by the Leica ADS40 sensor (an earlier generation of the Leica ADS series, with slightly different viewing angles and spectral configuration). Data from the forward-viewing, nadir-viewing, and backward-viewing linear arrays show the effects of relief displacement. The magnitude of displacement is greater in the forward-viewing image than in the backward-viewing image because the forward-viewing array is tilted further off nadir. The data shown in Figure 4.9 represent only a small portion of the full image dimensions. For this data acquisition the sensor was operated from a flying height of 1920 m above ground level. The original ground sampling distance for this data set was 20 cm.

4.6 GEOMETRIC CHARACTERISTICS OF ACROSS-TRACK SCANNER IMAGERY

The airborne thermal images illustrated in this discussion were all collected using across-track, or whiskbroom, scanning procedures. Not only are across-track scanners (multispectral and thermal) subject to altitude and attitude variations due to the continuous and dynamic nature of scanning, but also their images contain systematic geometric variations due to the geometry of across-track scanning. We discuss these sources of systematic and random geometric variations under separate headings below, although they occur simultaneously. Also, although we use thermal scanner images to illustrate the various geometric characteristics treated here, it should be emphasized that these characteristics hold for *all* across-track multispectral scanner images, not simply across-track thermal scanner images.

Spatial Resolution and Ground Coverage

Airborne across-track scanning systems are generally operated at altitudes in the range 300 to 12,000 m. Table 4.1 summarizes the spatial resolution and ground coverage that would result at various operating altitudes when using a system

TABLE 4.1 Ground Resolution at Nadir and Swath Width for Various Flying Heights of an Across-Track Scanner Having a 90° Total Field of View and a 2.5-mrad IFOV

Altitude	Flying Height above Ground (m)	Ground Resolution at Nadir (m)	Swath Width (m)
Low	300	0.75	600
Medium	6,000	15	12,000
High	12,000	30	24,000

having a 90° total field of view and a 2.5-mrad IFOV. The ground resolution at nadir is calculated from Eq. 4.1 ($D = H'\beta$). The swath width W can be calculated from

$$W = 2H' \tan \theta \tag{4.2}$$

where

W = swath width
H' = flying height above terrain
θ = one-half total field of view of scanner

Many of the geometric distortions characterizing across-track scanner imagery can be minimized by constraining one's analysis to the near-center portion of the imaging swath. Also, as we will discuss, several of these distortions can be compensated for mathematically. However, their effect is difficult to negate completely. As a consequence, across-track images are rarely used as a tool for precision mapping. Instead, data extracted from the images are normally registered to some base map when positional accuracy is required in the interpretation process.

Tangential-Scale Distortion

Unless it is geometrically rectified, across-track imagery manifests severe scale distortions in a direction perpendicular to the flight direction. The problem arises because a scanner mirror rotating at constant angular velocity does not result in a constant speed of the scanner's IFOV over the terrain. As shown in Figure 4.10, for any increment of time, the mirror sweeps through a constant incremental arc, $\Delta\theta$. Because the mirror rotates at a constant angular velocity, $\Delta\theta$ is the same at any scan angle θ. However, as the distance between the nadir and the ground resolution cell increases, the linear ground velocity of the resolution cell increases. Hence, the ground element, ΔX, covered per unit time increases with increasing distance from the nadir. This results in image scale compression at points away from the nadir as the ground spot covers a greater distance at its increasing ground speed. The resulting distortion is known as *tangential-scale distortion*. Note that it occurs only in the along-scan direction, perpendicular to the direction of flight. Image scale in the direction of flight is essentially constant.

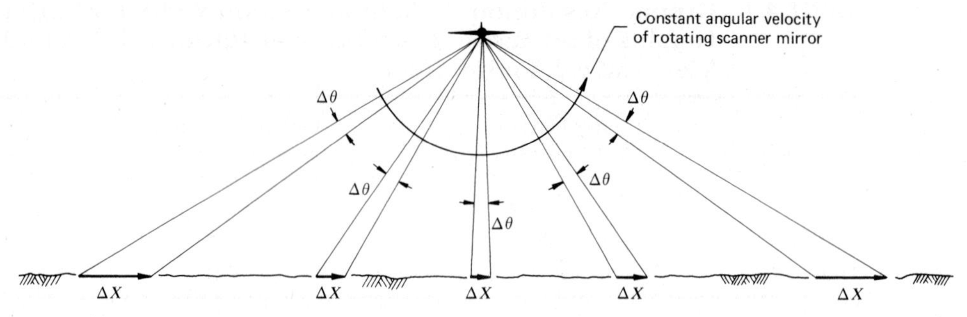

Figure 4.10 Source of tangential-scale distortion.

Figure 4.11 schematically illustrates the effect of tangential distortion. Shown in Figure 4.11*a* is a hypothetical vertical aerial photograph taken over flat terrain containing patterns of various forms. An unrectified across-track scanner image of the same area is shown in Figure 4.11*b*. Note that because of the constant longitudinal scale and varying lateral scale of the scanner imagery, objects do not maintain their proper shapes. Linear features—other than those parallel or normal to the scan lines—take on an S-shaped *sigmoid curvature*. Extreme compression of ground features characterizes the image near its edges. These effects are illustrated in Figure 4.12, which shows an aerial photograph and an across-track thermal scanner image of the same area. The flight line for the thermal image is vertical on the page. Note that the photograph and thermal image are reproduced with the same scale along the flight line but that the scale of the thermal image is compressed in a direction perpendicular to the flight line. Two diagonal roads that are straight on the aerial photograph take on a sigmoid curvature on the thermal image. Note also the light-toned water and trees on this nighttime thermal image (this phenomenon is discussed in Section 4.10).

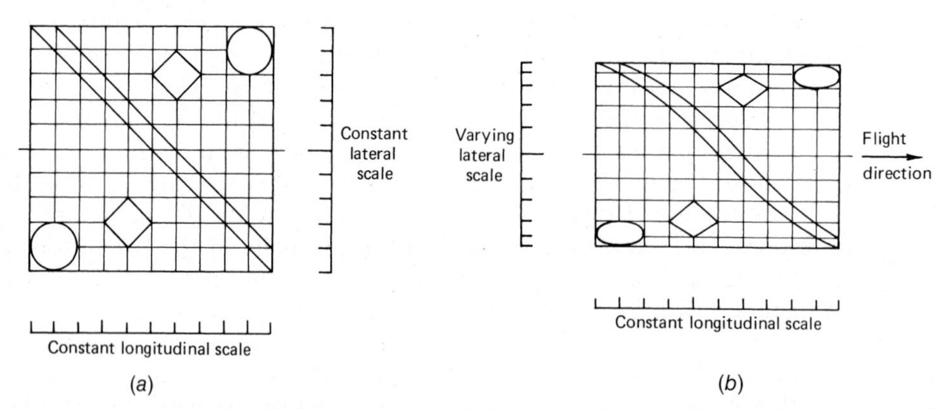

Figure 4.11 Tangential-scale distortion in unrectified across-track scanner imagery: (*a*) vertical aerial photograph; (*b*) across-track scanner imagery.

Figure 4.12 Comparison of aerial photograph and across-track thermal scanner image illustrating tangential distortion, Iowa County, WI: (*a*) panchromatic aerial photograph, 3000-m flying height; (*b*) nonrectified thermal image, 6:00 A.M., 300-m flying height. [(*a*) Courtesy USDA–ASCS. (*b*) Courtesy National Center for Atmospheric Research.]

Figure 4.13 further illustrates tangential-scale distortion. This scanner image shows a group of cylindrical oil storage tanks. The flight line was from left to right. Note how the scale becomes compressed at the top and bottom of the image, distorting the circular shape of the tank tops. Note also that the scanner views the sides as well as the tops of features located away from the flight line.

Figure 4.13 Across-track thermal scanner image illustrating tangential distortion, 100-m flying height. (Courtesy Texas Instruments, Inc.)

Tangential-scale distortion normally precludes useful interpretation near the edges of unrectified across-track scanner imagery. Likewise, geometric measurements made on unrectified scanner imagery must be corrected for this distortion. Figure 4.14 shows the elements involved in computing true ground positions from measurements made on a distorted image. On unrectified imagery, y coordinates will relate directly to *angular* dimensions, not to lineal dimensions. This results in the geometric relationship depicted in the figure, where the effective focal plane of image formation is shown as a curved surface below the aircraft. In order to determine the ground position Y_p corresponding to image point p, we must first compute θ_p from the relationship

$$\frac{y_p}{y_{\max}} = \frac{\theta_p}{\theta_{\max}}$$

Rearranging yields

$$\theta_p = \frac{y_p \theta_{\max}}{y_{\max}} \tag{4.3}$$

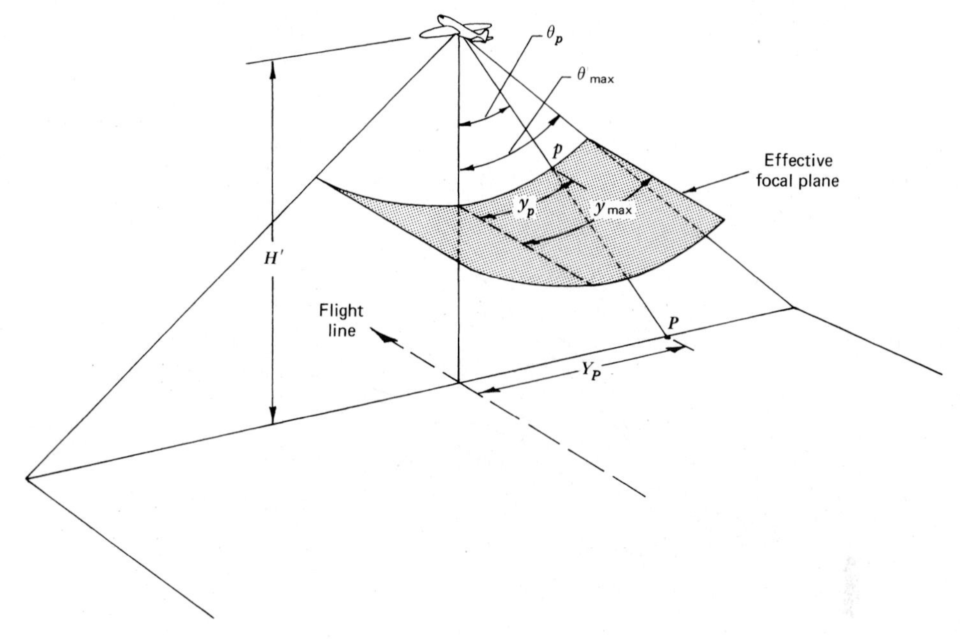

Figure 4.14 Tangential-scale distortion correction.

where

y_p = distance measured on image from nadir line to point p
y_{max} = distance from nadir line to edge of image
θ_{max} = one-half total field of view of scanner

Once θ_p has been computed, it may be trigonometrically related to ground distance Y_p by

$$Y_p = H' \tan \theta_p \tag{4.4}$$

When determining ground positions on unrectified imagery, the above process must be applied to each y coordinate measurement. Alternatively, the correction can be implemented electronically, or digitally, in the image recording process, resulting in *rectilinearized* images. In addition to permitting direct measurement of positions, rectilinearized imagery improves the ability to obtain useful interpretations in areas near the edge of images.

Resolution Cell Size Variations

Across-track scanners sense energy over ground resolution cells of continuously varying size. An increased cell size is obtained as the IFOV of a scanner moves outward from the flight nadir. The geometric elements defining the size of the ground resolution cell are shown in Figure 4.15. At the nadir line, the ground

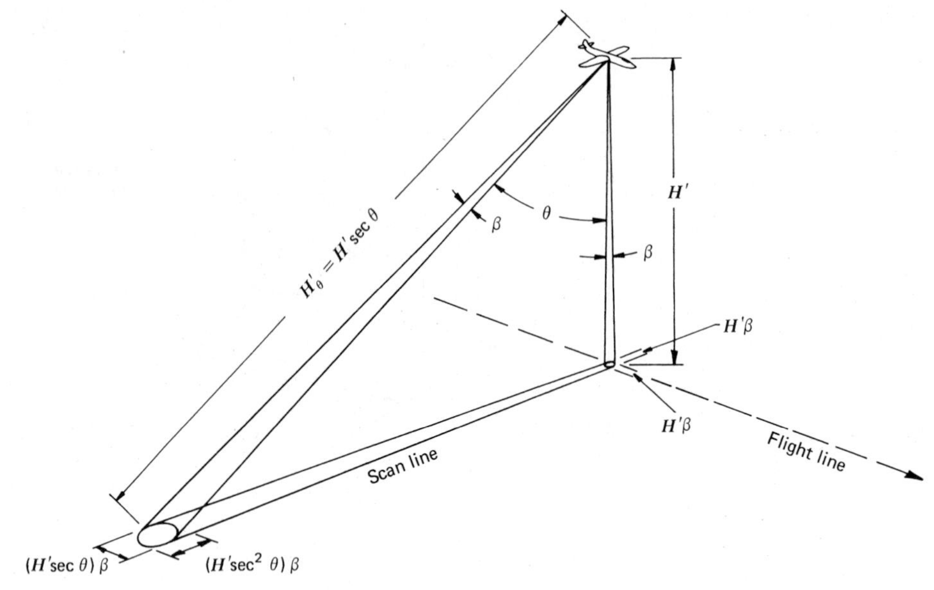

Figure 4.15 Resolution cell size variation.

resolution cell has a dimension of $H'\beta$. At a scan angle θ, the distance from the aircraft to the cell becomes $H'_\theta = H' \sec\theta$. Hence the size of the resolution cell increases. The cell has dimensions of $(H' \sec\theta)\beta$ in the direction of flight and $(H' \sec^2\theta)\beta$ in the direction of scanning. These are actually the *nominal* dimensions of the measurement cell. The *true* size and shape of a ground resolution cell are a function not only of β, H', and θ but also of the *response time* of a particular scanner's electronics. The response time is a measure of the time that a scanner takes to respond electronically to a change in ground reflected or emitted energy. With this added restriction, we see that optics control the resolution cell size in the direction of flight, while both optics and electronics can influence the cell size in the direction of scan.

Although it is rarely critical to know the precise degree of resolution cell size variation, it is important to realize the effect this variation has on the interpretability of the imagery at various scan angles. The scanner output at any point represents the integrated radiance from all features within the ground resolution cell. Because the cell increases in size near the edge of the image, only larger terrain features will completely fill the IFOV in these areas. When objects smaller than the area viewed by the IFOV are imaged, background features also contribute to the recorded signal. This is particularly important in thermal scanning applications where accurate temperature information is required. *For an object to be registered with the proper radiant temperature, its size must be larger than the ground resolution cell.* This effect may again limit the image analysis to the central portion of the image, even after rectilinearization is performed. However, an advantage of the changing size of the ground resolution cell is that it compensates for off-nadir radiometric falloff. If the ground resolution cell area were constant, the irradiance received by the scanner would decrease as $1/H'^2_\theta$. But since the ground resolution cell area increases as H'^2_θ, the irradiance falloff is precisely compensated and a consistent radiance is recorded over uniform surfaces.

One-Dimensional Relief Displacement

Figure 4.16 illustrates the nature of relief displacement characterizing across-track scanner images. Because all objects are viewed by the scanner only along "side-looking" scan lines, relief displacement occurs only in this single direction. (See also Figure 4.13.) An advantage to relief displacement is that it affords an opportunity to see a side view of objects. On the other hand, it can obscure the view of objects of interest. For example, a thermal mission might be planned to detect heat losses in steam lines in an urban setting. Tall buildings proximate to the objects of interest may completely obscure their view. In such cases it is often necessary to cover the study area twice, in perpendicular flight directions.

Figure 4.17 is a thermal image illustrating one-dimensional relief displacement. Note that the displacement of the tops of the tall buildings is greater with increasing distance from the nadir.

Flight Parameter Distortions

Because across-track scanner imagery is collected in a continuous fashion, it lacks the consistent relative orientation of image points found on instantaneously

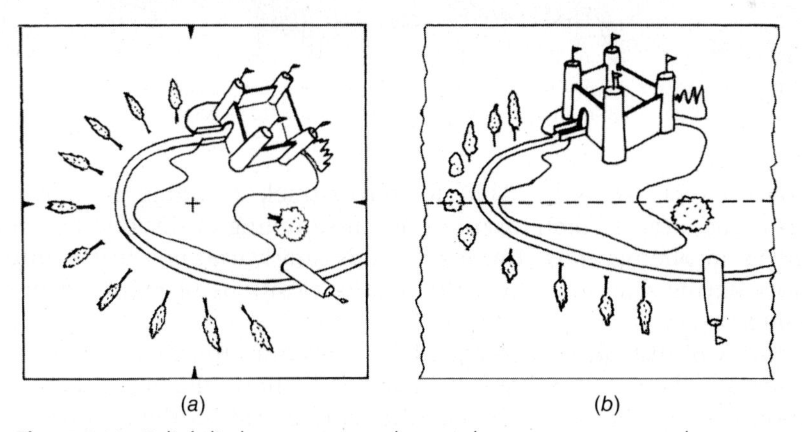

Figure 4.16 Relief displacement on a photograph versus an across-track scanner image. (*a*) In a vertical aerial photograph, vertical features are displaced radially from the principal point. (*b*) In an across-track scanner image, vertical features are displaced at right angles from the nadir line.

Figure 4.17 Across-track thermal scanner image illustrating one-dimensional relief displacement, San Francisco, CA, predawn, 1500-m flying height, 1 mrad IFOV. (Courtesy NASA.)

imaged aerial photographs. That is, across-track scanning is a dynamic continuous process rather than an intermittent sampling of discrete perspective projections as in photography. Because of this, any variations in the aircraft flight trajectory during scanning affect the relative positions of points recorded on the resulting imagery.

A variety of distortions associated with aircraft attitude (angular orientation) deviations are shown in Figure 4.18. The figure shows the effect that each type of distortion would have on the image of a square grid on the ground. This grid is shown in Figure 4.18*a*.

Figure 4.18*b* gives a sketch of an across-track scanner image acquired under constant aircraft altitude and attitude conditions. Only the tangential-scale

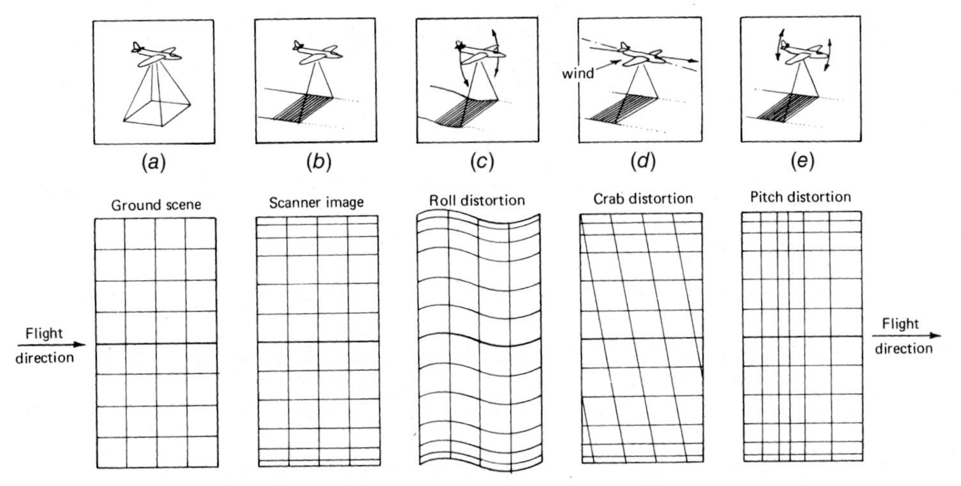

Figure 4.18 Across-track scanner imagery distortions induced by aircraft attitude deviations: (*a*) ground scene, (*b*) scanner image, (*c*) roll distortion, (*d*) crab distortion, (*e*) pitch distortion.

distortion is present in this case. In Figure 4.18*c* the effect of aircraft *roll* about its flight axis is shown. Roll causes the ground grid lines to be imaged at varying times in the mirror rotation cycle. Consequently, the image takes on a wavy appearance. When extreme crosswind is encountered during data acquisition, the axis of the aircraft must be oriented away from the flight axis slightly to counteract the wind. This condition is called *crab* and causes a skewed image (Figure 4.18*d*). Finally, as illustrated in Figure 4.18*e*, variations in aircraft *pitch* might cause local-scale changes in the direction of flight.

The effects of roll, crab, and, pitch are generally mitigated through the integration of a GPS/IMU with the scanner. Data from the GPS/IMU are used both in flight management and sensor control during data acquisition and in geometrically correcting the sensor data during post-processing.

4.7 GEOMETRIC CHARACTERISTICS OF ALONG-TRACK SCANNER IMAGERY

The geometric characteristics of along-track scanner images are quite different from those of across-track scanner images. Along-track scanners have no scanning mirror, and there is a fixed geometric relationship among the solid-state detector elements recording each scan line. As shown in Figure 4.19, the uniformly spaced detectors in the sensor array view uniformly spaced ground resolution elements. Unlike across-track scanners, along-track scanners are not subject to tangential scale distortion, so a vertical image acquired over flat terrain would have uniform scale in the cross-track direction.

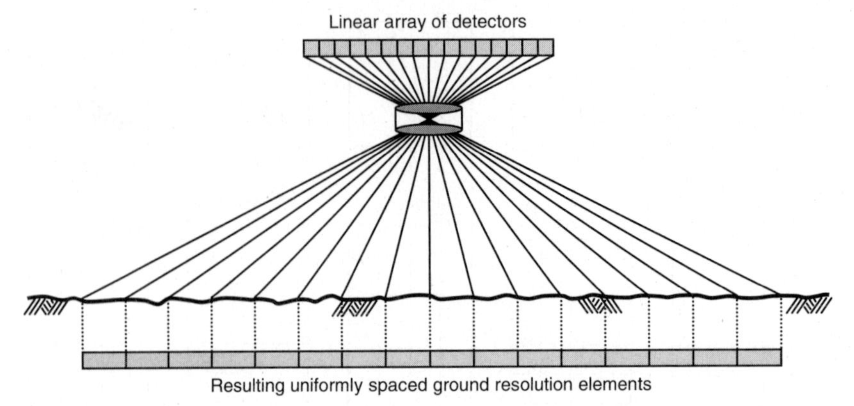

Figure 4.19 Geometry of one line of an along-track scanner image.

Like across-track scanners, along-track scanners manifest one-dimensional relief displacement, with objects above the terrain being displaced perpendicularly away from the nadir line. Figure 4.20 illustrates this effect. Objects along the nadir line are not displaced, while the degree of displacement increases toward the edge of each scan line. If the tops of the buildings in Figure 4.20 are all at the same elevation, the building rooftops will all have uniform scale (because of the absence of tangential distortion), but those farther from the nadir line will be displaced further (Figure 4.20*b*).

In essence, the geometry along each line of an along-track scanner image is similar to that of an aerial photograph. Line-to-line variation in imaging geometry is caused purely by any changes in the altitude or attitude (angular orientation) of the aircraft along a flight line. Often, on-board inertial measurement units and GPS systems are used to measure these variations and geometrically correct the data from along-track scanners.

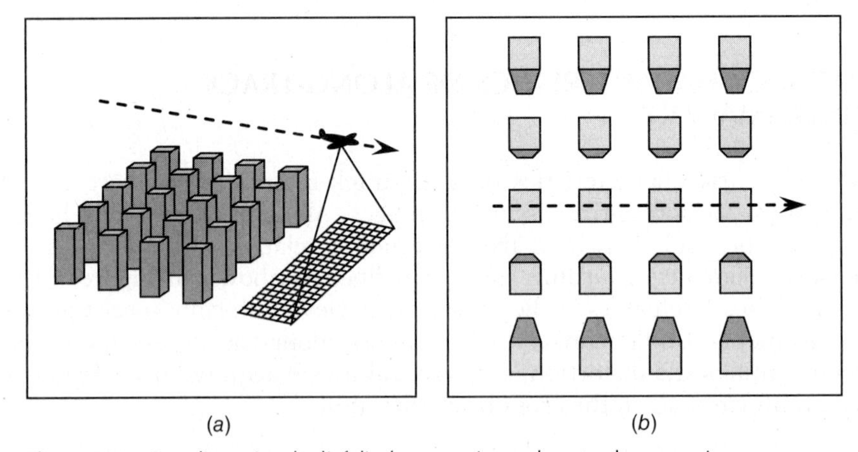

Figure 4.20 One-dimensional relief displacement in an along-track scanner image.

4.8 THERMAL IMAGING

As mentioned earlier, a *thermal imager* is merely a particular kind of multispectral scanner, namely, one whose detectors only sense in the thermal portion of the spectrum. As mentioned in Chapter 1 and discussed in more detail in the next section of this chapter, energy at these wavelengths is naturally emitted by objects as a function of their temperature. Thus, unlike most of the images discussed in previous sections of this book, thermal images are not dependent on reflected solar radiation, so these systems may be operated at any time of the day or night. Because our eyes are not sensitive to thermal infrared radiation, there is no "natural" way to represent a thermal image visually. In most cases, these images are represented in grayscale, with the brightness or darkness of each pixel determined by the intensity of thermal radiation measured over the corresponding ground resolution cell. The usual convention when looking at the earth's surface is to have higher radiant temperature areas displayed as lighter-toned image areas. For meteorological purposes, this convention is typically reversed so that clouds (cooler than the earth's surface) appear light toned. Occasionally, thermal images are displayed in other visual formats, such as using a rainbow of hues from blue (cool) to red or white (hot) to represent temperatures.

Due to atmospheric effects, thermal imaging systems are restricted to operating in either (or both) the 3- to 5-μm or 8- to 14-μm range of wavelengths (Section 1.3). *Quantum* or *photon* detectors are typically used for this purpose. These detectors are capable of very rapid (less than 1-μsec) response. They operate on the principle of direct interaction between photons of radiation incident on them and the energy levels of electrical charge carriers within the detector material. For maximum sensitivity, the detectors must be cooled to very low temperatures, to minimize their own thermal emissions. Often, the detector is surrounded by a *dewar* containing liquid nitrogen at 77 K. A dewar is a double-walled insulated vessel that acts like a thermos bottle to prevent the liquid coolant from boiling away at a rapid rate.

The spectral sensitivity ranges of three widely used types of photon are included in Table 4.2.

Thermal imagers became commercially available during the late 1960s. Early systems used an across-track scanning configuration, with typically only one or a few detectors; the data were detected electronically but recorded on film. Present-day thermal imaging systems may utilize along-track (linear array) sensors with over 1000 detectors, providing a high radiometric resolution and direct digital recording of the image data.

TABLE 4.2 Characteristics of Photon Detectors in Common Use

Type	Abbreviation	Useful Spectral Range (μm)
Mercury-doped germanium	Ge:Hg	3–14
Indium antimonide	InSb	3–5
Mercury cadmium telluride	HgCdTe (MCT), or "trimetal"	3–14

Figure 4.21 The TABI-1800 thermal imaging system. (Courtesy ITRES, Inc.)

Figure 4.21 shows the TABI-1800 sensor, an along-track airborne thermal imaging system developed by ITRES, Inc. This system has an array of 1800 mercury cadmium telluride (MCT) detectors with a radiometric resolution of 0.05°C. It measures thermal radiation in a single broad band covering the range of 3.7 to 4.8 μm. The system's total field of view is 40°, and its instantaneous field of view is 0.405 milliradians. When operated from altitudes of 250 to 3000 m above ground level, this yields imagery with a spatial resolution of approximately 10 cm to 1.25 m.

Figure 4.22 shows a TABI-1800 image of a gas-fired power station and former natural gas processing plant located near Balzac, Alberta, Canada. The image was acquired at 11:30 P.M. local time, long after sunset; the full-resolution version of this image has a spatial resolution of 15 cm. The brightness of each feature in this image is a function of the quantity of thermal radiation emitted, and as discussed in the next section of this chapter, this in turn is determined by both the temperature and the material composition of the feature. In this image, lighter tones are generally indicative of higher radiant temperatures. Thus, for example, the plume of hot gases being emitted by the smokestack at upper right appears bright white in the image, while another smokestack (at lower left) was not in use at the time and appears much darker. (Note also the contrary directions of relief displacement at these two smokestacks, indicating that the sensor's nadir line fell between them.) At several locations in the image, buried steam pipelines are visible as light-toned but less distinct linear features, such as the prominent L-shaped line near the center of the image. Although these pipelines are underground, they warm the surface enough to measurably increase the brightness of the thermal imagery. Finally, it should be noted that many of the very dark features in this image (including some metal rooftops and aboveground pipelines) are not necessarily particularly cold; instead, they appear dark due to the inherently low emissivity of bare metal, as discussed in the following section.

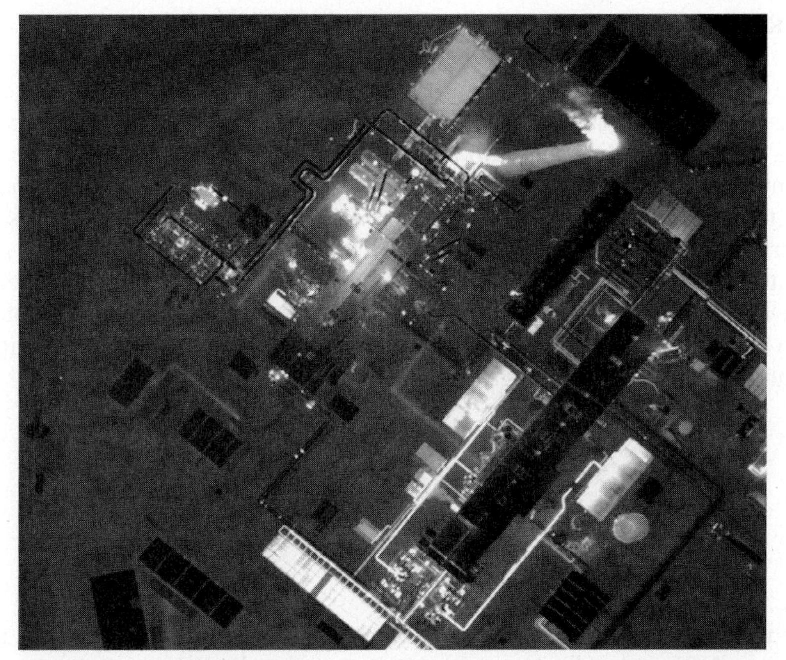

Figure 4.22 Thermal imagery from the TABI-1800 system, acquired over a natural gas processing facility and power plant near Balzac, Alberta, Canada. Light tones indicate higher radiant temperatures. (Courtesy ITRES, Inc.)

4.9 THERMAL RADIATION PRINCIPLES

A basic understanding of the nature of thermal radiation is necessary for the proper interpretation of a thermal image. In this section we review and extend some of the principles of blackbody radiation introduced in Section 1.2. We also treat how thermal radiation interacts with the atmosphere and various earth surface features.

Radiant Versus Kinetic Temperature

One normally thinks of temperature measurements as involving some measuring instrument being placed in contact with, or being immersed in, the body whose temperature is to be measured. When this is done, *kinetic temperature* is measured. Kinetic temperature is an "internal" manifestation of the average translational energy of the molecules constituting a body. In addition to this internal manifestation, objects radiate energy as a function of their temperature. This emitted energy is an "external" manifestation of an object's energy state. It is this external manifestation of an object's energy state that is remotely sensed using thermal scanning. The emitted energy is used to determine the *radiant temperature* of earth surface features. Later we describe how to relate kinetic and radiant temperature.

Blackbody Radiation

We have previously described the physics of electromagnetic radiation in accordance with the concepts of blackbody radiation (see Section 1.2). Recall that any object having a temperature greater than absolute zero (0 K, or −273°C) emits radiation whose intensity and spectral composition are a function of the material type involved and the temperature of the object under consideration. Figure 4.23 shows the spectral distribution of the energy radiated from the surface of a blackbody at various temperatures. All such blackbody curves have similar form, and their energy peaks shift toward shorter wavelengths with increases in temperature. As indicated earlier (Section 1.2), the relationship between the wavelength of peak spectral exitance and temperature for a blackbody is given by *Wien's displacement law*:

$$\lambda_m = \frac{A}{T} \tag{4.5}$$

where

λ_m = wavelength of maximum spectral radiant exitance, μm
A = 2898 μm K
T = temperature, K

The total radiant exitance coming from the surface of a blackbody at any given temperature is given by the area under its spectral radiant exitance curve. That is, if

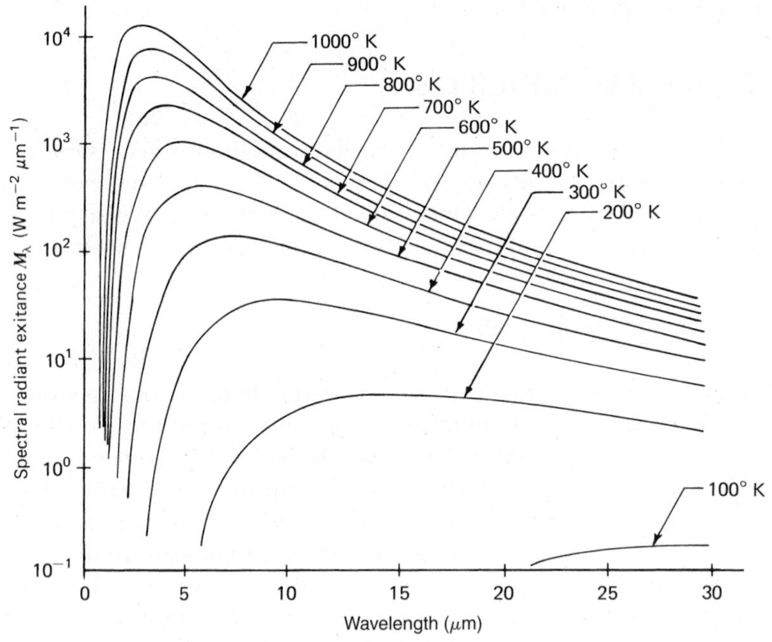

Figure 4.23 Spectral distribution of energy radiated from blackbodies of various temperatures.

a sensor were able to measure the radiant exitance from a blackbody at all wavelengths, the signal recorded would be proportional to the area under the blackbody radiation curve for the given temperature. This area is described mathematically by the *Stefan–Boltzmann law*, given in Eq. 1.4, and in expanded form here as:

$$M = \int_0^\infty M(\lambda)d\lambda = \sigma T^4 \tag{4.6}$$

where

M = total radiant exitance, $\mathrm{W\,m^{-2}}$
$M(\lambda)$ = spectral radiant exitance, $\mathrm{W\,m^{-2}\,\mu m^{-1}}$
σ = Stefan–Boltzmann constant, $5.6697 \times 10^{-8}\ \mathrm{W\,m^{-2}\,K^{-4}}$
T = temperature of blackbody, K

Equation 4.6 indicates that the total radiant exitance from the surface of a blackbody varies as the fourth power of absolute temperature. The remote measurement of radiant exitance M from a surface can therefore be used to infer the temperature T of the surface. In essence, it is this indirect approach to temperature measurement that is used in thermal sensing. Radiant exitance M is measured over a discrete wavelength range and used to find the radiant temperature of the radiating surface.

Radiation from Real Materials

While the concept of a blackbody is a convenient theoretical vehicle to describe radiation principles, real materials do not behave as blackbodies. Instead, all real materials emit only a fraction of the energy emitted from a blackbody at the equivalent temperature. The "emitting ability" of a real material, compared to that of a blackbody, is referred to as a material's *emissivity* ε.

Emissivity ε is a factor that describes how efficiently an object radiates energy compared to a blackbody. By definition,

$$\varepsilon(\lambda) = \frac{\text{radiant exitance of an object at a given temperature}}{\text{radiant exitance of a blackbody at the same temperature}} \tag{4.7}$$

Note that ε can have values between 0 and 1. As with reflectance, emissivity can vary with wavelength and viewing angle. Depending on the material, emissivity can also vary somewhat with temperature.

A *graybody* has an emissivity that is less than 1 but is constant at all wavelengths. At any given wavelength the radiant exitance from a graybody is a constant fraction of that of a blackbody. If the emissivity of an object varies with wavelength, the object is said to be a *selective radiator*. Figure 4.24 illustrates the comparative emissivities and spectral radiant exitances for a blackbody, a graybody (having an emissivity of 0.5), and a selective radiator.

Many materials radiate like blackbodies over certain wavelength intervals. For example, as shown in Figure 4.25*a*, water is very close (ε is 0.98 to 0.99) to

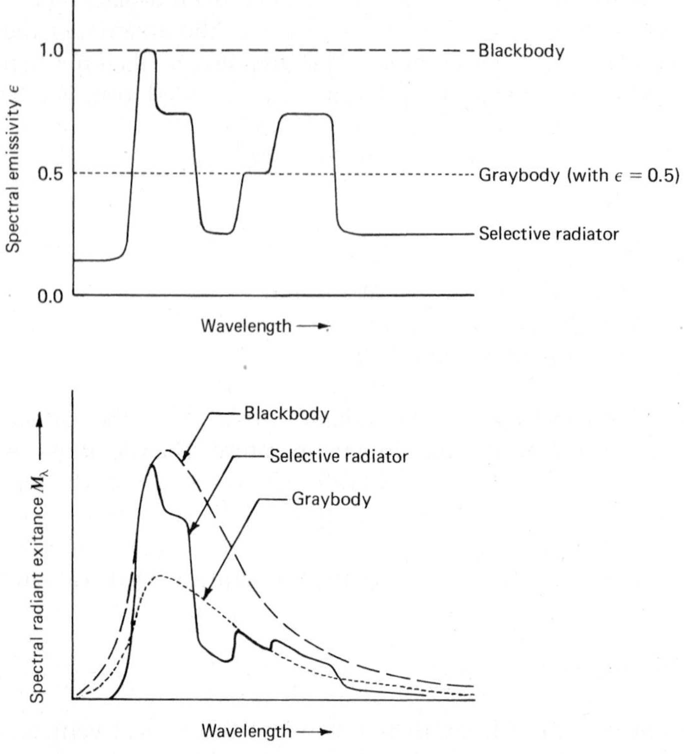

Figure 4.24 Spectral emissivities and radiant exitances for a blackbody, a graybody, and a selective radiator. (Adapted from Hudson, 1969.)

behaving as a blackbody radiator in the 6- to 14-μm range. Other materials, such as quartz, act as a selective radiator, with considerable variation in emissivity at different wavelengths in the 6- to 14-μm range (Figure 4.25*b*).

The 8- to 14-μm region of spectral radiant exitance is of particular interest because it not only includes an atmospheric window but also contains the peak energy emissions for most surface features. That is, the ambient temperature of earth surface features is normally about 300 K, at which temperature the peak emissions will occur at approximately 9.7 μm. For these reasons, most thermal sensing is performed in the 8- to 14-μm region of the spectrum. The emissivities of different objects vary greatly with material type in this range. However, for any given material type, emissivity is often considered constant in the 8- to 14-μm range when broadband sensors are being used. This means that within this spectral region materials are often treated as graybodies. However, close examination of emissivity versus wavelength for materials in this wavelength range shows that values can vary considerably with wavelength. Therefore, the within-band emissivities of materials sensed in the 10.5- to 11.5-μm range (National Oceanic and Atmospheric Administration [NOAA] AVHRR band 4) would not necessarily

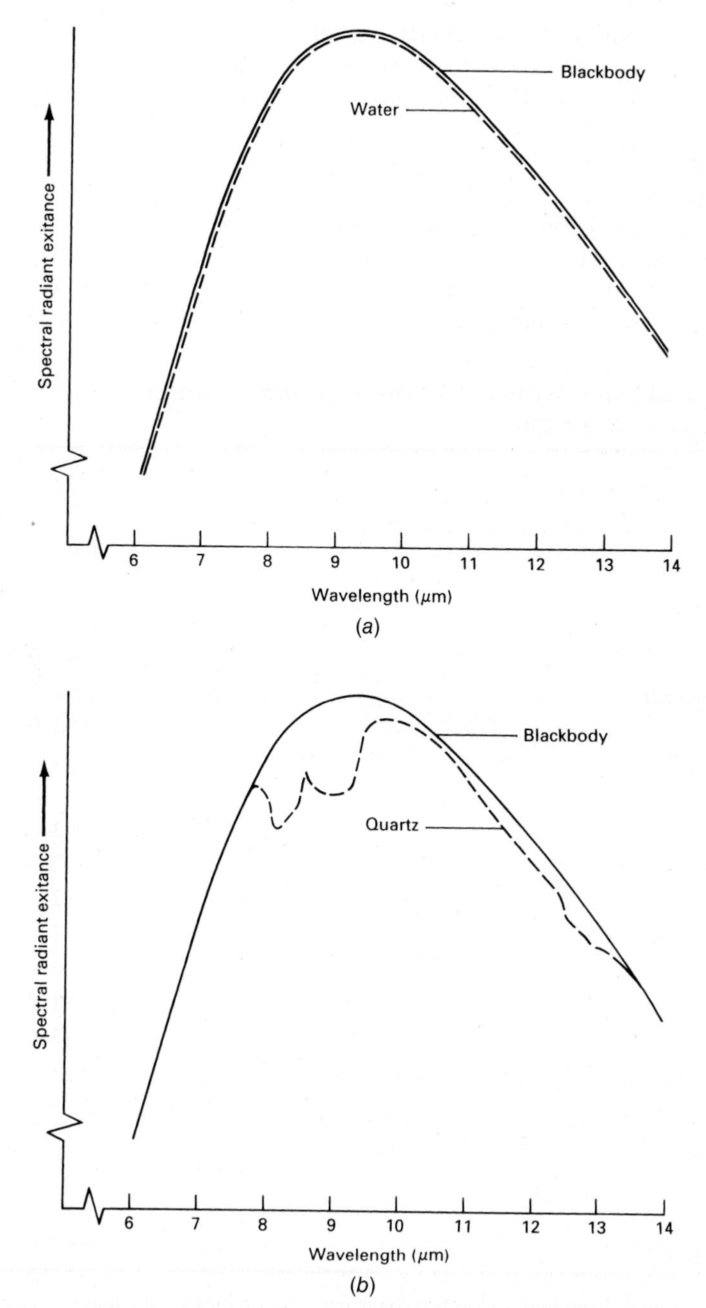

Figure 4.25 Comparison of spectral radiant exitances for (*a*) water versus a blackbody and (*b*) quartz versus a blackbody. (For additional curves, see Salisbury and D'Aria, 1992.)

be the same as the within-band emissivities of materials sensed in the 10.4- to 12.5-μm range (Landsat TM band 6). Furthermore, emissivities of materials vary with material condition. A soil that has an emissivity of 0.92 when dry could have an emissivity of 0.95 when wet (water-coated soil particles have an emissivity approaching that of water). Objects such as deciduous tree leaves can have a different emissivity when a single leaf is sensed (0.96) than when an entire tree crown is sensed (0.98). Table 4.3 indicates some typical values of emissivity over the 8- to 14-μm-wavelength range for various common materials.

It should be noted that as objects are heated above ambient temperature, their emissive radiation peaks shift to shorter wavelengths. In special-purpose

TABLE 4.3 Typical Emissivities of Various Common Materials over the Range of 8–14 μm

Material	Typical Average Emissivity ε over 8–14 μm[a]
Clear water	0.98–0.99
Wet snow	0.98–0.99
Human skin	0.97–0.99
Rough ice	0.97–0.98
Healthy green vegetation	0.96–0.99
Wet soil	0.95–0.98
Asphaltic concrete	0.94–0.97
Brick	0.93–0.94
Wood	0.93–0.94
Basaltic rock	0.92–0.96
Dry mineral soil	0.92–0.94
Portland cement concrete	0.92–0.94
Paint	0.90–0.96
Dry vegetation	0.88–0.94
Dry snow	0.85–0.90
Granitic rock	0.83–0.87
Glass	0.77–0.81
Sheet iron (rusted)	0.63–0.70
Polished metals	0.16–0.21
Aluminum foil	0.03–0.07
Highly polished gold	0.02–0.03

[a]Emissivity values (ordered from high to low) are typical average values for the materials listed over the range of 8–14 μm. Emissivities associated with only a portion of the 8- to 14-μm range can vary significantly from these values. Furthermore, emissivities for the materials listed can vary significantly, depending on the condition and arrangement of the materials (e.g., loose soil vs. compacted soil, individual tree leaves vs. tree crowns).

applications, such as forest fire mapping, systems operating in the 3- to 5-μm atmospheric window may be used. These systems offer improved definition of hot objects at the expense of the surrounding terrain at ambient temperature.

Atmospheric Effects

As is the case with all passive remote sensing systems, the atmosphere has a significant effect on the intensity and spectral composition of the energy recorded by a thermal system. As mentioned, atmospheric windows (Figure 4.26) influence the selection of the optimum spectral bands within which to measure thermal energy signals. Within a given window, the atmosphere intervening between a thermal sensor and the ground can increase or decrease the apparent level of radiation coming from the ground. The effect that the atmosphere has on a ground signal will depend on the degree of atmospheric absorption, scatter, and emission at the time and place of sensing.

Gases and suspended particles in the atmosphere may absorb radiation emitted from ground objects, resulting in a decrease in the energy reaching a thermal sensor. Ground signals can also be attenuated by scattering in the presence of suspended particles. On the other hand, gases and suspended particles in the atmosphere may emit radiation of their own, adding to the radiation sensed. Hence, atmospheric absorption and scattering tend to make the signals from ground objects appear colder than they are, and atmospheric emission tends to make ground objects appear warmer than they are. Depending on atmospheric conditions during imaging, one of these effects will outweigh the other. This will result in a

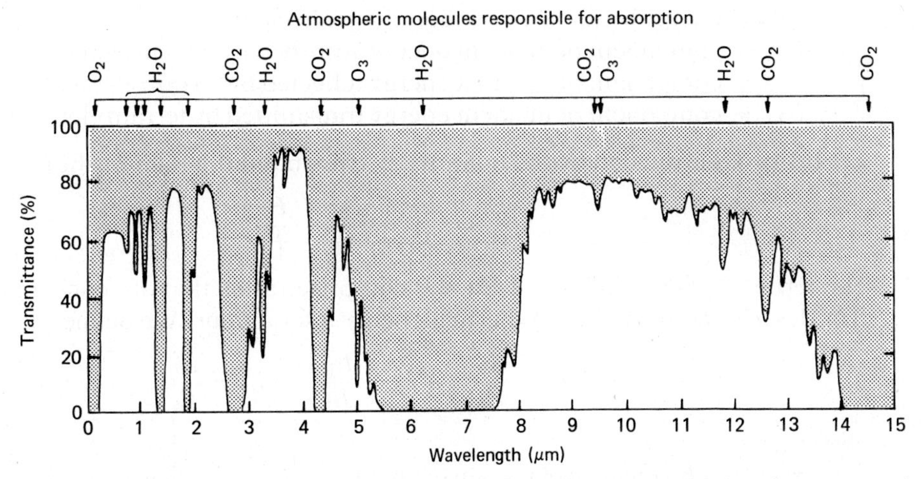

Figure 4.26 Atmospheric absorption of the wavelength range 0 to 15 μm. Note the presence of atmospheric windows in the thermal wavelength regions 3 to 5 μm and 8 to 14 μm. (Adapted from Hudson, 1969.)

biased sensor output. Both effects are directly related to the atmospheric path length, or distance, through which the radiation is sensed.

Thermal sensor measurements of temperature can be biased by as much as 2°C or more when acquired at altitudes as low as 300 m. Of course, meteorological conditions have a strong influence on the form and magnitude of the thermal atmospheric effects. Fog and clouds are essentially opaque to thermal radiation. Even on a clear day, aerosols can cause major modifications of signals sensed. Dust, carbon particles, smoke, and water droplets can all modify thermal measurements. These atmospheric constituents vary with site, altitude, time, and local weather conditions.

Atmospheric effects on radiant temperature measurements usually may not be ignored. The various strategies commonly employed to compensate for atmospheric effects are described later in this chapter. We now consider how thermal radiation interacts with ground objects.

Interaction of Thermal Radiation with Terrain Elements

In thermal sensing we are interested in the radiation emitted from terrain features. However, the energy radiated *from* an object usually is the result of energy incident *on* the feature. In Section 1.4 we introduced the basic notion that energy incident on the surface of a terrain element can be absorbed, reflected, or transmitted. In accordance with the principle of conservation of energy, we can state the relationship between incident energy and its disposition upon interaction with a terrain element as

$$E_{\mathrm{I}} = E_{\mathrm{A}} + E_{\mathrm{R}} + E_{\mathrm{T}} \tag{4.8}$$

where

E_{I} = energy incident on surface of terrain element
E_{A} = component of incident energy absorbed by terrain element
E_{R} = component of incident energy reflected by terrain element
E_{T} = component of incident energy transmitted by terrain element

If Eq. 4.8 is divided by the quantity E_{I}, we obtain the relationship

$$\frac{E_{\mathrm{I}}}{E_{\mathrm{I}}} = \frac{E_{\mathrm{A}}}{E_{\mathrm{I}}} + \frac{E_{\mathrm{R}}}{E_{\mathrm{I}}} + \frac{E_{\mathrm{T}}}{E_{\mathrm{I}}} \tag{4.9}$$

The terms on the right side of Eq. 4.9 comprise ratios that are convenient in further describing the nature of thermal energy interactions. We define

$$\alpha(\lambda) = \frac{E_{\mathrm{A}}}{E_{\mathrm{I}}} \quad \rho(\lambda) = \frac{E_{\mathrm{R}}}{E_{\mathrm{I}}} \quad \tau(\lambda) = \frac{E_{\mathrm{T}}}{E_{\mathrm{I}}} \tag{4.10}$$

where

$\alpha(\lambda)$ = *absorptance* of terrain element
$\rho(\lambda)$ = *reflectance* of terrain element
$\tau(\lambda)$ = *transmittance* of terrain element

We can now restate Eq. 4.8 in the form

$$\alpha(\lambda) + \rho(\lambda) + \tau(\lambda) = 1 \tag{4.11}$$

which defines the interrelationship among a terrain element's absorbing, reflecting, and transmitting properties.

Another ingredient necessary is the *Kirchhoff radiation law*. It states that the spectral emissivity of an object equals its spectral absorptance:

$$\varepsilon(\lambda) = \alpha(\lambda) \tag{4.12}$$

Paraphrased, "good absorbers are good emitters." While Kirchhoff's law is based on conditions of thermal equilibrium, the relationship holds true for most sensing conditions. Hence, if we apply it in Eq. 4.11, we may replace $\alpha(\lambda)$ with $\varepsilon(\lambda)$, resulting in

$$\varepsilon(\lambda) + \rho(\lambda) + \tau(\lambda) = 1 \tag{4.13}$$

Finally, in most remote sensing applications the objects we deal with are assumed to be opaque to thermal radiation. That is, $\tau(\lambda) = 0$, and it is therefore dropped from Eq. 4.13 such that

$$\varepsilon(\lambda) + \rho(\lambda) = 1 \tag{4.14}$$

Equation 4.14 demonstrates the direct relationship between an object's emissivity and its reflectance in the thermal region of the spectrum. The lower an object's reflectance, the higher its emissivity; the higher an object's reflectance, the lower its emissivity. For example, water has nearly negligible reflectance in the thermal spectrum. Therefore, its emissivity is essentially 1. In contrast, a material such as sheet metal is highly reflective of thermal energy, so it has an emissivity much less than 1.

The emissivity of an object has an important implication when measuring radiant temperatures. Recall that the Stefan–Boltzmann law, as stated in Eq. 4.6 $(M = \sigma T^4)$, applied to blackbody radiators. We can extend the blackbody radiation principles to real materials by reducing the radiant exitance M by the emissivity factor ε such that

$$M = \varepsilon\sigma T^4 \tag{4.15}$$

Equation 4.15 describes the interrelationship between the measured signal a thermal sensor "sees," M, and the parameters of temperature and emissivity. Note that because of emissivity differences, earth surface features can have the same temperature and yet have completely different radiant exitances.

The output from a thermal sensor is a measurement of the radiant temperature of an object, T_{rad}. Often, the user is interested in relating the radiant temperature of an object to its kinetic temperature, T_{kin}. If a sensor were to view a blackbody, T_{rad} would equal T_{kin}. For all real objects, however, we must account for the emissivity factor. Hence, the kinetic temperature of an object is related to its radiant temperature by

$$T_{\text{rad}} = \varepsilon^{1/4} T_{\text{kin}} \tag{4.16}$$

TABLE 4.4 Kinetic versus Radiant Temperature for Four Typical Material Types

		Kinetic Temperature T_{kin}		Radiant Temperature $T_{rad} = \varepsilon^{1/4} T_{kin}$	
Object	Emissivity ε	K	°C	K	°C
Blackbody	1.00	300	27	300.0	27.0
Vegetation	0.98	300	27	298.5	25.5
Wet soil	0.95	300	27	296.2	23.2
Dry soil	0.92	300	27	293.8	20.8

Equation 4.16 expresses the fact that for any given object the radiant temperature recorded by a remote sensor will always be less than the kinetic temperature of the object. This effect is illustrated in Table 4.4, which shows the kinetic versus the radiant temperatures for four objects having the same kinetic temperature but different emissivities. Note how kinetic temperatures are always underestimated if emissivity effects are not accounted for in analyzing thermal sensing data.

A final point to be made here is that *thermal sensors detect radiation from the surface (approximately the first 50 μm) of ground objects*. This radiation may or may not be indicative of the internal bulk temperature of an object. For example, on a day of low humidity, a water body having a high temperature will manifest evaporative cooling effects at its surface. Although the bulk temperature of the water body could be substantially warmer than that of its surface temperature, a thermal sensor would record only the surface temperature.

4.10 INTERPRETING THERMAL IMAGERY

Successful interpretations of thermal imagery have been made in many fields of application. These include such diverse tasks as determining rock type and structure, locating geologic faults, mapping soil type and soil moisture, locating irrigation canal leaks, determining the thermal characteristics of volcanoes, studying evapotranspiration from vegetation, locating cold-water springs, locating hot springs and geysers, determining the extent and characteristics of thermal plumes in lakes and rivers, studying the natural circulation patterns in water bodies, determining the extent of active forest fires, and locating subsurface fires in landfills or coal refuse piles.

Most thermal imaging operations, such as geologic and soil mapping, are qualitative in nature. In these cases, it is not usually necessary to know absolute ground temperatures and emissivities, but simply to study relative differences in the radiant temperatures within a scene. However, some thermal imaging operations require quantitative data analysis in order to determine absolute temperatures. An example would be the use of thermal remote sensing as an enforcement tool by a state department of natural resources to monitor surface water temperatures of the effluent from a nuclear power plant.

Various times of day can be utilized in thermal imaging studies. Many factors influence the selection of an optimum time or times for acquiring thermal data. Mission planning and image interpretation must take into consideration the effects of diurnal temperature variation. The importance of diurnal effects is shown in Figure 4.27, which illustrates the relative radiant temperatures of soils and rocks versus water during a typical 24-hr period. Note that just before dawn a quasi-equilibrium condition is reached where the slopes of the temperature curves for these materials are very small. After dawn, this equilibrium is upset and the materials warm up to a peak that is reached sometime after noon. Maximum scene contrast normally occurs at about this time and cooling takes place thereafter.

Temperature extremes and heating and cooling rates can often furnish significant information about the type and condition of an object. Note, for example, the temperature curve for water. It is distinctive for two reasons. First, its range of temperature is quite small compared to that of soils and rocks. Second, it reaches its maximum temperature an hour or two after the other materials. As a result, terrain temperatures are normally higher than water temperatures during the day and lower during the night. Shortly after dawn and near sunset, the curves for water and the other features intersect. These points are called thermal

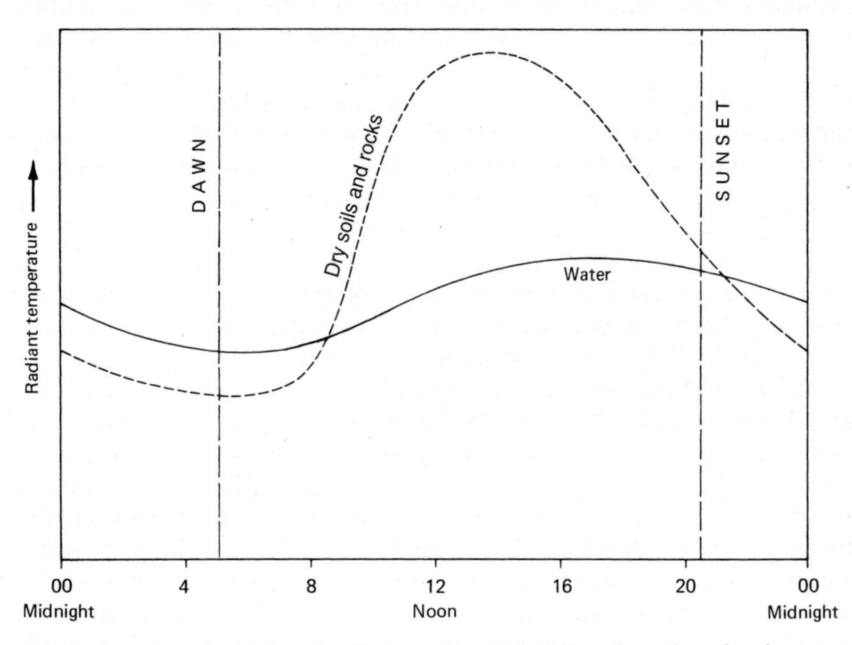

Figure 4.27 Generalized diurnal radiant temperature variations for soils and rocks versus water.

crossovers and indicate times at which no radiant temperature difference exists between two materials.

The extremes and rates of temperature variation of any earth surface material are determined, among other things, by the material's thermal conductivity, capacity, and inertia. *Thermal conductivity* is a measure of the rate at which heat passes through a material. For example, heat passes through metals much faster than through rocks. *Thermal capacity* determines how well a material stores heat. Water has a very high thermal capacity compared to other material types. *Thermal inertia* is a measure of the response of a material to temperature changes. It increases with an increase in material conductivity, capacity, and density. In general, materials with high thermal inertia have more uniform surface temperatures throughout the day and night than materials of low thermal inertia.

During the daytime, direct sunlight differentially heats objects according to their thermal characteristics and their sunlight absorption, principally in the visible and near-IR portion of the spectrum. Reflected sunlight can be significant in imagery utilizing the 3- to 5-μm band. Although reflected sunlight has virtually no direct effect on imagery utilizing the 8- to 14-μm band, daytime imagery contains thermal "shadows" in cool areas shaded from direct sunlight by objects such as trees, buildings, and some topographic features. Also, slopes receive differential heating according to their orientation. In the Northern Hemisphere, south-facing slopes receive more solar heating than north-facing slopes. Many geologists prefer "predawn" imagery for their work as this time of day provides the longest period of reasonably stable temperatures, and "shadow" effects and slope orientation effects are minimized. However, aircraft navigation over areas selected for thermal image acquisition is more difficult during periods of darkness, when ground features cannot be readily seen by the pilot. Other logistics also enter into the timing of thermal scanning missions. For example, scanning of effluents from power plant operations normally must be conducted during periods of peak power generation.

A number of thermal images are illustrated in the remainder of this section. In all cases, darker image tones represent cooler radiant temperatures and lighter image tones represent warmer radiant temperatures. This is the representation most commonly used in thermal images of earth surface features. As mentioned in Section 4.8, for meteorological applications the reverse representation is used to preserve the light-toned appearance of clouds.

Figure 4.28 illustrates the contrast between daytime (*a*) and nighttime (*b*) thermal images. The water in this scene (note the large lake at right and the small, lobed pond in lower center) appears cooler (darker) than its surroundings during the daytime and warmer (lighter) at night. The kinetic water temperature has changed little during the few hours of elapsed time between these images. However, the surrounding land areas have cooled considerably during the evening hours. Again, water normally appears cooler than its surroundings on daytime thermal images and warmer on nighttime thermal images, except for the case of open water surrounded by frozen or snow-covered ground where the water would appear warmer day and night. Trees can be seen in many places in

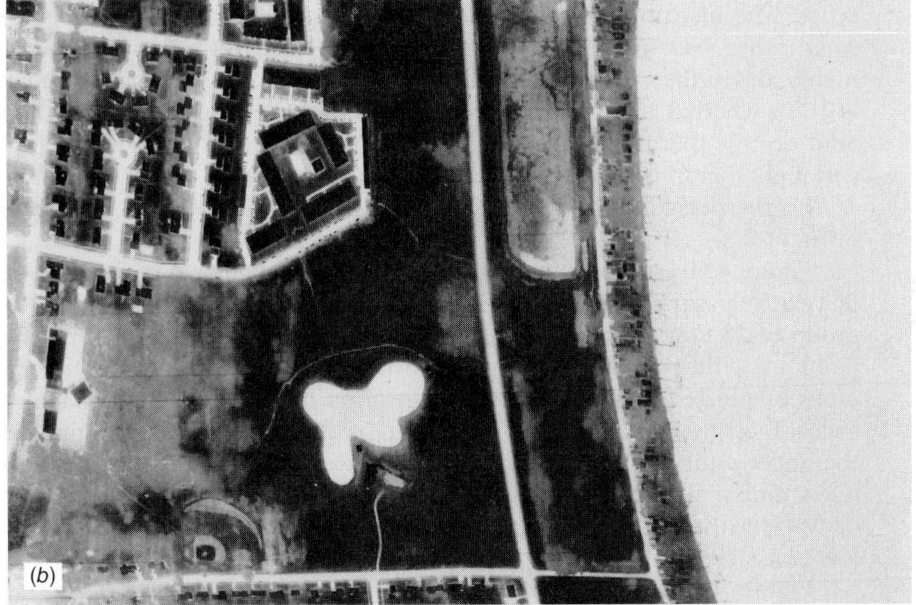

Figure 4.28 Daytime and nighttime thermal images, Middleton, WI: (*a*) 2:40 P.M., (*b*) 9:50 P.M.; 600 m flying height, 5 mrad IFOV. (Courtesy National Center for Atmospheric Research.)

these images (note the area above and to the right of the small pond). Trees generally appear cooler than their surroundings during the daytime and warmer at night. Tree shadows appear in many places in the daytime image (note the residential area at upper left) but are not noticeable in the nighttime image. Paved areas (streets and parking lots) appear relatively warm both day and night. The pavement surfaces heat up to temperatures higher than their surroundings during the daytime and lose heat relatively slowly at night, thus retaining a temperature higher than their surroundings.

Figure 4.29 is a daytime thermal image showing the former shoreline of glacial Lake Middleton, an ephemeral glacial lake that is now primarily agricultural fields. This was a small lake, about 800 ha in extent at its maximum. At its lowest level, the lake was only about 80 ha in size. The beach ridge associated with this lowest lake level is shown at B. The ridge is most evident at the lower right because the prevailing winds at the time of its formation were from the upper left. The ridge is a small feature, only 60 m wide and 0.5 to 1 m higher than the surrounding lakebed material. The beach ridge has a fine sandy loam surface soil 0.3 to 0.45 m thick underlain by deep sandy materials. The lakebed soils (A) are silt loam to a depth of at least 1.5 m and are seasonally wet with a groundwater table within 0.6 m of the ground surface in the early spring. At the time of this thermal image, most of the area shown here was covered with agricultural crops. The instrument sensed the radiant temperature of the vegetation over the soils rather than the bare soils themselves. Based on field radiometric measurements, the radiant temperature of the vegetation on the dry, sandy beach ridge soil is 16°C, whereas that over the wetter, siltier lakebed soil is 13°C. Although prominent on this thermal image, the beach ridge is often overlooked on panchromatic aerial photographs and is only partially mapped on a soil map of the area. Also seen on this thermal image are trees at C, bare soil at D, and closely mowed grass (a sod farm) at E.

Figure 4.30 contains two nighttime thermal images illustrating the detectability of relatively small features on large-scale imagery. In Figure 4.30*a* (9:50 P.M.), a group of 28 cows can be seen as white spots near the upper left. In Figure 4.30*b* (1:45 A.M.), they have moved near the bottom of the image. (Deer located in flat areas relatively free of obstructing vegetation have also been detected, with mixed success, on thermal imagery.) The large, rectangular, very dark-toned object near the upper right of Figure 4.30*a* is a storage building with a sheet metal roof having a low emissivity. Although the kinetic temperature of the roof may be as warm as, or warmer than, the surrounding ground, its radiant temperature is low due to its low emissivity.

Figure 4.31 illustrates how "shadows" in a daytime thermal image may be misleading. This high-resolution image was acquired using a 0.25-mrad IFOV, yielding a ground resolution cell size of approximately 0.3 m. Several helicopters are visible, parked near their hangers. The "thermal shadows" cast by these helicopters are not true shadows, because they appear dark due to their lower temperature rather than due to an obstruction of the source of illumination—the

Figure 4.29 Daytime thermal image. Middleton, WI, 9:40 A.M., 600 m flying height, 5 mrad IFOV. (Courtesy National Center for Atmospheric Research.)

Figure 4.30 Nighttime thermal images, Middleton, WI: (*a*) 9:50 P.M., (*b*) 1:45 A.M.; 600 m flying height, 5 mrad IFOV. (Courtesy National Center for Atmospheric Research.)

latter is not generally a factor in thermal remote sensing, because the objects and terrain in the image themselves are the source of illumination. Instead, the "thermal shadows" appear dark because the pavement beneath the helicopters is shaded from the sun and thus has a lower kinetic temperature. There may be a substantial time lag in the formation and dissipation of these thermal shadows. Note that in Figure 4.31 there are two such shadows left by helicopters that are not in their original parked positions. Two helicopters can also be seen with their blades running while sitting on the helipad in the lower right; these do not have shadows because the pavement beneath them has not been shaded long enough for its kinetic temperature to drop noticeably.

Figure 4.32 illustrates the heated cooling water from a coal-fired power plant discharging into Lake Michigan. This daytime thermal image shows that the plant's cooling system is recirculating its heated discharge water. Heated water initially flows to the right. Winds from the upper right at 5 m/sec cause the plume to double back on itself, eventually flowing into the intake channel. The ambient lake temperature is about 4°C. The surface water temperatures in the plume are 11°C near the submerged discharge and 6°C in the intake channel. On a late winter day with cold lake temperatures, such as shown here, the recirculating plume does not cause problems with the power plant operation. However, such an event

Figure 4.31 Daytime thermal image, Quantico, VA, 0.25 mrad IFOV. (Courtesy Goodrich ISR Systems.)

could cause operational problems during the summer because the intake water would be warmer than acceptable for the cooling system.

The use of aerial thermal scanning to study heat loss from buildings has been investigated in many cities. Figure 4.33 illustrates the typical form of imagery acquired in such studies. Note the striking differences among the radiant temperatures of various roofs as well as the temperature differences between various house roofs and garage/carport roofs of the same house. Such images are often

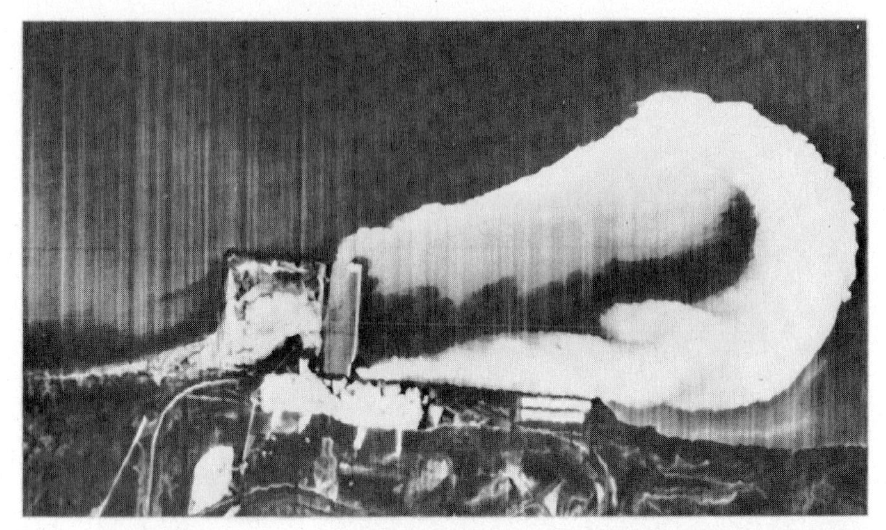

Figure 4.32 Daytime thermal image, Oak Creek Power Plant, WI, 1:50 P.M., 800 m flying height, 2.5 mrad IFOV. (Courtesy Wisconsin Department of Natural Resources/NASA.)

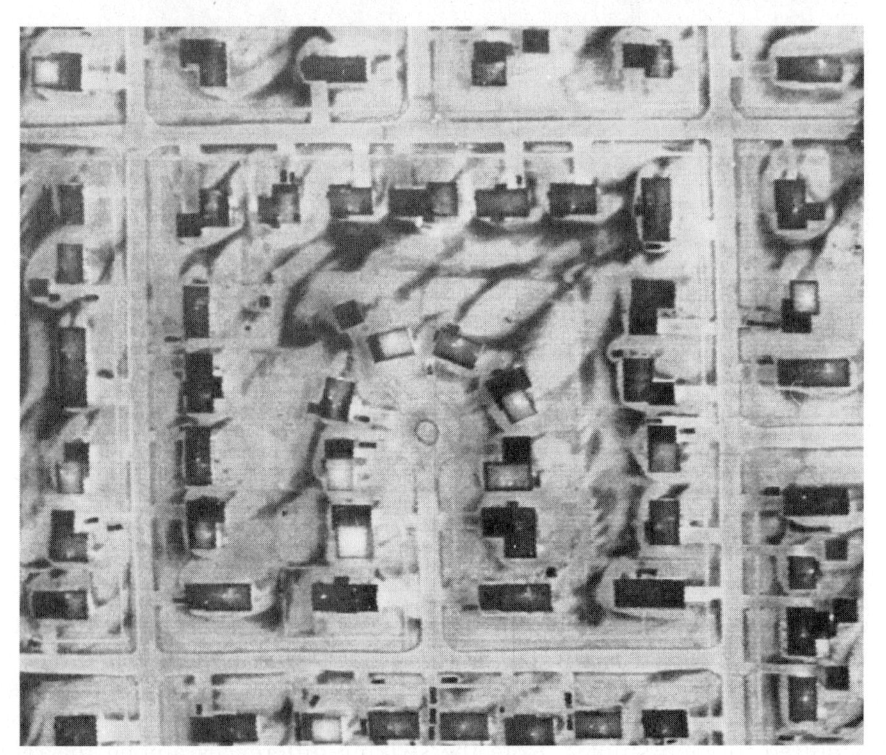

Figure 4.33 Nighttime thermal image depicting building heat loss in an Iowa City, approximately 2:00 A.M., snow-covered ground, air temperature approximately −4°C, 460 m flying height, 1 mrad IFOV. (Courtesy Iowa Utility Association.)

useful in assessing inadequate or damaged insulation and roofing materials. The dark-toned streaks aligned from upper right to lower left on the ground areas between houses result from the effects of wind on the snow-covered ground.

Although aerial thermal scanning can be used to estimate the amount of energy radiated from the roofs of buildings, *the emissivity of the roof surfaces must be known to determine the kinetic temperature of the roof surfaces*. With the exception of unpainted sheet metal roofs (Figure 4.30), which have a very low emissivity, roof emissivities vary from 0.88 to 0.94. (Painted metal surfaces assume the emissivity characteristics of the particular paint used.)

Thermal scanning of roofs to estimate heat loss is best accomplished on cold winter nights at least 6 to 8 hr after sunset, in order to minimize the effects of solar heating. Alternatively, surveys can be conducted on cold overcast winter days. In any case, roofs should be neither snow covered nor wet. Because of the side-looking characteristics of scanner images, the scanner vertically views the rooftops of buildings only directly beneath the plane. At the edge of the scan, it views the rooftops plus a portion of building sides. Roof pitch affects the temperature of roofs. A flat roof directly faces a night sky that is 20 to 30°C cooler than ambient air temperature and will therefore lose heat by radiation. Sloping roofs often receive radiation from surrounding buildings and trees, keeping their surfaces warmer than flat roofs. Attic ventilation characteristics must also be taken into account when analyzing roof heat loss.

When mapping heat loss by aerial thermal scanning, it must also be realized that heat loss from roofs constitutes only a portion of the heat lost from buildings, because heat is also lost through walls, doors, windows, and foundations. It is estimated that a house with comparable insulation levels in all areas loses about 10 to 15% of its heat through the roof. Homes with well-insulated walls and poorly insulated ceilings may lose more heat through the roof.

An alternative and supplement to aerial thermal scanning is the use of ground-based thermal scanning systems. Applications of these imaging systems range from detecting faulty electrical equipment (such as overheated transformers) to monitoring industrial processes to medical diagnosis. Depending on the detector used, the systems operate in the 3- to 5-μm or 8- to 14-μm wavelength bands. When operating in the range 3 to 5 μm, care must be taken not to include reflected sunlight in the scene.

Figure 4.34 shows a thermal image of the Zaca Fire near Santa Barbara, California. This fire burned nearly 1000 km^2 of wildlands during the summer of 2007. The image was acquired by the Autonomous Modular Sensor, a multispectral scanner operating in the visible, near-IR, mid-IR, and thermal spectral regions. The AMS is operated from NASA's Ikhana UAV (illustrated in Figure 1.24*a*). As discussed in Chapter 1, the use of UAVs as platforms for airborne remote sensing is becoming more common. The image shown in Figure 4.34 is from the AMS band 12 (which spans from 10.26 to 11.26 μm in this operating mode). The dark areas at upper right and extreme lower left represent unburned forest, which is kept relatively cool due to the shade from trees and the cooling effect of

Figure 4.34 Thermal IR image of the Zaca Fire near Santa Barbara, California, acquired by the Autonomous Modular Sensor (AMS) on NASA's Ikhana UAV. Compare to color images in Plate 9. (Courtesy NASA Airborne Science Program.)

evapotranspiration. Lighter-toned areas in the center of the image have been burned over, exposing the bare soil. The brightest patches in the lower right quadrant of the image represent the actively burning fire. Elsewhere in the image, recently burned steep hillsides facing toward the sun also appear relatively bright as their unvegetated slopes are heated by the sun.

Comparison of Figure 4.34 to Plate 9 shows the ability of long-wavelength thermal IR radiation to pass through smoke and haze. Plate 9(*a*) shows a true-color composite of visible bands on the Ikhana AMS (from the same image mosaic shown in Figure 4.34). Note that much of the area is largely or wholly obscured by smoke in this visible wavelength color composite image. Plate 9(*b*) shows a false-color composite with the thermal band in red and two mid IR bands in green and blue. In this false-color composite, the actively burning, recently burned, and unburned areas can be readily differentiated.

4.11 RADIOMETRIC CALIBRATION OF THERMAL IMAGES AND TEMPERATURE MAPPING

As mentioned previously, there are many applications of thermal remote sensing in which only *relative* differences in the brightness of various surface features need to be compared. In these cases, there is no need to quantify surface temperatures in the imagery. But in other cases it is important to be able to extract quantitative temperature measurements from the data. This requires some calibration process, in which the electronic signals measured by the sensor's detectors are converted to units of temperature. There are numerous approaches to calibration of thermal sensors, each with its own degree of accuracy and efficiency. What form of calibration is used in any given circumstance is a function of not only the equipment available for data acquisition and processing but also the requirements of the application at hand. We limit our discussion here to a general description of the two most commonly used calibration methods:

1. Internal blackbody source referencing
2. Air-to-ground correlation

As will become apparent in the following discussion, a major distinction between the two methods is that the first does not account for atmospheric effects but the second does. Other calibration methods are described by Schott (2007).

Internal Blackbody Source Referencing

Current generations of thermal imaging systems normally incorporate internal temperature references. These take the form of two blackbody radiation sources positioned so that they are viewed by the sensor during flight. The temperatures of these sources can be precisely controlled and are generally set at the "cold" and "hot" extremes of the ground scene to be monitored. By comparing the thermal signal recorded from the earth's surface to the radiation from these two precisely controlled blackbody calibration sources, a mathematical relationship can be established to calibrate the sensor.

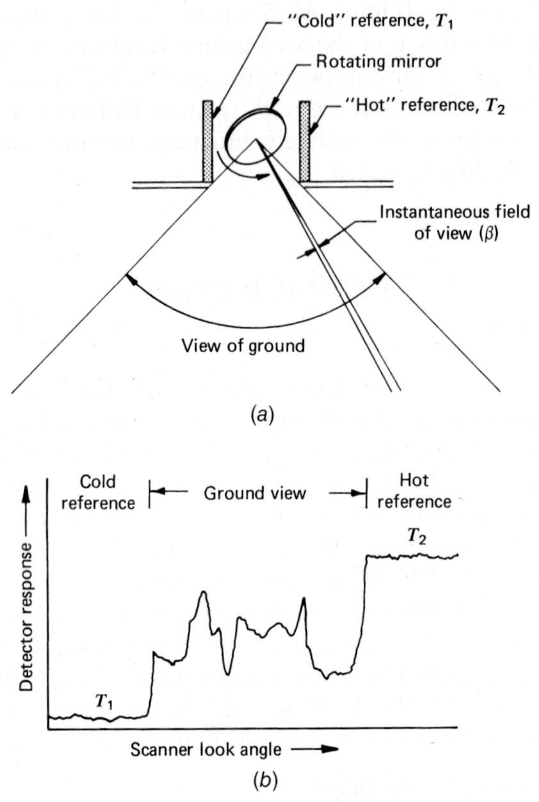

Figure 4.35 Internal blackbody source calibration in an across-track thermal scanner: (*a*) reference plate arrangement; (*b*) typical detector output for one scan line.

Figure 4.35 illustrates the configuration of an internally calibrated across-track thermal scanner, such as the ATLAS system discussed in Section 4.4. The arrangement of the reference sources, or "plates," relative to the field of view of the scanner is shown in Figure 4.35*a*. The detector signal typical of one scan line is illustrated in Figure 4.35*b*. Along each scan line, the scanner optics first view the cold-temperature reference plate (T_1), then sweep across the ground, and finally look at the hot reference plate (T_2). The scanner output at the two temperature plates is recorded along with the image data, and this process is repeated for each scan line. This provides a continuously updated reference by which the scanner output values can be related to absolute radiant temperature.

Along-track thermal imaging systems, such as the TABI-1800 instrument discussed in Section 4.8, also employ internal blackbody sources for radiometric calibration, although the designs differ from the across-track example shown in Figure 4.35. In either case, the internal source referencing procedure permits acceptable calibration accuracy for many applications. Actual versus predicted temperature discrepancies of less than 0.3°C are typical for missions flown at altitudes up to 600 m under clear, dry weather conditions. However, internal calibration still does not account for atmospheric effects. As indicated earlier (Section 4.9), under many prevailing mission conditions the atmosphere can bias scanner temperature measurements by as much as 2°C at altitudes as low as 300 m.

Air-to-Ground Correlation

Atmospheric effects can be accounted for in thermal image calibration by using empirical or theoretical atmospheric models. Theoretical atmospheric models use observations of various environmental parameters (such as temperature, pressure, and CO_2 concentration) in mathematical relationships that predict the effect the atmosphere will have on the signal sensed. Because of the complexity of measuring and modeling the factors that influence atmospheric effects, these effects are normally eliminated by correlating remotely sensed image data with actual surface measurements on an empirical basis.

Air-to-ground correlation is frequently employed in calibration of thermal water quality data, such as those acquired over heated water discharges. Surface temperature measurements are taken on the ground simultaneously with the passage of the aircraft. Thermometers, thermistors, or thermal radiometers operated from boats are commonly used for this purpose. Observations are typically made at points where temperatures are assumed to be constant over large areas. The corresponding scanner output value is then determined for each point of ground-based surface temperature measurement. A calibration curve is then constructed relating the scanned output values to the corresponding ground-based radiant temperature (Figure 4.36). Once the calibration relationship is defined (typically using statistical regression analysis procedures), it is used to estimate the temperature at all points on the scanner images where no ground data exist.

Figure 4.37 illustrates the type of thermal radiometer that can be used for air-to-ground correlation measurements. This particular instrument is an Everest Interscience handheld "infrared thermometer" that operates in the 8- to 14-μm range and displays radiant temperatures on a liquid-crystal display (LCD) panel. It reads temperatures in the range from -40 to $+100$°C, with a radiometric resolution of 0.1°C and an accuracy of 0.5°C. The instrument was designed as a plant stress monitor, and its LCD panel also displays dry-bulb air temperature and differential temperature (the difference between radiant temperature and air temperature).

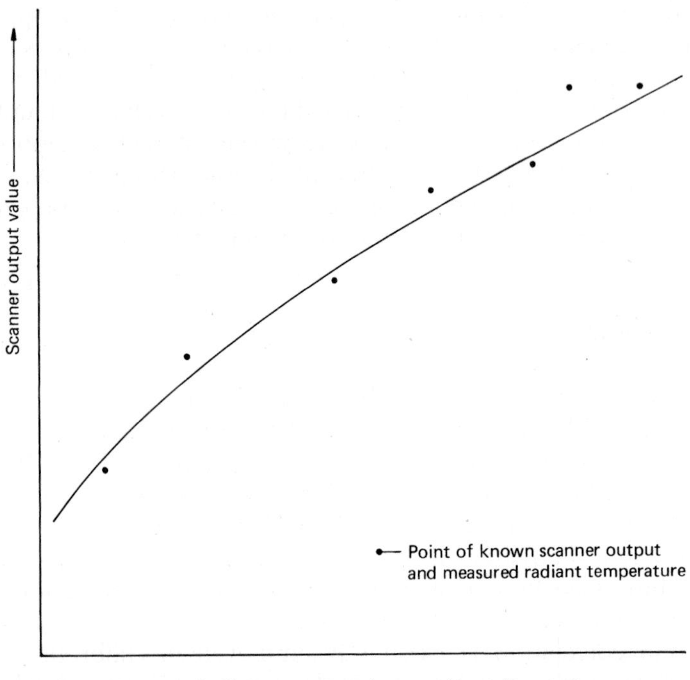

Figure 4.36 Sample calibration curve used to correlate scanner output with radiant temperature measured by radiometer.

Figure 4.37 Infrared thermometer (thermal radiometer). Spot radiant temperatures are read from LCD display panel. (Courtesy Everest Interscience, Inc.)

Temperature Mapping with Thermal Images

In many applications of thermal remote sensing techniques, it is of interest to prepare "maps" of surface temperature distributions. The digital data recorded by a thermal imager can be processed, analyzed, and displayed in a variety of ways. For example, consider image data for which a correlation has been developed to relate sensor output values to absolute ground temperatures. This calibration relationship can be applied to each point in the image data set, producing a matrix of absolute temperature values.

The precise form of a calibration relationship will vary with the temperature range in question, but for the sake of the example, we assume that a linear fit of the digital data to radiant exitance is appropriate. Under this assumption, a digital number, DN, recorded by a sensor, can be expressed by

$$\mathrm{DN} = A + B\varepsilon T^4 \tag{4.17}$$

where

A, B = system response parameters determined from sensor calibration procedures described earlier
ε = emissivity at point of measurement
T = kinetic temperature at point of measurement

Once A and B are determined, kinetic temperature T for any observed digital number DN is given by

$$T = \left(\frac{\mathrm{DN} - A}{B\varepsilon}\right)^{1/4} \tag{4.18}$$

The parameters A and B can be obtained from internal blackbody calibration, air-to-ground correlation, or any of the other calibration procedures. At a minimum, two corresponding temperature (T) and digital number (DN) values are needed to solve for the two unknowns A and B. Once parameters A and B are known, Eq. 4.18 can be used to determine the kinetic temperature for any ground point for which DN is observed *and* the emissivity is known. The calibrated data may be further processed and displayed in a number of different forms (e.g., isotherm maps, color-coded images, or GIS layers).

4.12 FLIR SYSTEMS

To this point in our discussion of aerial thermography, we have emphasized across-track and along-track imaging systems that view the terrain directly beneath an aircraft. *Forward-looking infrared (FLIR)* systems can be used to acquire oblique views of the terrain ahead of an aircraft. Some FLIR systems use cooled detectors similar to those discussed in Section 4.8, while others employ uncooled detectors, each of which functions as a *microbolometer*. This latter type of detector is thermally isolated from its surroundings and when exposed to thermal radiation its electrical resistance changes rapidly. In general, systems using cooled detectors are heavier and more expensive, but provide better radiometric resolution and faster imaging. Uncooled, microbolometer-based infrared imagers are generally lightweight and less expensive but do not provide as high precision in their temperature measurements. They also typically do not include the type of onboard blackbody-based calibration discussed in Section 4.11. Instead, these lightweight systems are typically brought to a laboratory for calibration using external sources on an as-needed basis (e.g., once per year).

Figure 4.38 shows two images acquired by FLIR systems. Figure 4.38*a* was acquired during the daytime over a storage tank facility. Conceptually, the FLIR system used to produce this image operates on the same basic principles as an across-track line scanning system. However, the mirror for the system points forward and optically sweeps the field of view of a linear array of thermal detectors across the scene of interest. Figure 4.38*b* was acquired using a helicopter-borne FLIR system. Thermal shadows can be seen next to the parked vehicles at the left side of the image, while the moving vehicle at the center of the image does not have a thermal shadow associated with it. Note also the heat emitted from the hood of those cars that have been operated most recently and the image of a person near the right side of the image.

Modern FLIR systems are extremely portable (typically weighing less than 30 kg) and can be operated on a wide variety of fixed-wing aircraft and helicopters as well as from ship- and ground-based platforms. Forward-looking imagery has been used extensively in military applications. Civilian use of FLIR is increasing in applications such as search and rescue operations, pollution control, firefighting, electrical transmission line maintenance, law enforcement activities, and nighttime vision systems for vehicles.

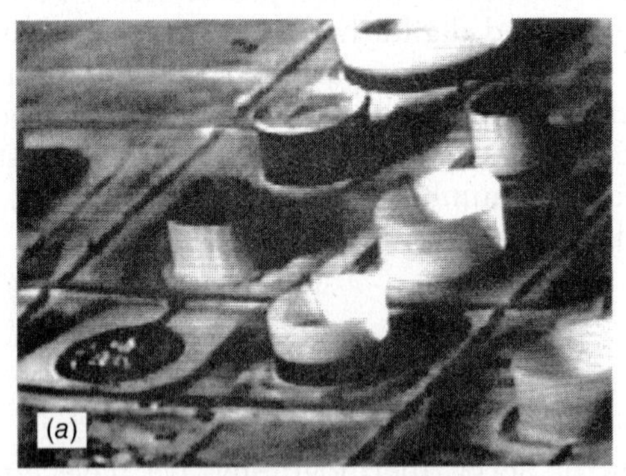

Figure 4.38 FLIR images. (*a*) Storage tank facility. Note level of liquid in each tank. (Courtesy of Raytheon Company, Inc.) (*b*) City street. Note thermal shadows adjacent to parked vehicles and the image of a person on the right side of the street. (Courtesy FLIR Systems, Inc.)

4.13 HYPERSPECTRAL SENSING

Hyperspectral sensors (sometimes referred to as *imaging spectrometers*) are instruments that acquire images in many, very narrow, contiguous spectral bands throughout the visible, near-IR, mid-IR, and thermal-IR portions of the spectrum. (They can employ across-track or along-track scanning or two-dimensional framing arrays.) These systems typically collect 100 or more bands of data, which enables the construction of an effectively continuous reflectance (or emittance in the case of thermal IR energy) spectrum for every pixel in the scene (Figure 4.39). These systems can discriminate among earth surface features that have diagnostic absorption and reflection characteristics over narrow wavelength intervals that are "lost" within the relatively coarse bandwidths of the various bands of conventional multispectral scanners. This concept is illustrated in Figure 4.40, which shows the laboratory-measured reflectance spectra for a number of common minerals over the wavelength range 2.0 to 2.5 μm. Note the diagnostic absorption features for these various material types over this spectral range. Also shown in this figure is the bandwidth of band 7 of the Landsat TM, ETM+, and OLI (Chapter 5). Whereas these sensors obtain only one data point corresponding to the integrated response over a spectral band 0.27 μm wide, a hyperspectral sensor is capable of obtaining many data points over this range using bands on the order of 0.01 μm wide. Thus, hyperspectral sensors can produce data of sufficient spectral resolution for direct identification of the materials, whereas the broader band TM cannot resolve these diagnostic spectral differences. Hence, while a

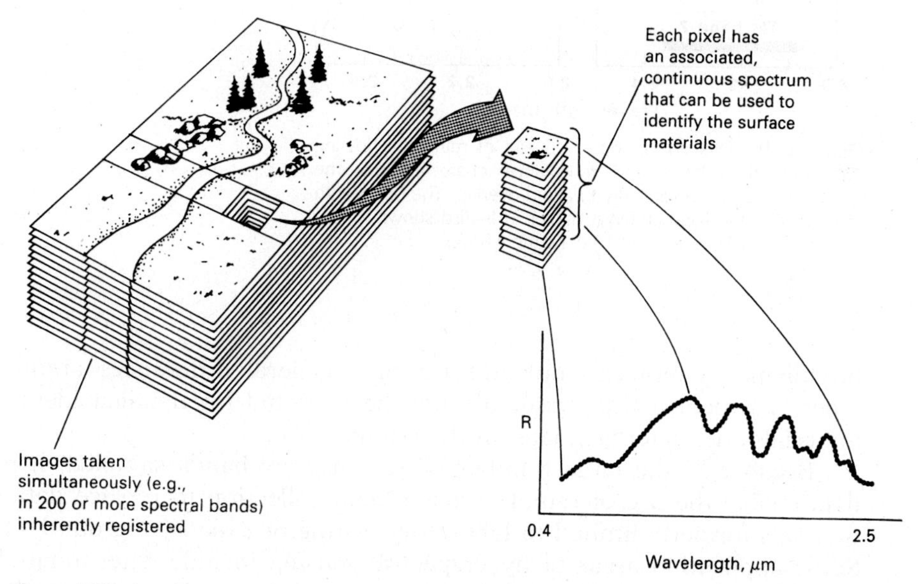

Figure 4.39 Imaging spectrometry concept. (Adapted from Vane, 1985.)

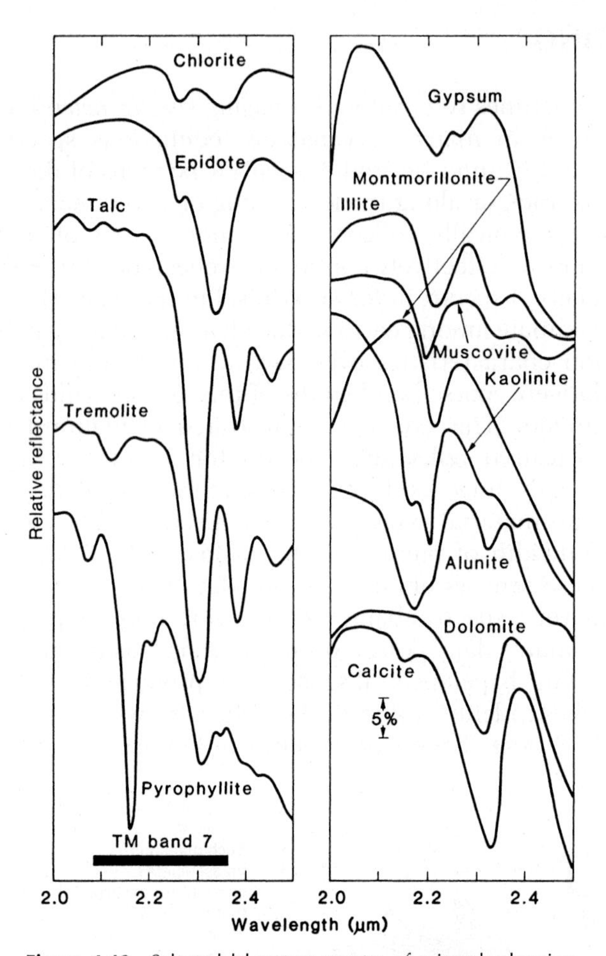

Figure 4.40 Selected laboratory spectra of minerals showing diagnostic absorptance and reflectance characteristics. The spectra are displaced vertically to avoid overlap. The bandwidth of band 7 of the Landsat TM (Chapter 5) is also shown. (From Goetz et al., 1985. Copyright 1985 by the AAAS.)

broadband system can only discriminate general differences among material types, a hyperspectral sensor affords the potential for detailed identification of materials and better estimates of their abundance.

Because of the large number of very narrow bands sampled, hyperspectral data enable the use of remote sensing data collection to replace data collection that was formerly limited to laboratory testing or expensive ground site surveys. Some application areas of hyperspectral sensing include determinations of surface mineralogy; water quality; bathymetry; soil type and erosion; vegetation type, plant stress, leaf water content, and canopy chemistry; crop type and condition; and snow and ice properties.

Extensive initial hyperspectral sensing research was conducted with data acquired by the *Airborne Imaging Spectrometer* (AIS). This system collected 128 bands of data, approximately 9.3 nm wide. In its "tree mode," the AIS collected data in contiguous bands between 0.4 and 1.2 μm; in its "rock mode," between 1.2 and 2.4 μm. The IFOV of the AIS was 1.9 mrad, and the system was typically operated from an altitude of 4200 m above the terrain. This yielded a narrow swath 32 pixels wide beneath the flight path for AIS-1 (64 pixels wide for AIS-2) with a ground pixel size of approximately 8 × 8 m.

Figure 4.41 shows an AIS image acquired over Van Nuys, California, on the first engineering test flight of the AIS. This image covers the area outlined in black lines on the corresponding photographic mosaic. Notable features include a

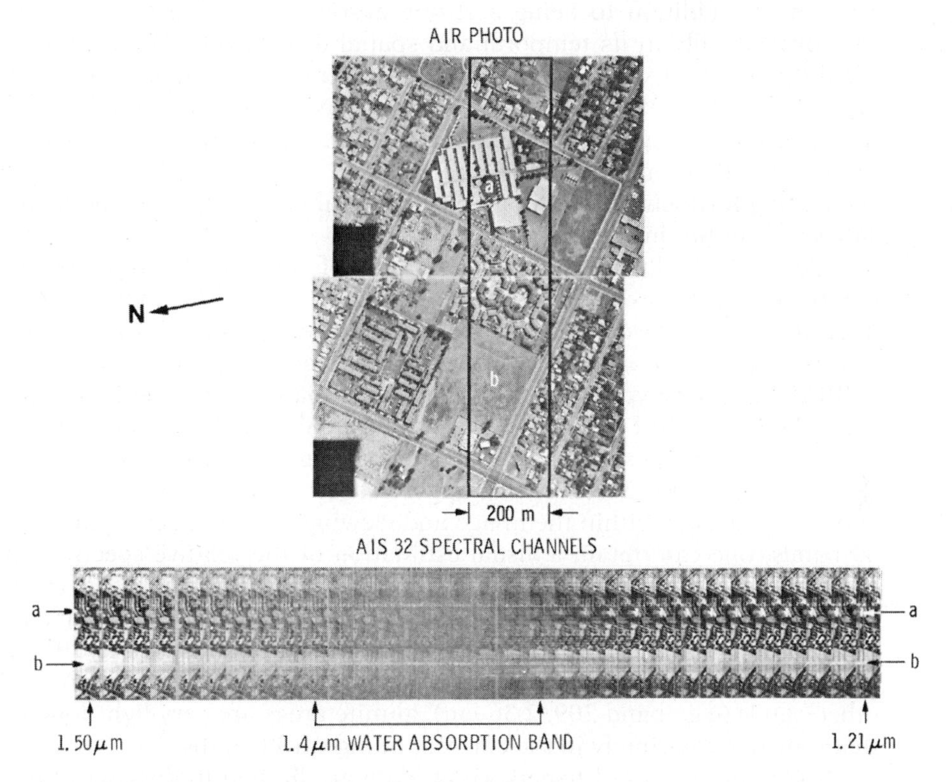

Figure 4.41 Airphoto mosaic and 32 spectral bands of AIS images covering a portion of the same area, Van Nuys, CA: (*a*) A school courtyard and (*b*) an open field are shown on the mosaic. The vertical black lines outline the coverage of the corresponding 32-pixel-wide AIS images taken in 32 spectral bands between 1.50 and 1.21 μm. The individual AIS images taken in 9.3-nm-wide, contiguous, spectral intervals are shown at the bottom. The spectral reflectance behavior of the well-watered courtyard and the open field are quite different and are primarily associated with differences in irrigation practices. (From Goetz et al., 1985. Copyright 1985 by the AAAS.)

field in the lower portion of the image, a condominium complex in the center, and a school in the upper half of the image. On this test flight, images were acquired in only 32 contiguous spectral bands in the region from 1.50 to 1.21 μm. A composite of the 32 AIS images, each in a different 9.3-nm-wide spectral band and each 32 pixels wide, is shown below the photographic mosaic. The most obvious feature in the AIS images is the loss of detail in the atmospheric water absorption band centered around 1.4 μm. However, there is some detail visible in the spectral images in this band. Details associated with reflectance variations are identified with arrows. For instance, the reflectance of a well-watered courtyard lawn inside the school grounds (location *a*) drops significantly beyond 1.4 μm in comparison with the reflectance of the unwatered field (location *b*).

Figure 4.41 also illustrates the pronounced effect water vapor can have on hyperspectral sensor data acquired near the 1.40-μm absorption band. Other atmospheric water vapor absorption bands occur at approximately 0.94, 1.14, and 1.88 μm. In addition to being a strong absorber, atmospheric water vapor can vary dramatically in its temporal and spatial distribution. Within a single scene, the distribution of water vapor can be very patchy and change on the time scale of minutes. Also, large variations in water vapor can result simply from changes in the atmospheric path length between the sensor and the ground due to changes in surface elevation. Given the strength and variability of water vapor effects, accounting for their influence on hyperspectral data collection and analysis is the subject of continuing research.

Figure 4.42 illustrates the appearance of 20 discrete hyperspectral sensor bands ranging from 1.98 to 2.36 μm used for the identification of minerals in a hydrothermally altered volcanic area at Cuprite, Nevada. These 20 bands represent a sequence of all odd-numbered bands between bands 171 and 209 of the AVIRIS airborne hyperspectral sensor. Broadband sensors such as the Landsat TM, ETM+, and OLI (Chapter 5) have been unable to discriminate clearly among various mineral types in this spectral region. Figure 4.42 displays the 20 bands in sequence, from band 171 at the upper left to band 209 at the lower right. By choosing one area within the image and viewing that area across the images of all 20 bands, one can obtain a visual estimation of the relative spectral behavior of the materials within the image. For example, labeled as *a* in band 189 (2.16 μm) are several dark spots near the center of the image that are occurrences of the mineral alunite. These alunite areas are darkest in bands 187 to 191 because of their absorption characteristics in wavelengths from 2.14 through 2.18 μm. In other bands (e.g., band 209, 2.36 μm), alunite areas are very light toned, meaning that they reflect strongly in the corresponding wavelengths.

Color composites of hyperspectral data are limited to displays of three bands at a time (with one band displayed as blue, one band as green, and one band as red). In an attempt to convey the spectral nature and complexity of hyperspectral images, hyperspectral data are commonly displayed in the manner shown in Plate 10. This plate shows two segments of imagery from the Hyperion hyperspectral sensor, an along-track sensor carried on the EO-1 satellite (described in

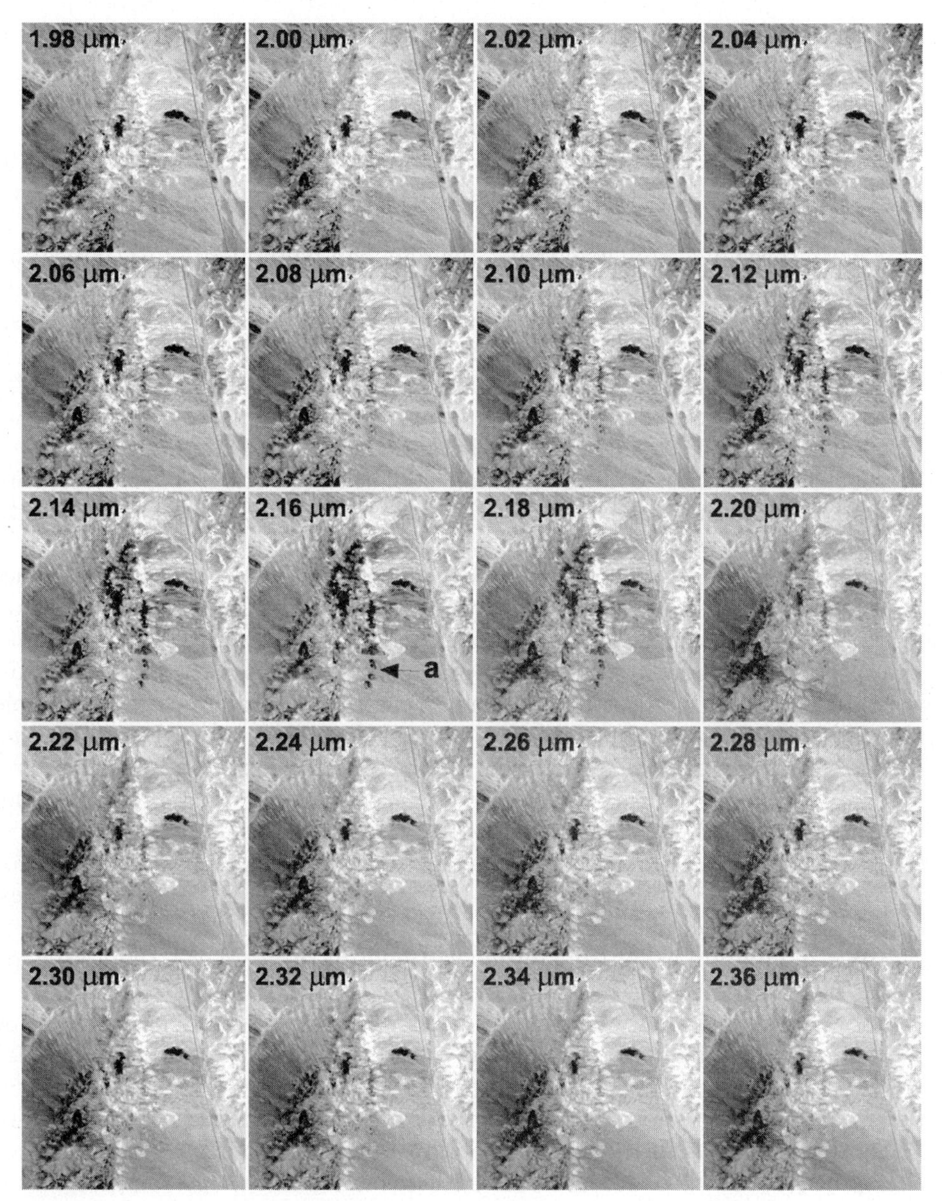

Figure 4.42 Images from bands 171 to 209 (1.98 to 2.36 μm) of the AVIRIS 224-channel imaging spectrometer, showing hydrothermically altered volcanic rocks at Cuprite, NV. Only odd-numbered channels are shown. (Courtesy NASA/JPL/Caltech.)

Chapter 5). The data shown in Plate 10 are two consecutive segments from a much longer image acquired over southern China. The northern of the two segments, on the left side of the plate, covers major portions of the industrial city of Shenzhen, while the southern of the two segments, on the right, shows a portion of Lantau Island, the largest island in Hong Kong. (There is a slight overlap between the southern edge of the left image and the northern end of the image at right.)

In these two isometric views, the data can be thought of as "cubes" having the dimensions of rows × columns × bands. The front of each cube is a color infrared composite, with the bands centered at 0.905, 0.661, and 0.560 μm displayed as red, green, and blue, respectively. The top and right side of each cube represent color-coded spectral radiance values corresponding to 183 spectral bands for each of the edge pixels along the top and right of the front of the cube. These spectral bands range from 0.47 μm, just behind the front of the cube, to 2.3 μm, at the back. In this color coding scheme, "cool" colors such as blue correspond to low spectral radiance values and "hot" colors such as red correspond to high spectral radiance values.

The shortest wavelengths shown on the sides of the cubes show generally elevated radiance values, due to Rayleigh scattering (Chapter 1). Forested areas in the mountains in both image cubes have relatively low radiance in the visible spectrum, as indicated by blue, cyan, and green hues. Urban areas, highways, and other artificial features tend to show much higher radiance in the visible region, with yellow and red hues present on the sides of the cubes. Beyond 0.7 μm, in the near IR, this pattern is reversed, with vegetated areas showing higher radiance levels than urban areas. Water in the bays and channels around Lantau Island shows relatively high radiance in the shortest wavelengths (0.47 to 0.55 μm) but the level of radiance decreases sharply in wavelengths approaching the near IR.

At various wavelengths in the near and mid IR, the sides of the cubes in Plate 10 show distinct black lines. These are the absorption bands, atmospheric "walls" where radiation is not transmitted through the atmosphere. Due to the low signal-to-noise ratio in these absorption regions, the Hyperion data processor does not always include the measured data in these bands (the data set is padded with values of 0 for all pixels in some of the absorption bands). This explains their featureless black appearance on the sides of the cubes in Plate 10.

Figure 4.43 shows representative examples of spectral signatures for various surface features in the EO-1 Hyperion image cubes from Plate 10. The detailed spectra for features such as trees, grasses and shrubs, and impervious surfaces in urban areas can be used to support many types of sophisticated quantitative analysis, from characterizing the health of vegetation to spectral mixture analysis (Chapter 7). Likewise, subtle differences in the spectra from pixels in the ocean, a reservoir (in the upper left corner of the left cube in Plate 10), and fishponds (in the center of Plate 10) can be used to diagnose the bio-optical properties of these varying water bodies. (Modeling optical properties of lakes from hyperspectral data is discussed in more detail later in this section.)

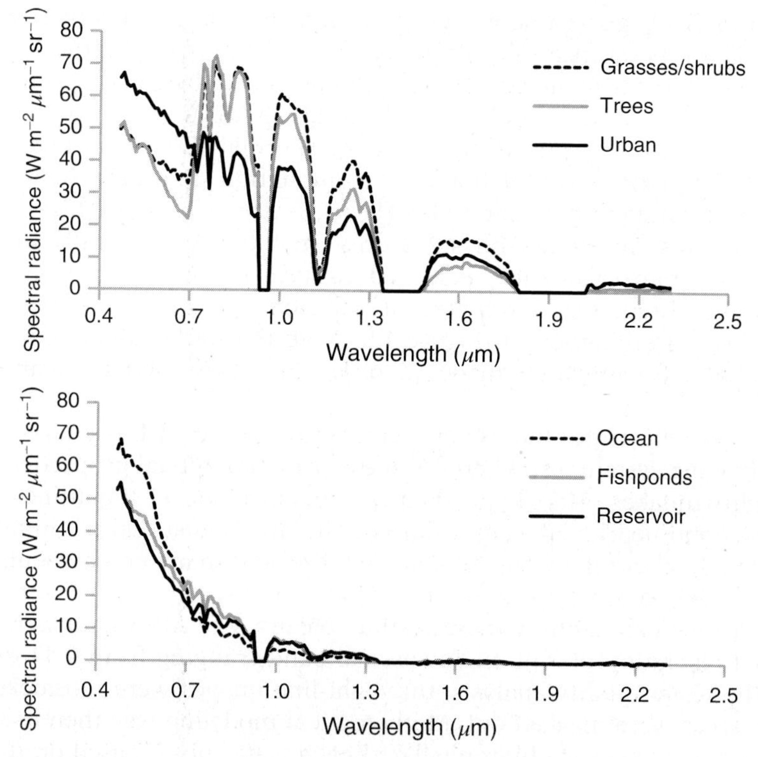

Figure 4.43 Spectral signatures of surface features in the EO-1 Hyperion images shown in Plate 10. Top: Terrestrial features. Bottom: Water bodies.

Substantial research has been undertaken to automate the analysis of the detailed spectral reflectance curves associated with individual pixels in hyperspectral data sets. Many successful approaches share the following general procedures (Kruse, 2012):

1. Atmospheric correction of the hyperspectral imagery to convert at-sensor spectral radiance values to surface reflectances.
2. Image enhancement procedures to reduce noise levels and to reduce the data volume.
3. Identification of "pure" spectra ("end-members") in the imagery, or acquisition of reference spectra from field spectroradiometry or spectral libraries.
4. Classification and/or mixture modeling of spectra from the imagery using the "end-members" and/or reference spectra from step 3.

These and other hyperspectral data analysis techniques are described in Chapter 7.

A number of government agencies and mapping/consulting companies maintain data sets of spectral reflectance curves. For example, the USGS maintains a "digital spectral library" that contains the reflectance spectra of more than 500 minerals, plants, and other materials. NASA's Jet Propulsion Laboratory maintains the Advanced Spaceborne Thermal Emission and Reflection Radiometer (ASTER) spectral library, a compilation of nearly 2000 spectra of natural and manufactured materials. The U.S. Army Topographic Engineering Center maintains "hyperspectral signatures" in the range of 0.400 to 2.500 μm for vegetation, minerals, soils, rock, and cultural features, based on laboratory and field measurements. Examples of selected spectra from some of these spectral libraries are shown in Figure 1.9. Note that atmospheric corrections to hyperspectral data must be made in order to make valid comparisons with library spectra.

Figure 4.44 illustrates the use of imagery from an AISA Eagle airborne hyperspectral imaging system, used here for measuring bio-optical properties related to water quality in lakes. AISA Eagle data were acquired on 11 flight-lines over Lake Minnetonka and nearby lakes in Minnesota in late August. Lake Minnetonka is a complex lake with a highly convoluted shoreline and many embayments, some of which are nearly separate from the rest of the lake. Water quality within the lake can vary dramatically among these distinct basins. The AISA data were collected at a spatial resolution of 3 m, in 86 spectral bands ranging from 0.44 to 0.96 μm. Prior to the water quality analysis, the flight-line images were mosaicked and all nonwater areas were masked out. Mathematical modeling was then used to compute the concentration of chlorophyll-*a* at each water pixel, based on the analysis of spectral reflectance and absorptance features (Gitelson et al., 2000). (Concentrations of phycocyanin, another algal pigment, and total suspended matter were also mapped.)

Figure 4.44*b* compares the spectral response from representative sites in the AISA imagery, identified by circles in Figure 4.44*a*. The spectral signature of Site 1 is typical of a eutrophic water body, with a modeled chlorophyll-*a* concentration over 120 μg/L. Its spectral response pattern is characterized by chlorophyll-induced absorption of blue wavelengths (below 0.5 μm) and red wavelengths (around 0.68 μm), a reflectance peak between 0.5 and 0.6 μm, and a second reflectance peak at 0.7 μm, where scattering from cellular structures is somewhat elevated and absorptance from both water and algal pigments is relatively low. The 0.7-μm reflectance peak is higher in this example than in any of the lake spectra shown in Figure 4.10, because of the extremely elevated chlorophyll levels at this site. In contrast, the spectral response at Site 2 is typical of relatively clear water, in this case having a modeled chlorophyll-*a* concentration below 8 μg/L.

Many airborne hyperspectral scanning instruments have been developed. Here, we will describe some of the more widely used instruments. Their key technical details are summarized in Table 4.5. In recent years, there has been a distinct trend toward the development of lightweight, portable hyperspectral instruments suitable for deployment on UAVs and other small platforms.

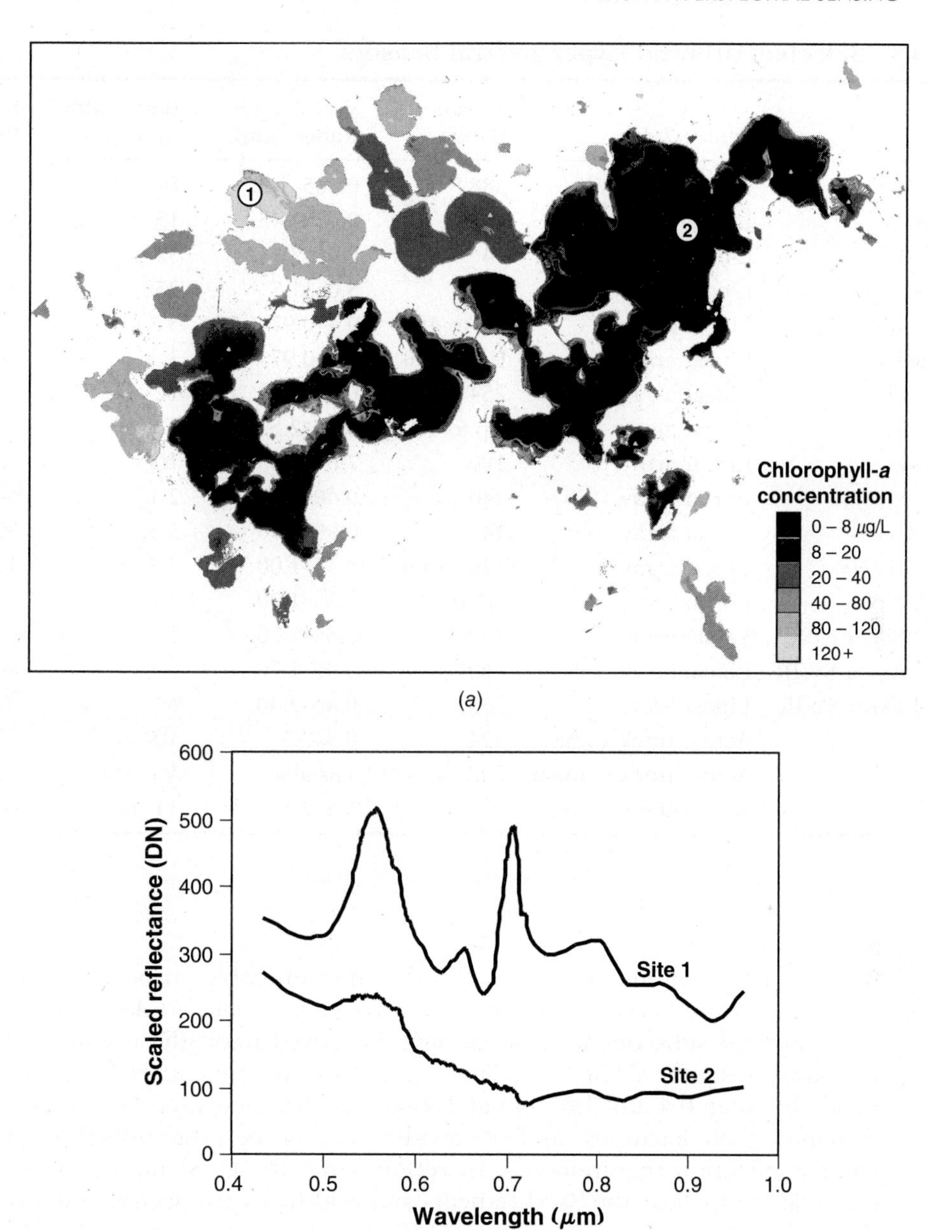

Figure 4.44 (*a*) Chlorophyll-*a* concentration in Lake Minnetonka, MN, from AISA Eagle airborne hyperspectral imagery, late August. (Courtesy University of Nebraska-Lincoln, Center for Advanced Land Management Information Technologies.) (*b*) Spectral response at two sites, indicated by circles in (*a*).

TABLE 4.5 Selected Airborne Hyperspectral Sensors

Sensor	Sensor Type	Sensor Bands	Wavelength Range (μm)	Bandwidth (nm)	Samples Per Line
CASI 1500	Linear array	288	0.365–1.05	Down to 1.9	1500
SASI 600	Linear array	100	0.95–2.45	15	600
MASI 600	Linear array	64	3.0–5.0	32	600
TASI 600	Linear array	32	8.0–11.5	250	600
AISA Eagle	Linear array	Up to 488	0.4–0.97	3.3	Up to 1024
AISA Eaglet	Linear array	Up to 410	0.4–0.97	3.3	1600
AISA Hawk	Linear array	254	0.97–2.5	12	320
AISA Fenix	Linear array	Up to 619	0.38–2.5	3.5–10	384
AISA Owl	Linear array	100	7.6–12.5	100	384
Pika II	Linear array	240	0.40–0.90	2.1	640
Pika NIR	Linear array	145	0.90–1.70	5.5	320
NovaSol visNIR	Linear array	120 to 180	0.38–1.00	3.3	1280
NovaSol Alpha-vis	Linear array	40 to 60	0.35–1.00	10	1280
NovaSol SWIR 640C	Linear array	170	0.85–1.70	5	640
NovaSol Alpha-SWIR	Linear array	160	0.90–1.70	5	640
NovaSol Extra-SWIR	Linear array	256	0.86–2.40	6	320
AVIRIS	Across-track scanner	224	0.4–2.5	10	677
HyMap	Across-track scanner	200	Variable	Variable	Variable
Probe-1	Across-track scanner	128	0.4–2.5	11–18	600

The first commercially developed, programmable airborne hyperspectral scanner was the *Compact Airborne Spectrographic Imager (CASI)*, available since 1989. Several subsequent versions have improved upon the original CASI, with the latest being the CASI 1500. This linear array system collects data in up to 288 bands between 0.4 and 1.05 μm at 1.9-nm spectral intervals. The precise number of bands, their locations, and bandwidths can all be programmed to suit particular application requirements. In recent years the CASI family of sensors has expanded to include the SASI (a near- and mid IR hyperspectral system) and two thermal-IR hyperspectral systems, MASI (in the 3 to 5 μm window) and TASI (8 to 11.5 μm). All of these instruments use linear arrays.

Another family of commercially developed, linear array hyperspectral systems consists of the *AISA Eagle*, *AISA Eaglet*, *AISA Hawk*, *AISA Fenix*, and *AISA Owl*, all manufactured by Specim. The Eagle and Eaglet cover the visible and near-infrared regions, with the Eaglet featuring a lightweight and compact design suitable for use on UAVs and similar small platforms. The Hawk operates in the near- and mid-IR regions, while the Owl covers the thermal IR. The AISA Fenix has over 600 spectral bands ranging from the upper end of the ultraviolet (0.38 μm) to the mid IR (2.5 μm).

Resonon manufactures several compact, lightweight hyperspectral sensors that can be used in a variety of settings, including the laboratory, manufacturing, ground-based imaging, and airborne platforms (including UAVs). The *Pika II* offers 240 bands in the visible and near IR, while the *Pika NIR* has 145 bands in the near IR and the lower range of the mid IR (up to 1.7 μm).

NovaSol has developed a line of small hyperspectral instruments (named *microHSI*), with weights ranging from 3.5 kg to less than 0.5 kg. This makes them suitable for operation on small UAVs, such as the helicopter-style vertical takeoff UAV shown in Figure 1.24*b*. The various sensors have 40 to 256 spectral bands from the ultraviolet (0.35 μm) to the mid IR (2.4 μm), with bandwidths from 3.3 to 10 nm.

The *Airborne Visible–Infrared Imaging Spectrometer (AVIRIS)* typically collects data in 224 bands that are approximately 9.6 nm wide in contiguous bands between 0.40 and 2.45 μm. When flown on NASA's ER-2 research aircraft at an altitude of 20 km, the swath width of this across-track scanner is approximately 10 km with a ground pixel resolution of approximately 20 m. An extensive archive of AVIRIS imagery from 1992 to the present is maintained by NASA.

The *HyMap (Hyperspectral Mapping)* system is an across-track airborne hyperspectral scanner built by Integrated Spectronics that senses in up to 200 bands. A variation of this scanner is the *Probe-1* built by Integrated Spectronics for ESSI (Earth Search Sciences Incorporated), which collects 128 bands of data in the 0.40- to 2.5-μm range.

The HyMap data shown in Plate 11 illustrate how it is possible to use airborne hyperspectral data to construct image-derived spectral reflectance curves for various minerals that are similar to laboratory-based spectral reflectance curves and to use these data to identify and plot the location of various minerals in an image. The area shown in (*a*) and (*b*) of Plate 11 is approximately 2.6 $\times$ 4.0 km in size and is located in the Mt. Fitton area of South Australia. This is an area of strongly folded sedimentary rocks (with some intrusive rocks). The HyMap data were collected in 128 narrow spectral bands over the range of 0.40 to 2.5 μm, with a ground resolution cell size of 5 m per pixel. Part (*a*) is a color-IR composite image using bands centered at 0.863, 0.665, and 0.557 μm, displayed as red, green, and blue, respectively. Part (*c*) shows laboratory-derived spectral reflectance curves for selected minerals occurring in the image. Part (*d*) shows spectral reflectance curves for these same minerals, based directly on the HyMap hyperspectral data. These curves were tentatively identified using visual comparison with the laboratory curves. Part (*b*) shows color overlays of these selected minerals on a grayscale image of the HyMap data. The colors in (*b*) correspond with those used in the spectral plots (*c*) and (*d*). Six different minerals are shown here. A preliminary analysis of the HyMap data without a priori information, and using only the 2.0- to 2.5-μm range, indicated the presence of over 15 distinct minerals.

4.14 CONCLUSION

In recent years, there has been a rapid increase in the number of new instruments developed for hyperspectral sensing. This technology holds the potential to provide a quantum jump in the quality of spectral data obtained about earth surface features. Hyperspectral sensing also provides a quantum jump in the quantity of data acquired. Visualization of these data is challenging, and the development of algorithms and methods for quantitative analysis of hyperspectral data remains a priority.

Research is also underway on the development of "ultraspectral" sensors that will provide data from thousands of very narrow spectral bands. This level of spectral sensitivity is seen as necessary for such applications as identifying specific materials and components of aerosols, gas plumes, and other effluents (Meigs et al., 2008).

In this chapter, we have treated the basic theory and operation of multispectral, thermal, and hyperspectral systems. We have illustrated selected examples of image interpretation, processing, and display, with an emphasis on data gathered using airborne systems. In Chapter 5, we discuss multispectral, thermal, and hyperspectral sensing from space platforms. In Chapter 7, we further describe how the data from these systems are processed digitally.

5 Earth Resource Satellites Operating in the Optical Spectrum

5.1 INTRODUCTION

Remote sensing from space is currently a worldwide, familiar, day-to-day activity that has numerous applications throughout science, government, and private industry. These applications range from global resource monitoring to such activities as land use planning, real estate development, natural disaster response, and vehicular navigation. Much of today's existing satellite remote sensing technology results directly or indirectly from the *Landsat* program, which was initiated in 1967 as the *Earth Resources Technology Satellite* (*ERTS*) program before being renamed. The Landsat program was the first formal civil research and development activity aimed at using satellites to monitor land resources on a global basis. The program was undertaken by NASA in cooperation with the U.S. Department of Interior. Landsat-1 was launched in 1972 and the program continues today, with the most recently launched satellite in the Landsat series being Landsat-8 (launched in 2013).

One of the primary factors that motivated the formation of the Landsat program was the initial development and success of meteorological satellites. Beginning with the first Television and Infrared Observation Satellite (TIROS-1) in 1960, the early weather satellites returned rather coarse views of cloud patterns and virtually indistinct images of the earth's surface. However, the field of satellite meteorology progressed quickly, and sensor refinements and improved data processing techniques led to more distinct images of both atmospheric and terrestrial features. Looking *through*, not just *at*, the earth's atmosphere had become possible. Also reinforcing the notion that systematic imaging of the earth's surface from space had great potential value in a range of scientific and practical applications were the photographs taken by astronauts during the space programs of the 1960s: Mercury, Gemini, and Apollo.

Having evolved substantially since the early days of the Landsat program, satellite remote sensing is now a highly international enterprise, with numerous countries and commercial firms developing and launching new systems on a regular basis. Global coverage of the earth is now readily available from multiple sources via the Internet. This has resulted in decisions being made with much more "spatial intelligence" than in any time in the past. It has also aided in conceiving of the earth as a system. Space remote sensing has brought a new dimension to understanding not only the natural wonders and processes operative on our planet but also the impacts of humankind on earth's fragile and interconnected resource base.

This chapter emphasizes satellite systems that operate within the *optical spectrum,* which extends from approximately 0.3 to 14 μm. This range includes UV, visible, near-, mid-, and thermal IR wavelengths. (It is termed the optical spectrum because lenses and mirrors can be used to refract and reflect such energy.) Substantial remote sensing from space is also performed using systems that operate in the *microwave* portion of the spectrum (approximately 1 mm to 1 m wavelength). Microwave remote sensing is the subject of Chapter 6.

The subject of remote sensing from space is a rapidly changing one, not only in terms of changes in technology but also in terms of changes in institutional arrangements. For example, many new commercial firms continue to be created while others are being merged or dissolved. Also, several satellite systems are being operated cooperatively by international consortia. Likewise, many countries have developed "dual use" systems that can be operated to serve both military and civilian purposes. Similarly, large single-satellite systems are being augmented or completely replaced by constellations of many smaller "minisats" (100 to 1000 kg), "microsats" (10 to 100 kg), or "nanosats" (1 to 10 kg). Smaller satellites can offer several potential advantages relative to their larger counterparts. They are generally less complex single-payload systems, reducing system engineering design and fabrication costs. Being small, they can often be flown as secondary payloads on other missions, reducing launch costs. Because of such cost savings, system operators can often afford to acquire more satellites, thereby increasing image coverage frequency and providing system redundancy. More frequent launches also provide the opportunity for more frequent introduction of new system technology. For example, many new satellites are nearly completely autonomous, or self-directed, in their operation and often are a part of a sensor web. A sensor web typically permits interactive communication and coordination among numerous in-situ sensors, satellite systems, and ground command resources.

In this dynamic environment, it is more important than ever that potential users of data from spaceborne remote sensing systems understand the basic design characteristics of these systems and the various trade-offs that determine whether a particular sensor will be suitable for a given application. Hence, the next section of this discussion presents an overview of the general characteristics common to all earth resource remote sensing satellite systems. We then examine a wide range of satellite systems, with reference to the ways their designs exemplify those fundamental characteristics.

Although we aim to be current in our coverage of spaceborne remote sensing systems, the rapid pace of change in this field suggests that some of the material in this chapter about certain specific systems will likely be dated rapidly.

We cannot hope to describe, or even list, all of the past, present, and planned future systems of potential interest to the readership of this book. Therefore, we focus our discussion primarily on representative samples of the major broad types of satellite systems available: moderate (4 to 80 m) resolution systems, high (less than 4 m) resolution systems, and hyperspectral sensors. We also provide brief descriptions of meteorological satellites, ocean monitoring satellites, the Earth Observing System, and space station remote sensing. Depending upon one's academic or work setting, one may wish to skip over certain of the sections in this chapter. We have written this discussion with this possibility in mind. Overall, the two-fold goal of this chapter is to present the reader with a sufficient background to understand the general range of options afforded by various past, present, and planned land observation satellites and to provide a frame of reference for consulting present and future online resources to obtain additional detail about any particular system.

5.2 GENERAL CHARACTERISTICS OF SATELLITE REMOTE SENSING SYSTEMS OPERATING IN THE OPTICAL SPECTRUM

At a detailed level, there are many technical design and operating characteristics that can differentiate one satellite remote sensing system from another in terms of suitability for a particular application. At the same time, there are some general characteristics that characterize all such systems. There are also many factors that a prospective remote sensing satellite data user must consider when selecting a particular system for a given application that deal more with data availability than the technical design of the system itself. In this section we introduce some basic terminology and concepts that will hopefully provide a general foundation for evaluating alternative systems for a given application.

Satellite Bus and Mission Subsystems

All earth monitoring remote sensing satellites incorporate two major subsystems: a *bus*, and a *mission system*. The bus is the carrying vehicle for the mission system, and the mission system is the sensor system payload itself. The bus forms the overall structure for the satellite and consists of a number of subsystems that support the mission system. One such subsystem provides electrical power from solar arrays that charge batteries when the satellite is on the sunlit portion of its orbit; the batteries provide power when the satellite is in the earth's shadow. Other subsystems provide such functions as on-board propulsion for orbit adjustment, attitude control, telemetry and tracking, thermal control, communications for satellite command and microwave transmission of sensor data when within range of a ground station, and data storage for delayed broadcasting if necessary. The bus is generally equipped with proven components that are as light weight as practical. A well designed bus can integrate a range of mission systems. For example, the bus employed for Landsat-1, -2, and -3 was a slightly modified version of that used for the Nimbus weather satellite.

Figure 5.1 illustrates the various buses and mission systems that have been used in the Landsat program over the past 40+ years. Shown in (*a*) is a schematic illustration of the observatory configuration employed by Landsat-1, -2, and -3. Part (*b*) is a similar depiction of the observatory configuration employed by Landsat-4 and -5. Part (*c*) is an artist's rendition of the observatory configuration for Landsat-7 (Landsat-6 failed at launch.) Part (*d*) is a laboratory photograph of Landsat-8 taken shortly after the integration of the mission system with the spacecraft bus. (Note that the solar array is stowed for launch in this picture.)

Satellite Orbit Design

One of the most fundamental characteristics of a satellite remote sensing system is its orbit. A satellite in orbit about a planet moves in an elliptical path with the planet at one of the foci of the ellipse. Important elements of the orbit include its *altitude, period, inclination,* and *equatorial crossing time*. For most earth observation satellites, the orbit is approximately circular, with altitudes more than 400 km above the earth's surface. The period of a satellite, or time to complete one orbit, is related to its altitude according to the formula (Elachi, 1987)

$$T_o = 2\pi(R_p + H')\sqrt{\frac{R_p + H'}{g_s R_p^2}} \tag{5.1}$$

where

T_o = orbital period, sec
R_p = planet radius, km (about 6380 km for earth)
H' = orbit altitude (above planet's surface), km
g_s = gravitational acceleration at planet's surface (0.00981 km/sec^2 for earth)

The inclination of a satellite's orbit refers to the angle at which it crosses the equator. An orbit with an inclination close to 90° is referred to as near polar because the

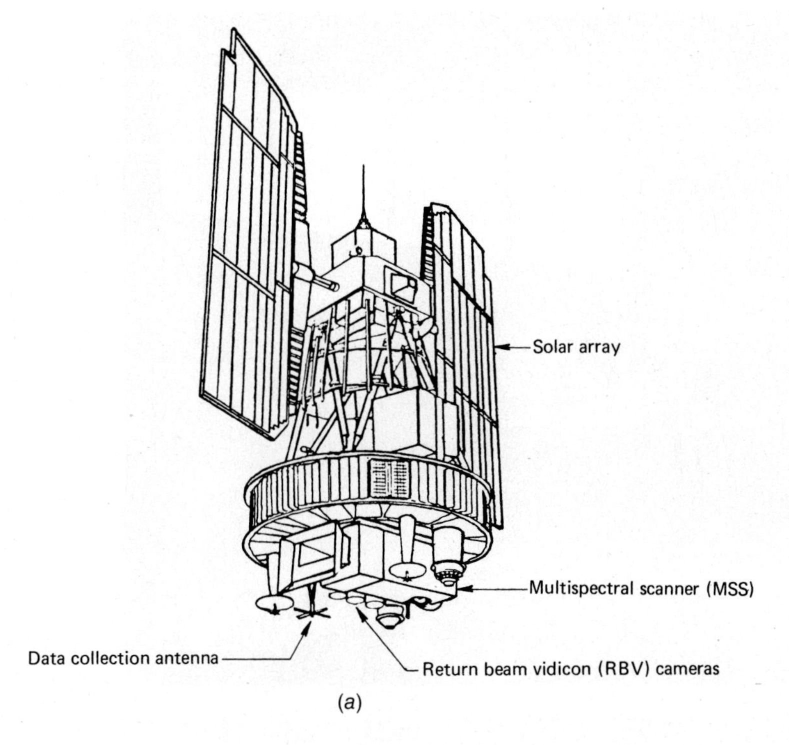

(*a*)

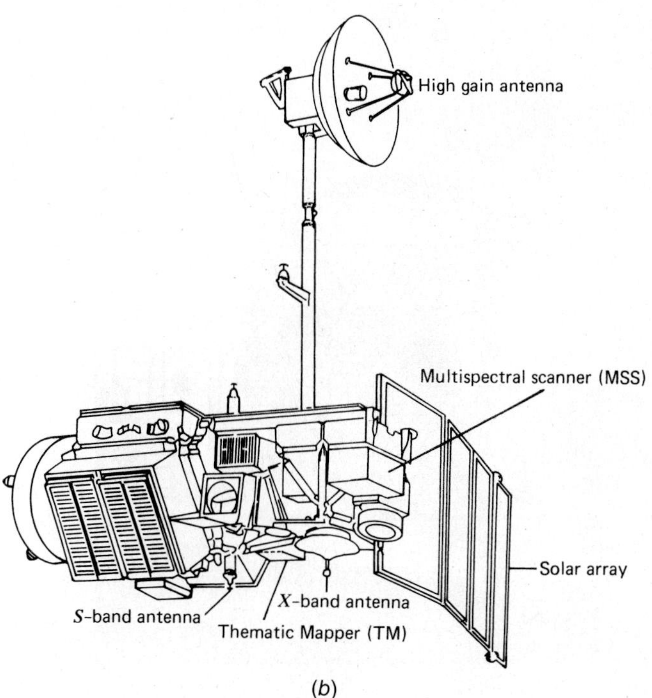

(*b*)

Figure 5.1 Satellite buses and sensor systems employed for: (*a*) Landsat-1, -2, and -3; (*b*) Landsat-4 and -5; (*c*) Landsat-7; (*d*) Landsat-8. (Adapted from NASA diagrams.)

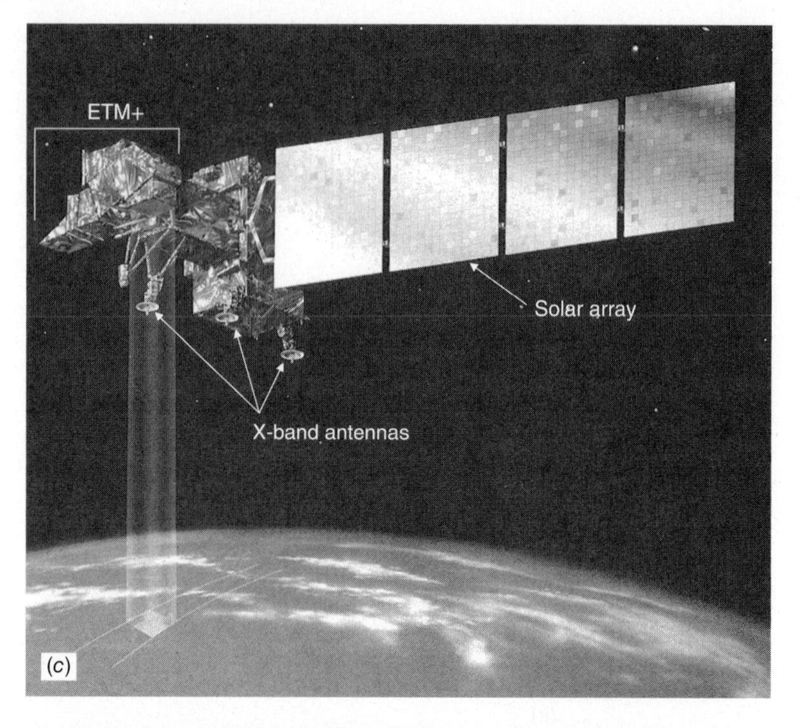

Figure 5.1 (*Continued*)

satellite will pass near the north and south poles on each orbit. An equatorial orbit, in which the spacecraft's ground track follows the line of the equator, has an inclination of 0°. Two special cases are *sun-synchronous* orbits and *geostationary* orbits. A sun-synchronous orbit results from a combination of orbital period and inclination such that the satellite keeps pace with the sun's westward progress as the earth rotates. Thus, the satellite always crosses the equator at precisely the same local sun time (the local clock time varies with position within each time zone). A geostationary orbit is an equatorial orbit at the altitude (approximately 36,000 km) that will produce an orbital period of exactly 24 hr. A geostationary satellite thus completes one orbit around the earth in the same amount of time needed for the earth to rotate once about its axis and remains in a constant relative position over the equator.

The above elements of a satellite's orbit are illustrated in Figure 5.2, for the Landsat-4, -5, -7, and -8 satellites. These satellites were launched into circular, near-polar orbits at an altitude of 705 km above the earth's surface. Solving Eq. 5.1 for this altitude yields an orbital period of about 98.9 min, which corresponds to approximately 14.5 orbits per day. The orbital inclination chosen for these satellites was 98.2°, resulting in a sun-synchronous orbit.

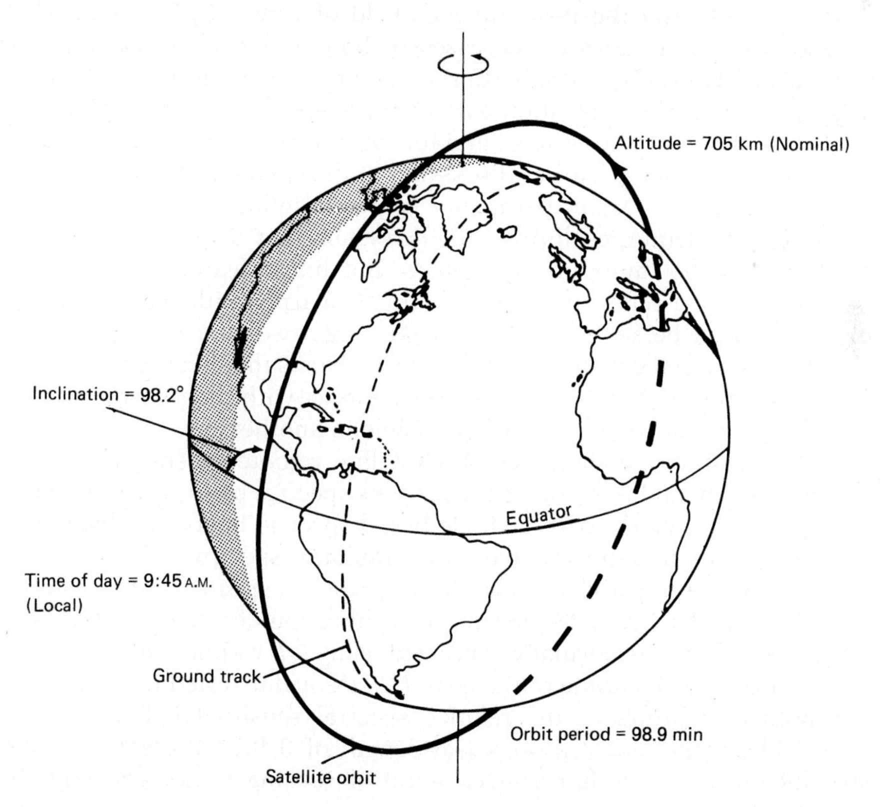

Figure 5.2 Sun-synchronous orbit of Landsat-4, -5, -7, and -8. (Adapted from NASA diagram.)

One point we should emphasize about sun-synchronous orbits is that they ensure repeatable illumination conditions during the specific seasons. This aids in mosaicking adjacent tracks of imagery and comparing annual changes in land cover and other ground conditions. While such orbits ensure *repeatable* illumination conditions, these conditions vary with location and season. That is, the sun's rays strike the earth at varying solar elevation angles as a function of both latitude and time. Likewise, the azimuth direction of solar illumination changes with season and latitude. In short, a sun-synchronous orbit does not compensate for changes in solar altitude, azimuth, or intensity. These factors are always changing and are compounded by variations in atmospheric conditions between scenes.

Sensor Design Parameters

A number of sensor design parameters influence the utility of any given satellite remote sensing system for various uses. A system's spatial resolution is certainly very important. The principles discussed in Chapter 4 that determine the spatial resolution for airborne sensors also apply to spaceborne systems. For example, at nadir the spatial resolution of an across-track scanner is determined by the orbital altitude and the instantaneous field of view (IFOV), while off nadir the ground resolution cell size increases in both the along-track and across-track dimensions. However, for spaceborne systems the curvature of the earth further degrades the spatial resolution for off-nadir viewing. This becomes particularly important for pointable systems and for systems with a wide total field of view.

As we noted in our earlier discussion of the spatial resolution of photographic systems (Section 2.7), this parameter is greatly influenced by scene contrast. For example, a moderate resolution satellite system may acquire data having a spatial resolution in the range of 30 to 80 m, but linear features as narrow as a few meters, having a reflectance that contrasts sharply with that of their surroundings, can often be seen on such images (e.g., two-lane roads, concrete bridges crossing water bodies). On the other hand, objects much larger than the GSD may not be apparent if they have a very low contrast with their surroundings, and features visible in one band may not be visible in another.

The spectral characteristics of a satellite remote sensing system include the number, width, and position of the sensor's spectral bands. The extreme cases are panchromatic sensors, with a single broad spectral band, and hyperspectral sensors, with a hundred or more narrow, contiguous spectral bands.

The relative sensitivity of each of the sensor's spectral bands as a function of wavelength is determined as part of the calibration of a sensor. The spectral sensitivity of each band is normally expressed using *full width, half maximum (FWHM)* terminology. The meaning of the term FWHM is illustrated in Figure 5.3, which is a hypothetical graph of the relative spectral sensitivity of a detector used to record blue light centered on a wavelength of 0.480 μm. Note that the relative spectral sensitivity within a given band of sensing is not constant. It is much

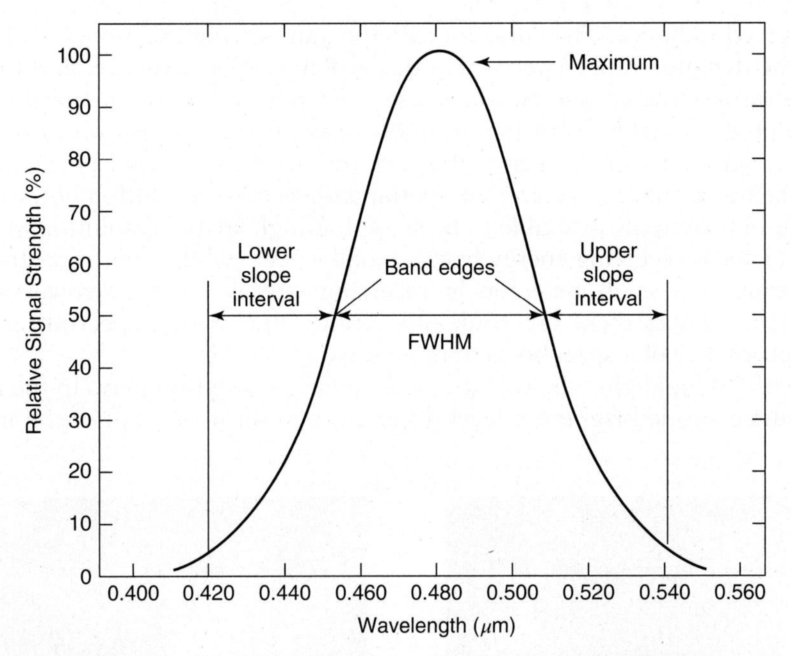

Figure 5.3 Full width, half maximum spectral bandwidth specification.

greater near the middle of the spectral sensitivity curve for each band. Accordingly, it would not be meaningful to express the entire width of the spectral sensitivity curve as the effective range of sensitivity in a band (because the sensitivity is so low at the tails of the spectral sensitivity curve). So, the practical sensitivity limits are expressed over the spectral bandwidth defined at the level where the sensor's sensitivity is at one-half (50%) of its maximum.

The points at which the FWHM intersects the spectral response curve in Figure 5.3 define the *lower band edge* (0.454 μm) and *upper band edge* (0.507 μm) of the spectral bandwidth. The interval between the band edges is normally used to specify the narrowest practical spectral range that is resolvable by a given sensor (even though there is some sensitivity outside of this range). The design or calibration specifications for a given sensor generally state the center wavelength location of each band, and its lower and upper band edges (and the allowable tolerances in these values). They may also describe the edge slope of the spectral sensitivity curve over a particular sensitivity range (often 1% to 50%, 5% to 50%, or both). In the case shown in Figure 5.3, a lower wavelength edge slope interval and an upper wavelength edge slope interval over the sensitivity range of 5% to 50% has been used to characterize the slope of the spectral response curve. In this example, both the lower and upper edge slope intervals are 0.034 μm wide.

The radiometric properties of a remote sensing system include the radiometric resolution, typically expressed as the number of data bits used to record the observed radiance. They also include the gain setting (Section 7.2) that determines the dynamic range over which the system will be sensitive, and the signal-to-noise ratio of the sensor. In some cases, the gain setting on one or more bands may be programmable from the ground to permit the acquisition of data under different conditions, such as over the dark ocean and the bright polar ice caps. It should be noted that increasing the spatial and spectral resolution both result in a decrease in the energy available to be sensed. A high spatial resolution means that each detector is receiving energy from a smaller area, while a high spectral resolution means that each detector is receiving energy in a narrower range of wavelengths. Thus, there are trade-offs among the spatial, spectral, and radiometric properties of a spaceborne remote sensing system.

Figure 5.4 highlights the role that radiometric resolution plays in the ability to discriminate subtle brightness level differences in an image. All of the images in

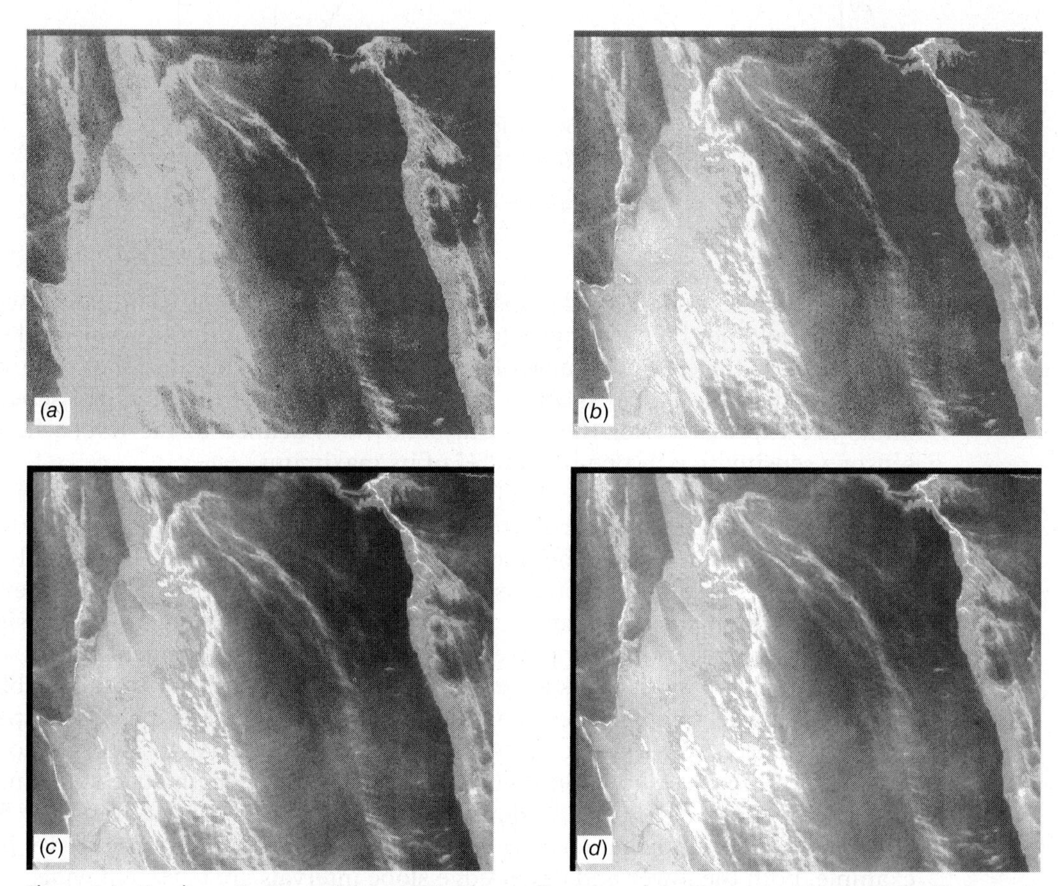

Figure 5.4 Panchromatic image (50-cm GSD) of a small portion of the Deepwater Horizon oil spill shown at multiple (simulated) radiometric resolutions: (*a*) 1-bit, 2 levels; (*b*) 2-bit, 4 levels; (*c*) 4-bit, 16 levels; (*d*) 8-bit, 256 levels. (Original data courtesy DigitalGlobe; processing courtesy University of Wisconsin-Madison SSEC.)

this illustration have the same spatial resolution, but the range of DNs used to encode each of them varies from 2 (1-bit), to 4 (2-bits), 16 (4-bits), and 256 (8-bits).

Some other factors to consider in evaluating a satellite system include its coverage area, revisit period (time between successive coverages), and off-nadir imaging capabilities. For sensors that cannot be steered to view off nadir, the swath width is determined by the orbital altitude and the sensor's total field of view, while the revisit period is determined by the orbital period and the swath width. For example, the Landsat-8 sensors and Terra/Aqua MODIS instruments are on satellite platforms that share the same orbital altitude and period, but MODIS has a much wider field of view. This gives it a much greater swath width, which in turn means that any given point on the earth will appear in MODIS imagery more frequently. If across-track pointing is used, both the swath width and revisit period will be affected. For certain time-sensitive applications, such as monitoring the effects of flooding, fires, and other natural disasters, a frequent revisit cycle is particularly important. Either across-track or along-track pointing can be used to collect stereoscopic imagery, which can assist in image interpretation and topographic analysis. The primary advantage of along-track pointing for stereoscopic image acquisition is that both images of the stereopair will be acquired within a very brief period of time, typically no more than seconds or minutes apart. Across-track stereoscopic images are often acquired several days apart (with the attendant changes in atmospheric and surface conditions between dates).

Yet another important factor to be considered when assessing a given sensor system is the spatial accuracy of the image data it produces. Spatial *accuracy* is not to be confused with spatial *resolution*. While the two are loosely related, they should not be equated. Spatial resolution simply refers to the ground dimensions of a pixel at ground level, irrespective of how accurately that pixel can be located in the real-world ground coordinate system. The absolute accuracy of a pixel location in the ground coordinate mapping system depends upon how well the sensor geometry can be modeled and the extent to which sensor data are processed to remove terrain displacements. While most images are georeferenced, they are not necessarily fully orthorectified (corrected for terrain displacements).

The absolute spatial accuracy of image data can be measured and expressed in various ways using many different statistical measures and mapping standards. The most recent of these mapping standards is the *National Standard for Spatial Data Accuracy* (NSSDA). This standard may be reviewed in detail online (including numerical examples) at www.fgdc.gov/standards. It is also treated in more detail in such references as Congalton and Green (2009).

The NSSDA specifies the estimation of the absolute spatial accuracy of image data by comparing image-derived ground coordinates for several image-identifiable check points to the ground coordinate of these features as measured by a more accurate, independent method (e.g., ground surveys). The discrepancy, or error, between the image-derived ground coordinate values (XYZ) and the more accurately measured values obtained at a minimum of 20 check points (GCPs) is estimated using *root mean square error* (RMSE). The standard can be applied to a range of data types. For data sets involving planimetric (XY) and elevation (Z) data, the horizontal and vertical accuracies are reported separately, in ground distance units. We limit this discussion to the determination of horizontal accuracy of image data. This involves the assessment of two-dimensional (X and Y), or "circular," errors about a point's true position. Thus, a horizontal accuracy computed according to the NSSDA standard specifies that 95% of all well-defined points in an image should fall within a radial distance of no more than the computed accuracy value from the point's true position. For example, if the computed accuracy for a given image is 6.2 m, only 5% of all well-defined points in the image would be located at a distance greater than 6.2 m from their true position.

To define how horizontal accuracy is computed according to the NSSDA standard:

let

$$\mathrm{RMSE}_x = \mathrm{sqrt}\left[\sum \left(x_{image,\,i} - x_{check,\,i}\right)^2 / n\right]$$

$$\mathrm{RMSE}_y = \mathrm{sqrt}\left[\sum \left(y_{image,\,i} - y_{check,\,i}\right)^2 / n\right]$$

where

$x_{image,\,i},\ y_{image,\,i}$ = the ground coordinates of the ith check point in the image
$x_{check,\,i},\ y_{check,\,i}$ = the ground coordinates of the ith check point in the control point datase
n = number of check points
i = an integer ranging from 1 to n

The horizontal error at any check point i is $\mathrm{sqrt}\left[\left(x_{image,\,i} - x_{check,\,i}\right)^2 + \left(y_{image,\,i} - y_{check,\,i}\right)^2\right]$.

The horizontal RMSE is

$$\mathrm{RMSE}_r = \mathrm{sqrt}\left[\mathrm{RMSE}_x^2 + \mathrm{RMSE}_y^2\right]$$

Finally, at the 95% confidence level,

$$\text{Horizontal Accuracy} = 1.7308 \times \mathrm{RMSE}_r \tag{5.2}$$

Note that the above equation assumes that all systematic errors have been eliminated to the fullest extent possible, that the x-component and y-component of the error are normally distributed and independent, and that RMSE_x and RMSE_y are equal. When RMSE_x and RMSE_y are not equal

$$\text{Horizontal Accuracy} = 1.2239 \times \left(\mathrm{RMSE}_x + \mathrm{RMSE}_y\right) \tag{5.3}$$

It is quite common to express the horizontal accuracy of a satellite image product using such terminology as CE 95. The "CE" stands for *circular error probable* and the "95" infers the use of a 95% confidence limit in the determination of accuracy. Most horizontal accuracy values for satellite image data employ either CE 95 or CE 90 specifications. The important point is that irrespective of how the spatial accuracy of an image product is specified, the potential user of such data must assess whether this accuracy is acceptable for the particular intended use.

Other System Considerations

Many other practical considerations determine the potential utility of any given satellite remote sensing system in any given application context. These include such factors as the procedures for tasking the satellite, data delivery options, data archiving, license/copyright restrictions, data formats (including data products derived from raw image data, e.g., vegetation index images), and data cost. These characteristics, many of which are often overlooked by prospective users of satellite imagery, have the potential to radically alter the usability of the imagery from any given system.

5.3 MODERATE RESOLUTION SYSTEMS

Launched on July 23, 1972, Landsat-1 was the first satellite specifically designed to acquire data about earth resources on a systematic, repetitive, long-term, moderate resolution, multispectral basis. Starting as a purely experimental endeavor, the Landsat program has evolved into an operational activity. From the program's inception, all Landsat data have been collected in accordance with an "open skies" principle, meaning there would be nondiscriminatory access to data collected anywhere in the world. The program continues to the present (2014) with the operation of Landsat-8, which was launched on February 11, 2013.

The second global scale, long-term, moderate resolution earth resource satellite monitoring program to come into existence was the *Système Pour l'Observation de la Terre*, or SPOT, program. SPOT-1 was launched on February 21, 1986, and the most recent satellite in the SPOT program, SPOT-6, was launched on September 9, 2012.

Because of the long-term global archives resulting from the Landsat and SPOT systems, we briefly summarize the operation of these "heritage" programs in the next two subsections of this discussion, respectively. This is followed by a general overview of the numerous other moderate resolution systems whose operations have commenced subsequent to the initiation of the Landsat and SPOT programs.

5.4 LANDSAT-1 TO -7

Of the eight Landsat satellites that have been launched to date, only one (Landsat-6) has failed to achieve orbit. Remarkably, even with a launch failure and several funding and administrative delays, Landsat-1, -2, -3, -4, -5, and -7 collectively acquired over 40 uninterrupted years of invaluable data. (Many refer to this circumstance as a stroke of good fortune combined with "over-engineering," given that the systems each had a design life requirement of only five years.) In the interest of brevity, we provide a collective historical summary of the salient characteristics of the Landsat-1 through Landsat-7 missions, followed by a separate discussion of Landsat-8.

Table 5.1 highlights the characteristics of the Landsat-1 through -7 missions. It should be noted that five different types of sensors have been included in various combinations on these missions. These are the *Return Beam Vidicon* (*RBV*), the *Multispectral Scanner* (*MSS*), the *Thematic Mapper* (*TM*), the *Enhanced Thematic Mapper* (*ETM*), and the *Enhanced Thematic Mapper Plus* (*ETM+*). Table 5.2 summarizes the spectral sensitivity and spatial resolution of each of these systems as included on the various missions.

The RBV sensors listed in Table 5.2 were television-like analog cameras that acquired a frame of imagery using a shutter device to record each image on a photosensitive surface in the focal plane of the camera. The surface was then scanned with an electron beam to produce an analog video signal that was subsequently used to print hardcopy negatives. Each RBV scene covered an earth surface area approximately 185 km × 185 km in size. Figure 5.5 is a sample Landsat-3 RBV image.

While not intended, RBV images were secondary sources of data in comparison to the MSS systems flown on Landsat-1, -2, and -3. Two factors contributed to this situation. First, RBV operations were plagued with various technical malfunctions. More importantly, the MSS systems were the first global monitoring systems capable of producing multispectral data in a digital format.

The MSS systems were across-track multispectral scanners that employed an oscillating (rather than spinning) mirror to scan back and forth in an east-west direction over a total field of view of approximately 11.56°. Six lines were scanned simultaneously in each of the reflective spectral bands. The scanning of six lines simultaneously dictated the use of six detectors per spectral band, but it reduced the rate of scanning mirror oscillation by a factor of six, resulting in improved system response characteristics. The mirror oscillated every 33 milliseconds, and the scanners only acquired data in the west-to-east direction of scanning. It took about 25 sec to acquire each scene. Like all Landsat scenes, the swath width of MSS images is approximately 185 km.

Figure 5.6 is a full-frame, band 5 (red) MSS scene covering a portion of central New York. Note that the image area is a parallelogram, not a square, because of the earth's rotation during the 25 sec it took for the satellite to travel from the top of the scene to the bottom. The U.S. archive of MSS scenes contains nearly 614,000 images. The worldwide database for MSS and RBV data contains some 1.3 million scenes.

TABLE 5.1 Characteristics of Landsat-1 to -7 Missions

Satellite	Launched	Decommissioned	RBV Bands	MSS Bands	TM Bands	Orbit
Landsat-1	July 23, 1972	January 6, 1978	1–3 (simultaneous images)	4–7	None	18 days/900 km
Landsat-2	January 22, 1975	February 25, 1982	1–3 (simultaneous images)	4–7	None	18 days/900 km
Landsat-3	March 5, 1978	March 31, 1983	A–D (one-band side-by-side images)	4–8[a]	None	18 days/900 km
Landsat-4	July 16, 1982[b]	June 15, 2001	None	1–4	1–7	16 days/705 km
Landsat-5	March 1, 1984[c]	November 18, 2011	None	1–4	1–7	16 days/705 km
Landsat-6	October 5, 1993	Failure upon launch	None	None	1–7 plus panchromatic band (ETM)	16 days/705 km
Landsat-7	April 15, 1999[d]	—	None	None	1–7 plus panchromatic band (ETM+)	16 days/705 km

[a]Band 8 (10.4–12.6 μm) failed shortly after launch.
[b]TM data transmission failed in August 1993.
[c]MSS powered off in August 1995; solar array drive problems began in November 2005; temporary power anomaly experienced in August 2009; TM operation suspended in November 2011; MSS re-activated for limited data collection in April 2012.
[d]Scan Line Corrector (SLC) malfunctioned on May 31, 2003.

TABLE 5.2 Sensors Used on Landsat-1 to -7 Missions

Sensor	Mission	Sensitivity (μm)	Resolution (m)
RBV	1, 2	0.475–0.575	80
		0.580–0.680	80
		0.690–0.830	80
	3	0.505–0.750	30
MSS	1–5	0.5–0.6	79/82[a]
		0.6–0.7	79/82[a]
		0.7–0.8	79/82[a]
		0.8–1.1	79/82[a]
	3	10.4–12.6[b]	240
TM	4, 5	0.45–0.52	30
		0.52–0.60	30
		0.63–0.69	30
		0.76–0.90	30
		1.55–1.75	30
		10.4–12.5	120
		2.08–2.35	30
ETM[c]	6	Above TM bands	30 (120 m thermal band)
		plus 0.50–0.90	15
ETM+	7	Above TM bands	30 (60 m thermal band)
		plus 0.50–0.90	15

[a]79 m for Landsat-1 to -3 and 82 m for Landsat-4 and -5.
[b]Failed shortly after launch (band 8 of Landsat-3).
[c]Landsat-6 launch failure.

It can be seen in Table 5.2 that the TM-class of instruments used in the Landsat-4 through -7 missions afforded substantial improvement in both their spatial and spectral resolutions in comparison to the RBV and MSS sensors used on the earlier missions. The TM-class of sensors also provided 8-bit (vs. 6-bit) radiometric resolution and improved geometric integrity compared to their predecessors. Table 5.3 lists the spectral bands utilized for the TM-class of instruments along with a brief summary of the intended practical applications of each.

Figure 5.7, a small portion of a Landsat scene, illustrates the comparative appearance of the seven Landsat TM bands. Here, the blue-green water of the lake, river, and ponds in the scene has moderate reflection in bands 1 and 2 (blue and green), a small amount of reflection in band 3 (red), and virtually no reflection in bands 4, 5, and 7 (near and mid IR). Reflection from roads and urban streets is highest in bands 1, 2, and 3 and least in band 4 (other cultural features

Figure 5.5 Landsat-3 RBV image, Cape Canaveral, FL. Scale 1:500,000. (Courtesy NASA.)

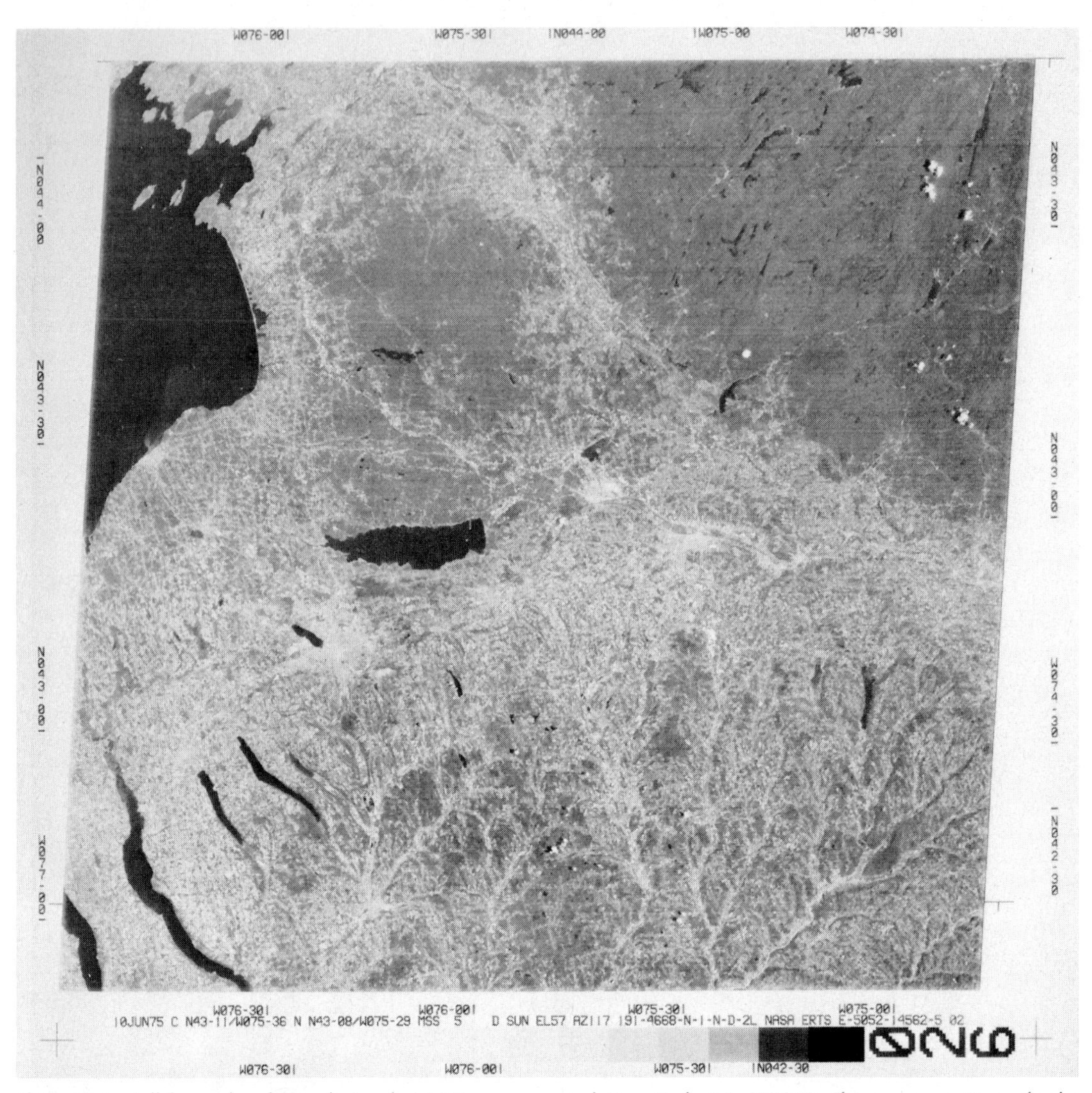

Figure 5.6 Full-frame, band 5 (red), Landsat MSS scene, central New York. 1:1,700,000. Shown are portions of Lake Ontario (upper left), Adirondack Mountains (upper right), and Finger Lakes region (lower left). (Courtesy NASA.)

such as new subdivisions, gravel pits, and quarries would have similar reflectances). Overall reflection from agricultural crops is highest in band 4 (near IR). Note also the high band 4 reflectance of the golf courses appearing to the right of the river and the lake. The distinct tonal lineations from upper right (northeast) to lower left (southwest) in these images are a legacy from the most recent glaciation of Wisconsin. Glacial ice movement from northeast to southwest left a terrain characterized by many drumlins and scoured bedrock hills. Present-day crop

TABLE 5.3 Thematic Mapper Spectral Bands

Band	Wavelength (μm)	Nominal Spectral Location	Principal Applications
1	0.45–0.52	Blue	Designed for water body penetration, making it useful for coastal water mapping. Also useful for soil/vegetation discrimination, forest-type mapping, and cultural feature identification.
2	0.52–0.60	Green	Designed to measure green reflectance peak of vegetation (Figure 1.9) for vegetation discrimination and vigor assessment. Also useful for cultural feature identification.
3	0.63–0.69	Red	Designed to sense in a chlorophyll absorption region (Figure 1.9) aiding in plant species differentiation. Also useful for cultural feature identification.
4	0.76–0.90	Near IR	Useful for determining vegetation types, vigor, and biomass content, for delineating water bodies, and for soil moisture discrimination.
5	1.55–1.75	Mid IR	Indicative of vegetation moisture content and soil moisture. Also useful for differentiation of snow from clouds.
6[a]	10.4–12.5	Thermal IR	Useful in vegetation stress analysis, soil moisture discrimination, and thermal mapping applications.
7[a]	2.08–2.35	Mid IR	Useful for discrimination of mineral and rock types. Also sensitive to vegetation moisture content.

[a]Bands 6 and 7 are out of wavelength sequence because band 7 was added to the TM late in the original system design process.

and soil moisture patterns reflect the alignment of this grooved terrain. The thermal band (band 6) has a less distinct appearance than the other bands because the ground resolution cell of this band is 120 m. It has an indistinct, rather than blocky, appearance because the data have been resampled into the 30-m format of the other bands. As would be expected on a summertime thermal image recorded during the daytime, the roads and urban areas have the highest radiant temperature, and the water bodies have the lowest radiant temperature.

Plate 12 shows six color composite images of the same area shown in Figure 5.7. Table 5.4 shows the color combinations used to generate each of these composites. Note that (*a*) is a "normal color" composite, (*b*) is a "color infrared" composite, and (*c*) to (*f*) are some of the many other "false color" combinations

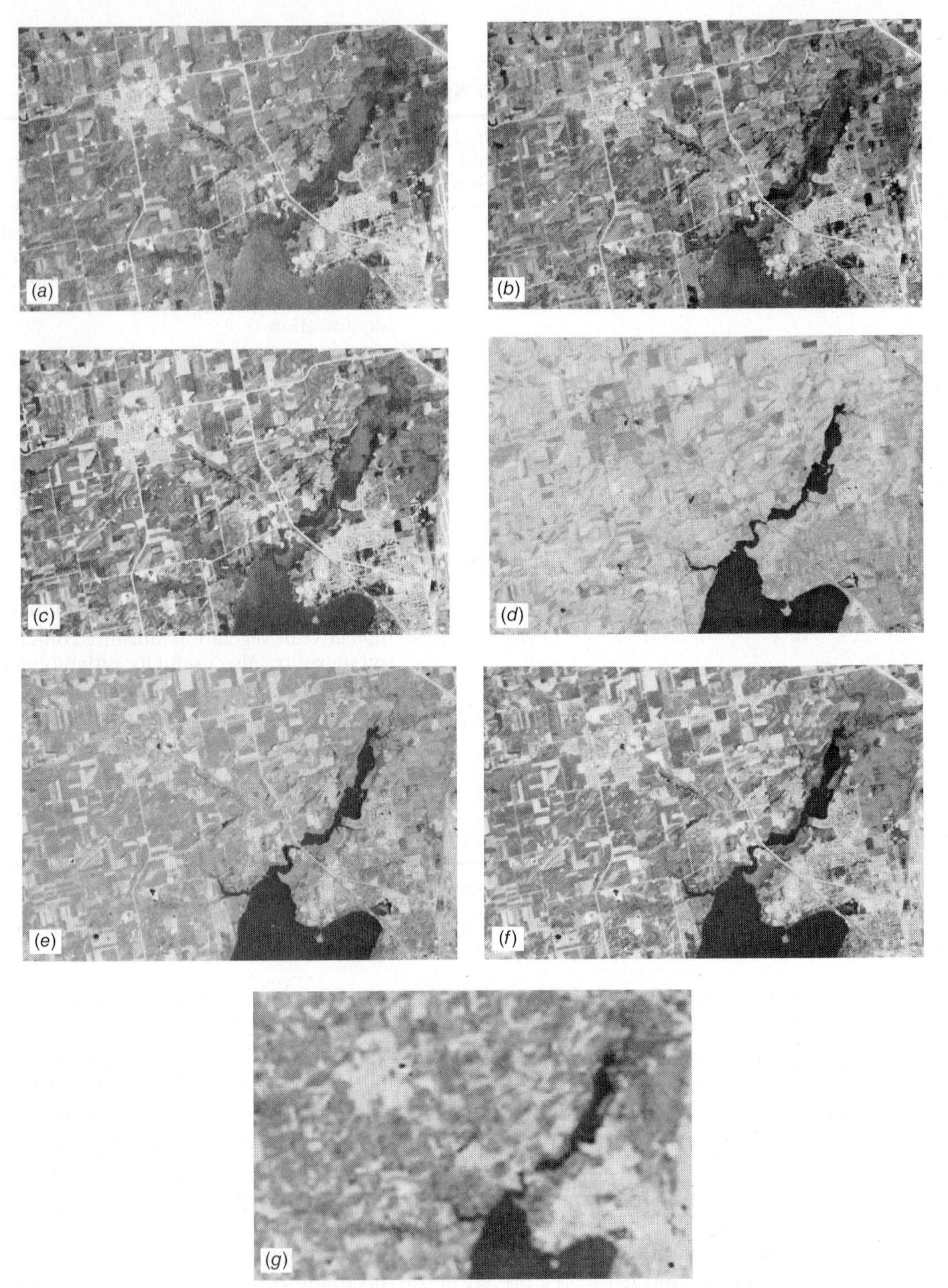

Figure 5.7 Individual Landsat TM bands, suburban Madison, WI, late August (scale 1:210,000): (*a*) band 1, 0.45 to 0.52 μm (blue); (*b*) band 2, 0.52 to 0.60 μm (green); (*c*) band 3, 0.63 to 0.69 μm (red); (*d*) band 4, 0.76 to 0.90 μm (near IR); (*e*) band 5, 1.55 to 1.75 μm (mid IR); (*f*) band 7, 2.08 to 2.35 μm (mid IR); (*g*) band 6, 10.4 to 12.5 μm (thermal IR). (Author-prepared figure.)

TABLE 5.4 TM Band–Color Combination Shown in Plate 12

	TM Band–Color Assignment in Composite		
Plate 12	Blue	Green	Red
(*a*)	1	2	3
(*b*)	2	3	4
(*c*)	3	4	5
(*d*)	3	4	7
(*e*)	3	5	7
(*f*)	4	5	7

that can be produced. A study at the USGS EROS (NOAA, 1984) showed an interpreter preference for several specific band–color combinations for various features. For the mapping of water sediment patterns, a normal color composite of bands 1, 2, and 3 (displayed as blue, green, and red) was preferred. For most other applications, such as mapping urban features and vegetation types, the combinations of (1) bands 2, 3, and 4 (color IR composite), (2) bands 3, 4, and 7, and (3) bands 3, 4, and 5 (all in the order blue, green, and red) were preferred. In general, vegetation discrimination is enhanced through the incorporation of data from one of the mid-IR bands (5 or 7). Combinations of any one visible (bands 1 to 3), the near-IR (band 4), and one mid-IR (band 5 or 7) band are also very useful. However, a great deal of personal preference is involved in band–color combinations for interpretive purposes, and for specific applications, other combinations could be optimum.

Figure 5.8 shows a Landsat TM band 6 (thermal) image of Green Bay and Lake Michigan (between the states of Wisconsin and Michigan). In this image, the land area has been masked out and is shown as black (using techniques described in Chapter 7). Based on a correlation with field observations of water surface temperature, the image data were "sliced" into six gray levels, with the darkest tones having a temperature less than 12°C, the brightest tones having a temperature greater than 20°C, and each of the four intermediate levels representing a 2°C range between 12 and 20°C.

Figure 5.9 illustrates a portion of the first image acquired by the Landsat-7 ETM+. This is a panchromatic band image depicting Sioux Falls, South Dakota, and vicinity. Features such as the airport (upper center), major roads, and new residential development (especially evident in lower right) are clearly discernible in this 15-m-resolution image. This band can be merged with the 30-m data from ETM+ bands 1 to 5 and 7 to produce "pan-sharpened" color images with essentially 15 m resolution (Section 7.6). Several other Landsat-7 ETM+ images are illustrated elsewhere in this book.

Figure 5.8 Landsat TM band 6 (thermal IR) image, Green Bay and Lake Michigan, Wisconsin–Michigan, mid-July. Scale 1:303,000. (Author-prepared figure.)

Whereas the MSS instruments collected data along six scan lines only when the IFOV was traversing in the west-to-east direction along a scan line, both the TM and the ETM+ acquired 16 scan lines of data during both the forward (east-to-west) and reverse (west-to-east) sweeps of its scan mirror. Because of this, a *scan line corrector* (*SLC*) was incorporated into the TM and ETM+ sensors. The function of the scan line corrector is illustrated in Figure 5.10. During each scan mirror sweep, the scan line corrector rotates the sensor line of sight backward along the satellite ground track to compensate for the forward motion of the spacecraft. This prevents the overlap and underlap of scan lines and produces straight scan lines that are perpendicular to the ground track.

Potential users of Landsat-7 data should be aware that the system's SLC failed on May 31, 2003. This resulted in the collection of "raw" images manifesting a

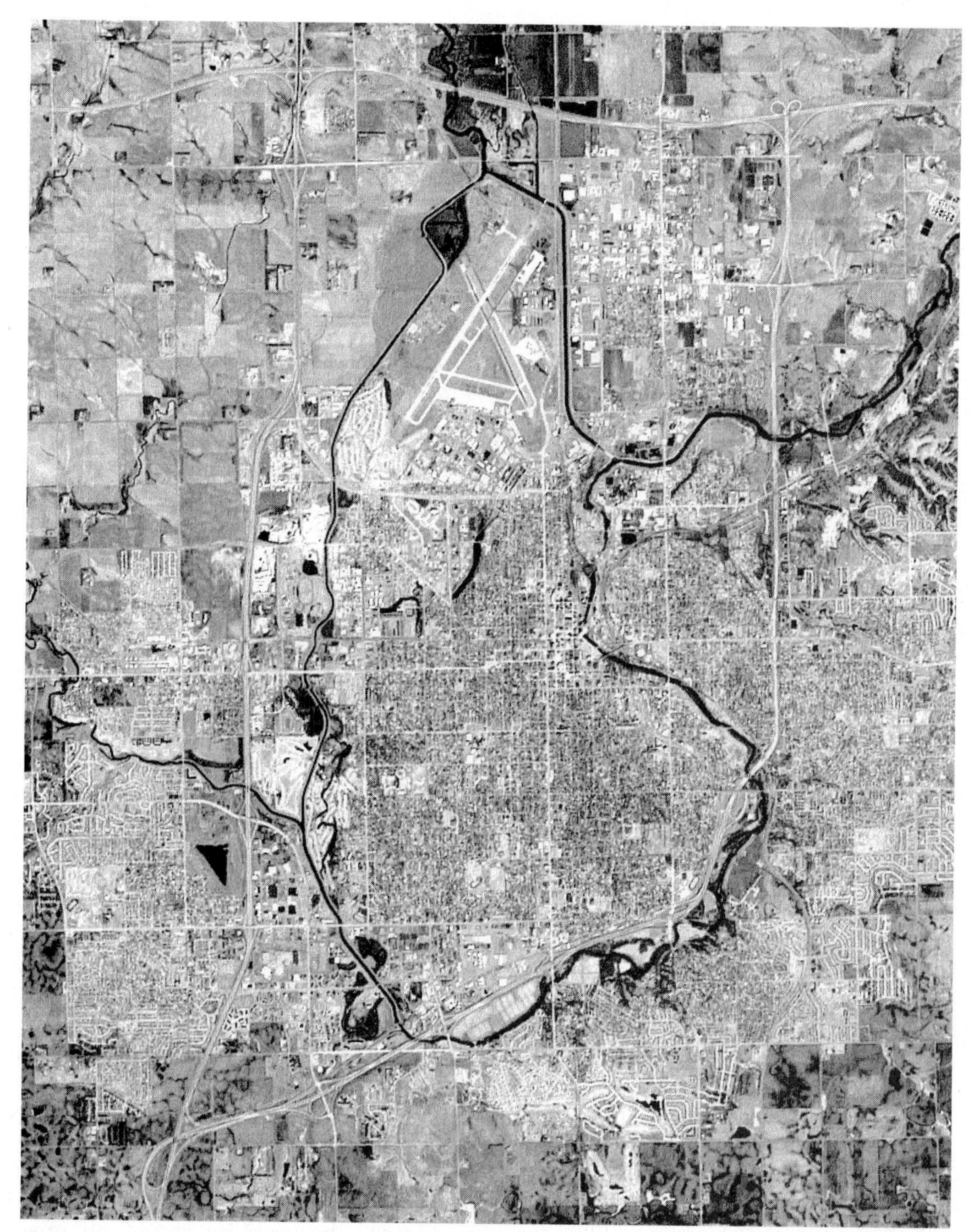

Figure 5.9 The first image acquired by the Landsat-7 ETM+. Panchromatic band (15 m resolution), Sioux Falls, SD, April 18, 1999. Scale 1:88,000. (Courtesy EROS Data Center.)

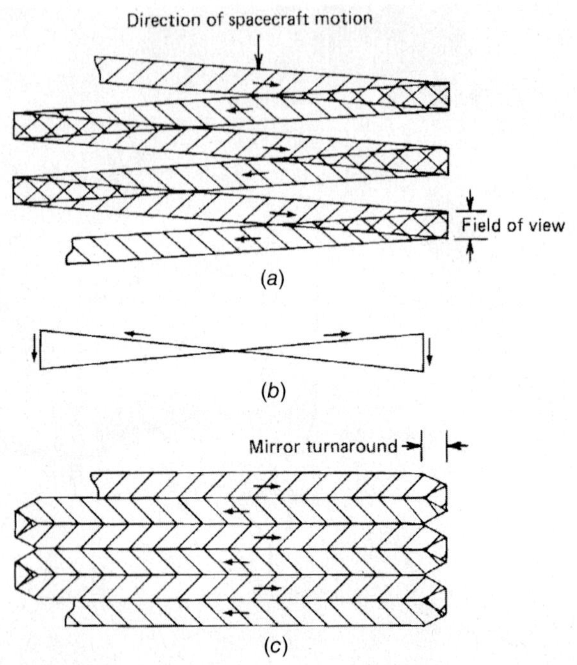

Figure 5.10 Schematic of TM scan line correction process: (*a*) uncompensated scan lines; (*b*) correction for satellite motion; (*c*) compensated scan lines. (Adapted from NASA diagram.)

zigzag pattern along the satellite ground track (Figure 5.10*a*). These raw images contain data gaps and duplications that are most pronounced along the eastward and westward edges of a scene and gradually diminish toward the center of the scene. The nature of this problem is illustrated in Figure 5.11, which shows a full-scene image acquired over California's Salton Sea area on September 17, 2003.

Figure 5.12 shows enlargements of an area located near the eastern edge of Figure 5.11. Figure 5.12*a* illustrates the duplications present near the edge of the scene prior to their removal, and Figure 5.12*b* shows the gaps in coverage subsequent to removal of duplicated pixels and application of geometric correction.

Several approaches have been developed to improve the utility of the "SLC-off" images collected since the scan line corrector failure developed. Among these are filling data gaps based on interpolating from neighboring scan lines, or using pixel values extracted from one or more alternative acquisition dates. Figure 5.12*c* illustrates the use of pixel values from a near-anniversary date, pre-SLC-failure image of the same scene (September 14, 2002) to fill the data gaps present in Figure 5.12*b*. One or more post-SLC-failure images can also be used to accomplish the gap-filling process. Clearly, the usability of SLC-off images, or image products derived from them, varies greatly based on the intended application of the data.

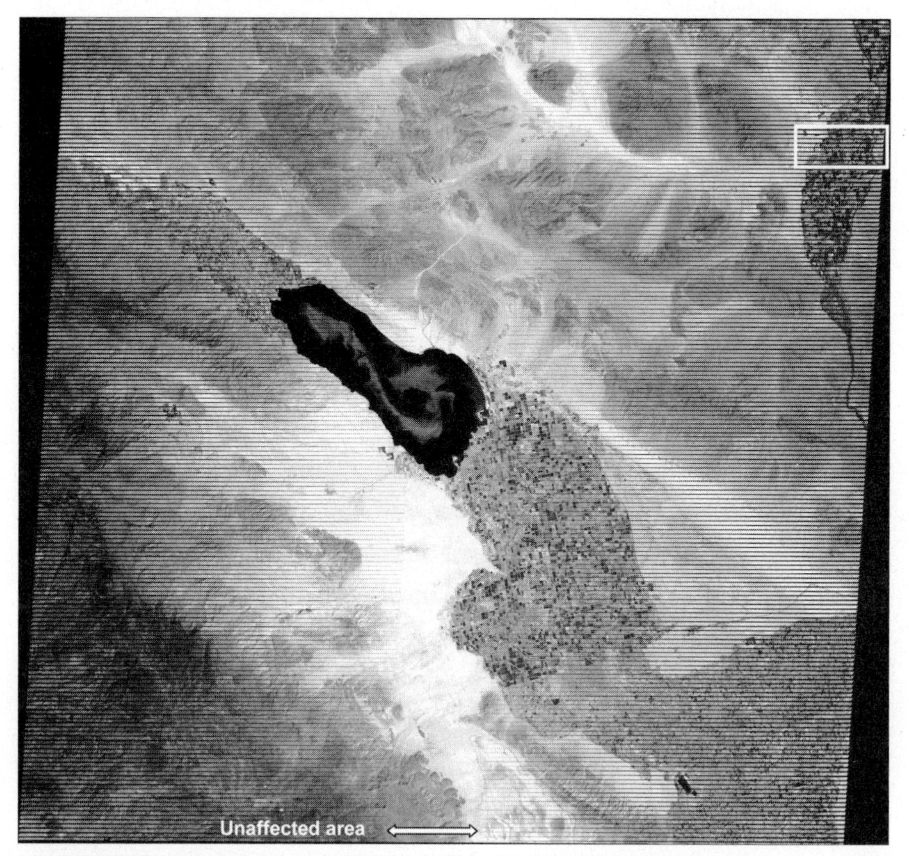

Figure 5.11 SLC-off, Landsat-7, band 3, full-scene image of California's Salton Sea and surrounding area (path 39, row 37) acquired on September 17, 2003. Level 1G (geometrically corrected). Data gap problems do not affect the center of the image, but are more pronounced toward the edges of the image. White rectangle shows area enlarged in Figure 5.12. (Courtesy USGS.)

The Landsat-1 through -7 missions combined to provide over 40 years of invaluable image coverage of the earth's land surface, coastal shallows, and coral reefs. These data now form a baseline for continued synoptic monitoring of the dynamics of global resources and processes at moderate spatial and temporal scales. At the same time, the Landsat-1 through Landsat-7 era was laden with a multitude of both administrative management and technical challenges. Several U.S. presidential administrations and sessions of the U.S. Congress debated the appropriate policy and operation of the U.S. land remote sensing program. Funding for the program and establishing the cost of the data resulting from the program were persistent issues (and continue to be so). Over time, the distribution of Landsat data in the United States has gone through four distinct phases: experimental, transitional, commercial, and governmental. Since 2008, the United

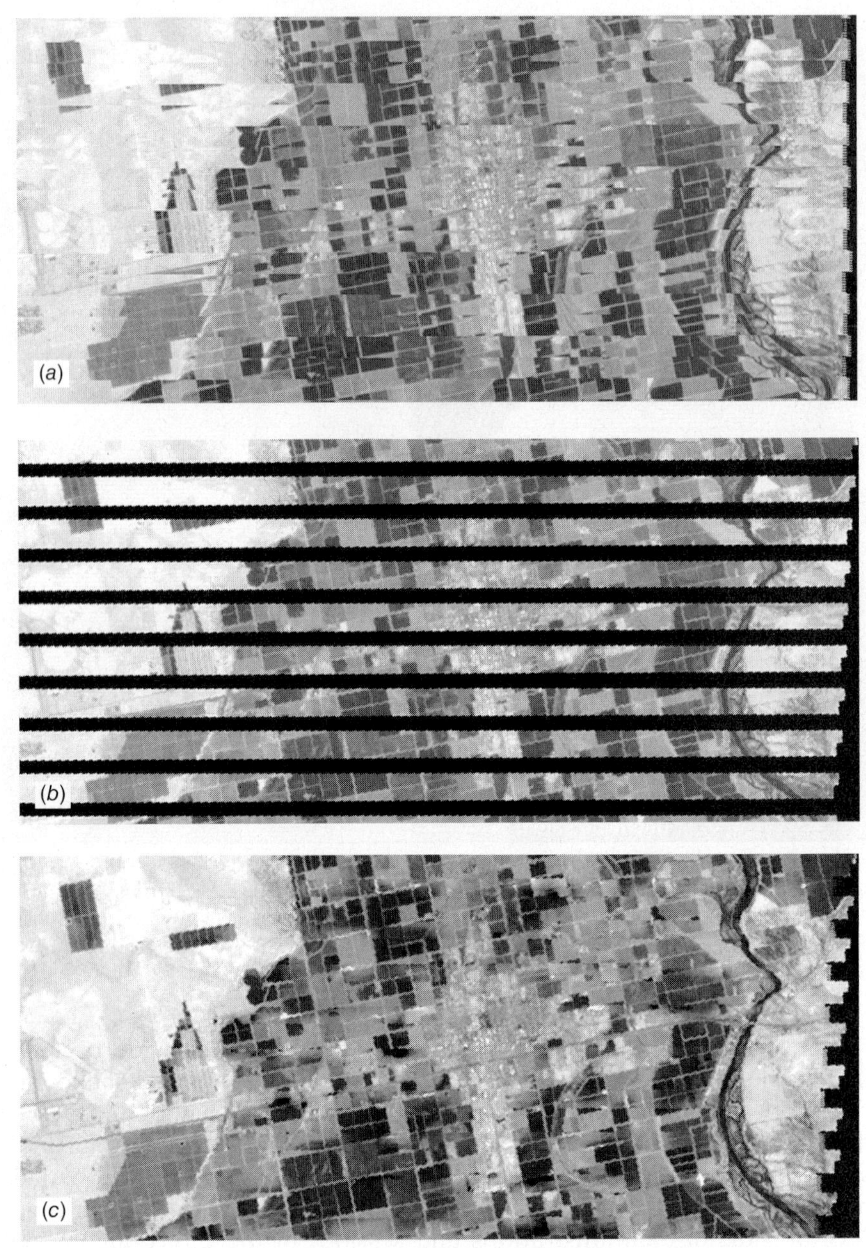

Figure 5.12 Enlargements of area shown in the white rectangle in Figure 5.11. (*a*) Pixel duplications present near the edge of the scene prior to elimination. (*b*) Data gaps present after duplication elimination and geometric correction. (*c*) Gap-filled image—gaps were filled with data from a near-anniversary date (September 14, 2002), pre-SLC-image of the same scene. (Courtesy USGS.)

States Geological Survey (USGS) has provided downloads of all archived Landsat scenes at no charge to any user through the *Earth Resources Observation and Science (EROS)* Center, located in Sioux Falls, South Dakota. Prior to the implementation of this policy, a maximum of approximately 25,000 Landsat images were sold in a year. Under the new policy, over 9 million images were downloaded in less than four years.

The USGS archive contains over 2.8 million images, and efforts are underway to consolidate nonredundant images that have been collected at international cooperator receiving stations. This initiative is expected to increase the total EROS archive by 3 to 4 million images, with 75% of the new images coming from the European Union, Australia, Canada, and China.

Further treatment of the history and technology of the Landsat-1 through -7 missions is beyond the scope of this discussion. (Additional information about these missions is available in such references as Lauer et al. (1997), PERS (2006), Wulder and Masek (2012), and previous editions of this book). Our focus in the remainder of this discussion is on summarizing the Landsat-8 mission.

5.5 LANDSAT-8

As previously mentioned, Landsat-8 was launched on February 11, 2013, from the Kennedy Space Center using an Atlas V rocket. During the prelaunch phase of the Landsat-8 mission, the program was called the *Landsat Data Continuity Mission (LDCM)*. Hence, "LDCM" and "Landsat-8" refer to the same mission and may be used interchangeably in other references or papers. As the LDCM moniker suggests, the primary purpose of the mission is to extend the collection of Landsat data on an operational basis into the future. NASA and USGS were the principal cooperators in the conception and operation of the mission. NASA was responsible primarily for the space segment of the mission, including spacecraft and sensor development, spacecraft launch, systems engineering, prelaunch calibration, and in-orbit checkout of the space segment. USGS led the development of the ground system and assumed the mission operation after completion of an on-orbit checkout period. Those operations include post-launch calibration, scheduling of data collection, as well as receiving, archiving, and distributing the resulting data.

The high level objectives for the mission are to (1) collect and archive moderate resolution (30 m GSD) reflective, multispectral image data permitting seasonal coverage of the global land mass for a period of no less than five years; (2) collect and archive moderate- to low-resolution (120 m) thermal multispectral image data, permitting seasonal coverage of the global land mass for a period of not less than three years; (3) ensure that the resulting data will be sufficiently consistent with the data from the previous Landsat missions in terms of acquisition geometry, coverage, spectral and spatial characteristics, calibration, output data quality, and data availability to permit assessment of land cover and land use change over multi-decadal periods; and (4) distribute standard data products to the general public on a nondiscriminatory basis and at a price no greater than the incremental cost of fulfilling a user request. (This latter goal is being reached by providing data distribution over the Internet at no cost to the user.)

Note that the minimum design life for the optical component of the Landsat-8 mission is five years and that for the thermal component is only three years. This difference results from the fact that the thermal system was approved considerably later than the optical one. The relaxation in design life was specified in order to expedite the development of the thermal component. The designers could save development time through selective redundancy in subsystem components as opposed to pursuing a more robust redundancy that would be dictated by a five-year design life. In any case, Landsat-8 includes enough fuel for at least 10 years of operation.

Landsat-8 was launched into a repetitive, near-circular, sun-synchronous, near polar orbit that is functionally equivalent to the orbits used for Landsats-4, -5, and -7. As was shown in Figure 5.2, this orbit has a nominal altitude of 705 km at the equator and has an inclination angle of 98.2° (8.2° from normal) with respect to the equator. The satellite crosses the equator on the north-to-south portion of each orbit at 10:00 A.M. mean local time (±15 min). Each orbit takes approximately 99 min, with just over 14.5 orbits being completed in a day. Due to earth rotation, the distance between ground tracks for consecutive orbits is approximately 2752 km at the equator (Figure 5.13).

The above orbit results in a 16-day repeat cycle for the satellite. As shown in Figure 5.14, the time interval between adjacent coverage tracks of the satellite is 7 days. The east-west swath width of each Landsat-8 scene continues to be 185 km, and the north-south dimension of each image is 180 km. All images are cataloged according to their location in the *Worldwide Reference System-2* (*WRS-2*). In this system each orbit within a cycle is designated as a path. Along these paths, the individual nominal sensor frame centers are designated as rows. Thus, a scene can be uniquely identified by specifying a path, row, and date. WRS-2 is made up of 233 paths, numbered 001 to 233, east to west, with path 001 crossing the equator at longitude 64°36′ W. The rows are numbered such that row 60 coincides with the equator on the orbit's descending node. Row 1 of each path starts at 80°47′N latitude. The WRS-2 system was also used for Landsat-4, -5, and -7. There is an eight-day offset between Landsat-7 and Landsat-8 coverage of each WRS-2 path. (Having a different orbital pattern, Landsat-1, -2, and -3, are cataloged using WRS-1. WRS-1 has 251 paths and the same number of rows as WRS-2.)

Landsat-8 incorporates two sensors, the *Operational Land Imager* (*OLI*) and the *Thermal Infrared Sensor* (*TIRS*). These two sensors have identical 15° fields of view and collect coincident datasets of the same ground area that are subsequently merged during ground processing into a single data product for each WRS-2 scene. The data are initially stored by an onboard solid-state recorder and then transmitted via an X-band data stream to several ground receiving stations. These receiving stations not only include the EROS Center in the United

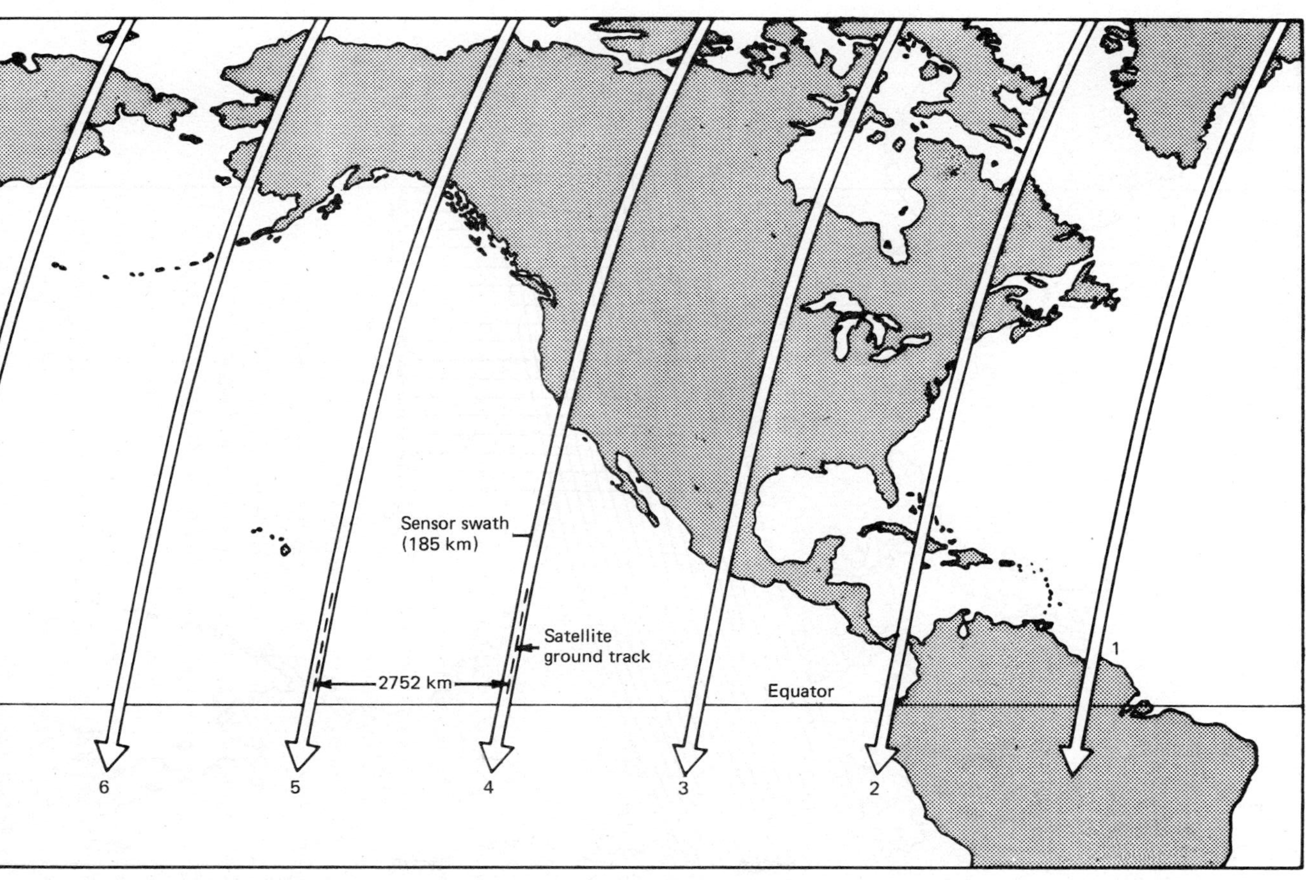

Figure 5.13 Spacing between adjacent Landsat-8 orbit tracks at the equator. The earth revolves 2752 km to the east at the equator between passes. (Adapted from NASA diagram.)

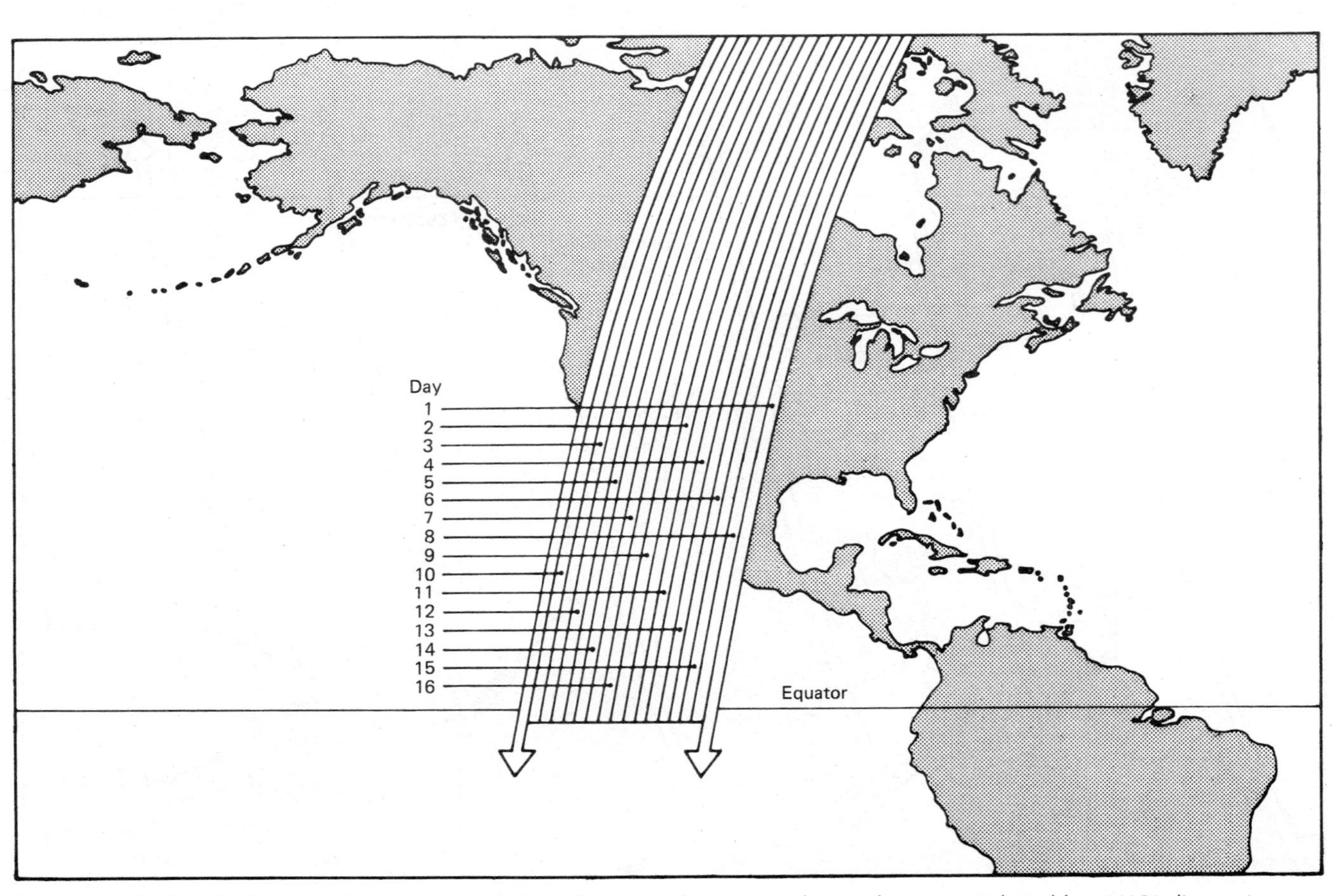

Figure 5.14 Timing of adjacent Landsat-8 coverage tracks. Adjacent swaths are imaged seven days apart. (Adapted from NASA diagram.)

States, but those operated by the network of sponsoring foreign governments cooperating in the Landsat program, referred to as International Cooperators (ICs). The various ground receiving stations forward the data to the EROS Center, which operates the *Data Processing and Archive System* (*DPAS*) for ingesting, processing, archiving, and disseminating all Landsat-8 data. The DPAS is designed to process an average of at least 400 scenes per day, within 24 hr of observation. The 12-bit data from both Landsat-8 sensors are typically processed to produce a "Level 1T" orthorectified (terrain corrected) GeoTIFF image product that is radiometrically corrected and co-registered to a UTM cartographic projection (WGS84 datum) or, in the case of polar scenes, registered to the Polar Stereographic projection. The OLI multispectral data are resampled to a 30 m GSD, and the panchromatic band is resampled to a 15 m GSD. The TIRS data are oversampled from the original 100 m sensor resolution (which exceeds the 120 m minimum requirement for the thermal bands) to a 30 m GSD for alignment with the OLI data. The specified spatial accuracy for the data is approximately 12 m CE 90.

Both the OLI (Figure 5.15*a*) and TIRS (Figure 5.15*b*) sensors employ linear array technology and pushbroom scanning, a first for the Landsat program. The primary advantages to the pushbroom technology are an improved signal-to-noise ratio due to longer detector dwell time, lack of moving parts, improved imaging platform stability, and consistent internal image geometry. The primary disadvantage relative to whiskbroom scanning is the need to cross-calibrate

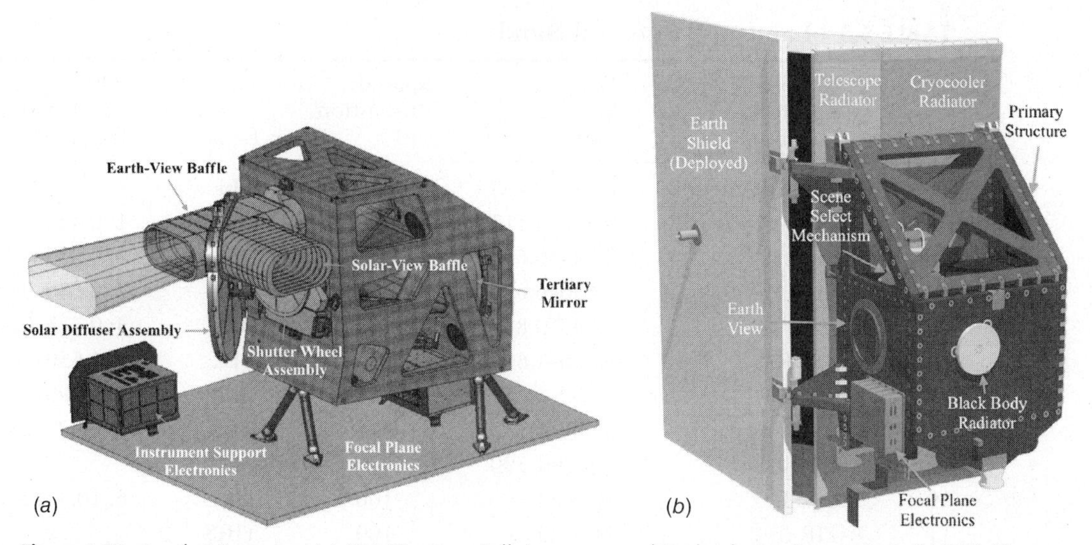

Figure 5.15 Landsat-8 sensors: (*a*) OLI (Courtesy Ball Aerospace and Technologies Corporation); (*b*) TIRS (Courtesy NASA). Reprinted from *Remote Sensing of Environment*, vol. 122, Irons, J.R., Dwyer, J.L., and J.A. Barsi, "The next Landsat satellite: The Landsat Data Continuity Mission," pp. 13 and 17 respectively, copyright 2012, with permission from Elsevier.

thousands of detectors to achieve spectral and radiometric uniformity across the focal plane of each instrument. In this regard, the OLI incorporates 6500 detectors per multispectral band and 13,000 detectors for the panchromatic band. These are distributed among 14 focal plane modules. The system also includes a covered "blind" band of detectors that is used to estimate detector bias during image acquisitions. Thus, the total number of detectors incorporated into the OLI is over 75,000.

TIRS images result from pushbroom scanning using three *quantum-well-infrared photodetector* (*QWIP*) arrays located in a cryogenically cooled focal plane. Each array is 640 detectors long in the cross-track direction with each detector providing a 100-m GSD. The arrays overlap slightly such that they produce an effective composite array of 1850 pixels to image a 185-km swath width. TIRS is the first spaceflight instrument to employ QWIP detector technology. The calibration of these detectors is accomplished by using a mirror controlled by a scene selector mechanism to alternate the field of view of the sensor between nadir (Earth), a blackbody calibration source, and a deep space view.

Table 5.5 summarizes the spectral bands resulting from compositing OLI and TIRS image data. In the interest of providing data consistency with the Landsat-7 ETM+ bands, several of the Landsat-8 reflective bands are identical to, or mimic closely, the ETM+ reflective bands. However, the widths of several of the OLI bands are refined in order to avoid or lessen the influence of various atmospheric absorption features present within the ETM+ bands. For example, OLI band 5 (0.845–0.885 μm) excludes a water vapor absorption feature located at 0.825 μm occurring near the

TABLE 5.5 Landsat-8 Spectral Bands

Band	Nominal Spectral Location	Band Width (μm)	Spatial Resolution (m)	Sensor	ETM+ Band Width Band: (μm)
1	Coastal/Aerosol	0.433–0.453	30	OLI	
2	Blue	0.450–0.515	30	OLI	1: 0.450–0.515
3	Green	0.525–0.600	30	OLI	2: 0.525–0.605
4	Red	0.630–0.680	30	OLI	3: 0.630–0.690
5	NIR	0.845–0.885	30	OLI	4: 0.775–0.900
6	SWIR 1	1.560–1.660	30	OLI	5: 1.550–1.750
7	SWIR 2	2.100–2.300	30	OLI	7: 2.090–2.350
8	Panchromatic	0.500–0.680	15	OLI	8: 0.520–0.900
9	Cirrus	1.360–1.390	30	OLI	
10	TIR 1	10.6–11.2	100	TIRS	6: 10.4–12.5
11	TIR 2	11.5–12.5	100	TIRS	

middle of the ETM+ near-infrared band (0.775–0.900 μm). Band 8 of the OLI, the panchromatic band, has also been narrowed relative to the ETM+ panchromatic band to improve the contrast between vegetated versus nonvegetated areas.

In addition to the above refinements in spectral band widths, the OLI includes two new bands. The first of these is band 1, a deep-blue band (0.433–0.453 μm) primarily to aid in monitoring ocean color in coastal areas as well as to estimate aerosol characteristics (e.g., reflectance, concentration, optical thickness). The second is band 9, a shortwave infrared band (1.360–1.390 μm) that contains a strong water absorption feature and facilitates the detection of cirrus clouds in OLI images. Cirrus clouds are thin wispy clouds that tend to be aligned in the same direction and generally do not completely cover the sky. Because they occur very high in the atmosphere, where it is very cold, they consist of ice crystals rather than water droplets. Being very thin, cirrus clouds do not reflect much energy in the visible portion of the spectrum. In band 9 imagery they appear bright relative to most land features. This can aid in assessing cirrus contamination in other bands.

The two TIRS bands present in Landsat-8 image data were added late in the mission design process primarily to aid in the measurement of water consumption over irrigated agricultural fields, especially in the semi-arid western United States. Efforts to obtain such data were advocated by several western state water management agencies, led by the Western States Water Council. In order to design and build TIRS on an accelerated schedule, a decision was made to specify a 120-m resolution for the thermal bands. (The actual final design features 100-m resolution.) While this represented a step back from the 60-m spatial resolution of the thermal data collected by Landsat-7 ETM+, the 120 m resolution was deemed sufficient for the water consumption application, which primarily involves monitoring fields that are irrigated by center-pivot irrigation systems, both in the U.S. Great Plains and in many other areas across the globe. These systems typically apply water over circular areas having diameters in the 400-m to 800-m range. The thermal data are also useful in many other applications besides measuring evapotranspiration and water consumption. These include mapping urban heat fluxes, monitoring heated water discharge plumes near power plants, and many others.

The particular bands used to acquire TIRS data were selected to enable correction of the thermal data for atmospheric effects using a "split window" algorithm. The basis for such algorithms is that (in the 10.0 μm–12.5 μm range) a reduction of surface emitted radiance due to atmospheric absorption is proportional to the difference of simultaneous at-sensor radiance measurements made at two different wavelengths (Caselles et al., 1998; Liang, 2004; Jiménez-Muñoz and Sobrino, 2007).

Overall, the Landsat-8 sensors and data are very well calibrated, which is essential to accurate detection and quantification of earth surface change over time. The integrity of Landsat-8 data not only results from extensive pre-launch calibration, but also from continuing on-orbit sensor calibration (augmented by ground-based measurements) throughout the life of the mission. This involves performing many different types of calibration operations over a range of nominal temporal scales. These include:

- **Twice every earth imaging orbit (approximately every 40 min).** During image acquisition, light enters the OLI optical system through an earth-view baffle (Figure 5.15*a*). Before and after each earth imaging interval a *shutter wheel assembly* is rotated on command into a position that forms a shutter over the OLI field of view and prevents light from entering the instrument. This provides a means of monitoring, and compensating for, the dark bias in the radiometric response of the system's detectors during imaging. The dark bias of the TIRS detectors is also determined twice each imaging orbit by first pointing the field of view of the TIRS instrument (Figure 5.15*b*) at deep space and then at an on-board blackbody of known and controlled temperature.
- **Once per day.** The OLI includes two *stimulation lamp assemblies*, each containing six small lamps inside a hemisphere. These lamps are capable of illuminating the OLI focal plane through the full OLI optical system with the instrument's shutter closed. One lamp assembly is used daily to monitor the stability of the OLI radiometric response to the known lamp illumination. The second lamp assembly is used at weekly (or longer) time intervals to aid in separating intrinsic changes in the intensity of the primary lamp assembly over time from changes in the radiometric response of the OLI sensor itself.
- **Once per week.** Solar-view OLI calibrations are nominally employed on a weekly basis to monitor the overall calibration of the sensor and to perform detector-to-detector normalization. These calibrations require a spacecraft maneuver to point the solar-view aperture toward the sun and employment of the three-position *solar diffuser wheel assembly*.

  When the OLI is being used for imaging, the solar diffuser wheel is rotated into a position that simply contains a hole that allows light to enter the sensor directly through the earth-view baffle. In either of the other two wheel positions, a solar diffuser panel is introduced to block the optical path through the earth-view baffle, the solar-view baffle is pointed at the sun, and the diffuser panel reflects solar illumination onto the OLI optical system. One position of the diffuser wheel introduces a "working" panel into the optical path, and the other exposes a "pristine" panel. The working panel is nominally employed on a weekly basis, and the pristine panel is exposed less frequently (approximately semiannually). The primary purpose of the pristine panel is to monitor changes in the spectral reflectance of the working panel due to frequent solar exposure.
- **Once per month.** A number of earth observation satellites, including the EO-1 system, have used the moon as a calibration source because the reflectance properties of the lunar surface are stable and can be modeled as a function of view and illumination geometries (Kieffer et al., 2003).

The OLI radiometric calibration is validated on a monthly basis by maneuvering the Landsat-8 spacecraft to image the moon near its full phase during the dark portion of the system's orbit.

- **Aperiodic.** As with many other satellite systems, including previous Landsat satellites, Landsat-8 collects image data over a range of surface calibration sites at irregular time intervals in order to further validate sensor calibration. These sites include such features as large lake areas of known surface temperature, dry lake playas for which ground observations of surface reflectance and atmospheric conditions are made at the time of overflight, and sites containing features for which highly accurate ground control point information exists.

Readers who wish to obtain additional information about the Landsat-8 mission in general, and its sensor payload in particular, are encouraged to consult Irons et al. (2012), the primary source upon which the above discussion is based.

Figure 5.16 illustrates several spectral bands associated with a subscene of Landsat-8 summertime imagery acquired over the diverse landscape of Juneau, Alaska, and vicinity. Four color composite images produced from the same data set are also shown in Plate 13. Juneau is the capital city of Alaska and is located within the Borough of Juneau. (Boroughs in Alaska are administrative areas similar to counties found elsewhere in the United States.) Juneau Borough has a population of approximately 32,000 and comprises approximately 8430 km^2.

The city of Juneau is located on the Gastineau Channel in the Alaskan Panhandle (SE Alaska) below steep mountains about 1100 m high and can be seen near the eastward end of the bridge crossing the channel in the lower right of this subscene. Atop these mountains is the Juneau Icefield, a large ice mass from which about 30 glaciers flow, including the Mendenhall Glacier (located outside of the coverage area of this subscene). The silt-laden meltwater from these glaciers flows via the Mendenhall River and into the Gulf of Alaska shown in the lower left of this subscene. Two of these glaciers (the Ptarmigan Glacier and the Lemon Creek Glacier) can be seen at the upper right of this subscene. Juneau Borough shares its eastern border with British Columbia, but it is landlocked, in that there are no roads across the mountains into British Columbia. Access to Juneau is by water and air. Vehicles are frequently ferried to and from the area. The airport serving Juneau can be seen just left of the center of this subscene.

The range of land cover types and features present in Figure 5.16 and Plate 13 helps to illustrate the relative utility of the various Landsat-8 spectral bands for differentiating among numerous landscape elements. For example, in addition to the airport itself, several other areas of urban and built-up land can be seen, including the developed area above the airport, roads and highways, downtown Juneau, and the urban area located across the channel from Juneau. These areas are best shown in the Landsat-8 visible bands included in (*a*), (*b*), (*c*), (*d*), and (*e*) of Figure 5.16 and the normal color composite, (*a*), of Plate 13.

Figure 5.16 Selected Landsat-8 OLI and TIR bands, Juneau, Alaska, and vicinity, August 1, 2013. Scale 1:190,000. (*a*) Band 8, 0.500–0.680 μm (panchromatic); (*b*) band 1, 0.433–0.453 μm (coastal aerosol); (*c*) band 2, 0.450–0.515 μm (blue); (*d*) band 3, 0.525–0.600 μm (green); (*e*) band 4, 0.630–0.680 μm (red); (*f*) band 5, 0.845–0.885 μm (near IR), (*g*) band 6, 1.560–1.660 μm (SWIR 1), (*h*) band 7, 2.100–2.300 μm (SWIR 2); (*i*) band 10, 10.6–11.2 μm (TIR 1). (See Table 5.5 for additional information about these bands.) (Author-prepared figure.)

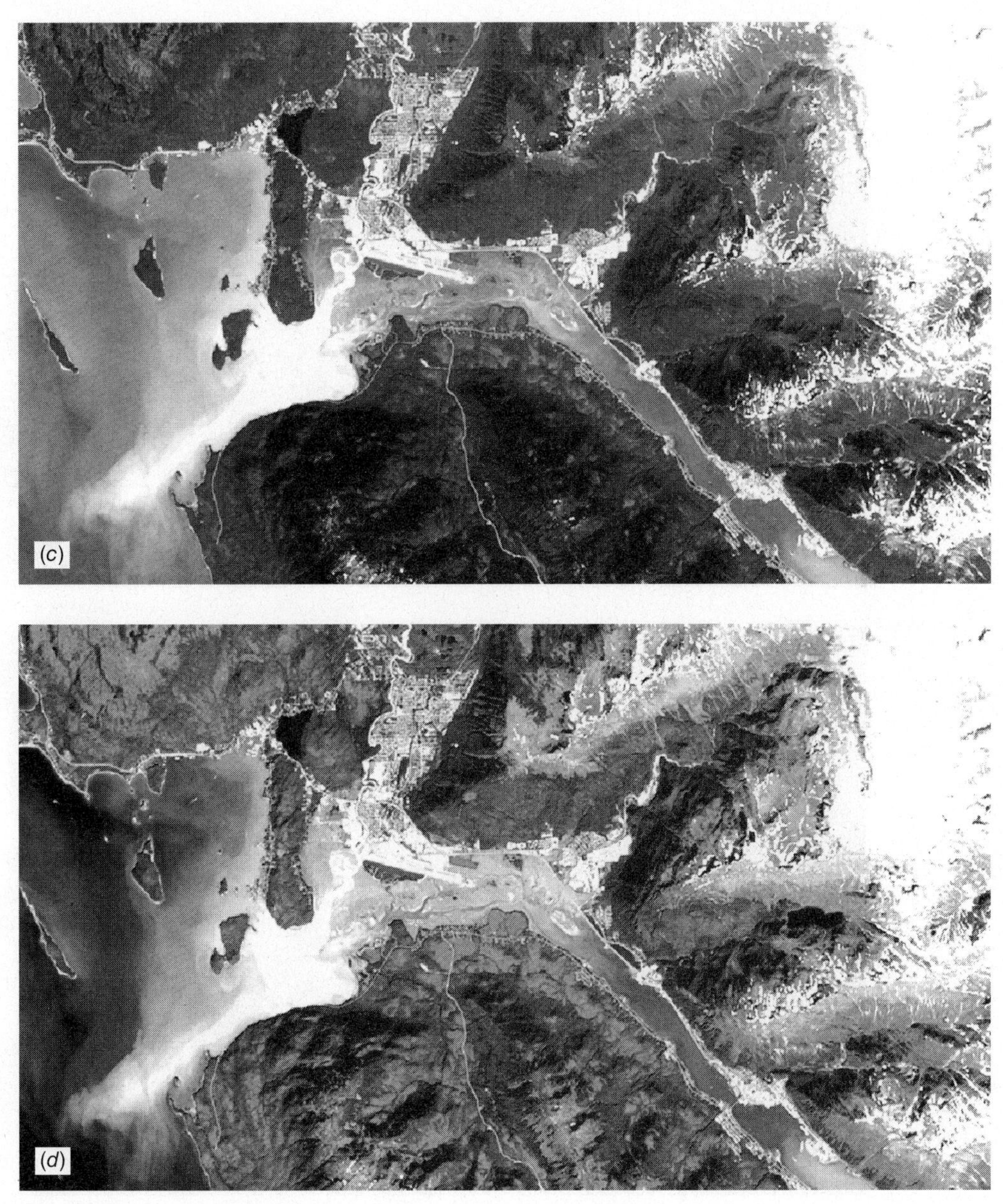

Figure 5.16 *(Continued)*

In the lower left portion of this subscene, the cold, silt-laden meltwater from the Mendenhall Glacier enters the saltwater of the Gulf of Alaska. The plume of silt-laden water is best seen in the visible bands of the image, (*a*)–(*e*) of Figure 5.16 and the Plate 13 color composites containing at least one visible band, (*a*)–(*c*). Also note that the appearance of the plume varies with the water

Figure 5.16 *(Continued)*

penetration (and resolution) of each band. In the thermal infrared band, the Mendenhall glacier meltwater is darker (colder) than the salt water into which it flows. (Waterways entering coastal areas in more temperate situations typically appear warmer than the receiving water.)

Figure 5.16 *(Continued)*

The snow and ice cover shown in the upper right of Figure 5.16, including the Ptarmigan and Lemon Creek Glaciers, is highly reflective in all the visible bands (*a*)–(*e*) as well as in the near-IR band (*f*). However, this cover type is highly absorbent in the shortwave-IR bands and appears very dark in bands (*g*) and (*h*). Snow

Figure 5.16 *(Continued)*

and ice is also distinguished from other cover types in Plate 13, especially in (*c*) and (*d*). Similarly, vegetated areas are best distinguished from other cover types in Plate 13.

5.6 FUTURE LANDSAT MISSIONS AND THE GLOBAL EARTH OBSERVATION SYSTEM OF SYSTEMS

As important and successful as the Landsat program has been over the course of its history, it has never enjoyed a long-term programmatic design or clear budgetary future. This fact is aptly stated in a recent National Research Council report titled *Landsat and Beyond: Sustaining and Enhancing the Nation's Land Imaging Program* (NRC, 2013). The preface for this report states:

> The nation's economy, security, and environmental vitality rely on routine observations of Earth's surface to understand changes to the landscape at local, regional, and global scales. The National Aeronautics and Space Administration (NASA) conceived and built the first Landsat satellites as a research activity. Over the years, Landsat missions have assumed an operational character, with a diverse set of users reliant on the continuing availability of Landsat imagery and derived data products. However, responsibility for funding, management, development, and operations of the Landsat series has changed hands numerous times, with responsibilities shifting among government agencies and private-sector entities. While the U.S. Department of the Interior's (DOI's) U.S. Geological Survey (USGS)

has established and maintained management of land remote sensing data acquisition, archiving, and dissemination, no clearly defined and sustainable land imaging program has yet to be created.

Thus, at this juncture (2015), the future prospects for the Landsat program are closely tied to continuing political discourse and budgetary uncertainty. As the above report stresses, "Landsat-8 has only a 5-year design life, and there is no assured successor. Landsat-9 is under discussion in the U.S. executive and congressional branches, but its configuration remains under debate. Prospects for missions beyond Landsat-9 are unclear, and the sharing of responsibilities with commercial and foreign contributors has not been articulated."

In addition to detailing and lamenting the chaotic history and ill-defined programmatic direction of the Landsat program, the NRC report describes several opportunities to be considered in order to move toward creation of a vibrant *Sustained and Enhanced Land Imaging Program* (*SELIP*) into the future. These range from changing the satellite acquisition and procurement process, to integrating Landsat designs more fully with the data sources generated by systems operated by international and commercial partners, widening the Landsat swath width to reduce revisit time, to employing constellations of small satellites as an augmentation or replacement for one satellite with a full suite of sensors. Under any of these technological scenarios, it is clear that cost will be a major consideration. The life-cycle costs for every Landsat mission since that of Landsat-4 have been approximately $1 billion, when adjusted to 2012 dollars. Such costs are a clear impediment to program sustainability.

One development that offers hope for the future of the Landsat program is the formulation and release of the first-ever *National Plan for Civil Earth Observations*. This plan was released by the White House Office of Science & Technology Policy (OSTP, 2014). It establishes priorities and related actions needed to continue and enhance U.S. earth observations of all types dealing with monitoring earth's land surfaces, oceans, and atmosphere. The report highlights the importance of a range of civil earth observations from space in this context. With respect to land imaging, it specifically charges NASA and USGS with the joint implementation of a 25-year program for sustained land imaging that results in future data that are "fully compatible with the 42-year record of Landsat observations." The plan also stresses the role of non-optical sensing capabilities to supplement optical imaging, and the need for international cooperation.

Irrespective of its future programmatic and technological form, a vibrant SELIP will continue to be central to the *U.S. Group on Earth Observations* (*USGEO*). USGEO was established in 2005 under OSTP's Committee on Environmental and Natural Resources to lead federal efforts to achieve the goal of an integrated earth observation system at the national level. USGEO includes representatives from 17 federal agencies and the Executive Office of the President. USGEO is a founding member and important contributor to the international

Group on Earth Observations (*GEO*). GEO includes more than 85 member countries, the European Commission, and over 65 other participating organizations and is developing the *Global Earth Observation System of Systems* (*GEOSS*).

The basic premise for the development of GEOSS is that many remote sensing, in-situ, surface-based, and ocean-based earth observation systems exist across the globe, but these systems are usually single purpose in nature. "GEOSS recognizes that no matter how effective and efficient all of our single-purpose Earth observation systems may be, their value multiplies when they work in synergy" (www.usgeo.gov). GEOSS focuses on facilitating such synergy on an international and interdisciplinary basis by engendering the linkage and interoperability of observation and information systems for monitoring and forecasting changes in the global environment. The aim is to encourage policy development based on science derived from more than "snapshot assessments." Nine "social benefit areas" form the current focus of GEOSS activities: agriculture, biodiversity, climate, disasters, ecosystems, energy, human health, water, and weather. Further information about these efforts can be found at www.earthobservations.org.

5.7 SPOT-1 TO -5

The French government decided to undertake the development of the SPOT satellite program in 1978. Shortly thereafter, Sweden and Belgium agreed to participate in the program. Conceived and designed by the French Centre National d'Etudes Spatiales (CNES), SPOT developed into a large-scale international program with ground receiving stations and data distribution outlets located in nearly 40 countries with over 30 million images collected since the beginning of the program. The first satellite in the program, SPOT-1, was launched from the Kourou Launch Range in French Guiana on February 21, 1986, onboard an Ariane launch vehicle. This satellite began a new era in space remote sensing, for it was the first earth resource satellite system to include a linear array sensor and employ pushbroom scanning techniques. It also was the first system to have pointable optics. This enabled side-to-side off-nadir viewing capabilities, and it afforded full-scene stereoscopic imaging from two different satellite tracks permitting coverage of the same area.

Table 5.6 summarizes the sensors used on the SPOT-1 to -5 missions. All of these earlier systems were launched by ESA Ariane rockets and involved some form of government funding. However, the long-range goal for the SPOT program was to make it a wholly commercially sponsored enterprise. This goal was reached during the development of SPOT-6 and -7, which form the subject of the next section of this discussion.

Like the Landsat satellites, SPOT satellites employ a circular, near-polar, sun-synchronous orbit. The nominal orbit has an altitude of 832 km and an inclination of 98.7°. The satellites descend across the equator at 10:30 A.M. local solar

TABLE 5.6 Sensors Used on SPOT-1 to -5 Missions

Mission	Launched	High Res. Instruments	Spectral Bands	Vegetation Instrument	Vegetation Spectral Bands
1–3	February 21, 1986 January 21, 1990 September 25, 1993	2 HRVs	1 Pan (10 m) Green (20 m) Red (20 m) NIR (20 m)	—	—
4	March 23, 1998	2 HRVIRs	1 Pan (10 m) Green (20 m) Red (20 m) NIR (20 m) Mid IR (20 m)	Yes	Blue (1000 m) Red (1000 m) NIR (1000 m) Mid IR (1000 m)
5	May 3, 2002	2 HRGs	2 Pan (5 m) Green (10 m) Red (10 m) NIR (10 m) Mid IR (20 m)	Yes	Blue (1000 m) Red (1000 m) NIR (1000 m) Mid IR (1000 m)

time, with somewhat later crossings in northern latitudes and somewhat earlier crossings in southern latitudes. For example, SPOT crosses areas at a latitude of 40°N at approximately 11:00 A.M. and areas at a latitude of 40°S at 10:00 A.M.

The orbit pattern for SPOT repeats every 26 days. This means any given point on the earth can be imaged using the same viewing angle at this frequency. However, the pointable optics of the system enable off-nadir viewing during satellite passes separated alternatively by one and four (and occasionally five) days, depending on the latitude of the area viewed (Figure 5.17). For example, during the 26-day period separating two successive satellite passes directly over a point located at the equator, seven viewing opportunities exist (day D and days $D + 5$, $+10$, $+11$, $+15$, $+16$, and $+21$). For a point located at a latitude of 45°, a total of 11 viewing opportunities exist (day D and days $D + 1$, $+5$, $+6$, $+10$, $+11$, $+15$, $+16$, $+20$, $+21$, and $+25$). This "revisit" capability is important in two respects. First, it increases the potential frequency of coverage of areas where cloud cover is a problem. Second, it provides an opportunity for viewing a given area at frequencies ranging from successive days to several days to a few weeks. Several application areas, particularly within agriculture and forestry, require repeated observations over these types of time frames.

The sensor payload for SPOT-1, -2, and -3 consisted of two identical *high resolution visible* (HRV) imaging systems and auxiliary magnetic tape recorders. Each HRV operated in either of two modes of sensing: (1) a 10-m-resolution panchromatic (black-and-white) mode over the range 0.51 to 0.73 μm or (2) a 20-m-resolution multispectral (color IR) mode over the ranges 0.50 to 0.59 μm, 0.61 to 0.68 μm, and 0.79 to 0.89 μm. The 10-m data were collected by a 6000-element CCD array, and the 20-m data were collected by three 3000-element CCD arrays. All data were effectively encoded over 8 bits. Each instrument's field of view was 4.13°, such that the ground swath of each HRV scene was 60-km wide if collected under nadir viewing conditions.

The first element in the optical system for each HRV was a plane mirror that could be rotated to either side by ground command through an angle of ±27° (in 45 steps of 0.6° each). This allowed each instrument to image any point within a strip extending 475 km to either side of the satellite ground track (Figure 5.18). The size of the actual ground swath covered naturally varied with the pointing angle employed. At the 27° maximum value, the swath width for each instrument was 80 km. When the two instruments were pointed so as to cover adjacent image fields at nadir, the total swath width was 117 km and the two fields overlapped by 3 km (Figure 5.19). While each HRV instrument was capable of collecting panchromatic and multispectral data simultaneously, resulting in four data streams, only two data streams could be transmitted at one time. Thus, either panchromatic or multispectral data could be transmitted over a 117-km-wide swath, but not both simultaneously. Figure 5.20, which shows Baltimore, Maryland's Inner Harbor, illustrates the spatial resolution of SPOT-1, -2, and -3 panchromatic imagery.

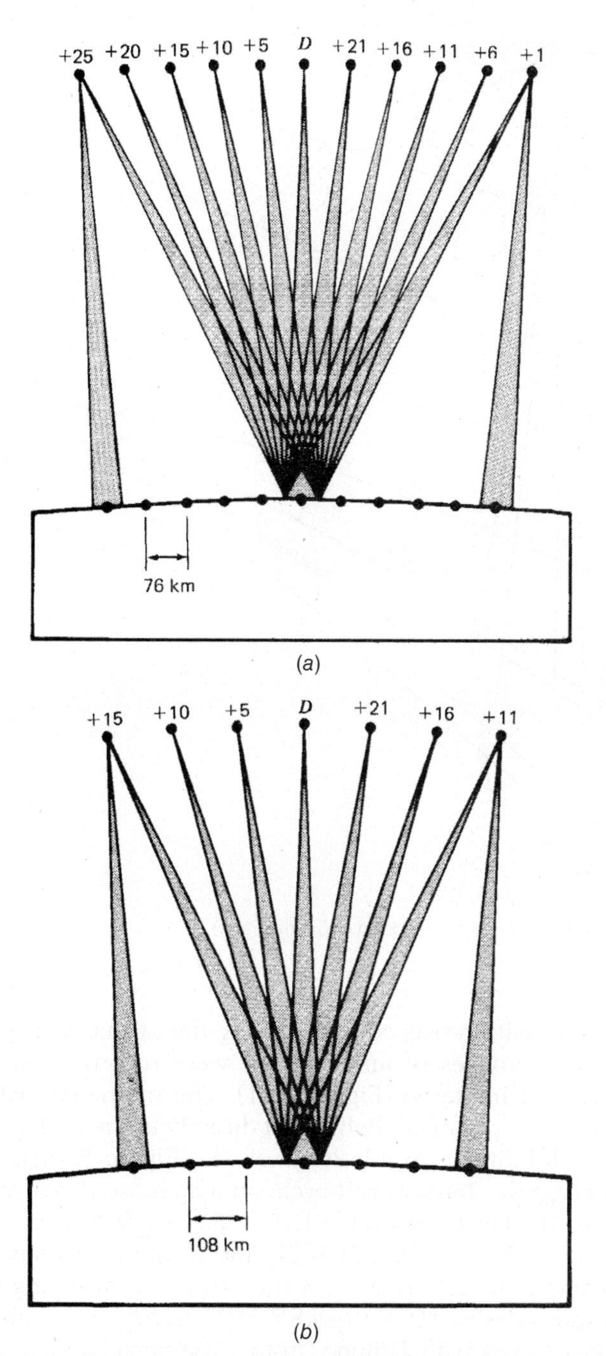

Figure 5.17 SPOT revisit pattern: (*a*) latitude 45°; (*b*) latitude 0°. (Adapted from CNES diagram.)

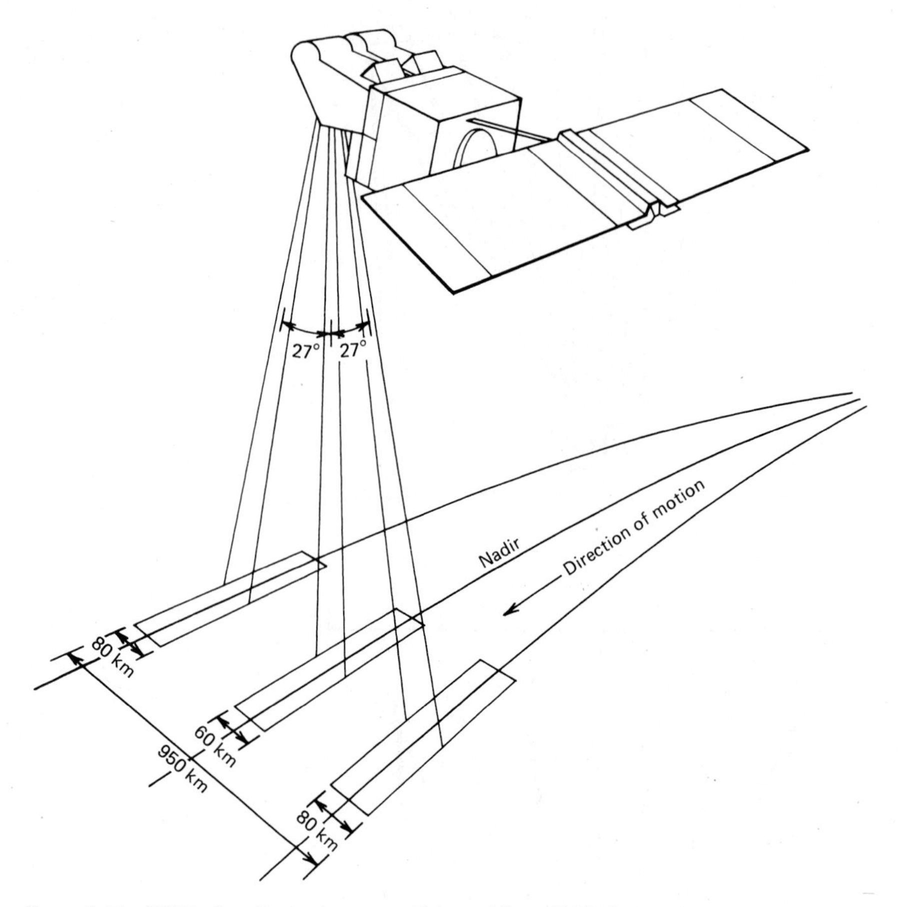

Figure 5.18 SPOT off-nadir viewing range. (Adapted from CNES diagram.)

Stereoscopic image acquisition was collected using the off-nadir viewing capability of the HRV. That is, images of an area that were recorded on different satellite tracks can be viewed in stereo (Figure 5.21). The frequency with which stereoscopic coverage could be obtained, being tied directly to the revisit schedule for the satellite, varied with latitude. At a latitude of 45° (Figure 5.17*a*), there are six possible occasions during the 26-day orbit cycle on which *successive-day* stereo coverage could be obtained (day D with $D + 1$, $D + 5$ with $D + 6$, $D + 10$ with $D + 11$, $D + 15$ with $D + 16$, $D + 20$ with $D + 21$, and $D + 25$ with day D of the next orbit cycle). At the equator (Figure 5.17*b*), only two stereoviewing opportunities on successive days were possible ($D + 10$ with $D + 11$ and $D + 15$ with $D + 16$). The base–height ratio also varied with latitude, from approximately 0.50 at 45° to approximately 0.75° at the equator. (When it was not necessary to acquire stereoscopic

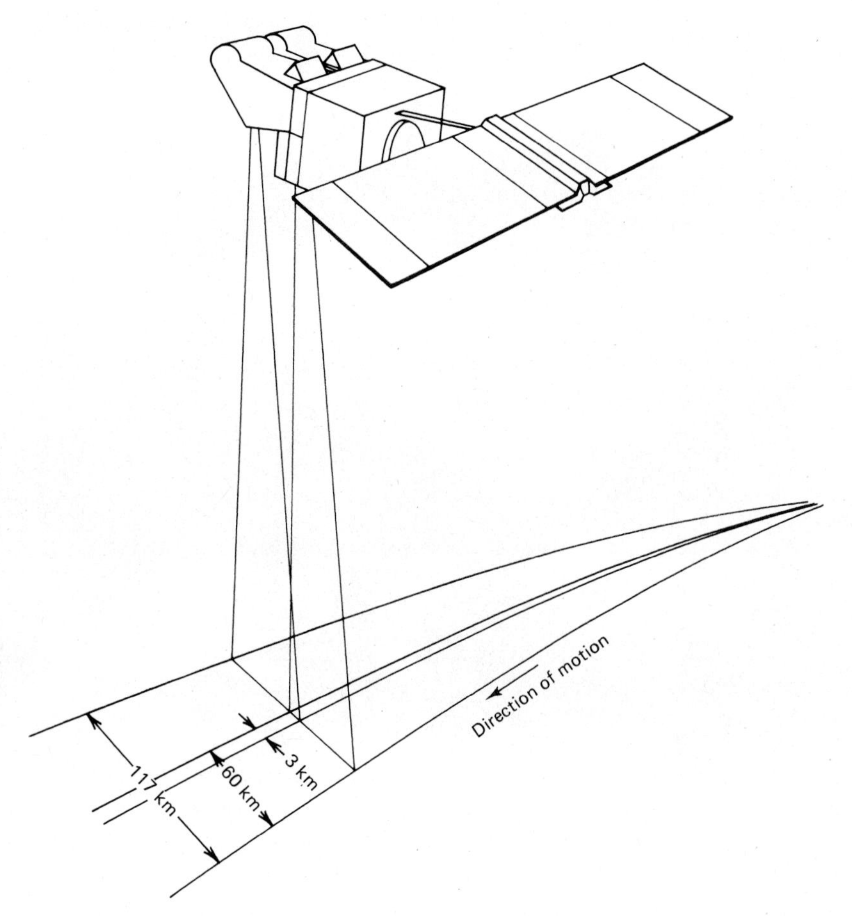

Figure 5.19 SPOT ground coverage with HRVs recording adjacent swaths. (Adapted from CNES diagram.)

coverage on a successive-day basis, the range of possible viewing opportunities and viewing geometries greatly increased.)

Figure 5.22 illustrates a stereopair of images acquired as a preliminary "engineering data set" on the third and fourth days of operation of SPOT-1. In these images of Libya, an eroded plateau is seen at the top and center, and alluvial fans are shown at the bottom. Some vertical streaking can be seen, especially in the left-hand image. These are a consequence of the fact that the individual detectors of the CCDs were not fully calibrated at the time of this early acquisition.

Using the parallax resulting when SPOT data are acquired from two different orbit tracks, perspective views of a scene can be generated. Figure 5.23 is a perspective view of the Albertville area in the French Alps produced in this manner.

Figure 5.20 SPOT-1 panchromatic image, Baltimore, MD, Inner Harbor, late August. Scale 1:70,000. (© CNES 1986. Distribution Airbus Defence and Space (DS)/Spot Image.)

Such views can also be produced by processing data from a single image using DEM data for the same coverage area (as discussed in Chapter 7).

While SPOT-1, -2, and -3 were effectively identical, SPOT-4 was designed to provide long-term continuity of data collection but (like Landsat) with successive improvements in the technical capability and performance of the sensing systems involved. The primary imaging systems on board this satellite were the *high resolution visible and infrared* (*HRVIR*) sensors and the *Vegetation* instrument. The *HRVIR* system included two identical sensors capable of imaging a total nadir swath width of 120 km. Among the major improvements in the system was the addition of a 20-m-resolution band in the mid-IR portion of the spectrum (between 1.58 and 1.75 μm). This band improved the capability of the data for such applications as vegetation monitoring, mineral discrimination, and soil moisture mapping. Furthermore, mixed 20- and 10-m data sets were coregistered onboard instead of during ground processing. This was accomplished by replacing the panchromatic band of SPOT-1, -2, and -3 (0.49 to 0.73 μm) with the "red"

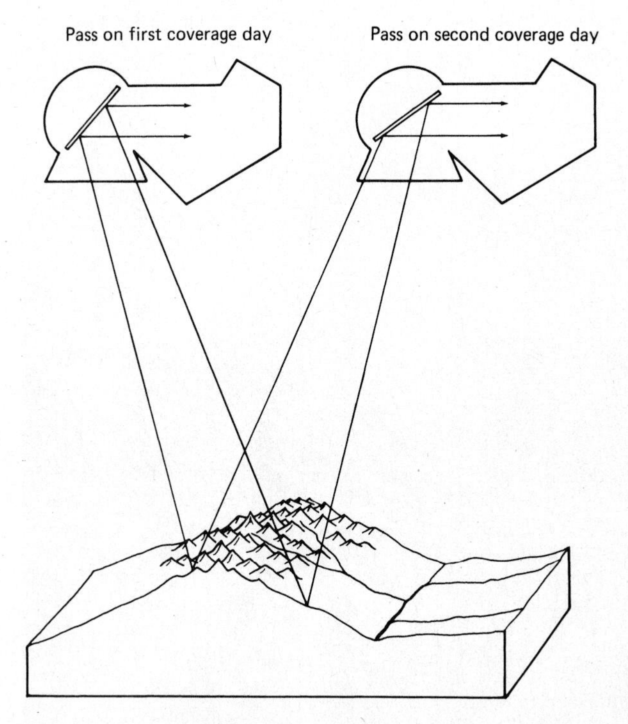

Figure 5.21 SPOT stereoscopic imaging capability. (Adapted from CNES diagram.)

band from these systems (0.61 to 0.68 μm). This band was used to produce both 10-m black-and-white images and 20-m multispectral data.

Another major addition to SPOT-4's payload was the Vegetation instrument. While designed primarily for vegetation monitoring, the data from this system are useful in a range of applications where frequent, large-area coverage is important. (We illustrate several of these applications in our discussion of meteorological and ocean-monitoring satellites.) The Vegetation instrument employed push-broom scanning to provide a very wide angle image swath 2250 km wide with a spatial resolution at nadir of approximately 1 km. It also provided global coverage on a daily basis. (A few zones near the equator were covered every other day.) As shown in Table 5.7, the Vegetation instrument used three of the same spectral bands as the HRVIR system, namely, the red, near-IR, and mid-IR bands, but it also incorporated a blue band (0.43 to 0.47 μm) for oceanographic applications.

Flown in combination, the Vegetation instrument and the HRVIR system afforded the opportunity for coincident sampling of comparable large-area, coarse-resolution data and small-area, fine-resolution data. Comparison of the data from both systems is facilitated by the use of the same geometric reference

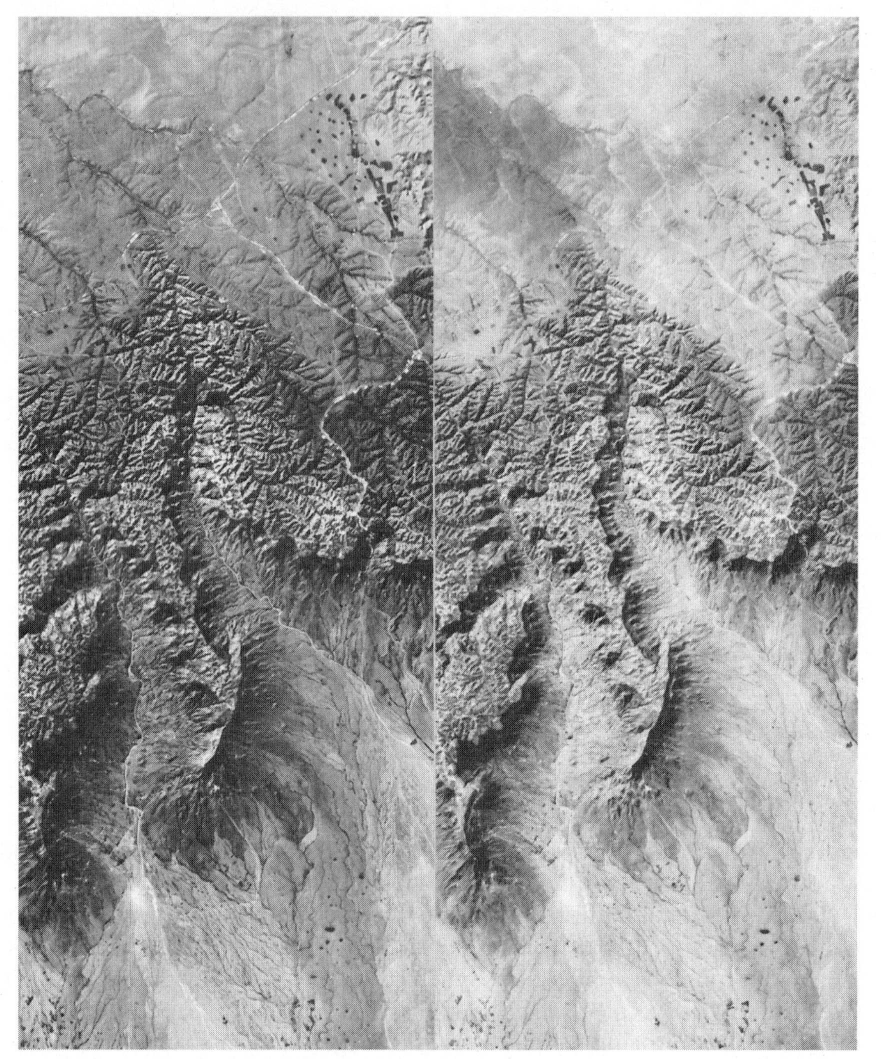

Figure 5.22 SPOT-1 panchromatic image, stereopair, February 24 and 25, 1986, Libya. (© CNES 1986. Distribution Airbus DS/Spot Image.)

system for both instruments and the spectral bands common to the two systems (Table 5.7). Given the flexibility of the combined system, data users could tailor sampling strategies to combine the benefits of Vegetation's revisit capability and the HRVIR system's high spatial resolution to validate the lower resolution data. Applications of the combined data sets range from regional forecasting of crop yield to monitoring forest cover and observing long-term environmental change.

Plate 14 shows four images obtained with the SPOT Vegetation instrument. In (*a*), the red tones associated with the healthy green vegetation in western Europe

Figure 5.23 Perspective view of the Albertville area in the French Alps, generated entirely from SPOT data, on successive days in late July. (© CNES 1993. Distribution Airbus DS/Spot Image.)

TABLE 5.7 Comparison of Spectral Bands Used by the SPOT-4 HRVIRs and the Vegetation Instrument

Spectral Band (μm)	Nominal Spectral Location	HRVIR	Vegetation Instrument
0.43–0.47	Blue	—	Yes
0.50–0.59	Green	Yes	—
0.61–0.68	Red	Yes	Yes
0.79–0.89	Near IR	Yes	Yes
1.58–1.75	Mid IR	Yes	Yes

contrast sharply with the yellow-brown colors of the North African desert. Snow cover in the Alps is also readily apparent in this image. In (*b*), most of the area around the Red Sea does not support much vegetation, but the Nile River floodplain and delta stand out in red tones near the upper left of the image. In (*c*) much of the land area in this portion of Southeast Asia is a bright red color associated with lush tropical vegetation. Plate 14*d* is a combination of red, near-IR, and mid-IR bands, in contrast to the blue, red, and near-IR combination used for (*a*), (*b*), and (*c*). Here the resulting colors are quite different than seen in the previous color band combinations. This image covers portions of the U.S. Pacific Northwest and western Canada. Mountainous areas with significant snow cover and bare rock are a distinct pink color. The areas with the most lush vegetation are depicted in brown tones in areas of Washington and Oregon west of the Cascade Mountains and on Vancouver Island, British Columbia. (The brown colors result from high reflection in the near-IR and mid-IR bands.) The white streaks over the ocean area at upper left are aircraft contrails. The bright white area in the lower left part of the image is coastal fog.

The SPOT-5 satellite incorporates two *high resolution geometric* (*HRG*) instruments, a single *high resolution stereoscopic* (*HRS*) instrument, and a Vegetation instrument similar to that on SPOT-4. The HRG systems are designed to provide high spatial resolution, with either 2.5- or 5-m-resolution panchromatic imagery, 10-m resolution in the green, red, and near-IR multispectral bands, and 20-m resolution in the mid-IR band. The panchromatic band has a spectral range similar to that of the HRV on SPOT-1, -2, and -3 (0.48 to 0.71 μm), rather than the high resolution red band that was adopted for the SPOT-4 HRVIR. The high resolution mode in the panchromatic band is achieved by using two linear arrays of CCDs, offset horizontally within the focal plane by one-half the width of a CCD. Each linear array has a nominal ground sampling distance of 5 m. During processing at the ground station, data from the two linear arrays are interlaced and interpolated, resulting in a single panchromatic image with 2.5-m resolution.

Figure 5.24 shows a portion of a SPOT-5 HRG image collected over Stockholm, Sweden. This image was among the first acquired after SPOT-5's launch in May, 2002. The 10-m red band is shown in (*a*), and 2.5-m panchromatic imagery for the central city is shown in (*b*). Stockholm is situated along a waterway between Lake Mälaren, on the left side of (*a*), and an inlet of the Baltic Sea. Between the city and the sea lies an archipelago of some 24,000 islands, only some of which are visible at the right side of this image. The Old City area includes the royal palace, located near the north end of the island of Stadsholmen, in the center of image (*b*). At 2.5-m resolution, numerous bridges, docks, and boats can be seen along the waterways of the city.

The HRS instrument incorporates fore-and-aft stereo collection of panchromatic imagery, to facilitate the preparation of orthorectified products and digital elevation models (DEMs) at a resolution of 10 m. It is also possible to acquire across-track stereoscopic imagery from the twin HRG instruments, which are pointable to $\pm 31°$ off nadir. Table 5.8 summarizes the spatial and spectral characteristics of the HRG and HRS instruments on SPOT-5.

Figure 5.24 SPOT-5 HRG imagery of Stockholm, Sweden, May, 2002: (*a*) Band 2, 10 m resolution (red); (*b*) panchromatic band, 2.5 m resolution. (© CNES 2002. Distribution Airbus DS/Spot Image.)

TABLE 5.8 Sensors Carried on SPOT-5

Sensor	Spectral Band (μm)	Spatial Resolution (m)	Swath Width (km)	Stereoscopic Coverage
HRG	Pan: 0.48–0.71	2.5 or 5[a]	60–80	Cross-track pointing to ±31.06°
	B1: 0.50–0.59	10		
	B2: 0.61–0.68	10		
	B3: 0.78–0.89	10		
	B4: 1.58–1.75	20		
HRS	Pan: 0.49–0.69	5–10[b]	120	Along-track stereo coverage
Vegetation 2[c]	B0: 0.45–0.52	1000	2250	
	B2: 0.61–0.68	1000		
	B3: 0.78–0.89	1000		
	B4: 1.58–1.75	1000		

[a]The HRG panchromatic band has two linear arrays with 5 m spatial resolution that can be combined to yield 2.5 m resolution.
[b]The HRS panchromatic band has a spatial resolution of 10 m in the along-track direction and 5 m in the across-track direction.
[c]For consistency with the band numbering on other SPOT sensors, the Vegetation instrument's blue band is numbered 0 and the red band is numbered 2; there is no band 1.

5.8 SPOT-6 AND -7

As previously mentioned, SPOT-6 and -7 are completely commercially funded systems that ushered in a new era of development and operation for the SPOT program beginning in 2009. These systems were initially operated by Astrium Geoinformation Services, and since 2013, have been managed by *Airbus Defence & Space*. The goal for the SPOT-6 and -7 missions is to provide continuity of SPOT data until the 2024 timeframe. Although these systems fit our definition of "high resolution" satellites, we treat them here to provide a comprehensive summary of all of the SPOT satellites operated to date in one location in this discussion.

SPOT-6 and -7 are the first "twins" in the SPOT family, and are separated (phased) by 180° in the same orbit. This provides for an atleast-daily global revisit capability. SPOT-6 was launched on September 9, 2010, by the *Indian Space Research Organization* (*ISRO*) using its *Polar Satellite Launch Vehicle* (*PSLV*). SPOT-7 was launched in the same manner on June 30, 2014. (SPOT-6 and -7 are operated on the same orbit as the high resolution Pleiades 1A and 1B satellites, listed in Table 5.16. When these four satellites are operated as a constellation, intraday revisit of any point on earth is possible.)

The SPOT-6 and -7 systems have been designed to afford improved image acquisition, tasking, and data delivery relative to their predecessors. The nominal acquisition mode provides image coverage that is 60 km wide across-track and

600 km long in the north-south direction. However, several smaller images can be collected in multi-strip format in a single path, and non-north-south acquisitions are also possible. Fore-and-aft stereopair and stereo triplet image collection capabilities are also available. This provides for viewing angles between two consecutive images of either 15° or 20° and a B/H ratio between 0.27 and 0.40.

Each satellite can also be retasked every four hours, providing reactivity to changing weather and short-notice user requests for applications such as emergency response. Rapid data availability is accomplished through online delivery. Current data products include orthorectified images having a location accuracy of 10 m CE90 where reference DEMs exist.

The imaging sensor onboard the SPOT-6 and -7 systems is a pushbroom instrument called the New *AstroSat Optical Modular Instrument* (*NAOMI*). It was designed and developed to be configured in many different ways for various other missions. The SPOT 6/7 configuration results in the simultaneous collection of 12-bit data in a 1.5-m panchromatic band and four 6-m multispectral bands. The spectral sensitivity of each of these five bands is given in Table 5.9.

Figure 5.25 shows the island of Bora Bora in French Polynesia, a major international tourist destination. Bora Bora was formed about four million years ago after volcanoes erupted from the seabed. As the volcano sank back into the Pacific Ocean, a ring of coral reefs (atolls) formed to mark the former coastline. A lagoon also formed between the reef and the island around this time. The island is about 39 km^2 in size, with a permanent population of about 9000 people. Shown in (*a*) is a subscene of SPOT-6 data imaged on September 12, 2012, just three days after the launch of the satellite. This image shows the island and its surrounding coral reef, with large lagoons between the reef and the island. A color composite of the 6-m spatial resolution original blue, green, and red band data is shown here as a grayscale rendering. Shown in (*b*) is an enlargement of an area on the northern part of the reef that includes the island's airport. Tourist transportation from the airport to the main island is by boat. Shown in (*c*) and (*d*) is an area just inside the northeast part of the reef that shows two sets of "over-water bungalows." These popular bungalows are fitted with a glass floor through which guests can observe the marine life below. Other such bungalow hotels dot the lagoon as well.

TABLE 5.9 Spot-6 and -7 NAOMI Spectral Bands and Spatial Resolutions

Nominal Spectral Location	Band Width (μm)	Spatial Resolution (m)
Panchromatic	0.450–0.745	1.5
Blue	0.450–0.520	6
Green	0.530–0.590	6
Red	0.625–0.695	6
Near Infrared	0.760–0.890	6

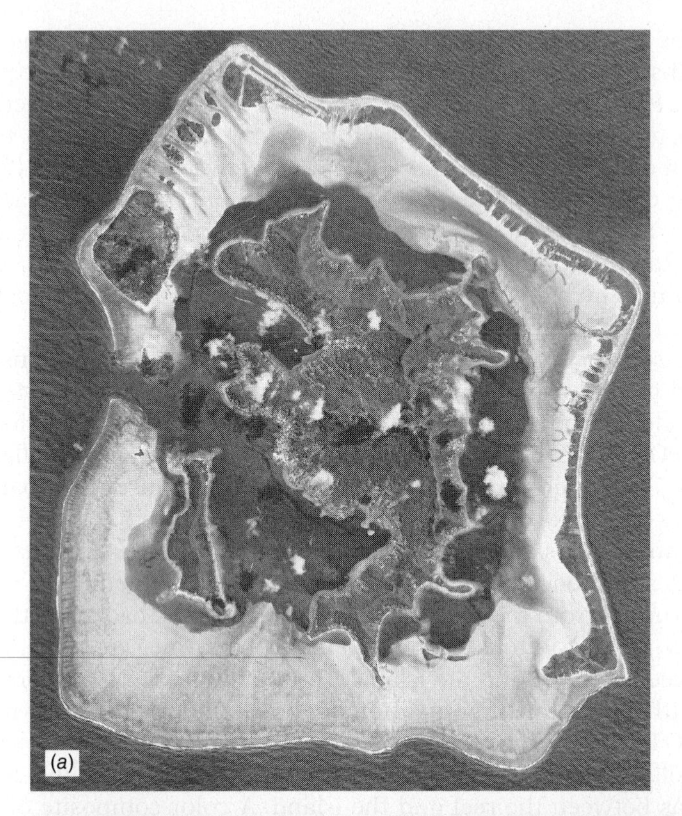

Figure 5.25 Images of the island of Bora Bora in French Polynesia: (*a*) Subscene of SPOT-6 data of the island of Bora Bora, imaged on September 12, 2012, showing the island and its surrounding coral reef, with large lagoons between the reef and the island. The spatial resolution of the original blue, green, and red band data shown here in grayscale is 6 m. (*b*) Enlargement of an area on the northern part of the reef that includes the island's airport. (c) Enlargement of an area just inside the northeast part of the reef that shows "over-water bungalows." (© Airbus DS/Spot Image 2012.) (*d*) Low altitude oblique airphoto of the bungalows. (Courtesy Moeava de Rosemont – Atea Data Aerial Imaging.)

Figure 5.25 *(Continued)*

5.9 EVOLUTION OF OTHER MODERATE RESOLUTION SYSTEMS

Due to their historical long-term use and widespread availability, Landsat and SPOT data have been emphasized in this discussion. However, more than 100 other earth observation systems operating in the optical spectrum have either provided historical data, are continuing to collect data, or are planned for the near future. Clearly, describing each of these systems is beyond the scope of this discussion. Our aim is to simply characterize the general evolution of these systems and underscore that remote sensing from space has become a major international activity having profound scientific, social, and commercial significance.

The evolution of land-oriented optical remote sensing satellite systems has been characterized by three general time periods: (1) the Landsat and SPOT "heritage" period when these two systems completely dominated civilian remote sensing from space, (2) the immediate "follow-on" period (between approximately 1988 and 1999) when several other moderate resolution systems came into existence, and (3) the period since 1999, wherein both "high resolution" and moderate resolution systems have served as complementary data sources. This latter period has also included the development of space borne hyperspectral sensing systems.

In next two sections of this discussion we summarize briefly the major "follow-on" moderate resolution systems launched prior to 1999, and those launched since 1999. Thereafter, we treat high resolution and hyperspectral systems. In the context of this discussion, we emphasize that the use of the terminology "moderate resolution" refers to systems having a spatial resolution of 4 to 60 m, and "high resolution" refers to systems having a spatial resolution less than 4 m. Several systems have different spatial resolutions among their various spectral bands. We classify these systems as moderate or high resolution based upon their highest or best resolution in any spectral band (often a panchromatic band). It should also be noted that the 4-m boundary we establish between moderate and high resolution is an arbitrary one. As technology changes, this boundary will likely shift toward lower values. Furthermore, from a user's perspective, what constitutes a moderate versus high resolution data source depends upon the application.

5.10 MODERATE RESOLUTION SYSTEMS LAUNCHED PRIOR TO 1999

Shortly after the launch of SPOT-1, the Republic of India began operating its series of moderate resolution *Indian Remote Sensing* (*IRS*) satellites with the launch of IRS-1A in 1988. Table 5.10 summarizes the characteristics of the early satellites in

TABLE 5.10 Summary of IRS-1A through IRS-1D Satellite Characteristics

Satellite	Launch Year	Sensor	Spatial Resolution (m)	Spectral Band (μm)	Swath Width (km)
IRS-1A	1988	LISS-I	72.5	0.45–0.52	148
				0.52–0.59	
				0.62–0.68	
				0.77–0.86	
		LISS-II	36.25	0.45–0.52	146
				0.52–0.59	
				0.62–0.68	
				0.77–0.86	
IRS-1B	1991	Same as IRS-1A			
IRS-1C	1995	Pan	5.8	0.50–0.75	70
		LISS-III	23	0.52–0.59	142
				0.62–0.68	
				0.77–0.86	
			70	1.55–1.70	148
		WiFS	188	0.62–0.68	774
			188	0.77–0.86	774
IRS-1D	1997	Same as IRS-1C			

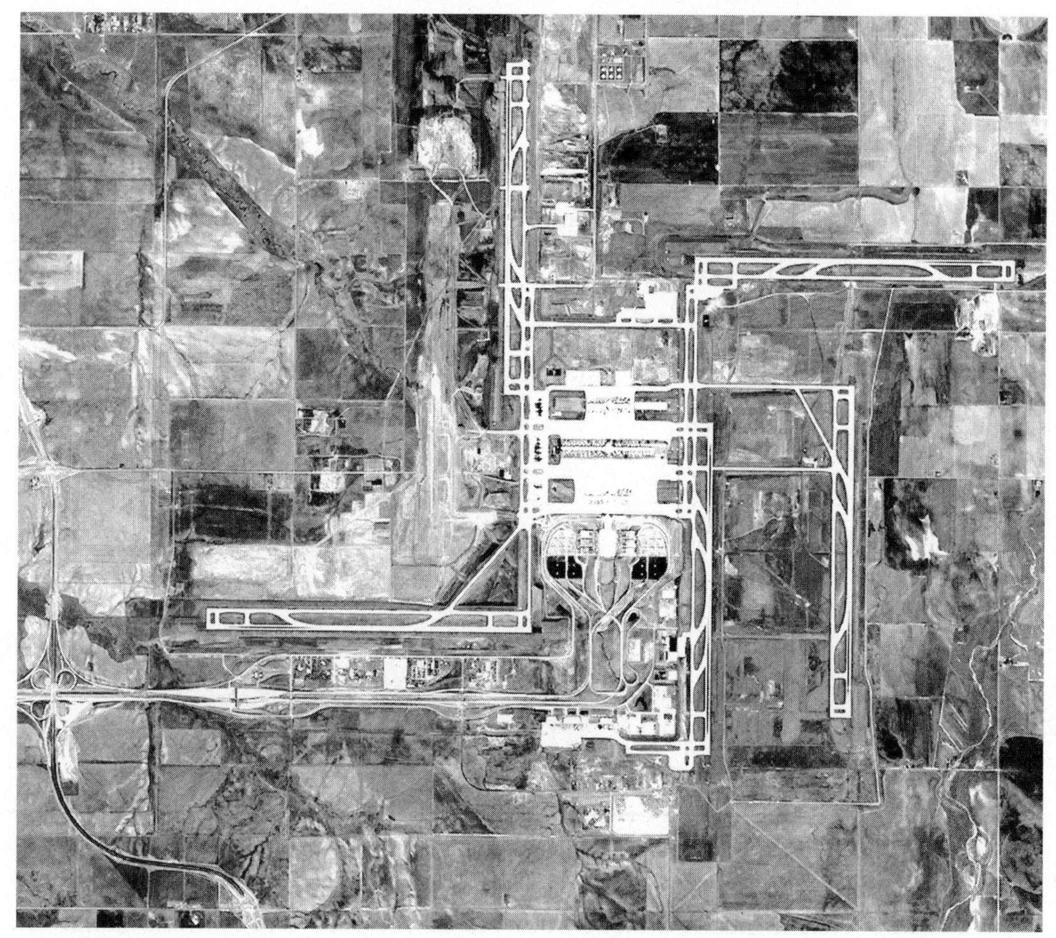

Figure 5.26 IRS panchromatic image (5.8-m resolution), Denver International Airport. Scale 1:85,000. (Courtesy of EOTec LLC.)

this series, IRS-1A through IRS-1D. Figure 5.26 is a portion of a LISS-III panchromatic image that includes the Denver International Airport and illustrates the level of detail available with the 5.8-m data. (As will be highlighted in the next two sections of this discussion, India has continued since the operation of its initial systems to be a major participant in the evolution of satellite remote sensing.)

There were several other early entrants into the enterprise of designing and operating satellite systems that had earth resource management capabilities. Among these was the former Soviet Union, with the launch of two *RESURS-01* satellites in 1985 and 1988. The RESURS program was then continued by Russia, with launches of RESURS-01-3 in 1994 and RESURS-01-4 in 1998. The latter two were the first in the series from which data were readily available outside the former Soviet Union (Table 5.11).

Japan launched its first land remote sensing satellite (*JERS-1*) in 1992. While designed primarily as a radar remote sensing mission (see Chapter 6), JERS-1 also carried the Optical Sensor (OPS), an along-track multispectral sensor. The

TABLE 5.11 Summary of RESURS-O1-3 and RESURS-O1-4 Satellite Characteristics

			Spatial Resolution (m)			
Satellite	Launch Year	Sensor	Along Track	Across Track	Spectral Band (μm)	Swath Width (km)
RESURS-O1-3	1994	MSU-E	35	45	0.5–0.6	45
					0.6–0.7	
					0.8–0.9	
		MSU-SK	140	185	0.5–0.6	600
					0.6–0.7	
					0.7–0.8	
					0.8–1.0	
			560	740	10.4–12.6	
RESURS-O1-4	1998	MSU-E	33	29	0.5–0.6	58
					0.6–0.7	
					0.8–0.9	
		MSU-SK	130	170	0.5–0.6	710
					0.6–0.7	
					0.7–0.8	
					0.8–1.0	
			520	680	10.4–12.6	

OPS system included along-track stereoscopic coverage in one band, for topographic mapping purposes.

In 1996, Japan also launched the *Advanced Earth Observing Satellite* (*ADEOS*), which included as one of its sensors the *Advanced Visible and Near Infrared Radiometer* (*AVNIR*). This system included one 8-m resolution band and four 16-m resolution bands, as well as a cross-track pointing capability for stereoscopic data acquisition.

Japan also cooperated with the United States in the development of the *Advanced Spaceborne Thermal Emission and Reflection Radiometer* (*ASTER*), which was launched in 1999 aboard the *Terra* satellite. We discuss this system in Section 5.19 under the heading of the Earth Observing System (EOS).

5.11 MODERATE RESOLUTION SYSTEMS LAUNCHED SINCE 1999

Table 5.12 summarizes a broad array of selected moderate resolution systems that have been launched since 1999 (and those planned for launch by 2016). Space clearly precludes our ability to describe the details of each of these systems

TABLE 5.12 Selected Moderate Resolution Satellites Launched or Planned for Launch since 1999

Year Launched	Satellite	Country	Pan. Res. (m)	MS Res. (m)	Swath Width (km)
1999	CBERS-1	China/Brazil	20, 80	20, 80, 260	113, 120, 890
	KOMPSAT-1	Korea	6.6	—	17
2000	MTI	US	—	5, 20	12
	EO-1	US	10	30	37
2001	Proba	ESA	8	18, 36	14
2002	DMC AlSat-1	Algeria	—	32	650
2003	CBERS-2	China/Brazil	20, 80	20, 80, 260	113, 120, 890
	DMC BilSat	Turkey	12	26	24, 52, 650
	DMC NigeriaSat-1	Nigeria	—	32	650
	UK DMC-1	UK	—	32	650
	ResourceSat-1	India	6	6, 24, 56	23, 141, 740
2004	ThaiPhat	Thailand	—	36	600
2005	MONITOR-E -1	Russia	8	20	94, 160
	DMC Beijing-1	China	4	32	650
2007	CBERS-2B[a]	China/Brazil	20	20, 80, 260	113, 120, 890
2008	RapidEye-A,B,C,D,E	German Co.	—	6.5	77
	X-Sat	Singapore	—	10	50
	HJ-1A, HJ-1B	China	—	30	720
2009	DMC Deimos-1	Spain	—	22	650
	UK DMC-2	UK	—	22	650
2011	DNC NigeriaSat-X	Nigeria	—	22	650
	ResourceSat-2	India	6	6, 24, 56	70, 140, 740
2013	Landsat-8	US	15	30, 100	185
	Proba-V	Belgium/ESA	—	100	2250
	CBERS-3[b]	China/Brazil	5	10, 20, 64	60, 120, 866
2015	VENμS*	Israel/France	—	5	28
	CBERS-4[c] *	China/Brazil	5	10, 20, 60	60, 120, 866
	SENTINEL-2A	ESA		10, 20, 60	290
	ResourceSat-2A*	India		5.8, 24, 55	70, 140, 740
2016	SENTINEL-2B*	ESA		10, 20, 60	290

[a]Also included a panchromatic High Resolution Camera having 2.7-m resolution and a 27-km swath width.
[b]CBERS-3 failed to launch properly on December 9, 2013.
[c]Also includes a multispectral instrument having 40-m resolution in visible, near-, and mid-IR bands, and an 80-m thermal band with a 120-km swath width.
*Indicates system planned for launch after the time of this writing (2014).

(e.g., orbit characteristics, sensor design and performance, tasking, and data availability, format options, and cost). Readers who wish to obtain such information are encouraged to consult various online resources that provide such detail. These resources include, but are not limited to:

- Commercial or governmental internet sites specifically devoted to treatment of a given mission or system.
- The ESA Earth Observation Portal's Earth Observation Missions Database (www.eoportal.org).
- The University of Twente, Faculty of Geo-Information Science and Earth Observation, ITC, Database of Satellites and Sensors (www.itc.nl/research/products/sensordb/searchsat.aspx).

Several of the satellite systems listed in Table 5.12 played a particular role as initial test-beds for new concepts or the design of systems that would follow them. For example, the *Earth Observing-1* (*EO-1*) mission was launched, as part of NASA's *New Millennium Program* (*NMP*). The purpose of the NMP was to develop and test new technology in space missions in general. The EO-1 mission validated new developments in lightweight materials, detector arrays, spectrometers, communication, power generation, propulsion, and data storage. It set the stage for the continued development of earth observing systems having significant increases in performance while also having reduced cost, volume, and mass. It also demonstrated the power of "autonomous," or self-directed, satellite operations. Such operations ranged from onboard feature detection to modify image acquisition tasking decisions, to employing the satellite as part of a *sensor web*.

A sensor web typically permits interactive communication and coordination among numerous in-situ sensors, satellite systems, and ground command resources. Sensor webs are particularly useful for monitoring transient events, such as wildfires, floods, and volcanoes. In such cases, a combination of satellites might be used to form an ad hoc constellation that enables collaborative autonomous image acquisitions triggered by detection of the event. The software for planning, scheduling, and controlling this collaborative data collection is both space- and ground-based and typically incorporates modeling and artificial intelligence capabilities. Such approaches are central to the concept of the GEOSS discussed earlier in Section 5.6.

The EO-1 mission was originally designed to be a one-year technology validation and demonstration in support of the design of follow-on systems to Landsat-7. This baseline phase of the mission was so successful that after a year the program was continued, with the support of funding from USGS, as an extended mission. Three land-imaging sensors were part of the mission: the *Atmospheric Corrector* (*AC*) and *Hyperion*, both of which are discussed below in the context of hyperspectral satellite systems, and the *Advanced Land Imager* (*ALI*).

The ALI served to validate several of the design characteristics that were eventually incorporated into the Landsat-8 OLI. For example, the ALI employed pushbroom scanning, a panchromatic band that did not extend into the near IR, a coastal/aerosol band identical to that present on the OLI (band 1), and a band that closely approximated the near-IR band present on the OLI (band 5).

The entire concept of small, low-cost, and highly capable earth-imaging "microsatellite" systems was advanced with the 2001 launch of the European Space Agency's *Project for On-board Autonomy* (*Proba*). Like EO-1, Proba was designed primarily as a one-year technology demonstration mission. The satellite was only 60 × 60 × 80 cm in size and weighed only 94 kg. Its operation was essentially autonomous, meaning that everyday tasks such as guidance, control, scheduling, computation and control of sensor pointing, and payload resource management were performed with minimal human intervention. The primary earth-imagers onboard included the *High Resolution Camera* (*HRC*) and the *Compact High Resolution Imaging Spectrometer* (*CHRIS*). The HRC was a panchromatic system yielding 8-m-resolution images. We discuss the CHRIS system under hyperspectral systems.

The year 2002 marked the first launch, of Algeria's *DMC AlSat-1*, in a series of microsatellites comprising the *Disaster Monitoring Constellation* (*DMC*). The DMC is an international program conceived by Surrey Satellite Technology Limited (SSTL) of the United Kingdom. The constellation is coordinated by DMC International Imaging Ltd (DMCii), a subsidiary of SSTL. The goal of the program is to engender international collaboration in the launch of a constellation of satellites specifically designed for disaster monitoring (and useful for a range of other applications). While each satellite in the constellation is built and owned by an individual country, they are launched into the same orbit and operated cooperatively. In this way, participants in the program can realize the benefits of a constellation of satellites in orbit simultaneously while maintaining independent ownership and low cost.

While the design of the various DMC satellites is not absolutely identical, all of the earlier satellites in the constellation provided moderate (22-, 26-, or 32-m) resolution data using spectral bands located in the green, red, and near-IR (equivalent to bands 2, 3, and 4 of the Landsat ETM+). With a swath width in excess of 600 km, each satellite provides global coverage at least every four days. Four satellites flying in the same orbit can therefore provide global coverage on a daily basis. As shown in Table 5.12, moderate resolution systems launched into the DMC after AlSat-1 included those operated by Turkey, Nigeria, the UK, China, and Spain. Turkey's BilSat-1 included a 12-m-resolution panchromatic imager and an 8 band multispectral system. China's Beijing-1 also incorporated a 4-m-resolution panchromatic imager.

The *IRS ResourceSat-1* and *-2* listed in Table 5.12 carry three imaging systems similar to those on IRS-1C and IRS-1D but with improved spatial resolution. The general characteristics of these sensors are included in Table 5.13. Although these satellites carry a highly capable complement of sensors, one possible limitation of these systems for some applications (e.g., water quality studies) is the lack of data collection in the blue portion of the spectrum.

TABLE 5.13 Summary of IRS ResourceSat-1 and -2 Satellite Characteristics

Sensor	Spatial Resolution (m)	Spectral Band (μm)	Swath Width (km)
LISS-IV	6	0.52–0.59	70 (Any single-band mode)
		0.62–0.68	23 (Multispectral mode)
		0.77–0.86	
LISS-III	24	0.52–0.59	141
		0.62–0.68	
		0.77–0.86	
		1.55–1.70	
AWiFS	56 (nadir)	Same as LISS-III	740

The German system listed in Table 5.12 consists of a constellation of five identical, 150-kg minisatellites developed by RapidEye Inc. A single launch was used to place all of the satellites comprising the constellation into the same orbit plane. The satellites are evenly spaced in a sun-synchronous orbit at an altitude of approximately 630 km and an inclination of 98°. The systems can be pointed to provide a revisit time of one day between ±70° latitude on a global basis.

The *RapidEye Earth Imaging System* (*REIS*) is a five-band multispectral imager that employs linear array pushbroom scanning. The system was designed primarily to serve the agricultural and cartographic markets of Europe and the United States. However, the market for the data is actually global and extends into many other application areas. Table 5.14 lists the spectral bands employed by each REIS.

Both the VENμS and Sentinel missions listed in Table 5.12 support the European *Copernicus* program, previously known as *Global Monitoring for Environment and Security* (*GMES*). This is a joint initiative led by the European Commission (EC) in partnership with ESA and the European Environment Agency (EEA). The overall aim of the program is to provide improved environmental and security information to policymakers, the business community, and general

TABLE 5.14 REIS Spectral Bands

Band	Wavelength (μm)	Nominal Spectral Location	Spatial Resolution (m)
1	0.440–0.510	Blue	6.5
2	0.520–0.590	Green	6.5
3	0.630–0.685	Red	6.5
4	0.690–0.730	Red edge	6.5
5	0.760–0.850	Near IR	6.5

public through the integration of satellite observations, in-situ data, and modeling. The broad information services provided by Copernicus support the primary areas of land management, monitoring the marine and atmospheric environments, emergency management, security, and climate change. Copernicus is Europe's main contribution to the Global Earth Observation System of Systems (GEOSS).

VENμS stands for Vegetation and Environment monitoring on a New Micro-Satellite. It is a research demonstration mission developed jointly by the Israel Space Agency (ISA) and CNES. The science-oriented component of the mission is focused primarily on vegetation monitoring, in addition to water color characterization. Every two days the satellite will cover 50 well-characterized sites representative of the earth's main inland and coastal ecosystems. This will employ a 12-band imager having a 5.3-m spatial resolution and 28-km swath width. The sensing bands vary from 15 to 50 nm in width and are located in the visible and near IR.

The Sentinel 2A and 2B systems are identical and represent one of five families of Sentinel missions being undertaken in support of Copernicus. The Sentinel-1 missions are polar-orbiting radar missions focused both on land and ocean applications. (These missions are described in Section 6.13.) The Sentinel-2 missions are described below. Sentinel-3 involves polar-orbiting multi-instrument missions to provide high-accuracy optical, radar, and altimetry data for marine and land applications. Sentinel-4 and -5 will provide data for atmospheric monitoring from geostationary and polar orbit, respectively. (Sentinel-5 Precursor is a system that will be launched prior to Sentinel-5 in order to bridge the data gap created when the SCIAMACHY spectrometer on board the Envisat satellite failed in 2012. This instrument mapped the concentration of trace gases and aerosols in the troposphere and stratosphere.)

As shown in Table 5.12, the Sentinel-2A satellite is scheduled for launch during 2015. In support of Copernicus, the data from this system (and Sentinel-2B) will support operational generation of mapping products of two major types: generic land cover, land use, and change detection; and geophysical variables such as leaf area index, leaf water content, and leaf chlorophyll content. Each satellite will occupy a sun-synchronous orbit at an altitude of 786 km, an inclination of 98.5°, and a descending node local overpass time of 10:30 A.M. (identical to SPOT and very close to that of Landsat) . With its 290-m swath width, each satellite will afford a global 10-day revisit cycle at the equator (and a 5-day revisit time when both satellites are operational with 180° orbit phasing). The satellites will systematically acquire data over land and coastal areas from 56°S to 84°N latitude. Data will also be collected over islands larger than 100 km^2, EU islands, all others less than 20 km from the coastline, the entire Mediterranean Sea, all inland water bodies, and closed seas. The system can also be operated in an emergency lateral pointing mode, providing access to any point within its global coverage area within one to two days.

Sentinel-2 data will be collected by the *Multispectral Imager* (*MSI*), which incorporates pushbroom scanning. Table 5.15 lists the spectral bands in which MSI data are collected. The collective spectral range of the 13 spectral bands involved span the approximate range of 0.4–2.4 μm (visible, near IR, and SWIR). The data have 12-bit quantization. As shown, four bands are acquired at a spatial resolution of 10 m, six bands have a resolution of 20 m, and three bands have a resolution of 60 m. The 60-m bands are used solely for data calibration or correction. The bands indicated with a "*" in Table 5.15 contribute primarily to the mission objectives of land cover classification, vegetation condition assessment, snow/ice/cloud differentiation, and mineral discrimination. Some bands (2, 8, 8_a, and 12) support both the calibration and correction process, as well as the primary mapping process. It should also be noted that three relatively narrow bands (5, 6, and 7) are located within the "red edge" portion of the spectrum to specifically facilitate the differentiation of vegetation types and conditions.

It should also be noted in Table 5.12 that the "V" associated with the Proba-V mission stands for "Vegetation." The Proba-V mission, initiated by ESA and the Belgian Federal Science Policy Office (Belspo) was designed to extend and

TABLE 5.15 Sentinel-2 MSI Spectral Bands

Band	Nominal Spectral Location	Center λ (μm)	Band Width (μm)	Description or Use	Spatial Resolution (m)
1	Visible	0.433	0.020	Blue [Aerosols Correction]	60
2*	Visible	0.490	0.065	Blue [Aerosols Correction]	10
3*	Visible	0.560	0.035	Green	10
4*	Visible	0.665	0.030	Red	10
5*	NIR	0.705	0.015	Red Edge	20
6*	NIR	0.740	0.015	Red Edge	20
7*	NIR	0.775	0.020	Red Edge	20
8*	NIR	0.842	0.115	[Water Vapor Correction]	10
8_a*	NIR	0.865	0.020	[Water Vapor Correction]	20
9	NIR	0.940	0.020	[Water Vapor Correction]	60
10	SWIR	1.375	0.030	[Cirrus Detection]	60
11*	SWIR	1.610	0.090	Snow/Ice/Cloud	20
12*	SWIR	2.190	0.180	Snow/Ice/Cloud & [Aerosols Correction]	20

*Indicates bands that contribute primarily to the mission objectives of land cover classification, vegetation condition assessment, snow/ice/cloud differentiation, and mineral discrimination.
[ ] Indicates bands used solely or secondarily for calibration or correction.

improve upon the 15-year collection of Vegetation-type data acquired by SPOT-4 and SPOT-5.

The Proba-V satellite (launched May 7, 2013) is smaller than a cubic meter and has a total mass of 140 kg. The sensor onboard has a mass of only 40 kg and requires only 35 W of power for its operation. The system employs pushbroom scanning using three earth-viewing telescopes and contiguous focal plane arrays in a compact "three-mirror anastigmat" (TMA) design. In this manner, the sensor can record over a 102° field of view. Each TMA records four spectral bands that were chosen to be functionally equivalent to the SPOT Vegetation sensor bands. These include: blue (0.44 to 0.49 μm), red (0.61 to 0.70 μm), near-IR (0.77 to 0.91 μm), and SWIR (1.56 to 1.63 μm).

Proba-V is in a sun-synchronous orbit at an altitude of 820 km having a 98.73° inclination. This results in a local descending crossing time of 10:30 to 11:30 A.M. The swath width of the system is approximately 2250 km, and the daily latitude coverage is from 75°N to 35°N and 35°S to 56°S, providing coverage of most land areas every day, and complete land coverage every two days.

The resolution of the Proba-V sensor is 100 m at nadir and 350 m at image edge. Typical data products include daily and 10-day composites at one-third and 1 km for the visible and near-IR bands, and two-thirds and 1 km for the SWIR band. Plate 15 illustrates the first uncalibrated global mosaic of vegetation data collected by the system while it was still being commissioned during mid-2013. Given the coverage, temporal, spatial, and spectral resolution of the Proba-V data, their applications extend beyond the realm of vegetation monitoring (e.g., tracking of extreme weather, monitoring inland waters, among many others).

5.12 HIGH RESOLUTION SYSTEMS

As stated earlier, the era of civil high resolution (<4 m) satellite systems began in 1999. Table 5.16 lists 30 high resolution systems that have either been launched or are planned for launch by 2015. It can be seen in this table that contrary to the case with moderate resolution systems, many operators of high resolution sensors have been, and continue to be, commercial firms. In great part, this owes to the broad commercial market for higher resolution data in comparison to moderate resolution data. The applications for high resolution data range from infrastructure planning and management to commercial and military intelligence, disaster response, water resource management, forest management, agricultural management, urban planning and management, site design and assessment, and land use and transportation planning, among numerous others.

The first four high resolution systems launched were commercially developed and operated. These included three U.S. commercial systems (*IKONOS, QuickBird, and OrbView-3*), and one Israeli commercial system (*EROS-A*). All of the initial U.S. systems featured functionally equivalent spectral bands: one panchromatic (0.45–0.90 μm) and four multispectral (1: 0.45–0.52 μm), (2: 0.52–0.60 μm),

TABLE 5.16 Selected High Resolution Satellites Launched or Planned for Launch since 1999

Year Launched	Satellite	Country	Pan. Res. (m)	MS Res. (m)	Swath Width (km)
1999	IKONOS	US (commercial)	1	4	11
2000	EROS A	Israel (commercial)	1.9	—	14
2001	QuickBird	US (commercial)	0.61	2.44	16
2003	OrbView-3	US (commercial)	1	4	8
2004	FormoSat-2	Taiwan	2	8	24
2005	Cartosat 1	India	2.5	—	30
	TopSat	UK	2.5	5	10, 15
2006	ALOS	Japan	2.5	10	35, 70
	EROS B	Israel (commercial)	0.7	—	7
	Resurs DK-1	Russia	1	3	28
	KOMPSAT-2	Korea	1	4	15
2007	Cartosat 2	India	0.8	—	10
	WorldView-1	US (commercial)	0.45 (0.5)[a]	—	16
	THEOS	Thailand	2	15	22, 90
2008	CartoSat-2A	India	0.8	—	10
	GeoEye-1	US (commercial)	0.41 (0.5)[a]	1.64 (2)[b]	15
2009	WorldView-2	US (commercial)	0.46 (0.5)[a]	1.84 (2)[b]	16
	EROS C	Israel (commercial)	0.7	2.80	11
2010	CartoSat-2B	India	0.8	—	10
2011	DMC NigeriaSat-2	Nigeria	2.5	5,32	20, 20, 320
	Pleiades-1	France	0.7	2	20
2012	KOMPSAT-3	Korea	0.7	2.8	17
	Pleiades-2	France	0.7	2	20
2013	GeoEye-2*	US (Commercial)	0.34 (0.5)[a]	1.4 (2)[b]	14
2014	CartoSat-1A*	India	1.25	2.5	60
	DMC-3 (A,B,C)*	China (Commercial)	1	4	23, 23
	CartoSat-2C*	India	0.8	2	10
	WorldView-3	US (Commercial)	0.31 (0.4)[a]	1.24 (1.6)[b], 3.7 (7.5)[c]	13
	KOMPSAT-3A*	Korea	0.7	2.8	17
2015	CartoSat-3*	India	0.3	—	15

[a]Panchromatic data initially resampled to 0.4 m for non-US-government data users. Available to all users at full resolution as of February 21, 2015.
[b]Multispectral data (at wavelengths shorter than SWIR) initially resampled to 1.6 m for non-US-government data users.
[c]SWIR data initially resampled to 7.5 m for non-US-government data users.
*Indicates system planned for launch after the time of this writing (2014). Several systems will likely be launched later than indicated here. GeoEye-2 is being used as a ground spare to be launched when it is needed to meet increased user demand or to replace an on-orbit DigitalGlobe satellite.

(3: 0.63–0.69 μm), and (4: 0.76–0.90 μm). The EROS-A system included a panchromatic band only (0.50–0.90 μm).

After an unsuccessful launch of the first IKONOS satellite, a second IKONOS (coming from the Greek word for "image") became the first high resolution system to be launched successfully. The system employed linear array technology to collect 1 m panchromatic and 4 m multispectral data. The panchromatic and multispectral bands could be combined to produce "pan-sharpened" multispectral imagery (with an effective resolution of 1 m). The system also was highly maneuverable and employed "body pointing" of the entire spacecraft to permit data acquisition in any direction within 45° of nadir.

The first IKONOS image, collected on September 30, 1999, is shown in Figure 5.27*a*. Enlargements of selected portions of the image, including a runway at Ronald Reagan Washington National Airport (*b*) and the Washington Monument (*c*), illustrate the level of detail that can be seen in 1-m-resolution panchromatic imagery. At the time this image was acquired, the Washington Monument was undergoing renovation, and scaffolding can be seen on the side facing the sensor. The monument also shows the effects of relief displacement in this image.

Plate 16 shows a small portion of an IKONOS image acquired during early summer over an agricultural area in eastern Wisconsin. Included are a false-color composite of IKONOS multispectral bands 2, 3, and 4 at original 4-m resolution; the panchromatic band at 1-m resolution; and a merged image showing the results of the pan-sharpening process (Section 7.6). The sharpened image has the spectral information present in the original multispectral data, but with an effective spatial resolution of 1 m.

For several years, the highest resolution satellite imagery available to the public was provided by QuickBird. Figure 5.28 shows a panchromatic image (61-cm resolution) of the Taj Mahal of India, collected by QuickBird on February 15, 2002. The complex was built by the Muslim Emperor Shah Jahan as a mausoleum for his wife, queen Mumtaz Mahal. The height of the minarets at the four corners of the main structure is indicated by the length of their shadows. In addition, the peaks of the domes appear to be offset from the centers of the domes, due to relief displacement in this image.

Plate 17 illustrates the use of pan-sharpening techniques to produce color QuickBird images with an effective 61 cm resolution. The image shows a true-color composite derived from the merger of data from the QuickBird panchromatic band and multispectral bands 1, 2, and 3 shown as blue, green, and red, respectively. It was acquired on May 10, 2002, over the Burchardkai Container Terminal at the Port of Hamburg. The outlines and colors of individual cargo containers can be readily differentiated at the resolution of this image.

As shown in Table 5.16, the resolution of the QuickBird system was superseded by that of WorldView-1 (in 2007), GeoEye-1 (in 2008), and Worldview-2 (in 2009). All three of these systems resulted in data resampled to 0.5 m for non-U.S.-governmental users. A third generation of high resolution systems began with the

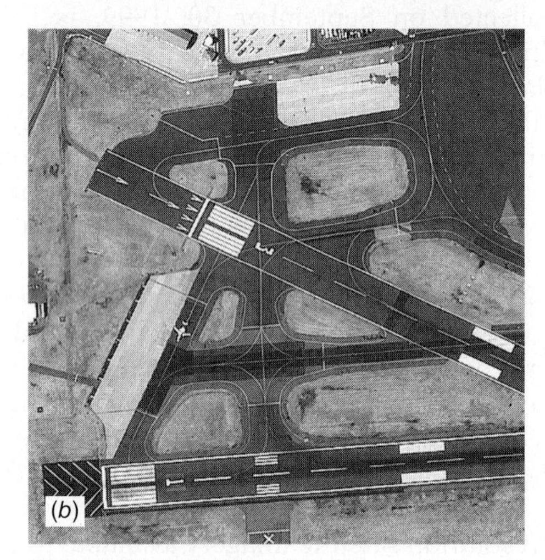

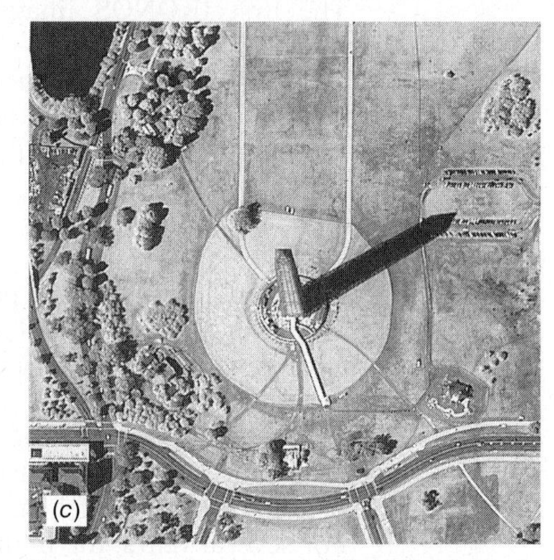

Figure 5.27 First IKONOS image, acquired on September 30, 1999, over Washington, DC: (*a*) Panchromatic image; (*b*) enlargement showing runway at Ronald Reagan Washington National Airport; (*c*) enlargement showing Washington Monument (with scaffolding present during renovation activity). (Courtesy DigitalGlobe.)

2014 launch of WorldView-3, to be followed by GeoEye-2 and CartoSat-3. These systems will collect panchromatic data at a native resolution of 0.3 to 0.34 m.

To say that high resolution satellite remote sensing has been a growing and rapidly changing enterprise since the launch of IKONOS in 1999 is an understatement. This sector has witnessed a broad array of not only technical, but also institutional, changes. In the United States, for example, IKONOS was originally operated by Space Imaging Corporation. OrbView-3 was operated by ORBIMAGE, and QuickBird was operated by DigitalGlobe, Inc. In January 2006, ORBIMAGE acquired Space Imaging and began operating under the brand name GeoEye. That corporate entity operated OrbView-3 and IKONOS and launched GeoEye-1 in 2008. Construction of GeoEye-1 was supported under a contract awarded by the National Geospatial-Intelligence Agency (NGA) under the agency's *NextView* program. The aim of that program was to

Figure 5.28 QuickBird panchromatic image of the Taj Mahal, India, 61-cm resolution. (Courtesy DigitalGlobe.)

provide U.S. commercial satellite operators the financial support needed to develop next-generation commercial imaging systems and develop nongovernment markets for the resulting data and services. The NextView program also supported the development of DigitalGlobe's WorldView-1 system. (DigitalGlobe funded WorldView-2 independently.)

Upon the completion of the NextView program, both GeoEye and DigitalGlobe were supported under NGA's EnhancedView program, begun in 2010.

However, budget cuts slated for GeoEye's portion of this funding during 2012 led to the eventual purchase of GeoEye by DigitalGlobe in early 2013. The merged companies continue to operate under the DigitalGlobe brand name.

DigitalGlobe's initial combined satellite constellation consisted of five satellites: IKONOS, QuickBird, WorldView-1, GeoEye-1, and WorldView-2. During 2013, the firm completed construction of GeoEye-2 as a ground spare to serve either as a replacement for the other on-orbit satellites or to meet increased customer demand for image data. In addition, WorldView-3 was launched mid-2014.

Commercial high resolution satellite firms are clearly not limited to those in the United States. The international scope of this activity is ever-broadening. At the same time, government-operated programs have become numerous on an international basis (as shown by the countries listed in Table 5.16). But one example of such international activity is the DMC-3 Constellation to be launched in late 2014 or early 2015 by DMCii.

GeoEye-1 is able to collect 0.41-m-resolution panchromatic data and 1.64-m multispectral data. However, panchromatic data produced at this resolution are only available to approved U.S. government users. Imagery supplied to commercial customers is resampled to 0.5 m, the highest resolution permitted under the system's operating license.

The orbit for GeoEye-1 is sun-synchronous, with an altitude of 684 km, a 98° inclination, and a 10:30 A.M. equatorial crossing time. The nominal swath width for the system is 15 km, but the system can be pointed as much as 60° from nadir in any direction. This permits the generation of a broad range of monoscopic and stereoscopic image products as well as DEMs. The geolocation accuracy of the data without the use of ground control points is 3 m or less. The revisit period for the system is three days or less, but depends upon latitude and look angle. Table 5.17 summarizes the spectral bands in which the 11-bit GeoEye-1 data are collected.

The imaging system onboard WorldView-1 provides panchromatic data only. The spectral range over which these data are collected is 0.40 μm to 0.90 μm. The spatial resolution of the system at nadir is 0.45 m, but is resampled to 0.5 m for non–U.S.-government data users.

TABLE 5.17 GeoEye-1 Spectral Bands

Band (μm)	Nominal Spectral Location	Spatial Resolution (m)
0.450–0.800	Panchromatic	0.41
0.450–0.510	Blue	1.64
0.520–0.600	Green	1.64
0.655–0.690	Red	1.64
0.780–0.920	Near IR	1.64

The orbit of WorldView-1 is sun-synchronous, with an altitude of 450 km, an inclination of 97.2°, and a 10:30 A.M. equatorial crossing time. The nominal swath width for the system is 16 km at nadir, and the system can be body-pointed as much as 40°, affording a 775-km total viewing field. Its average revisit period is 1.7 days at 1-m GSD or less.

WorldView-2 was launched in 2009. Panchromatic data from this system are supplied to non-U.S. government users at a spatial resolution of 0.5 m. The panchromatic data are collected over the range of 0.450 to 0.800 μm. The system also provides 8-band multispectral data at a resolution of 1.85 m at nadir. The nominal spectral regions for the multispectral bands include: Coastal (0.400 to 0.450 μm), Blue (0.450 to 0.510 μm), Green (0.510 to 0.580 μm), Yellow (0.585 to 0.625 μm), Red (0.630 to 0.690 μm), Red Edge (0.705 to 0.745 μm), Near IR1 (0.770 to 0.895 μm), and Near IR2 (0.860 to 1.040 μm).

The orbit of WorldView-2 is sun-synchronous, with an altitude of 770 km. This higher orbit (compared to the 450 km of WorldView-1) affords a wider total field of view (16.4 km) and an average revisit period of just over one day. Like WorldView-1, WorldView-2 has a 10:30 A.M. equatorial crossing time and collects data with 11-bit digitization.

WorldView-3 is termed by some as a "super-spectral" satellite. The system not only collects 0.31-m panchromatic and 1.24-m multispectral data in the same spectral bands as WorldView-2, but it also collects data in eight SWIR bands (at 3.7 m) and 12 "*CAVIS*" bands (at 30 m). The SWIR bands were chosen to improve penetration of haze, fog, smog, dust, smoke, mist, and cirrus. The CAVIS bands are used to correct for the effects of *c*louds, *a*erosols, *v*apors, *i*ce, and *s*now. All panchromatic and multispectral bands, other than the SWIR bands, are collected at 11 bits; the SWIR bands are collected at 14 bits. Table 5.18 summarizes the SWIR and CAVIS spectral bands collected by the system.

The orbit for WorldView-3 is sun-synchronous, with an altitude of 617 km, and a 1:30 P.M. equatorial crossing time. The revisit frequency at 40°N latitude is less than a day for the collection of 1 m GSD data. The predicted spatial accuracy of the data without ground control is less than 3.5 m CE90. As with its predecessors, WorldView-3 employs control moment gyroscopes (CMGs), which permit the satellite to reorient over a desired collection area in 4–5 sec, compared to the 35–45 sec for traditional attitude control technology.

As shown in Table 5.16, the years 2014 and 2015 will include the launch of several other high resolution systems—including India's continuation of its *CartoSat* series and China's launch of its DMC-3 constellation. The design of the DMC-3 mission consists of a constellation of three identical satellites launched together and equally spaced in the same orbit. With the ability to point up to 45° off nadir, the constellation affords a daily revisit capability. The constellation represents a new international business model for satellite development in that it is owned and operated by DMCii, but its capacity is leased by a Chinese company called 21 AT (standing for Twenty First Century Aerospace Technology Company, Ltd).

TABLE 5.18 Summary of the WorldView-3 SWIR and CAVIS Spectral Bands

SWIR Bands	Band Width (μm)	CAVIS Bands	Band Width (μm)
SWIR1	1.195–1.225	Desert Clouds	0.405–0.420
SWIR2	1.550–1.590	Aerosol-1	0.459–0.509
SWIR3	1.640–1.680	Green	0.525–0.585
SWIR4	1.710–1.750	Aerosol-2	0.620–0.670
SWIR5	2.145–2.185	Water-1	0.845–0.885
SWIR6	2.185–2.225	Water-2	0.897–0.927
SWIR7	2.235–2.285	Water-3	0.930–0.965
SWIR8	2.295–2.365	NDVI-SWIR	1.220–1.252
		Cirrus	1.350–1.410
		Snow	1.620–1.680
		Aerosol-3	2.105–2.245
		Aerosol-3[a]	2.105–2.245

[a]Two identical Aerosol-3 detectors are located on different ends of the focal plane in order to provide parallax to estimate cloud heights.

Space precludes additional discussion of past, present, and future high resolution satellite systems. The reader is cautioned that the plans for many future systems are subject to change, and given the broadening demand for high resolution data, several additional systems will be developed over time. Current information is best obtained by consulting relevant Internet sites.

5.13 HYPERSPECTRAL SATELLITE SYSTEMS

Whereas the development of moderate and high resolution *multispectral* satellite systems has been a relatively rapid and broad ranging activity, the development of *hyperspectral* satellite systems has been a longer-term activity involving relatively few systems at present. Satellite systems that collect image data in numerous, narrow, contiguous spectral bands are complex and costly. They also generate tremendous volumes of data that complicate image data collection, transmission, storage, and processing.

Several early attempts to develop or launch satellite hyperspectral systems ended in failure or program cancellation, leading to decades of unrealized potential for hyperspectral remote sensing from space. But tremendous increases in technological capability are changing this circumstance. Based on the proven success of the first few hyperspectral systems launched, the planned launch of various second-generation systems, and the increased availability and performance of hyperspectral image analysis software, the utilization of spaceborne hyperspectral imaging will likely grow substantially in the next few years.

Among the earliest of hyperspectral satellite sensors to be launched successfully were the Hyperion and AC systems carried on the previously described EO-1 spacecraft, and the CHRIS sensor included in the previously described Proba mission. The EO-1 Hyperion instrument nominally provides 242 spectral bands of data over the 0.36- to 2.6-μm range, each of which has a width of 0.010 to 0.011 μm. Some of the bands, particularly those at the lower and upper ends of the range, exhibit a poor signal-to-noise ratio. As a result, during Level 1 processing, only 198 of the 242 bands are calibrated; radiometric values in the remaining bands are set to 0 for most data products. Hyperion consists of two distinct linear array sensors, one with 70 bands in the UV, visible, and near IR, and the other with 172 bands in the near IR and mid IR (with some overlap in the spectral sensitivity of the two arrays). The spatial resolution of this experimental sensor is 30 m, and the swath width is 7.5 km. Data from the Hyperion system are distributed by the USGS. (We illustrated Hyperion data earlier in Plate 10.)

The EO-1 AC (also referred to as the *LEISA AC or LAC*) is a hyperspectral imager of coarse spatial resolution covering the 0.85- to 1.5-μm-wavelength range. It was designed to correct imagery from other sensors for atmospheric variability due primarily to water vapor and aerosols. The AC has a spatial resolution of 250 m at nadir. Acquisition of AC data was ended after the first year of the EO-1 mission, although Hyperion data continue to be collected as of the time of this writing (2014).

CHRIS is the primary sensor aboard the Proba satellite. It is a small (79 cm $\times$ 26 cm $\times$ 20 cm) and lightweight (14 kg) instrument capable of imaging in up to 62, nearly contiguous, narrow spectral bands at a resolution of 34 m. It can also be operated at a spatial resolution of 17 m. In the higher resolution mode, a typical nadir image is 13 km $\times$ 13 km in size and consists of data from 18 spectral bands. The system operates in the visible/near-infrared range (400 to 1050 nm) with a minimum spectral sampling interval ranging between 1.25 (at 400 nm) and 11 nm (at 1050 nm). The system can also be pointed in both the along- and across-track directions. As many as five different images of an area can be obtained at viewing angles of $-55°$, $-36°$, $0°$, $36°$, and $55°$ in the along-track direction within a time period of 2.5 min. In this manner, the bidirectional reflectance properties of features can be assessed. CHRIS had a one-year design life, but it has operated for more than nine years and its data have served more than 300 scientific groups in more than 50 countries. Applications of the data have ranged from land surface monitoring to coastal zone studies and aerosol monitoring.

Several near-term and future hyperspectral satellite systems are in various stages of planning and development. Representative of such systems is the Italian Space Agency's (ASI) *PRISMA* (*Hyperspectral Precursor and Application Mission*) to be launched in 2015. The PRISMA sensor incorporates both a hyperspectral (30-m GSD) and panchromatic (5-m GSD) instrument that share the same collecting optics. The system incorporates pushbroom scanning and a 30-km swath width. The hyperspectral imager collects data using two spectrometers: a visible/near IR (VNIR) spectrometer collecting 66 bands of data over a spectral range of 0.4–1.01 μm, and a SWIR spectrometer collecting 171 bands of data over a spectral range of 0.92–2.50 μm. The bandwidth for each band is 10 nm. The panchromatic band operates over the spectral range of 0.40–0.70 μm. The satellite employs a sun-synchronous orbit at an altitude of approximately 770 km, an inclination of 98.19°, with a 10:30 A.M. equatorial crossing time. Data collection will focus on an area of interest including all of Europe and the Mediterranean region.

Designed primarily as an interdisciplinary, scientific global-scale hyperspectral satellite mission is the German *Environmental Mapping and Analysis Program (EnMAP)*. The overall goal of EnMAP is to measure and model the dynamics of earth's terrestrial and aquatic ecosystems in support of addressing major environmental problems related to human activity and climate change (Kaufmann et al., 2012). The focus is on global observation of ecological parameters based upon monitoring a broad array of biophysical, biochemical, and geochemical variables. The aim is to increase understanding of biospheric/geospheric processes in the context of resource sustainability. The mission is managed by the Space Agency of the German Aerospace Center (DLR) under the scientific lead of the German Research Centre for Geosciences (GFZ). The launch of the EnMAP satellite is planned for 2015.

The EnMAP hyperspectral imager employs pushbroom scanning in a total of 244 spectral bands over the range of 0.42 to 2.45 μm. In the visible and near IR (0.42 to 1.00 μm), 89 bands are used to collect data with a spectral sampling interval of 6.5 nm and an 8.1-nm bandwidth. In the SWIR (0.9 to 2.45 μm), 155 bands collect data with a sampling interval of 10 nm and a 12.5-nm bandwidth. The GSD for all data is 30 m, with a nominal spatial accuracy of less than 100 m. Data are recorded at 14-bit radiometric resolution, with a radiometric calibration accuracy of better than 5%.

The EnMAP satellite occupies a sun-synchronous orbit, at an altitude of 643 km, an inclination of 97.96°, with an 11:00 A.M. equatorial crossing time. The swath width is 30 km at nadir, and the system can be pointed across-track up to 30°. The nominal revisit time for the system is 23 days for nadir-pointing images and as short as four days when 30° pointing is employed. Thus, the system affords relatively high spatial resolution, very high spectral resolution, and relatively high temporal resolution capabilities. Because of these qualities, it is likely that the data from this scientific mission will have numerous operation and commercial applications as well.

The *Hyperspectral Imager for the Coastal Ocean* (*HICO*) is a hyperspectral sensor mounted on the International Space Station (ISS). Installed on the ISS in 2009, HICO was originally sponsored by the U.S. Office of Naval Research, before being transferred to NASA's ISS Program in mid-2013. The sensor has 87 functional spectral bands of 5.7-nm width located in the range from 0.4 to 0.9 μm, an optimal range for most aquatic remote sensing applications. Its nominal spatial resolution is 90 m when viewing at nadir. One limitation of HICO is that its geographic coverage is constrained by the 52° orbital inclination of the ISS, so there is no coverage over high-latitude regions. (Remote sensing from the ISS is discussed further in Section 5.20.)

Several other hyperspectral satellite systems are planned for the future. For example, the Italian Space Agency (ISI) and the Israeli Space Agency (ISA) are designing a hyperspectral instrument for the *Spaceborne Hyperspectral Applicative Land an Ocean Mission* (*SHALOM*). The design of the system is being driven by potential applications of the resulting data for such purposes as environmental monitoring and risk management. At the same time, NASA is in the design phase of a hyperspectral mission called HyspIRI. The instruments to be flown on this mission are planned to include a visible to short wave infrared (VSWIR) imaging spectrometer and a thermal scanner (TIR). The imaging spectrometer will record the 0.380- to 2.50-μm range in 10-nm bands. The TIR will record eight bands of data between 3 to 12 μm. Both instruments will have 60-m spatial resolution. The satellite will employ a 626-km sun-synchronous orbit with a 10:30 A.M. equatorial crossing time. The revisit period for the visible spectrometer will be 19 days; that of the TIR will be five days. The anticipated launch date for the satellite is some time after 2022.

5.14 METEOROLOGICAL SATELLITES FREQUENTLY APPLIED TO EARTH SURFACE FEATURE OBSERVATION

Designed specifically to assist in weather prediction and climate monitoring, *meteorological satellites*, or *metsats*, generally incorporate sensors that have very coarse spatial resolution compared to land-oriented systems. On the other hand, metsats afford the advantages of global coverage at very high temporal resolution. Accordingly, metsat data have been shown to be useful in natural resource applications where frequent, large-area mapping is required and fine detail is not. Apart from the advantage of depicting large areas at high temporal resolution, the coarse spatial resolution of metsats also greatly reduces the volume of data to be processed for a particular application.

Metsats can employ either a polar-orbiting or geostationary orbit design. The operation of such satellites is a very broad-scale international activity involving the United States, Europe, India, China, Russia, Japan, and South Korea, among others. In Europe, the launch and operation of metsats is coordinated by the *European Organization for the Exploitation of Meteorological Satellites* (*EUMETSAT*). Over 30 member and cooperating states participate in EUMETSAT. We make no attempt to treat all of the metsat systems in operation and planned for the future. Space limits our discussion to three representative metsat series that have been, or currently are, operated by the United States. The first is the recently retired *Polar-Orbiting Environmental Satellite* (*POES*) series that was managed by the National Oceanic and Atmospheric Administration (NOAA). The second is the *Joint Polar Satellite System* (*JPSS*), which is replacing the POES series. The third program we discuss is the NOAA *Geostationary Operational Environmental Satellite* (*GOES*) series. All three of these programs incorporate a range of meteorological sensors. We treat only the salient characteristics of those sensors used most often in land (and ocean) remote sensing applications.

5.15 NOAA POES SATELLITES

Several generations of satellites have been flown in the NOAA POES series. Previous editions of this book have discussed earlier generations of these systems, which date back to 1978. Here we treat the final generation of the POES series. These include the NOAA-15 to NOAA-19 missions, all of which included the third-generation of a sensor called the *Advanced Very High Resolution Radiometer* (*AVHRR/3*). (These satellites were launched on May 13, 1998; September 21, 2000; June 24, 2002; May 20, 2005, and February 6, 2009, respectively.) NOAA-15, -17, and -19 had daytime north-to-south equatorial crossing times of 7:30 A.M. and 10:00 A.M., respectively. NOAA-16 and -18 had a nighttime north-to-south equatorial crossing time of approximately 2:00 A.M. The satellites orbited at altitudes ranging from 830 to 870 km above the earth with an orbit period of about 102 min and an inclination of 98.7°. Table 5.19 lists the basic characteristics and primary applications of the AVHRR/3's six spectral bands (only five of which were transmitted to the ground at any time). Figure 5.29 shows the 2600-km swath width characteristic of the system. Coverage was acquired at a ground resolution of 1.1 km at nadir. This resolution naturally

TABLE 5.19 AVHRR/3 Spectral Bands

Band	Wavelength (μm)	Nominal Spectral Location	Typical Use
1	0.58–0.68	Red	Daytime cloud, snow, ice, and vegetation mapping
2	0.725–1.00	Near IR	Daytime land/water boundaries, snow, ice and vegetation mapping
3A	1.58–1.64	Mid IR	Snow/ice detection
3B	3.55–3.93	Thermal	Day/night cloud and surface temperature mapping
4	10.30–11.30	Thermal	Day/night cloud and surface temperature mapping
5	11.5–12.50	Thermal	Sea surface temperature, cloud mapping, correction for atmospheric water vapor path radiance

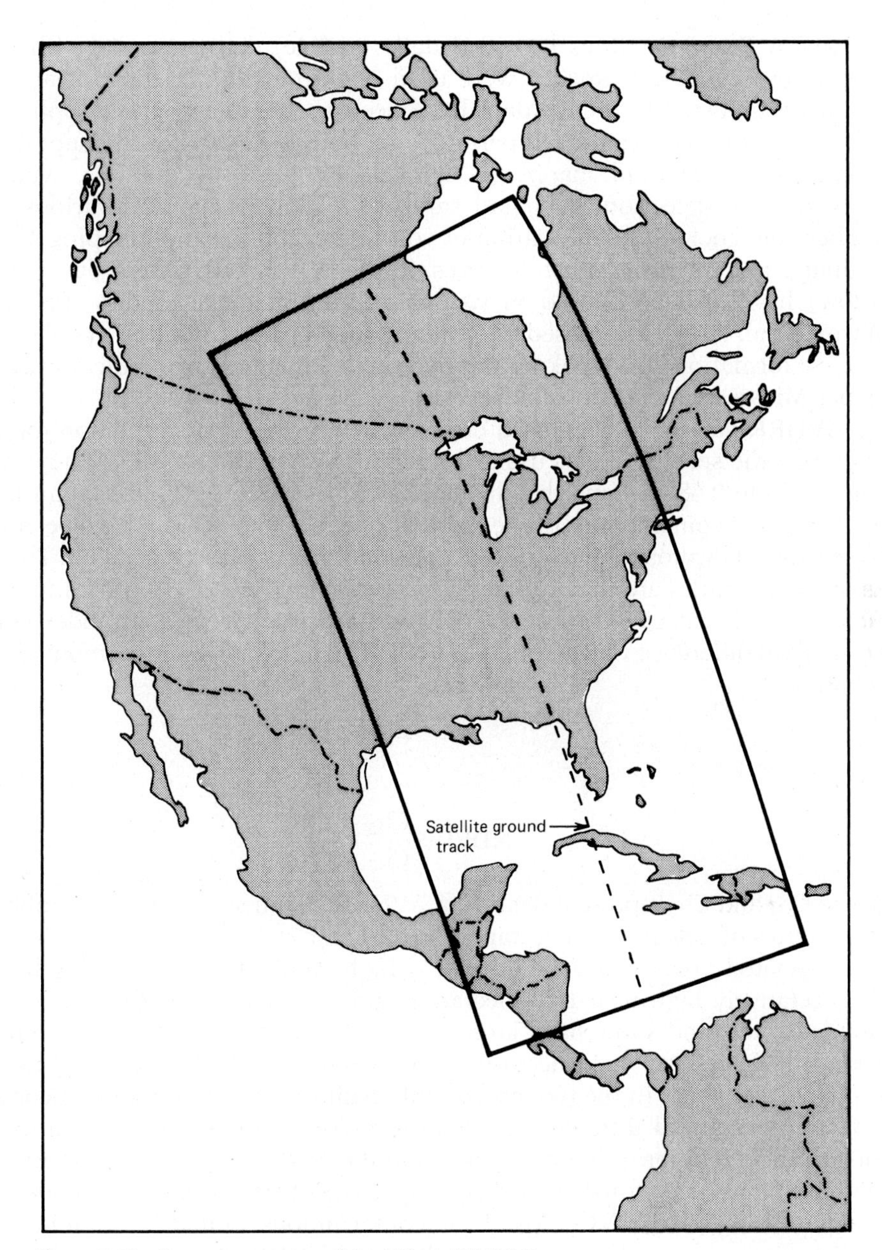

Figure 5.29 Example coverage of the NOAA AVHRR/3.

becomes coarser with increases in the viewing angle off nadir. NOAA received AVHRR/3 data at full resolution and archived them in two different forms. Selected data were recorded at full resolution, referred to as local area coverage (LAC) data. All of the data were sampled down to a nominal resolution of 4 km, referred to as global area coverage (GAC) data.

NOAA POES satellites provided daily (visible) and twice-daily (thermal IR) global coverage. Such images were used operationally in a host of applications requiring timely data. POES data have been used extensively in applications as varied as surface water temperature mapping, snow cover mapping, flood monitoring, vegetation mapping, regional soil moisture analysis, wildfire fuel mapping, fire detection, dust and sandstorm monitoring, and various geologic applications including observation of volcanic eruptions and mapping of regional drainage and physiographic features. Both POES and GOES data (described below) have also been used in regional climate change studies. For example, Wynne et al. (1998) have used 14 years of such data to monitor the phenology of lake ice formation and breakup as a potential climate change indicator in the U.S. Upper Midwest and portions of Canada.

AVHRR data have been used extensively for large-area vegetation monitoring. Typically, the spectral bands used for this purpose have been the channel 1 visible band (0.58 to 0.68 μm) and the channel 2 near-IR band (0.725 to 1.00 μm). Various mathematical combinations of the AVHRR channel 1 and 2 data have been found to be sensitive indicators of the presence and condition of green vegetation. These mathematical quantities are thus referred to as *vegetation indices*. Two such indices have been routinely calculated from AVHRR data—a simple vegetation index (VI) and a normalized difference vegetation index (NDVI). These indices are computed from the equations

$$\mathrm{VI} = \mathrm{Ch}_2 - \mathrm{Ch}_1 \tag{5.4}$$

and

$$\mathrm{NDVI} = \frac{\mathrm{Ch}_2 - \mathrm{Ch}_1}{\mathrm{Ch}_2 + \mathrm{Ch}_1} \tag{5.5}$$

where Ch_1 and Ch_2 represent data from AVHRR channels 1 and 2, preferably expressed in terms of radiance or reflectance (see Chapter 7).

Vegetated areas will generally yield high values for either index because of their relatively high near-IR reflectance and low visible reflectance. In contrast, clouds, water, and snow have larger visible reflectance than near-IR reflectance. Thus, these features yield negative index values. Rock and bare soil areas have similar reflectances in the two bands and result in vegetation indices near zero.

The normalized difference vegetation index is preferred to the simple index for global vegetation monitoring because the NDVI helps compensate for changing illumination conditions, surface slope, aspect, and other extraneous factors.

Plate 18 shows NDVI maps of the conterminous United States for six 2-week periods during a single year prepared from AVHRR data. The NDVI values for vegetation range from a low of 0.05 to a high of 0.66+ in these maps and are displayed in various colors (see key at bottom of plate). Clouds, snow, and bright nonvegetated surfaces have NDVI values of less than zero. The NDVI selected for each pixel is the greatest value on any day during the 14-day period (the highest NDVI value is assumed to represent the maximum vegetation "greenness" during the period). This process eliminates clouds from the composite (except in areas that are cloudy for all 14 days).

Scientists associated with the EROS have also produced numerous global NDVI composites to meet the needs of the international community (NOAA GVI, 2006). In these efforts, data collected at 29 international receiving stations, in addition to that recorded by NOAA, were acquired (starting April 1, 1992) on a daily global basis. Several tens of thousands of AVHRR images have been archived for this program and numerous 10-day maximum NDVI composites have been produced (Eidenshink and Faundeen, 1994).

Numerous investigators have related the NDVI to several vegetation phenomena. These phenomena have ranged from vegetation seasonal dynamics at global and continental scales to tropical forest clearance, leaf area index measurement, biomass estimation, percentage ground cover determination, and photosynthetically active radiation estimation. In turn, these vegetation attributes are used in various models to study photosynthesis, carbon budgets, water balance, and related processes.

Notwithstanding the widespread use of the AVHRR NDVI, it should be pointed out that a number of factors can influence NDVI observations that are unrelated to vegetation conditions. Among these factors are variability in incident solar radiation, radiometric response characteristics of the sensor, atmospheric effects (including the influence of aerosols), and off-nadir viewing effects. Also, the AVHRR scans over ±55° off nadir (compared to ±7.7° for Landsat). This causes a substantial change not only in the size of the ground resolution cell along an AVHRR scan line but also in the angles and distances over which the atmosphere and earth surface features are viewed. Normalizing for such effects is the subject of continuing research. (We further discuss vegetation indices in Section 7.19.)

5.16 JPSS SATELLITES

The JPSS represents the second generation of U.S. operational polar orbiting metsats. The JPSS program builds upon the 40 years of partnership between NOAA and NASA, with NASA acting as the satellite design and acquisition agent and NOAA responsible for managing and operating the resulting satellites. There are three satellites in the JPSS program: the *Suomi National Polar-orbiting Partnership* (*SNPP*), *JPSS-1*, and *JPSS-2*. The SNPP (often referred to as the Suomi NPP) was launched on October 28, 2011, the JPSS-1 has a planned launch date of 2017, and the anticipated launch date for JPSS-2 is 2022.

The SNPP is named in honor of Verner E. Suomi, a meteorologist at the University of Wisconsin–Madison, and often recognized as the "Father of Satellite Meteorology." The satellite is providing on-orbit testing and validation of sensors, algorithms, ground operations, and data processing systems that will be used in the operation of JPSS-1 and -2. The SNPP is in a sun-synchronous orbit at an altitude of 824 km, with an inclination of 98.74° and an equatorial crossing time of 10:30 A.M. There are five instruments on board the satellite. Four of these are designed to measure such meteorological parameters as atmospheric temperature and moisture, ozone concentration, and the space and time of the earth's radiation budget. Of primary interest for land (and ocean) applications is the fifth instrument, the *Visible Infrared Imaging Radiometer Suite* (*VIIRS*). VIIRS incorporates the radiometric accuracy of the NPOS AVHRR along with improved spectral and spatial resolution.

As shown in Table 5.20, VIIRS includes 22 spectral bands. Five of these bands are referred to as "Imagery" resolution bands (I1-I5) and have a spatial resolution of 0.375 km at nadir (0.8 km at scan edge). Sixteen of the remaining 17 "Moderate" resolution bands (M1–M16) have a resolution of 0.75 km at nadir and 1.6 km at scan edge. The 17th Moderate resolution band (DNB) is a broad (0.5–0.9 μm)

TABLE 5.20 VIIRS Spectral Bands

Band	Primary Driving Parameter(s)	Band Width (μm)	Nominal Spectral Region	Nadir Spatial Resolution (km)	Edge Spatial Resolution (km)
I1	Vis Imagery/NDVI	0.600–0.680	Red	0.375	0.8
I2	Land Imagery/NDVI	0.846–0.885	NIR	0.375	0.8
I3	Snow/Ice	1.580–1.640	SWIR	0.375	0.8
I4	Cloud Imagery	3.550–3.930	TIR	0.375	0.8
I5	Cloud Imagery	10.50–12.40	TIR	0.375	0.8
M1	Ocean Color/Aerosol	0.402–0.422	Blue	0.75	0.8
M2	Ocean Color/Aerosol	0.436–0.454	Blue	0.75	1.6
M3	Ocean Color/Aerosol	0.478–0.498	Blue	0.75	1.6
M4	Ocean Color/Aerosol	0.545–0.565	Green	0.75	1.6
M5	Ocean Color/Aerosol	0.662–0.682	Red	0.75	1.6
M6	Atmospheric Correction	0.739–0.754	NIR	0.75	1.6
M7	Ocean Color/Aerosol	0.846–0.885	NIR	0.75	1.6
M8	Cloud Particle/Snow Grain Size	1.230–1.250	NIR	0.75	1.6
M9	Cirrus Clouds	1.371–1.386	SWIR	0.75	1.6
M10	Snow Fraction	1.580–1.640	SWIR	0.75	1.6
M11	Clouds/Aerosol	2.225–2.275	SWIR	0.75	1.6
M12	SST	3.660–3.840	TIR	0.75	1.6
M13	SST/Fire Detection	3.973–4.128	TIR	0.75	1.6
M14	Cloud Top	8.400–8.700	TIR	0.75	1.6
M15	SST	10.263–11.263	TIR	0.75	1.6
M16	SST	11.538–12.488	TIR	0.75	1.6
DNB	Day/Night Band	0.5–0.9	VIS/NIR	0.75	0.75 Full Width

"day-night band," which can be used to collect nighttime visible/near-IR image data at a resolution of 0.75 km across the entire scan width. The VIIRS system employs a whiskbroom scanner that has a total field of view of 112°, resulting in a 3000-km swath width, which affords daily global image coverage. The swath width and coverage location of each spectral band are individually programmable (providing improved resolution views of selected coverage areas near nadir).

Figure 5.30 illustrates three composite images of VIIRS day-night band data collected over nine days during April 2012 and 13 days in October 2012. It

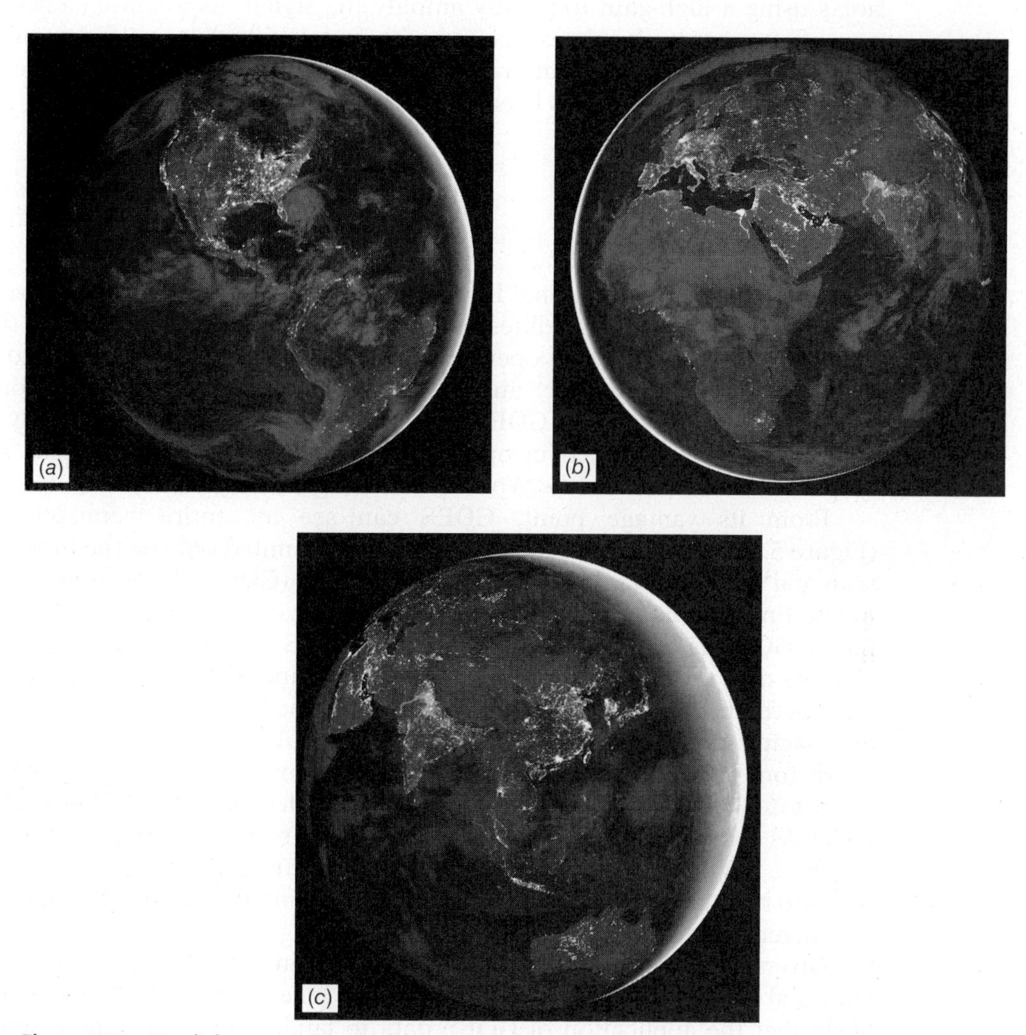

Figure 5.30 City lights composite images acquired by the Suomi NPP VIIRS in April and October 2012. (*a*) North and South America; (*b*) Europe, Africa, and western portions of Asia; (*c*) eastern portions of Asia, and Australia. (Courtesy NOAA, NASA, DOD.)

required 312 orbits and 2.5 terabytes of data to obtain cloud-free imagery of the entire continental land surfaces and islands of the earth. Figure 5.30*a* includes North and South America; Figure 5.30*b* includes Europe, Africa, and western portions of Asia. Note the distinct appearance of the Nile River Valley and Delta in this image. This area represents only 5% of the land area of Egypt, but includes nearly 97% of that country's population. Figure 5.30*c* includes eastern portions of Asia and Australia.

The VIIRS sensor carefully "tunes" its gain on a pixel-by-pixel basis to record very bright pixels using a low-gain to avoid oversaturation and to record very dim areas using a high-gain to greatly amplify the signal. As a consequence, the system often records signals associated with features such as wildfires, oil and gas flares, auroras, volcanoes, and reflected moonlight. In the case of Figure 5.30, the light from such features has been removed to emphasize the city lights.

5.17 GOES SATELLITES

The Geostationary Operational Environmental Satellites (GOES) are part of a network of meteorological satellites located in geostationary orbit around the globe. The United States normally operates two such systems, typically located at 75°W longitude (and the equator) and at 135°W longitude (and the equator). These systems are referred to as GOES-EAST and GOES-WEST, respectively. Similar systems have been placed in operation by several other countries as part of a corporate venture within the World Meteorological Organization.

From its vantage point, GOES can see an entire hemispherical disk (Figure 5.31). The repeat frequency is therefore limited only by the time it takes to scan and relay an image. The first GOES satellite (GOES-1) was launched in 1975. At the time of this writing (2014), the most recent launch within the series was that of GOES-15, in 2010. The design of the GOES imaging systems has seen many changes over the years. The next system to be launched (designated GOES-R until it is successfully operating on-orbit) will mark the beginning of a new generation of imaging capabilities. Whereas GOES-8 to GOES-15 employed five spectral bands (one visible and four thermal) at resolutions of 4 to 8 km, the GOES-R series of satellites will feature a 16-band imager called the *Advanced Baseline Imager* (*ABI*). As can be seen in Table 5.21, the ABI will incorporate two visible, one near-IR, three SWIR, and 10 thermal bands at resolutions from 0.5 to 2 km. The ABI will also be capable of full disk scanning every 5 min versus the 25 min required by the current generation of GOES satellites.

Given the major improvements of the ABI in terms of spectral, spatial, and temporal resolution relative to the current generation of GOES satellites, it is likely that the application of GOES data to land and ocean remote sensing will increase substantially over the next few years. GOES-R is scheduled for launch in 2015, to be followed by GOES-S in 2017, GOES-T in 2019, and GOES-U in 2024.

Figure 5.31 GOES visible band (0.55 to 0.7 μm) image of hemispherical disk including North and South America, early September. A hurricane is clearly discernible in the Gulf of Mexico. (Courtesy NOAA/National Environmental Satellite, Data, and Information Service.)

5.18 OCEAN MONITORING SATELLITES

The oceans, which cover more than 70% of the earth's surface, have important influences on global weather and climate. Like the atmosphere, oceans are characterized by circulation dynamics that are neither steady nor smooth, but rather highly variable over a broad range of spatial and temporal scales. These dynamics, coupled with the sheer size of oceans, make them difficult to monitor solely from boats, buoys, drifters, or even airborne platforms. Hence, satellite observations are extremely useful for extrapolating from point or small area observations to provide spatially detailed synoptic views of the oceans over large areas and extended time periods.

In the case of ocean monitoring satellite systems operating in the optical portion of the spectrum, thermal infrared radiance and water color are the main parameters that are used to assist in the mapping of surface temperature,

TABLE 5.21 ABI Spectral Bands

Band	Primary Driving Parameter(s)	Band Width (μm)	Nominal Spectral Region	Nadir Spatial Resolution (km)
1	Daytime Aerosol Over Land/ Coastal Water Mapping	0.45–0.49	Blue	1
2	Daytime Clouds/Fog Insolation/Winds	0.59–0.69	Red	0.5
3	Daytime Vegetation/ Aerosol Over Water/Winds	0.846–0.885	NIR	1
4	Daytime Cirrus Clouds	1.371–1.386	SWIR	2
5	Daytime Cloud-Top Phase And Particle Size/Snow	1.58–1.64	SWIR	1
6	Daytime Land/Cloud Properties/Vegetation/Snow	2.225–2.275	SWIR	2
7	Surface and Clouds/Fog At Night/Fire/Winds	3.80–4.00	TIR	2
8	High-Level Atmospheric Water Vapor/Winds/Rain	5.77–6.6	TIR	2
9	Mid-Level Atmospheric Water Vapor/Winds/Rain	6.75–7.15	TIR	2
10	Low-Level Atmospheric Water Vapor/Winds/SO_2	7.24–7.44	TIR	2
11	Total Water For Stability/ Cloud Phase/Dust/SO_2/Rain	8.3–8.7	TIR	2
12	Total Ozone/Turbulence/ Winds	9.42–9.8	TIR	2
13	Surface and Clouds	10.1–10.6	TIR	2
14	Imagery/SST/Clouds/ Rain	10.8–11.6	TIR	2
15	Total Water/Ash/SST	11.8–12.8	TIR	2
16	Air Temperature/Cloud Heights and Amounts	13.0–13.6	TIR	2

chlorophyll, suspended solids, currents, and *gelbstoffe* (yellow substance), particularly in near-shore and coastal waters. In the case of radar and lidar satellite systems (discussed in Chapter 6), observations might focus on such measurements as surface topography, currents, wave fields, and wind stress. Both active and passive systems are used extensively to monitor sea ice. Also, many useful ocean observations are often made by land-oriented satellite systems and metsats.

The field of satellite oceanography was greatly enhanced by the launch of two early-generation ocean monitoring systems in 1978. These were the Seasat and Nimbus-7 satellites. Seasat (Section 6.11) carried several instruments dedicated to ocean monitoring that operated in the microwave portion of the spectrum. Nimbus-7 carried the *Coastal Zone Color Scanner* (*CZCS*), which was employed on a limited-coverage "proof-of-concept" basis. The CZCS was designed specifically to measure the color and temperature of the coastal zones of the oceans. The CZCS ceased operation in mid-1986. Follow-on systems included Japan's *Marine Observations Satellite* (*MOS-1*) launched in 1987 and MOS-1b launched in 1990. Additional information about these early-generation systems can be found in earlier editions of this book.

Another major source of ocean-oriented remote sensing data was the *Sea-viewing Wide-Field-of-View Sensor* (*SeaWiFS*). SeaWiFS incorporated an eight-channel across-track scanner operating over the range of 0.402 to 0.885 μm (see Table 5.22). This system was designed primarily for the study of ocean biogeochemistry and was a joint venture between NASA and private industry.

Under a procurement arrangement known as a "data buy," NASA contracted with Orbital Sciences Corporation (OSC) to build, launch, and operate SeaWiFS on the OSC OrbView-2 satellite to meet NASA's science requirements for ocean-monitoring data. OSC (now part of DigitalGlobe) retained the rights to market the resulting data for commercial applications. This was the first time that industry led the development of an entire mission. The system was launched on August 1, 1997, and produced data until December 11, 2010.

From a scientific perspective, SeaWiFS data have been used for the study of such phenomena as ocean primary production and phytoplankton processes; cycles of carbon, sulfur, and nitrogen; and ocean influences on physical climate, including heat storage in the upper ocean and marine aerosol formation.

TABLE 5.22 Major Characteristics of the SeaWiFS System

Spectral bands	402–422 nm
	433–453 nm
	480–500 nm
	500–520 nm
	545–565 nm
	660–680 nm
	745–785 nm
	845–885 nm
Ground resolution	1.1 km
Swath width	2800 km

Commercial applications have ranged from fishing to navigation, weather prediction, and agriculture.

SeaWiFS produced two major types of data. *Local Area Coverage* (*LAC*) data with 1.13-km nadir resolution were broadcast directly to receiving stations and *Global Area Coverage* (*GAC*) data (subsampled on board every fourth line, every fourth pixel) were recorded on board the OrbView-2 spacecraft for subsequent transmission. The system was designed to obtain full global coverage (GAC) data every two days. The LAC data were collected less frequently on a research priority basis. A 705-km orbit, with a southbound equatorial crossing time of 12:00 noon and a ±58.3° scan angle, characterized the system. This provided a scan swath width of approximately 2800 km.

Although originally designed for ocean observation, SeaWiFS has provided a unique opportunity to study land and atmospheric processes as well. In its first year of operation alone, SeaWiFS provided new scientific insights into a number of phenomena. Among others, these include the El Niño and La Niña climate processes; a range of natural disasters, including fires in Florida, Canada, Indonesia, Mexico, and Russia; floods in China; dust storms in the Sahara and Gobi Deserts; hurricanes in numerous locations; and unprecedented phytoplankton blooms. More than 800 scientists representing 35 countries accessed the data in the first year of system operation, and more than 50 ground stations throughout the world began receiving the data. In a sense, with its ability to observe ocean, land, and atmospheric processes simultaneously, SeaWIFS acted as a precursor to the series of instruments comprising the EOS.

Plate 19 is a composite of SeaWiFS GAC data for an 11-month time period showing chlorophyll *a* concentrations in the ocean areas and normalized difference vegetation index values over the land areas.

The Korean KOMPSAT-1 satellite, launched in 1999, incorporated a global ocean color monitoring system called the Ocean Scanning Multi-spectral Imager (OSMI). The OSMI was a whiskbroom-type scanner that operated in up to six spectral bands (in the 0.4–0.9 μm range). The center and bandwidth of each band could be controlled through ground command. The nadir resolution of the system was 850 m, and the swath width of the system was 800 km.

Launched March 1, 2002 by the European Space Agency, Envisat-1 included 10 optical and radar instruments. Among these were the *Advanced Synthetic Aperture Radar* (*ASAR*), discussed in Section 6.13, and the *Medium Resolution Imaging Spectrometer* (*MERIS*). The primary mission of MERIS was the measurement of sea color in ocean and coastal areas (with additional use in atmospheric and land observation).

MERIS was a pushbroom instrument having a 1150-km-wide swath that was divided into five segments for data acquisition. Five identical sensor arrays collected data side by side with a slight overlap between adjacent swaths. Data were collected at 300-m nadir resolution over coastal zones and land areas. Open-ocean data were also collected at a reduced resolution of 1200 m through onboard combinations of 4×4 adjacent pixels across and along track.

MERIS had the ability to record data in as many as 15 bands. This system was also programmable by ground command such that the number, location, and width of each spectral band could be varied within the 0.4- to 1.05-μm range. MERIS allowed global coverage every three days.

Readers interested in further information about ocean monitoring satellites and their applications are encouraged to consult any of the myriad of texts and references on the subject. Another source of timely information about ocean color sensors in particular is the *International Ocean Colour Coordinating Group* (*IOCCG*). This organization maintains an Internet-based listing of historical, current, and scheduled ocean color sensors at www.ioccg.org.

5.19 EARTH OBSERVING SYSTEM

The Earth Observing System (*EOS*) is one of the primary components of a NASA-initiated concept originally referred to as *Mission to Planet Earth* (*MTPE*). Over time, this concept has been renamed or recast as the *Earth Science Enterprise* (*ESE*) and the *Earth-Sun Missions*. Currently (2014), EOS is a major component of the Earth Science Division of NASA's Science Mission Directorate and includes numerous satellite missions and related research programs. The overall focus of EOS is on improved understanding of the earth as an integrated system and its response to natural and human-induced changes. The aim is to improve the prediction of climate, weather, and natural hazards in support of both scientific research and national policymaking.

The first operational satellite system to be included in the EOS program structure was Landsat-7. Since its inception, the program has not only included operational systems, but also systems under development, and planned programs for the future. Clearly, programs of this magnitude, cost, and complexity are subject to change. Also, the EOS program includes numerous platforms and sensors that are outside the realm of land-oriented remote sensing. We make no attempt to describe the overall program. Rather, we limit our attention to the first two EOS-dedicated platforms, the *Terra* and *Aqua* spacecraft. Sometimes described as the "flagship" of EOS, Terra was launched on December 18, 1999. It was followed on May 4, 2002, by Aqua. Both of these platforms are complex systems with multiple remote sensing instruments. Five sensors are included on Terra:

ASTER: Advanced Spaceborne Thermal Emission and Reflection Radiometer
CERES: Clouds and the Earth's Radiant Energy System
MISR: Multi-Angle Imaging Spectro-Radiometer
MODIS: Moderate Resolution Imaging Spectro-Radiometer
MOPITT: Measurements of Pollution in the Troposphere

Aqua carries six instruments, two of which (MODIS and CERES) are also present on Terra. The remaining four instruments on Aqua include:

AMSR/E: Advanced Microwave Scanning Radiometer-EOS
AMSU: Advanced Microwave Sounding Unit
AIRS: Atmospheric Infrared Sounder
HSB: Humidity Sounder for Brazil

Table 5.23 summarizes the salient characteristics and applications associated with each of the instruments on Terra and Aqua.

The intent in the design of Terra and Aqua is to provide a suite of highly synergistic instruments on each platform. For example, four of the five

TABLE 5.23 Sensors Carried on Terra and Aqua

Sensor	Terra/ Aqua	General Characteristics	Primary Applications
ASTER	Terra	Three scanners operating in the visible and near, mid, and thermal IR, 15–90 m resolution, along-track stereo	Study vegetation, rock types, clouds, volcanoes; produce DEMs; provide high resolution data for overall mission requirements
MISR	Terra	Four-channel CCD arrays providing nine separate view angles	Provide multiangle views of earth surface features, data on clouds and atmospheric aerosols, and correction for atmospheric effects to data from ASTER and MODIS
MOPITT	Terra	Three channel near-IR scanner	Measure carbon monoxide and methane in the atmospheric column
CERES	Both	Two broadband scanners	Measure radiant flux at top of the atmosphere to monitor earth's total radiation energy balance
MODIS	Both	Thirty-six-channel imaging spectrometer; 250 m to 1 km resolution	Useful for multiple land and ocean applications, cloud cover, cloud properties
AIRS	Aqua	Hyperspectral sensor with 2378 channels, 2- to 14-km spatial resolution	Measure atmospheric temperature and humidity, cloud properties, and radiative energy flux
AMSR/E	Aqua	Twelve-channel microwave radiometer, 6.9 to 89 GHz	Measure precipitation, land surface wetness and snow cover, sea surface characteristics, cloud properties
AMSU	Aqua	Fifteen-channel microwave radiometer, 50 to 89 GHz	Measure atmospheric temperature and humidity
HSB	Aqua	Five-channel microwave radiometer, 150 to 183 MHz	Measure atmospheric humidity

instruments on Terra are used in a complementary manner to obtain data about cloud properties. Similarly, measurements made by one instrument (e.g., MISR) can be used to atmospherically correct the data for another (e.g., MODIS). In composite, each spacecraft provides detailed measurements contributing to a number of interrelated scientific objectives.

Both Terra and Aqua are quite massive for earth-observing satellites. Terra is approximately 6.8 m long and 3.5 m in diameter and weighs 5190 kg, while Aqua is 16.7 m × 8.0 m × 4.8 m and weighs 2934 kg. The satellites are in near-polar, sun-synchronous orbits at 705 km altitude, and their orbits fit those of the WRS-2 numbering system developed for Landsat-4, -5, -7, and -8. Terra has a 10:30 A.M. equatorial crossing time (descending), a time chosen to minimize cloud cover. Aqua has a 1:30 P.M. equatorial crossing time (ascending). Among the instruments carried on Terra and Aqua, the three of greatest interest to the readership of this book are the MODIS, ASTER, and MISR instruments.

MODIS

Flown on both the Terra and Aqua satellites, MODIS is a sensor that is intended to provide comprehensive data about land, ocean, and atmospheric processes simultaneously. Its design is rooted in various earlier sensors, or "heritage instruments," such as the AVHRR and CZCS. However, MODIS is a highly improved successor to these earlier systems. MODIS not only provides two-day repeat global coverage with greater spatial resolution (250, 500, or 1000 m, depending on wavelength) than the AVHRR, but it also collects data in 36 carefully chosen spectral bands (Table 5.24), with 12-bit radiometric sensitivity. In addition, MODIS data are characterized by improved geometric rectification and radiometric calibration. Band-to-band registration for all 36 MODIS channels is specified to be 0.1 pixel or better. The 20 reflected solar bands are absolutely calibrated radiometrically with an accuracy of 5% or better. The calibrated accuracy of the 16 thermal bands is specified to be 1% or better. These stringent calibration standards are a consequence of the EOS requirement for a long-term continuous series of observations aimed at documenting subtle changes in global climate. This data set must not be sensor dependent; hence, the emphasis on sensor calibration.

The total field of view of MODIS is ±55°, providing a swath width of 2330 km. A large variety of atmospheric, oceanic, and land surface data products can be derived from MODIS data. Among the current atmospheric products are aerosol and cloud properties and water vapor and temperature profiles. Representative ocean products include sea surface temperature and chlorophyll concentrations. Land surface products include, among others: surface reflectance, land surface temperature and emissivity, land cover/change, vegetation indices, thermal anomalies/fire, leaf area index/fraction of photosynthetically active radiation, net primary vegetation production, BRDF/albedo, and vegetation conversion.

TABLE 5.24 MODIS Spectral Bands

Primary Use	Band	Bandwidth	Resolution (m)
Land/cloud	1	620–670 nm	250
boundaries	2	841–876 nm	250
Land/cloud	3	459–479 nm	500
properties	4	545–565 nm	500
	5	1230–1250 nm	500
	6	1628–1652 nm	500
	7	2105–2155 nm	500
Ocean color/	8	405–420 nm	1000
phytoplankton/	9	438–448 nm	1000
biogeochemistry	10	483–493 nm	1000
	11	526–536 nm	1000
	12	546–556 nm	1000
	13	662–672 nm	1000
	14	673–683 nm	1000
	15	743–753 nm	1000
	16	862–877 nm	1000
Atmospheric	17	890–920 nm	1000
water vapor	18	931–941 nm	1000
	19	915–965 nm	1000
Surface/cloud	20	3.660–3.840 μm	1000
temperature	21[a]	3.929–3.989 μm	1000
	22	3.929–3.989 μm	1000
	23	4.020–4.080 μm	1000
Atmospheric	24	4.433–4.498 μm	1000
temperature	25	4.482–4.549 μm	1000
Cirrus clouds	26[b]	1.360–1.390 μm	1000
Water vapor	27	6.538–6.895 μm	1000
	28	7.175–7.475 μm	1000
	29	8.400–8.700 μm	1000
Ozone	30	9.580–9.880 μm	1000
Surface/cloud	31	10.780–11.280 μm	1000
temperature	32	11.770–12.270 μm	1000
Cloud top	33	13.185–13.485 μm	1000
altitude	34	13.485–13.758 μm	1000
	35	13.785–14.085 μm	1000
	36	14.085–14.385 μm	1000

[a]Bands 21 and 22 are similar, but band 21 saturates at 500 K versus 328 K.
[b]Wavelength out of sequence due to change in sensor design.

Plate 20 shows a MODIS-based sea surface temperature map of the western Atlantic, including a portion of the Gulf Stream. Land areas and clouds have been masked out. Cold water, approximately 7°C, is shown in purple, with blue, green, yellow, and red representing increasingly warm water, up to approximately 22°C. The warm current of the Gulf Stream stands out clearly in deep red colors.

MODIS data are particularly useful for understanding complex regional-scale systems and their dynamics. This is illustrated in Plate 21, which depicts a synoptic view of water surface temperatures in all five of the Laurentian Great Lakes simultaneously. With the temporal frequency of MODIS data collection, understanding the dynamics of such phenomena over a range of timescales is possible (cloud cover permitting).

ASTER

ASTER is an imaging instrument that is a cooperative effort between NASA and Japan's Ministry of International Trade and Industry. In a sense, ASTER serves as a "zoom lens" for the other instruments aboard Terra in that it has the highest spatial resolution of any of them. ASTER consists of three separate instrument subsystems, each operating in a different spectral region, using a separate optical system, and built by a different Japanese company. These subsystems are the *Visible and Near Infrared* (*VNIR*), the *Short Wave Infrared* (*SWIR*), and the *Thermal Infrared* (*TIR*), respectively. Table 5.25 indicates the basic characteristics of each of these subsystems.

The VNIR subsystem incorporates three spectral bands that have 15 m ground resolution. The instrument consists of two telescopes—one nadir looking with a three-band CCD detector and the other backward looking (27.7° off nadir) with a single-band (band 3) detector. This configuration enables along-track stereoscopic image acquisition in band 3 with a base-height ratio of 0.6. This permits the construction of DEMs from the stereo data with a vertical accuracy of approximately 7 to 50 m. Cross-track pointing to 24° on either side of the orbit path is accomplished by rotating the entire telescope assembly.

The SWIR subsystem, which failed in April 2008, operated in six spectral bands through a single, nadir-pointing telescope that provided 30-m resolution. Cross-track pointing to 8.55° was accomplished through the use of a pointing mirror.

The TIR subsystem operates in five bands in the thermal IR region with a resolution of 90 m. Unlike the other instrument subsystems, it incorporates a whiskbroom scanning mirror. Each band uses 10 detectors, and the scanning mirror functions both for scanning and cross-track pointing to 8.55°.

All ASTER bands cover the same 60-km imaging swath with a pointing capability in the cross-track direction to cover ±116 km from nadir. This means that any point on the globe is accessible at least once every 16 days with full 14-band spectral coverage from the VNIR, SWIR (until April 2008), and TIR. In that the VNIR subsystem had a larger pointing capability, it could collect data up to ±318 km from nadir, with an average revisit period of four days at the equator.

TABLE 5.25 ASTER Instrument Characteristics

Characteristic	VNIR	SWIR[b]	TIR
Spectral range	Band 1: 0.52–0.60 μm, nadir looking	Band 4: 1.600–1.700 μm	Band 10: 8.125–8.475 μm
	Band 2: 0.63–0.69 μm, nadir looking	Band 5: 2.145–2.185 μm	Band 11: 8.475–8.825 μm
	Band 3: 0.76–0.86[a] μm, nadir looking	Band 6: 2.185–2.225 μm	Band 12: 8.925–9.275 μm
	Band 3: 0.76–0.86[a] μm, backward looking	Band 7: 2.235–2.285 μm	Band 13: 10.25–10.95 μm
		Band 8: 2.295–2.365 μm	Band 14: 10.95–11.65 μm
		Band 9: 2.360–2.430 μm	
Ground resolution (m)	15	30	90
Cross-track pointing (deg)	±24	±8.55	±8.55
Cross-track pointing (km)	±318	±116	±116
Swath width (km)	60	60	60
Quantization (bits)	8	8	12

[a]Stereoscopic imaging subsystem.
[b]SWIR system failed in April 2008.

Three individual bands of an ASTER data set for a 13.3 × 54.8-km area in Death Valley, California, are shown as Figure 5.32. Note that this area represents only a portion of the full ASTER swath width. Plate 22 shows three color composites of ASTER data for the same area as shown in Figure 5.32. Plate 22*a* is a color composite of three VNIR (visible and near-IR) bands, part (*b*) shows three SWIR (mid-IR) bands, and part (*c*) shows three TIR (thermal IR) bands. Various geological and vegetation differences are displayed in the three parts of this plate. For example, the bright red areas in (*a*) are patches of vegetation on the Furnace Creek alluvial fan that reflect more highly in the near-IR than in the green or red parts of the spectrum. These are seen as blue in (*b*), but are difficult to discern in (*c*), which shows thermal bands with 90-m resolution as contrasted with the 15 m resolution of (*a*) and the 30-m resolution of (*b*). A striking difference in geological materials is seen in (*c*), where surfaces containing materials high in the mineral quartz have a red color.

Because the nadir-looking and backward-looking band 3 sensors can obtain data of the same area from two different angles, stereo ASTER data can be used to produce DEMs. Various bands of ASTER image data can then be "draped" over

Figure 5.32 ASTER data, Death Valley, CA, early March: (*a*) band 2 (red); (*b*) band 5 (mid IR); (*c*) band 13 (thermal IR). Scale 1:300,000. (Author-prepared figure.)

Figure 5.33 ASTER data draped over a digital elevation model, Death Valley, CA (view to the north). Black-and-white copy of a color composite derived from thermal IR bands. (Courtesy NASA/JPL.)

the DEMs to produce perspective views of an area, as illustrated in Figure 5.33. (The area shown in Figure 5.33 roughly corresponds with that in Figure 5.32 and Plate 22.) Such perspective views can help image analysts to obtain an overview of the area under study. For example, in this figure, bedrock mountains are located at left (west) and right (east). The broad, very light-toned area through the center of the image is the floor of Death Valley. Between the mountains and the valley floor, many alluvial fans (essentially continuous on the west side of the valley floor) can be seen.

MISR

The MISR instrument differs from most other remote sensing systems in that it has nine sensors viewing the earth simultaneously at different angles. Each sensor consists of a set of linear arrays providing coverage in four spectral bands (blue, green, red, and near IR). One sensor is oriented toward nadir (0°); four sensors are viewing forward at angles of 26.1°, 45.6°, 60.0°, and 70.5°; and four are viewing backward at similar angles. The system has a swath width of 360 km, and a spatial resolution of 250 m in the nadir-viewing sensor and 275 m in the off-nadir sensors.

The MISR viewing angles were selected to address several different objectives. The nadir-viewing sensor provides imagery with the minimum interference from the atmosphere. The two sensors at 26.1° off nadir were selected to provide stereoscopic coverage for measurement of cloud-top altitudes, among other purposes. The 45.6° sensors provide information about aerosols in the atmosphere, which are of great importance for studies of global change and the earth's radiation budget. The 60.0° sensors minimize the effects of directional differences in the scattering of light from clouds and can be used to estimate the hemispherical reflectance of land surface features. Finally, the 70.5° sensors were chosen to provide the most oblique angle that could be achieved within practical limitations, in order to maximize the impact of off-nadir scattering phenomena.

Analysis of MISR imagery contributes to studies of the earth's climate and helps monitor the global distribution of particulate pollution and different types of haze. For example, Plate 23 shows a series of MISR images of the eastern United States, acquired during a single orbit (during early March) at four different angles. The area shown ranges from Lakes Ontario and Erie to northern Georgia and covers a portion of the Appalachian Mountains. The eastern end of Lake Erie was covered by ice on this date and thus appears bright in these images. Plate 23*a* was acquired at an angle of 0° (nadir-viewing), while (*b*), (*c*), and (*d*) were acquired at forward-viewing angles of 45.6°, 60.0°, and 70.5°, respectively. Areas of extensive haze over the Appalachian Mountains are almost invisible in the 0° image but have a major effect on the 70.5° image, due to the increased atmospheric path length. The images in (*c*) and (*d*) appear brighter due to increased path radiance, but also have a distinct color shift toward the blue end of the spectrum. This is a result of the fact that Rayleigh scattering in the atmosphere is inversely proportional to the fourth power of the wavelength, as discussed in Chapter 1.

5.20 SPACE STATION REMOTE SENSING

Astronauts have used handheld cameras and other sensing systems to image the earth since the Mercury missions of the early 1960s. The first American space workshop, Skylab, was used to acquire numerous forms of remote sensing data in 1973. The Russian MIR space station was fitted with a dedicated remote sensing module called PRIRODA that was launched in 1996. With contributions from 12 nations, the PRIRODA mission consisted of a broad variety of different sensors, including numerous optical systems as well as both active and passive microwave equipment.

The *International Space Station* (*ISS*) has provided an ideal platform for continuing the tradition of earth observation from human-tended spacecraft. The orbital coverage of the ISS extends from 52°N to 52°S latitude. It orbits the earth every 90 min at an altitude of 400 km, and the revisit period over any given point at the equator (within 9° off nadir) is 32 hr. Hence, approximately 75% of the land surface of the globe, most of the world's coastlines, 100% of the rapidly changing tropics, and 95% of the world's population can be observed from this platform.

Over time, two designs of earth-viewing windows have been used to facilitate the collection of earth observation imagery from the ISS. The first was a 50-cm-diameter, high-optical-quality window installed in the Destiny laboratory module, which was activated in 2001. Since 2010, images have been acquired from within the *Cupola Observation Module*, shown in Figure 5.34. The Cupola is a dome-shaped module that affords a 360° view outside of the ISS. It is approximately

Figure 5.34 NASA astronaut Chris Cassidy using a 400-mm lens on a digital still camera to photograph a target of opportunity on Earth from the ISS Cupola Observation Module. (Courtesy NASA.)

2 m in diameter and 1.5-m tall and incorporates six side windows and a nadir viewing window approximately 79 cm in diameter. As of August 1, 2013, ISS crew members had taken over 1,200,000 frames of imagery using film and digital cameras and other sensors.

Figure 5.35 is an example of an ISS astronaut photograph taken in response to a short-lived natural event. This image was taken on May 23, 2006, and shows the ash plume of an eruption of the Cleveland Volcano. ISS astronauts were the first to observe the eruption, and alerted the Alaska Volcano Observatory. The eruption lasted for only six hours. Cleveland Volcano is a strato volcano with a summit elevation of about 1730 meters and is the second most active volcano in Alaska. The magma that feeds eruptions of ash and lava flows from Cleveland Volcano is generated by

Figure 5.35 Eruption of Cleveland Volcano, Alaska, photographed from the ISS on May 23, 2006. (Courtesy Image Science and Analysis Laboratory, NASA-Johnson Space Center.)

subduction of the Pacific Plate beneath the North American Plate. As one tectonic plate subducts beneath another, melting of materials above and within the subducting plate produces magma that can eventually move to the surface and erupt through a vent. This is the same process responsible for eruptions of the Cascade Volcanoes of the northwest United States, such as Mt. Ranier and Mt. St. Helens.

Figure 5.36 shows a bright ring of lights surrounding San Francisco Bay at night and includes the cities of San Francisco (center-left), Oakland (across from San Francisco, on the east side of the bay), and San Jose (south of the bay). This photograph was with a Nikon camera by an ISS astronaut on the evening of October 21, 2013. The more densely populated areas and the major roads are the brightest areas in the image. Five of the bay's primary bridges are visible in this photograph.

As discussed in Section 5.13, the HICO sensor has also been operated from the ISS since 2009, collecting hyperspectral imagery at 90 m resolution, primarily for coastal and inland aquatic remote sensing research. Images coming down from the ISS on a daily basis are added to an ever-growing (more than 1,700,000-image) database of views of the earth acquired by astronauts since the early years of the U.S. space program. This online database is maintained at the NASA Johnson Space Center (NASA eol, 2013). Data collected from the ISS are used for applications ranging from validating other global terrestrial data sets to real-time monitoring of ephemeral events and human-induced global change. Real-time human oversight of the acquisition of such data complements greatly the capabilities of the various satellite systems discussed in this chapter.

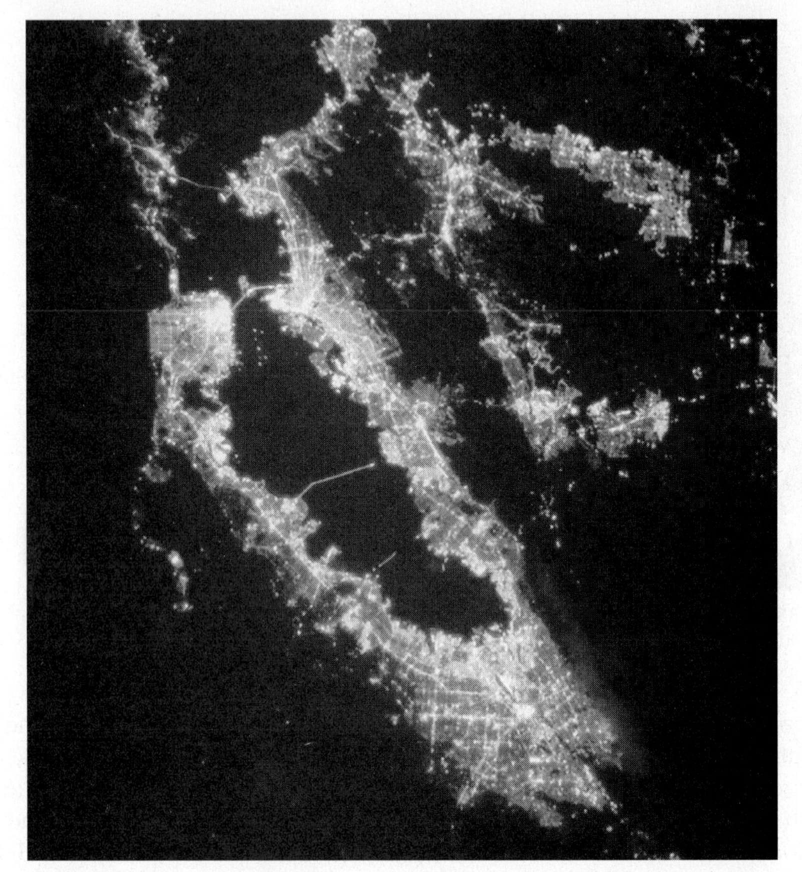

Figure 5.36 City lights of San Francisco, California, and vicinity photographed from the ISS on the evening of September 23, 2013. The camera used to acquire this photograph was a Nikon D35 digital camera fitted with a 50-mm focal length lens. (Courtesy Image Science and Analysis Laboratory, NASA-Johnson Space Center.)

5.21 SPACE DEBRIS

One of our objectives in presenting this chapter has been to highlight the large number of past, present, and planned earth observing satellite missions. However, the increasing number of these and other types of satellites and space missions has led to serious concerns about the development of unsustainable amounts and spatial densities of *space debris*. Space debris is also known as *space junk* or *space waste*. This debris consists of objects that have outlived their usefulness and now orbit around the earth. Examples range from spent rocket stages to inoperable satellites, fragments from collisions of objects in space, and dust. Depending on altitude, such debris can remain in space for tens, hundreds, thousands, or even millions of years (at very high altitudes).

As shown in Figure 5.37, space debris is most concentrated in two primary debris fields. These are a ring of objects in *Geosynchronous Orbit* (*GEO*) and a

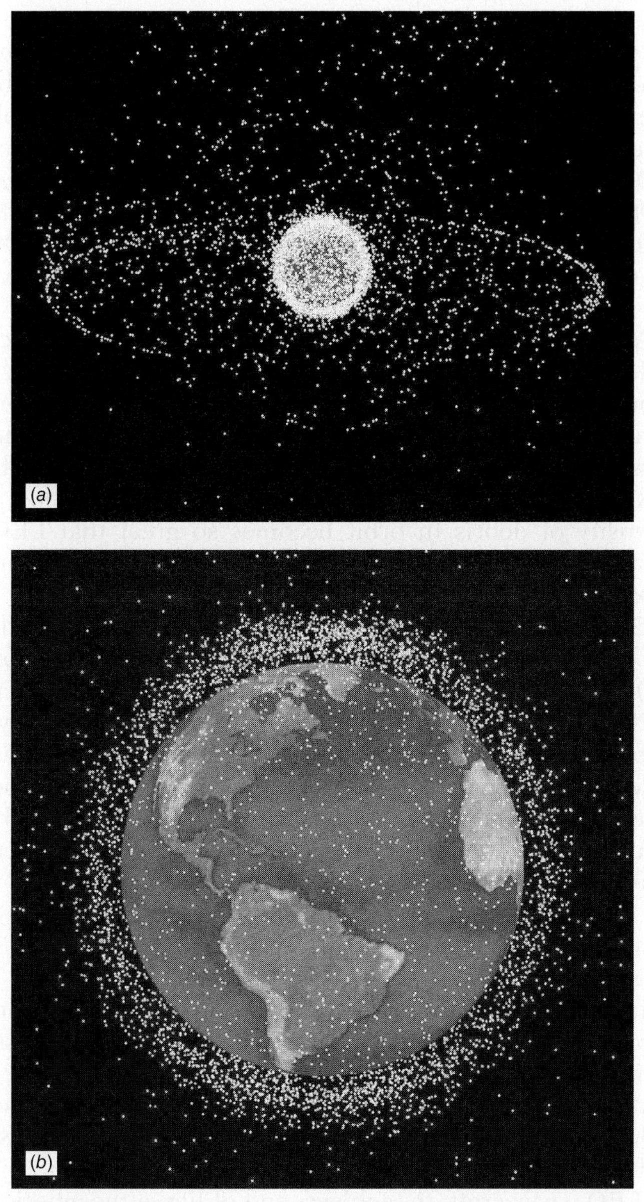

Figure 5.37 Space debris populations. (*a*) Oblique view from outside of geosynchronous orbit showing two primary debris fields (the ring of objects in geosynchronous orbit and the cloud of debris in low earth orbit); (*b*) enlarged view of debris in low earth orbit. (Courtesy NASA.)

cloud of debris in *Low Earth Orbit* (*LEO*). LEO consists of orbits with altitudes below 2000 km (where most polar-orbiting satellites operate). Orbital debris can travel at relative orbital velocities of 15.5 km/sec in LEO and can easily damage or destroy operational satellites and spacecraft if their orbits overlap and cause a collision. The NASA Orbital Debris Program Office estimates that there are more than 21,000 objects larger than 10 cm in diameter in LEO. Only approximately 1000 of these are operational spacecraft. The remainder consists of objects that no longer serve any useful purpose. About 64% of routinely tracked objects are fragments from breakups caused by explosions and collisions of satellites or rocket bodies. LEO contains approximately 500,000 objects between 1 and 10 cm in size and over 100 million debris particles smaller than 1 cm in size. Maximum spatial concentrations of space debris occur at altitudes of 800 to 1000 km and near 1400 km. Spatial densities in GEO and near the orbits of navigational satellites are smaller by two to three orders of magnitude.

In 1978, NASA scientist Donald J. Kessler proposed a possible scenario in which the debris density in LEO could eventually be high enough that collisions between objects could cause a collisional cascade wherein each collision generating space debris increases the likelihood of further collisions, a "domino effect" (Kessler and Cour-Palais, 1978). This scenario is frequently called the *Kessler syndrome* or the *Kessler effect*. The implications of such a condition would be profound if the density of debris in orbit becomes so great that LEO satellite operations and space exploration would become extremely difficult and expensive, *if not unfeasible,* if LEO becomes impassable.

In the United States, a 2011 National Research Council study concluded that LEO orbital debris is at the "tipping point," or threshold, of causing a Kessler syndrome (NRC, 2011). More recently, the findings from the 6th European Conference on Space Debris (2013) included such conclusions as those expressed by the head of ESA's Space Debris Office, Heiner Klinkrad: "Our understanding of the growing space debris problem can be compared with our understanding of the need to address Earth's changing climate some 20 years ago." He further stated that "While measures against further debris creation and actively de-orbiting defunct satellites are technically demanding and potentially costly, there is no alternative to protect space as a valuable resource for our critical satellite infrastructure."

Space debris reduction is dictated simultaneously by economics, service to citizens, resource stewardship, and science. In that context, satellite operators worldwide, including those designing and flying earth resource, telecommunication, broadcast, weather and climate-monitoring, and navigation missions, are now actively focusing on measures to control space debris. For example, ESA has many initiatives underway in the area of space debris mitigation. These include, but are not limited to, prevention of in-orbit explosions, removal of mass (5 to 10 objects per year) from regions with high object densities and long orbital lifetimes, reorbiting LEO spacecraft to achieve end-of-life atmospheric re-entry, reorbiting into orbits at altitudes greater than 2000 km, reorbiting spacecraft operating near the geosynchronous ring into a "graveyard orbit" approximately 300 km above the GEO ring. Several technologies are being developed to approach, capture, and deorbit various types of debris. Clearly, mitigation of space debris will be the subject of substantial research and development over the next several years. Successful implementation of such endeavors will continue to be contingent upon substantial international cooperation (www.iadc-online.org).

6 MICROWAVE AND LIDAR SENSING

6.1 INTRODUCTION

An increasing amount of valuable environmental and resource information is being derived from sensors that operate in the *microwave* portion of the electromagnetic spectrum. In the context of the sensors we have discussed thus far, microwaves are not "micro" at all. That is, the microwave portion of the spectrum includes wavelengths within the approximate range of 1 mm to 1 m. Thus, the longest microwaves are about 2,500,000 times longer than the shortest light waves!

There are two distinctive features that characterize microwave energy from a remote sensing standpoint:

1. Microwaves are capable of penetrating the atmosphere under virtually all conditions. Depending on the wavelengths involved, microwave energy can "see through" haze, light rain and snow, clouds, and smoke.
2. Microwave reflections or emissions from earth materials bear no direct relationship to their counterparts in the visible or thermal portions of the spectrum. For example, surfaces that appear "rough" in the visible portion of the spectrum may be "smooth" as seen by microwaves. In general, microwave

responses afford us a markedly different "view" of the environment—one far removed from the views experienced by sensing light or heat.

In this chapter we discuss both airborne and spaceborne, as well as *active* and *passive*, microwave sensing systems. Recall that the term "active" refers to a sensor that supplies its own source of energy or illumination. *Radar* is an active microwave sensor, and it is the major focus of attention in this chapter. In recent years there has been a great increase in the availability of spaceborne radar imagery, most notably from constellations of new, high-resolution imaging radar satellites. To a lesser extent, we also treat the passive counterpart to radar, the *microwave radiometer*. This device responds to the extremely low levels of microwave energy that are naturally emitted and/or reflected from ambient sources (such as the sun) by terrain features.

We conclude this chapter with a discussion of *lidar* remote sensing. Like radar, lidar sensors are active remote sensing systems. However, they use pulses of laser light, rather than microwave energy, to illuminate the terrain. Over the past decade, the use of airborne lidar systems (primarily for topographic mapping) has rapidly expanded. As with radar, the outlook for applications of such systems is an extremely promising one.

6.2 RADAR DEVELOPMENT

The word *radar* originated as an acronym for *radio detection and ranging*. As its name implies, radar was developed as a means of using radio waves to detect the presence of objects and to determine their distance and sometimes their angular position. The process entails transmitting short bursts, or pulses, of microwave energy in the direction of interest and recording the strength and origin of "echoes" or "reflections" received from objects within the system's field of view.

Radar systems may or may not produce images, and they may be ground based or mounted in aircraft or spacecraft. A common form of nonimaging radar is the type used to measure vehicle speeds. These systems are termed *Doppler radar* systems because they utilize Doppler frequency shifts in the transmitted and returned signals to determine an object's velocity. Doppler frequency shifts are a function of the relative velocities of a sensing system and a reflector. For example, we perceive Doppler shifts in sound waves as a change in pitch, as in the case of a passing car horn or train whistle. The Doppler shift principle is often used in analyzing the data generated from imaging radar systems.

Another common form of radar is the *plan position indicator (PPI)* system. These systems have a circular display screen on which a radial sweep indicates the position of radar "echoes." Essentially, a PPI radar provides a continuously updated plan-view map of objects surrounding its rotating antenna. These systems are common in weather forecasting, air traffic control, and navigation applications. However, PPI systems are not appropriate for most remote sensing applications because they have rather poor spatial resolution.

Airborne and spaceborne radar remote sensing systems are collectively referred to as *imaging radar*. Imaging radar systems employ an antenna fixed below the aircraft (or spacecraft) and pointed to the side. Thus, such systems were originally termed *side-looking radar (SLR)*, or *side-looking airborne radar (SLAR)* in the case of airborne systems. Modern imaging radar systems use advanced data processing methods (described in Section 6.4) and are referred to as *synthetic aperture radar (SAR)* systems. Regardless of the terminology used to identify them, imaging radar systems produce continuous strips of imagery depicting extensive ground areas that parallel the platform's flight line.

Imaging radar was first developed for military reconnaissance purposes in the late 1940s. It became an ideal military reconnaissance system, not only because it affords nearly an all-weather operating capability, but also because it is an active, day-or-night imaging system. The military genesis of imaging radar has had two general impacts on its subsequent application to civilian remote sensing uses. First, there was a time lag between military development, declassification, and civilian application. Less obvious, but nonetheless important, is the fact that military radar systems were developed to look at military targets. Terrain features that "cluttered" radar imagery and masked objects of military importance were naturally not of interest in original system designs. However, with military declassification and improvement in nonmilitary capabilities, imaging radar has evolved into a powerful tool for acquiring natural resource data.

In the years since the first edition of this book was published, the science and technology of radar remote sensing have undergone dramatic progress. Today, at least eight major satellite radar systems provide imagery day and night, worldwide to monitor the planet's oceans, land, and ice caps. These data support operations ranging from agricultural forecasting to natural disaster response. Scientists have learned a great deal about the ways that radar signals interact with different environments and are using this understanding to develop new applications of radar technology.

Some of the earliest broad-scale applications of radar remote sensing occurred in regions where persistent cloud cover limits the acquisition of imagery in the optical portion of the spectrum. The first such large-scale project for mapping terrain with side-looking airborne radar was a complete survey of the Darien province of Panama. This survey was undertaken in 1967 and resulted in images used to produce a mosaic of a 20,000-km^2 ground area. Prior to that time, this region had never been photographed or mapped in its entirety because of persistent (nearly perpetual) cloud cover. The success of the Panama radar mapping project led to the application of radar remote sensing throughout the world.

In 1971, a radar survey was begun in Venezuela that resulted in the mapping of nearly 500,000 km^2 of land. This project resulted in improvements in the accuracy of the location of the boundaries of Venezuela with its neighboring countries. It also permitted a systematic inventory and mapping of the country's water resources, including the discovery of the previously unknown source of several major rivers. Likewise, improved geologic maps of the country were produced.

Also beginning in 1971 was Project Radam (standing for *Radar of the Amazon*), a reconnaissance survey of the Amazon and the adjacent Brazilian northeast. At that time, this was the largest radar mapping project ever undertaken. By the end of 1976, more than 160 radar mosaic sheets covering an area in excess of

8,500,000 km^2 had been completed. Scientists used these radar mosaics as base maps in a host of studies, including geologic analysis, timber inventory, transportation route location, and mineral exploration. Large deposits of important minerals were discovered after intensive analysis was made of newly discovered features shown by radar. Mapping of previously uncharted volcanic cones, and even large rivers, resulted from this project. In such remote and cloud-covered areas of the world, radar imagery is a prime source of inventory information about potential mineral resources, forestry and range resources, water supplies, transportation routes, and sites suitable for agriculture. Such information is essential to planning sustainable development in such ecologically sensitive areas.

Along with the humid tropics, the Arctic and Antarctic regions present some of the most challenging conditions for optical remote sensing. At high latitudes, where cloud cover is frequent and the sun is low in the sky or absent for much of the year, imaging radar is used to track the extent and movement of sea ice to ensure the safety of shipping lanes. On the circumpolar land masses, radar is used to measure the velocity of glaciers and to map surface features in areas that may only rarely, if ever, be visited by humans. Figure 6.1 shows the first-ever comprehensive image mosaic of Antarctica, produced in 1997 by the Radarsat Antarctic Mapping Program (RAMP; Jezek, 2003). In this image, the seemingly featureless expanse of Antarctica turns out to be far more variable than previously realized,

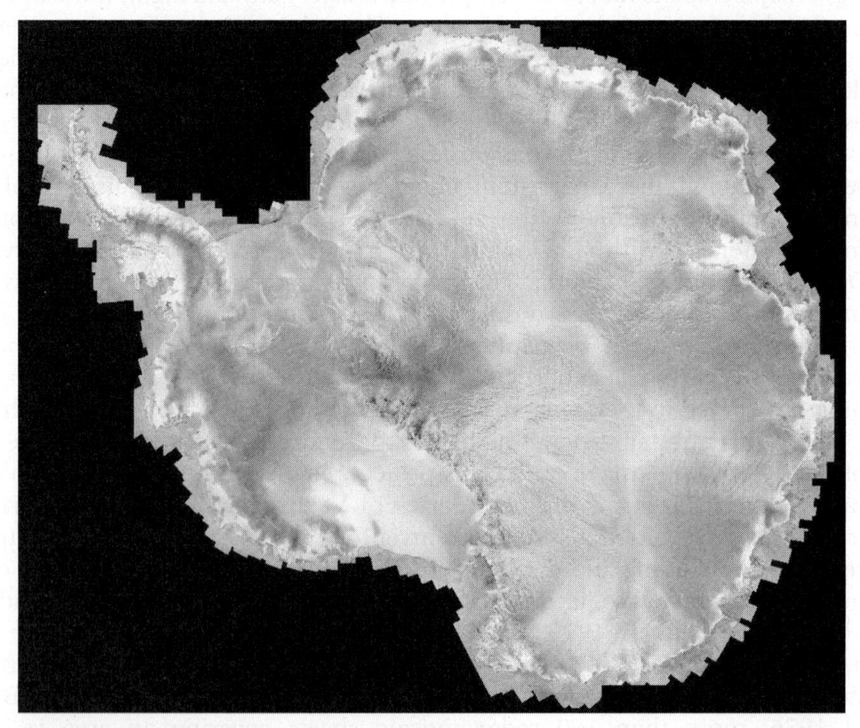

Figure 6.1 Radarsat mosaic of the entire continent of Antarctica, derived from over 3150 individual radar images. (Courtesy Canadian Space Agency.)

with extensive areas of higher and lower radar reflectivity associated with differing conditions of snow deposition and surface melt; the presence of features such as crevasses, snow dune fields, and ice shelves; and varying roughness of the subglacial terrain. In addition to this image mosaic, Radarsat imagery has been processed using the technique of radar interferometry (Section 6.9) to produce a digital elevation model of the whole continent, while repeated image acquisitions in subsequent years by Radarsat-2 have enabled measurement of the velocity of ice flow likewise across the entirety of the continent.

Numerous other applications of radar imagery have been demonstrated in the areas of geologic mapping, mineral exploration, flood inundation mapping, forestry, agriculture, and urban and regional planning. Over the oceans, radar imagery has been used extensively to determine wind, wave, and ice conditions and to map the extent and movement of oil spills. Finally, the sophisticated data processing methods used in radar interferometry are now being widely employed for topographic mapping and for measuring changes in topography caused by factors ranging from the movements of magma within volcanoes to the gradual subsidence of the surface in areas of groundwater extraction.

Radar remote sensing from space began with the launch of *Seasat* in 1978 and continued with the Shuttle Imaging Radar (SIR) and Soviet Cosmos experiments in the 1980s. The "operational" era of spaceborne radar remote sensing began in the early 1990s with the *Almaz-1, ERS-1, JERS-1,* and *Radarsat-1* systems launched by the former Soviet Union, European Space Agency, Japan, and Canada, respectively. In the new century, radar satellites have proliferated, culminating with the development of multi-satellite "constellations" such as Germany's TerraSAR-X/TanDEM-X pair, Italy's COSMO-SkyMed suite of satellites, and other multi-satellite missions planned by Canadian and European space agencies. The field of spaceborne radar remote sensing thus continues to be characterized by rapid technological advances, an expanding range of sources of data, and a high level of international participation.

6.3 IMAGING RADAR SYSTEM OPERATION

The basic operating principle of an imaging radar system is shown in Figure 6.2. Microwave energy is transmitted from an antenna in very short bursts or pulses. These high energy pulses are emitted over a time period on the order of microseconds (10^{-6} sec). In Figure 6.2, the propagation of one pulse is shown by indicating the wavefront locations at successive increments of time. Beginning with the solid lines (labeled 1 through 10), the transmitted pulse moves radially outward from the aircraft in a constrained (or narrow) beam. Shortly after time 6, the pulse reaches the house, and a reflected wave (dashed line) is shown beginning at time 7. At time 12, this return signal reaches the antenna and is registered at that time on the antenna response graph (Figure 6.2*b*). At time 9, the transmitted wavefront is reflected off the tree, and this "echo" reaches the antenna at time 17. Because the tree is less reflective of radar waves than the house, a weaker response is recorded in Figure 6.2*b*.

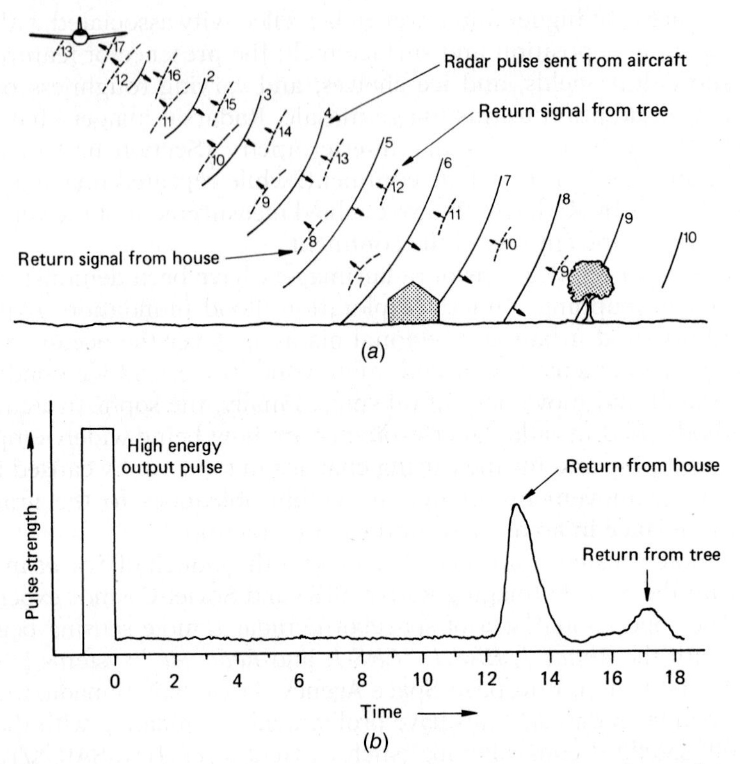

Figure 6.2 Operating principle of imaging radar: (*a*) propagation of one radar pulse (indicating the wavefront location at time intervals 1–17); (*b*) resulting antenna return.

By electronically measuring the return time of signal echoes, the range, or distance, between the transmitter and reflecting objects may be determined. Because the energy propagates in air at approximately the velocity of light c, the slant range, $\overline{SR}$, to any given object is given by

$$\overline{SR} = \frac{ct}{2} \tag{6.1}$$

where

$\overline{SR}$ = slant range (direct distance between transmitter and object)
c = speed of light (3×10^8 m/sec)
t = time between pulse transmission and echo reception

(Note that the factor 2 enters into the equation because the time is measured for the pulse to travel the distance both to and from the target, or twice the range.) This principle of determining distance by electronically measuring the transmission–echo time is central to imaging radar systems.

One manner in which radar images are created is illustrated in Figure 6.3. As the aircraft advances, the antenna (1) is continuously repositioned along the flight

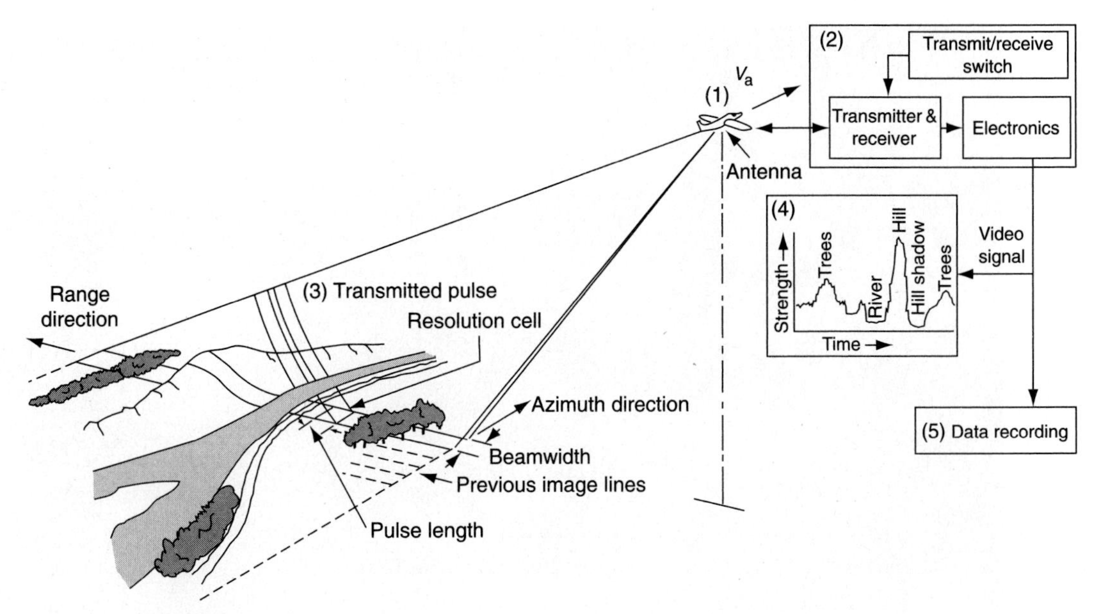

Figure 6.3 Imaging radar system operation. (Adapted from Lewis, 1976.)

direction at the aircraft velocity V_a. The antenna is switched from a transmitter to a receiver mode by a synchronizer switch (2). A portion of each transmitted pulse (3) is returned (as an echo) from terrain features occurring along a single antenna beamwidth. Shown in (4) is the return signal from one line of data. Return signals (echoes) are received by the airborne antenna, processed, and then recorded (5). Spaceborne systems operate on the same general principle.

It is essential to note that radar systems do not directly measure the reflectance of the surface. Instead, they record the intensity of the radiation that is *backscattered* by the surface—that is, the fraction of the incident energy that is reflected directly backward toward the sensor. Consider, for example, a very dark area within a radar image. One possible explanation for this pattern would be that the surface has very low reflectance at radar wavelengths. On the other hand, it is also possible that the surface has a very high reflectance, but that it is smooth enough to act as a specular reflector (Chapter 1). Both of these cases would produce low levels of measured backscatter, but for very different reasons.

There are several ways of quantifying and representing the strength of the returning pulse measured by the radar antenna. One commonly used radiometric representation is signal *power*, either directly or in a log-transformed version with units of *decibels (dB)*. The visual appearance of radar images is often improved by converting them to *magnitude* format, in which pixel values are scaled in direct proportion to the square root of power. With digital imagery, it is straightforward to convert among these (and other) image formats. Appendix C discusses these concepts and terms in more detail.

Figure 6.4 shows an example of a high-resolution spaceborne radar image, acquired by the TerraSAR-X radar satellite over the Three Gorges Dam on the

Figure 6.4 TerraSAR-X image of the Three Gorges Dam on the Yangtze River, China (scale 1:170,000). Imagery acquired in October at 1 m to 5 m spatial resolution. (Copyright: DLR e.V. 2009, Distribution Airbus DS/Infoterra GmbH.)

Yangtze River in China. The dam is 181 m high and spans a length of 2300 m. In terms of installed capacity, it is the largest electrical generating station in the world, and it also serves to reduce the risk of flooding on the lower Yangtze. In this image, the river appears as a dark band running from left to right (west to east) across the center of the image. Its predominantly dark tone is due to the fact that the smooth surface of the water acts as a specular reflector, producing very little radar backscatter and a correspondingly low amplitude of the returning radar signal. The dam itself, and other surrounding structures, appear relatively bright due to their geometric configuration and the materials used in their construction; likewise, ships and barges on the river reflect strongly back toward the radar antenna and appear as bright points or rectangles. The land surface north and south of the river appears quite rough, because of the irregular mountainous terrain. As will be noted in Section 6.5 and elsewhere in this chapter, the side-looking geometry of radar images tends to emphasize topography by giving slopes facing toward the sensor a bright appearance while slopes facing away are darker, with especially prominent shadows in steep terrain. In Figure 6.4, the prominent rightward-pointing shape of the hills and mountains is an indicator that the TerraSAR-X satellite was viewing this area from the right (east) when the image was acquired.

The effectiveness of radar imagery in "illuminating" the underlying topography is further demonstrated in Figure 6.5, a satellite radar image of an area of folded sedimentary rocks in the Appalachian Mountains of Pennsylvania. Here, the "sidelighting" nature of radar images obtained as illustrated in Figure 6.3 is

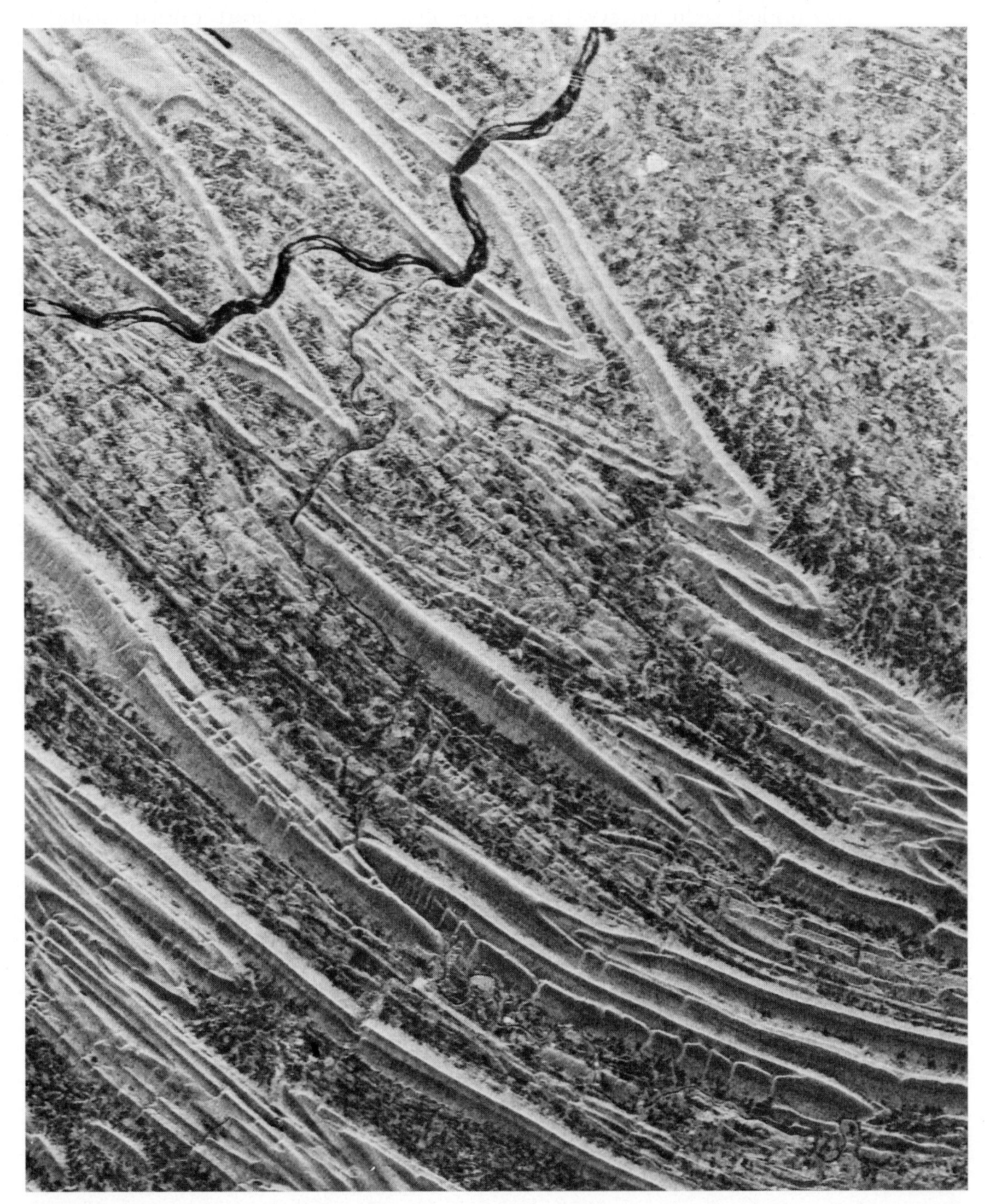

Figure 6.5 Seasat radar image of Appalachian Mountains of Pennsylvania, L band, midsummer. Scale 1:575,000. (Courtesy NASA/JPL/Caltech.)

apparent. In Figure 6.5, the signals from the radar system in the spacecraft were transmitted toward the bottom of the page, and the signals received by the radar system are those reflected back toward the top of the page. Note that, as mentioned above, the topographic slopes of the linear hills and valleys associated with the folded sedimentary rocks that face the spacecraft return strong signals, whereas the flatter areas and slopes facing away from the spacecraft return weaker signals. Note also that, as with the Yangtze River in Figure 6.4, the river seen at upper left of Figure 6.5 is dark toned because of specular reflection away from the sensor. (Section 6.8 describes earth surface feature characteristics influencing radar returns in some detail.)

Figure 6.6 illustrates the nomenclature typically used to describe the geometry of radar data collection. A radar system's *look angle* is the angle from

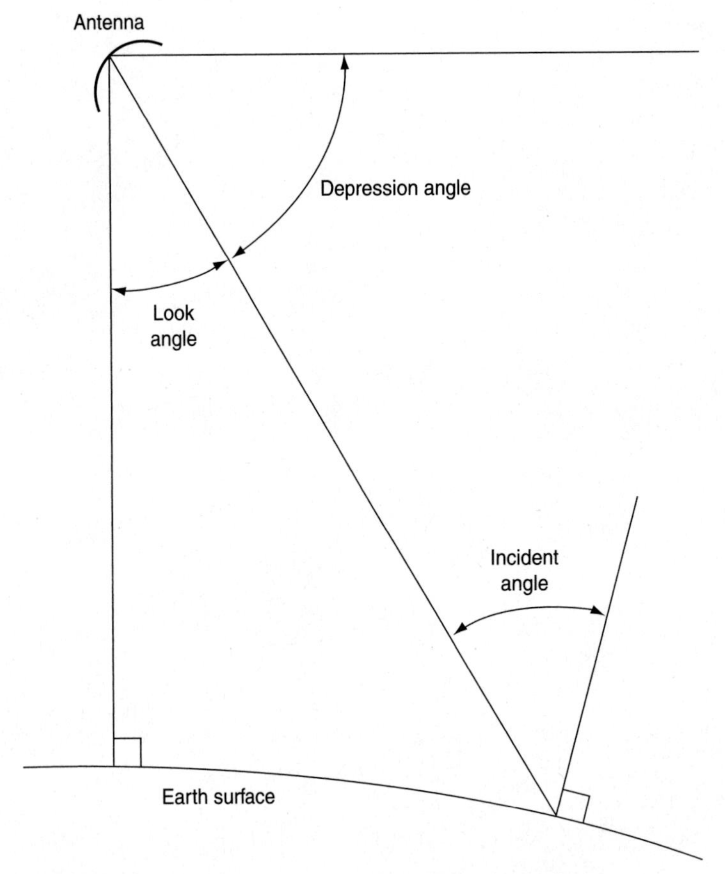

Figure 6.6 Nomenclature for the geometry of radar data collection.

nadir to a point of interest on the ground. The complement of the look angle is called the *depression angle*. The *incident angle* is the angle between the incident radar beam at the ground and the normal to the earth's surface at the point of incidence. (The incident angle is often referred to using the grammatically questionable term "incidence angle" in the literature. The two terms are synonymous.) In the case of airborne imaging over flat terrain, the incident angle is approximately equal to the look angle. In radar imaging from space, the incident angle is slightly greater than the look angle due to earth curvature. The *local incident angle* is the angle between the incident radar beam at the ground and the normal to the ground surface at the point of incidence. The incident angle and the local incident angle are the same only in the case of level terrain.

The ground resolution cell size of an imaging radar system is controlled by two independent sensing system parameters: *pulse length* and *antenna beamwidth*. The pulse length of the radar signal is determined by the length of time that the antenna emits its burst of energy. As can be seen in Figure 6.7, the signal pulse length dictates the spatial resolution in the direction of energy propagation. This direction is referred to as the *range* direction. The width of the antenna beam determines the resolution cell size in the flight, or *azimuth*, direction. We consider each of these elements controlling radar spatial resolution separately.

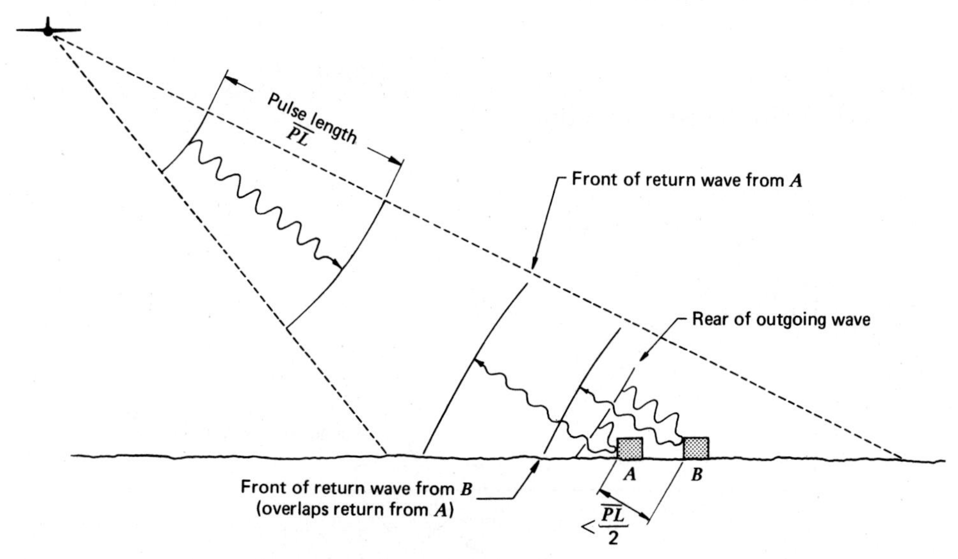

Figure 6.7 Dependence of range resolution on pulse length.

Range Resolution

For a radar system to image separately two ground features that are close to each other in the range direction, it is necessary for the reflected signals from all parts of the two objects to be received separately by the antenna. Any overlap in time between the signals from two objects will cause their images to be blurred together. This concept is illustrated in Figure 6.7. Here a pulse of length $\overline{PL}$ (determined by the duration of the pulse transmission) has been transmitted toward buildings A and B. Note that the slant-range distance (the direct sensor-to-target distance) between the buildings is less than $\overline{PL}/2$. Because of this, the pulse has had time to travel to B and have its echo return to A while the end of the pulse at A continues to be reflected. Consequently, the two signals are overlapped and will be imaged as one large object extending from building A to building B. If the slant-range distance between A and B were anything greater than $\overline{PL}/2$, the two signals would be received separately, resulting in two separate image responses. Thus, the slant-range resolution of an imaging radar system is independent of the distance from the aircraft and is equal to half the transmitted pulse length.

Although the *slant-range* resolution of an imaging radar system does not change with distance from the aircraft, the corresponding *ground-range* resolution does. As shown in Figure 6.8, the ground resolution in the range direction varies inversely with the cosine of the depression angle. This means that the ground-range resolution becomes smaller with increases in the slant-range distance.

Accounting for the depression angle effect, the ground resolution in the range direction R_r is found from

$$R_r = \frac{c\tau}{2\cos\theta_d} \tag{6.2}$$

where τ is the pulse duration.

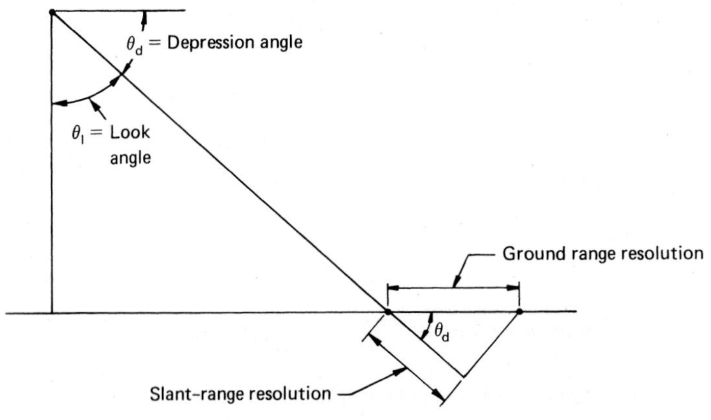

Figure 6.8 Relationship between slant-range resolution and ground-range resolution.

EXAMPLE 6.1

A given imaging radar system transmits pulses over a duration of 0.1 μsec. Find the range resolution of the system at a depression angle of 45°.

Solution

From Eq. 6.2

$$R_r = \frac{(3 \times 10^8 \text{ m/sec})(0.1 \times 10^{-6} \text{ sec})}{2 \times 0.707} = 21 \text{ m}$$

Azimuth Resolution

Early imaging radar systems were significantly limited in their resolution in the azimuth dimension. With the development of a set of advanced data processing techniques known as *synthetic aperture processing*, it became possible to greatly improve the azimuth resolution of these systems. We begin this discussion of azimuth resolution by considering the inherent resolution of a simple radar without synthetic aperture processing, and then move on to discuss the great improvement in resolution for synthetic aperture radar systems.

As shown in Figure 6.9, the resolution of an imaging radar system in the azimuth direction, R_a, is determined by the angular *beamwidth* β of the antenna and the slant range $\overline{SR}$. As the antenna beam "fans out" with increasing distance from the aircraft, the azimuth resolution deteriorates. Objects at points A and B in

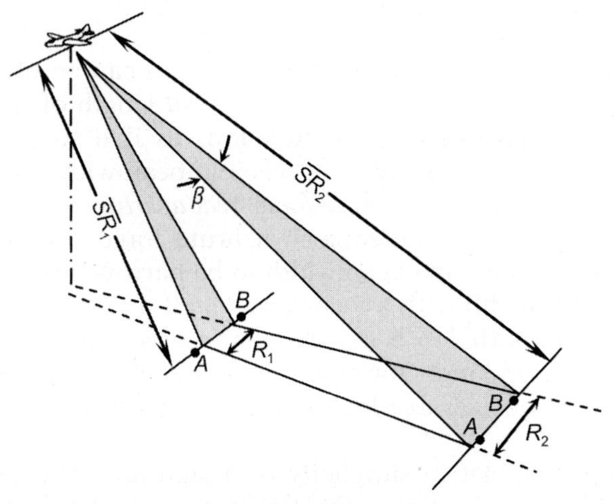

Figure 6.9 Dependence of azimuth resolution (R_a) on antenna beamwidth (β) and slant range ($\overline{SR}$).

Figure 6.8 would be resolved (imaged separately) at $\overline{SR}_1$ but not at $\overline{SR}_2$. That is, at distance $\overline{SR}_1$, A and B result in separate return signals. At distance $\overline{SR}_2$, A and B would be in the beam simultaneously and would not be resolved.

Azimuth resolution R_a is given by

$$R_a = \overline{SR} \cdot \beta \tag{6.3}$$

EXAMPLE 6.2

A given imaging radar system has a 1.8-mrad antenna beamwidth. Determine the azimuth resolution of the system at slant ranges of 6 and 12 km.

Solution

From Eq. 6.3

$$R_{a6\,km} = (6 \times 10^3 \text{ m})(1.8 \times 10^{-3}) = 10.8 \text{ m}$$

and

$$R_{a12\,km} = (12 \times 10^3 \text{ m})(1.8 \times 10^{-3}) = 21.6 \text{ m}$$

The beamwidth of the antenna of an imaging radar system is directly proportional to the wavelength of the transmitted pulses λ and inversely proportional to the length of the antenna $\overline{AL}$. That is,

$$\beta = \frac{\lambda}{\overline{AL}} \tag{6.4}$$

For any given wavelength, the effective antenna beamwidth can be narrowed by one of two different means: (1) by controlling the *physical* length of the antenna or (2) by synthesizing a *virtual* antenna that functions as if it were of greater length than the physical antenna. Those systems wherein beamwidth is controlled by the physical antenna length are called *brute force, real aperture,* or *noncoherent* radars. As expressed by Eq. 6.4, the antenna in a brute force system must be many wavelengths long for the antenna beamwidth to be narrow. For example, to achieve even a 10-mrad beamwidth with a 5-cm-wavelength radar, a 5-m antenna is required $[(5 \times 10^{-2}\text{ m})/(10 \times 10^{-3}) = 5 \text{ m}]$. To obtain a resolution of 2 mrad, we would need an antenna 25 m long! Obviously, antenna length requirements of brute force systems present considerable logistical problems if fine resolution at long range is the objective.

Brute force systems enjoy relative simplicity of design and data processing. Because of resolution problems, however, their operation is often restricted to relatively short range, low altitude operation, and the use of relatively short wavelengths. These restrictions are unfortunate because short-range, low altitude operation limits the area of coverage obtained by the system, and short wavelengths experience more atmospheric attenuation and dispersion.

6.4 SYNTHETIC APERTURE RADAR

The deficiencies of brute force operation are overcome in *synthetic aperture radar (SAR)* systems. These systems employ a short physical antenna, but through modified data recording and processing techniques, they synthesize the effect of a very long antenna. The result of this mode of operation is a very narrow effective antenna beamwidth, even at far ranges, without requiring a physically long antenna or a short operating wavelength.

At the detailed level, the operation of SAR systems is quite complex. Suffice it to say here that these systems operate on the principle of using the sensor motion along track to transform a single physically short antenna into an array of such antennas that can be linked together mathematically as part of the data recording and processing procedures (Elachi, 1987). This concept is shown in Figure 6.10. The "real" antenna is shown in several successive positions along the flight line. These successive positions are treated mathematically (or electronically) as if they were simply successive elements of a single, long synthetic antenna. Points on the ground at near range are viewed by proportionately fewer antenna elements than those at far range, meaning effective antenna length increases with range. This

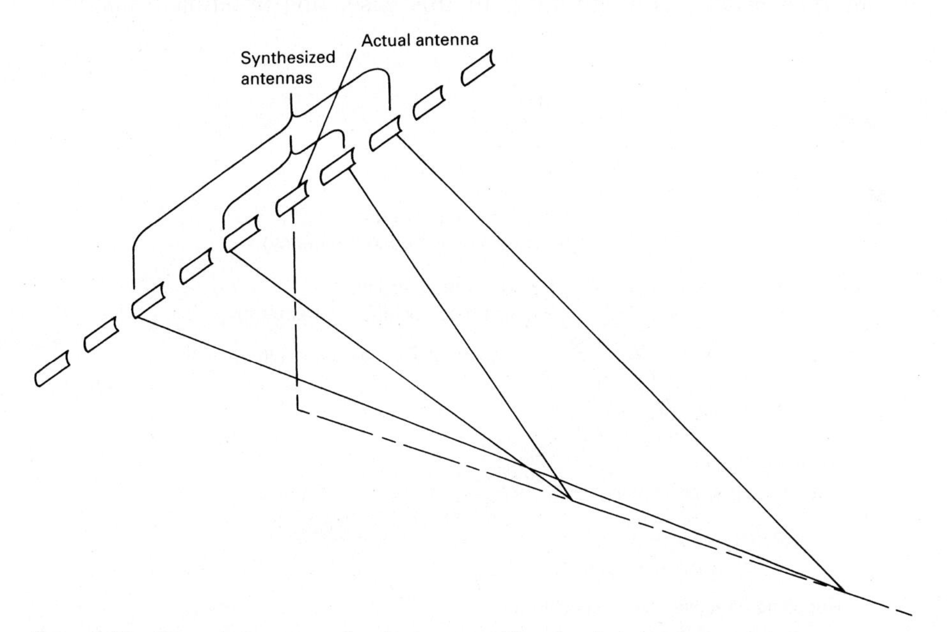

Figure 6.10 Concept of an array of real antenna positions forming a synthetic aperture.

results in essentially constant azimuth resolution irrespective of range. Through this process, antennas as long as several kilometers can be synthesized with spaceborne synthetic aperture systems.

Figure 6.11 illustrates yet another approach to explaining how synthetic aperture systems operate, namely discriminating only the near-center return signals from the real antenna beamwidth by detecting Doppler frequency shifts. Recall that a Doppler shift is a change in wave frequency as a function of the relative velocities of a transmitter and a reflector. Within the wide antenna beam, returns from features in the area ahead of the aircraft will have upshifted (higher) frequencies resulting from the Doppler effect. Conversely, returns from the area behind the aircraft will have downshifted (lower) frequencies. Returns from features near the centerline of the beamwidth will experience little or no frequency shift. By processing the return signals according to their Doppler shifts, a very small effective beamwidth can be generated.

Figure 6.12 illustrates the variation with distance of the ground resolution cell size of real aperture systems (*a*) versus synthetic aperture systems (*b*). Note that as the distance from the aircraft increases, the azimuth resolution size increases with real aperture systems and remains constant with synthetic aperture systems and that the ground range resolution size decreases with both systems.

The discussion to this point has assumed that a single physical antenna acts as both transmitter and receiver. This configuration is referred to as *monostatic* SAR and is the most commonly used arrangement for imaging radar systems. As shown in Figure 6.13, however, it is also possible to operate two radar antennas in a *bistatic* configuration. In this case, one antenna transmits while both

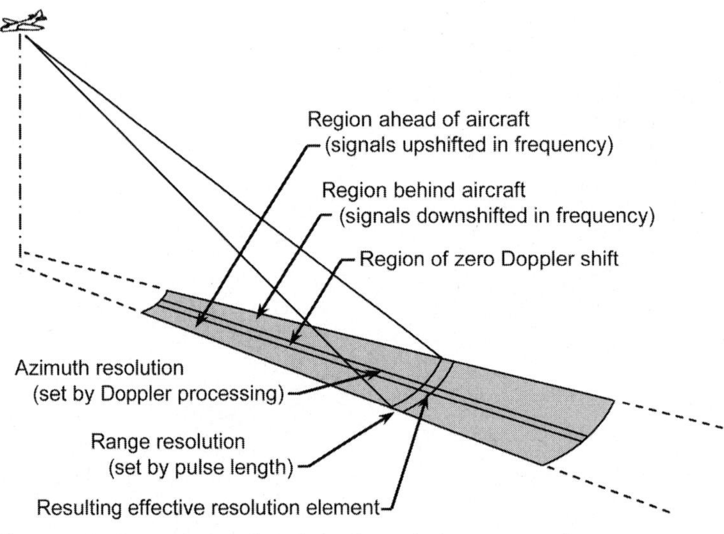

Figure 6.11 Determinants of resolution in synthetic aperture radar.

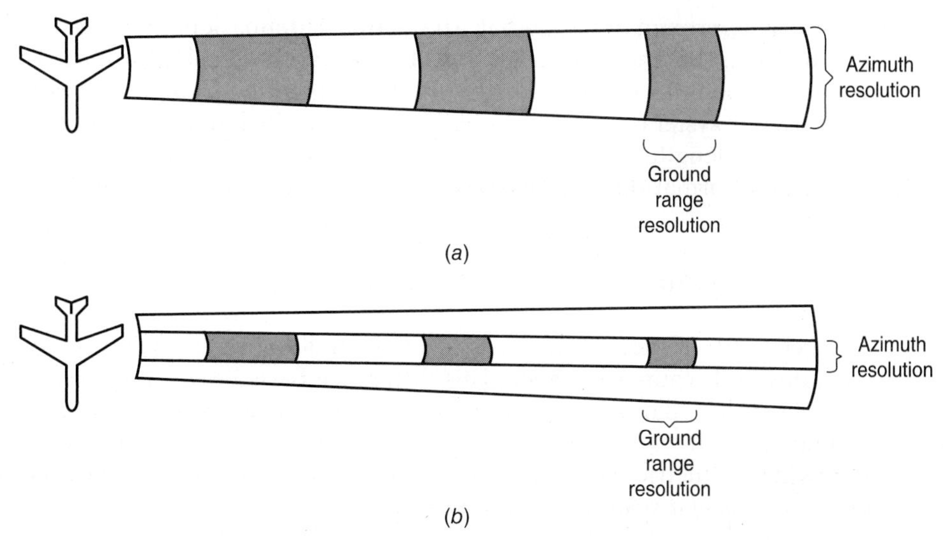

Figure 6.12 Variation with distance of spatial resolution of real aperture (*a*) versus synthetic aperture (*b*) imaging radar systems.

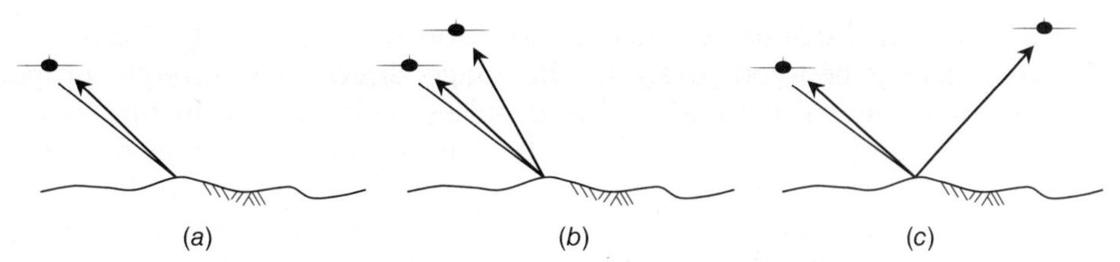

Figure 6.13 Monostatic (*a*) versus bistatic (*b*) and bistatic forward-scattering (*c*) radar imaging.

antennas receive signals reflected from the surface. The bistatic configuration shown in 6.13*b* is particularly useful for *SAR interferometry* (Section 6.9), a technique in which radar images are acquired over a given area from two slightly different angles. Figure 6.13*c* shows a variant of bistatic radar in which the passive (receiving-only) antenna is located on the opposite side of the area being imaged, so it records radar signals that are forward-scattered rather than backscattered. The Shuttle Radar Topography Mission (Section 6.19) and the TerraSAR-X/TanDEM-X satellite constellation (Section 6.16) are examples of systems that have used bistatic SAR imaging.

SAR systems can also be divided into *unfocused* and *focused* systems. Again the details of these systems are beyond our immediate concern. The interesting point about these systems is that the theoretical resolution of *unfocused* systems is a function of wavelength and range, not antenna length. The theoretical resolution of a *focused* system is a function of antenna length, regardless of range or wavelength. More particularly, the resolution of a focused synthetic aperture system is approximately one-half the actual antenna length. That is, the *shorter* the antenna, the finer the resolution. In theory, the resolution for a 1-m antenna would be about 0.5 m, whether the system is operated from an aircraft or a spacecraft! Radar system design is replete with trade-offs among operating range, resolution, wavelength, antenna size, and overall system complexity. (For additional technical information on SAR systems, see Raney, 1998.)

6.5 GEOMETRIC CHARACTERISTICS OF RADAR IMAGERY

The geometry of radar imagery is fundamentally different from that of both photography and scanner imagery. This difference basically results because radar is a *distance* rather than an *angle* measuring system. The influences this has on image geometry are many and varied. Here we limit our discussion to treatment of the following geometric elements of radar image acquisition and interpretation: *scale distortion, relief displacement*, and *parallax*.

Slant-Range Scale Distortion

Radar images can be recorded in two geometric formats. In a *slant-range* format, the spacing between pixels in the range direction is directly proportional to the time interval between received pulses. This interval in turn is proportional to the slant-range distance from the sensor to the object being imaged, rather than to the horizontal ground distance from the nadir line to the object. This has the effect of compressing the image scale at near range and expanding it at far range. By contrast, in a *ground-range* format, image pixels are spaced in direct proportion to their distance along a theoretical flat ground surface.

Figure 6.14 illustrates the characteristics of slant-range and ground-range image formats. The symbols A, B, and C represent objects of equal size that are equally separated in the near, middle, and far range. The respective ground ranges to the points are $\overline{GR_A}$, $\overline{GR_B}$, and $\overline{GR_C}$. Based directly on the signal return time, the slant-range image shows unequal distances between the features as well as unequal widths for the features. The result is a varying image scale that is at a minimum in the near range and progresses hyperbolically to a maximum at the far range. Therefore, on a slant-range presentation, object width $A_1 < B_1 < C_1$ and distance $\overline{AB} < \overline{BC}$. Applying a hyperbolic correction, a ground-range image of essentially constant scale can be formed with width $A = B = C$ and distance $\overline{AB} = \overline{BC}$. For a given swath width, the change in scale across an image decreases with increasing flying height. Thus, satellite systems have less scale change across an image than do airborne systems.

Obviously, the scale distortions inherent in slant-range imagery preclude its direct use for accurate planimetric mapping. However, approximate ground range

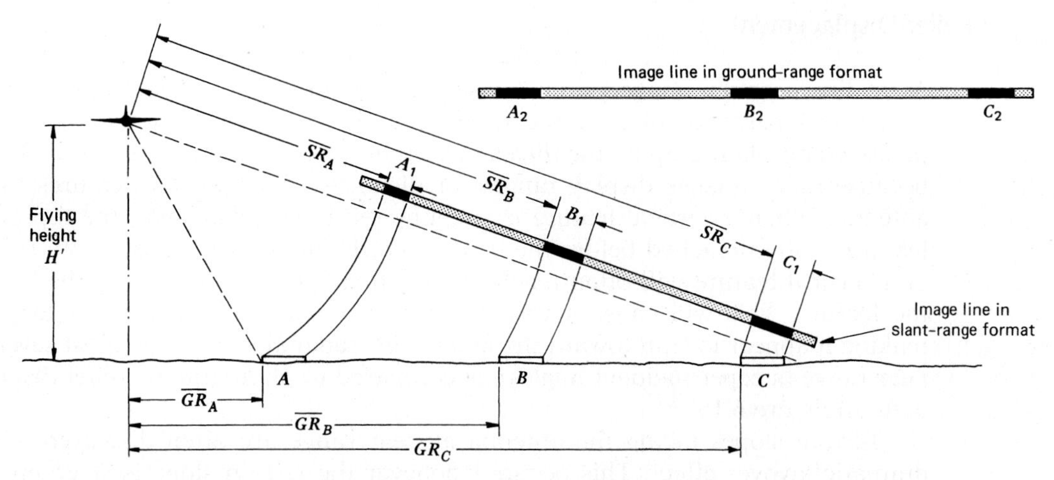

Figure 6.14 Slant-range versus ground-range image format. (Adapted from Lewis, 1976.)

$\overline{GR}$ can be derived from slant range $\overline{SR}$ and flying height H', under the assumption of flat terrain. From Figure 6.14 it can be seen that

$$\overline{SR}^2 = H'^2 + \overline{GR}^2$$

so

$$\overline{GR} = \left(\overline{SR}^2 - H'^2\right)^{1/2} \tag{6.5}$$

Therefore, a ground-range distance can be calculated from an image slant-range distance if the flying height is known. The assumption of flat terrain should be noted, however, and it should be pointed out that flight parameters also affect both range and azimuth scales. The range scale will vary with changes in aircraft altitude, and the azimuth scale will be dependent on precise synchronization between the aircraft ground speed and the data recording system.

Maintaining consistent scale in the collection and recording of radar imagery is a complex task. Whereas scale in the range (or across-track) direction is determined by the speed of light, scale in the azimuth (or along-track) direction is determined by the speed of the aircraft or spacecraft. To reconcile and equalize these independent scales, strict control of data collection parameters is needed. In most airborne systems, this is provided by a GPS and *inertial navigation and control system*. This device guides the aircraft at the appropriate flying height along the proper course. Angular sensors measure aircraft roll, crab, and pitch and maintain a constant angle of the antenna beam with respect to the line of flight. Inertial systems also provide the output necessary to synchronize the data recording with the aircraft ground speed. Spaceborne systems provide a more stable flight platform.

Relief Displacement

As in line scanner imagery, relief displacement in radar images is one dimensional and perpendicular to the flight line. However, in contrast to scanner imagery and photography, the *direction* of relief displacement is reversed. This is because radar images display ranges, or distances, from terrain features to the antenna. When a vertical feature is encountered by a radar pulse, the top of the feature is often reached before the base. Accordingly, return signals from the top of a vertical feature will often reach the antenna before returns from the base of the feature. This will cause a vertical feature to "lay over" the closer features, making it appear to lean toward the nadir. This radar *layover effect*, most severe at near range (steeper incident angles), is compared to photographic relief displacement in Figure 6.15.

Terrain slopes facing the antenna at near range are often displayed with a dramatic layover effect. This occurs whenever the terrain slope is steep enough such that the top of the slope is imaged before the bottom of the slope. This condition is met by pyramid 1 in Figure 6.16. This pyramid is located at the near-range side of the image, and the radar signals arrive at a very steep incident angle (see Section 6.3 for the terminology used to describe radar imaging geometry). Radar signals reflected from the top of this pyramid will be received earlier than those reflected from its base (Figure 6.16*b*), causing layover (Figure 6.16*c*). From Figure 6.16, it can be seen that layover is most extreme at short ranges and steep incident angles.

When the terrain is flatter, and/or the incident angle is less steep, layover does not occur. That is, the radar pulse reaches the base of the feature before the top. Nonetheless, the slopes of the surfaces still will not be presented in their true size and shape. As shown in Figure 6.16*c*, in the case of pyramid 4 (located at far range), the slope facing toward the radar antenna is compressed slightly. This

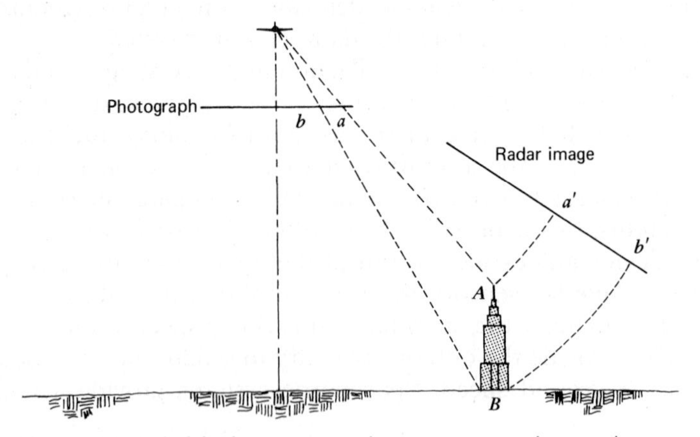

Figure 6.15 Relief displacement on radar images versus photographs.

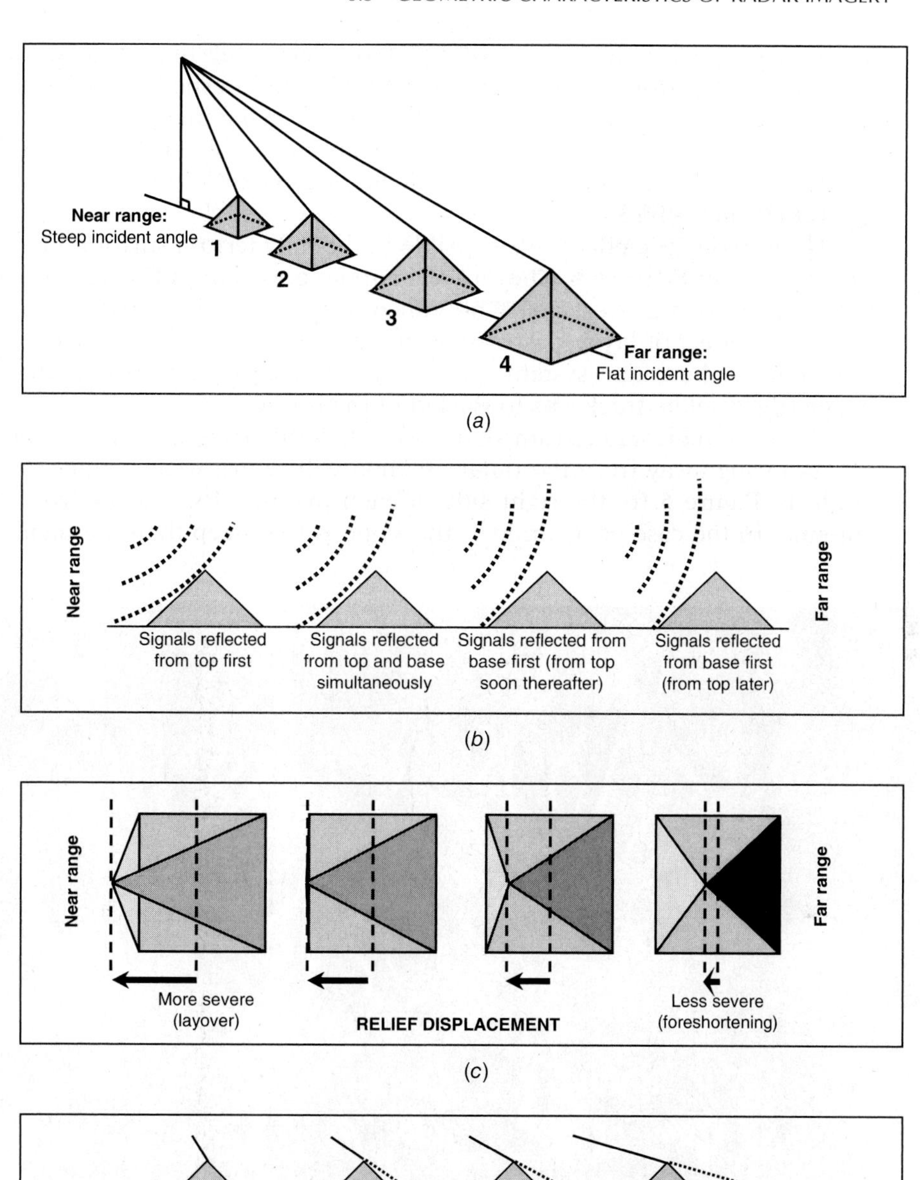

(d)

Figure 6.16 Relief displacement and shadowing in radar images. (*a*) Relationship between range and incident angle. (*b*) Relationship between terrain slope and wavefront of incident radiation. (*c*) Appearance of resulting image, showing brightness and geometric characteristics. Note differences in severity of relief displacement, ranging from layover (pyramid 1) to slight foreshortening (pyramid 4). (*d*) Increased length of shadows at flat incident angles. Note absence of shadows at pyramid 1, and lengthy shadow at pyramid 4.

foreshortening effect is a form of relief displacement that is less severe than layover. Moving from far range to near range, the foreshortening becomes more extreme, until the top and base of the feature are being imaged simultaneously, at which point the slope of the feature has been foreshortened to zero length. At still shorter range distances (or steeper incident angles), foreshortening is replaced by layover (Figure 6.16*c*).

These geometric effects were visible in the hilly terrain surrounding the Three Gorges Dam in Figure 6.4. They are even more apparent in Figure 6.17, a satellite radar image of the west coast of Vancouver Island, British Columbia. In this image, the effect of layover is very prominent because of the relatively steep look direction of the radar system and the mountainous terrain contained in this image (the satellite track was to the left of the image).

The look angle and terrain slope also affect the phenomenon of *radar shadow*. Slopes facing away from the radar antenna will return weak signals or no signal at all. In Figure 6.16, the right side of each pyramid faces away from the radar antenna. In the case of pyramid 1, the slope is less steep than the incident angle,

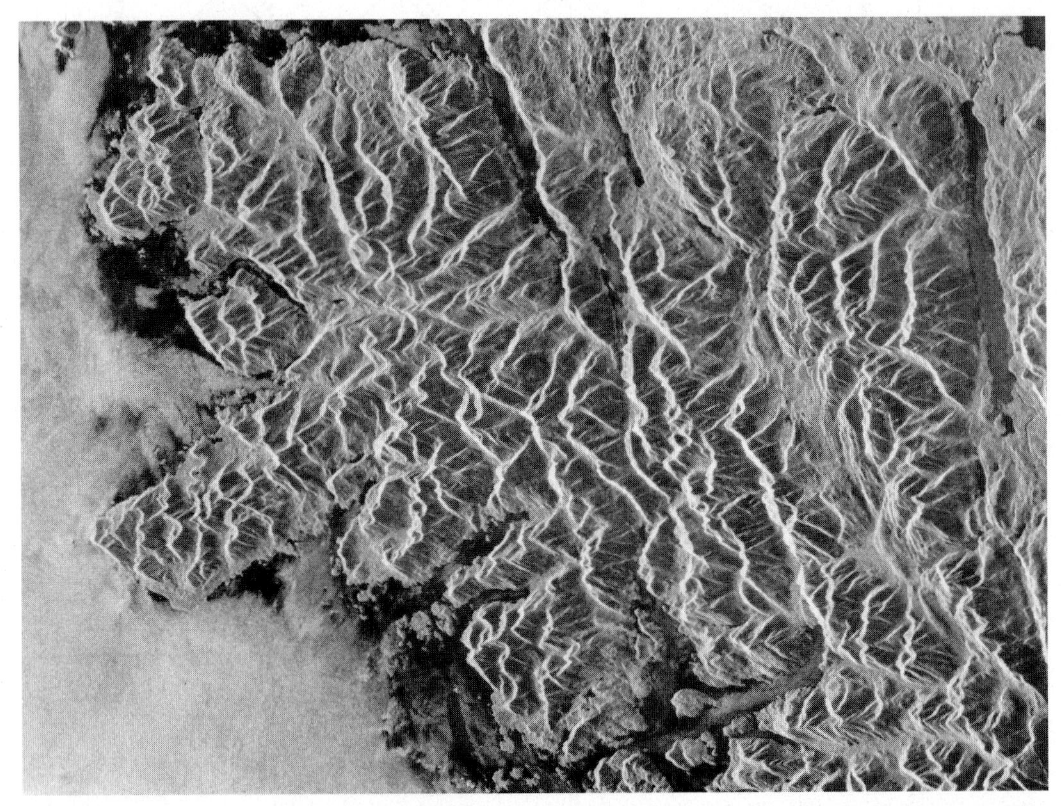

Figure 6.17 ERS-1 radar image, C band, Vancouver Island, British Columbia, midsummer. Scale 1:625,000. (ESA, Courtesy Canada Centre for Remote Sensing.)

and thus the right side will still be illuminated by the radar pulse. This illumination, however, will be very slight, and the resulting return signals will be weak, causing this slope to appear relatively dark on the image. In the case of pyramid 4, located at far range, the slope of its right side is steeper than the incident angle of the radar pulse. Thus, it will receive no illumination at all and will appear completely black. In fact, this pyramid will cast a radar shadow across an area extending further in the far-range direction (Figure 6.16*d*). The length of this shadowed area increases with the range distance, because of the corresponding flattening of the incident angle.

In summary, we can see that there is a trade-off between relief displacement and shadowing. Radar images acquired at steep incident angles will have severe foreshortening and layover, but little shadowing. Images acquired at flatter incident angles will have less relief displacement, but more of the imagery will be obscured by shadows. The effects of shadowing in radar images are discussed further in Section 6.8.

Parallax

When an object is imaged from two different flight lines, differential relief displacements cause image parallax on radar images. This allows images to be viewed stereoscopically. Stereo radar images can be obtained by acquiring data from flight lines that view the terrain feature from opposite sides (Figure 6.18*a*). However, because the radar sidelighting will be reversed on the two images in the stereopair, stereoscopic viewing is somewhat difficult using this technique.

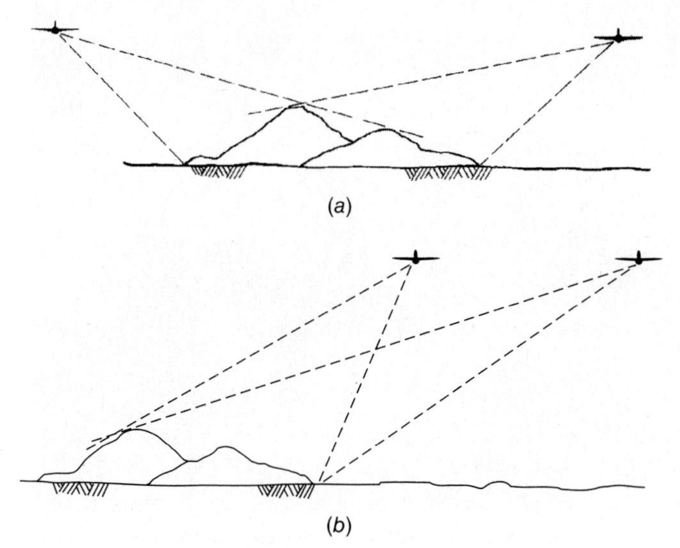

Figure 6.18 Flight orientations to produce parallax on radar images: (*a*) opposite-side configuration; (*b*) same-side configuration.

Accordingly, stereo radar imagery is often acquired from two flight lines at the same altitude on the same side of the terrain feature. In this case, the direction of illumination and the sidelighting effects will be similar on both images (Figure 6.18*b*). It is also possible to acquire stereo radar imagery in the same-side configuration by using different flying heights on the same flight line and, therefore, varying the antenna look angle.

Figure 6.19 shows a stereopair of spaceborne radar images of volcanic terrain in Chile that were acquired from two laterally offset flight lines at the same altitude on the same side of the volcano. This resulted in the two different look angles (45° and 54°) that were used for data collection. Although the stereo convergence angle is relatively small (9°), the stereo perception of the imagery is excellent because of the ruggedness of the terrain. The volcano near the bottom of the figure is Michinmahuida volcano; it rises 2400 m above the surrounding terrain. The snow-covered slopes of this volcano appear dark toned, because of absorption of the radar signal by the snow.

Figure 6.19 Shuttle imaging radar stereopair (SIR-B), Michinmahuida Volcano, Chiloe Province, Chile. Scale 1:350,000. The data for this stereopair were collected at two incident angles from the same altitude with same-side illumination. (Courtesy NASA/JPL/Caltech.)

In addition to providing a stereoscopic view, image parallax may be measured and used to compute approximate feature heights. As with aerial photography, parallax is determined by measuring mutual image displacements on the two images forming a stereomodel. Such measurements are part of the science of *radargrammetry*, a field beyond the scope of our interest in this text. (For further information on radargrammetry, see Leberl, 1990, 1998.)

6.6 TRANSMISSION CHARACTERISTICS OF RADAR SIGNALS

The two primary factors influencing the transmission characteristics of the signals from any given radar system are the wavelength and the polarization of the energy pulse used. Table 6.1 lists the common wavelength bands used in pulse transmission. The letter codes for the various bands (e.g., K, X, L) were originally selected arbitrarily to ensure military security during the early stages of radar development. They have continued in use as a matter of convenience, and various authorities designate the various bands in slightly different wavelength ranges.

Naturally, the wavelength of a radar signal determines the extent to which it is attenuated and/or dispersed by the atmosphere. Serious atmospheric effects on radar signals are confined to the shorter operating wavelengths (less than about 4 cm). Even at these wavelengths, under most operating conditions the atmosphere only slightly attenuates the signal. As one would anticipate, attenuation generally increases as operating wavelength decreases, and the influence of clouds and rain is variable. Whereas radar signals are relatively unaffected by clouds, echoes from heavy precipitation can be considerable. Precipitation echoes are proportional, for a single drop, to the quantity D^6/λ^4, where D is the drop diameter. With the use of short wavelengths, radar reflection from water droplets is substantial enough to be used in PPI systems to distinguish regions of precipitation. For example, rain and clouds can affect radar signal returns when the radar

TABLE 6.1 Radar Band Designations

Band Designation	Wavelength λ (cm)	Frequency $\nu = c\lambda^{-1}$ [MHz (10^6 cycles sec^{-1})]
K_a	0.75–1.1	40,000–26,500
K	1.1–1.67	26,500–18,000
K_u	1.67–2.4	18,000–12,500
X	2.4–3.75	12,500–8000
C	3.75–7.5	8000–4000
S	7.5–15	4000–2000
L	15–30	2000–1000
P	30–100	1000–300

wavelength is 2 cm or less. At the same time, the effect of rain is minimal with wavelengths of operation greater than 4 cm. With K- and X-band radar, rain may attenuate or scatter radar signals significantly.

Figure 6.20 illustrates an unusual shadow effect created by severe rainfall activity and demonstrates the wavelength dependence of radar systems. In this X-band airborne radar image, the bright "cloudlike" features are due to backscatter from rainfall. The dark regions "behind" these features (especially well seen at lower right) can be created by one of two mechanisms. One explanation is that they are "hidden" because the rain completely attenuates the incident energy, preventing illumination of the ground surface. Alternatively, a portion of the

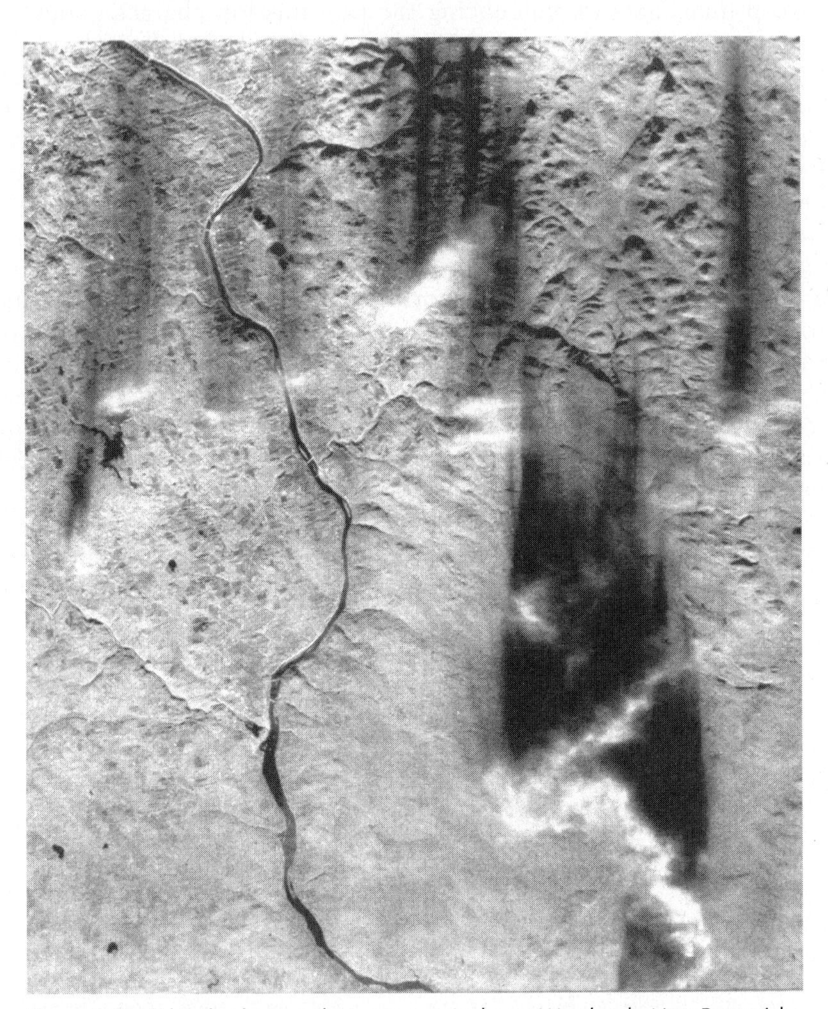

Figure 6.20 X-band airborne radar image acquired near Woodstock, New Brunswick, Canada, illustrating an unusual shadow effect created by severe rainfall activity and radar signal attenuation. (Courtesy Agriculture and Agri-Food Canada, Fredericton, NB, Canada.)

energy penetrates the rain during radar signal transmission but is completely attenuated after backscatter. This is termed *two-way attenuation*. In either case, much less energy is returned to the receiving antenna. When the signal is not completely attenuated, some backscatter is received (upper left).

Another effect of rainfall on radar images is that rain occurring at the time of data acquisition can change the physical and dielectric properties of the surface soil and vegetation, thus affecting backscatter. The effects on backscatter from moisture in or on the soil, plants, and other surface features will be discussed in Section 6.8.

Irrespective of wavelength, radar signals can be transmitted and/or received in different modes of *polarization*. The polarization of an electromagnetic wave describes the geometric plane in which its electrical field is oscillating. With multipolarization radar systems, the signal can be filtered in such a way that its electrical wave vibrations are restricted to a single plane perpendicular to the direction of wave propagation. (Unpolarized energy vibrates in all directions perpendicular to that of propagation.) Typically, radar signals are transmitted in a plane of polarization that is either parallel to the antenna axis (*horizontal polarization, H*) or perpendicular to that axis (*vertical polarization, V*), as shown in Figure 6.21. Likewise, the radar antenna may be set to receive only signals with a specified polarization. This results in four typical polarization combinations (*HH*, *VV*, *HV*, and *VH*), where the first letter indicates the transmitted polarization and the second indicates the received polarization. The *HH* and *VV* cases are referred to as *like-polarized* or *co-polarized* signals, while *HV* and *VH* are referred to as being *cross-polarized*.

Many objects will respond to polarized incident radiation by reflecting signals that are partially polarized and partially depolarized. In this case, the degree of polarization is defined as the power of the polarized component divided by the total reflected power. Measuring the polarization properties of an object's reflectance pattern is referred to as *polarimetry*. Clearly, polarimetry requires the ability to measure reflected energy at multiple polarizations. One common configuration for polarimetric radars is to have two channels, one operating in *H* mode and one

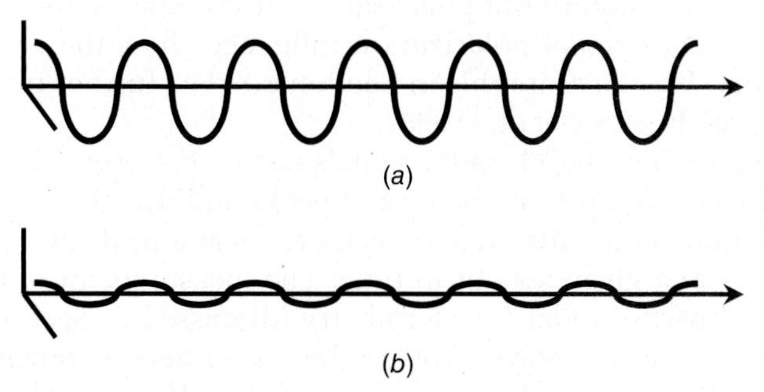

Figure 6.21 Polarization of radar signals. (*a*) Vertically polarized wave. (*b*) Horizontally polarized wave.

in *V*. When the first channel transmits a pulse, both channels are used to collect returning signals, resulting in *HH* and *HV* measurements. In the next instant, the second channel transmits, and the two record the ensuing *VH* and *VV* returns.

Any system that measures more than one polarization (e.g., *HH* and *HV*) is referred to as a "polarization diversity" or *multipolarization* radar. A system that measures all four of these orthogonal polarization states (*HH*, *HV*, *VH*, and *VV*) is referred to as being "fully polarimetric" or *quadrature polarized*. This is a particularly useful design, because it provides more insight into the physics of the interaction between the radar signals and the surface being imaged. For example, it may be possible to distinguish single-bounce from double-bounce scattering, to differentiate surface and volume scattering, and to measure various physical properties of the surface.

Quadrature polarization also provides sufficient information to calculate the theoretical backscatter properties of the surface in any plane of polarization, not just the *H* and *V* planes. In fact, the *H* and *V* polarizations are actually only two special cases (albeit the most commonly used in remote sensing) of a much more complex set of possible polarization states. In theory, any polarization state can be described using two parameters, *ellipticity* (χ) and *orientation* or *inclination angle* (Ψ). Ellipticity refers to the tendency of a wave to rotate its plane of polarization around the axis in which it is propagating. It ranges from $-45°$ to $+45°$. Waves with an ellipticity of 0° are said to be linearly polarized, those with positive ellipticity are said to be left-handed, and those with negative ellipticity are said to be right-handed. In the extreme cases, polarizations of $+45°$ and $-45°$ are referred to as left-circular and right-circular, respectively.

The orientation angle of a wave's polarization can range from 0° to 180°. Technically, a horizontally polarized wave is one with an ellipticity of 0° and an orientation of either 0° or 180°. A vertically polarized wave has an ellipticity of 0° and an orientation of 90°.

When a quadrature-polarized radar system collects *HH*, *HV*, *VH*, and *VV* measurements, it is possible to use matrix transformations to compute the theoretical response that would be expected from the surface when transmitting in any combination of (χ_T, Ψ_T) and receiving in any combination of (χ_R, Ψ_R).

Because various objects modify the polarization of the energy they reflect to varying degrees, the mode of signal polarization influences how the objects look on the resulting imagery. We illustrate this in Section 6.8. For further information on radar polarization, see Boerner et al. (1998).

Note that, as shown in Figure 6.21, radar signals can be represented as sinusoidal waves that cycle through a repeating pattern of peaks and troughs. The *phase* of a wave refers to its position along this cycle—whether it is at a peak, at a trough, or somewhere in between—at a given instant in time. The measurement of radar signal phase is an essential part of radar interferometry (discussed in Section 6.9). In addition, the phase of radar waves plays a role in the phenomenon known as radar image speckle, which is discussed in the next section of this chapter.

For long wavelengths (P band) at high altitude (greater than 500 km), the ionosphere can have significant effects on the transmission of radar signals. These effects occur in at least two ways. First, passage through the ionosphere can result in a propagation delay, which leads to errors in the measurement of the slant range. Second, there is a phenomenon known as *Faraday rotation*, whereby the plane of polarization is rotated somewhat, in direct proportion to the amount of ionospheric activity from the planet's magnetic field. These factors can cause significant problems for polarimetric, long wavelength, orbital SAR systems. (For further information on these effects, see Curlander and McDonough, 1991.)

6.7 OTHER RADAR IMAGE CHARACTERISTICS

Other characteristics that affect the appearance of radar images are *radar image speckle* and *radar image range brightness variation*. These factors are described below.

Radar Image Speckle

All radar images contain some degree of *speckle*, a seemingly random pattern of brighter and darker pixels in the image. Radar pulses are transmitted coherently, such that the transmitted waves are oscillating in phase with one another. However, Figure 6.22 shows that the waves backscattered from within a single ground resolution cell (or pixel) on the earth's surface will travel slightly different distances from the antenna to the surface and back. This difference in distance means that the returning waves from within a single pixel may be in phase or out of phase by varying degrees when received by the sensor. Where the returning waves are in phase with one another, the intensity of the resulting combined signal will be amplified by *constructive interference*. At the opposite extreme, where returning waves from within a single pixel are at completely opposite phases (that is, when one wave is at the peak of its cycle and another is at the trough), they will tend to cancel each other out, reducing the intensity of the combined signal (this is known as *destructive interference*). Constructive and destructive interference produce a seemingly random pattern of brighter and darker pixels in radar images, giving them the distinctly grainy appearance known as speckle.

For example, part 3 of Figure 6.22 shows a grid of 24 pixels representing a ground area with a uniformly dark linear feature (perhaps a smooth level road) crossing a uniformly lighter-toned background. Because of the effect of speckle, the resulting radar image will show pseudo-random variations in the apparent backscatter from every pixel in the image. This makes it more difficult to identify and differentiate features within the imagery. Speckle is often described imprecisely as "random noise," but it is important to realize that the seemingly random

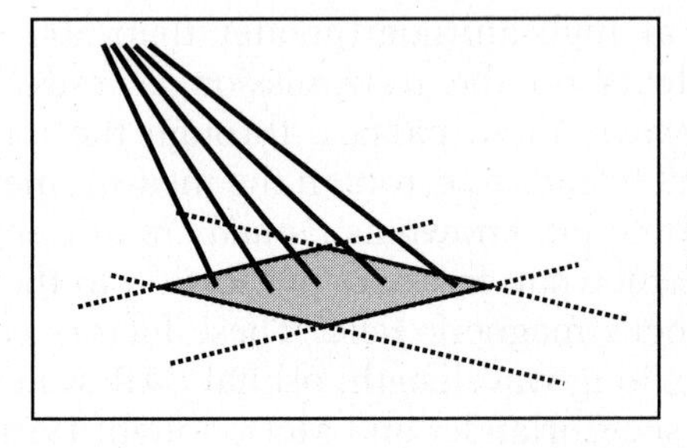

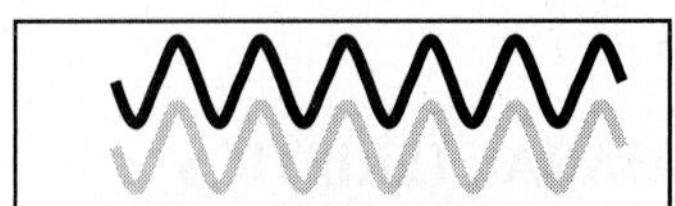

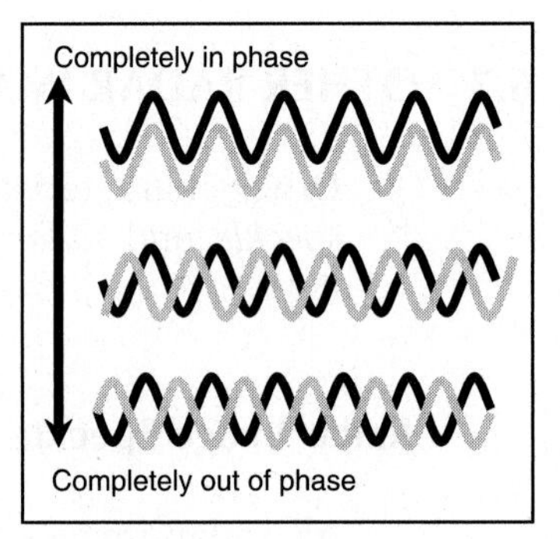

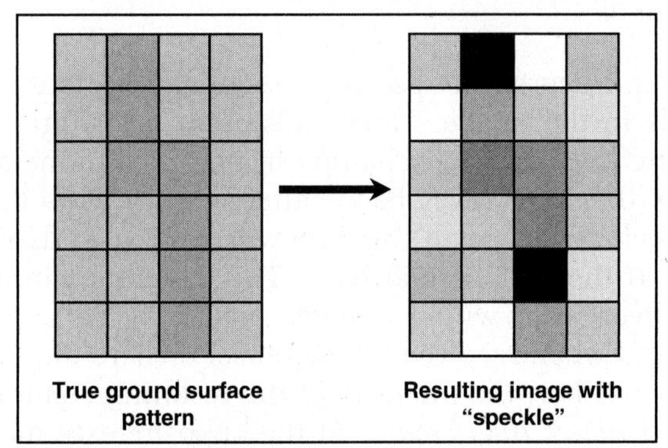

Figure 6.22 Speckle formation in radar images.

variations in backscatter are a direct result of the sub-pixel-scale geometry of the radar illumination conditions. Thus, if two images are acquired from the same position, with the same wavelength and polarization, and with the same ground surface conditions, the speckle pattern in the two images will be highly correlated. In fact, this principle is employed as part of the process of radar interferometry (Section 6.9).

Speckle can be reduced through the application of image processing techniques, such as averaging neighboring pixel values or by special filtering and averaging techniques, but it cannot be completely eliminated. One technique useful for reducing speckle is *multiple-look processing*. In this procedure, several independent images of the same area, produced by using different portions of the synthetic aperture, are averaged together to produce a smoother image. The number of statistically independent images being averaged is called the *number of looks*, and the amount of speckle is inversely proportional to the square root of this value. Given that the input data characteristics are held constant, the size of the resolution cell of the output image is directly proportional to the number of looks. For example, a four-look image would have a resolution cell four times larger than a one-look image and a speckle standard deviation one-half that of a one-look image. Hence, both the number of looks and the resolution of a system contribute to the overall quality of a radar image. For further information on number of looks versus other image characteristics, such as speckle and resolution, see Raney (1998) and Lewis, Henderson, and Holcomb (1998).

Figure 6.23 illustrates the appearance of radar images of the same scene with different numbers of looks. The amount of speckle is much less with four looks (*b*) than with one look (*a*), and even less with 16 looks (*c*). These images were specially processed such that the image resolution is the same for all three parts of the figure; 16 times as much data were required to produce the image in (*c*) as the image in (*a*).

Radar Image Range Brightness Variation

Synthetic aperture radar images often contain a systematic gradient in image brightness across the image in the range direction. This is principally caused by two geometric factors. First, the size of the ground resolution cell decreases from near range to far range, reducing the strength of the return signal. Second, and more significantly, backscatter is inversely related to the local incident angle (i.e., as the local incident angle increases, backscatter decreases), which is in turn related to the distance in the range direction. As a result, radar images will tend to become darker with increasing range. This effect is typically more severe for airborne radar systems than for spaceborne systems because the range of look angles is larger for the airborne systems with a lower flying height (for the same swath width). To some degree, mathematical models can be used to compensate for this effect, resulting in images without visible range brightness illumination effects. Some SAR systems (e.g., SIR-C) correct for the first of these geometric factors (decreasing ground resolution cell size), but not the second (increasing local incident angle), the effects of which are more complex.

Figure 6.24 is an airborne SAR image with a difference in look angle from near to far range of about 14°. Figure 6.24*a* has no compensation for range-related brightness variation, while a simple empirical model has been used to compensate for this effect in Figure 6.24*b*.

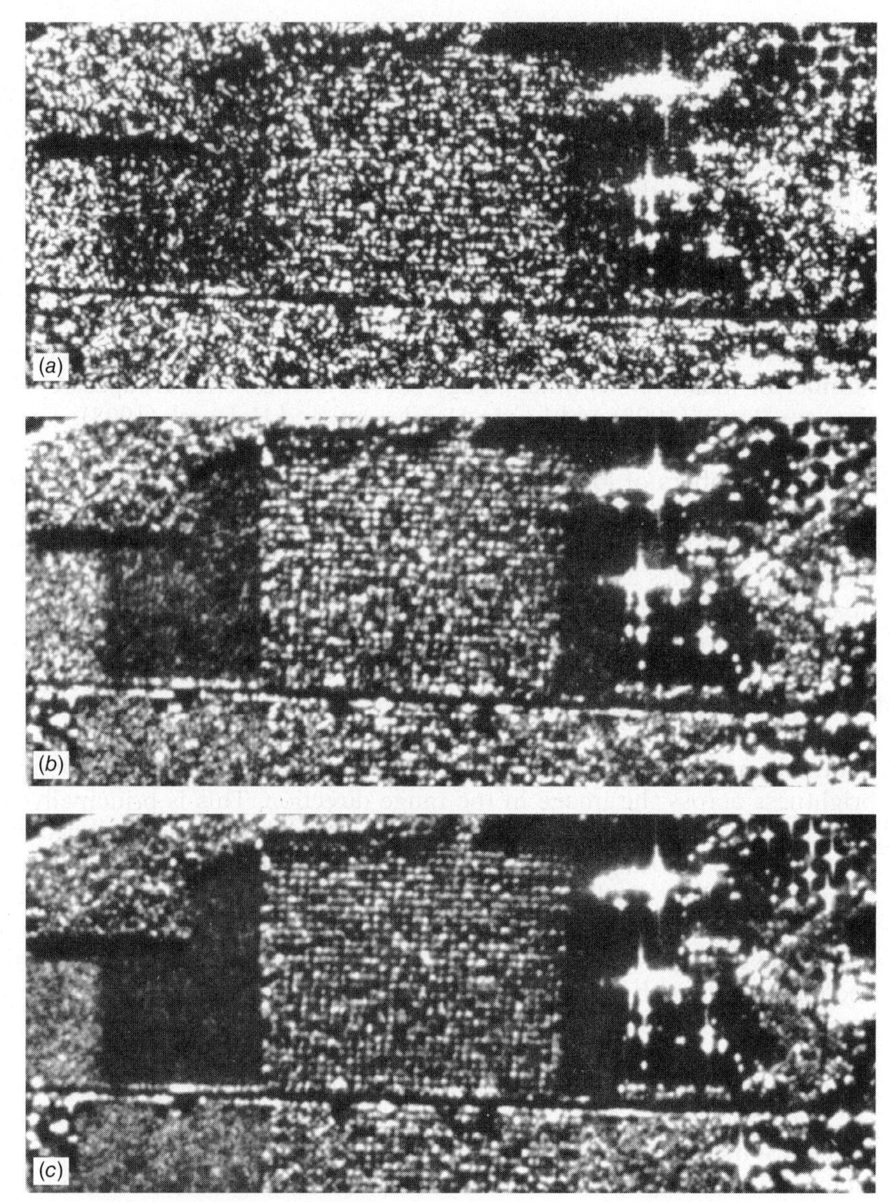

Figure 6.23 An example of multilook processing and its effect on image speckle: (*a*) one look; (*b*) four looks; (*c*) 16 looks. X-band airborne SAR radar image. Note that speckle decreases as the number of looks increases. These images were specially processed such that the image resolution is the same for all three parts of the figure; 16 times as much data were required to produce the image in (*c*) as the image in (*a*). (From American Society for Photogrammetry and Remote Sensing, 1998. Images copyright © John Wiley & Sons, Inc.)

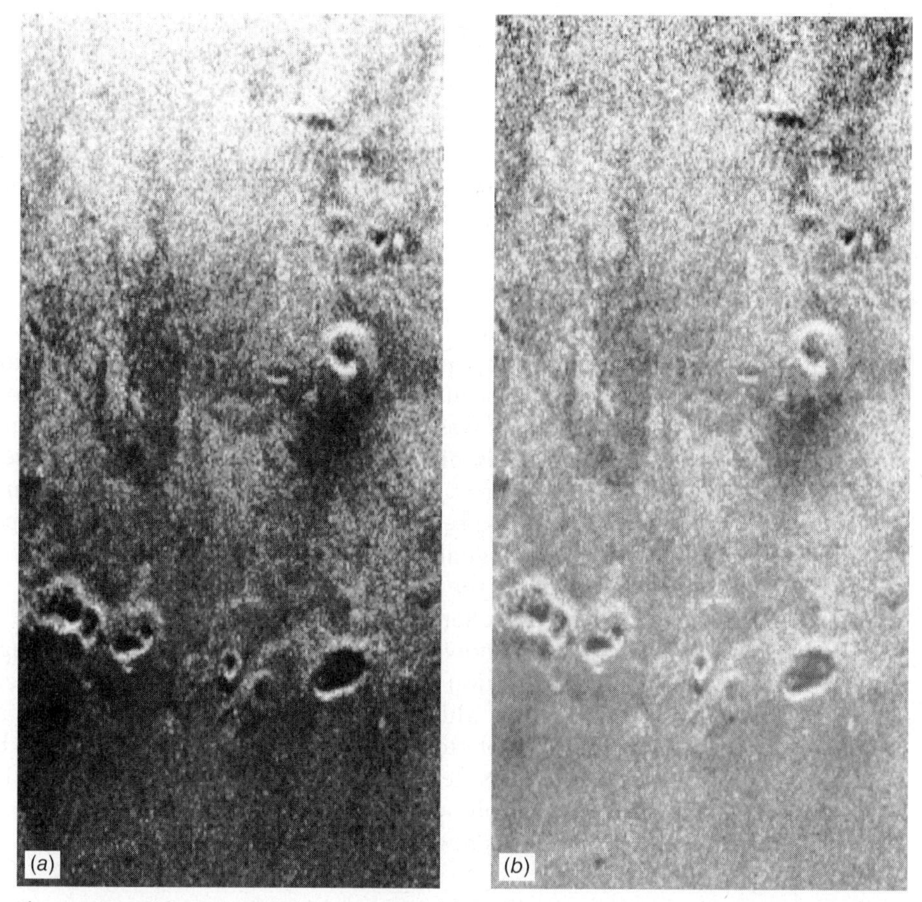

Figure 6.24 Airborne SAR radar images, Hualalai volcano, Hawaii: (*a*) without compensation for range brightness falloff; (*b*) with compensation for range brightness falloff. The difference in look angle from near range (top of the image) to far range (bottom of the image) is about 14°. (Courtesy NASA/JPL/Caltech.)

6.8 RADAR IMAGE INTERPRETATION

Radar image interpretation has been successful in many fields of application. These include, for example, mapping major rock units and surficial materials, mapping geologic structure (folds, faults, and joints), mapping vegetation types (natural vegetation and crops), determining sea ice types, and mapping surface drainage features (streams and lakes).

Because of its sidelighted character, radar imagery superficially resembles aerial photography taken under low sun angle conditions. However, a host of earth surface feature characteristics work together with the wavelength, incident angle, and polarization of radar signals to determine the intensity of radar returns from various objects. These factors are many, varied, and complex. Although several theoretical models have been developed to describe how various objects reflect radar energy, most practical knowledge on the subject has been derived from empirical observation. It has been found that the primary factors influencing objects' return signal intensity are their geometric and electrical characteristics; these are described below. The effects of radar signal polarization are illustrated, and radar wave interactions with soil, vegetation, water and ice, and urban areas are also described.

Geometric Characteristics

Again, one of the most readily apparent features of radar imagery is its side-lighted character. This arises through variations in the relative sensor/terrain geometry for differing terrain orientations, as was illustrated in Figure 6.16. Variations in local incident angle result in relatively high returns from slopes facing the sensor and relatively low returns, or no returns, from slopes facing away from the sensor.

In Figure 6.25, the return-strength-versus-time graph has been positioned over the terrain such that the signals can be correlated with the feature that produced them. Above the graph is the corresponding image line, in which the signal strength has been converted schematically to brightness values. The response from this radar pulse initially shows a high return from the slope facing the sensor. This is followed by a duration of no return signal from areas blocked from illumination by the radar wave. This radar shadow is completely black and sharply defined, unlike shadows in photography that are weakly illuminated by energy scattered by the atmosphere. Note that radar shadows can be seen in several radar images in this chapter. Following the shadow, a relatively weak response is recorded from the terrain that is not oriented toward the sensor.

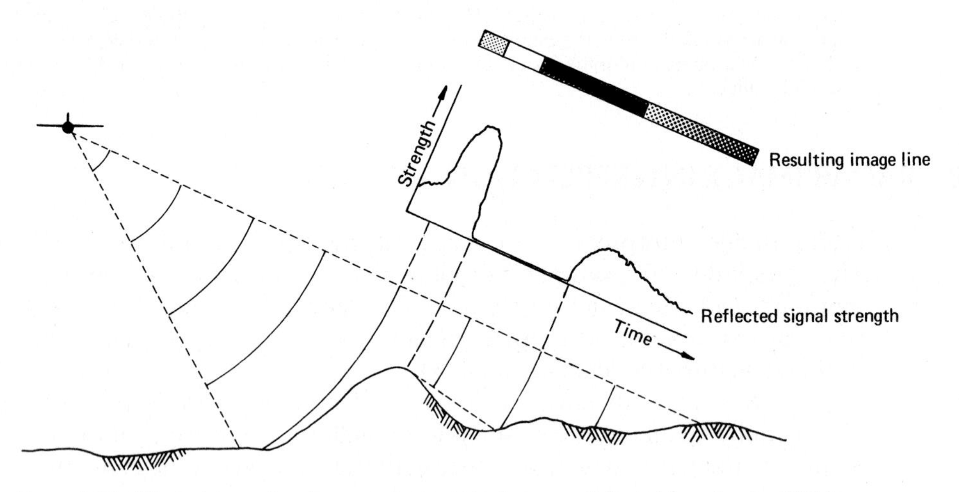

Figure 6.25 Effect of sensor/terrain geometry on radar imagery. (Adapted from Lewis, 1976.)

Radar backscatter and shadow areas are affected by different surface properties over a range of local incident angles. As a generalization, for local incident angles of 0° to 30°, radar backscatter is dominated by topographic slope. For angles of 30° to 70°, surface roughness dominates. For angles greater than 70°, radar shadows dominate the image.

Figure 6.26 illustrates radar reflection from surfaces of varying roughness and geometry. The *Rayleigh criterion* states that surfaces can be considered "rough" and act as diffuse reflectors (Figure 6.26*a*) if the root-mean-square (rms) height of the surface variations exceeds one-eighth of the wavelength of sensing ($\lambda/8$) divided by the cosine of the local incident angle (Sabins, 1997). Such surfaces scatter incident energy in all directions and return a significant portion of the incident energy to the radar antenna. Surfaces are considered "smooth" by the Rayleigh criterion and act as specular reflectors (Figure 6.26*b*) when their rms height variation is less than approximately $\lambda/8$ divided by the cosine of the local incident angle. Such surfaces reflect most of the energy away from the sensor, resulting in a very low return signal.

The Rayleigh criterion does not consider that there can be a category of surface relief that is intermediate between definitely rough and definitely smooth surfaces. A *modified Rayleigh criterion* is used to typify such situations. This criterion considers rough surfaces to be those where the rms height is greater than $\lambda/4.4$ divided by the cosine of the local incident angle and smooth when the rms height variation is less than $\lambda/25$ divided by the cosine of the local incident angle (Sabins, 1997). Intermediate values are considered to have intermediate roughnesses. Table 6.2 lists the surface height variations that can be considered smooth, intermediate, and rough for several radar bands for local incident angles of 20°, 45°, and 70°. (Values for other wavelength bands and/or incident angles can be calculated from the information given above.)

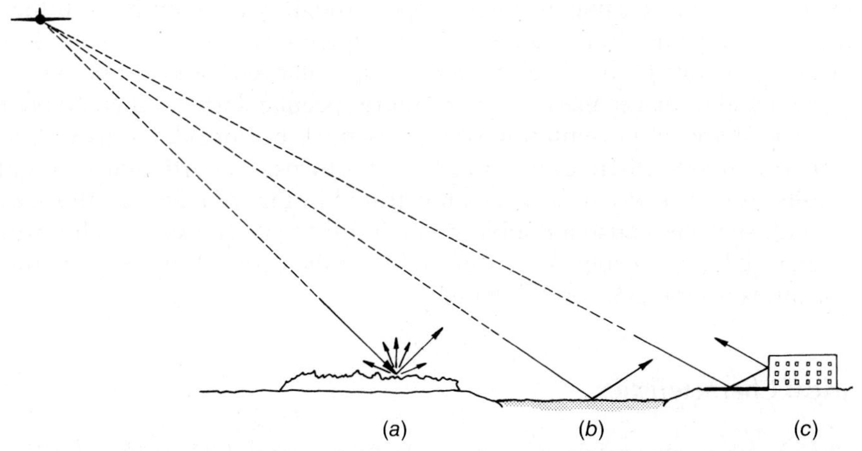

Figure 6.26 Radar reflection from various surfaces: (*a*) diffuse reflector, (*b*) specular reflector, (*c*) corner reflector.

TABLE 6.2 Definition of Synthetic Aperture Radar Roughness Categories for Three Local Incident Angles[a]

	Root-Mean-Square Surface Height Variation (cm)		
Roughness Category	K_a Band ($\lambda = 0.86$ cm)	X Band ($\lambda = 3.2$ cm)	L Band ($\lambda = 23.5$ cm)
(a) Local incident angle of 20°			
Smooth	<0.04	<0.14	<1.00
Intermediate	0.04–0.21	0.14–0.77	1.00–5.68
Rough	>0.21	>0.77	>5.68
(b) Local incident angle of 45°			
Smooth	<0.05	<0.18	<1.33
Intermediate	0.05–0.28	0.18–1.03	1.33–7.55
Rough	>0.28	>1.03	>7.55
(c) Local incident angle of 70°			
Smooth	<0.10	<0.37	<2.75
Intermediate	0.10–0.57	0.37–2.13	2.75–15.6
Rough	>0.57	>2.13	>15.6

[a]The table is based on a modified Rayleigh criterion.
Source: Adapted from Sabins, 1997.

Figure 6.27 graphically illustrates how the amount of diffuse versus specular reflection for a given surface roughness varies with wavelength, and Table 6.3 describes how rough various surfaces appear to radar pulses of various wavelengths using the modified Rayleigh criterion described above. It should be noted that some features, such as cornfields, might appear rough when seen in both the visible and the microwave portion of the spectrum. Other surfaces, such as roadways, may be diffuse reflectors in the visible region but specular reflectors of microwave energy. In general, radar images manifest many more specular surfaces than do photographs.

The shape and orientation of objects must be considered as well as their surface roughness when evaluating radar returns. A particularly bright response results from a *corner reflector,* as illustrated in Figure 6.26*c*. In this case, adjacent smooth surfaces cause a double reflection that yields a very high return. Because corner reflectors generally cover only small areas of the scene, they typically appear as bright spots on the image.

Electrical Characteristics

The electrical characteristics of terrain features work closely with their geometric characteristics to determine the intensity of radar returns. One measure of an

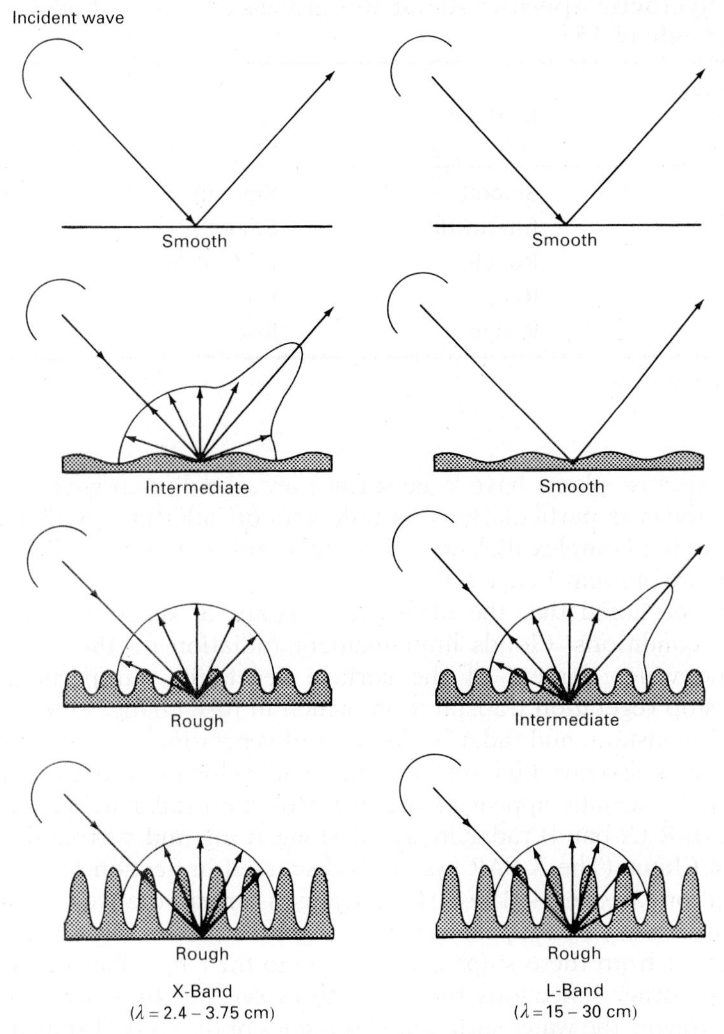

Figure 6.27 X-band and L-band radar reflection from surfaces of varying roughness. (Modified from diagram by Environmental Research Institute of Michigan.)

object's electrical character is the *complex dielectric constant*. This parameter is an indication of the reflectivity and conductivity of various materials.

In the microwave region of the spectrum, most natural materials have a dielectric constant in the range 3 to 8 when dry. On the other hand, water has a dielectric constant of approximately 80. Thus, the presence of moisture in either soil or vegetation can significantly increase radar reflectivity. In fact, changes in radar signal strength from one material to another are often linked to changes in moisture content much more closely than they are to changes in the materials

TABLE 6.3 Synthetic Aperture Radar Roughness at a Local Incident Angle of 45°

Root-Mean-Square Surface Height Variation (cm)	K_a Band ($\lambda = 0.86$ cm)	X Band ($\lambda = 3.2$ cm)	L Band ($\lambda = 23.5$ cm)
0.05	Smooth	Smooth	Smooth
0.10	Intermediate	Smooth	Smooth
0.5	Rough	Intermediate	Smooth
1.5	Rough	Rough	Intermediate
10.0	Rough	Rough	Rough

Source: Adapted from Sabins, 1997.

themselves. Because plants have large surface areas and often have a high moisture content, they are particularly good reflectors of radar energy. Plant canopies with their varying complex dielectric constants and their microrelief often dominate the texture of radar image tones.

It should be noted that the dielectric constant of vegetation changes with atmospheric conditions. Clouds limit incident radiation on the earth's surface, changing the water content of the surface vegetation. In particular, clouds decrease or stop vegetation transpiration, which in turn changes the water potential, dielectric constant, and radar backscatter of vegetation.

Metal objects also give high returns, and metal vehicles, bridges, silos, railroad tracks, and poles usually appear as bright features on radar images. Figure 6.28 shows an X-SAR (X-band) radar image of Hong Kong and surrounding areas in southeastern China. (The X-SAR system is discussed in Section 6.12.) The image was acquired on October 4, 1994. Hong Kong is among the world's busiest seaports, and numerous ships appear as small, bright features in the imagery. The very high backscatter from these ships is partly due to their metallic composition and partly due to corner reflections from structures on the ships, as well as corner reflections involving the water surface and the sides of the ships. Urban areas in this image also show very high backscatter, because of the presence of large buildings that likewise act as corner reflectors. The sea surface acts as a specular reflector and appears dark. Note also the effects of layover (relief displacement) in the steep, mountainous terrain on Hong Kong island at the right side of the image.

Some particularly bright features in Figure 6.28, such as ships and large buildings, show anomalous cross-shaped patterns radiating outward from the central points. These artifacts occur when the backscatter from an object (generally a metallic corner reflector) is high enough to exceed the dynamic range of the radar system, saturating the antenna's electronics. These "side-lobe" patterns are frequently observed whenever large, angular metallic objects such as bridges, ships, and offshore oil rigs are imaged against a backdrop of dark, smooth water.

Figure 6.28 X-SAR radar image of Hong Kong, China. Scale 1:170,000. (Courtesy DLR and NASA/JPL/Caltech.)

Effect of Polarization

Figure 6.29 illustrates the effect of signal polarization on radar images. The figure shows a pair of L-band radar images from SIR-C (Section 6.12), covering part of the island of Sumatra in Indonesia. Most of the area shown here consists of relatively undisturbed tropical rainforest, interspersed with large tracts that have been cleared for palm oil plantations. The images were acquired in October 1994. Newly cleared areas (where the rainforest cover was removed within five years before the image acquisition) appear as bright polygons in the *HH*-polarized image. Older clearings where planted palm trees are growing are not easily distinguished in the *HH* image, but appear much darker in the *HV* image. In the lower right corner of the area, a chain of lakes located in coastal marshes appear dark in both images due to specular reflection. In general, the choice of polarization(s) used in acquiring radar imagery will depend on the type of landscape features being studied. The most detailed information about surface materials would be provided by fully polarimetric ("quad-pol") radar systems offering all four polarization bands. However, radar system design involves trade-offs among the spatial resolution, extent of coverage, and number of polarizations available. Where fully polarimetric data are not available, dual-polarization systems

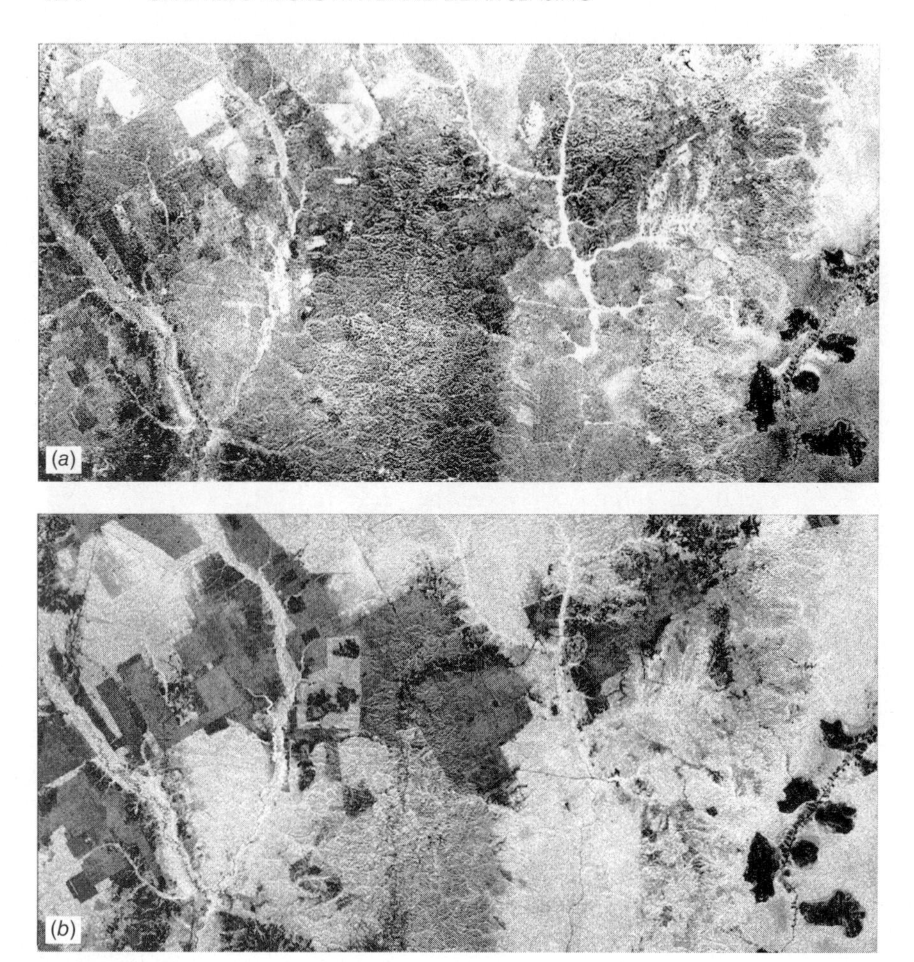

Figure 6.29 SIR-C radar images of Sumatra, Indonesia. (*a*) L-*HH*. (*b*) L-*HV*. (Courtesy NASA/JPL/Caltech.)

including both a like- and a cross-polarized band will often yield a great deal of information about surface features.

Plate 24 shows a pair of polarimetric L-band radar images acquired by NASA's UAVSAR, a radar system mounted on an uninhabited aerial vehicle (UAV; see Section 1.8). In these images, the *HH* polarization is shown in red, the *HV* polarization in green, and the *VV* polarization in blue. In *(a)*, Iceland's Hofsjokull glacier appears in green and magenta along the upper portion of the image. These colors indicate different scattering mechanisms that, in turn, can be used to draw inferences about surface conditions on the glacier. The green hue results from depolarized scattering (including relatively greater *HV* scattering) in the ablation zone, where the surface is rougher and melt is occurring. At higher elevations, the glacier's surface is smoother and the radar response is dominated by *HH* and *VV* backscatter.

Surrounding Hofsjokull, many familiar glacial landforms can be seen, including moraines, outwash plains, braided streams, ice-margin lakes, and other features.

Plate 24*b* shows a UAVSAR image of the area surrounding the Bahia de San Lorenzo, an arm of the Gulf of Fonseca extending into Honduras. The bay is surrounded by an intricate network of natural drainage channels flowing through mangrove swamps. In 1999, the mangroves in this region were designated as a "wetland of international importance" under the Ramsar Convention. Prominent hills at the lower left and upper right appear bright green, due to the increased cross-polarized response from trees growing on slopes that are too steep to farm. At lower elevations, agricultural fields occur on dryer land and aquaculture ponds intermingle with the mangrove swamps. Green patches within the swamps (with relatively higher *HV* backscatter) support more large woody biomass while the red, magenta, and blue areas are generally cleared, with the relative proportions of *HH* and *VV* backscatter determined by the surface roughness, the type of vegetation present, and the radar incident angle.

Soil Response

Because the dielectric constant for water is at least 10 times that for dry soil, the presence of water in the top few centimeters of bare (unvegetated) soil can be detected in radar imagery. Soil moisture and surface wetness conditions become particularly apparent at longer wavelengths. Soil moisture normally limits the penetration of radar waves to depths of a few centimeters. However, signal penetrations of several meters have been observed under extremely dry soil conditions with L-band radar.

Figure 6.30 shows a comparison of Landsat TM (*a*) and spaceborne synthetic aperture radar (*b*) imagery of the Sahara Desert near Safsaf Oasis in southern Egypt. This area is largely covered by a thin layer of windblown sand, which obscures many of the underlying bedrock and drainage features. Field studies in the area have shown that L-band (23 cm) radar signals can penetrate up to 2 m through this sand, providing imagery of subsurface geologic features. The dark, braided patterns in (*b*) represent former drainage channels from an ancient river valley, now filled with sand more than 2 m deep. While some of these channels are believed to date back tens of millions of years, others most likely formed during intervals within the past half-million years when the region experienced a wetter climate. Archaeologists working in this area have found stone tools used by early humans more than 100,000 years ago. Other features visible in the radar imagery primarily relate to bedrock structures, which include sedimentary rocks, gneisses, and other rock types. Very few of these features are visible in the Landsat imagery, due to the obscuring sand cover.

Vegetation Response

Radar waves interact with a vegetation canopy as a group of volume scatterers composed of a large number of discrete plant components (leaves, stems, stalks,

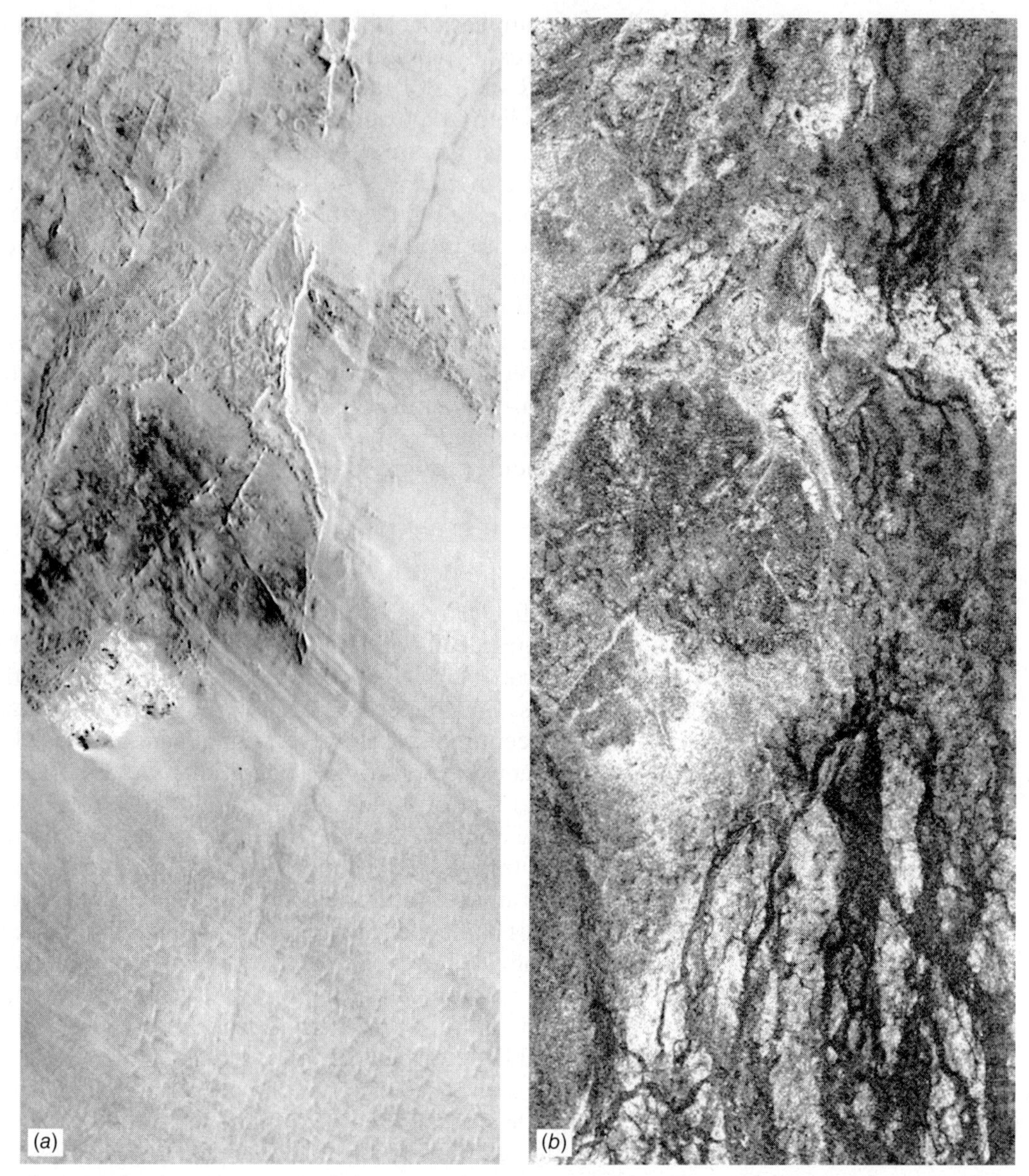

Figure 6.30 Sahara Desert near Safsaf Oasis, southern Egypt: (*a*) Landsat TM image; (*b*) SIR-C image, L band, *HH* polarization, 45° incident angle. North is to the upper left. Scale 1:170,000. (Courtesy NASA/JPL/Caltech.)

limbs, etc.). In turn, the vegetation canopy is underlain by soil that may cause surface scattering of the energy that penetrates the vegetation canopy. When the radar wavelengths approximate the mean size of plant components, volume scattering is strong, and if the plant canopy is dense, there will be strong backscatter from the vegetation. In general, shorter wavelengths (2 to 6 cm) are best for sensing crop canopies (corn, soybeans, wheat, etc.) and tree leaves. At these wavelengths, volume scattering predominates and surface scattering from the underlying soil is minimal. Longer wavelengths (10 to 30 cm) are best for sensing tree trunks and limbs.

In addition to plant size and radar wavelength, many other factors affect radar backscatter from vegetation. Recall that vegetation with a high moisture content returns more energy than dry vegetation. Also, more energy is returned from crops having their rows aligned in the azimuth direction than from those aligned in the range direction of radar sensing.

Figure 6.31 shows a pair of L-band radar images of an agricultural area located near Winnipeg, Alberta. The images were acquired from NASA's UAVSAR (mentioned previously in the discussion of Plate 24) on June 17 (6.31*a*) and July 17 (6.31*b*). The light-toned features are agricultural fields with higher soil moisture and/or crops with a higher moisture content than the darker-toned areas. Circular features represent center-pivot irrigation systems. The numerous differences in radar brightness between corresponding fields in 6.31*a* and 6.31*b* are primarily due to changes in plant growth and soil moisture during the one-month interval between image acquisitions. Healthy plants have a high water content, and thus a high dielectric constant, which in turn increases the reflectivity of the crop surface. That is, leaves with a high moisture content reflect radar waves more strongly than dry leaves, bare soil, or other features. Likewise, the vertical structure of a crop canopy increases the radar backscatter relative to the specular reflections that would typify a bare, smooth field.

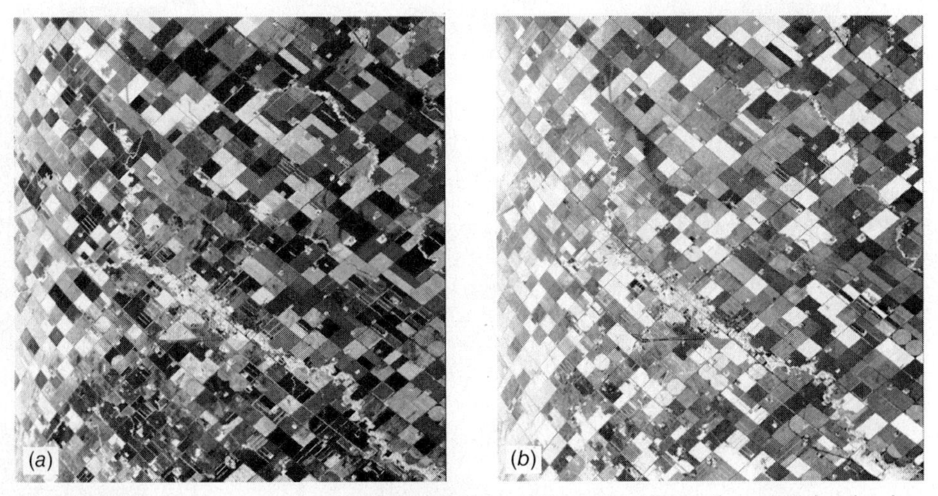

Figure 6.31 Radar imagery of agricultural crops near Winnepeg, Alberta, from an L-band radar system operated on an uninhabited aerial vehicle (UAV). (*a*) June 17. (*b*) July 17. (Courtesy NASA/JPL/Caltech.)

The lighter-toned linear feature wandering through the images in Figure 6.31 from upper left to lower right is a watercourse whose banks are lined with trees, shrubs, and other riparian vegetation. The brightness of this feature is due both to the increased roughness of the vegetation present and to its increased moisture content. Often, vegetated areas that are flooded or are adjacent to standing water can cause a corner reflector effect. Each stalk of vegetation forms a right angle with the calm water. Combined, these can produce a bright radar return, which is a useful indicator of water standing beneath a vegetation canopy (see also Figure 6.34).

Figure 6.32 illustrates the effect of wavelength on the appearance of airborne SAR images. Here, the scene is imaged with three different wavelengths. Most crop types reflect differently in all three wavelength bands, with generally lighter tones in the C band and darker tones in the P band. Many crop types in this image could be identified by comparing the relative amounts of backscatter in the three different bands.

Figure 6.33 shows a C-band (*a*) and an L-band (*b*) image of an area in northern Wisconsin that is mostly forested, containing many lakes. Because of specular reflection from their smooth surfaces, the lakes appear dark throughout both images. A tornado scar can be seen as a dark-toned linear feature running through

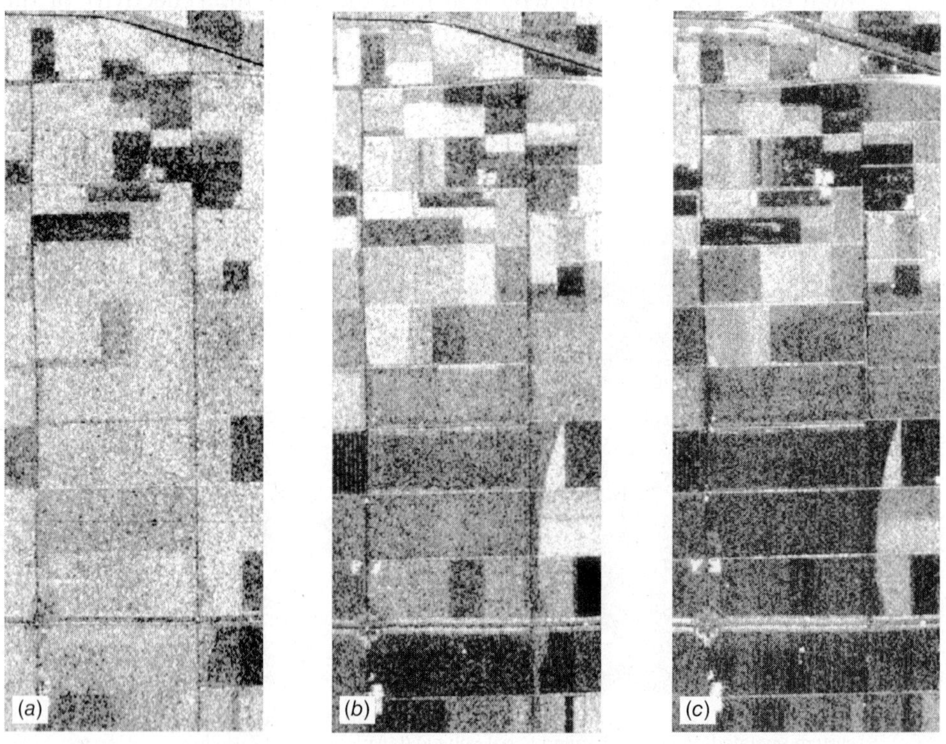

Figure 6.32 Airborne SAR images of an agricultural area in the Netherlands: (*a*) C band (3.75–7.5 cm); (*b*) L band (15–30 cm); (*c*) P band (30–100 cm). *HH* polarization. (From American Society for Photogrammetry and Remote Sensing, 1998. Images copyright © John Wiley & Sons, Inc.)

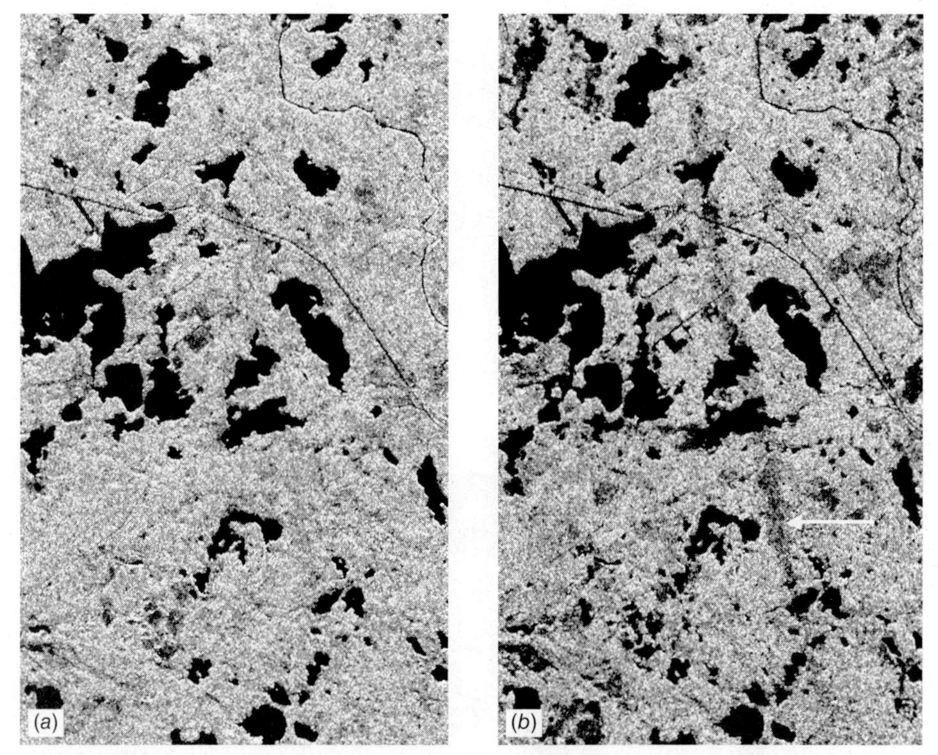

Figure 6.33 SIR-C images of a forested area in northern Wisconsin: (*a*) C-band image; (*b*) L-band image. Scale 1:150,000. Note the dark-toned lakes throughout the image and the tornado scar that is visible only in the L-band image. (Courtesy NASA/JPL/Caltech and UW-Madison Environmental Remote Sensing Center.)

the center of Figure 6.33*b*, from upper left to lower right. The tornado occurred 10 years before the date of this image. It destroyed many buildings and felled most of the trees in its path. After the tornado damage, timber salvage operations removed most of the fallen trees, and young trees were established. At the time of acquisition of these spaceborne radar images, the young growing trees in the area of the tornado scar appear rough enough in the C-band (6-cm) image that they blend in with the larger trees in the surrounding forested area. In the L-band (24-cm) image, they appear smoother than the surrounding forested area and the tornado scar can be seen as a dark-toned linear feature.

Incident angle also has a significant effect on radar backscatter from vegetation. Figure 6.34 shows spaceborne SAR images of a forested area in northern Florida and further illustrates the effect of radar imaging at multiple incident angles on the interpretability of radar images. The terrain is flat, with a mean elevation of 45 m. Sandy soils overlay weathering limestone; lakes are sinkhole lakes. Various land cover types can be identified in Figure 6.34*b* by their tone, texture, and shape. Water bodies (W) have a dark tone and smooth texture. Clear-cut areas (C) have a dark tone with a faint mottled texture and rectangular to angular

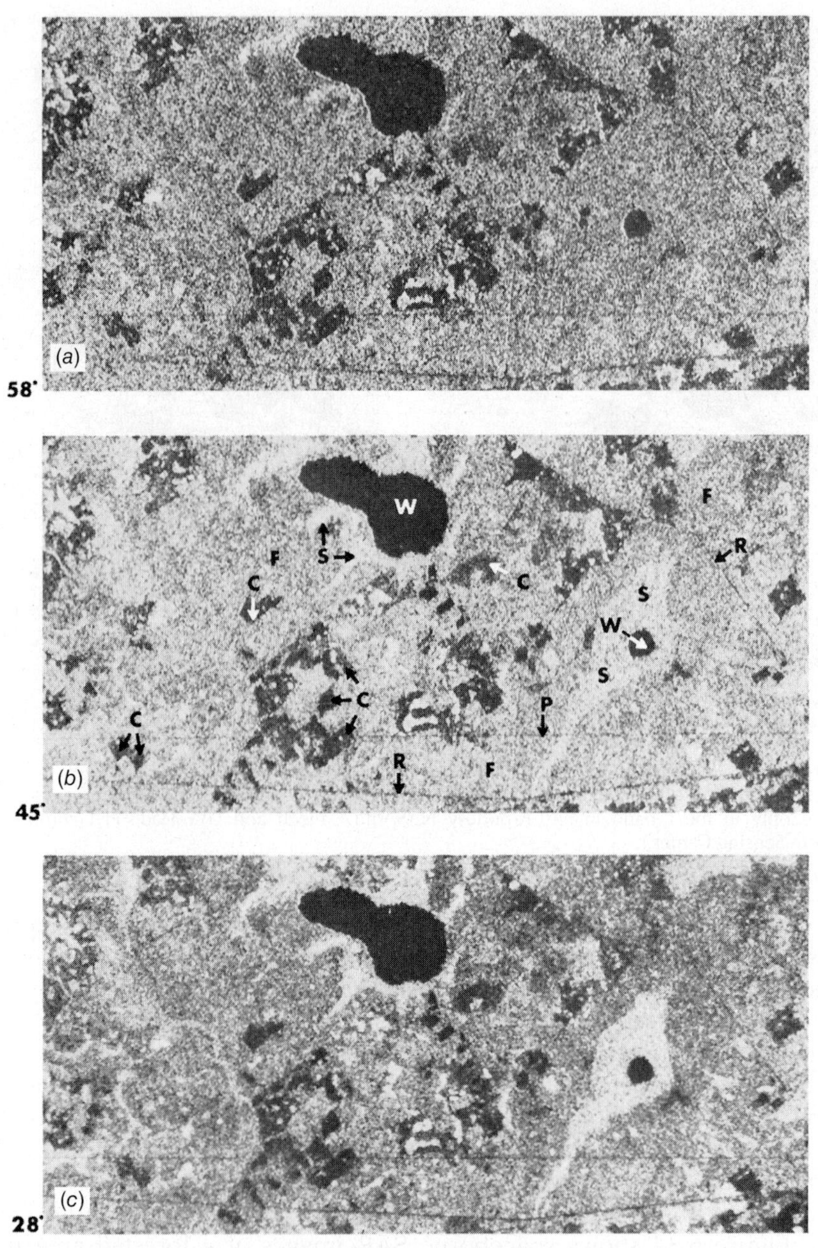

Figure 6.34 SIR-B images, northern Florida, L band (scale 1:190,000): (*a*) 58° incident angle, October 9; (*b*) 45° incident angle, October 10; (*c*) 28° incident angle, October 11. C = clear-cut area; F = pine forest; P = powerline right-of-way; R = road; S = cypress-tupelo swamp; W = open water. (Courtesy Department of Forestry and Natural Resources, Purdue University, and NASA/JPL/Caltech.)

shapes. The powerline right-of-way (P) and roads (R) appear as dark-toned, narrow, linear swaths. Pine forest (F), which covers the majority of this image, has a medium tone with a mottled texture. Cypress-tupelo swamps (S), which consist mainly of deciduous species, have a light tone and a mottled texture. However, the relative tones of the forested areas vary considerably with incident angle. For example, the cypress-tupelo swamp areas are dark toned at an incident angle of 58° and cannot be visually distinguished from the pine forest. These same swamps are somewhat lighter toned than the pine forest at an incident angle of 45° and much lighter toned than the pine forest at an incident angle of 28°. The very high radar return from these swamps on the 28° image is believed to be caused by specular reflection from the standing water in these areas acting in combination with reflection from the tree trunks, resulting in a complex corner reflector effect (Hoffer, Mueller, and Lozano-Garcia, 1985). This effect is more pronounced at an incident angle of 28° than at larger incident angles because the penetration of radar waves through the forest canopy is greater at the smaller angle.

Water and Ice Response

Smooth water surfaces act as specular reflectors of radar waves and yield no returns to the antenna, but rough water surfaces return radar signals of varying strengths. Experiments conducted with the Seasat radar system (L-band system with look angles of 20° to 26°, as described later in Section 6.11) showed that waves with a wavelength greater than 100 m could be detected when wave heights were greater than about 1 m and surface wind speeds exceeded about 2 m/sec (Fu and Holt, 1982). It was also found that waves moving in the range direction (moving toward or away from the radar system) could be detected more readily than waves moving in the azimuth direction.

Radar images from space have revealed interesting patterns that have been shown to correlate with ocean bottom configurations. Figure 6.35*a* is a spaceborne SAR image of the English Channel near the Strait of Dover. Here, the channel is characterized by tidal variations of up to 7 m and reversing tidal currents with velocities at times over 1.5 m/sec. Also, there are extensive sand bars on both sides of the strait and along the coasts of France and England. The sand bars in the channel are long, narrow ridges from 10 to 30 m in depth, with some shallower than 5 m. Together with the high volume of ship traffic, these sand bars make navigation in the channel hazardous. By comparing this image with Figure 6.35*b*, it can be seen that the surface patterns on the radar image follow closely the sand bar patterns present in the area. Tidal currents at the time of image acquisition were 0.5 to 1.0 m/sec, generally in a northeast-to-southwest direction. The more prominent patterns are visible over bars 20 m or less in depth.

Radar backscatter from ice is dependent on the dielectric properties and spatial distribution of the ice. In addition, such factors as ice age, surface roughness, internal geometry, temperature, and snow cover also affect radar backscatter.

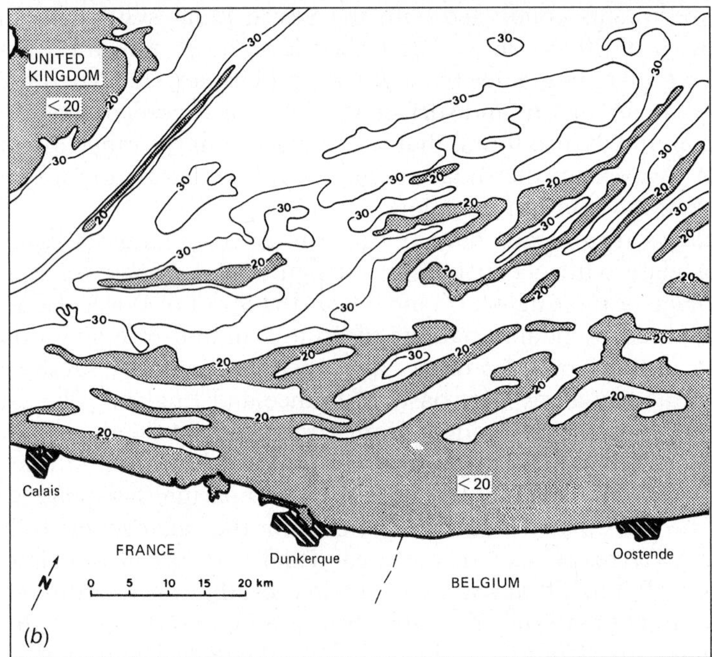

Figure 6.35 English Channel near the Strait of Dover: (*a*) Seasat-1 SAR image, L band, midsummer; (*b*) map showing ocean bottom contours in meters. (Courtesy NASA/JPL/Caltech.)

X- and C-band radar systems have proven useful in determining ice types and, by inference, ice thickness. L-band radar is useful for showing the total extent of ice, but it is often not capable of discriminating ice type and thickness.

Urban Area Response

As illustrated in Figure 6.36, urban areas typically appear light toned in radar images because of their many corner reflectors.

Figure 6.37, an airborne SAR image of Sun City, Arizona, illustrates the effect of urban building orientation on radar reflection. The "corner reflection" from buildings located on the part of the circular street system where the wide faces of the houses (front and rear) face the direction from which the radar waves have

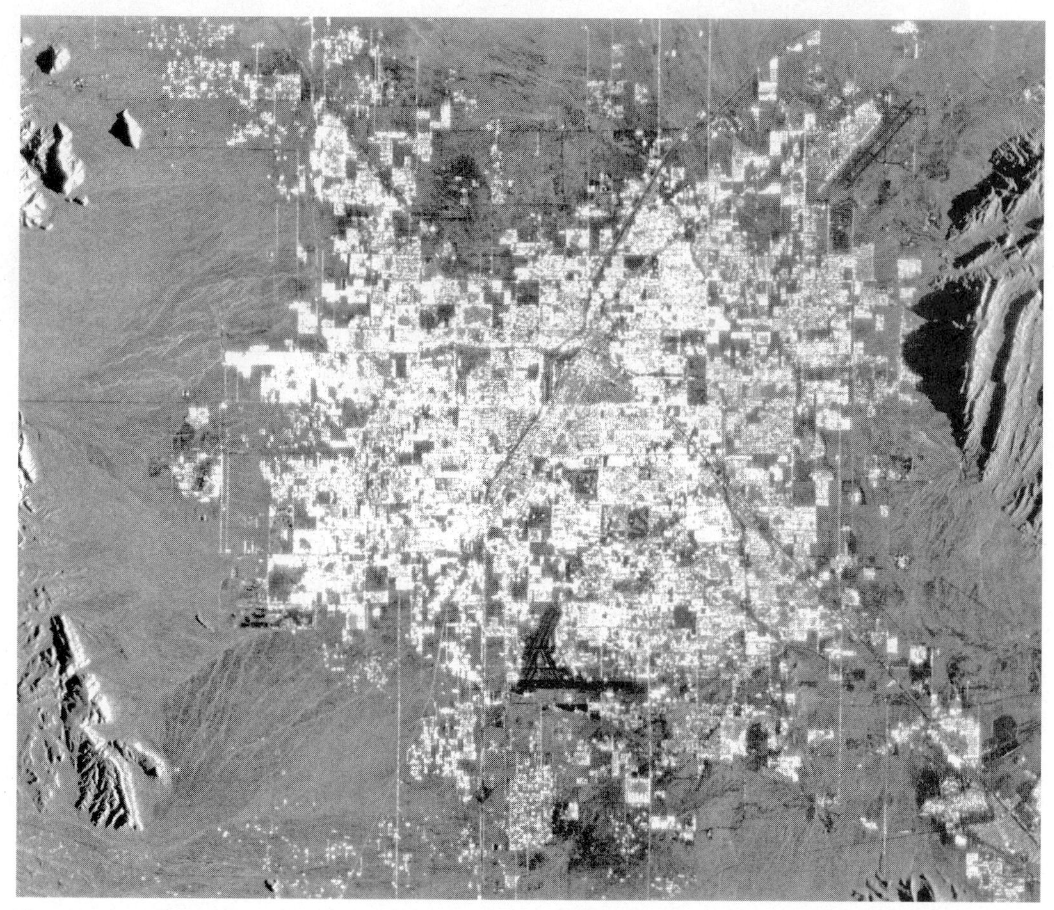

Figure 6.36 Airborne SAR image, Las Vegas, NV, X band, *HH* polarization. North is to the top, the look direction is from the right. Scale 1:250,000. (From American Society for Photogrammetry and Remote Sensing, 1998. Images copyright © John Wiley & Sons, Inc.)

Figure 6.37 Airborne SAR image, Sun City, AZ, X band. The look direction is from the top of the image. Scale 1:28,000. (From American Society for Photogrammetry and Remote Sensing, 1998. Images copyright © John Wiley & Sons, Inc.)

originated provides the strongest radar returns. At right angles to this direction, there is again a relatively strong radar return where the sides of the houses face the direction from which the radar waves have originated. This effect is sometimes called the *cardinal effect,* a term that has survived from the early days of radar remote sensing. At that time, it was noted that reflections from urban areas, often laid out according to the cardinal directions of a compass, caused significantly larger returns when the linear features were illuminated at an angle orthogonal to their orientation, hence the name cardinal effect (Raney, 1998). Other earth surface features also respond with a similar effect. For example, the orientation of row crops affects their response, as described earlier in this section, and the orientation of ocean waves also strongly affects their response.

Summary

In summary, as a generalization, larger radar return signals are received from slopes facing the aircraft, rough objects, objects with a high moisture content, metal objects, and urban and other built-up areas (resulting from corner reflections). Surfaces acting as diffuse reflectors return a weak to moderate signal and may often have considerable image texture. Low returns are received from surfaces acting as specular reflectors, such as smooth water, pavements, and playas (dry lakebeds). No return is received from radar "shadow" areas.

6.9 INTERFEROMETRIC RADAR

As discussed in Section 6.5, the presence of differential relief displacement in overlapping radar images acquired from different flight lines produces image parallax. This is analogous to the parallax present in aerial photographs or electro-optical scanner data. Just as photogrammetry can be used to measure surface topography and feature heights in optical images, radargrammetry can be used to make similar measurements in radar images. In recent years, much attention has been paid to an alternative method for topographic mapping with radar. *Imaging radar interferometry* is based on analysis of the phase of the radar signals as received by two antennas located at different positions in space. (The concept of the phase of a radar signal was introduced in Section 6.6, and it is discussed in more detail in Appendix C.) As shown in Figure 6.38, the radar signals returning from a single point P on the earth's surface will travel slant-range distances r_1 and r_2 to antennas A_1 and A_2, respectively. The difference between lengths r_1 and r_2 will result in the signals being out of phase by some phase difference ϕ, ranging

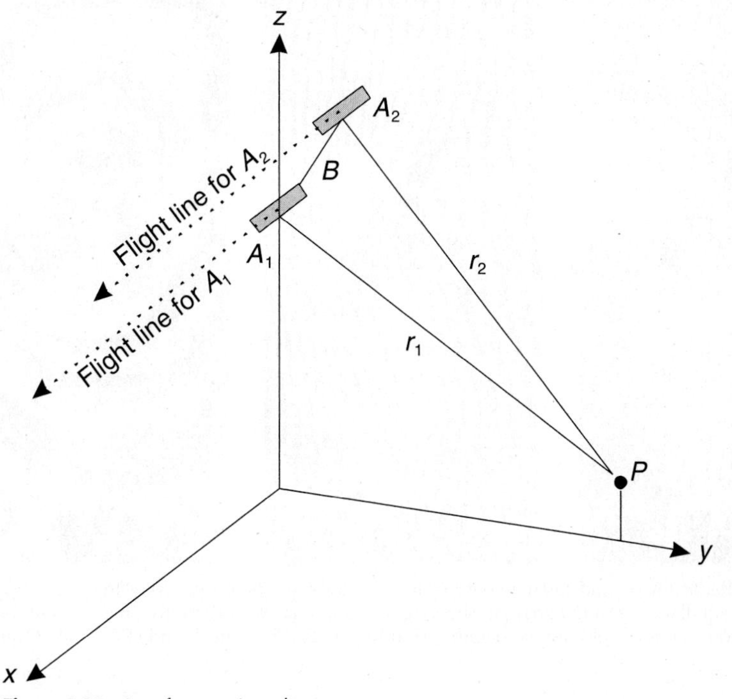

Figure 6.38 Interferometric radar geometry.

from 0 to 2π radians. If the geometry of the *interferometric baseline* (B) is known with a high degree of accuracy, this phase difference can be used to compute the elevation of point P.

Figure 6.39 illustrates the interferometric determination of earth's surface elevations. Figure 6.38*a* is a SAR image of a large volcano. Figure 6.38*b* shows an *interferogram*, which displays the phase difference values for each pixel of an interferometric radar data set. The resulting interference pattern consists of a series of stripes, or *fringes*, that represent differences in surface height and sensor position. When the effect of sensor position is removed, a *flattened interferogram* is produced in which each fringe corresponds to a particular elevation range (Figure 6.39*c*).

There are several different approaches to collecting interferometric radar data. In the simplest case, referred to as *single-pass interferometry*, two antennas are located on a single aircraft or satellite platform. One antenna acts as both a transmitter and receiver, while the second antenna acts only as a receiver. In this bistatic case, as shown in Figure 6.38, the interferometric baseline is the physical

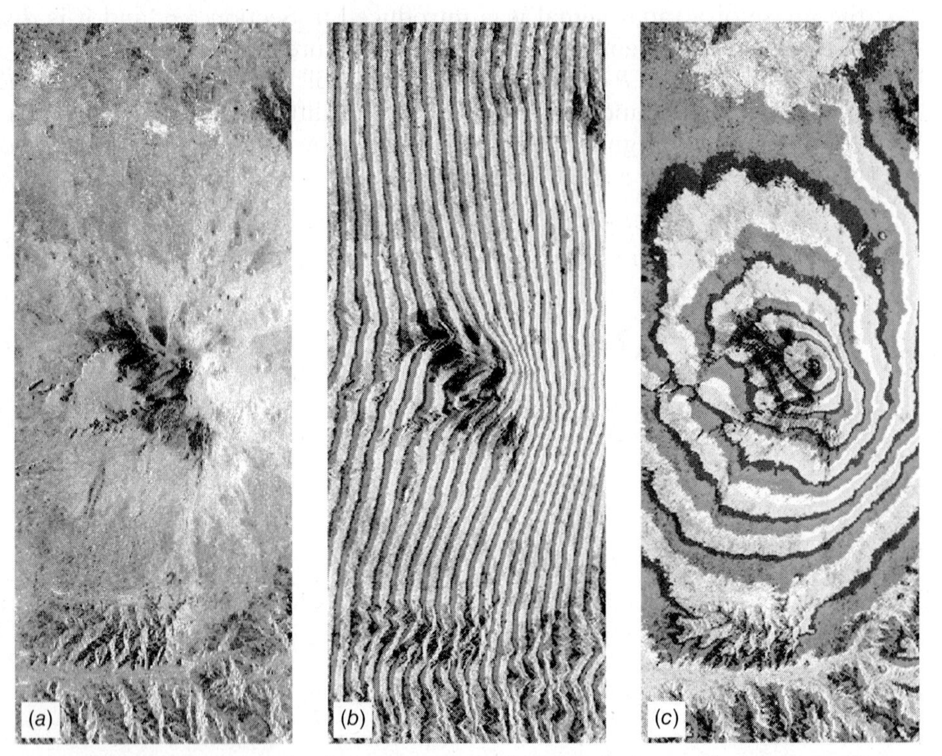

Figure 6.39 Radar image and radar interferograms: (*a*) SIR-C radar image, Mt. Etna, Italy; (*b*) raw interferogram; (*c*) flattened interferogram showing elevation ranges. Parts (b) and (c) are black-and-white reproductions of color interferograms. (Courtesy NASA/JPL/Caltech and EROS Data Center.)

distance between the two antennas. Alternatively, in *repeat-pass interferometry*, an aircraft or satellite with only a single radar antenna makes two or more passes over the area of interest, with the antenna acting as both a transmitter and receiver on each pass. The interferometric baseline is then the distance between the two flight lines or orbital tracks. It is generally desirable to have the sensor pass as close as possible to its initial position, to keep this baseline small. For airborne repeat-pass interferometry, the flight lines should generally be separated by no more than tens of meters, while for spaceborne systems this distance can be as much as hundreds or thousands of meters.

In repeat-pass interferometry, the position and orientation of objects on the surface may change substantially between passes, particularly if the passes are separated by an interval of days or weeks. This results in a situation known as *temporal decorrelation* in which precise phase matching between the two signals is degraded. For example, in a forested area the individual leaf elements at the top of the canopy may change position due to wind action over the course of a single day. For a short wavelength system such as an X-band SAR, which is highly sensitive to individual leaves and other small features, this decorrelation may limit the use of repeat-pass interferometry substantially. In more arid landscapes, where vegetation is sparse, temporal decorrelation will be less of a problem. Likewise, longer wavelength interferometric radar systems will tend to be less affected by temporal decorrelation than will systems using shorter wavelengths. The use of single-pass interferometry avoids the problem of surface change between passes, so that decorrelation is not a problem.

The effects of decorrelation in short-wavelength repeat-pass radar interferometry are illustrated in Figure 6.40. C- and L-band radar images were acquired (with *VV* polarization) over a site in northwestern Wisconsin on October 8 and 10, 1994 by the SIR-C radar system. Repeat-pass interferograms were created by combining the images from the two dates, and the magnitude and phase-difference data were extracted using the principles discussed in Appendix C. In the magnitude images (Figure 6.40*a* and *c*), nonforested areas such as pastures and clear-cuts show up distinctly as darker patches, particularly in the longer wavelength L band. In the C-band phase data (Figure 6.40*b*), there are relatively clear fringe patterns in some parts of the imagery—primarily the nonforested areas—but elsewhere the fringes disappear, due to decorrelation between the two dates. In contrast, the L-band phase data (Figure 6.40*d*) show clear, well-correlated fringes throughout the scene. (Note that the spacing of the fringes in the L-band image is much wider than the spacing of the C-band fringes, because the L-band wavelength is four times as long.) The difference in the clarity of the fringe patterns in the C-band and L-band images is due to the fact that the short-wavelength band is more sensitive to changes in the position of small features such as individual leaves and branches in the top of the forest canopy. Ultimately, in this case, temporal decorrelation prevented construction of a digital elevation model (DEM) from the C-band interferometric data, but the L-band data were successfully used to derive a DEM.

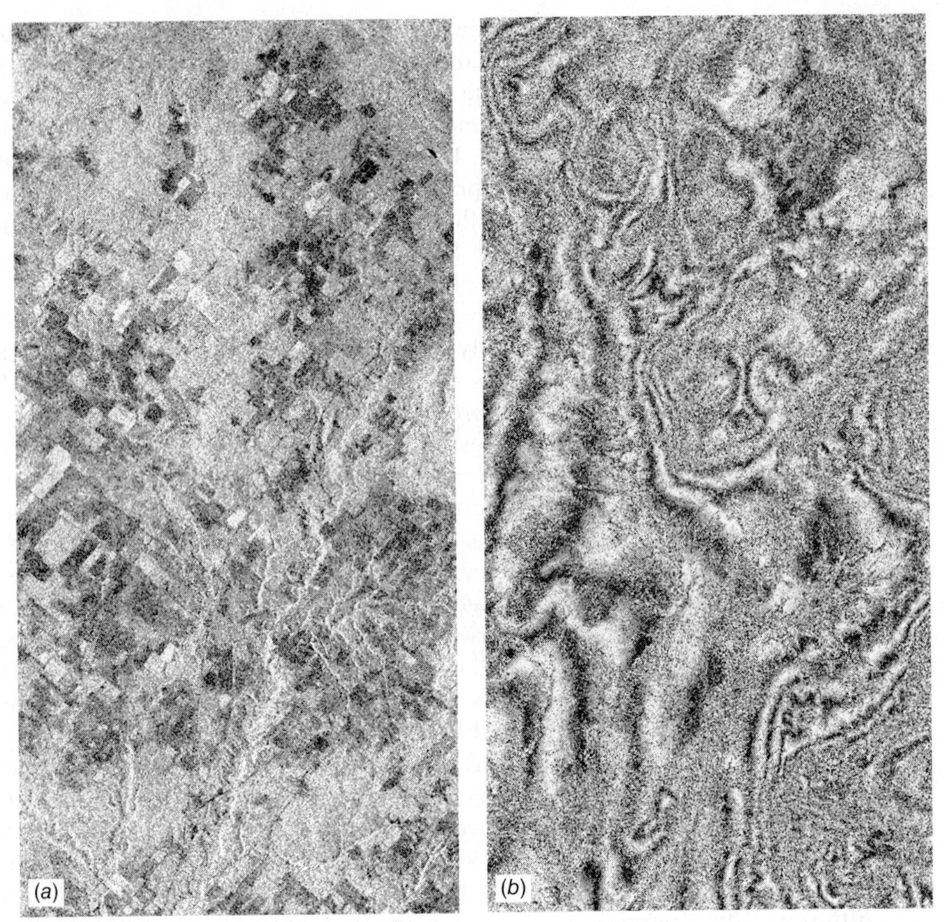

Figure 6.40 C- and L-band interferometric radar images, northwestern Wisconsin. (*a*) C-band, magnitude. (*b*) C-band, phase. (*c*) L-band, magnitude. (*d*) L-band, phase. (Author-prepared figure.)

In some cases, repeat-pass interferometry can actually be used to study surface changes that have occurred between the two passes. In addition to the "before" and "after" images, this approach—known as *differential interferometry*—also requires prior knowledge about the underlying topography. This can be in the form of an existing DEM, but errors in the DEM will lead to incorrect estimates of surface change. A better approach is to acquire one interferometric image pair from the period before the surface change occurs, for use in combination with a third image acquired after the change. In either case, the phase difference between the before and after images can be corrected to account for topography, with the residual phase differences then representing changes in the position of features on the surface. If the interferometric correlation between the two images is high, these changes can be accurately measured to within a small

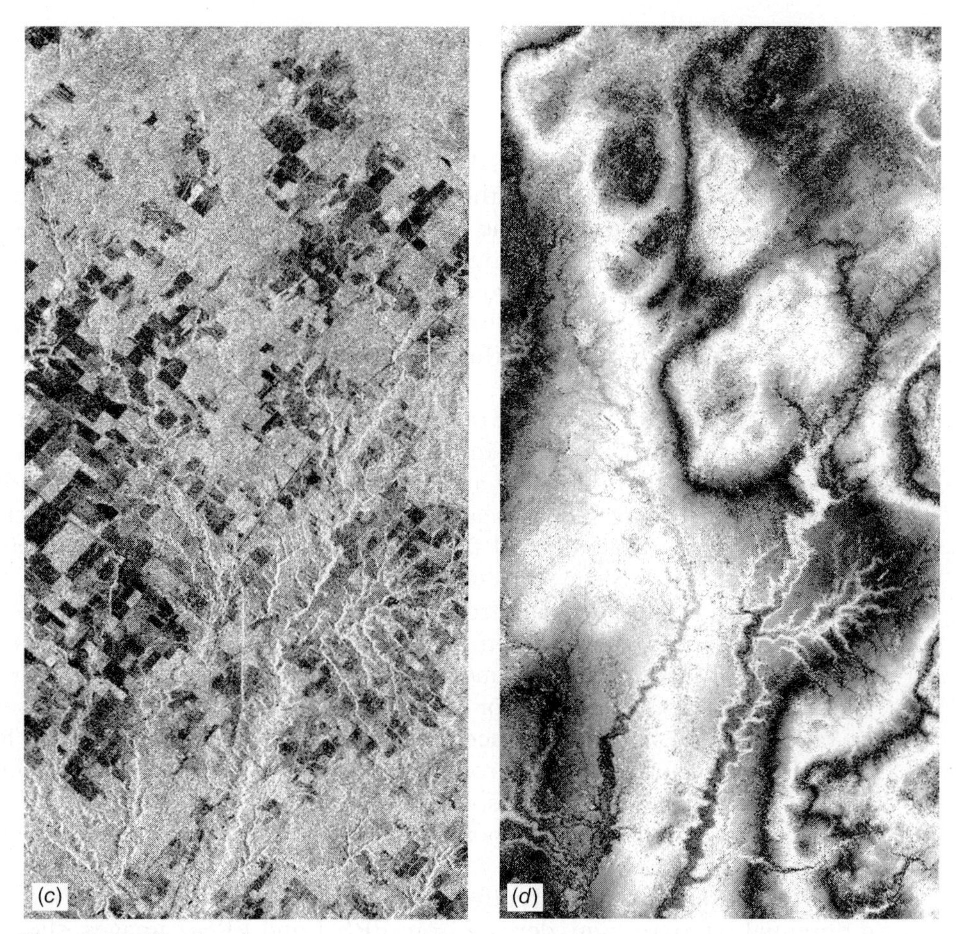

Figure 6.40 *(Continued)*

fraction of the radar system's wavelength—often to less than 1 cm. With a single pair of images, surface changes are measured only as line-of-sight displacements, meaning that only the degree to which a point moved toward or away from the radar look direction can be measured. If two sets of interferometric image pairs are available from different look directions, such as from the ascending and descending segments of a satellite's orbit, the two-dimensional movement of the surface can be derived.

This approach works best for changes that affect large areas in a spatially correlated manner, such as the entire surface of a glacier moving downhill, as opposed to changes that occur in a spatially disjointed manner, such as the growth of trees in a forest. Plate 25 provides three examples of applications of differential interferometry. In Plate 25*a* this technique is used to assess surface deformation during the Tōhoku earthquake that struck Japan on March 11, 2011. This magnitude 9.0 earthquake and the tsunami that followed it caused extensive destruction in the northeastern coastal region of the main Japanese island of Honshu. The radar interferograms shown in this plate were produced using data from Japan's ALOS PALSAR satellite radar system (Section 6.14), acquired both before and after the earthquake. As shown in the scale bar at bottom, each color cycle from one cyan "fringe" to the next represents approximately 12 cm of surface movement in the radar's line-of-sight direction (additional movement may have occurred perpendicular to this axis). Scientists at the Japan Aerospace Exploration Agency (JAXA) have analyzed these data and estimated that up to 4 m of surface displacement occurred in the most affected parts of this area.

Plate 25*b* shows an example of the use of radar interferometry for monitoring slower changes in topography, in this case the continuing ground uplift caused by magma accumulation below a volcano in the central Oregon Cascade Range. Scientists from the USGS Cascades Volcano Observatory, in connection with other agencies, have confirmed the slow uplift of a broad area centered about 5 km west of South Sister volcano. The radar interferogram shown in Plate 25*b* was produced using radar data from the European Space Agency's ERS satellites. In this repeat-pass interferogram, each full color band from blue to red represents about 2.8 cm of ground movement in the direction of the radar satellite. (No information is available for the uncolored areas, where forest vegetation, or other factors, hinders the acquisition of useful radar data.) The four concentric bands show that the ground surface moved toward the satellite by as much as 10 cm between August 1996 and October 2000. Surface uplift caused by magma accumulation at depth can be a precursor to volcanic activity at the earth's surface.

A third example of differential interferometry is illustrated in Plate 25*c*. This differential interferogram, derived from ERS-1 and ERS-2 images, shows surface displacement between April 1992 and December 1997 in the area around Las Vegas, Nevada. For most of the past century, pumping of groundwater from an underground aquifer for domestic and commercial consumption has caused land subsidence at a rate of several centimeters per year in Las Vegas, with significant damage to the city's infrastructure. In recent years, artificial recharge of groundwater has been used in an attempt to reduce subsidence. Analysis of the interferometric radar imagery in combination with geologic maps shows that the spatial extent of subsidence is controlled by geologic structures (faults, indicated by white lines in Plate 25*c*) and sediment composition (clay thickness). The maximum detected subsidence during the 1992 to 1997 period was 19 cm. Other potential applications of differential radar interferometry include monitoring the movement of glaciers and ice sheets, measuring displacement across faults after earthquakes, and detecting land subsidence due to oil extraction, mining, and other activities.

6.10 RADAR REMOTE SENSING FROM SPACE

The earliest civilian (non-classified) spaceborne imaging radar missions were the experimental spaceborne systems Seasat-1 (1978) and three Shuttle Imaging Radar systems (SIR-A, SIR-B, and SIR-C) that orbited for short periods between 1981 and 1994. The SIR-C antenna was employed again in February 2000 for the Shuttle Radar Topography Mission (SRTM), a brief but highly productive operational program to map global topography using radar interferometry.

In the 1990s the first true "operational" (non-experimental) radar remote sensing satellites were developed. During the four-year period from 1991 to 1995, radar satellites were launched by the former Soviet Union, the European Space Agency, and the national space agencies of Japan and Canada. Since then, the number of radar satellite systems has increased dramatically, culminating with a shift toward multi-satellite "constellations" to provide rapid global coverage and tandem single-pass interferometry. These multi-satellite missions include Germany's TerraSAR-X/TanDEM-X pair, the Italian constellation of four COSMO-SkyMed satellites, the European Space Agency's forthcoming pair of Sentinel-1 satellites, and the planned Radarsat Constellation Mission.

The advantages of spaceborne radar systems are obvious. Because radar is an active sensor that can gather data both day and night, radar images may be acquired during both the south-to-north (ascending) and north-to-south (descending) portions of a satellite's orbit. This is in contrast to electro-optical remote sensing systems that normally only acquire imagery on the daylight portion of each orbit. Spaceborne radar imagery can also be collected where clouds, haze, and other atmospheric conditions would prevent the acquisition of electro-optical imagery. Thus, spaceborne radar systems are ideally suited for applications where imagery can be dependably acquired wherever and whenever it is needed.

In general, radar images acquired at small incident angles (less than 30°) emphasize variations in surface slope, although geometric distortions due to layover and foreshortening in mountainous regions can be severe. Images with large incident angles have reduced geometric distortion and emphasize variations in surface roughness, although radar shadows increase.

A limitation in the use of airborne radar imagery is the large change in incident angle across the image swath. In these circumstances, it is often difficult to distinguish differences in backscatter caused by variations in incident angle from those actually related to the structure and composition of the surface materials present in an image. Spaceborne radar images overcome this problem because they have only small variations in incident angle. This makes their interpretation less difficult.

Beginning with SIR-C in 1994, most spaceborne radar systems have made use of advanced beam steering techniques to collect data in three broad categories of imaging modes. The basic configuration, often called *Stripmap* mode, represents the imaging process that has been discussed throughout the previous sections of this chapter. For wide-area coverage, the *ScanSAR* imaging mode involves electronically steering the radar beam back and forth in the range direction. In effect, the beam illuminates two or more adjacent swaths in alternation, with the far-range side of the first swath being contiguous with the near-range side of the second swath, and so on. The multiple swaths are then processed to form a single, wide image. The disadvantage of ScanSAR mode is that the spatial resolution is reduced.

An additional imaging configuration, referred to as *Spotlight*, involves steering the radar beam in azimuth rather than in range, in order to dwell on a given site for a longer period of time. As the satellite approaches the target area, the beam is directed slightly forward of the angle at which it is normally transmitted; then, while the satellite moves past, the beam swings back to continue covering the target area. Through an extension of the synthetic aperture principle, this Spotlight mode allows a finer resolution to be achieved by acquiring more "looks" over the target area from a longer segment of the orbit path. This increase in resolution comes at the expense of continuous coverage because while the antenna is focusing on the target area it is missing the opportunity to image other portions of the ground swath. Many of the latest high-resolution radar satellites employ Spotlight mode to achieve resolutions on the order of 1 to 3 m.

The SIR-C radar mission provided the first tests of both ScanSAR and Spotlight modes from space. Figure 6.41 shows the first ScanSAR image from space,

Figure 6.41 SIR-C "ScanSAR" image of the Weddell Sea off Antarctica, October. Scale 1:2,500,000. (Courtesy NASA/JPL/Caltech.)

acquired over the Weddell Sea off Antarctica on October 5, 1994. The dimensions of the image are 240 km by 320 km. The upper left half of the image shows an area of the open sea, with a uniform gray tone. First-year seasonal pack ice (0.5 to 0.8 m thick) occupies the lower right corner of the image. In between these areas, in the lower left and center of the image, there are two large oceanic circulation features or eddies, each approximately 50 km wide and rotating in a clockwise direction. Very dark areas within and adjacent to the eddies are newly formed ice, whose smooth surface acts as a specular reflector. This type of spaceborne Scan-SAR imagery is an important resource for applications such as tracking the extent of sea ice and movements of icebergs over large areas, as an aid to shipping and navigation in high-latitude oceans.

6.11 SEASAT-1 AND THE SHUTTLE IMAGING RADAR MISSIONS

Seasat-1 was the first of a proposed series of satellites oriented toward oceanographic research. The Seasat-1 satellite was launched on June 27, 1978, into an 800-km near-polar orbit. Approximately 95% of the earth's oceans were to be covered by the system. Unfortunately, prime power system failure 99 days after launch limited the image data produced by the satellite.

Seasat-1 employed a spaceborne L-band (23.5-cm) SAR system with *HH* polarization. It was designed to generate imagery across a 100-km swath with a look angle of 20° to 26° and four-look 25 m resolution in both range and azimuth. Table 6.4 summarizes these characteristics (as well as those of the SIR systems).

Although the primary rationale for placing the imaging radar system on board Seasat-1 was its potential for monitoring the global surface wave field and polar sea ice conditions, the resultant images of the oceans revealed a much wider spectrum of oceanic and atmospheric phenomena, including internal waves, current

TABLE 6.4 Characteristics of Major Experimental Synthetic Aperture Radar Systems

Characteristic	Seasat-1	SIR-A	SIR-B	SIR-C
Launch date	June 1978	November 1981	October 1984	April 1994 October 1994
Length of mission	99 days	3 days	8 days	10 days
Nominal altitude, km	800	260	225	225
Wavelength band	L band	L band	L band	X band (X-SAR) C and L bands (SIR-C)
Polarization	*HH*	*HH*	*HH*	*HH, HV, VV, VH* (X band *HH* only)
Look angle	20–26° (fixed)	47–53° (fixed)	15–60° (variable)	15–60° (variable)
Swath width, km	100	40	10–60	15–90
Azimuth resolution, m	25	40	25	25
Range resolution, m	25	40	15–45	15–45

boundaries, eddies, fronts, bathymetric features, storms, rainfalls, and windrows. Seasat-1 was also operated over the world's land areas, and many excellent images illustrating applications to geology, water resources, land cover mapping, agricultural assessment, and other land-related uses were obtained. Figures 6.5 and 6.35 (described previously) are examples of Seasat-1 imagery.

Following the successful but brief Seasat-1 mission, a number of early spaceborne radar experiments were conducted using radar systems operated from the Space Shuttle—SIR-A in 1981, SIR-B in 1984, and two SIR-C missions in 1994. Table 6.4 summarizes the characteristics of these three systems.

SIR-A

The SIR-A experiments were conducted from the Space Shuttle during November 1981. This was the second flight of the Space Shuttle, and the first scientific payload ever flown on the Shuttle. The SIR-A system possessed many of the same characteristics as the radar system onboard Seasat-1, most notably its long-wavelength L-band (23.5-cm) antenna with *HH* polarization. The principal differences between these two were SIR-A's larger look angle (47° to 53°), narrower swath, and slightly coarser resolution (40 m). SIR-A obtained imagery over 10 million km^2 of the earth's surface and acquired radar images of many tropical, arid, and mountainous regions for the first time.

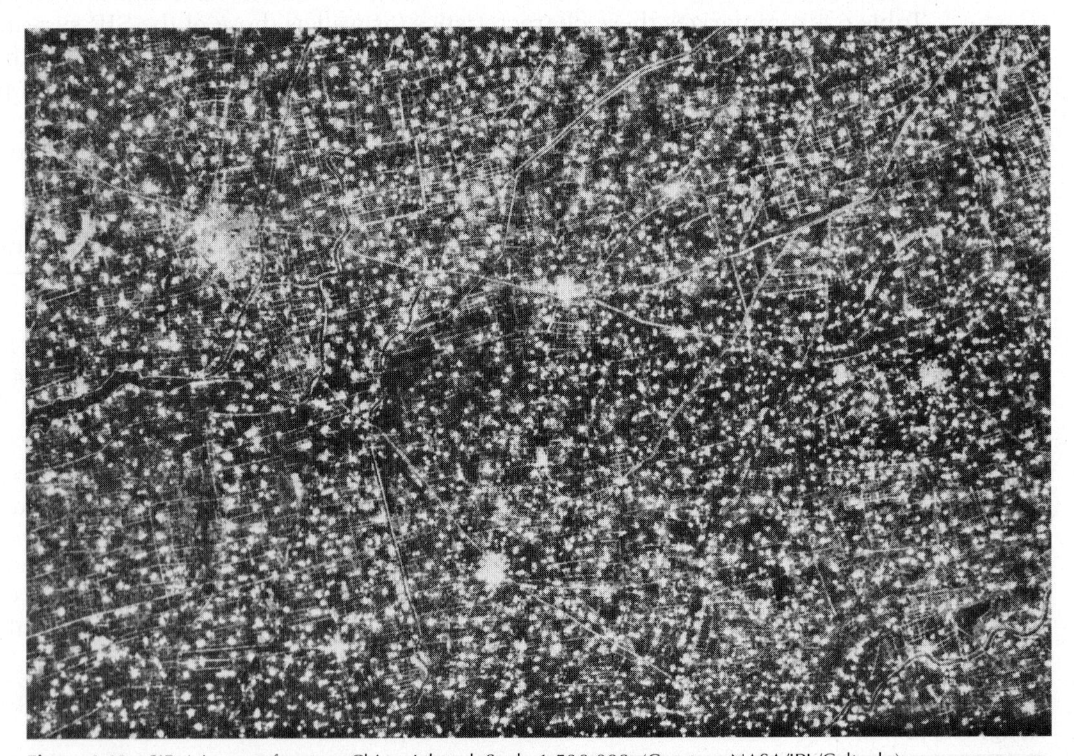

Figure 6.42 SIR-A image of eastern China, L band. Scale 1:530,000. (Courtesy NASA/JPL/Caltech.)

Figure 6.42 is a SIR-A image showing villages, roads, and cultivated fields in eastern China. Each of the hundreds of white spots on this image is a village. The agricultural crops common in this area are winter wheat, kaoliang, corn, and millet. The dark linear and winding features with white lines on each side are rivers and drainageways located between levees.

SIR-B

The SIR-B experiments were conducted from the Space Shuttle during October 1984. Again, an L-band system with *HH* polarization was used. The principal difference between the SIR-A and SIR-B radar systems is that SIR-B was equipped with an antenna that could be tilted mechanically to beam its radar signals toward the earth at varying look angles (ranging from 15° to 60°). This provided the opportunity for scientific studies aimed at assessing the effect of various incident angles on radar returns. In addition, it provided the opportunity for the acquisition of stereo radar images.

Figure 6.43 shows SIR-B images of Mt. Shasta, a 4300-m-high volcano in northern California. These images illustrate the effect of incident angle on elevation displacement. In Figure 6.43*a*, having an incident angle of 60°, the peak of the volcano is imaged near its center. In Figure 6.43*b*, having an incident angle of 30°, the peak is imaged near the top of the figure (the look direction was from top to bottom in this figure). Several light-toned tongues of lava can be seen on the flanks of this strato volcano. The surface of the young lava flow seen at upper left in this radar image consists of unvegetated angular chunks of basalt $\frac{1}{3}$ to 1 m in size that present a very rough surface to the L-band radar waves. The other, somewhat darker toned lava flows on the flanks of Mt. Shasta are older, have weathered more, and are more vegetated.

SIR-C

SIR-C missions were conducted in April and October 1994. SIR-C was designed to explore multiple-wavelength radar sensing from space. Onboard SIR-C were L-band (23.5-cm) and C-band (5.8-cm) systems from NASA and an X-band (3.1-cm) system (known as X-SAR) from a consortium of the German Space Agency (DARA) and the Italian Space Agency (ASI). The multiple wavelength bands available allowed scientists to view the earth in up to three wavelength bands, either individually or in combination. SIR-C look angles were variable in one-degree increments from 15° to 60°, and four polarizations were available (*HH*, *HV*, *VV*, and *VH*).

The scientific emphasis in selecting sites for SIR-C image acquisition was on studying five basic themes—oceans, ecosystems, hydrology, geology, and rain and clouds. Ocean characteristics studied included large surface and internal waves, wind motion at the ocean surface, and ocean current motion, as well as sea ice characteristics and distribution. Ecosystem characteristics studied included land use, vegetation type and extent, and the effects of fires, flooding, and clear cutting.

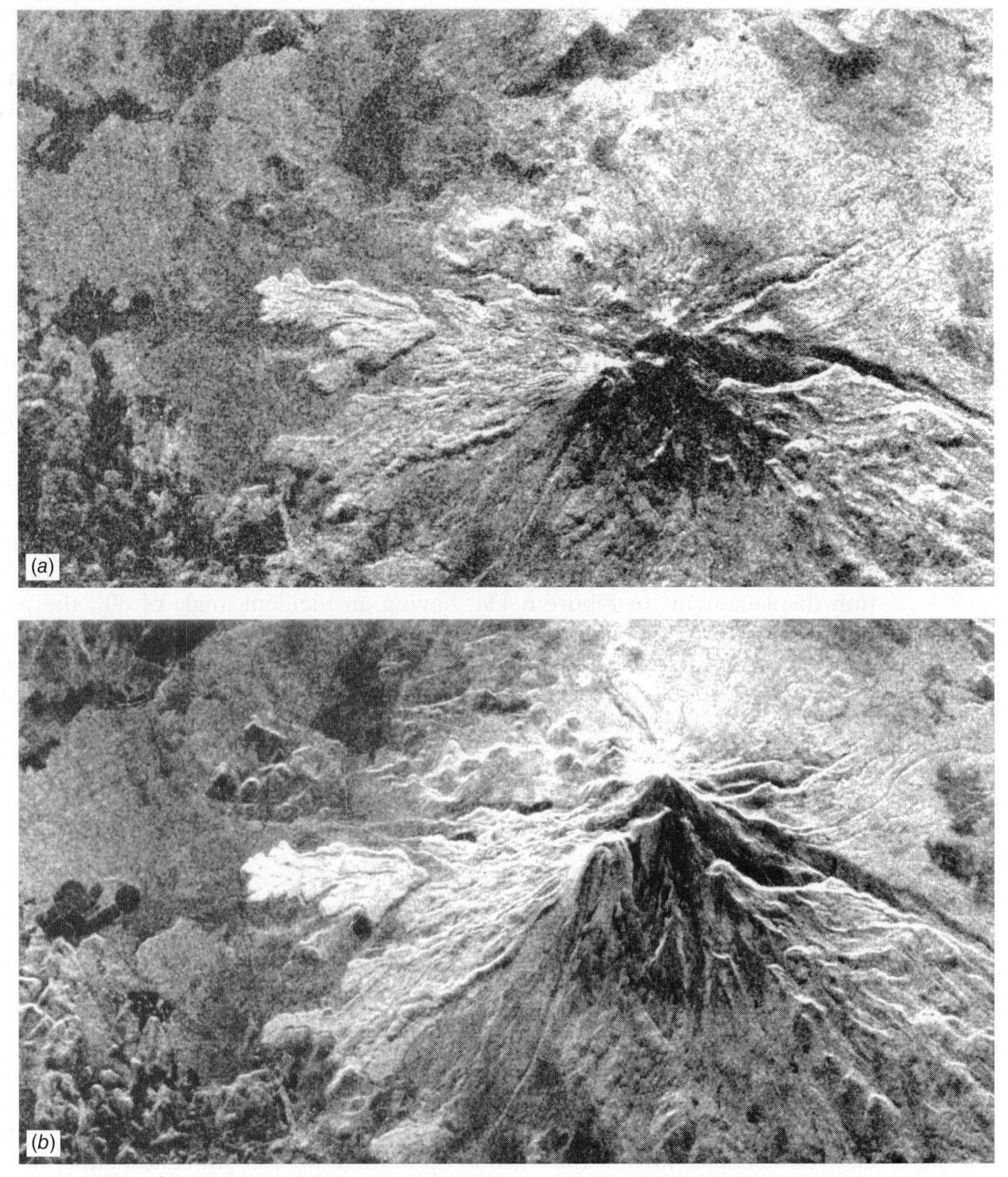

Figure 6.43 SIR-B images, Mt. Shasta, CA, L band, mid-fall (scale 1:240,000): (*a*) 60° incident angle, (*b*) 30° incident angle. Note the severe layover of the mountain top in (*b*). (Courtesy NASA/JPL/Caltech.)

Hydrologic studies focused on water and wetland conditions, soil moisture patterns, and snow and glacier cover. Geologic applications included mapping geologic structures (including those buried under dry sand), studying soil erosion, transportation and deposition, and monitoring active volcanoes. Also under study was the attenuation of radar signals at X-band and C-band wavelengths by rain and clouds.

The multi-wavelength, multi-polarization design of SIR-C provides many more options for viewing and analyzing the imagery. For visual interpretation, three different wavelength–polarization combinations are used to produce a composite image, with one wavelength/polarization combination displayed as blue, one as green, and one as red. If the different wavelength–polarization images show backscatter from different features with different intensities, then these features are displayed with different colors. This can be seen in Plate 26, which shows a volcano-dominated landscape in central Africa; parts of Rwanda, Uganda, and the Democratic Republic of the Congo (formerly Zaire) are each present in this image. In this image, C-band data with *HH* polarization are displayed as blue, C-band data with *HV* polarization are displayed as green, and L-band data with *HV* polarization are displayed as red. The volcano at top center is Karisimba, 4500 m high. The green band on the lower slopes of Karisimba volcano, to the right of its peak, is an area of bamboo forest, one of the world's few remaining natural habitats for mountain gorillas. Just right of the center of the image is Nyiragongo volcano, an active volcano 3465 m high. The lower portion of the image is dominated by Nyamuragira volcano, 3053 m high, and the many lava flows (purple in this image) that issue from its flanks.

Plate 27 shows SIR-C imagery of a portion of Yellowstone National Park in Wyoming. Yellowstone was the world's first national park and is known for its geological features, including geysers and hot springs. The park and the surrounding region also provide habitat for populations of grizzly bears, elk, and bison. In 1988, massive forest fires burned across some 3200 km^2 within the park. The intensity of the burn varied widely, leaving a complex mosaic of heavily burned, lightly burned, and unburned areas that will continue to dominate the park's landscape for decades. The effects of these fires can be clearly seen in the L_{VH}-band SIR-C image (*a*), acquired on October 2, 1994. Unburned lodgepole pine forest returned a strong response in the L_{VH}-band and thus appear relatively bright in this image. Areas of increasing burn intensity have proportionately less above-ground forest biomass present; these areas produced less L_{VH} backscatter and appear darker. Yellowstone Lake, near the bottom of the image, appears black due to specular reflection and negligible L_{VH} backscatter.

Plate 27*b* shows a map of above-ground forest biomass, derived from the SIR-C data shown in (*a*) and from field measurements by the Yellowstone National Biological Survey. Colors in the map indicate the amount of biomass, ranging from brown (less than 4 tons per hectare) to dark green (nonburned forest with a biomass of greater than 35 tons per hectare). Rivers and lakes are shown in blue. The ability of long-wavelength and cross-polarized radar systems to estimate forest biomass may provide a valuable tool for natural resource managers and scientists, whether in the wake of natural disasters such as fires and windstorms, or in the course of routine forest inventory operations.

Unlike the subsequent Shuttle Radar Topography Mission (Section 6.19), which collected data over the majority of the earth's land surface, SIR-C data were only collected over specific areas of research interest. Furthermore, due to the experimental nature of the SIR-C system, not all of the data from the two missions were processed to the full-resolution "precision" level. Those images that were processed (prior to the termination of the program in mid-2005) are now archived at the USGS EROS Data Center, and can be ordered online through EarthExplorer.

6.12 ALMAZ-1

The Soviet Union (just prior to its dissolution) became the first country to operate an earth-orbiting radar system on a commercial basis with the launch of *Almaz-1* on March 31, 1991. Almaz-1 returned to earth on October 17, 1992, after operating for about 18 months. Other Almaz missions were planned prior to the dissolution of the Soviet Union but were subsequently canceled.

The primary sensor on board Almaz-1 was a SAR system operating in the S-band spectral region (10 cm wavelength) with *HH* polarization. The look angle for the system could be varied by rolling the satellite. The look angle range of 20° to 70° was divided into a standard range of 32° to 50° and two experimental ranges of 20° to 32° and 50° to 70°. The effective spatial resolution varied from 10 to 30 m, depending on the range and azimuth of the area imaged. The data swaths were approximately 350-km wide. Onboard tape recorders were used to record all data until they were transmitted in digital form to a ground receiving station.

6.13 ERS, ENVISAT, AND SENTINEL-1

The European Space Agency (ESA) launched its first remote sensing satellite, *ERS-1*, on July 17, 1991, and its successor *ERS-2* on April 21, 1995. Both had a projected life span of at least three years; ERS-1 was retired from service on March 10, 2000, and ERS-2 ended its mission on September 5, 2011. The characteristics of both systems were essentially the same. They were positioned in sun-synchronous orbits at an inclination of 98.5° and a nominal altitude of 785 km. During the 1995 to 2000 period, a particular focus of the two satellites was tandem operation for repeat-pass radar interferometry.

ERS-1 and ERS-2 carried three principal sensors: (1) a C-band active microwave instrumentation (AMI) module, (2) a Ku-band radar altimeter, and (3) an along-track scanning radiometer. We limit this discussion to the AMI, which consisted of a C-band, *VV*-polarized SAR system plus a non-imaging microwave scatterometer. The ERS SAR had a fixed (and relatively steep) look angle of 23° and a four-look resolution of approximately 30 m. The choice of wavelength, polarization, and look angle were selected primarily to support imaging of the oceans, although ERS-1 and -2 images have also been used for many land applications.

Figure 6.17 (described previously) is an example of an ERS-1 image. Plates 25*b* and 25*c*, described in Section 6.9, illustrate the use of ERS-1 and ERS-2 in tandem for differential radar interferometry.

On March 1, 2002, ESA launched *Envisat*. This large satellite platform carries a number of instruments, including the ocean monitoring system MERIS (Section 5.18), and an advanced imaging radar system. Following the launch, Envisat was maneuvered into an orbit matching that of ERS-2, just 30 min ahead of ERS-2 and covering the same ground track to within 1 km. Envisat operated for 10 years before contact was lost with the satellite on April 8, 2012.

Envisat's SAR instrument, the *Advanced Synthetic Aperture Radar (ASAR)* system, represented a substantial advance over that of its ERS predecessors. Like the ERS-1 and -2 SARs, Envisat's ASAR operated in the C band. However, ASAR had multiple imaging modes offering various combinations of swath width, resolution, look angle, and polarization. In its normal Image Mode, ASAR generated four-look high resolution (30 m) images at either *HH* or *VV* polarization, with swath widths ranging from 58 to 109 km and look angles ranging from 14° to 45°. Other ASAR modes were based on the ScanSAR technique discussed in Section 6.10. These provided coarser-resolution (150 to 1000 m) imagery over a 405-km swath with either *HH* or *VV* polarization. The final ASAR mode, Alternating Polarization Mode, represented a modified ScanSAR technique that alternates between polarizations over a single swath, rather than alternating between near- and far-range swaths. This mode provides 30-m-resolution dual-polarization imagery, with one of three polarization combinations (*HH* and *VV, VV and VH,* or *HH* and *HV*). Thus, Envisat was the first operational radar satellite to offer multiple polarization radar imagery.

Figure 6.44 is an Envisat ASAR image showing an oil slick in the Atlantic Ocean off the coast of Spain. On November 13, 2002, the oil tanker *Prestige* was located off the west coast of Spain when its hull ruptured. The tanker had been damaged in a storm and eventually sank on November 19. The image in Figure 6.44 was acquired on November 17, four days after the initial incident, and shows the damaged ship as a bright point (inset) at the head of an extensive oil slick. Over 3.8 million liters of fuel oil spilled from the tanker during this incident. Oil films have a dampening effect on waves, and the smoother, oil-coated water acts as more of a specular reflector than the surrounding water, thus appearing darker.

Just as Envisat represented an improvement over the earlier ERS satellite radars, the ESA's Sentinel-1A and -1B satellites will build upon the ERS/Envisat heritage while incorporating several advances. The first Sentinel-1 satellite was launched on April 3, 2014, and the second is planned for launch in 2016. Working together, these two imaging radar satellites will provide C-band, single- or dual-polarized imagery with coverage of mid- to high-latitudes every one to three days, with data delivery over the internet within an hour of image acquisition. This rapid delivery of data is designed to support operations such as ship tracking and sea ice monitoring, detection and mapping of oil spills, response to natural disasters such as floods and wildfires, and other time-critical applications.

The imaging modes for the Sentinal-1 SAR systems include a strip-map mode with 5-m resolution and 80-km swath; an extra-wide swath mode offering 40-m resolution data over a 400-km swath; and a single-pass interferometric mode

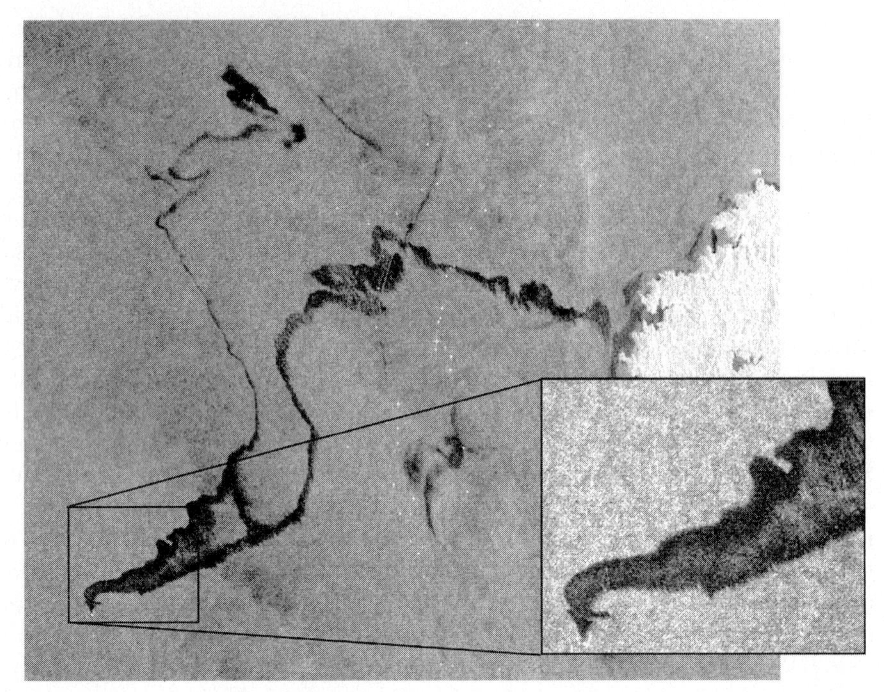

Figure 6.44 Envisat ASAR image showing an oil slick in the Atlantic Ocean off the coast of Spain. (Courtesy ESA.)

using both satellites in tandem to acquire interferometric data over a 250-km swath at 20-m resolution. As we will see in the following sections, this approach of using a "constellation" of multiple radar satellites orbiting in formation is being widely adopted to facilitate both the rapid acquisition of data worldwide and the collection of interferometric data via tandem pairs of satellites.

6.14 JERS-1, ALOS, AND ALOS-2

Developed by the National Space Development Agency of Japan (NASDA), the *JERS-1* satellite was launched on February 11, 1992, and operated until October 12, 1998. It included both a four-band optical sensor (OPS) and an L-band (23-cm) SAR operating with *HH* polarization. The radar system had a three-look ground resolution of 18 m and covered a swath width of 75 km at a look angle of 35°.

On January 24, 2006, NASDA launched its *Advanced Land Observing Satellite (ALOS)*. This spacecraft operated until May 12, 2011. Among the systems onboard this satellite was a *Phased Array L-band Synthetic Aperture Radar (PALSAR)* system. The PALSAR instrument was designed to be an improved follow-on to

Japan's earlier L-band radar satellite system, JERS-1. The ALOS PALSAR had a cross-track pointing capability over a range of incident angles from 8° to 60°. In its fine resolution mode, PALSAR collected either single- or dual-polarization imagery. In its ScanSAR mode, PALSAR covered a large area with a resolution of 100 m. Finally, PALSAR also had a fully polarimetric mode, in which it collected imagery in all four linear polarizations (*HH*, *HV*, *VH*, and *VV*).

Figure 6.45 shows a PALSAR image of Nagoya, a major Japanese port city. The image was acquired on April 21, 2006. Numerous structures associated with the port facilities can be seen in the image, along with bridges, ships, and waterways. The city's airport, Chubu Centrair International Airport, is located on an artificial island at the left side of the image. Plate 25*a* discussed previously provides an example of the use of ALOS PALSAR data for differential interferometry.

Following on the successes of ALOS, Japan's national space agency launched *ALOS-2* in May 2014. The radar system on ALOS-2, PALSAR-2, shares many of the characteristics of its predecessor, but with improved spatial resolution and the ability to image on either the left or right side of its orbit track.

Table 6.5 lists the characteristics of ALOS-2's radar imaging modes. Currently, ALOS-2 is the only radar satellite operating at the relatively long L-band wavelength range (all other current and near-future proposed imaging radar satellites operate at the shorter C- or X-bands). With its predecessors JERS-1 and ALOS-1 also having employed this longer wavelength, Japan's NASDA appears to be filling a critical niche not met by any other spaceborne radar systems since the brief Seasat-1 and SIR missions of 1978–1994.

Figure 6.45 ALOS PALSAR image of Nagoya, Japan, April. Scale 1:315,000. Black-and-white copy of a color multipolarization L-band image. (Courtesy NASDA.)

TABLE 6.5 ALOS-2 PALSAR-2 Imaging Modes

Mode	Polarization(s)	Resolution, m	Swath, km
Spotlight	Single (*HH* or *VV*)	1–3	25
Stripmap (ultrafine)	Single or dual	3	50
Stripmap (high sensitive)	Single, dual, or quad	6	50
Stripmap (fine)	Single, dual, or quad	10	70
ScanSAR (nominal)	Single or dual	100	350
ScanSAR (wide)	Single or dual	60	490

6.15 RADARSAT

Radarsat-1, launched on November 28, 1995, was the first Canadian remote sensing satellite. It was developed by the Canadian Space Agency in cooperation with the United States, provincial governments, and the private sector. Canada was responsible for the design, control, and operations of the overall system, while NASA provided the launch services. Radarsat-1 long outlived its expected lifetime of five years, finally ceasing operations on March 29, 2013. Among its major accomplishments was the production of the first complete, high-resolution map of Antarctica, as discussed at the beginning of this chapter.

The Radarsat-1 SAR was a C-band (5.6-cm) system with *HH* polarization. In contrast to the ERS and JERS-1 systems that preceded it, Radarsat-1's SAR could be operated in a variety of beam selection modes providing various swath widths, resolutions, and look angles. Virtually all subsequent spaceborne radars have adopted this flexible approach.

Radarsat-1 was followed by *Radarsat-2*, launched on December 14, 2007. It too employs a C-band system, but offers *HH*, *VV*, *HV*, and *VH* polarization options. It has beam selection modes similar to Radarsat-1 but with an increased number of modes available. These modes provide swath widths from 10 to 500 km, look angles from 10° to 60°, resolutions varying from 1 to 100 m, and number of looks varying from 1 to 10.

The orbit for Radarsat-2 is sun synchronous and at an altitude of 798 km and inclination of 98.6°. The orbit period is 100.7 min and the repeat cycle is 24 days. The radar antenna can be operated in either right-looking or left-looking configurations, providing one-day repeat coverage over the high Arctic and approximately three-day repeat coverage at midlatitudes.

Table 6.6 and Figure 6.46 summarize the modes in which the system operates.

Rather than following Radarsat-2 with another single spacecraft, the Canadian Space Agency is developing plans for the *Radarsat Constellation Mission*, consisting of at least three identical systems to be launched beginning in 2018, with the possibility of adding another three satellites in the future. Like Radarsat-1 and -2,

TABLE 6.6 Radarsat-2 Imaging Modes

Mode	Polarization Options	Swath Width, km	Look Angle	Resolution,[a] m
Standard	Single, dual	100	20–49°	27
Wide	Single, dual	150–165	20–45°	27
Fine	Single, dual	45	30–50°	8
Extended high	Single, dual	75	49–60°	24
Extended low	Single, dual	170	10–23°	34
ScanSAR narrow	Single, dual	305	20–47°	50
ScanSAR wide	Single, dual	510	20–49°	100
Fine quad-pol	Quad	25	18–49°	10
Standard quad-pol	Quad	25	18–49°	27
Ultra-fine	Single	20	20–49°	3
Spotlight	Single	18	20–49°	1
Multilook fine	Single	50	37–48°	10

[a]Resolution values are approximate. Azimuth and range resolution values differ, and range resolution varies with range (which in turn varies with look angle).

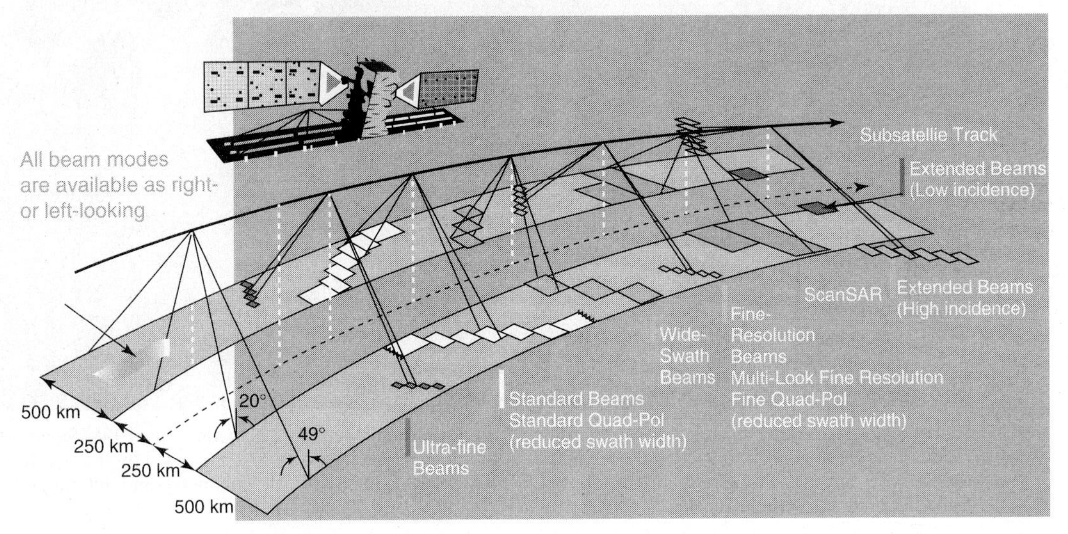

Figure 6.46 Radarsat-2 imaging modes.

the satellites in the Radarsat Constellation Mission would operate at the C-band, in a variety of geometric and polarimetric modes. One new operating mode will be a "low noise" configuration for detecting oil slicks and flat sea ice on the ocean surface, both of which can act as specular reflectors that return only very weak signals

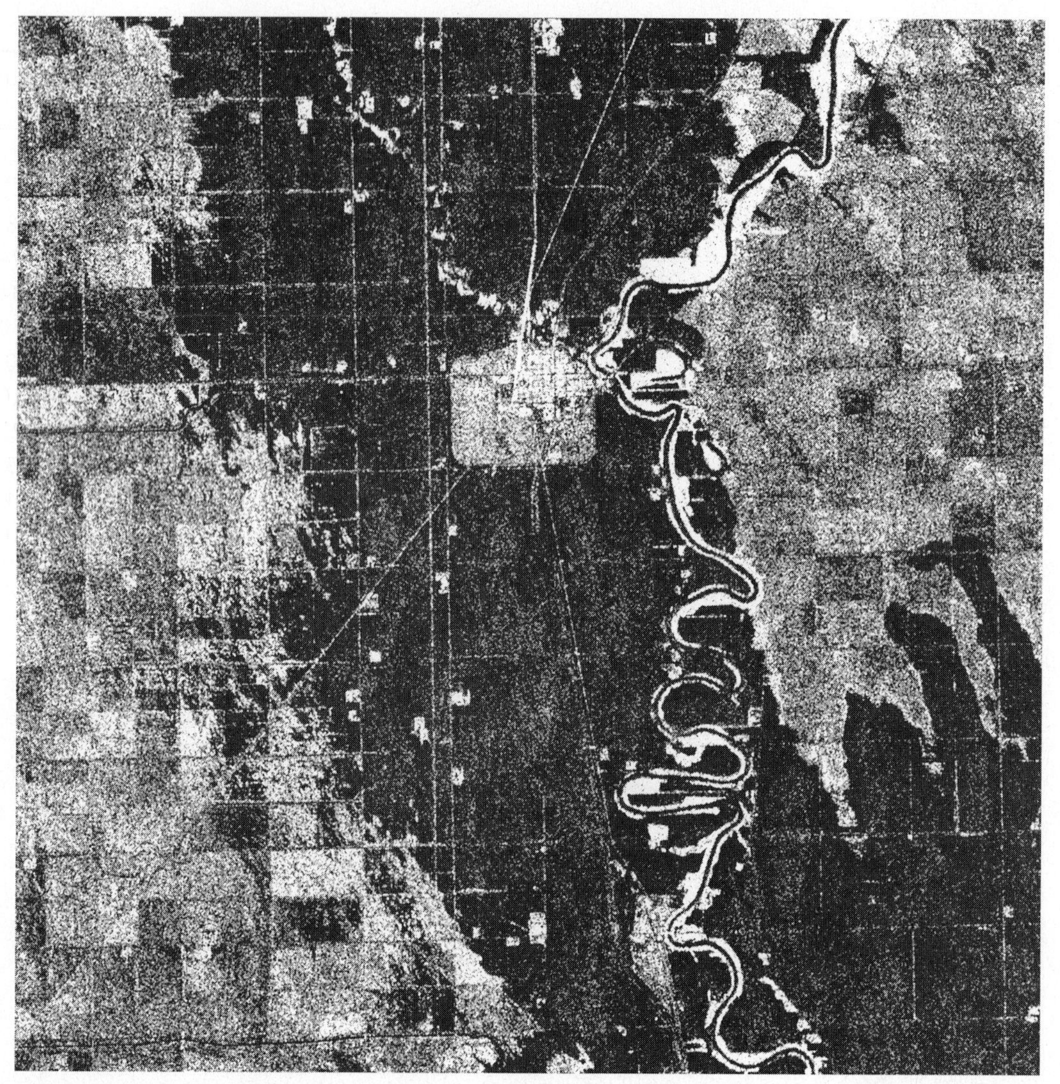

Figure 6.47 Radarsat-1 image showing flooding of the Red River, Manitoba, Canada, May 1996. Standard beam mode, incident angle 30° to 37°. Scale 1:135,000. (Canadian Space Agency 1996. Received by the Canada Centre for Remote Sensing. Processed and distributed by RADARSAT International. Enhanced and interpreted by the Canada Centre for Remote Sensing.)

to the radar sensor. With three to six identical satellites flying in formation, the revisit period will be brief, providing one-day revisit opportunities independent of cloud cover or time of day/night over much of the globe. The brief revisit interval between satellites will also facilitate the use of repeat-pass interferometry by reducing the time lag between acquisitions and correspondingly reducing the risk of temporal decorrelation.

The primary applications for which Radarsat-1 and -2 were designed, and which are guiding the design of the Radarsat Constellation Mission, include ice reconnaissance, coastal surveillance, land cover mapping, and agricultural and forestry monitoring. The near-real-time monitoring of sea ice is important for reducing the navigational risks of Arctic ships. Other uses include disaster monitoring (e.g., oil spill detection, landslide identification, flood monitoring), snow distribution mapping, wave forecasting, ship surveillance in offshore economic zones, and measurement of soil moisture. The different operating modes of the system allow both broad monitoring programs to be conducted as well as more detailed investigations using the fine resolution mode.

Figure 6.47 shows flooding of the Red River, Manitoba, Canada, in May 1996. The broad, dark area from lower right to the top of the image is smooth, standing water. The lighter-toned areas to the left and upper right are higher, nonflooded, ground. Where standing water is present under trees or bushes, corner reflection takes place and the area appears very light toned. This is especially evident near the Red River. (See also Figure 6.33 for an example of this effect.) The town of Morris can be identified as a light-toned rectangle in the flooded area. The town is protected by a levee and, as a result, has not flooded. Other, smaller areas that have not flooded (but are surrounded by water) can also be seen in this image.

6.16 TERRASAR-X, TANDEM-X, AND PAZ

On June 15, 2007, the German Aerospace Center launched *TerraSAR-X*, a new X-band radar satellite that provides imagery on demand at resolutions as fine as 1 m. A virtually identical twin satellite known as *TanDEM-X* was launched three years later, on June 21, 2010. The two systems are operated in tandem, in a bistatic configuration (Section 6.4) where one system transmits radar signals and both record the amplitude and phase of the backscattered response, effectively forming a large X-band single-pass interferometric radar. This provides on-demand topographic mapping capability anywhere on earth at 2 m accuracy, significantly better than anything previously available from space.

For imaging purposes, the two satellites each offer a variety of modes. In the standard stripmap mode, imagery is collected at approximately 3-m resolution, over a swath width of 30 km (for single polarization data) or 15 km (dual polarization). Several Spotlight modes provide 1- to 3-m resolution imagery over a 5-km to 10-km swath, again with the choice of single or dual polarization. Finally, in ScanSAR mode the instruments cover a 100-km-wide swath at a spatial resolution of 3 m in range by 18 m in azimuth.

In their tandem configuration, TerraSAR-X and TanDEM-X follow orbit tracks separated by approximately 200 to 500 m. The interferometric data collected in this configuration are being used to develop a globally uniform topographic data set with 12.5-m by 12.5-m pixel spacing. Elevations in this DEM have better than 2-m relative and 10-m absolute accuracy. When complete, it will provide a replacement for the near-global but coarser resolution topographic data from the Shuttle Radar Topography Mission, discussed in Section 6.19. The initial data collection for this new global DEM took just over one year. The results began to be made available in 2014, following a second year of data collection to improve the accuracy in areas of rough terrain.

A third satellite based on the TerraSAR-X design, and flying in the same orbit, is planned for launch in 2015. This system, known as *PAZ* ("peace" in Spanish) will further extend the opportunities for data collection by this constellation of X-band SARs, leading to very frequent revisit opportunities and more interferometric image pairs for mapping the dynamic topography of the earth's surface.

Figure 6.48 shows an example of TerraSAR-X imagery acquired in its High Resolution Spotlight mode over a copper mine in Chuquicamata, Chile. The mine, located in the Atacama Desert of northern Chile, is the largest copper mine in the world by volume. The imagery consists of a mosaic of data from two separate orbit tracks, with ground range resolutions ranging from 1.04 m to 1.17 m. The large whorled pattern at upper left, with relatively dark texture, is the open pit mine, with an access road spiraling into it. The bright dots within this pattern are vehicles working in the mine. Other features in this image include buildings, roads, and tailing piles associated with the mine operations.

Another High Resolution Spotlight mode image from TerraSAR-X is shown in Figure 6.49. This image covers a portion of Charles de Gaulle Airport, the largest airport in France, located outside Paris. The architectural design of Terminal 1, located left of the center of the image, has been compared to the shape of an octopus. As in this example, radar images of airports tend to have high contrast, due to the close proximity of large flat surfaces (which appear dark, due to specular

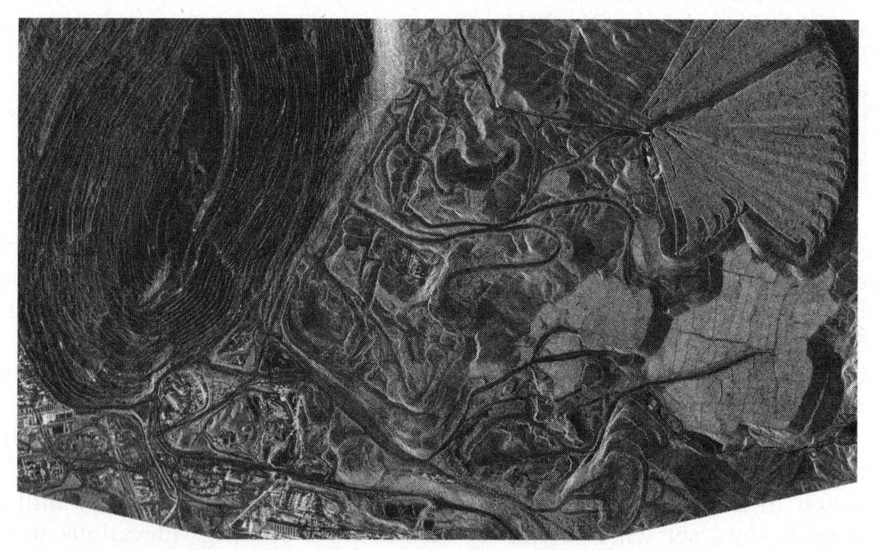

Figure 6.48 High Resolution Spotlight imagery from TerraSAR-X, over a copper mine in Chuquicamata, Chile (scale 1:44,000). Full resolution of this image ranges from 1.04 m to 1.17 m. (Copyright: DLR e.V. 2009, Distribution Airbus DS/Infoterra GmbH.)

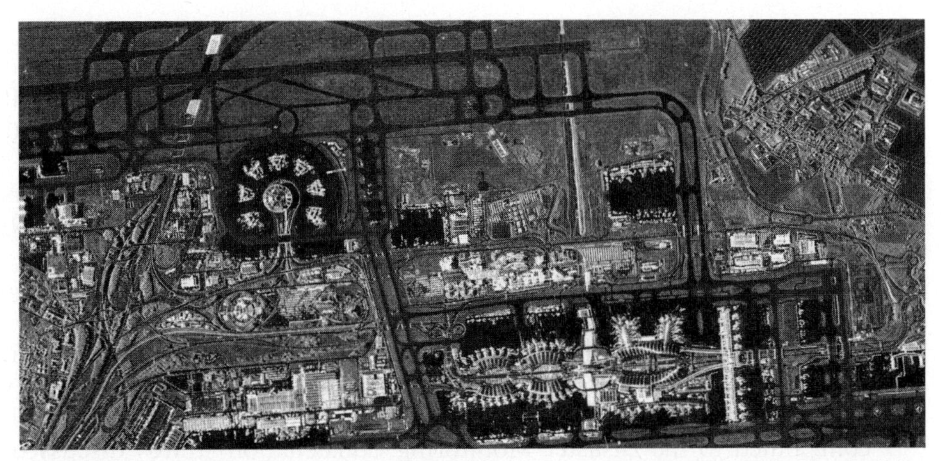

Figure 6.49 Charles de Gaulle Airport, France, shown in a TerraSAR-X High Resolution Spotlight image (scale 1:67,000). Full resolution of this image is 2.4 m. (Copyright: DLR e.V. 2009, Distribution Airbus DS/Infoterra GmbH.)

reflection) and angular, metallic structures dominated by high-intensity corner reflections (which are very bright in tone).

6.17 THE COSMO-SKYMED CONSTELLATION

While 2007 saw the launch of the first satellite in the TerraSAR-X family, it also featured the launch of the first two satellites in the Italian Space Agency's *COSMO-SkyMed* constellation, on June 8 and December 9 of that year. Two additional COSMO-SkyMed satellites were launched on October 25, 2008, and November 5, 2010. Like the German TerraSAR-X satellites, COSMO-SkyMed operates in the X-band. The satellites in this constellation share a common orbit plane, with an altitude of 620 km. The individual satellites have a 16-day repeat cycle but are spaced several days apart, providing multi-satellite revisit periods ranging from one day to several days; however, due to the cross-track pointability of the SAR antennas, most locations can be imaged every 12 hours.

Like the other recent SAR systems discussed here, COSMO-SkyMed offers a variety of imaging modes, ranging from the 1-m resolution Spotlight mode (covering a 10-km by 10-km area) to several ScanSAR modes covering a 100-km to 200-km swath at single-look resolutions of 16 m to 30 m. The system's emphasis is on rapid tasking and frequent image acquisitions for applications ranging from disaster response to defense and security. It can also be used for repeat-pass interferometry with temporal baselines as short as one day.

A second generation of COSMO-SkyMed satellites is currently being planned for launch beginning in 2017. If carried out as planned, this should ensure the continued availability of these data as the first generation of satellites in the constellation reaches the end of their design life.

6.18 OTHER HIGH-RESOLUTION SPACEBORNE RADAR SYSTEMS

As described in Chapter 5, the Indian Space Research Organization (ISRO) has devoted significant resources to developing and launching electro-optical satellites operating in the visible and infrared regions of the spectrum. At the same time, ISRO has not neglected the field of radar remote sensing. The first civilian SAR satellite built by ISRO was launched on April 26, 2012. Named *RISAT-1*, it features a C-band SAR with single, dual, and quad polarization options, and resolution modes ranging from 1 m to 50 m. It is primarily intended for agricultural and natural-resource monitoring applications.

Another agency with a long record of successful optical satellite launches that is now moving into the field of radar remote sensing is the UK's Surrey Satellite Technology Ltd. (SSTL), whose typically lightweight and nimble satellites have contributed to the Disaster Monitoring Constellation (DMC) discussed in Chapters 5 and 8. SSTL is currently preparing its first NOVASAR-S satellite, an S-band radar system that the company hopes will be the first of a series of radar satellites. While radar systems are typically massive in size and in power requirements, NOVASAR-S would be less than a quarter of the size of India's RISAT-1. This, in turn, would lead to greatly reduced launch costs, making a multi-satellite constellation more affordable.

If launched as planned, NOVASAR-S would be the first S-band radar satellite since Almaz-1 ceased operations more than two decades ago. Compared to the extensive body of research on the scattering behavior of shorter and longer wavelength C- and L-band radar systems, there have been relatively few studies at this intermediate wavelength. The NOVASAR-S would be able to collect up to three of the four linear polarization combinations (*HH, HV, VH,* and *VV*) during any given image acquisition, an improvement over the dual-polarization mode but not offering the theoretically complete measurement of polarimetric scattering that would be provided by a fully quad-pol system.

Given that many weather-related natural disasters (including tropical storms, floods, and tornadoes) are often accompanied by heavy cloud cover, the addition of one or more radar satellites to the DMC would certainly strengthen that constellation's role in disaster monitoring and response. The availability of an S-band radar system to complement the other X-, C-, and L-band radar satellites currently operational or planned for launch would also likely be beneficial for agricultural and natural resources management applications, although more research is needed in this area.

At the opposite extreme from the numerous short-wavelength, high-resolution SAR systems described in the preceding sections, NASA's planned *Soil Moisture Active Passive* (SMAP) mission will include a long-wavelength L-band radar with a very coarse spatial resolution and a wide field of view (swath width 1000 km). While the single-look resolution for SMAP's radar will vary widely across this wide swath, its data will be distributed with multilook processing resampled to a uniform 1-km grid.

In combination with a passive microwave radiometer that shares the same platform, SMAP's radar will help measure soil moisture and moisture state (frozen or liquid). These data are intended to contribute to systematic global monitoring of the water cycle, climate, and water exchanges among the soil column, ecosystems, and the atmosphere. It will also support forecasting of weather, floods/droughts, agricultural productivity, and other processes that involve soil moisture and soil state.

6.19 SHUTTLE RADAR TOPOGRAPHY MISSION

The *Shuttle Radar Topography Mission (SRTM)* was a joint project of the National Imagery and Mapping Agency (NIMA) and NASA to map the world in three dimensions. During a single Space Shuttle mission on February 11 to 22, 2000, SRTM collected single-pass radar interferometry data covering 119.51 million km^2 of the earth's surface, including over 99.9% of the land area between 60° N and 56° S latitude. This represents approximately 80% of the total land surface worldwide and is home to nearly 95% of the world's population.

The C-band and X-band antennas from the 1994 SIR-C/X-SAR shuttle missions (Section 6.11) were used for data collection. To provide an interferometric baseline suitable for data acquisition from space, a 60-m-long rigid mast was extended when the shuttle was in orbit, with a second pair of C-band and X-band antennas located at the end of the mast. The primary antennas in the shuttle's payload bay were used to send and receive data, while the outboard antennas on the mast operated only in receiving mode. The extendible mast can be seen during testing prior to launch in Figure 6.50. In Figure 6.50*a* most of the mast is stowed within the canister visible in the background. In Figure 6.50*b* the mast is extended to its full length. The two outboard antennas were not installed on the mast at the time of these tests; during the mission they were mounted on the triangular plate visible at the end of the mast in Figure 6.50. An artist's rendition of the shuttle in orbit during SRTM is provided in Figure 6.51. This illustration shows the position and orientation of the various SRTM components, including the main antenna inside the shuttle's payload bay; the canister for storage of the mast, the mast itself, and the outboard antennas at the end of the mast.

The system collected 12 terabytes of raw data during the 11-day mission, a volume of data that would fill over 15,000 CD-ROMs. Processing this volume of data took two years to complete. The elevation data are being distributed by the U.S. Geological Survey. The SRTM processor produces digital elevation models with a pixel spacing of 1 arcsecond of latitude and longitude (about 30 m[a]). The absolute horizontal and vertical accuracy of the data are better than 20 and 16 m,

[a]Prior to 2014, the data for areas outside the U.S. were aggregated to 3 arcseconds (about 90 m).

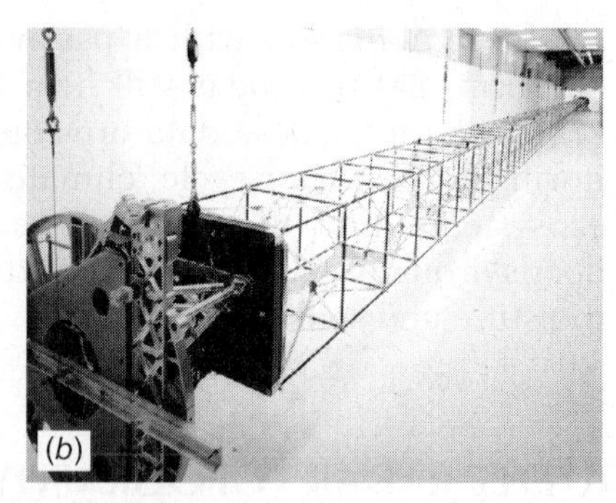

Figure 6.50 The SRTM extendable mast during prelaunch testing: (*a*) view of the mast emerging from the canister in which it is stowed during launch and landing; (*b*) the mast fully extended. (Courtesy NASA/JPL/Caltech.)

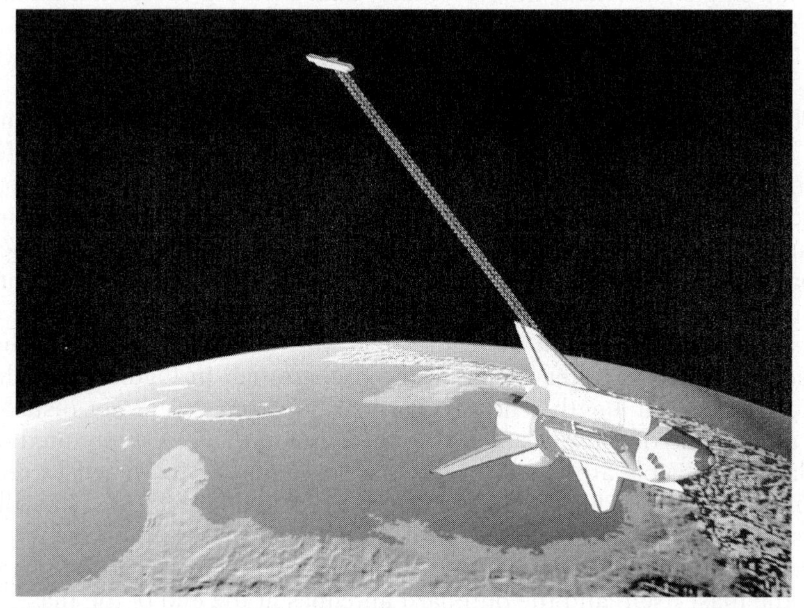

Figure 6.51 Artist's rendition of the shuttle in orbit during SRTM, showing the positions of the main antenna inside the payload bay, the canister, the mast, and the outboard antennas. (Courtesy NASA/JPL/Caltech.)

respectively. In addition to the elevation data, the SRTM processor produces orthorectified radar image products and maps showing the expected level of error in the elevation model.

Figure 6.52 shows a perspective view of a DEM for the Los Angeles metropolitan area. The DEM was derived from interferometric analysis of SRTM imagery,

Figure 6.52 Perspective view of a DEM of the Los Angeles area, derived from SRTM interferometric radar data. A Landsat-7 ETM+ image has been draped over the DEM to show land cover patterns. (Courtesy NASA/JPL/Caltech.)

and a Landsat-7 ETM+ image was draped over the DEM. This view is dominated by the San Gabriel Mountains, with Santa Monica and the Pacific Ocean in the lower right and the San Fernando Valley to the left.

Technical problems during the shuttle mission caused 50,000 km^2 of the targeted land area to be omitted by SRTM. These omitted areas represent less than 0.01% of the land area intended for coverage. All the omitted areas were located within the United States, where topographic data are already available from other sources.

As the first large-scale effort to utilize single-pass radar interferometry for topographic mapping from space, the project has proved to be highly successful. The resulting topographic data and radar imagery represent a unique and highly valuable resource for geospatial applications. However, with the anticipated near future availability of globally consistent, higher-resolution spaceborne radar topographic data such as the database now being compiled from TerraSAR-X and TanDEM-X interferometry (Section 6.16), the groundbreaking SRTM data set will likely come to be seen as an important early stepping stone on the path to a world where continually updated topographic data are available at increasingly high resolution across the whole earth. Figure 6.53 compares the resolution of the 12.5-m resolution globally consistent digital elevation data set from TerraSAR-X/ TanDEM-X to the pre-2014 3-arcsecond (90-m) resolution SRTM elevation data, for a site near Las Vegas, NV. Just as a decade ago the SRTM project represented a dramatic improvement in the consistency and availability of digital elevation data worldwide, the high-resolution DEMs currently being produced from

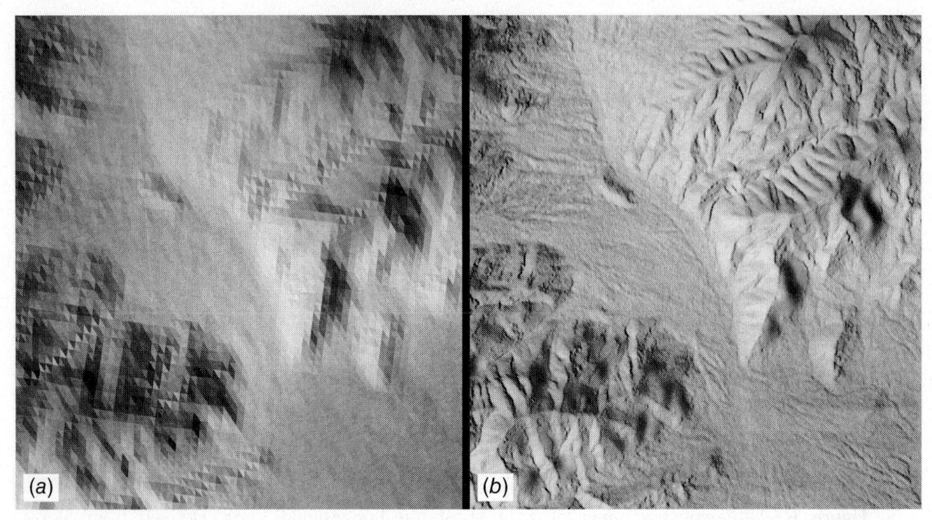

Figure 6.53 Comparison of digital elevation data from (*a*) the 3-arcsecond (90-m) resolution SRTM data set to (*b*) the 12.5-m resolution elevation data currently being collected worldwide by TerraSAR-X and TanDEM-X. (Courtesy German Aerospace Center – DLR.)

spaceborne radar satellites will represent a similar quantum jump in resolution and detail over the 3-arcsecond SRTM data set.

6.20 SPACEBORNE RADAR SYSTEM SUMMARY

The proliferation of new radar satellites and multi-satellite constellations, and the diversity of operating modes offered by all of these systems, render the task of

TABLE 6.7 Characteristics of Major Past Operational Spaceborne SAR Systems

Years in operation	Satellite	Country	Radar band	Pol. mode	Look angle	Resolution, m
1991–1992	Almaz-1	Soviet Union	S	*HH*	20–70°	10–30
1991–2000	ERS-1	ESA	C	*VV*	23°	30
1992–1998	JERS-1	Japan	L	*HH*	35°	18
1995–2011	ERS-2	ESA	C	*VV*	23°	30
1995–2013	Radarsat-1	Canada	C	*HH*	10–60°	8–100
2002–2012	Envisat	ESA	C	Dual	14–45°	30–1000
2006–2011	ALOS	Japan	L	Quad	10–51°	10–100

TABLE 6.8 Characteristics of Major Current Operational Spaceborne SAR Systems

Year of launch	Satellite	Country	Radar band	Pol. mode	Look angle	Resolution, m
2007	TerraSAR-X	Germany	X	Dual	15–60°	1–18
2007	COSMO-SkyMed 1	Italy	X	Quad	20–60°	1–100
2007	COSMO-SkyMed 2	Italy	X	Quad	20–60°	1–100
2007	Radarsat-2	Canada	C	Quad	10–60°	1–100
2008	COSMO-SkyMed 3	Italy	X	Quad	20–60°	1–100
2010	TanDEM-X	Germany	X	Dual	15–60°	1–18
2010	COSMO-SkyMed 4	Italy	X	Quad	20–60°	1–100
2012	RISAT-1	India	C	Quad	12–50°	1–50
2014	ALOS-2	Japan	L	Quad	10–60°	1–100
2014	Sentinel 1A	ESA	C	Dual	20–47°	5–40

TABLE 6.9 Characteristics of Major Planned Future Spaceborne SAR Systems

Expected launch	Satellite	Country	Radar band	Pol. mode	Look angle	Resolution, m
2015	SEOSAR/Paz	Spain	X	Dual	15–60°	1–18
2015	NOVASAR-S	UK	S	Tri[a]	15–70°	6–30
2015	SMAP	US	L	Tri[b]	35–50°	1000+
2016	Sentinel 1B	ESA	C	Dual	20–47°	5–40
2017	COSMO-SkyMed 2nd Generation-1	Italy	X	Quad	20–60°	1–100
2017	COSMO-SkyMed 2nd Generation-2	Italy	X	Quad	20–60°	1–100
2018	Radarsat Constellation 1, 2, 3	Canada	C	Quad	10–60°	3–100

[a]While NOVASAR-S could collect data in any of the four polarization combinations *HH, HV, VH,* and *VV*, it could record at most three of these polarizations during any given image acquisition.
[b]The SMAP radar will have fixed *HH, HV,* and *VV* polarizations.

summarizing the diversity of spaceborne radar sensors increasingly difficult. Tables 6.7, 6.8, and 6.9 are presented to assist in this process by compiling the essential characteristics of past, current, and planned future spaceborne radars. Note that there has been a general trend toward increasing design sophistication, including multiple polarizations, multiple look angles, and multiple combinations of resolution and swath width. However, none of these systems operates at more than one wavelength band—a fact that testifies to the technical challenges involved in designing spaceborne multi-wavelength radar systems.

6.21 RADAR ALTIMETRY

The radar systems described in the preceding sections of this chapter all are side-looking instruments, designed to collect image data on a pixel-by-pixel basis across a wide spatial area. Another class of radar remote sensing systems, however, is designed to look directly downward and measure the precise distance from the radar antenna to the earth's surface. These sensors are referred to as radar altimeters, and while they do not normally produce image data per se, the spatial data they collect on the earth's oceans, lakes, ice sheets, land surface, and seafloor are widely used in the geosciences and in practical applications ranging from water resources management to monitoring and forecasting the behavior of the El Nino/Southern Oscillation.

The basic principle behind radar altimetry is quite simple: The radar antenna transmits a microwave pulse downward to the surface and measures the elapsed time for the returning pulse. This in turn can be used to calculate the distance from the antenna to the surface. If the position of the antenna is known accurately, the elevation of the surface can be determined with a similar degree of accuracy. While this appears straightforward, in practice there are several challenges to the design and operation of radar altimeters.

Many applications of radar altimetry require centimeter-scale resolution in the vertical dimension. To ensure this, the transmitted pulse duration must be very brief, on the order of nanoseconds. This in turn would require an unfeasibly large power for the transmitted signal. Radar altimeters (and many imaging radars as well) circumvent this problem by using sophisticated signal processing methods, such as a pulse compression approach that involves modulating the signal (referred to as "chirp").

During the signal's transit from the antenna to the surface, it fans outward, such that for spaceborne radar altimeters the normal diameter of the pulse's footprint on the surface may be 5 to 10 km or more. The shape of the returning signal recorded by the antenna (referred to as the *waveform*) is a composite of thousands of individual echoes from scattering nodes within this broad footprint. The roughness of the surface will affect the waveform, such that as the surface roughness increases, it becomes more challenging to identify a specific elevation for the altimeter's footprint. Figure 6.54 illustrates the interaction between a transmitted pulse from a radar altimeter and an ideal, flat surface. At time $t = 1$, the pulse has not yet intersected the surface. At $t = 2$, only the very center of the wavefront is being scattered back, producing a weak but rapidly increasing signal. At $t = 3$, a larger area is being illuminated, and because this area of illumination is still directly beneath the sensor, it sends back a strong echo. At $t = 4$ and $t = 5$, the

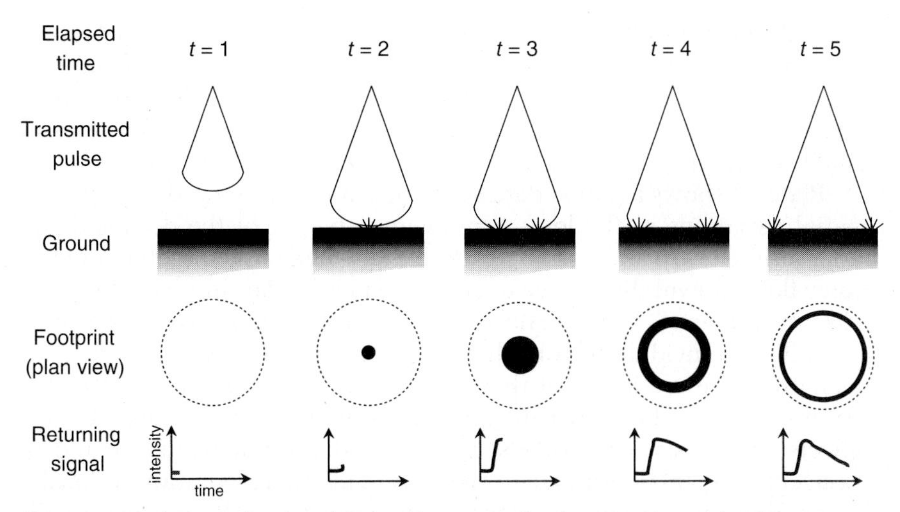

Figure 6.54 Schematic diagram of the signal transmitted and received by a radar altimeter.

wavefront is expanding outward away from the center of the altimeter's footprint, and the returned signal consequently becomes weaker. This produces the characteristic waveform of a radar altimeter, with a steeply rising leading edge followed by a gradually tapering trailing edge. Over a more complex surface with a rougher texture, the front of the waveform will be less steep, the peak will be less obvious, and the trailing edge will be noisier (Figure 6.55).

Since the early 1990s spaceborne radar altimeters have been operated from a variety of platforms, including the ERS-1, ERS-2, and Envisat satellites (Section 6.13); the *Topex/Poseidon* mission (1992–2005) and its successors *Jason-1* (2001–present) and *OSTM/Jason-2* (2008–present), all three of which are joint efforts of the United States and France); and others. Of particular note, the ESA's

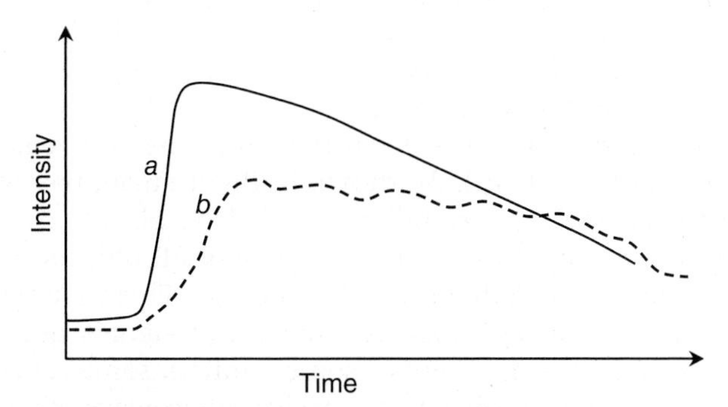

Figure 6.55 Returning pulses from radar altimeter measurements over a flatter surface (solid line *a*) and a rougher surface (dashed line *b*).

radar altimetry satellite *Cryosat-2* was successfully launched on April 8, 2010. The Cryosat-2 mission is primarily focused on monitoring the seasonal and long-term dynamics of the earth's land ice and sea ice, including the large polar ice sheets of Greenland and Antarctica, smaller ice caps and glaciers elsewhere, and the seasonally expanding and contracting Arctic and Antarctic sea ice.

Plate 29 shows a global data set of ocean bathymetry, derived from analysis of spatial variations in the long-term average height of the oceans as measured by radar altimeters, primarily Topex/Poseidon. While one might assume that because water flows downhill, the ocean surface must be flat, in reality there are irregularities in "sea level" at both short and long time scales. Sea level temporarily rises and falls in individual regions due to changes in local atmospheric pressure, wind strength and direction, and the dynamics of ocean currents. Over long periods, sea level tends to be higher in some areas than others, because the ocean surface reflects the presence of ridges, valleys, and abyssal plains on the ocean floor. (Conceptually, this is similar to the finer-scale representation of local bathymetry in the English channel as seen in the Seasat-1 SAR imagery in Figure 6.35.) Again, the radar altimeters do not see through the ocean directly; instead, they map the surficial expression of broad-scale submarine features deep below the surface.

6.22 PASSIVE MICROWAVE SENSING

Operating in the same spectral domain as radar, passive microwave systems yield yet another "look" at the environment—one quite different from that of radar. Being passive, these systems do not supply their own illumination but rather sense the naturally available microwave energy within their field of view. They operate in much the same manner as thermal radiometers and scanners. In fact, passive microwave sensing principles and sensing instrumentation parallel those of thermal sensing in many respects. As with thermal sensing, blackbody radiation theory is central to the conceptual understanding of passive microwave sensing. Again as in thermal sensing, passive microwave sensors exist in the form of both radiometers and scanners. However, passive microwave sensors incorporate antennas rather than photon detection elements.

Most passive microwave systems operate in the same spectral region as the shorter wavelength radar (out to 30 cm). As shown in Figure 6.56, passive microwave sensors operate in the low energy tail of the 300 K blackbody radiation curve typifying terrestrial features. In this spectral region, all objects in the natural environment emit microwave radiation, albeit faintly. This includes terrain elements and the atmosphere. In fact, passive microwave signals are generally composed of a number of source components—some emitted, some reflected, and some transmitted. Over any given object, a passive microwave signal might include (1) an emitted component related to the surface temperature and material attributes of the object, (2) an emitted component coming from the atmosphere,

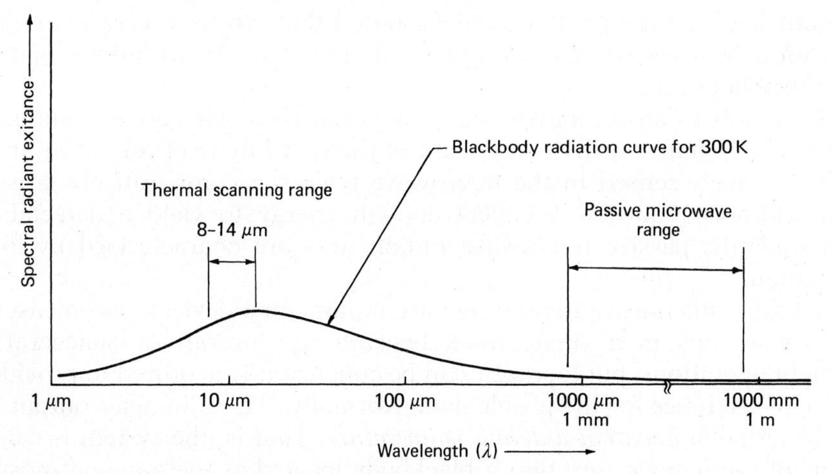

Figure 6.56 Comparison of spectral regions used for thermal versus passive microwave sensing.

(3) a surface-reflected component from sunlight and skylight, and (4) a transmitted component having a subsurface origin. In short, the intensity of remotely sensed passive microwave radiation over any given object is dependent not only on the object's temperature and the incident radiation but also on the emittance, reflectance, and transmittance properties of the object. These properties in turn are influenced by the object's surface electrical, chemical, and textural characteristics, its bulk configuration and shape, and the angle from which it is viewed.

Because of the variety of its possible sources and its extremely weak magnitude, the signal obtained from various ground areas is "noisy" compared to that provided by cameras, scanners, or radars. The interpretation of this signal is thus much more complex than that of the other sensors discussed. In spite of the difficulties, passive microwave systems are widely used in applications as diverse as measuring atmospheric temperature profiles, assessing snow water content, tracking the extent of sea ice, and analyzing subsurface variations in soil, water, and mineral composition.

Microwave Radiometers

The basic configuration of a typical microwave radiometer system is straightforward. Scene energy is collected at the antenna. A microwave switch permits rapid, alternate sampling between the antenna signal and a calibration temperature reference signal. The low strength antenna signal is amplified and compared with that of the internal reference signal. The difference between the antenna signal and the reference signal is electronically detected and input to some mode of readout and recording. (It should be noted that we have greatly simplified the operation of a microwave radiometer and that many variations of the design described here exist.)

Common to all radiometer designs is the trade-off between antenna beamwidth and system sensitivity. Because of the very low levels of radiation available to be passively sensed in the microwave region, a comparatively large antenna beamwidth is required to collect enough energy to yield a detectable signal. Consequently, passive microwave radiometers are characterized by low spatial resolution.

Profiling microwave radiometers are nonimaging devices that measure microwave emissions in a single track beneath the aircraft or spacecraft. During daylight operations, photography can be concurrently acquired to provide a visual frame of reference for the profile data. Normally, the radiometer output is expressed in terms of *apparent antenna temperature*. That is, the system is calibrated in terms of the temperature that a blackbody located at the antenna must reach to radiate the same energy as was actually collected from the ground scene.

Imaging Microwave Radiometers

Conceptually, an *imaging microwave radiometer* operates on the same principle as a profiling (nonimaging) radiometer, except that the antenna's field of view is rotated across the line of flight. Both airborne and spaceborne imaging microwave radiometers have been developed. As with radar systems, these passive microwave systems can be operated day or night, under virtually all weather conditions. They can also be designed to sense in multiple spectral bands within the microwave portion of the spectrum. By using spectral bands located in atmospheric "windows," passive microwave systems can be used to measure the properties of the earth's surface. For meteorological applications, other spectral bands may be positioned to measure atmospheric temperature profiles and to determine the atmospheric distribution of water and ozone. Like radar systems, imaging microwave radiometers can also be designed to measure the polarization of emitted microwaves (Kim, 2009).

Figure 6.57 shows three segments of a strip of imagery acquired with an airborne scanning passive microwave radiometer, or scanner. The image covers a transect running from Coalinga, California, visible at the western (left) end, to Tulare Lake (dry, now in agriculture) at the eastern (right) end in California's San Joaquin Valley. (Note that the image has the tonal and geometric appearance of thermal scanner imagery. However, in this image bright areas are radiometrically "cold" and dark areas are "warm.") Agricultural fields are visible along the length of the transect. The striping in several of the fields is due to irrigation. The darker fields are natural vegetation or dry bare soil. Radiance measurements made using this type of data have been found to relate quite systematically to the moisture content of the top 50 mm of the soil.

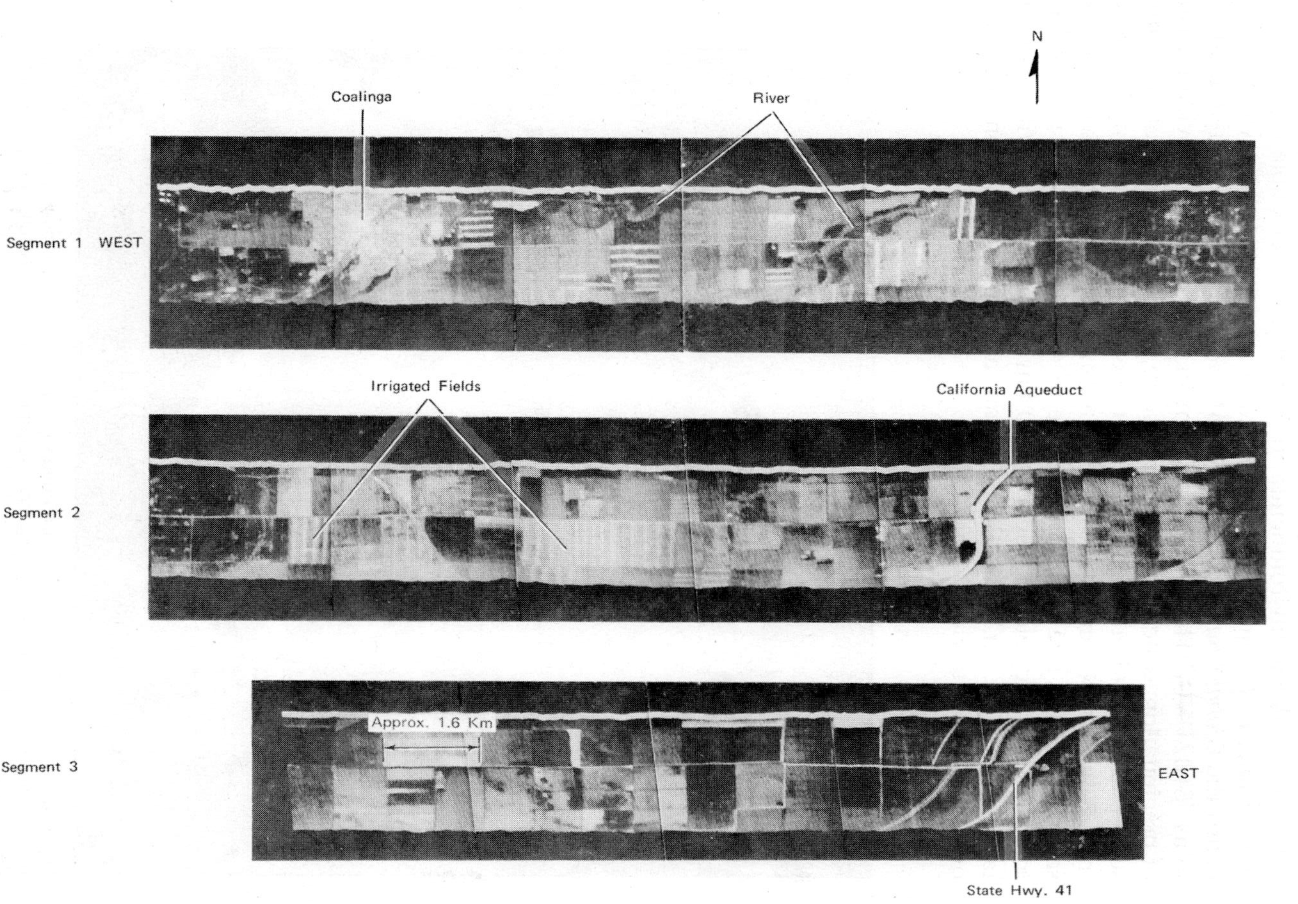

Figure 6.57 Passive microwave image transect, Coalinga to Tulare Lake, CA, midsummer, 760 m flying height. (Courtesy Geography Remote Sensing Unit, University of California–Santa Barbara, and Naval Weapons Center, China Lake, CA.)

The coarse resolution of spaceborne passive microwave sensors does not preclude their value for synoptic measurement of the earth's surface and atmospheric properties. In fact, such systems represent an important resource for coarse-scale global environmental monitoring. One example of such a system is the *Advanced Microwave Scanning Radiometer 2 (AMSR2),* which is carried on board the *Global Change Observation Mission-1* (GCOM-1) satellite, also referred to as "SHIZUKU." The AMSR2 instrument is designed to provide measurements of precipitation, water vapor, soil moisture, snow cover, sea surface temperature, sea ice extent, and other parameters. It operates at six wavelength bands ranging from 3.3 mm to 4.3 cm (89 to 7 GHz). A 2-m diameter antenna on the instrument scans across the line of flight at a rate of 40 revolutions per minute, covering a total across-track field of view of 1450 km on the surface. With a sampling distance of 5 to 10 km, AMSR2 images are appropriate for monitoring conditions over continental to global scales.

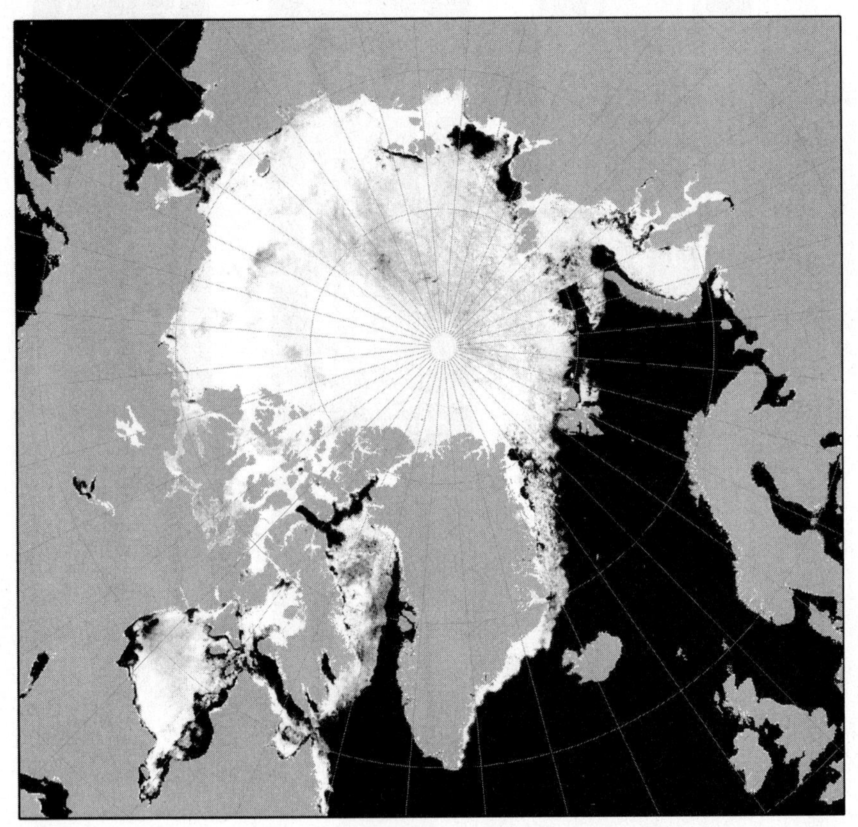

Figure 6.58 Map showing sea ice distribution in the Arctic Ocean, derived from AMSR2 passive microwave radiometer imagery, June. (Courtesy JAXA and Institute of Environmental Physics, University of Bremen.)

Figure 6.58 shows a map of sea ice in the Arctic Ocean, derived from AMSR2 imagery acquired on June 10, 2013. Northern Greenland and the Canadian Arctic archipelago are located at the lower center and left of this image, with Hudson Bay in the far lower left corner, while the northern coast of Siberia and its offshore islands are at the upper center and upper right. The methods used to derive sea ice concentration from AMSR2 microwave radiometry data are based on those described in Spreen et al. (2008) for an earlier generation of imaging microwave radiometer, AMSR-E.

In this map, the brightest tones represent 100% cover of sea ice, while darker tones represent lower sea ice concentrations. In the Arctic, some ice persists from year to year in the central part of the basin, while the outer extent of the ice expands and contracts with the seasons. The extent of ice cover in the Arctic has been gradually decreasing since 1979, and recent years have seen particularly striking declines in sea ice in this region. Some climate models suggest that the Arctic Ocean will become nearly or completely free of multiyear ice later in this century, while extrapolation of trends from the past three decades suggests that the disappearance of multiyear ice could happen much earlier. Sea ice plays an important role in the global climate cycle, modulating transfers of salinity within the ocean and transfers of heat and water vapor from the ocean to the atmosphere. Thus, efforts to monitor its extent are an essential component of research programs in global climate variability and climate change.

6.23 BASIC PRINCIPLES OF LIDAR

Lidar (which stands for *light detection and ranging*), like radar, is an active remote sensing technique. This technology involves transmitting pulses of laser light toward the ground and measuring the time of pulse return. The return time for each pulse back to the sensor is processed to calculate the distances between the sensor and the various surfaces present on (or above) the ground.

The use of lidar for accurate determination of terrain elevations began in the late 1970s. Initial systems were profiling devices that obtained elevation data only directly under the path of an aircraft. These initial laser terrain systems were complex and not necessarily suited for cost-effective terrain data acquisition over large areas, so their utilization was limited. (Among the primary limitations was the fact that neither airborne GPS nor Inertial Measurement Units, or IMUs, were yet available for accurate georeferencing of the raw laser data.) One of the more successful early applications of lidar was the determination of accurate water depths. In this situation the first reflected return records the water surface, closely followed by a weaker return from the bottom of the water body. The depth of the water can then be calculated from the differential travel time of the pulse returns (Figure 6.59).

The advantages of using lidar to supplement or replace traditional photogrammetric methodologies for terrain and surface feature mapping stimulated

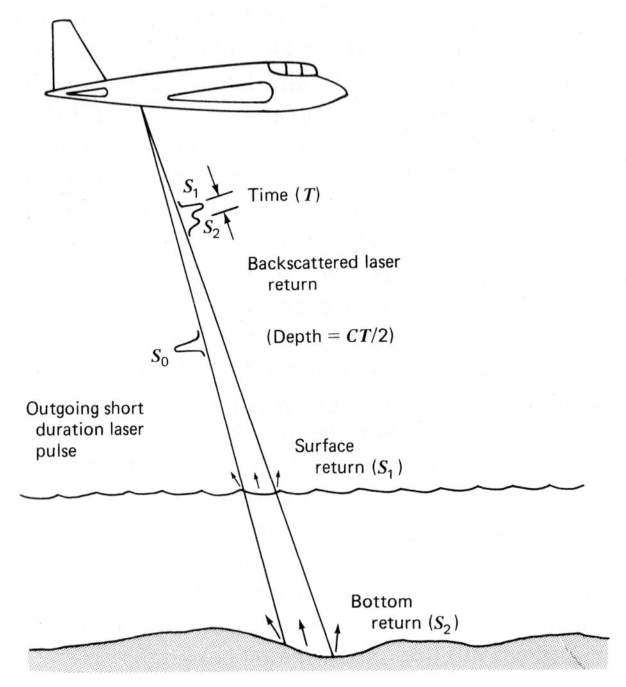

Figure 6.59 Principle of lidar bathymetry. (Adapted from Measures, 1984.)

the development of high-performance scanning systems. Among their advantages, these systems afford the opportunity to collect terrain data about steep slopes and shadowed areas (such as the Grand Canyon) and inaccessible areas (such as large mud flats and ocean jetties).

Modern lidar acquisition begins with an aircraft equipped with high-precision GPS (for X, Y, Z sensor location), an IMU (for measuring the angular orientation of the sensor with respect to the ground), a rapidly pulsing (10,000 to over 100,000 pulses/sec) laser, a highly accurate clock, substantial onboard computer support, reliable electronics, and robust data storage. Flight planning for lidar acquisition requires special considerations. Flights are conducted using a digital flight plan without ground visibility, often at night. Individual flight lines are planned with sufficient overlap (30 to 50%) to assure that data gaps do not occur in steep terrain. Areas with dense vegetation cover or steep terrain usually require a narrow field of view, so that most of the lidar pulses are pointing nearly straight down. These areas also typically require a higher density of laser pulses per square meter. This density (sometimes also expressed as a distance between pulses, or *post spacing*) is derived from the altitude and speed of the aircraft, the scanning angle, and the scan rate. Figure 6.60 portrays the operation of an airborne lidar scanning system.

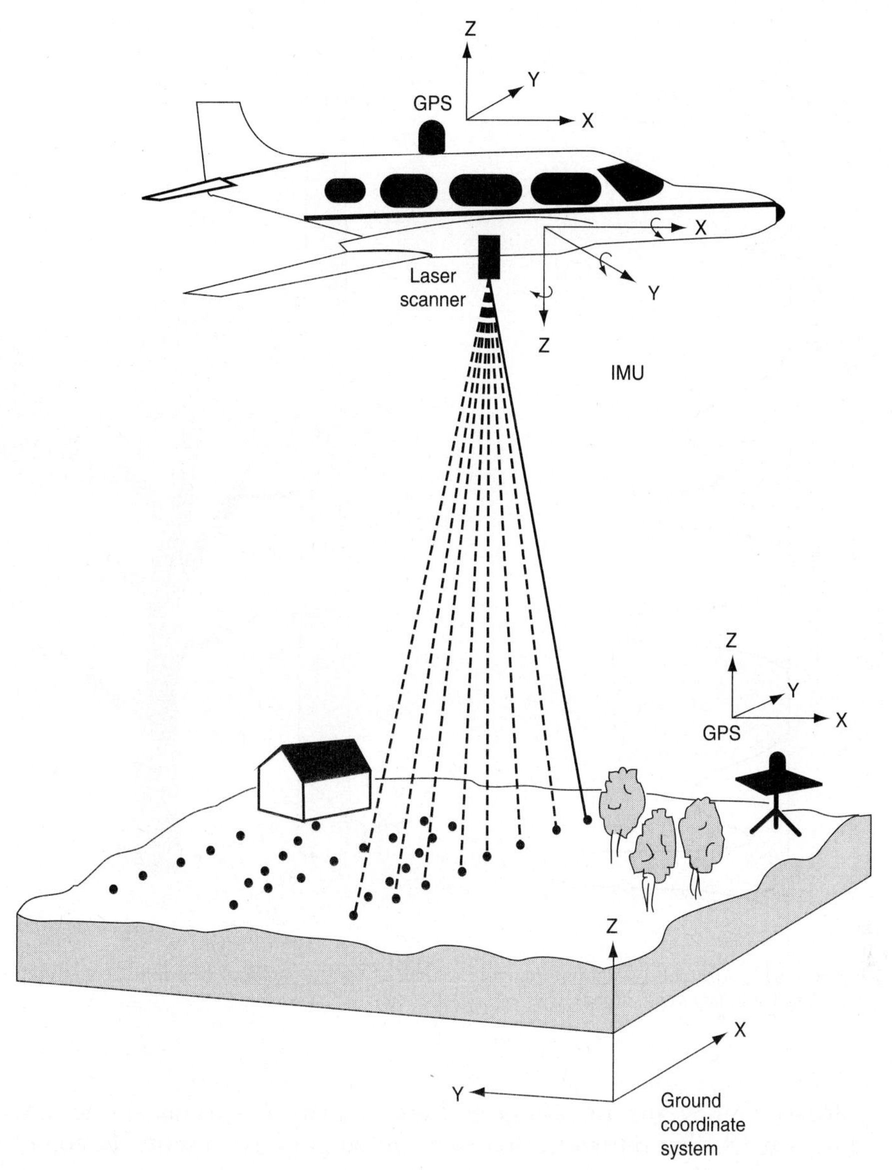

Figure 6.60 Components of an airborne scanning lidar system. (Courtesy EarthData International and Spencer B. Gross, Inc.)

In addition to rapid pulsing, modern systems are able to record five or more returns per pulse, which permits these systems to discriminate not only such features as a forest canopy and bare ground but also surfaces in between (such as the intermediate forest structure and understory). Figure 6.61 illustrates a theoretical pulse emitted from a lidar system and traveling along a line of sight through

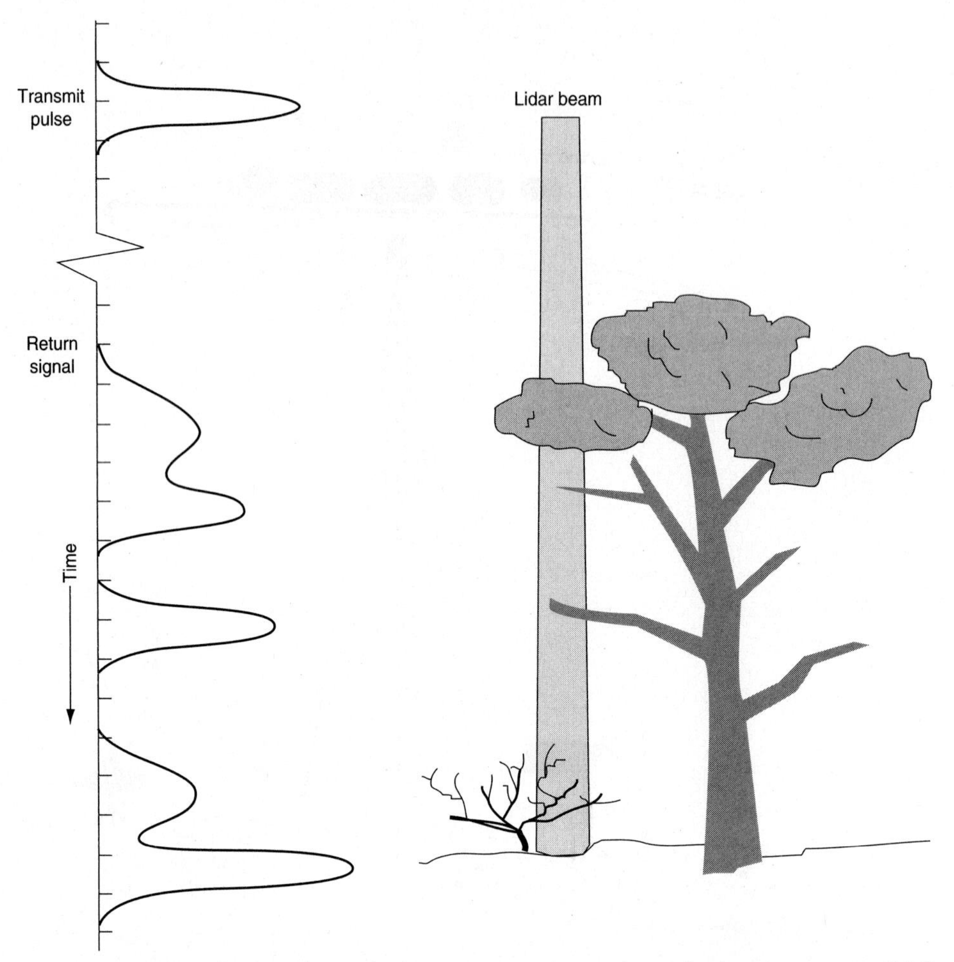

Figure 6.61 Lidar pulse recording multiple returns as various surfaces of a forest canopy are "hit." (Courtesy EarthData International.)

a forest canopy and recording multiple returns as various surfaces are "hit." In this case, the first return from a transmitted pulse represents the top of the forest canopy at a given location. The last return may represent the ground surface, if there are sufficient gaps in the canopy for portions of the transmitted pulse to reach the ground and return, or a point above the ground surface if the canopy is too dense. In urban areas, unobstructed surfaces such as building roofs will produce a single return, while areas with trees present will produce multiple returns from the tree canopy and ground surface.

Depending on the surface complexity (e.g., variable vegetation heights, terrain changes) and other mission parameters, lidar data sets can be remarkably large.

For coarser-resolution, large-area mapping, a lidar acquisition may produce 0.5 to 2 points per square meter (500,000 to 2 million points per square km), while detailed mapping of small areas where dense vegetation is present may require 10 to 50 (or more) points per square meter (10 to 50 million points per square km) from multiple-return lidar systems.

As with any airborne GPS activity, the lidar system requires one or more surveyed ground base stations to be established in or near the project area for differential correction of the airborne GPS data. In addition, a calibrated alignment process for the GPS position of the sensor and the orientation parameters is required to assure and verify the accuracy of the lidar data sets.

Lidar data are typically stored and distributed as a LAS file, a digital format sponsored by the American Society for Photogrammetry and Remote Sensing (ASPRS) and now the industry standard. For ease of use, the data from overlapping lidar flight lines should typically be merged into a seamless mosaic and then divided into numerous individual tiles, each stored in its own LAS file. (This is recommended due to the computational intensity of many 3D lidar visualization and processing methods, and the fact that the LAS file format does not include the detailed internal indexing needed to rapidly extract the subset of points needed at any one time.)

6.24 LIDAR DATA ANALYSIS AND APPLICATIONS

Raw lidar data collected by the sensor are processed using GPS differential correction and filtered for noise removal, then prepared as a file of X, Y, Z points. These can be visualized as a *point cloud* (Section 1.5 and Chapter 3) with the reflecting point for each returning pulse represented as a point in three-dimensional space. Figure 6.62 shows a lidar point cloud for a stand of needle-leaved evergreen trees in the Sierra Nevadas. The overall shape of each individual tree crown can be discerned within the point cloud. Before further processing or analysis, these lidar data points are usually classified to isolate different types of features and surfaces. The simplest classification would categorize points as "ground" or "non-ground," but it is preferable to provide more detailed categories such as ground, buildings, water, low and high vegetation classes, and noise. Typically, this process involves a mixture of automated classification and manual editing. Once the lidar points have been classified, they can be used to create a variety of derived products, ranging from DEMs and other three-dimensional models to high-resolution contour maps. A distinct advantage to lidar is that all the data are georeferenced from inception, making them inherently compatible with GIS applications.

Figure 6.63 shows a small portion of an airborne lidar data set collected over the coast of West Maui, Hawaii. The lidar data were collected using a Leica Geosystems ALS-40 Airborne Laser Scanner from an altitude of 762 m, over a 25° field of view. The scan rate was 20,000 pulses per second, and the lidar ground

Figure 6.62 Lidar point cloud over a forest in the Sierra Nevada, California. (Courtesy of Qinghua Guo and Jacob Flanagan, UC Merced.)

points were spaced 2 m apart. The position and orientation of the system was measured in-flight using GPS and an IMU, with a stationary GPS located at a nearby airport for differential correction (see Chapter 1). The resulting elevation measurements have a vertical accuracy of 16 cm root mean square (RMS) error.

Many small topographic features can be identified in the lidar data shown in Figure 6.63. The left (west) side of the lidar swath includes a narrow strip of the ocean surface, on which individual waves are visible. A narrow section of land between the shoreline and the Honoapiilani coastal highway shows many small streets, houses, and trees. Across the highway, lands formerly used for growing sugar cane extend up-slope, interrupted by narrow, steep-sided gullies. At the northern end of the image, numerous features associated with a golf course at Kapalua can be identified. The inset (with dimensions of 1000 m by 2200 m) shows an airport runway, numerous small structures, and individual trees and shrubs located in one of the large gullies.

Plate 28 shows an airborne lidar image of the coastline around Reid State Park in Sagadahoc County, Maine. The lidar data were acquired in May, with a spacing of 2 m between lidar points on the surface. The data were used to create a

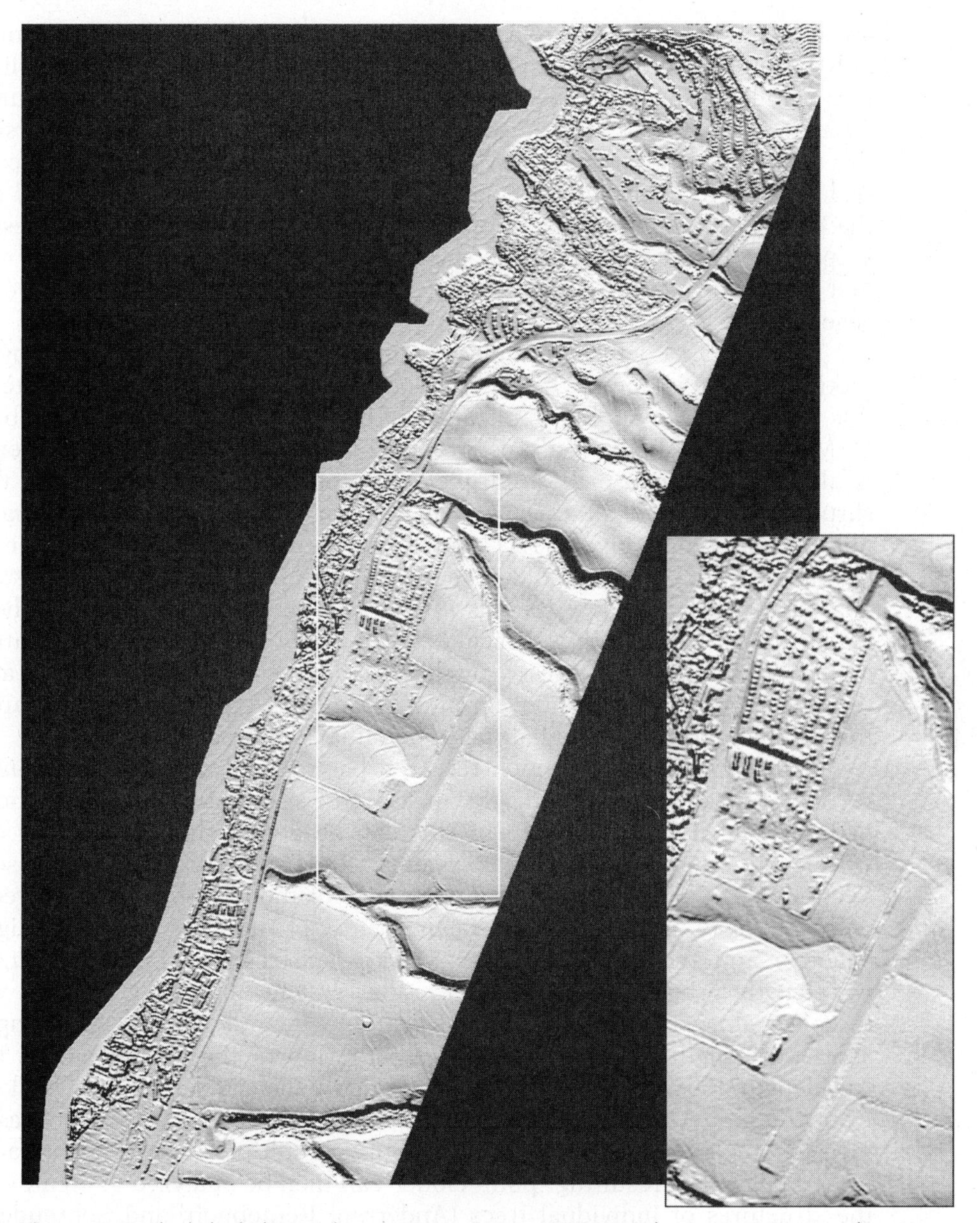

Figure 6.63 Shaded-relief DEM from Leica Geosystems ALS-40 Airborne Laser Scanner data, West Maui, Hawaii. Scale 1:48,000 (main image), 1:24,000 (inset). (Courtesy NOAA Coastal Services Center.)

DEM, which is represented here in shaded relief and with color-coded elevations. (See Chapter 1 for a discussion of methods for representing topographic data.) The nearshore waters of the Gulf of Maine appear blue in this visualization. Low-lying land areas are represented in green. In this area, many such low-lying areas are salt marshes, which have a uniform, flat appearance in this DEM, interrupted only by subtle but distinctly visible networks of tidal channels. A long linear feature parallel to the shoreline at the right side of the image is a line of dunes along a barrier spit. A road and causeway can be seen crossing the salt marsh behind this line of dunes. Other roads, as well as buildings and other features, can be seen elsewhere in the lidar visualization. At elevations above the tidal marshes, the DEM has a much rougher appearance, indicative of the heavily forested landscape. Like Figure 6.63, Plate 28 shows a digital surface model (DSM) where the first lidar return from each point is used to represent the top of the canopy in vegetated areas.

The ability to detect multiple returning pulses from a plant canopy and the ground surface is one of the most revolutionary aspects of lidar remote sensing. For topographic mapping applications, this allows the creation of a bare-earth DEM by using only the last pulse received from each point. On the other hand, it is also possible to use the multiple returning echoes to characterize the trees, shrubs, and other vegetation present within the "footprint" of each transmitted pulse. Plate 1 and Figure 1.21, both discussed previously in Chapter 1, show examples of the separation of aboveground vegetation from the terrain surface via analysis of early and late returns in spatially dense lidar data sets. The canopy height model in Plate 1*c* is typical of what might be used in forest management applications to monitor forest growth, estimate the volume of timber in a stand or the storage of carbon in aboveground woody biomass, and create maps of features ranging from wildlife habitat to potential wildfire fuel loadings.

At fine spatial scales, much research has focused on using horizontally and vertically detailed canopy height measurements to develop geometric models of individual tree crown shapes. This can be done using either lidar or multiwavelength interferometric radar systems. (Radar interferometry was discussed in Section 6.9.) Figure 6.64 shows a side view of models of individual trees, along with remotely sensed point-measurements of the tree crowns (light-toned spheres) and the ground surface (darker spheres). In Figure 6.64*a*, a dual-wavelength airborne interferometric radar system known as GeoSAR was used to collect imagery at the X-band (3 cm) and P-band (85 cm) wavelengths. The short-wavelength X-band is sensitive to the top of the canopy, while the long-wavelength P-band primarily passes through the canopy and is backscattered from the ground surface. In Figure 6.64*b*, an airborne lidar sensor was used to measure the same stand of trees, recording multiple returns from each transmitted pulse. The resulting "point cloud" can then be analyzed to create models of the structures of individual trees (Andersen, Reutebuch, and Schreuder, 2002). Collecting lidar data or interferometric radar imagery over the same stand on multiple dates permits estimation of tree growth rates.

Another current research topic in lidar is *full-waveform analysis* (Mallet and Bretar, 2009). Whereas "traditional" lidar systems may record one or more discrete returns from each transmitted pulse, full-waveform lidars digitize the continuous returning signal at a uniform sampling frequency. In essence, instead

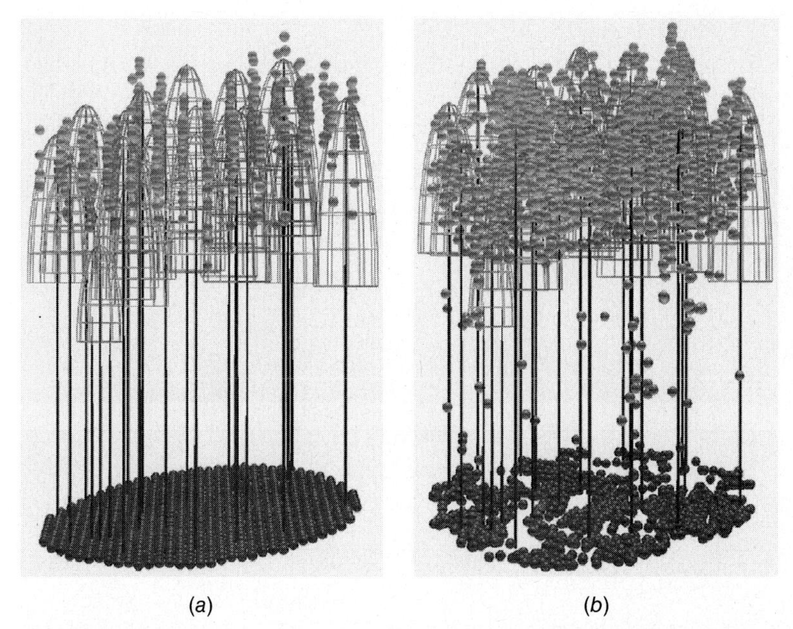

Figure 6.64 Models of tree structures at the Capitol Forest site, Washington State. (*a*) From interferometric radar. (*b*) From lidar. (Courtesy Robert McGaughey, USDA Forest Service PNW Research Station.)

of recording the timing of four or five peaks of the returning wave, these systems record the entire shape of the wave itself. Typically, a full-waveform lidar analysis involves sampling the shape of the lidar pulse at 1 nanosecond (ns) intervals, with the signal amplitude at each sampling interval recorded as an 8-bit digital number. The result of this process is a digital lidar waveform that can be considered as analogous to the digital spectral signature from a hyperspectral system. By analyzing this waveform, it is possible to more fully characterize the physical properties of the medium (e.g., a plant canopy) through which the transmitted signal passes on its way to the ground surface. Full-waveform lidar data can also be used for more efficient automated classification of types of lidar point returns.

One example of such a system is NASA's *Experimental Advanced Airborne Research Lidar (EAARL)*. When acquired from an altitude of 300 m, EAARL lidar spots have a 20-cm diameter on the surface. The returning wave from each spot is digitized at a resolution of 1 ns, corresponding to a vertical precision of 14 cm in air or 11 cm in water. This system has been used in coastal and marine mapping surveys, for studying emergent coastal vegetation communities and coral reefs (Wright and Brock, 2002).

A third area of active research is on the use of lidar data for detecting and quantifying changes in the three-dimensional position, shape, and volume of objects or surfaces. By acquiring repeat lidar data sets over a target area at multiple points in

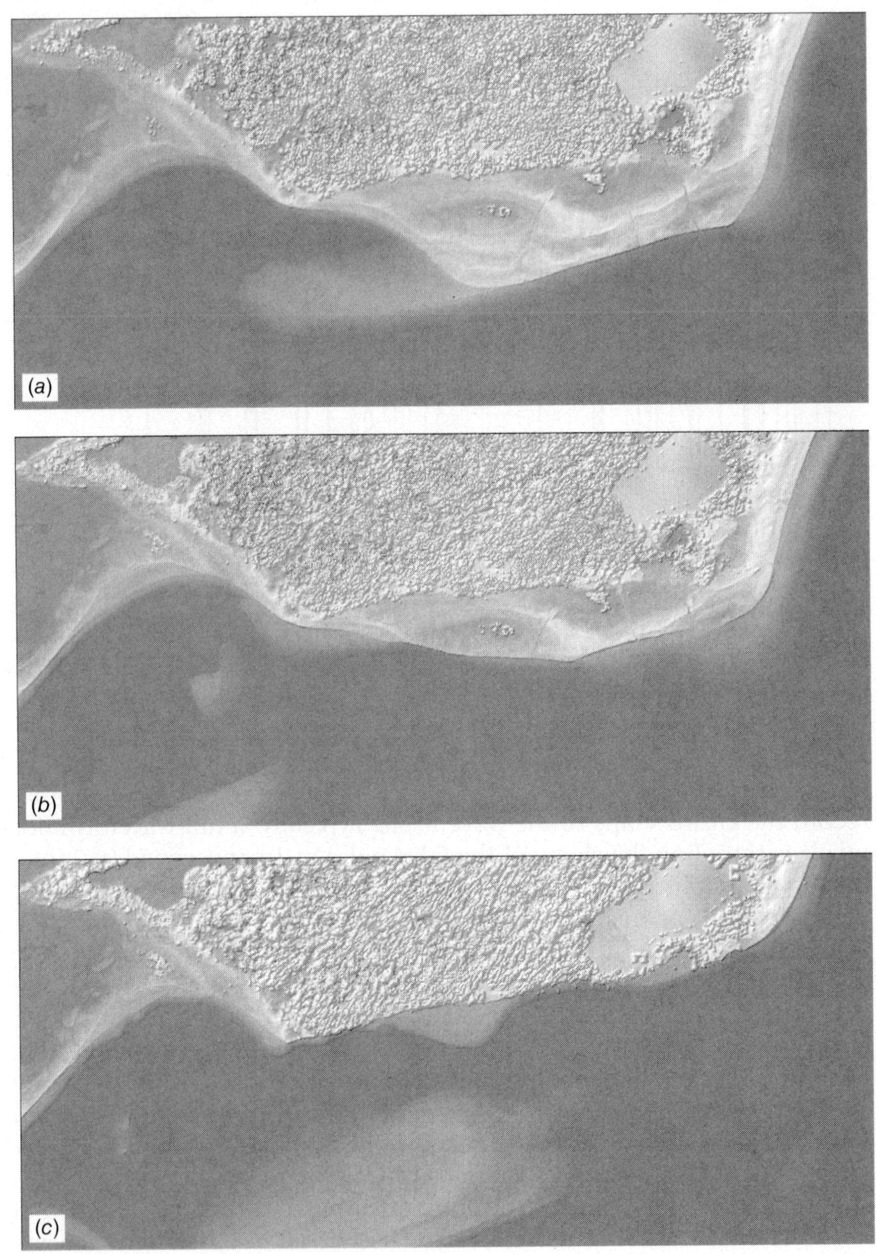

Figure 6.65 Lidar monitoring of coastal erosion and sand bar migration at the mouth of the Morse River, Popham Beach, Maine. Shaded relief renderings of DEMs from airborne lidar data. (*a*) May 5, 2004. (*b*) June 6, 2007. (*c*) May 24, 2010. (Scale 1:12,000). (Data courtesy NOAA Coastal Services Center.)

time, fine-scale topographic changes such as erosion, stream channel migration, changes in snow and ice mass, and other phenomena can be recorded. Figure 6.65 shows a series of three DEMs for a coastal site at the mouth of the Morse River in Maine (near the location of Plate 28) where processes such as mass wasting, channel migration, and sandbar movement are all occurring. By comparing the DEMs from 2004, 2007, and 2010, it is possible to calculate the volume of material transported to or away from each part of the site.

Ground-based lidar systems have also been developed, for use in characterizing the three-dimensional shape of complex objects such as buildings and bridges and the fine-scale "microtopography" of landscape sites. One such system, the Leica HDS6000, can record up to 500,000 points per second, through a complete 360° by 310° scanning angle. The positional accuracy of its measurements is better than 6 mm at a distance of 10 m.

Modern lidar systems have the capability to capture reflectance data from the returning pulses, in addition to the three-dimensional coordinates of the returns. Like the strength of radar returns, the intensity of lidar "echoes" varies with the wavelength of the source energy and the composition of the material returning the incoming signal. For example, commercial lidar systems frequently utilize a

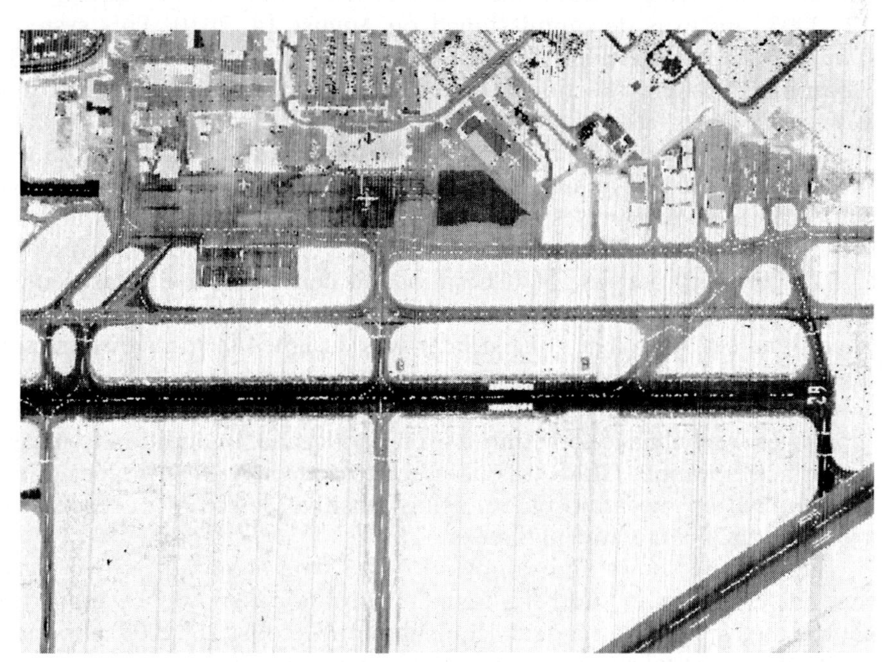

Figure 6.66 Lidar intensity image of a portion of the Lewiston/Nez Perce County Airport, Idaho. The data were acquired by a Leica Geosystems ALS 40 Airborne Laser Scanner at 24,000 pulses per second and 2-m post spacing. (Courtesy of i-TEN Associates.)

1.064-μm near-infrared wavelength pulse. At this wavelength, the reflectance of snow is 70 to 90%, mixed forest cover reflectance is 50 to 60%, and black asphalt has a reflectance near 5%. The lidar measurements of these reflectance percentage values are referred to as *lidar intensity* and may be processed to produce a georeferenced raster image. The spatial resolution of an intensity image is determined by the lidar point spacing, while the radiometric characteristics of the image are influenced by the shape of the laser footprint and the scan angle of the pulse (e.g., increasing scan angles increase the amount of canopy viewed in a forested area). These intensity images are useful for identification of broad land cover types and can serve as ancillary data for post-processing.

Figure 6.66 is an example of a lidar intensity image. The image covers a portion of the Lewiston/Nez Perce County Airport, Idaho. It was flown with a Leica Geosystems ALS 40 Airborne Laser Scanner at 24,000 pulses per second and 2-m post spacing.

6.25 SPACEBORNE LIDAR

The first free-flying satellite lidar system was launched successfully on January 12, 2003, and was decommissioned on August 14, 2010. This system, the *Ice, Cloud, and Land Elevation Satellite (ICESat)*, was a component of NASA's Earth Observing System (Section 5.19. ICESat carried a single instrument, the *Geoscience Laser Altimeter System (GLAS)*, which included lidars operating in the near infrared and visible, at 1.064 and 0.532 μm, respectively. GLAS transmitted 40 times per second, at an eye-safe level of intensity, and measured the timing of the received return pulses to within 1 ns. The laser footprints at ground level were approximately 70 m in diameter and were spaced 170 m apart along-track.

The primary purpose of ICESat was to collect precise measurements of the mass balance of polar ice sheets and to study how the earth's climate affects ice sheets and sea level. The ICESat lidar was designed to measure very small changes in ice sheets when data are averaged over large areas. This permits scientists to determine the contribution of the ice sheets of Greenland and Antarctica to global sea level change to within 0.1 cm per decade. In addition to measurements of surface elevation, GLAS was designed to collect vertical profiles of clouds and aerosols within the atmosphere, using sensitive detectors to record backscatter from the 0.532-μm-wavelength laser.

Figure 6.67 shows an example of GLAS data acquired on two passes over a lake in the desert in southern Egypt. Figure 6.67*a* shows the individual GLAS-illuminated spots (from a data acquisition on October 26, 2003) superimposed on a MODIS image taken on the same date. Figure 6.67*b* shows a cross section along the GLAS transect. In this cross section, the solid line represents the topography of the surface as obtained from a Shuttle Radar Topography Mission (SRTM) DEM, acquired on February 2000 before this lake basin was filled. Thus, this line represents the bathymetry of the lake, or the lake bottom elevations along the

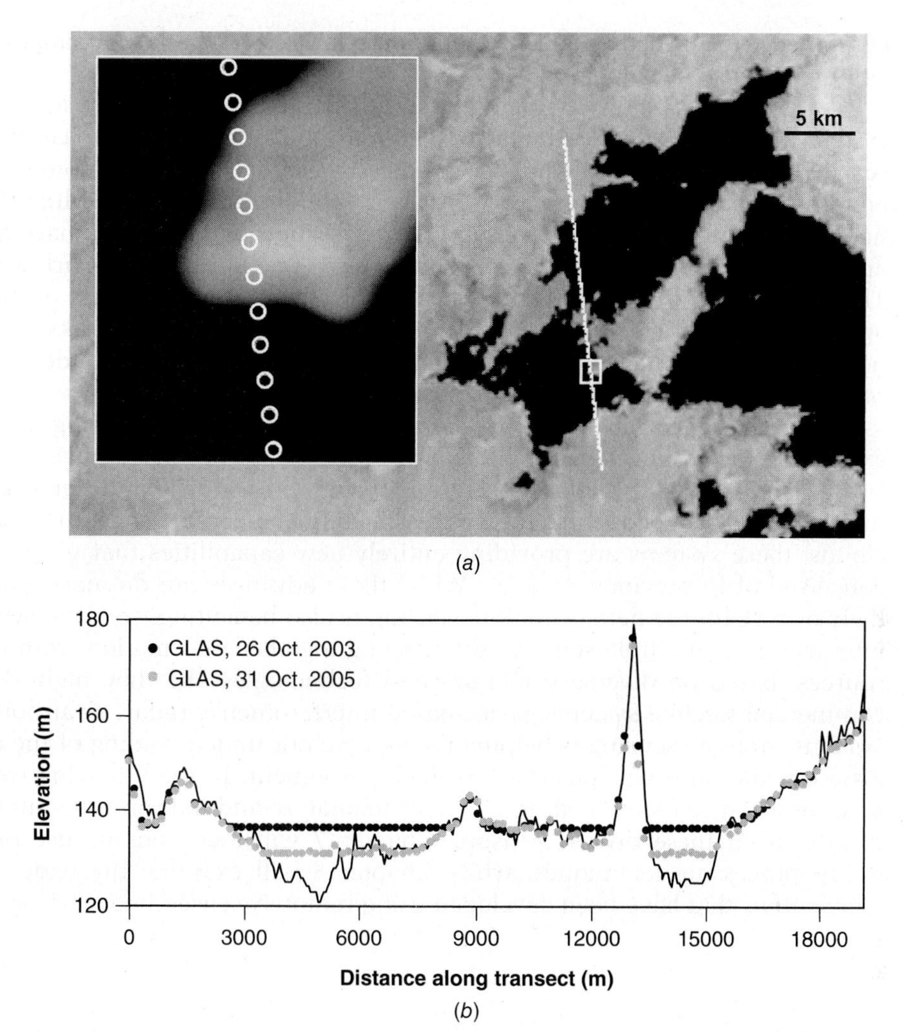

Figure 6.67 ICESat GLAS data over southern Egypt. (*a*) GLAS illuminated spots superimposed on MODIS imagery, October 26, 2003. (*b*) Cross section showing two dates of GLAS measurements (points), along with surface elevations from SRTM DEM data. (From Chipman and Lillesand, 2007; courtesy Taylor & Francis, *International Journal of Remote Sensing*, http://www.tandf.co.uk/journals.)

transect. The dark points in Figure 6.67*b* represent the GLAS elevation measurements from October 26, 2003. Where those points fall on land, they correspond closely to the ground surface elevation obtained from the SRTM DEM. Where they extend across the lake, they show the water level on that date. For comparison, the gray points in Figure 6.67*b* are GLAS elevation measurements from October 31, 2005. During the intervening years, the GLAS measurements indicate that the lake level dropped by 5.2 m (Chipman and Lillesand, 2007). When combined with measurement of the change in area of the lake, changes in lake volume can also be calculated.

A follow-on mission, ICESat-2, is currently planned for launch in 2016. This system will continue its predecessor's studies of polar and montane ice sheets and sea ice and will also contribute to studies of global vegetation biomass. It will employ a visible (green) laser with a wavelength of 0.532 μm, but while ICESat-1 had a pulse rate of 40 transmitted pulses per second, ICESat-2 will have a rate of approximately 10,000 pulses per second, closer to the rate of an airborne lidar. The laser will also be split into six beams, grouped into three pairs each 3.3 km apart. Along the track, the measurement spacing will be 70 cm, thanks to the high pulse rate. This configuration should allow greatly improved spatial detail in ICESat-2's observations of global ice and ecosystems.

As we have seen in this chapter, radar and lidar remote sensing are rapidly expanding the frontiers of remote sensing. With applications as diverse as measuring landscape subsidence from groundwater extraction (via differential radar interferometry) to modeling the shapes of individual tree crowns in lidar point clouds, these systems are providing entirely new capabilities that were not even conceived of in previous decades. While these advances are dramatic enough in their own right, the field of remote sensing is also benefiting from the synergistic interactions of multiple sensors. By bringing together information from multiple sources, based on diverse principles and technologies—be they high-resolution commercial satellite systems, space-based interferometric radar, or airborne lidar systems—remote sensing is helping us improve our understanding of the earth as a system and our own "position" in the environment. In the following two chapters, we will discuss many different ways to analyze and interpret the information coming from these diverse sensors. Chapter 7 will focus on the use of digital image processing techniques, while Chapter 8 will examine the wide array of applications that have been developed using remotely sensed imagery.

7 DIGITAL IMAGE ANALYSIS

7.1 INTRODUCTION

Digital image analysis refers to the manipulation of digital images (Section 1.5) with the aid of a computer. This could range from an amateur photographer using freely available software to adjust the contrast and brightness of pictures from her or his digital camera to a team of scientists using neural-network classification to map mineral types in an airborne hyperspectral image. In this chapter, we will begin with simple and widely used methods for enhancing digital images, correcting errors, and generally improving image quality prior to further visual interpretation or digital analysis. Many of these techniques are broadly applicable to a wide range of types of remotely sensed data. We then proceed to cover more advanced and specialized techniques. The enhancement, processing, and analysis of digital images comprise an extremely broad subject, often involving procedures that can be mathematically complex. For most topics in this chapter, we focus on the concepts and principles involved, without delving too deeply into the mathematics and algorithms that are used to implement these digital analysis methods.

The use of computers for digital processing and analysis began in the 1960s with early studies of airborne multispectral scanner data and digitized aerial photographs. However, it was not until the launch of Landsat-1 in 1972 that digital image data became widely available for land remote sensing applications. At that time, not only was the theory and practice of digital image processing in its infancy, but also the cost of digital computers was very high, and their computational efficiency was very low by modern standards. Today, access to low-cost, efficient computer hardware and software is commonplace, and the sources of digital image data are many and varied. These sources range from commercial and governmental earth resource satellite systems, to the meteorological satellites, to airborne scanner data, to airborne digital camera data, to image data generated by photogrammetric scanners and other high resolution digitizing systems. All of these forms of data can be processed and analyzed using the techniques described in this chapter.

The central idea behind digital image processing is quite simple. One or more images are loaded into a computer. The computer is programmed to perform calculations using an equation, or series of equations, that take pixel values from the raw image as input. In most cases, the output will be a new digital image whose pixel values are the result of those calculations. This output image may be displayed or recorded in pictorial format or may itself be further manipulated by additional software. The possible forms of digital image manipulation are seemingly infinite. However, virtually all these procedures may be categorized into one (or more) of the following seven broad types of computer-assisted operations:

1. **Image preprocessing.** These operations aim to correct distorted or degraded image data to create a more faithful representation of the original scene and to improve an image's utility for further manipulation later on. This typically involves the initial processing of raw image data to eliminate noise present in the data, to calibrate the data radiometrically, to correct for geometric distortions, and to expand or contract the extent of an image via mosaicking or subsetting. These procedures are often termed *preprocessing* operations because they normally precede further manipulation and analysis of the image data to extract specific information. We briefly discuss these procedures in Section 7.2.
2. **Image enhancement.** These procedures are applied to image data in order to more effectively render the data for subsequent interpretation. In many cases, image enhancement involves techniques for heightening the visual distinctions among features in a scene, ultimately increasing the amount of information that can be interpreted from the data. We summarize the various broad approaches to enhancement in Section 7.3. In Section 7.4, we treat specific procedures that manipulate the contrast of an image (level slicing and contrast stretching). In Section 7.5, we discuss spatial feature manipulation (spatial filtering, convolution, edge enhancement, and Fourier analysis). In Section 7.6, we consider enhancements involving multiple spectral bands of imagery (spectral ratioing, principal

and canonical components, vegetation components, and intensity–hue–saturation color space transformations).

3. **Image classification.** The objective of image classification is to replace visual interpretation of image data with quantitative techniques for automating the identification of features in a scene. This normally involves the analysis of multiple bands of image data (typically multispectral, multitemporal, polarimetric, or other sources of complementary information) and the application of statistically based decision rules for determining the land cover identity of each pixel in an image. When these decision rules are based solely on the spectral radiances observed in the data, we refer to the classification process as *spectral pattern recognition*. In contrast, the decision rules may be based on the geometric shapes, sizes, and patterns present in the image data. These procedures fall into the domain of *spatial pattern recognition*. Hybrid methods, in which both spatial and spectral patterns are used for classification, are increasingly common. In any case, the intent of the classification process is to categorize all pixels in a digital image into one of several land cover classes, or "themes." These categorized data may then be used to produce *thematic maps* of the land cover present in an image and/or produce summary statistics on the areas covered by each land cover type. Due to their importance, image classification procedures comprise the subject of more than one-third of the material in this chapter (Sections 7.7 to 7.17). We emphasize "supervised," "unsupervised," and "hybrid" approaches to spectrally based image classification before introducing more specialized topics such as object-based classification, classification of mixed pixels, and the use of neural networks in the classification process. Finally, we describe various procedures for assessing and reporting the accuracy of image classification results.
4. **Analysis of change over time.** Many remote sensing projects involve the analysis of two or more images from different points in time, to determine the extent and nature of changes over time. In Section 7.18, we explore "change detection" methods to identify specific areas of change. With the proliferation of temporally rich data sets that offer dozens or hundreds of images of a given area over a period of weeks or years, there is also a new interest in time-series analysis of remotely sensed imagery (Section 7.19), such as the analysis of seasonal cycles, interannual variability, and long-term trends in dense multitemporal data sets.
5. **Data fusion and GIS integration.** These procedures are used to combine image data for a given geographic area with other geographically referenced data sets for the same area. These other data sets might simply consist of image data generated on other dates by the same sensor or by other remote sensing systems. Frequently, the intent of data merging is to combine remotely sensed data with other ancillary sources of information

in the context of a GIS. For example, image data are often combined with soil, topographic, ownership, zoning, and assessment information. We discuss data fusion in Section 7.20.

6. **Hyperspectral image analysis.** Virtually all of the image processing principles introduced in this chapter in the context of multispectral image analysis may be extended directly to the analysis of hyperspectral data. However, the basic nature and sheer volume of hyperspectral data sets is such that various image processing procedures have been developed to analyze such data specifically. We introduce these procedures in Section 7.21.
7. **Biophysical modeling.** The objective of biophysical modeling (Section 7.22) is to relate quantitatively the digital data recorded by a remote sensing system to biophysical features and phenomena measured on the ground. For example, remotely sensed data might be used to estimate such varied parameters as crop yield, pollution concentration, or water depth.

We have made the above subdivisions of the topic of digital image processing to provide the reader with a conceptual roadmap for studying this chapter. Although we treat each of these procedures as distinct operations, they are often used in combination, as part of a chain of operations beginning with a raw image and ending with some highly transformed output product. For example, the restoration process of noise removal can often be considered an enhancement procedure. Likewise, certain enhancement procedures (such as principal components analysis) can be used not only to enhance the data but also to improve the efficiency of classification operations. In a similar vein, data merging can be used in image classification in order to improve classification accuracy. Hence, the boundaries between the various operations we discuss separately here are not well defined in practice.

7.2 PREPROCESSING OF IMAGES

In nearly every case, there are certain preprocessing operations that are performed on the raw data of an image prior to its use in any further enhancement, interpretation, or analysis. Some of these operations are designed to correct flaws in the data, while others make the data more amenable to further processing. In this section we will discuss these preprocessing operations under the general headings of noise removal, radiometric correction, geometric correction, and image subsetting and mosaicking. Some of these operations may be performed by the data provider, before the imagery is provided to the analysts who will be interpreting it. In other cases, the users may need to perform one or more of these preprocessing steps themselves.

Noise Removal

Image noise is any unwanted disturbance in image data that is due to limitations in the sensing, signal digitization, or data recording process. The potential sources of noise range from periodic drift or malfunction of a detector, to electronic interference between sensor components, to intermittent "hiccups" in the data transmission and recording sequence. Noise can degrade or totally mask the true radiometric information content of a digital image. Hence, noise removal usually precedes any subsequent enhancement or classification of the image data. The objective is to restore an image to as close an approximation of the original scene as possible.

The nature of noise correction required in any given situation depends upon whether the noise is systematic (periodic), random, or some combination of the two. For example, multispectral sensors that sweep multiple scan lines simultaneously often produce data containing systematic *striping* or *banding*. This stems from variations in the response of the individual detectors used within each band. Such problems were particularly prevalent in the collection of early Landsat data. While the multiple detectors used for each band were carefully calibrated and matched prior to launch, the radiometric response of one or more tended to drift over time, resulting in relatively higher or lower values along every sixth line in the image data. In this case valid data are present in the defective lines, but they must be normalized with respect to their neighboring observations.

Several *destriping* procedures have been developed to deal with the type of problem described above. One method is to compile a set of histograms for the image—one for each detector involved in a given band. These histograms are then compared in terms of their descriptive statistics (mean, median, standard deviation, and so on) to identify radiometric differences or malfunctions among the detectors. An empirical correction model can then be calculated to adjust the histograms for the malfunctioning detectors' lines to resemble those for the normal data lines. This adjustment factor is applied to each pixel in the problem lines while the others are not altered (Figure 7.1).

Another line-oriented noise problem sometimes encountered in digital data is *line drop*. In this situation, a number of adjacent pixels along a line (or an entire line) may contain spurious DNs, often values of 0 or "no data." This problem is normally addressed by replacing the defective DNs with the average of the values for the pixels occurring in the lines just above and below (Figure 7.2).

Random noise problems in digital data are handled somewhat differently than striping or line drops. This type of noise is characterized by nonsystematic variations in gray levels from pixel to pixel called *bit errors* or *salt and pepper noise*. Such noise is often referred to as being "spiky" in character, and it causes images to have a "salt and pepper" or "snowy" appearance with anomalously bright and/or dark pixels scattered across the imagery.

Figure 7.1 Destriping algorithm illustration: (*a*) original image manifesting striping with a six-line frequency; (*b*) restored image resulting from applying histogram algorithm. (Author-prepared figure.)

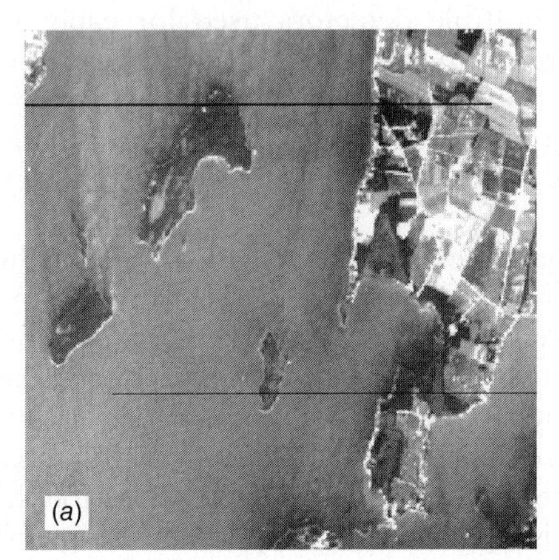

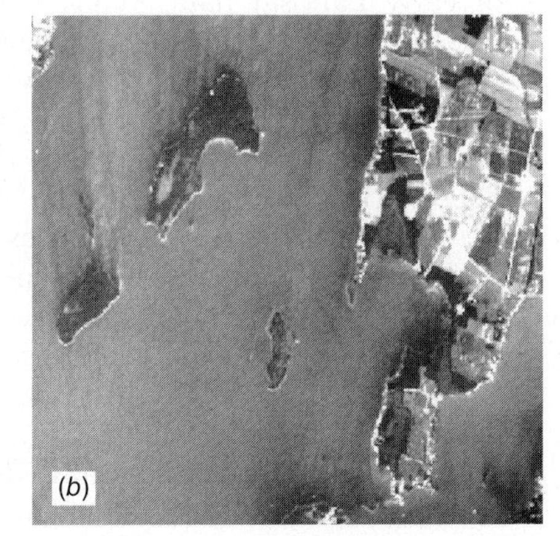

Figure 7.2 Line drop correction: (*a*) original image containing two line drops; (*b*) restored image resulting from averaging pixel values above and below defective line. (Author-prepared figure.)

Bit errors are handled by recognizing that noise values normally change much more abruptly than true image values. Thus, noise can be identified by comparing each pixel in an image with its neighbors. If the difference between a given pixel value and its surrounding values exceeds an analyst-specified threshold, the pixel is assumed to contain noise. The noisy pixel value can then be replaced by the average of its neighboring values. Moving neighborhoods or windows of 3×3 or 5×5 pixels are typically used in such procedures. Figure 7.3 illustrates the results of applying a noise reduction algorithm to an image with salt-and-pepper noise. (The use of moving windows for filtering an image, whether for noise reduction or other purposes, is discussed further in Section 7.5.)

Figure 7.3 Result of applying noise reduction algorithm: (*a*) original image data with noise-induced "salt-and-pepper" appearance; (*b*) image resulting from application of noise reduction algorithm. (Author-prepared figure.)

Radiometric Correction

Assuming that any striping, line drops, or other noise has been removed, the radiance measured by any given system over a given object is influenced by such factors as changes in scene illumination, atmospheric conditions, viewing geometry, and instrument response characteristics. Some of these effects, such as viewing geometry variations, are greater in the case of airborne data collection than in satellite image acquisition. Also, the need to perform correction for any or all of these influences depends directly upon the particular application at hand.

Over the course of the year, there are systematic, seasonal changes in the intensity of solar irradiance incident on the earth's surface. If remotely sensed images taken at different times of the year are being compared, it is usually necessary to apply a *sun elevation correction* and an *earth–sun distance correction*. The sun elevation correction accounts for the seasonal position of the sun relative to the earth (Figure 7.4). Through this process, image data acquired under different solar illumination angles are normalized by calculating pixel brightness values assuming the sun was at the zenith on each date of sensing. The correction is usually applied by dividing each pixel value in a scene by the sine of the solar elevation angle (or cosine of the solar zenith angle) for the particular time and location of imaging.

The earth–sun distance correction is applied to normalize for the seasonal changes in the distance between the earth and the sun. The earth–sun distance is usually expressed in astronomical units. (An astronomical unit is equivalent to the mean distance between the earth and the sun, approximately 149.6×10^6 km.) The irradiance from the sun decreases as the square of the earth–sun distance.

Ignoring atmospheric effects, the combined influence of solar zenith angle and earth–sun distance on the irradiance incident on the earth's surface can be expressed as

$$E = \frac{E_0 \cos \theta_0}{d^2} \tag{7.1}$$

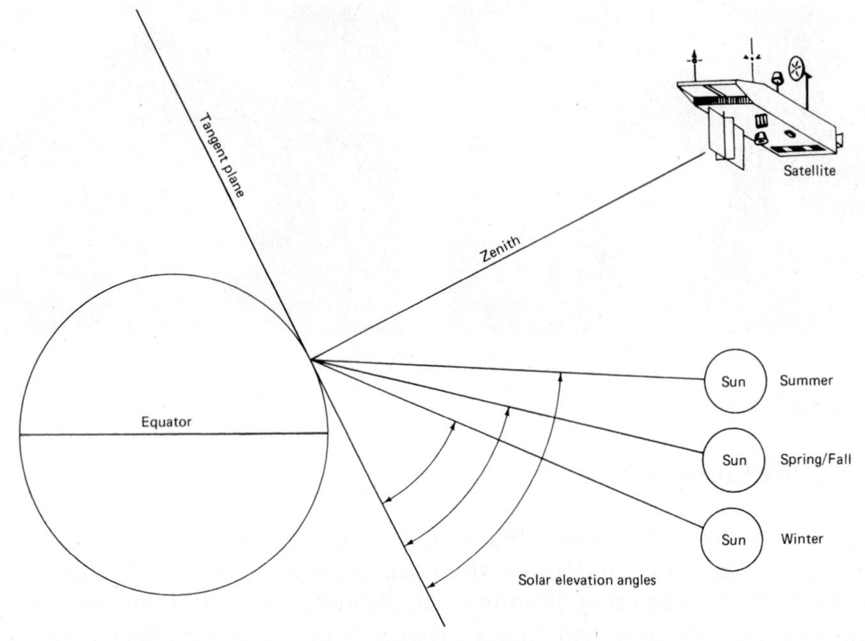

Figure 7.4 Effects of seasonal change on solar elevation angle. (The solar zenith angle is equal to 90° minus the solar elevation angle.)

where

E = normalized solar irradiance
E_0 = solar irradiance at mean earth–sun distance
θ_0 = sun's angle from the zenith
d = earth–sun distance, in astronomical units

(Information on the solar elevation angle and earth–sun distance for a given scene is normally part of the ancillary data supplied with the digital data.)

As initially discussed in Chapter 1, the influence of solar illumination variation is compounded by atmospheric effects. The atmosphere affects the radiance measured at any point in the scene in two contradictory ways. First, it attenuates (reduces) the energy illuminating a ground object. Second, it acts as a reflector itself, adding a scattered, extraneous "path radiance" to the signal detected by a sensor. Thus, the composite signal observed at any given pixel location can be expressed by

$$L_{\text{tot}} = \frac{\rho E T}{\pi} + L_{\text{p}} \tag{7.2}$$

where

L_{tot} = total spectral radiance measured by sensor
ρ = reflectance of object
E = irradiance on object
T = transmission of atmosphere
L_{p} = path radiance

(All of the quantities in Eq. 7.2 depend on wavelength.)

Only the first term in Eq. 7.2 contains valid information about ground reflectance. The second term represents the scattered path radiance, which introduces "haze" in the imagery and reduces image contrast. (Recall that scattering is wavelength dependent, with shorter wavelengths normally manifesting greater scattering effects.) *Haze compensation* procedures are designed to minimize the influence of path radiance effects. One means of haze compensation in multispectral data is to observe the radiance recorded over target areas of essentially zero reflectance. For example, the reflectance of deep clear water is essentially zero in the near-infrared region of the spectrum. Therefore, any signal observed over such an area represents the path radiance, and this value can be subtracted from all pixels in that band. This process is referred to as *dark object subtraction*.

For convenience, haze compensation routines are often applied uniformly throughout a scene. This may or may not be valid, depending on the uniformity of the atmosphere over a scene. When extreme viewing angles are involved in image acquisition, it is often necessary to compensate for the influence of varying the atmospheric path length through which the scene is recording. In such cases, off-nadir pixel values are usually normalized to their nadir equivalents. (Plate 23 illustrates the dependence of haze on viewing angle, using MISR imagery acquired at different angles.)

More advanced methods have been developed for *atmospheric correction* of optical and thermal images, when simple haze-removal techniques like dark object subtraction are insufficient. These algorithms are broadly divided into those based on empirical correction using spectral data from the imagery itself and those using radiative transfer methods to model atmospheric scattering and absorption from physical principles. In many cases, these algorithms may require information about local atmospheric conditions at the time of image acquisition.

When spectral data from more than one image need to be compared, but there is not sufficient information available for a complete atmospheric correction process, an alternative is *radiometric normalization* of the images. This process involves adjusting the brightness values of one or more secondary images to match a single base image. The images must at least partially overlap, and the overlap area must contain several temporally stable targets, features whose true surface reflectance is assumed to be constant over time. Typically, the analyst identifies a set of these targets covering a range of brightness values, then uses a statistical method such as linear regression to establish a model relating the brightness values in each secondary image to the corresponding brightness values in the base image. Each secondary image is then normalized using its own regression model.

The radiometric normalization process is illustrated in Figure 7.5 for a base image and two secondary images that show brightness variations due to both land cover change and extraneous factors (atmospheric conditions, solar illumination, or changes in sensor calibration). Pixels in image A are lighter-toned than the corresponding pixels in the base image, perhaps due to atmospheric haze, while pixels in image B are darker overall due to a lower sun angle. Once the

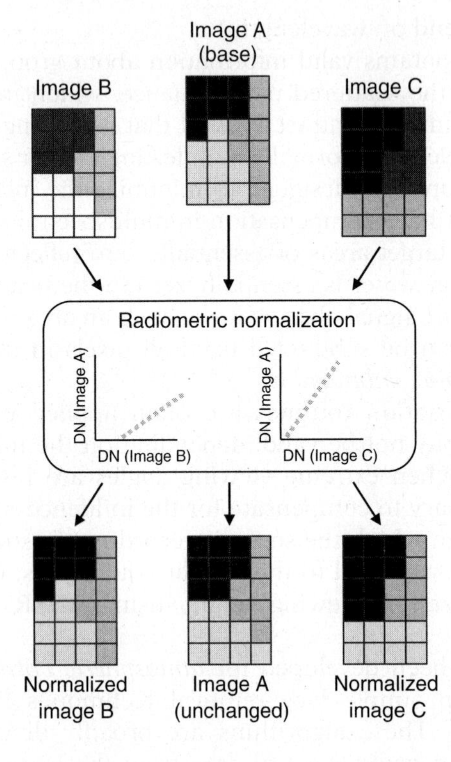

Figure 7.5 Radiometric normalization of three images of the same area. Top: Original images. Middle: Linear regression models used to normalize images B and C to match the radiometric range of image A. Bottom: Normalized images.

images have been normalized, the brightness values match more closely; any residual differences in brightness then correspond to actual land cover changes.

Another radiometric data processing activity involved in many quantitative applications of digital image data is *conversion of DNs to absolute radiance (or reflectance) values*. This operation accounts for the exact form of the A-to-D response functions for a given sensor and is essential in applications where measurement of absolute radiances is required. For example, such conversions are necessary when changes in the absolute reflectance of objects are to be measured over time using different sensors (e.g., the TM on Landsat-5 versus the OLI on Landsat-8). Likewise, such conversions are important in the development of mathematical models that physically relate image radiance or reflectance data to quantitative ground measurements (e.g., water quality measurements).

Normally, detectors and data systems are designed to produce a linear response to incident spectral radiance. Each spectral band of the sensor has its own linear response function, and its characteristics are monitored using onboard calibration lamps (and temperature references for thermal bands). The response function converts the raw electrical signal associated with a given level of spectral radiance into a scaled integer (DN). Inverting this response function allows the image analyst to convert DNs back into spectral radiance values.

In many cases (e.g., for most Landsat-8 OLI images provided by the U.S. Geological Survey), the model coefficients needed to do this conversion are provided as a multiplicative *gain* and an additive *intercept* (or *offset*), for each spectral band of the image. A given band can then be converted to spectral radiance using equation 7.3:

$$L = G * \mathrm{DN} + B \tag{7.3}$$

where

L = spectral radiance (over the spectral bandwidth of the channel)
G = slope of calibration function (channel gain)
DN = digital number value recorded
B = intercept of calibration function (channel offset)

Alternatively, the scaling factors may be provided in the form of a minimum and maximum spectral radiance for each band, corresponding to a minimum and maximum DN. Typically, the minimum DN will be 0 or 1, and the maximum will be determined by the number of bits used (e.g., 255 for 8-bit data or 1023 for 10-bit data). In this case, equation 7.4 is used to convert a single band of the image from DNs to spectral radiance:

$$L = \left(\frac{L_{\mathrm{MAX}} - L_{\mathrm{MIN}}}{\mathrm{DN}_{\mathrm{MAX}} - \mathrm{DN}_{\mathrm{MIN}}}\right)(\mathrm{DN} - \mathrm{DN}_{\mathrm{MIN}}) + L_{\mathrm{MIN}} \tag{7.4}$$

where

L = spectral radiance (over the spectral bandwidth of the channel)
$L_{\mathrm{MAX}}, L_{\mathrm{MIN}}$ = maximum and minimum spectral radiance
$\mathrm{DN}_{\mathrm{MAX}}, \mathrm{DN}_{\mathrm{MIN}}$ = maximum and minimum scaled DN
DN = digital number value recorded

Often the L_{MAX} and L_{MIN} values published for a given sensor are expressed in units of mW $\mathrm{cm}^{-2}\,\mathrm{sr}^{-1}\,\mu\mathrm{m}^{-1}$. That is, the values are often specified in terms of radiance per unit wavelength. To estimate the total within-band radiance in such cases, the value obtained from Eq. 7.4 must be multiplied by the width of the spectral band under consideration. Hence, a precise estimate of within-band radiance requires detailed knowledge of the spectral response curves for each band. The terms and concepts used to describe the radiometric units of an image are discussed further in online Appendix A.

Geometric Correction

Raw digital images usually contain geometric distortions so significant that they cannot be used directly as a map base without subsequent processing. The sources of these distortions range from variations in the altitude, attitude, and velocity of the sensor platform to factors such as panoramic distortion, earth curvature, atmospheric refraction, relief displacement, and nonlinearities in the sweep of a sensor's IFOV. The intent of geometric correction is to compensate for the distortions introduced by these factors so that the corrected image will have the highest practical geometric integrity. Increasingly, some or all of the necessary geometric corrections are performed automatically by image data providers. In other cases, however, the image analyst may need to use some of the methods described here to geometrically correct any distortions in the imagery and transform them into a standard coordinate system.

Systematic distortions are well understood and easily corrected by applying formulas derived by modeling the sources of the distortions mathematically. For example, a highly systematic source of distortion involved in multispectral scanning from satellite altitudes is the eastward rotation of the earth beneath the satellite during imaging. This causes each optical sweep of the scanner to cover an area slightly to the west of the previous sweep. This is known as *skew distortion*. The process of *deskewing* the resulting imagery involves offsetting each successive scan line slightly to the west. The skewed-parallelogram appearance of satellite multispectral scanner data is a result of this correction.

Random distortions and residual unknown systematic distortions are corrected by analyzing well-distributed ground control points (GCPs) occurring in an image. As with their counterparts on aerial photographs, GCPs are features of known ground location that can be accurately located on the digital imagery. Some features that might make good control points are highway intersections and distinct shoreline features. In the correction process numerous GCPs are located both in terms of their two-image coordinates (column, row numbers) on the distorted image and in terms of their ground coordinates (typically measured from a map or GPS located in the field, in terms of UTM coordinates or latitude and longitude). These values are then submitted to a least squares regression analysis to determine coefficients for two *coordinate transformation equations* that can be used to interrelate the geometrically correct (map) coordinates and the distorted-image coordinates. (Appendix B describes one of the more common forms of coordinate transformation, the *affine transformation*.) Once the coefficients for these equations are determined, the distorted-image coordinates for any map position can be precisely estimated. Expressing this in mathematic notation,

$$x = f_1(X, Y) \qquad y = f_2(X, Y) \tag{7.5}$$

where

(x, y) = distorted-image coordinates (column, row)
(X, Y) = correct (map) coordinates
f_1, f_2 = transformation functions

Intuitively, it might seem as though these equations are stated backward! That is, they specify how to determine the distorted-image positions corresponding to correct, or undistorted, map positions. But that is exactly what is done during the geometric correction process. We first define an undistorted output matrix of "empty" map cells and then fill in each cell with the gray level of the corresponding pixel, or pixels, in the distorted image. This process is illustrated in Figure 7.6. This diagram shows the geometrically correct output matrix of cells (solid lines) superimposed over the original, distorted matrix of image pixels (dashed lines). After producing the transformation function, a process called *resampling* is used to determine the pixel values to fill into the output matrix from the original image matrix. This process is performed using the following operations:

1. The coordinates of each element in the undistorted output matrix are transformed to determine their corresponding location in the original input (distorted-image) matrix.
2. In general, a cell in the output matrix will not directly overlay a pixel in the input matrix. Accordingly, the intensity value or digital number (DN) eventually assigned to a cell in the output matrix is determined on the basis of the pixel values that surround its transformed position in the original input matrix.

A number of different resampling schemes can be used to assign the appropriate DN to an output cell or pixel. To illustrate this, consider the shaded output pixel shown in Figure 7.6. The DN for this pixel could be assigned simply on the basis of the DN of the closest pixel in the input matrix, disregarding the slight offset. In our example, the DN of the input pixel labeled *a* would be transferred to the

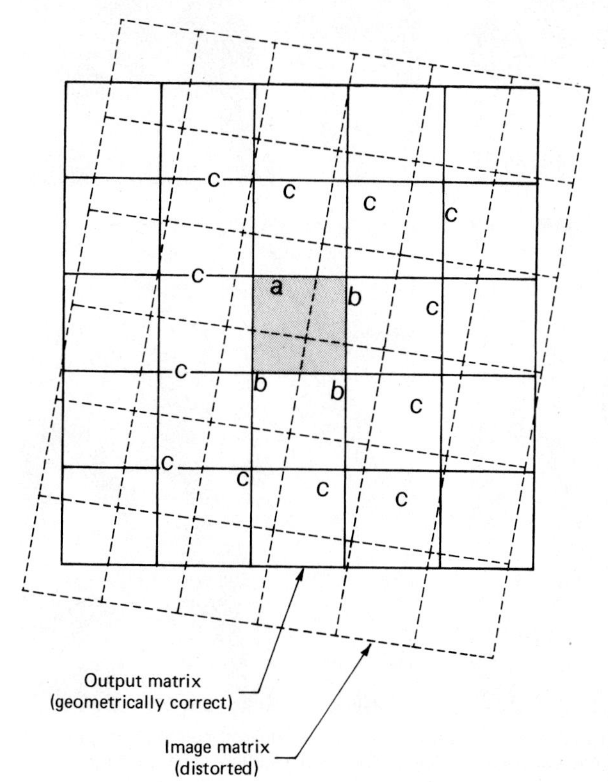

Figure 7.6 Matrix of geometrically correct output pixels superimposed on matrix of original, distorted input pixels.

shaded output pixel. This approach is called *nearest neighbor* resampling. It offers the advantage of computational simplicity and avoids having to alter the original input pixel values. However, features in the output matrix may be offset spatially by up to one-half pixel. This can cause a disjointed appearance in the output image product. Figure 7.7*b* is an example of a nearest neighbor resampled Landsat TM image. Figure 7.7*a* shows the original, distorted image.

More sophisticated methods of resampling evaluate the values of several pixels surrounding a given pixel in the input image to establish a "synthetic" DN to be assigned to its corresponding pixel in the output image. The *bilinear*

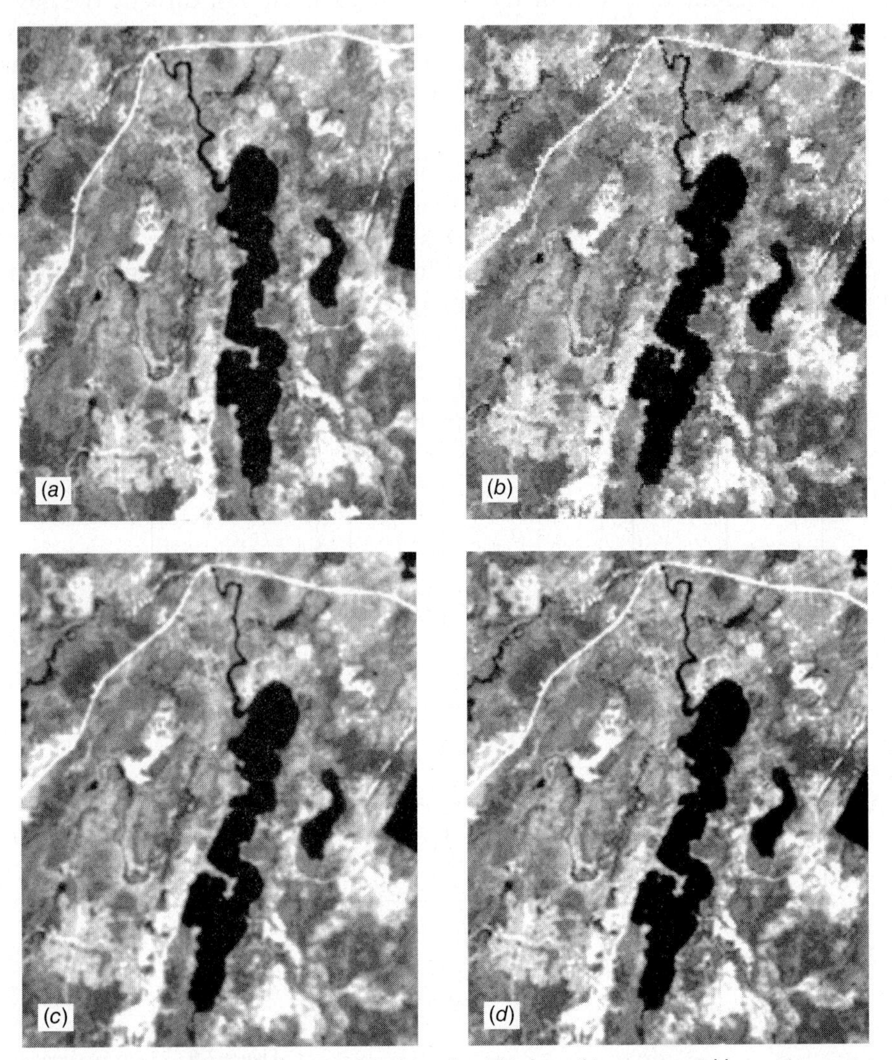

Figure 7.7 Resampling results: (*a*) original Landsat TM data; (*b*) nearest neighbor assignment; (*c*) bilinear interpolation; (*d*) cubic convolution. Scale 1:100,000. (Author-prepared figure.)

interpolation technique takes a distance-weighted average of the DNs of the four nearest pixels (labeled *a* and *b* in the distorted-image matrix in Figure 7.6). This process is simply the two-dimensional equivalent to linear interpolation. As shown in Figure 7.7*c*, this technique generates a smoother-appearing resampled image. However, because the process alters the gray levels of the original image, problems may be encountered in subsequent spectral pattern recognition analyses of the data. (Because of this, resampling is often performed after, rather than prior to, image classification procedures.)

An improved restoration of the image is provided by the *bicubic interpolation* or *cubic convolution* method of resampling. In this approach, the transferred synthetic pixel values are determined by evaluating the block of 16 pixels in the input matrix that surrounds each output pixel (labeled *a*, *b*, and *c* in Figure 7.6). Cubic convolution resampling (Figure 7.7*d*) avoids the disjointed appearance of the nearest neighbor method and provides a slightly sharper image than the bilinear interpolation method. (Again, this method alters the original image gray levels to some extent; other types of resampling can be employed to minimize this effect.)

As we discuss later, resampling techniques are important in several digital processing operations besides the geometric correction of raw images. For example, resampling is used to overlay or register multiple dates of imagery. It is also used to register images of differing resolution. Also, resampling procedures are used extensively to register image data and other sources of data in GISs. Appendix B contains additional details about the implementation of the various resampling procedures discussed in this section.

Many—but by no means all—sources of remotely sensed imagery are now provided with an even higher level of geometric correction, referred to as *terrain correction*. In this case, the imagery has been transformed into a map coordinate system via a process that uses a digital elevation model (DEM) to compensate for the effects of relief displacement and sensor tilt, in accordance with the photogrammetric principles discussed in Chapter 3. In essence, terrain correction produces an "orthoimage" that can be used as a map for geometric measurement and for overlay with other geographic data sources. In areas of steep slopes and large differences in elevation, the quality of the terrain correction results will depend on the accuracy and resolution of the DEM used in the correction process.

Subsetting, layer stacking, and mosaicking

A final step in the image preprocessing sequence often involves *subsetting* the image (s) to reduce the data volume, *layer stacking* to combine multiple separate bands or layers in a single image, and/or *mosaicking* multiple images to cover a broader area. Subsetting may be used to reduce the spatial extent of an image, cropping the image to cover only the specific area of interest, and it may also involve selecting only certain spectral bands. Layer stacking is often used when individual spectral bands are provided in separate files, but it can also be used to combine two or more different images (perhaps from different dates or different sensors).

If multiple images with different spatial extents are required to cover an area of interest, a mosaic may be created using data from each image to cover its own portion of the mosaic area. Care is required in this process, to ensure that there are not radiometric artifacts (e.g., abrupt changes in image brightness) at the boundaries between images in the mosaic. Where the images are acquired with widely disparate viewing angles, sophisticated processing may be required to avoid these kinds of artifacts caused by atmospheric haze and surface bidirectional reflectance factors. Often, if further analysis will be required (e.g., image classification), the analyst will choose to postpone producing the mosaic, instead performing the other image analysis procedures on each individual image, before finally mosaicking the end results after all processing and analysis steps.

7.3 IMAGE ENHANCEMENT

As previously mentioned, the primary goal of image enhancement is to improve the visual interpretability of an image by increasing the apparent distinction between the features in the scene. The range of possible image enhancement and display options available to the image analyst is virtually limitless. Most enhancement techniques may be categorized as either point or neighborhood operations. *Point operations* modify the brightness value of each pixel in an image data set independently. *Neighborhood operations* modify the value of each pixel based on neighboring brightness values. Either form of enhancement can be performed on single-band (monochrome) images or on the individual components of multi-image composites. The resulting images may also be recorded or displayed in black and white or in color. Choosing the appropriate enhancement(s) for any particular application is an art and often a matter of personal preference.

Enhancement operations are normally applied to image data after the appropriate preprocessing steps from Section 7.2 have been performed. Noise removal, in particular, is an important precursor to most enhancements. Without it, the image interpreter is left with the prospect of analyzing enhanced noise!

Below, we discuss the most commonly applied digital enhancement techniques. These techniques can be categorized as *contrast manipulation*, *spatial feature manipulation*, or *multi-image manipulation*. Within these broad categories, we treat the following:

1. **Contrast manipulation.** Gray-level thresholding, level slicing, and contrast stretching.
2. **Spatial feature manipulation.** Spatial filtering, edge enhancement, and Fourier analysis.
3. **Multi-image manipulation.** Multispectral band ratioing and differencing, vegetation and other indices, principal components, canonical components, vegetation components, intensity–hue–saturation (IHS) and other color space transformations, and decorrelation stretching.

7.4 CONTRAST MANIPULATION

Gray-Level Thresholding

Gray-level thresholding is used to *segment* an input image into two classes—one for those pixels having values below an analyst-defined gray level and one for those above this value. Below, we illustrate the use of thresholding to prepare a *binary mask* for an image. Such masks are used to segment an image into two classes so that additional processing can then be applied to each class independently.

Shown in Figure 7.8*a* is a Landsat-8 OLI band 4 (visible red band) image of the coastline of New Zealand's South Island. The image displays a broad range of gray levels over both land and water. Let us assume that we wish to show the brightness variations in this band in the water areas only. Because many of the gray levels for land and water overlap in this band, it would be impossible to separate these two classes using a threshold set in this band. This is not the case in the OLI band 5 (near-infrared band) shown in Figure 7.8*b*. The histogram of DNs for the band 5 image (Figure 7.8*c*) shows that water strongly absorbs the incident energy in this near-infrared band (low DNs), while the land areas are highly reflective (high DNs). Note that the DNs on this histogram exceed the "raw" 12-bit range provided by Landsat OLI because they have been scaled to 16-bit integers. A threshold set at DN = 6000 permits separation of these two classes in the band 5 data. This binary classification can then be applied to the band 4 data to enable display of brightness variations in only the water areas. This is illustrated in Figure 7.8*d*. In this image, the band 4 land pixel values have all been set to 0 (black) based on their classification in the band 5 binary mask. The band 4 water pixel values have been preserved and show an enhanced representation of variability in suspended sediment along the shoreline and in rivers and ponds.

Level Slicing

Level slicing is an enhancement technique whereby the DNs distributed along the x axis of an image histogram are divided into a series of analyst-specified intervals or "slices." All of the DNs falling within a given interval in the input image are then displayed at a single DN in the output image. Consequently, if six different slices are established, the output image contains only six different gray levels. The result looks something like a contour map, except that the areas between boundaries are occupied by pixels displayed at the same DN. Each level can also be shown as a single color.

Figure 7.8*e* illustrates the application of level slicing to the "water" portion of the scene illustrated in Figure 7.8*d*. Here, Landsat-8 OLI band 4 data have been level sliced into multiple levels in those areas previously determined to be water from the band 5 binary mask.

Level slicing is used extensively in the display of thermal infrared images in order to show discrete temperature ranges coded by gray level or color. (See Figure 5.8.)

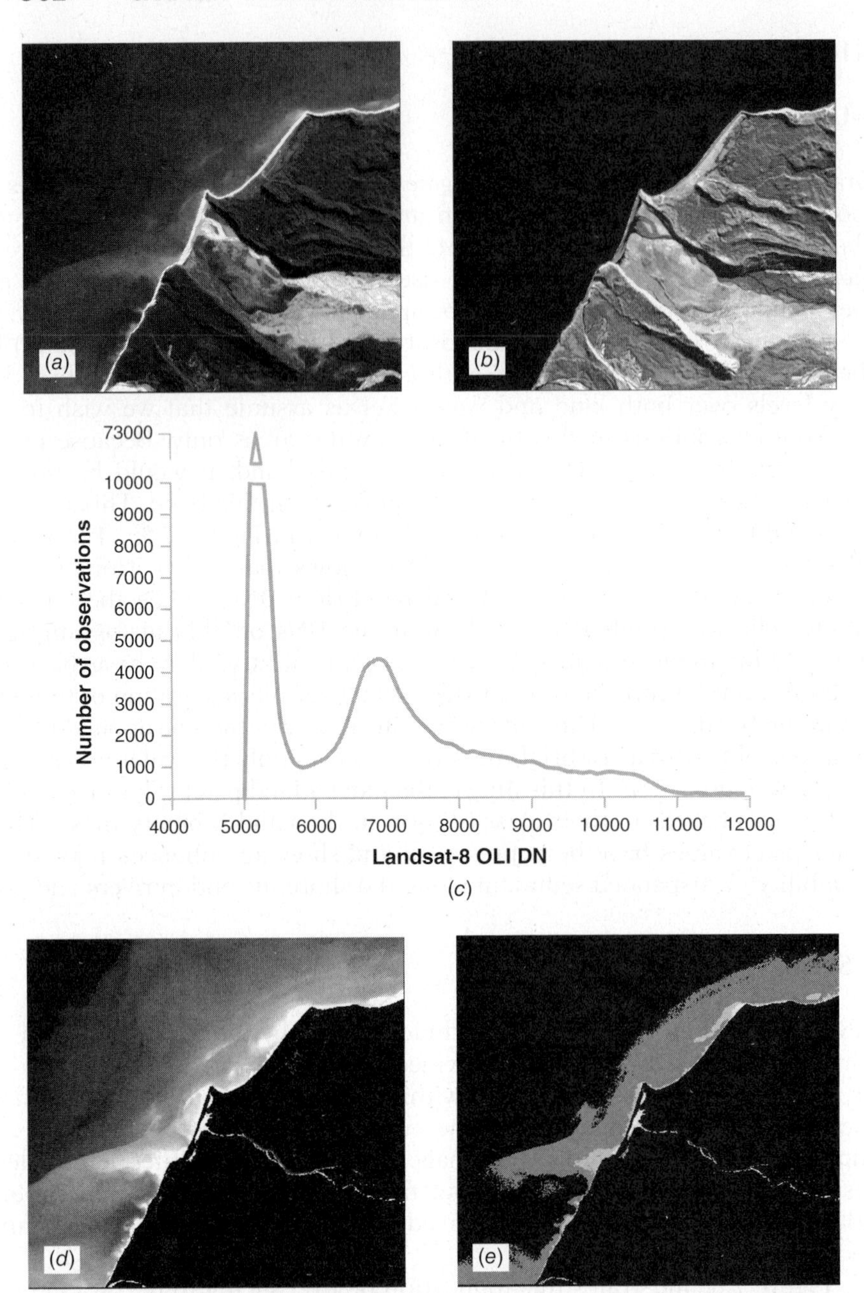

Figure 7.8 Gray-level thresholding for binary image segmentation: (*a*) original Landsat-8 OLI band 4 (red band) image containing continuous distribution of gray tones; (*b*) OLI band 5 (near infrared band) image; (*c*) OLI band 5 histogram, with the original 12-bit OLI radiometric range scaled to 16-bit DNs; (*d*) OLI band 4 brightness variation in water areas only; (*e*) level slicing operation applied to OLI band 4 data in areas determined to be water. (Author-prepared figure.)

Contrast Stretching

Image display and recording devices often operate over a range of 256 brightness levels (the maximum number represented in 8-bit computer encoding). Sensor data in a single image rarely cover this exact range—they may utilize only a small part of this 8-bit range (in low-contrast areas) or may cover a much wider range (in high-contrast areas imaged by sensors with more than 8-bit radiometric resolution). Hence, the intent of contrast stretching is to alter the range of brightness values present in an input image so as to optimally utilize the full 8-bit range of display values. The result is an output image that is designed to accentuate the contrast between features of interest to the image analyst.

To illustrate the contrast stretch process, consider a hypothetical sensing system whose image output levels can vary from 0 to 255. Figure 7.9*a* illustrates a histogram of brightness levels recorded in one spectral band over a scene. Assume that our hypothetical output device (e.g., computer monitor) is also capable of displaying 256 gray levels (0 to 255). Note that the histogram shows scene brightness values occurring only in the limited range of 60 to 158. If we were to use these image values directly in our display device (Figure 7.9*b*), we would be using only a small portion of the full range of possible display levels. Display levels 0 to 59 and 159 to 255 would not be utilized. Consequently, the tonal information in the scene would be compressed into a small range of display values, reducing the interpreter's ability to discriminate radiometric detail.

A more expressive display would result if we were to expand the range of image levels present in the scene (60 to 158) to fill the range of display values (0 to 255). In Figure 7.9*c*, the range of image values has been uniformly expanded to fill the total range of the output device. This uniform expansion is called a *linear stretch*. Subtle variations in input image data values would now be displayed in output tones that would be more readily distinguished by the interpreter. Light tonal areas would appear lighter and dark areas would appear darker.

In our example, the linear stretch would be applied to each pixel in the image using the algorithm

$$\mathrm{DN}' = \left(\frac{\mathrm{DN} - \mathrm{MIN}}{\mathrm{MAX} - \mathrm{MIN}}\right)255 \tag{7.6}$$

where

- DN' = digital number assigned to pixel in output image
- DN = original digital number of pixel in input image
- MIN = minimum value of input image, to be assigned a value of 0 in the output image (60 in our example)
- MAX = maximum value of input image, to be assigned a value of 255 in the output image (158 in our example).

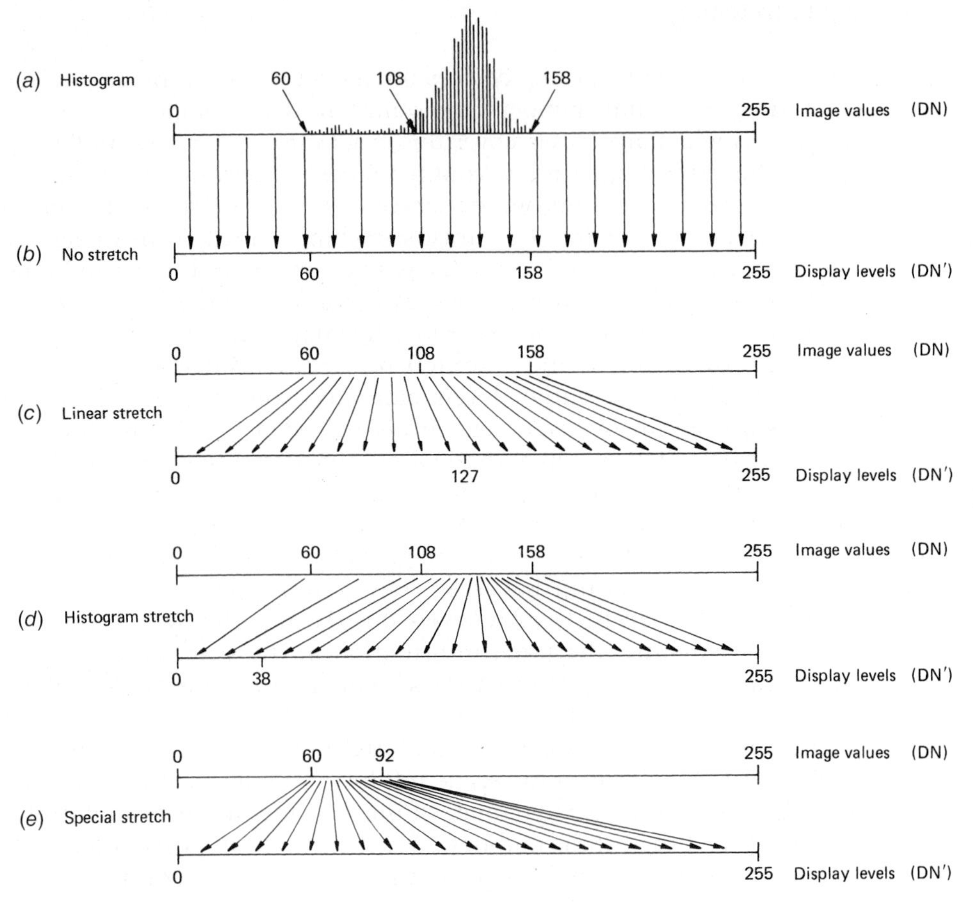

Figure 7.9 Principle of contrast stretch enhancement.

One drawback of the linear stretch is that it assigns as many display levels to the rarely occurring image values as it does to the frequently occurring values. For example, as shown in Figure 7.9*c*, half of the dynamic range of the output device (0 to 127) would be reserved for the small number of pixels having image values in the range 60 to 108. The bulk of the image data (values 109 to 158) are confined to half the output display levels (128 to 255). Although better than the direct display in (*b*), the linear stretch would still not provide the most expressive display of the data.

To improve on the above situation, a *histogram-equalized stretch* can be applied. In this approach, image values are assigned to the display levels on the basis of their frequency of occurrence. As shown in Figure 7.9*d*, more display values (and hence more radiometric detail) are assigned to the frequently occurring portion of the histogram. The image value range of 109 to 158 is now stretched over a large portion of the display levels (39 to 255). A smaller portion (0 to 38) is reserved for the infrequently occurring image values of 60 to 108.

For special analyses, specific features may be analyzed in greater radiometric detail by assigning the display range exclusively to a particular range of image values. For example, if water features were represented by a narrow range of values in a scene, characteristics in the water features could be enhanced by stretching this small range to the full display range. As shown in Figure 7.9*e*, the output range is devoted entirely to the small range of image values between 60 and 92. On the stretched display, minute tonal variations in the water range would be greatly exaggerated. The brighter land features, on the other hand, would be "washed out" by being displayed at a single, bright white level (255).

The visual effect of applying a contrast stretch algorithm is illustrated in Figure 7.10. An original Landsat-8 OLI image covering part of the Nile Delta in Egypt is shown in (*a*). The city of Cairo lies close to the apex of the delta on the right edge of the scene. Because of the wide range of image values present in this scene, the original image shows little radiometric detail. That is, features of similar brightness are virtually indistinguishable.

In Figure 7.10*b*, the brightness range of the desert area has been linearly stretched to fill the dynamic range of the output display. Patterns that were indistinguishable in the low contrast original are now readily apparent in this product. An interpreter wishing to analyze features in the desert region would be able to extract far more information from this display.

Because it reserves all display levels for the bright areas, the desert enhancement shows no radiometric detail in the darker irrigated delta region, which is displayed as black. If an interpreter were interested in analyzing a feature in this area, a different stretch could be applied, resulting in a display as shown in Figure 7.10*c*. Here, the display levels are devoted solely to the range of values present in the delta region. This rendering of the original image enhances brightness differences in the heavily populated and intensively cultivated delta, at the expense of all information in the bright desert area. Population centers stand out vividly in this display, and brightness differences among crop types are accentuated.

The contrast stretching examples we have illustrated represent only a small subset of the range of possible transformations that can be applied to image data. For example, nonlinear stretches such as sinusoidal transformations can be applied to image data to enhance subtle differences within "homogeneous" features such as forest stands or volcanic flows. Also, we have illustrated only monochromatic stretching procedures. Enhanced color images can be prepared by applying these procedures to separate bands of image data independently and then combining the results into a composite display.

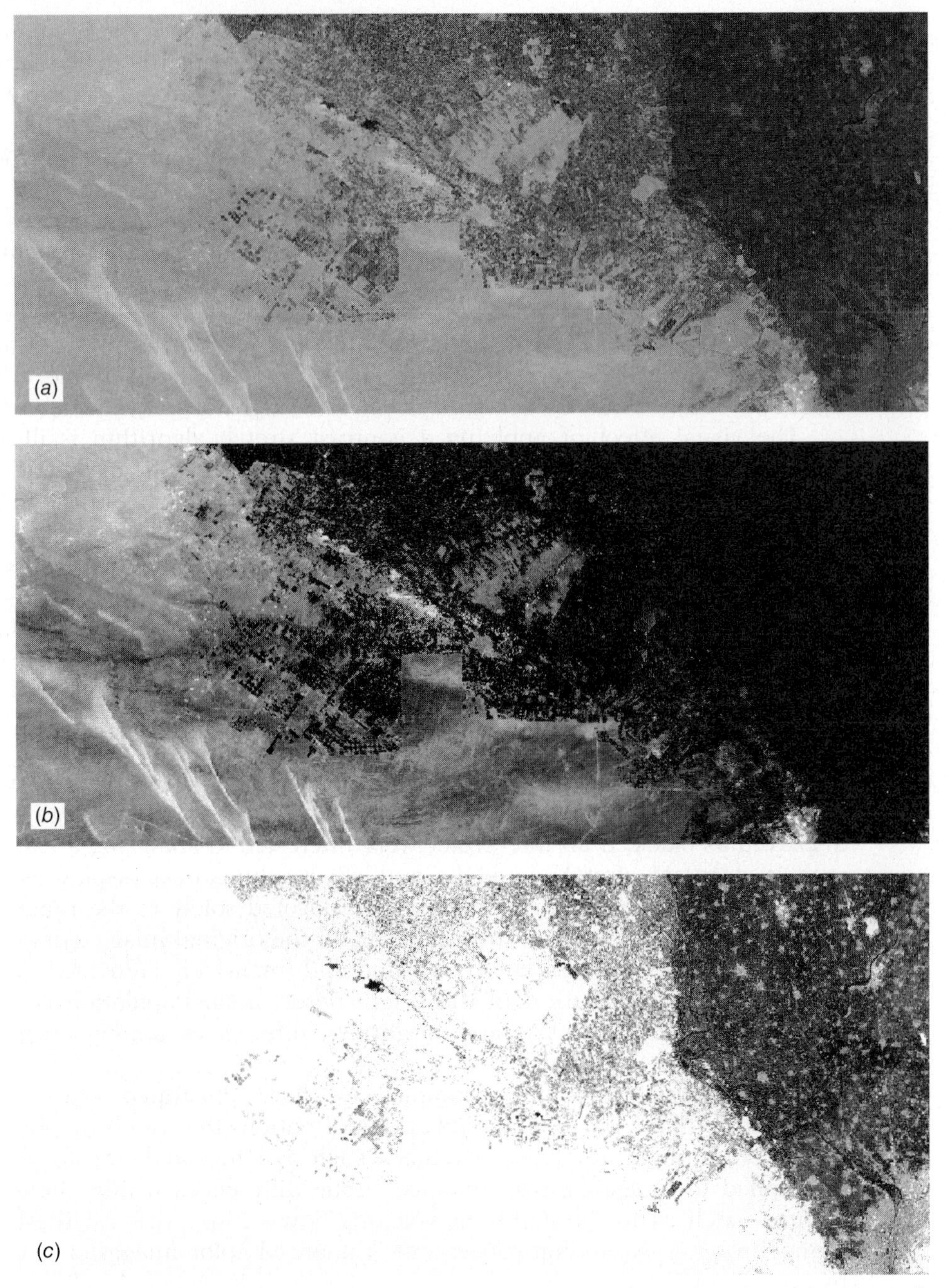

Figure 7.10 Effect of contrast stretching Landsat-8 OLI data acquired over the Nile Delta: (*a*) original image; (*b*) stretch that enhances contrast in bright image areas; (*c*) stretch that enhances contrast in dark image areas. (Author-prepared figure.)

7.5 SPATIAL FEATURE MANIPULATION

Spatial Filtering

In contrast to spectral filters, which serve to block or pass energy over various spectral ranges, spatial filters emphasize or deemphasize image data of various *spatial frequencies*. Spatial frequency refers to the "roughness" of the tonal variations occurring in an image. Image areas of high spatial frequency are tonally "rough." That is, the gray levels in these areas change abruptly over a relatively small number of pixels (e.g., across roads or field borders). "Smooth" image areas are those of low spatial frequency, where gray levels vary only gradually over a relatively large number of pixels (e.g., large agricultural fields or water bodies). *Low-pass filters* are designed to emphasize low frequency features (large-area changes in brightness) and deemphasize the high frequency components of an image (local detail). *High-pass filters* do just the reverse. They emphasize the detailed high frequency components of an image and deemphasize the more general low frequency information.

Spatial filtering is a "neighborhood" operation in that pixel values in an original image are modified on the basis of the gray levels of neighboring pixels. For example, a simple low-pass filter may be implemented by passing a moving window throughout an original image and creating a second image whose DN at each pixel corresponds to the neighborhood average within the moving window at each of its positions in the original image. Assuming a 3×3-pixel window is used, the center pixel's DN in the new (filtered) image would be the average value of the 9 pixels in the original image contained in the window at that point. Among other applications, low-pass filters are very useful for reducing random noise (see the discussion of noise removal in Section 7.2).

A simple high-pass filter may be implemented by subtracting a low-pass filtered image (pixel by pixel) from the original, unprocessed image. Figure 7.11 illustrates the visual effect of applying this process to an image. The original image is shown in Figure 7.11*a*. Figure 7.11*b* shows the low frequency component image, and Figure 7.11*c* illustrates the high frequency component image. Note that the low frequency component image (*b*) reduces deviations from the neighborhood average, which smooths or blurs the detail in the original image, reduces the gray-level range, but emphasizes the large-area brightness regimes of the original image. The high frequency component image (*c*) enhances the spatial detail in the image at the expense of the large-area brightness information. Both images have been contrast stretched. (Such stretching is typically required because spatial filtering reduces the gray-level range present in an image.)

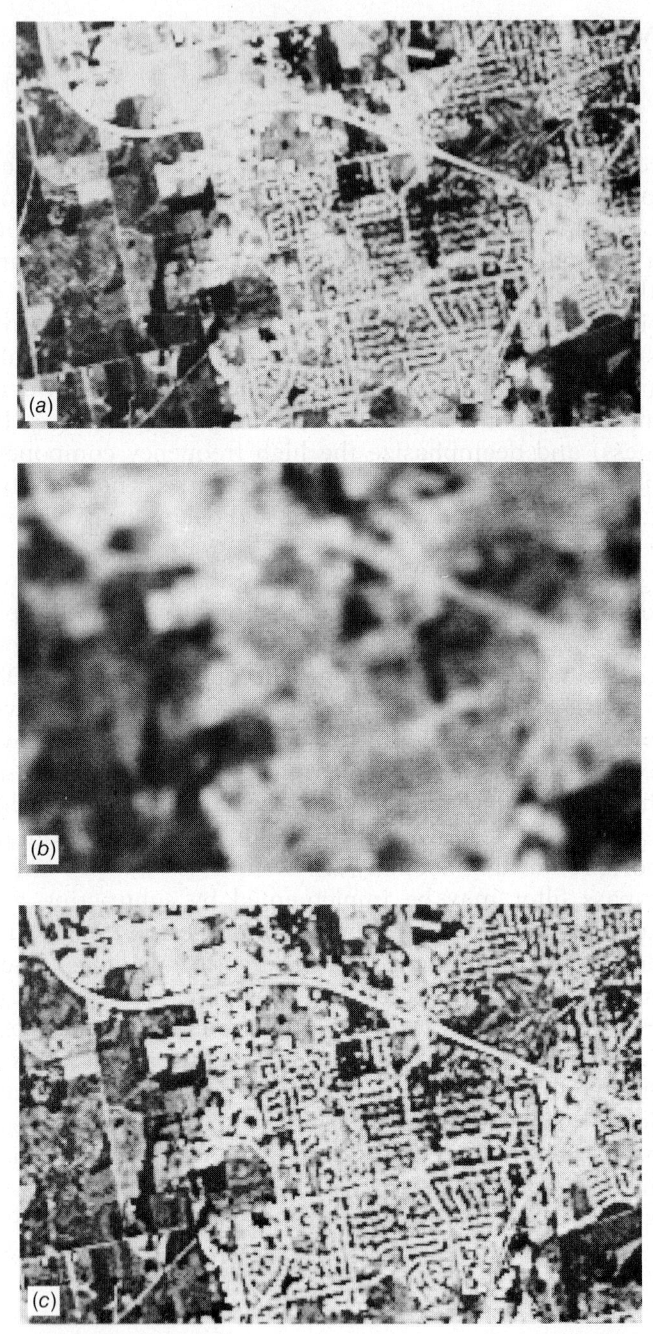

Figure 7.11 Effect of spatial filtering Landsat TM data: (*a*) original image; (*b*) low frequency component image; (*c*) high frequency component image. (Author-prepared figure.)

Convolution

Spatial filtering is but one special application of the generic image processing operation called *convolution*. Convolving an image involves the following procedures:

1. A moving window is established that contains an array of coefficients or weighting factors. Such arrays are referred to as *masks, operators,* or *kernels,* and they are normally an odd number of pixels in size (e.g., 3×3, 5×5, 7×7).
2. The kernel is moved throughout the original image, and the DN at the center of the kernel in a second (convoluted) output image is obtained by multiplying each coefficient in the kernel by the corresponding DN in the original image and adding all the resulting products. This operation is performed for each pixel in the original image.

Figure 7.12 illustrates a 3×3-pixel kernel with all of its coefficients equal to $\frac{1}{9}$. Convolving an image with this kernel would result in simply averaging the values

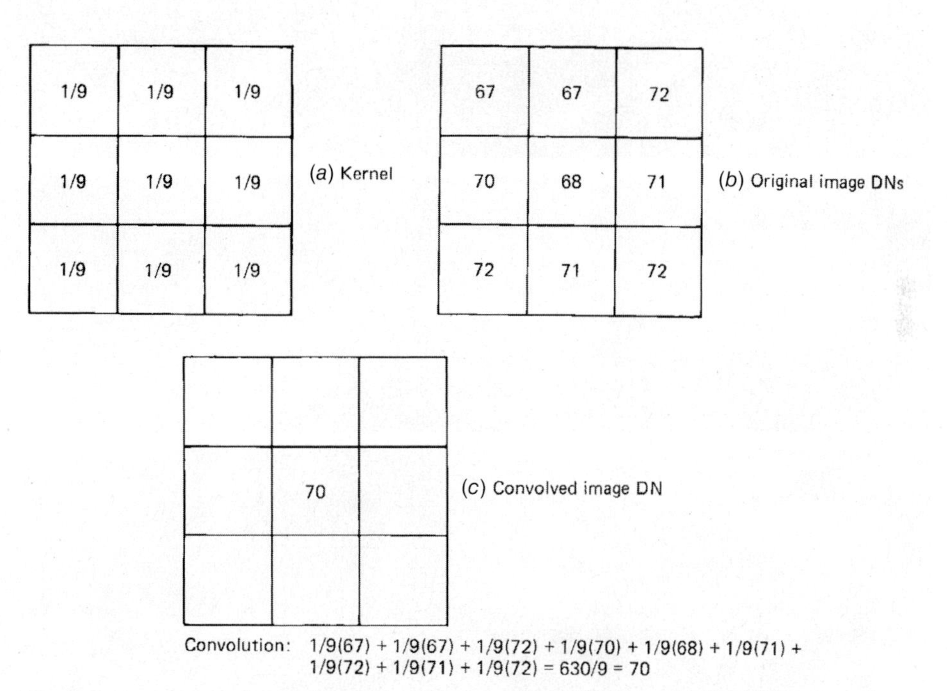

Figure 7.12 Concept of convolution. Shown is a 3×3-pixel kernel with all coefficients equal to $\frac{1}{9}$. The central pixel in the convolved image (in this case) contains the average of the DNs within the kernel.

in the moving window. This is the procedure that was used to prepare the low frequency enhancement shown in Figure 7.11*b*. However, images emphasizing other spatial frequencies may be prepared by simply altering the kernel coefficients used to perform the convolution. Figure 7.13 shows three successively lower frequency enhancements (*b*, *c*, and *d*) that have been derived from the same original data set (*a*).

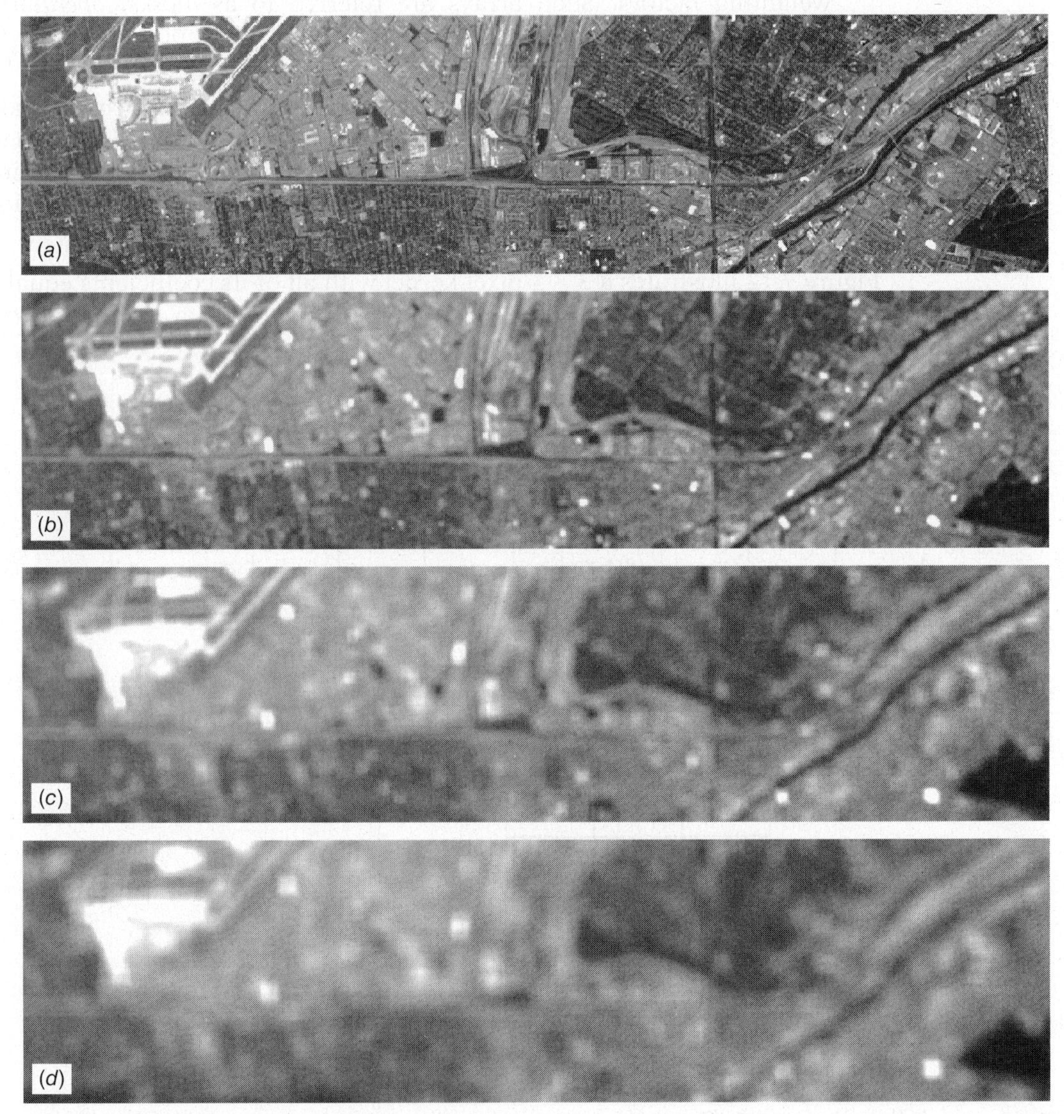

Figure 7.13 Frequency components of an image resulting from varying the kernel used for convolution: (*a*) original image; (*b–d*) successively lower frequency enhancements. (Author-prepared figure.)

The influence convolution may have on an image depends directly upon the size of the kernel used and the values of the coefficients contained within the kernel. The range of kernel sizes and weighting schemes is limitless. For example, by selecting the appropriate coefficients, one can center-weight kernels, make them of uniform weight, or shape them in accordance with a particular statistical model (such as a Gaussian distribution). In short, convolution is a generic image processing operation that has numerous applications in addition to spatial filtering. (Recall the use of "cubic convolution" as a resampling procedure.)

Edge Enhancement

We have seen that high frequency component images emphasize the spatial detail in digital images. That is, these images exaggerate local contrast and are superior to unenhanced original images for portraying linear features or edges in the image data. However, high frequency component images do not preserve the low frequency brightness information contained in original images. Edge-enhanced images attempt to preserve both local contrast and low frequency brightness information. They are produced by "adding back" all or a portion of the gray values in an original image to a high frequency component image of the same scene. Thus, edge enhancement is typically implemented in three steps:

1. A high frequency component image is produced containing the edge information. The kernel size used to produce this image is chosen based on the roughness of the image. "Rough" images suggest small filter sizes (e.g., 3×3 pixels), whereas large sizes (e.g., 9×9 pixels) are used with "smooth" images.
2. All or a fraction of the gray level in each pixel of the original scene is added back to the high frequency component image. (The proportion of the original gray levels to be added back may be chosen by the image analyst.)
3. The composite image is contrast stretched. This results in an image containing local contrast enhancement of high frequency features that also preserves the low frequency brightness information contained in the scene.

Directional first differencing is another enhancement technique aimed at emphasizing edges in image data. It is a procedure that systematically compares each pixel in an image to one of its immediately adjacent neighbors and displays the difference in terms of the gray levels of an output image. This process is mathematically akin to determining the first derivative of gray levels with respect to a given direction. The direction used can be horizontal, vertical, or diagonal. In Figure 7.14, a horizontal first difference at pixel A would result from subtracting the DN in pixel H from that in pixel A. A vertical first difference would result from

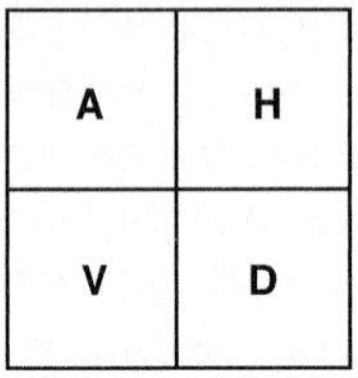

A	H
V	D

Horizontal first difference = $DN_A - DN_H$

Vertical first difference = $DN_A - DN_V$

Diagonal first difference = $DN_A - DN_D$

Figure 7.14 Primary pixel (A) and reference pixels (H, V, and D) used in horizontal, vertical, and diagonal first differencing, respectively.

subtracting the DN at pixel V from that in pixel A; a diagonal first difference would result from subtracting the DN at pixel D from that in pixel A.

It should be noted that first differences can be either positive or negative, so a constant such as the display value median (127 for 8-bit data) is normally added to the difference for display purposes. Furthermore, because pixel-to-pixel differences are often very small, the data in the enhanced image often span a very narrow range about the display value median and a contrast stretch must be applied to the output image.

First-difference images emphasize those edges normal to the direction of differencing and deemphasize those parallel to the direction of differencing. For example, in a horizontal first-difference image, vertical edges will result in large pixel-to-pixel changes in gray level. On the other hand, the vertical first differences for these same edges would be relatively small (perhaps zero). This effect is illustrated in Figure 7.15 where vertical features in the original image (*a*) are emphasized in the horizontal first-difference image (*b*). Horizontal features in the original image are highlighted in the vertical first-difference image (*c*). Features emphasized by the right and left diagonal first differences are shown in (*d*) and (*e*) respectively.

One popular nondirectional filter commonly used in edge enhancement is the *Laplacian filter*. This filter highlights all edges in an image, but it subdues the appearance of homogeneous or smoothly varying areas (Figure 7.15*f*). In many cases, the output from a Laplacian filter, or other nondirectional edge detector, is merged with (or added back to) the original image, as previously discussed (Figure 7.16).

Fourier Analysis

The spatial feature manipulations we have discussed thus far are implemented in the *spatial domain*—the (x,y) coordinate space of images. An alternative coordinate space that can be used for image analysis is the *frequency domain*. In this approach, an image is separated into its various spatial frequency components through

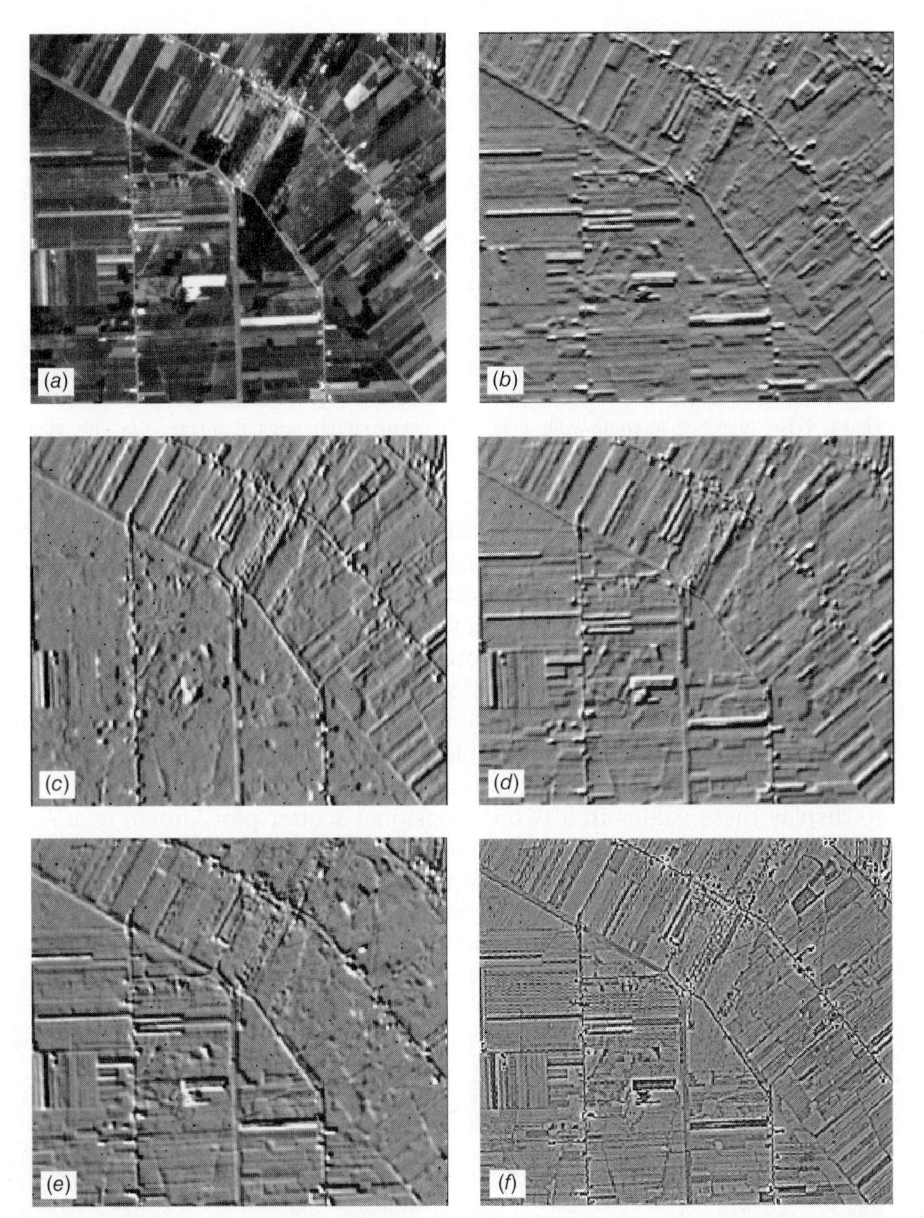

Figure 7.15 Edge enhancement through directional first differencing: (*a*) original image; (*b*) horizontal first difference; (*c*) vertical first difference; (*d*) right diagonal first difference; (*e*) left diagonal first difference; (*f*) Laplacian edge detector. (Author-prepared figure.)

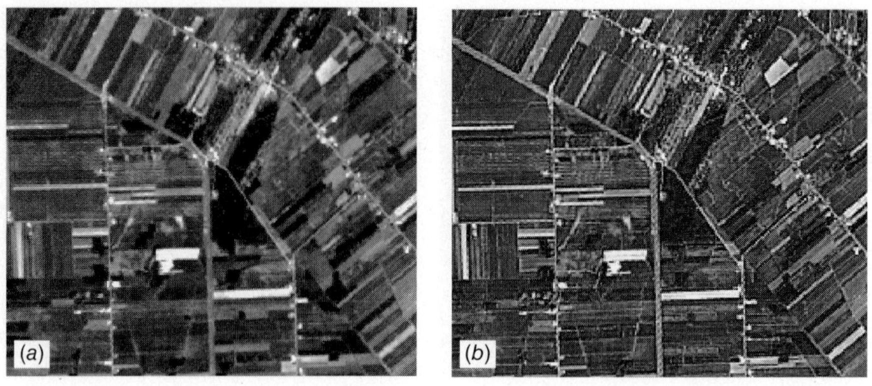

Figure 7.16 Image sharpening with edge enhancement: (*a*) original image; (*b*) with edge enhancement added back to original image. (Author-prepared figure.)

application of a mathematical operation known as the *Fourier transform*. A quantitative description of how Fourier transforms are computed is beyond the scope of this discussion. Conceptually, this operation amounts to fitting a continuous function through the discrete DN values if they were plotted along each row and column in an image. The "peaks and valleys" along any given row or column can be described mathematically by a combination of sine and cosine waves with various amplitudes, frequencies, and phases. A Fourier transform results from the calculation of the amplitude and phase for each possible spatial frequency in an image.

After an image is separated into its component spatial frequencies, it is possible to display these values in a two-dimensional scatter plot known as a *Fourier spectrum*. Figure 7.17 illustrates a digital image in (*a*) and its Fourier spectrum in (*b*). The lower frequencies in the scene are plotted at the center of the spectrum and

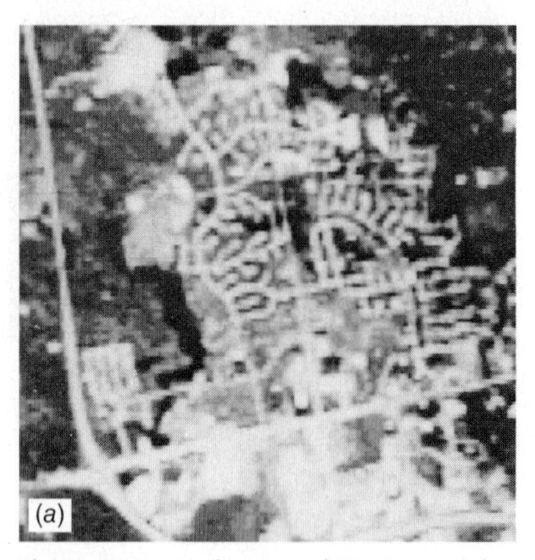

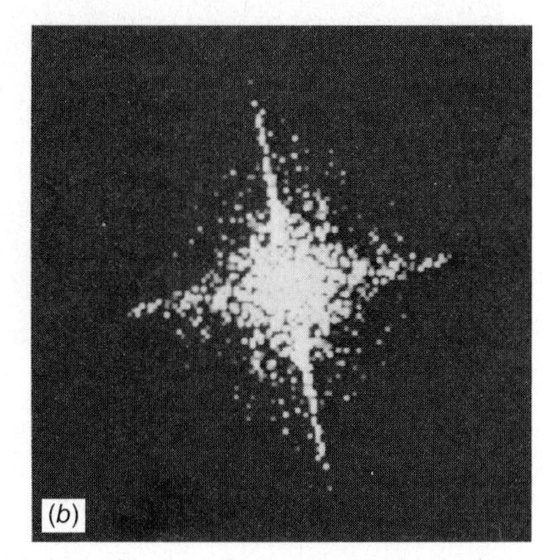

Figure 7.17 Application of Fourier transform: (*a*) original scene; (*b*) Fourier spectrum of (*a*). (Author-prepared figure.)

progressively higher frequencies are plotted outward. Features trending horizontally in the original image result in vertical components in the Fourier spectrum; features aligned vertically in the original image result in horizontal components in the Fourier spectrum.

If the Fourier spectrum of an image is known, it is possible to regenerate the original image through the application of an *inverse Fourier transform*. This operation is simply the mathematical reversal of the Fourier transform. Hence, the Fourier spectrum of an image can be used to assist in a number of image processing operations. For example, spatial filtering can be accomplished by applying a filter directly on the Fourier spectrum and then performing an inverse transform. This is illustrated in Figure 7.18. In Figure 7.18*a*, a circular high frequency

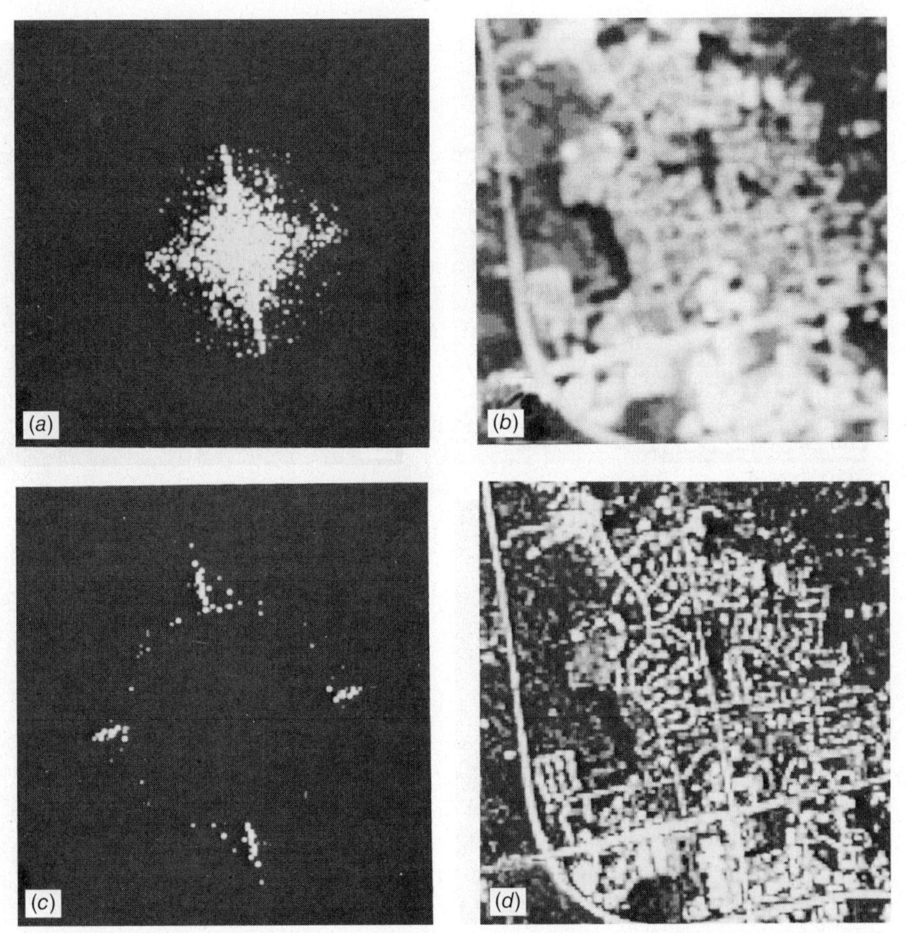

Figure 7.18 Spatial filtering in the frequency domain: (*a*) high frequency blocking filter; (*b*) inverse transform of (*a*); (*c*) low frequency blocking filter; (*d*) inverse transform of (*c*). (Author-prepared figure.)

blocking filter has been applied to the Fourier spectrum shown previously in Figure 7.17*b*. Note that this image is a low-pass filtered version of the original scene. Figures 7.18*c* and *d* illustrate the application of a circular low frequency blocking filter (*c*) to produce a high-pass filtered enhancement (*d*).

Figure 7.19 illustrates another common application of Fourier analysis—the elimination of image noise. Shown in Figure 7.19*a* is an airborne multispectral scanner image containing substantial noise. The Fourier spectrum for the image is shown in Figure 7.19*b*. Note that the noise pattern, which occurs in a horizontal direction in the original scene, appears as a band of frequencies trending in the vertical direction in the Fourier spectrum. In Figure 7.19*c* a vertical *wedge block filter* has been applied to the spectrum. This filter passes the lower

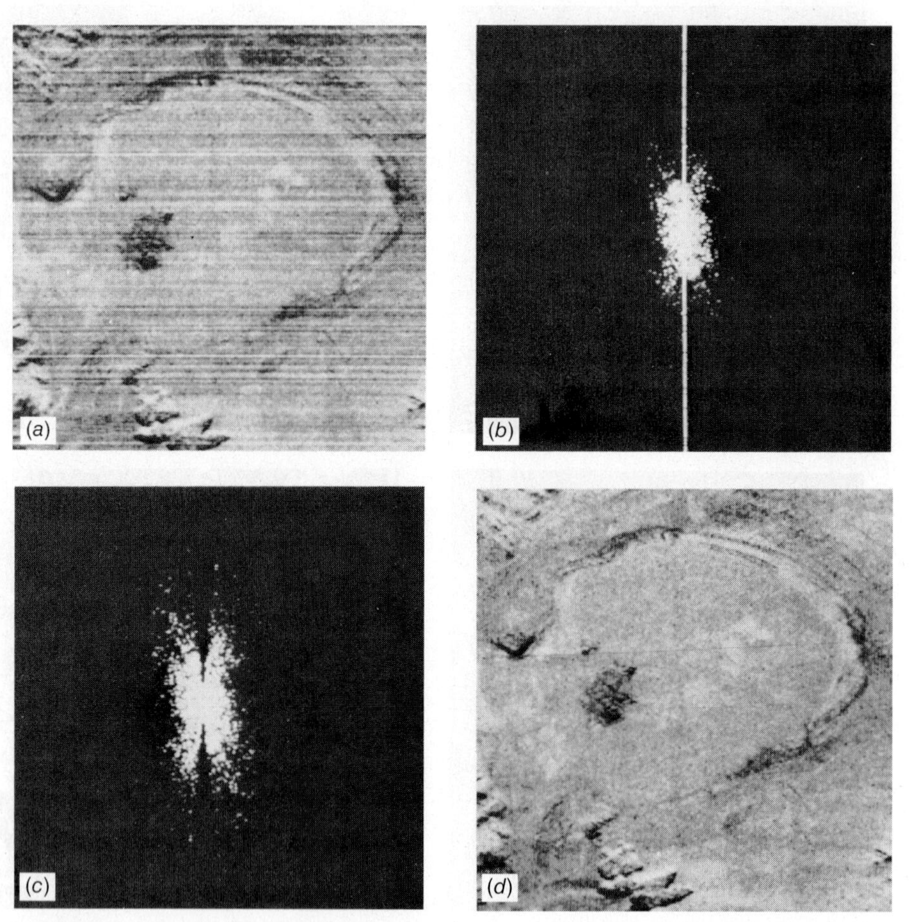

Figure 7.19 Noise elimination in the frequency domain. (*a*) Airborne multispectral scanner image containing noise. (Courtesy NASA.) (*b*) Fourier spectrum of (*a*). (*c*) Wedge block filter. (*d*) Inverse transform of (*c*). (Author-prepared figure.)

frequency components of the image but blocks the high frequency components of the original image trending in the horizontal direction. Figure 7.19*d* shows the inverse transform of (*c*). Note how effectively this operation eliminates the noise inherent in the original image.

7.6 MULTI-IMAGE MANIPULATION

Spectral Ratioing

Ratio images are enhancements resulting from the division of DN values in one spectral band by the corresponding values in another band. A major advantage of ratio images is that they convey the spectral or color characteristics of image features, regardless of variations in scene illumination conditions. This concept is illustrated in Figure 7.20, which depicts two different land cover types (deciduous and coniferous trees) occurring on both the sunlit and shadowed sides of a ridge line. The DNs observed for each cover type are substantially lower in the shadowed area than in the sunlit area. However, the ratio values for each cover type are nearly identical, irrespective of the illumination condition. Hence, a ratioed image of the scene effectively compensates for the brightness variation caused by the varying topography and emphasizes the color content of the data.

Ratioed images are often useful for discriminating subtle *spectral* variations in a scene that are masked by the *brightness* variations in images from individual spectral bands or in standard color composites. This enhanced discrimination is due to the fact that ratioed images clearly portray the variations in the *slopes* of the

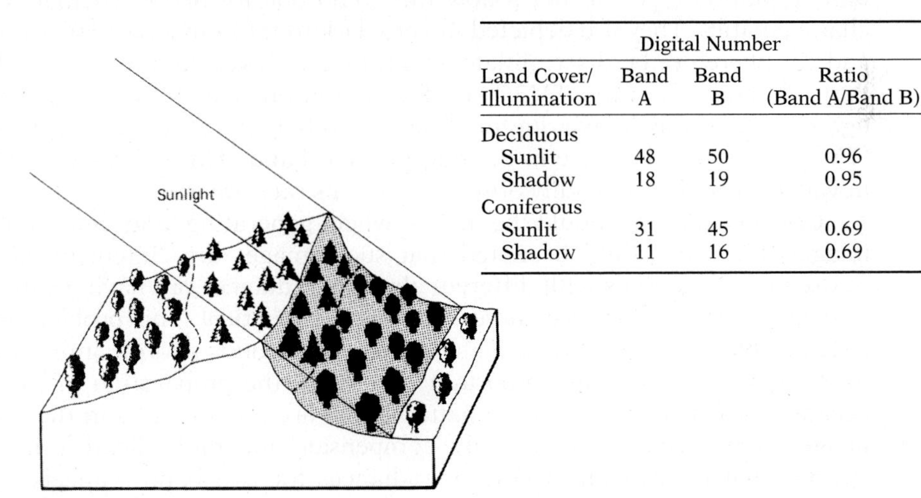

Land Cover/ Illumination	Digital Number Band A	Band B	Ratio (Band A/Band B)
Deciduous			
Sunlit	48	50	0.96
Shadow	18	19	0.95
Coniferous			
Sunlit	31	45	0.69
Shadow	11	16	0.69

Figure 7.20 Reduction of scene illumination effects through spectral ratioing. (Adapted from Sabins, 1997.)

spectral reflectance curves between the two bands involved, regardless of the absolute reflectance values observed in the bands. These slopes are typically quite different for various material types in certain bands of sensing. For example, the near-infrared-to-red ratio for healthy vegetation is normally very high. That for stressed vegetation is typically lower (as near-infrared reflectance decreases and the red reflectance increases). Thus a near-infrared-to-red (or red-to-near-infrared) ratioed image might be very useful for differentiating between areas of the stressed and nonstressed vegetation. This type of ratio has also been employed extensively in vegetation indices aimed at quantifying relative vegetation greenness and biomass.

Obviously, the utility of any given spectral ratio depends upon the particular reflectance characteristics of the features involved and the application at hand. The form and number of ratio combinations available to the image analyst also varies depending upon the source of the digital data. The number of possible ratios that can be developed from n bands of data is $n(n-1)$. Thus, for the six nonthermal bands of Landsat TM or ETM+ data there are $6(6-1)$, or 30, possible combinations.

Figure 7.21 illustrates four representative ratio images generated from TM data. These images depict higher ratio values in brighter tones. Shown in (*a*) is the ratio TM1/TM2. Because these two bands are highly correlated for this scene, the ratio image has low contrast. In (*b*) the ratio TM3/TM4 is depicted so that features such as water and roads, which reflect highly in the red band (TM3) and little in the near-infrared band (TM4), are shown in lighter tones. Features such as vegetation appear in darker tones because of their relatively low reflectance in the red band (TM3) and high reflectance in the near infrared (TM4). In (*c*) the ratio TM5/TM2 is shown. Here, vegetation generally appears in light tones because of its relatively high reflectance in the mid-infrared band (TM5) and its comparatively lower reflectance in the green band (TM2). However, note that certain vegetation types do not follow this trend due to their particular reflectance characteristics. They are depicted in very dark tones in this particular ratio image and can therefore be discriminated from the other vegetation types in the scene. Part (*d*) shows the ratio TM3/TM7. Roads and other cultural features appear in lighter tone in this image due to their relatively high reflectance in the red band (TM3) and low reflectance in the mid-infrared band (TM7). Similarly, differences in water turbidity are readily observable in this ratio image.

Certain caution should be taken when generating and interpreting ratio images. First, it should be noted that such images are "intensity blind." That is, dissimilar materials with different absolute radiances but having similar slopes of their spectral reflectance curves may appear identical. This problem is particularly troublesome when these materials are contiguous and of similar image texture. Noise removal is an important prelude to the preparation of ratio images because ratioing enhances noise patterns that are uncorrelated in the component images. Furthermore, ratios only compensate for multiplicative illumination effects. That is, division of DNs or radiances for two bands cancels only those

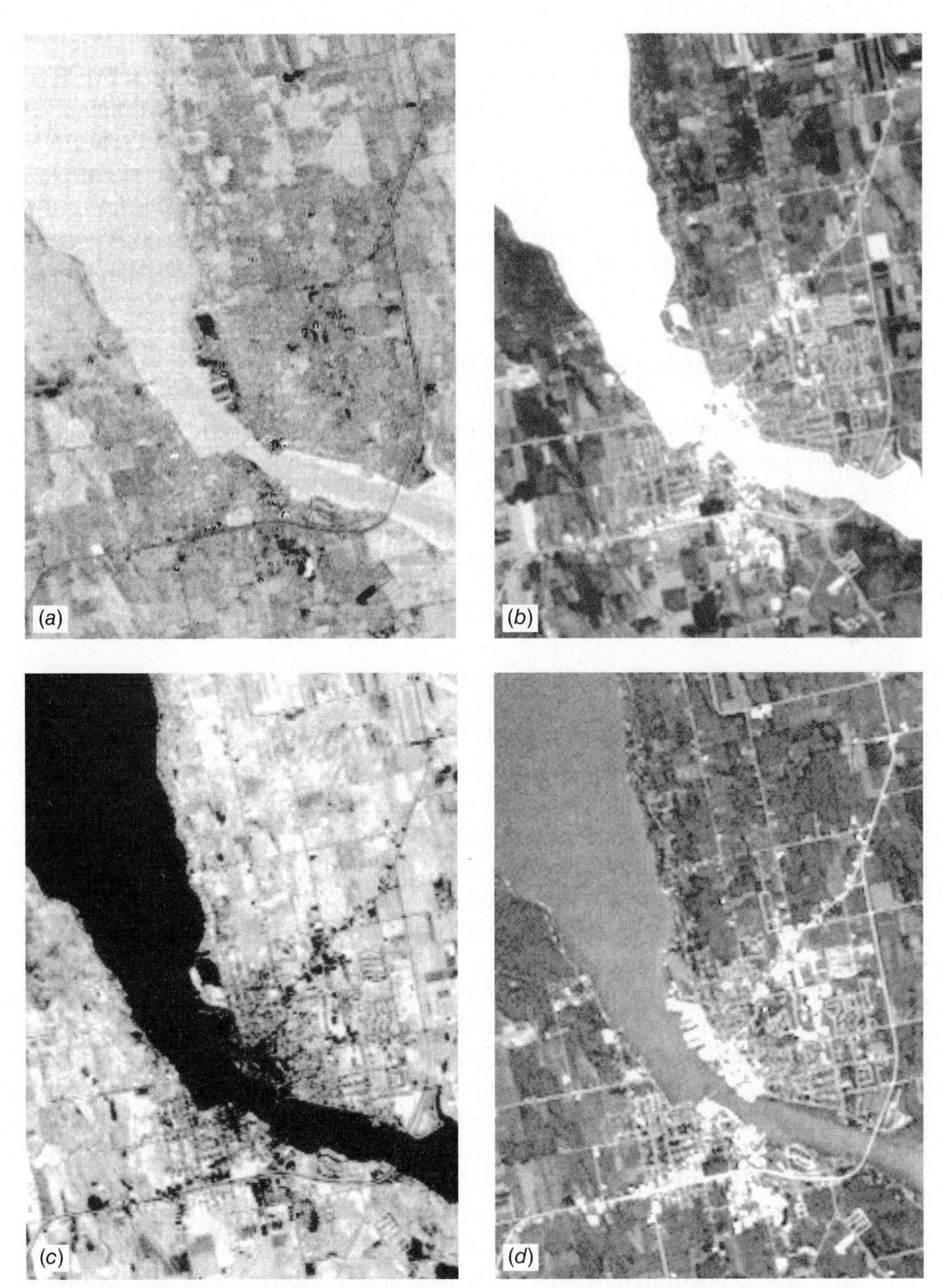

Figure 7.21 Ratioed images derived from midsummer Landsat TM data, near Sturgeon Bay, WI (higher ratio values are displayed in brighter image tones): (*a*) TM1/TM2; (*b*) TM3/TM4; (*c*) TM5/TM2; (*d*) TM3/TM7. (Author-prepared figure.)

factors that are operative equally in the bands and not those that are additive. For example, atmospheric haze is an additive factor that might need to be removed prior to ratioing to yield acceptable results. It should also be noted that ratios can "blow up" mathematically (be undefined) if the band in the denominator has a DN of zero. At the same time, ratios less than 1 are common and rounding to integer values will compress much of the ratio data into gray level values of 0 or 1. Hence, it is important to store ratio values as floating-point numbers rather than integers, or to scale the results of ratio computations over a wider range.

Normalized Difference Ratios and Other Indices

Several of the limitations of simple ratios are addressed by the use of *normalized difference indices* and other mathematical variants. One of the most widely used normalized difference indices is the Normalized Difference Vegetation Index (NDVI), discussed in Chapter 5 and illustrated in Plate 18. This index is based on the difference of reflectance in the near-infrared and red bands:

$$\mathrm{NDVI} = \frac{\rho_{\mathrm{NIR}} - \rho_{\mathrm{RED}}}{\rho_{\mathrm{NIR}} + \rho_{\mathrm{RED}}} \tag{7.7}$$

where ρ_{NIR} and ρ_{RED} are the spectral reflectance in the sensor's near-infrared and red bands, respectively. Clearly, high NDVI values will result from the combination of a high reflectance in the near infrared and lower reflectance in the red band. This combination is typical of the spectral "signature" of vegetation, as discussed in Chapter 1. Non-vegetated areas, including bare soil, open water, snow/ice, and most construction materials, will have much lower NDVI values.

Figure 7.22 shows examples of NDVI data derived from two dates of Landsat imagery at a site in China's far western Xinjiang region. Bright pixels correspond to densely vegetated land, primarily irrigated agricultural fields growing cotton and other crops; these areas have NDVI values greater than 0.2. The darkest portions of the NDVI images (with NDVI values below 0) are nonvegetated, including open water in some river channels and areas of bare sand in the desert. Intermediate gray tones represent areas of natural vegetation, primarily *Populus euphratica* woodlands, various types of shrublands, and other riparian plant communities found along current and former river channels. (Further examples of NDVI data for this same area, from MODIS rather than Landsat, are shown in Figures 7.58 and 7.59 and are discussed in Section 7.19.)

NDVI is widely used for vegetation monitoring and assessment. It has been shown to be well correlated not only with crop biomass accumulation, but also with leaf chlorophyll levels, leaf area index values, and the photosynthetically active radiation absorbed by a crop canopy. However, in some conditions, NDVI becomes less effective, and thus a variety of alternative vegetation indices have been proposed to replace or supplement it. The *Soil Adjusted Vegetation Index*

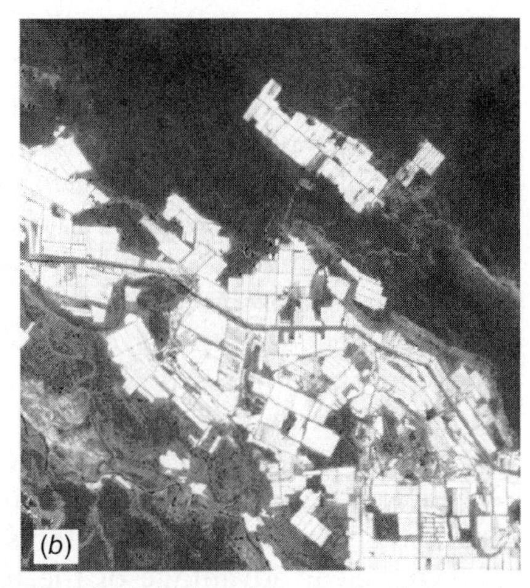

Figure 7.22 Normalized Difference Vegetation Index (NDVI) images from Landsat imagery in (*a*) 2000 and (*b*) 2011, Xinjiang, China. Brightest patches are irrigated agricultural fields; intermediate gray tones are riparian vegetation along current or former river channels; dark areas are sparsely vegetated or desert. (Author-prepared figure.)

(SAVI; Huete, 1988) is designed to compensate for a soil background in sparsely vegetated areas:

$$\text{SAVI} = \left(\frac{\rho_{\text{NIR}} - \rho_{\text{RED}}}{\rho_{\text{NIR}} + \rho_{\text{RED}} + L}\right)(1 + L) \tag{7.8}$$

where L is a correction factor between 0 and 1, with a default value of 0.5 (as L approaches 0, SAVI becomes close to NDVI). At the opposite extreme, when vegetation cover reaches moderate to high levels, the *green normalized difference vegetation index* (GNDVI) may be a more reliable indicator of crop conditions. The GNDVI is identical in form to the NDVI except that the green band is substituted for the red band. Similarly, the *wide dynamic range vegetation index* (WDRVI) has been shown to have improved sensitivity to moderate-to-high levels of photosynthetic green biomass (Viña, Henebry, and Gitelson, 2004).

With the availability of MODIS data on a global basis, several new vegetation indices have been proposed. For example, the *enhanced vegetation index* (EVI) has been developed as a modified NDVI with an adjustment factor to minimize soil background influences and a blue band correction of red band data to lessen atmospheric scattering. (An example of EVI data from MODIS is presented in Plate 32 and discussed in Section 7.19.) Likewise, the *MERIS terrestrial chlorophyll index* (*MTCI*) (Dash and Curran, 2004) has been developed to optimize the measurement of chlorophyll using the standard band settings of the MERIS sensor. Literally dozens of other vegetation indices have been proposed, and as new sensors provide access to spectral bands that were not previously included on spaceborne remote sensing platforms, it is likely that further entries in this list will be developed in the future.

Beyond vegetation, other mathematical indices have been described in which combinations of spectral bands are used to represent different surface feature types. For example, the *Normalized Difference Snow Index* (NDSI; Hall et al., 1998) is based on bands in the green and mid-infrared regions, such as Landsat TM/ETM+ bands 2 and 5, or Landsat-8 bands 3 and 6. Because snow reflects strongly in the visible spectrum but absorbs radiation at 1.6 μm in the mid-IR, the normalized difference of these two bands emphasizes snow cover. Figure 7.23 shows raw imagery from the green and mid-infrared bands (7.23*a* and *b*, respectively) for a site in the Andes Mountains of Peru. Snow and ice appear bright in the green band and quite dark in the mid-infrared. In the resulting NDSI image (*c*), snow and ice cover are highlighted. (Open water also has a high value in NDSI, but can easily be differentiated from snow in the visible region, where water is dark while snow is bright.)

One clear advantage of the normalized difference indices is that, unlike the simple ratios described previously, they are constrained to fall within the range of -1 to $+1$. Thus, they avoid some of the scaling difficulties presented by the simple ratios.

Principal and Canonical Components

Extensive interband correlation is a problem frequently encountered in the analysis of multispectral image data. That is, images generated by digital data from various wavelength bands often appear similar and convey essentially the same information. Principal and canonical component transformations are two techniques designed to reduce such redundancy in multispectral data. These transformations may be applied either as an enhancement operation prior to visual interpretation of the data or as a preprocessing procedure prior to automated classification of the data. If employed in the latter context, the transformations generally increase the computational efficiency of the classification process because both principal and canonical component analyses may result in a reduction in the dimensionality of the original data set. Stated differently, the purpose of these procedures is to compress all of the information contained in an original n-band data set into fewer than n "new bands." The new bands are then used in lieu of the original data.

A detailed description of the statistical procedures used to derive principal and canonical component transformations is beyond the scope of this discussion. However, the concepts involved may be expressed graphically by considering a two-band image data set such as that shown in Figure 7.24. In (*a*), a random sample of pixels has been plotted on a scatter diagram according to their gray levels as originally recorded in bands A and B. Superimposed on the band A–band B axis system are two new axes (axes I and II) that are rotated with respect to the

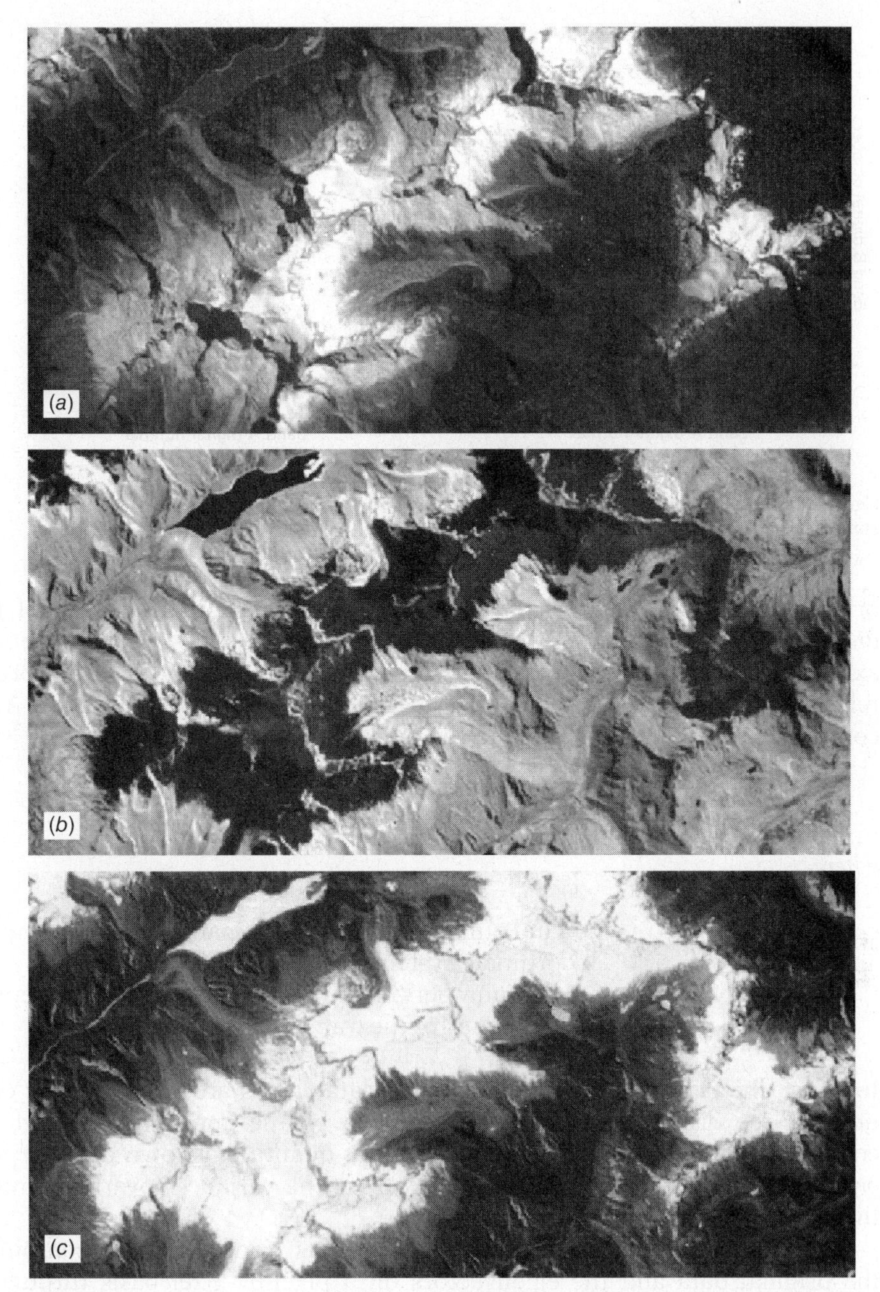

Figure 7.23 Landsat-7 ETM+ imagery and derived Normalized Difference Snow Index (NDSI) images of glaciers and snowfields in the Andes mountains, Peru. (*a*) ETM+ band 2 (green). (*b*) ETM+ band 5 (mid-infrared). (*c*) NDSI. (Author-prepared figure.)

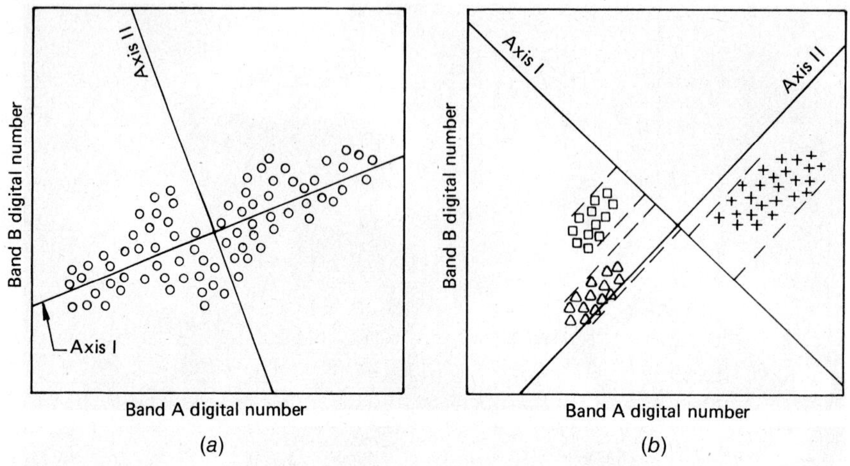

Figure 7.24 Rotated coordinate axes used in (*a*) principal component and (*b*) canonical component transformations.

original measurement axes and that have their origin at the mean of the data distribution. Axis I defines the direction of the *first principal component,* and axis II defines the direction of the *second principal component*. The form of the relationship necessary to transform a data value in the original band A–band B coordinate system into its value in the new axis I–axis II system is

$$\begin{aligned} \mathrm{DN_I} &= a_{11}\mathrm{DN_A} + a_{12}\mathrm{DN_B} \\ \mathrm{DN_{II}} &= a_{21}\mathrm{DN_A} + a_{22}\mathrm{DN_B} \end{aligned} \tag{7.9}$$

where

$\mathrm{DN_I}, \mathrm{DN_{II}}$ = digital numbers in new (principal component image) coordinate system

$\mathrm{DN_A}, \mathrm{DN_B}$ = digital numbers in old (original) coordinate system

$a_{11}, a_{12}, a_{21}, a_{22}$ = coefficients for the transformation

In short, the principal component image data values are simply linear combinations of the original data values multiplied by the appropriate transformation coefficients. These coefficients are statistical quantities known as *eigenvectors* or *principal components*. They are derived from the variance/covariance matrix for the original image data.

Hence, a principal component image results from the linear combination of the original data and the eigenvectors on a pixel-by-pixel basis throughout the image. Often, the resulting principal component *image* is loosely referred to as simply a *principal component*. This is theoretically incorrect in that the eigenvalues themselves are the principal components, but we will sometimes not make this distinction elsewhere in this book. It should be noted in Figure 7.24*a* that the data along the direction of the first principal component (axis I) have a greater variance or dynamic range than the data plotted against either of the original axes (bands A and B). The data along the second principal component direction have far less variance. This is characteristic of all principal component images. In general, the first principal component image (PC1) includes the largest percentage of the total scene variance and succeeding component images (PC2, PC3, . . . , PCn) each contain a decreasing percentage of the scene variance. Furthermore, because successive components are chosen to be orthogonal to all previous ones, the data they contain are uncorrelated.

Principal component enhancements are generated by displaying contrast-stretched images of the transformed pixel values. We illustrate the nature of these displays by considering the Landsat MSS images shown in Figure 7.25. This figure depicts the four MSS bands of a scene covering the Sahl al Matran area, Saudi Arabia. Figure 7.26 shows the principal component images for this scene. Some areas of geologic interest labeled in Figure 7.25 are (A) alluvial material in a dry stream valley, (B) flat-lying quaternary and tertiary basalts, and (C) granite and granodiorite intrusion.

Note that in Figure 7.26, PC1 expresses the majority (97.6%) of the variance in the original data set. Furthermore, PC1 and PC2 explain virtually all of the variance in the scene (99.4%). This compression of image information in the first two principal component images of Landsat MSS data is typical. Because of this, we refer to the *intrinsic dimensionality* of Landsat MSS data as being effectively 2. Also frequently encountered with Landsat MSS data, the PC4 image for this scene contains virtually no information and tends to depict little more than system noise. However, note that both PC2 and PC3 illustrate certain features that were obscured by the more dominant patterns shown in PC1. For example, a semi-circular feature (labeled C in Figure 7.25) is clearly defined in the upper right portion of the PC2 and PC3 images (appearing bright and dark, respectively). This feature was masked by more dominant patterns both in the PC1 image and in all bands of the original data. Also, its tonal reversal in PC2 and PC3 illustrates the lack of correlation between these images.

Figure 7.27 illustrates the use of principal component analysis of MODIS data to enhance and track the extent of a massive smoke plume resulting from a fire at a tire recycling facility located in Watertown, Wisconsin. (Section 8.13 includes an oblique airphoto and Landsat TM image of this fire.) The Aqua satellite passed over this fire just over four hours after it began. By this time, the plume of smoke extended 150 km to the southeast, stretching across the city of Milwaukee, Wisconsin, and over central Lake Michigan. Note that it is very difficult to discriminate the extent of the plume extending over Lake Michigan in the "raw images" shown in Figure 7.27*a* and the enlargement of this figure shown in Figure 7.27*b*. However, the full extent of the plume is very apparent in Figure 7.27*c*, which helps increase the contrast between the plume and the underlying land and water surfaces. Figure 7.27*c* was generated from the third principal component of MODIS bands 1–5 and 8.

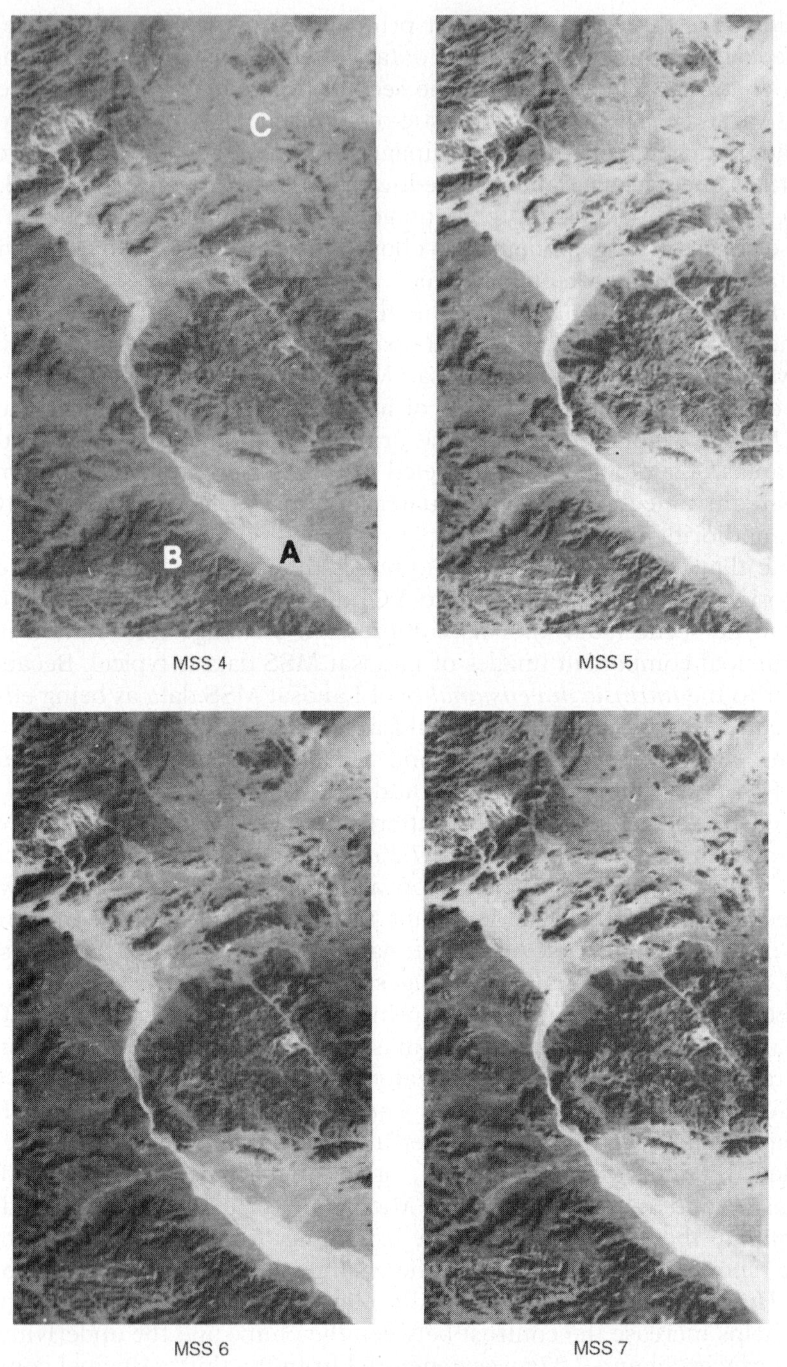

Figure 7.25 Four MSS bands covering the Sahl al Matran area of Saudi Arabia. Note the redundancy of information in these original image displays. (Courtesy NASA.)

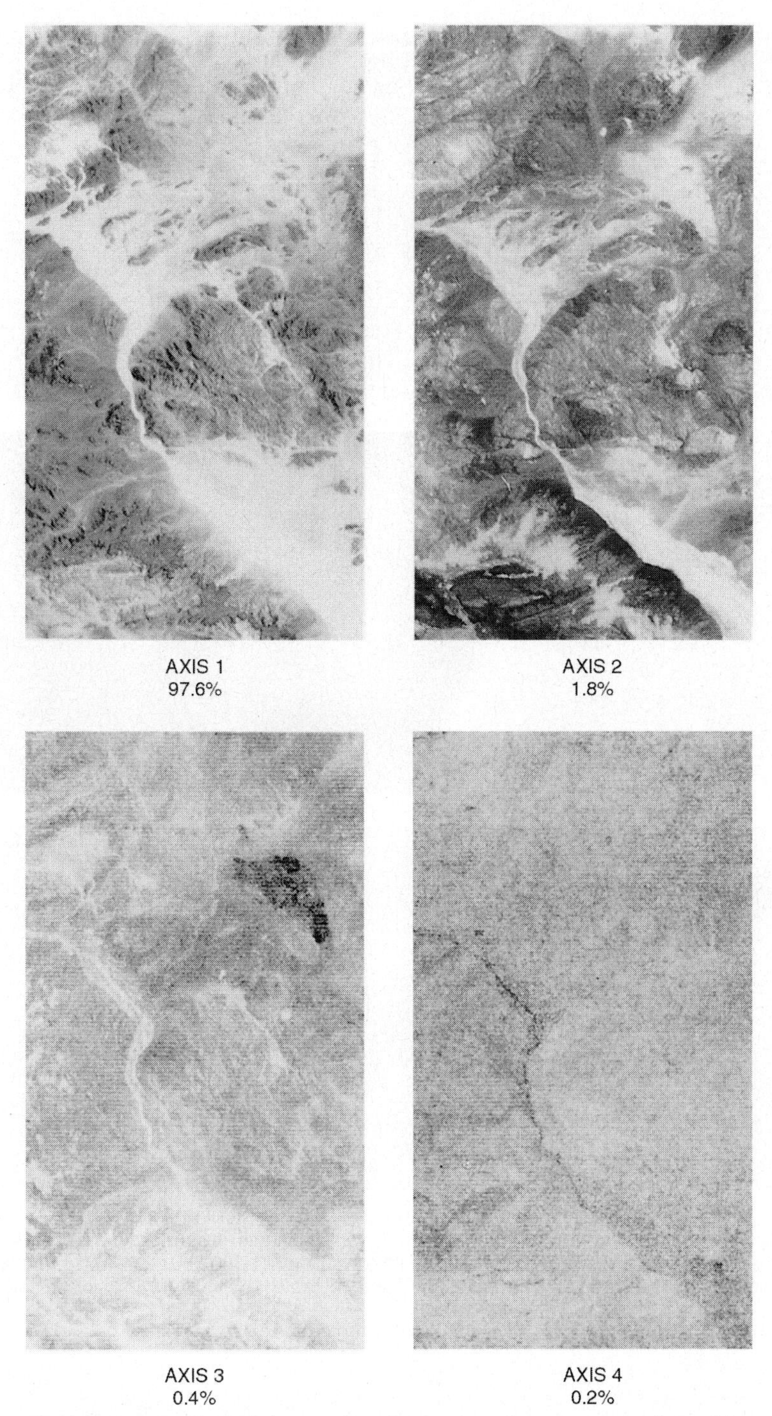

Figure 7.26 Transformed data resulting from principal component analysis of the MSS data shown in Figure 7.25. The percentage of scene variance contained in each axis is indicated. (Courtesy NASA.)

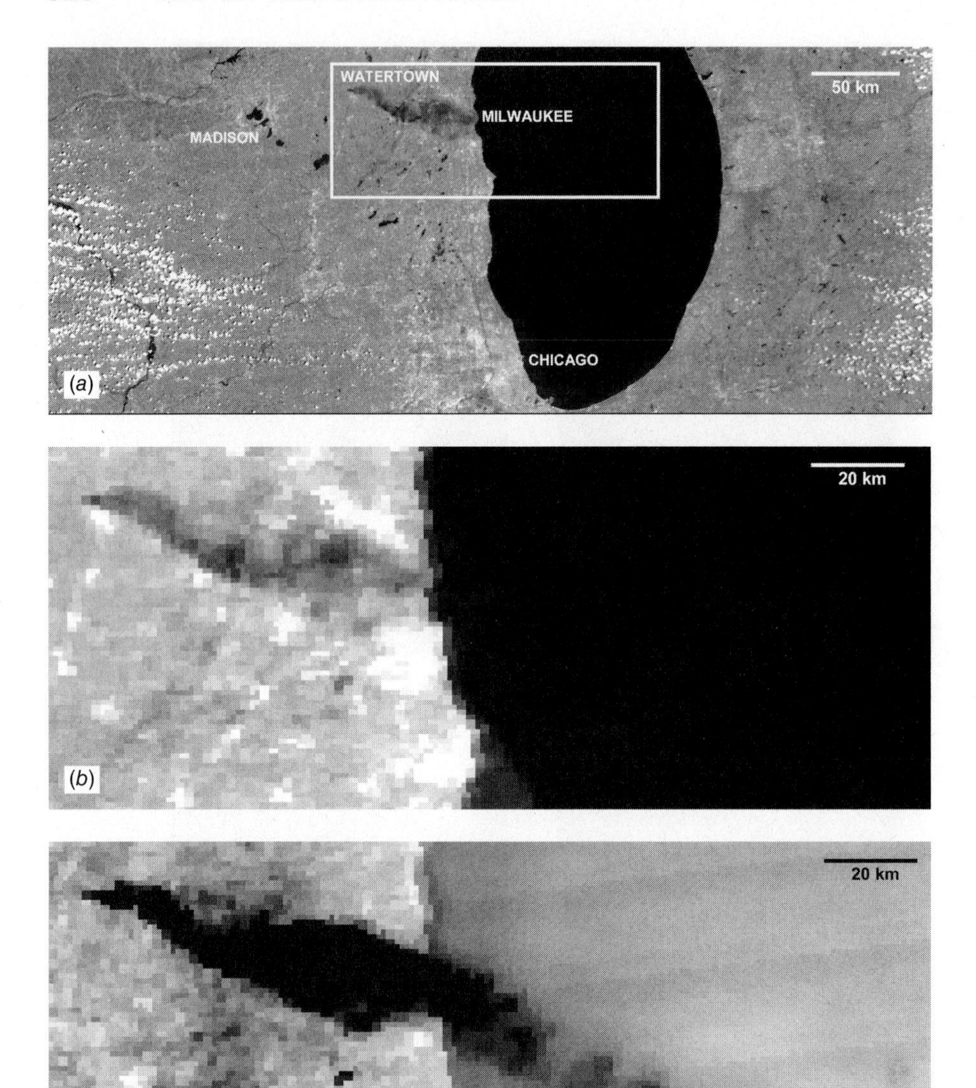

Figure 7.27 Use of principal component analysis to increase the contrast of a smoke plume with underlying land and water areas. (*a*) Portion of raw MODIS regional image. Scale 1:4,000,000. (*b*) Enlargement of plume area in MODIS band 4 image. Scale 1:1,500,000. (*c*) Third principal component image derived from MODIS bands 1–5 and 8. Scale 1:1,500,000. (Author-prepared figure.)

As in the case of ratio images, principal component images can be analyzed as separate black-and-white images (as illustrated here), or any three component images may be combined to form a color composite. If used in an image classification process, principal component data are normally treated in the classification algorithm simply as if they were original data. However, the number of components used is normally reduced to the intrinsic dimensionality of the data, thereby making the image classification process much more efficient by reducing the amount of computation required. (For example, the visible, near-infrared, and mid-infrared bands on Landsat TM, ETM+, and OLI can often be reduced to just three principal component images for classification purposes.)

Principal component enhancement techniques are particularly appropriate where little prior information concerning a scene is available. *Canonical component analysis*, also referred to as multiple discriminant analysis, may be more appropriate when information about particular features of interest is known. Recall that the principal component axes shown in Figure 7.24*a* were located on the basis of a random, undifferentiated sample of image pixel values. In Figure 7.24*b*, the pixel values shown are derived from image areas containing three different analyst-defined feature types (the feature types are represented by the symbols Δ, □, and +). The canonical component axes in this figure (axes I and II) have been located to maximize the *separability* of these classes while minimizing the variance within each class. For example, the axes have been positioned in this figure such that the three feature types can be discriminated solely on the basis of the first canonical component images (CC1) values located along axis I. Canonical component images not only improve classification efficiency but also can improve classification accuracy for the identified features due to the increased spectral separability of classes.

Vegetation Components

In addition to the vegetation indices described previously, various other forms of linear data transformations have been developed for vegetation monitoring, with differing sensors and vegetation conditions dictating different transformations. The *"tasseled cap" transformation* produces a set of *vegetation components* useful for agricultural crop monitoring. The majority of information is contained in two or three components that are directly related to physical scene characteristics. Brightness, the first component, is a weighted sum of all bands and is defined in the direction of the principal variation in soil reflectance. The second component, greenness, is approximately orthogonal to brightness and is a contrast between the near-infrared and visible bands. Greenness is strongly related to the amount of green vegetation present in the scene. A third component, called wetness, relates to canopy and soil moisture.

Vegetation components can be used in many ways. For example, Bauer and Wilson (2005) have related the greenness component of the tasseled cap transformation of Landsat TM data to the amount of green vegetation as a measure of impervious surface area in urban or developed areas (roads, parking lots, rooftops, etc.). In this case, the greenness component is used to indicate the relative *lack* of vegetation. Although the majority of TM pixels in an urban area are mixtures of two or more cover types, the method affords a means of estimating the fraction of each pixel that is impervious. The approach involves converting the six reflective bands of TM data to greenness and then establishing a polynomial regression model of the relationship between greenness and the percent of impervious area at the individual pixel level. The regression is established by carefully measuring the amount of impervious surfaces in approximately 50 selected sample areas from higher resolution data that are registered to the Landsat imagery. A range of kinds and amounts of impervious areas are used in this process. Once the regression model is developed, it is applied to all of the pixels classified as urban or developed in the study area.

Figures 7.28 and 7.29 illustrate the application of this procedure to a Landsat-5 TM image acquired over Woodbury, Minnesota, a rapidly growing suburb east of St. Paul. Shown in Figures 7.28*a* and 7.28*b* are the original raw TM band 3 (red) and TM band 4 (near-IR) images for the scene, respectively. The tasseled cap relative brightness, greenness, and wetness images are shown in parts (*c*), (*d*), and (*e*), respectively. The regression relating the percent impervious area of a pixel to its greenness value is shown in (*f*).

Figure 7.29 illustrates the application of regressions developed from both a 1986 image and the 2002 image shown here. Note the dramatic increase in the amount of impervious surface area over this time period. Such information is extremely valuable in support of developing "best management practices" for dealing with increased storm water runoff velocities and volumes, as well as minimizing pollution from a given area. Within Minnesota, this approach has been employed at local to regional and statewide scales. Impervious surface (and forest canopy density) data for the United States as a whole have also been generated as part of the National Land Cover Database (NLDC) program (Homer et al., 2004).

Intensity–Hue–Saturation Color Space Transformation

Digital images are typically displayed as additive color composites using the three primary colors: red, green, and blue (RGB). Figure 7.30 illustrates the interrelation among the RGB components of a typical color display device (such as a color monitor). Shown in this figure is the RGB *color cube*, which is defined by the brightness levels of each of the three primary colors. For a display with 8-bit-per-pixel data encoding, the range of possible DNs for each color component is 0 to 255. Hence, there are 256^3 (or 16,777,216) possible combinations of red, green, and blue DNs that can be displayed by such a device. Every pixel in a composited display may be represented by a three-dimensional coordinate position somewhere

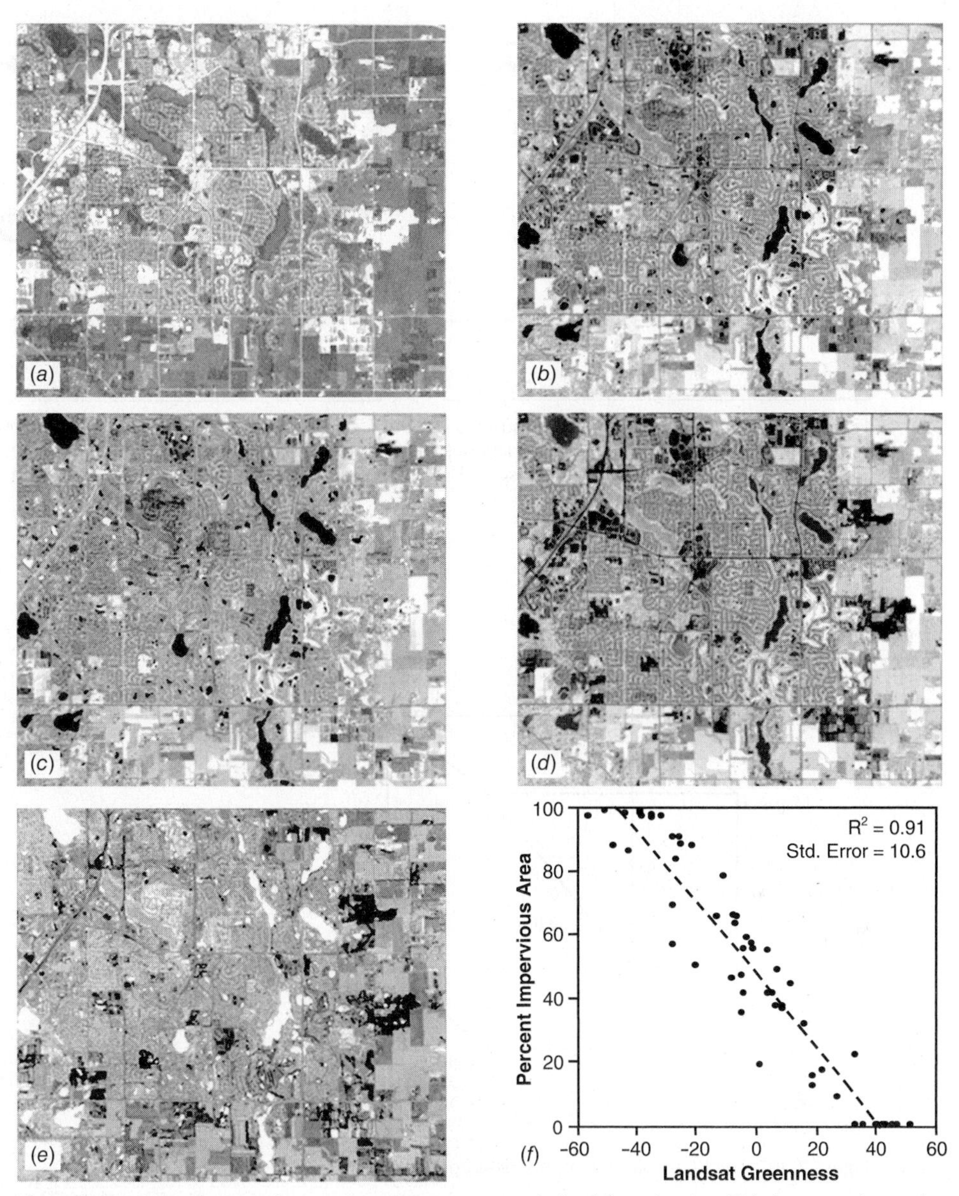

Figure 7.28 Use of tasseled cap greenness component derived from Landsat TM data to estimate the percentage of each pixel that is impervious in an image of Woodbury, MN. (*a*) Original TM band 3 (red) image. (*b*) Original TM band 4 (near-IR) image. (*c*) Brightness image. (*d*) Greenness image. (*e*) Wetness image. (*f*) Regression relationship relating percent impervious area of a pixel to greenness. (Courtesy University of Minnesota, Remote Sensing and Geospatial Analysis Laboratory, and the North American Lake Management Society.)

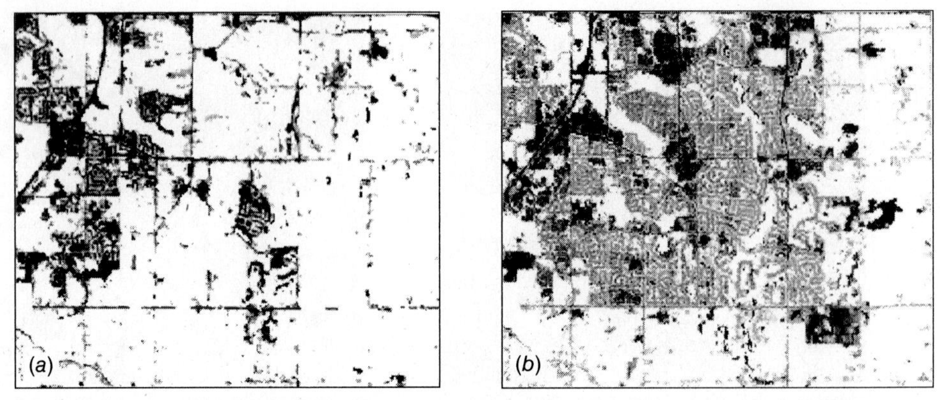

Figure 7.29 Classification of imperviousness for the City of Woodbury, MN. (*a*) 1986 impervious surface area 789 ha (8.5%) (*b*) 2002 impervious surface area 1796 ha (19.4%). (Courtesy University of Minnesota, Remote Sensing and Geospatial Analysis Laboratory, and the North American Lake Management Society.)

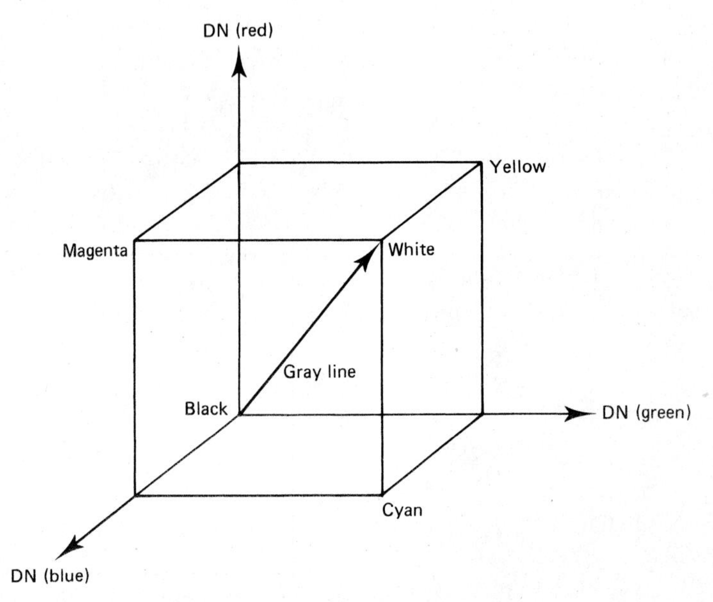

Figure 7.30 The RGB color cube. (Adapted from Schowengerdt, 1997.)

within the color cube. The line from the origin of the cube to the opposite corner is known as the *gray line* because DNs that lie on this line have equal components of red, green, and blue.

The RGB displays are used extensively in digital processing to display normal color, false color infrared, and arbitrary color composites. For example, a normal color composite may be displayed by assigning TM or ETM+ bands 1, 2, and 3 to the blue, green, and red components, respectively. A false color infrared composite results when bands 2, 3, and 4 are assigned to these respective components. Arbitrary color composites result when other bands or color assignments are used. Color composites may be contrast stretched on a RGB display by manipulating the contrast in each of the three display channels (using a separate lookup table for each of the three color components).

An alternative to describing colors by their RGB components is the use of the *intensity–hue–saturation* (*IHS*) system. *Intensity* relates to the total brightness of a color. *Hue* refers to the dominant or average wavelength of light contributing to a color. *Saturation* specifies the purity of color relative to gray. For example, pastel colors such as pink have low saturation compared to such high saturation colors as crimson. Transforming RGB components into IHS components *before* processing may provide more control over color enhancements.

Figure 7.31 shows one (of several) means of transforming RGB components into IHS components. This particular approach is called the *hexcone model,* and it

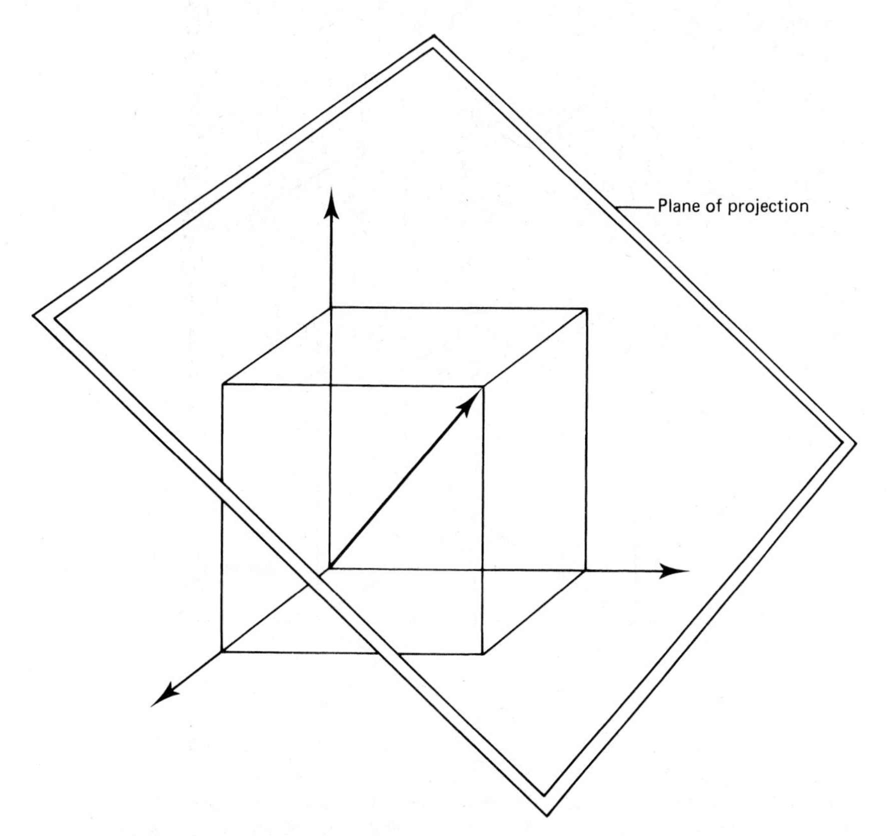

Figure 7.31 Planar projection of the RGB color cube. A series of such projections results when progressively smaller subcubes are considered between white and black.

involves the *projection* of the RGB color cube onto a plane that is perpendicular to the gray line and tangent to the cube at the corner farthest from the origin. The resulting projection is a hexagon. If the plane of projection is moved from white to black along the gray line, successively smaller color *subcubes* are projected and a series of hexagons of decreasing size result. The hexagon at white is the largest and the hexagon at black degenerates to a point. The series of hexagons developed in this manner define a solid called the *hexcone* (Figure 7.32*a*).

In the hexcone model intensity is defined by the distance along the gray line from black to any given hexagonal projection. Hue and saturation are defined at a

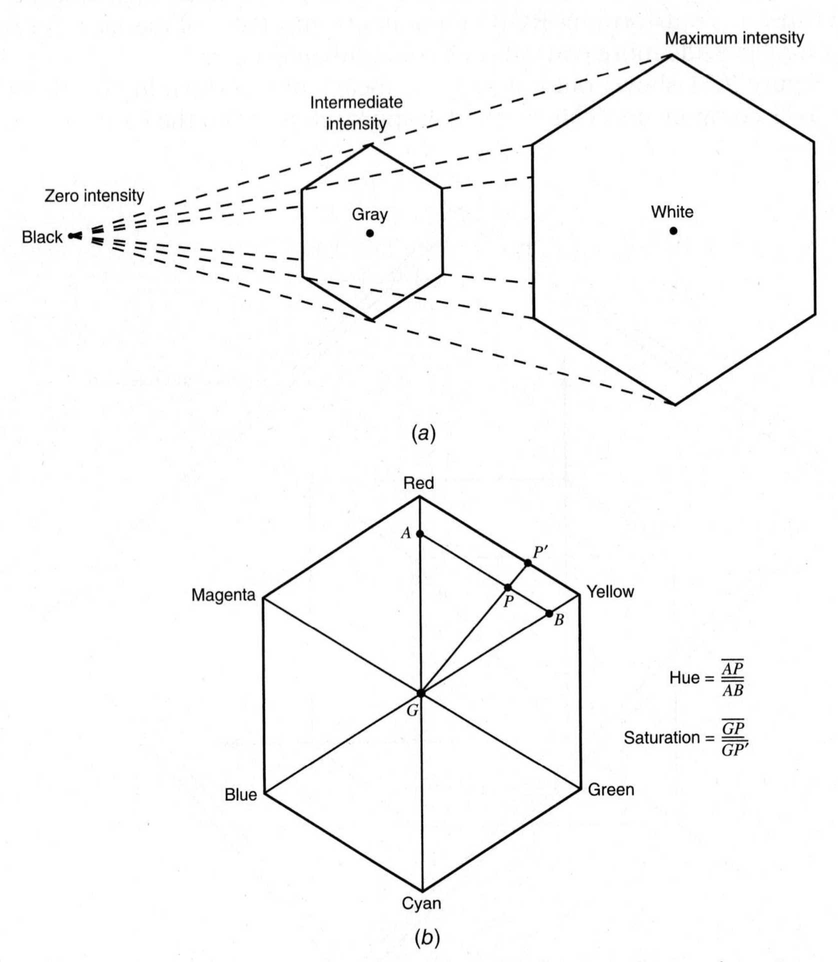

Figure 7.32 Hexcone color model. (*a*) Generation of the hexcone. The size of any given hexagon is determined by pixel intensity. (*b*) Definition of hue and saturation components for a pixel value, *P*, having a typical, nonzero intensity. (Adapted from Schowengerdt, 1997.)

given intensity, within the appropriate hexagon (Figure 7.32*b*). Hue is expressed by the angle around the hexagon, and saturation is defined by the distance from the gray point at the center of the hexagon. The farther a point lies away from the gray point, the more saturated the color. (In Figure 7.32*b*, linear distances are used to define hue and saturation, thereby avoiding computations involving trigonometric functions.)

At this point we have established the basis upon which any pixel in the RGB color space can be transformed into its IHS counterpart. Such transformations are often useful as an intermediate step in image enhancement. This is illustrated in Figure 7.33. In this figure the original RGB components are shown transformed first into their corresponding IHS components. The IHS components are then manipulated to enhance the desired characteristics of the image. Finally, these modified IHS components are transformed back to the RGB system for final display.

Among the advantages of IHS enhancement operations is the ability to vary each IHS component independently, without affecting the others. For example, a contrast stretch can be applied to the intensity component of an image, and the hue and saturation of the pixels in the enhanced image will not be changed (because they typically are in RGB contrast stretches). The IHS approach may also be used to display spatially registered data of varying spatial resolution. For example, high resolution data from one source may be displayed as the intensity component, and low resolution data from another source may be displayed as the hue and saturation components. For example, IHS procedures are often used to merge same-sensor, multiresolution data sets (e.g., 0.46-m panchromatic and 1.84-m multispectral WorldView-2 data, 1.5-m panchromatic and 6-m multispectral SPOT-6 data, or 15-m panchromatic and 30-m multispectral Landsat ETM+ or OLI data). The result is a composite image having the spatial resolution of the higher resolution data and the color characteristics of the original coarser resolution data. These techniques can also be used to merge data from different sensing systems (e.g., digital orthophotos with satellite data).

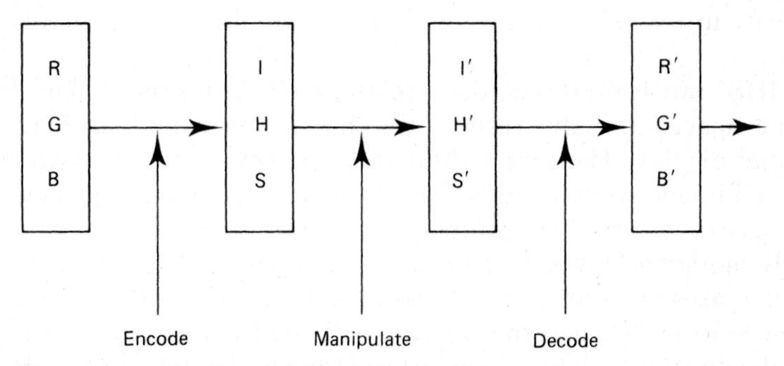

Figure 7.33 IHS/RGB encoding and decoding for interactive image manipulation. (Adapted from Schowengerdt, 1997.)

One caution to be noted in using IHS transformations to merge multi-resolution data is that direct substitution of the panchromatic data for the intensity component may not always produce the best final product in terms of the color balance of the merged product faithfully replicating the color balance of the original multispectral data. In such cases, alternative spectral merging algorithms, similar to IHS, can be used. One such approach was used in the production of Plate 16 (Chapter 5). This plate illustrates the merger of IKONOS 1-m-resolution panchromatic data and 4-m-resolution multispectral data. The result is a composite having the spatial resolution of the 1-m panchromatic data and the color characteristics of the original 4-m multispectral data.

While these types of resolution-enhancement techniques are not generally used for images being subjected to quantitative analysis, they have been shown to be very useful as an aid to visual image interpretation. In fact, merged panchromatic and multispectral imagery has become a standard product offered by many distributors of high resolution satellite imagery. The development and application of methods for merging images using IHS and other transformations are the subject of continuing research.

Decorrelation Stretching

Decorrelation stretching is a form of multi-image manipulation that is particularly useful when displaying multispectral data that are highly correlated. Data from multispectral or hyperspectral sensors that include multiple bands in the same region of the spectrum often fall into this category. Traditional contrast stretching of highly correlated data as R, G, and B displays normally only expands the range of intensities; it does little to expand the range of colors displayed, and the stretched image still contains only pastel hues. For example, no areas in a highly correlated image are likely to have high DNs in the red display channel but low values in the green and blue (which would produce a pure red). Instead, the reddest areas are merely a reddish-gray. To circumvent this problem, decorrelation stretching involves exaggeration of the least correlated information in an image primarily in terms of saturation, with minimal change in image intensity and hue.

As with IHS transformations, decorrelation stretching is applied in a transformed image space, and the results are then transformed back to the RGB system for final display. The major difference in decorrelation stretching is that the transformed image space used is that of the original image's principal components. The successive principal components of the original image are stretched independently along the respective principal component axes (Figure 7.24*a*). By definition, these axes are statistically independent of one another so the net effect of the stretch is to emphasize the poorly correlated components of the original data. When the stretched data are then transformed back to the RGB system, a display having increased color saturation results. There is usually little difference in the perceived hues and intensities due to enhancement. This makes interpretation of the enhanced image straightforward, with the decorrelated information exaggerated primarily in terms of saturation. Previously pastel hues become much more saturated.

Because decorrelation stretching is based on principal component analysis, it is readily extended to any number of image channels. Recall that the IHS procedure is applied to only three channels at a time.

7.7 IMAGE CLASSIFICATION

The overall objective of image classification procedures is to automatically categorize all pixels in an image into land cover classes or themes. Often this is done using *spectral patterns*; that is, pixels that share similar combinations of spectral reflectance or emissivity are grouped together in classes that are assumed to represent particular categories of surface features. No attention is paid to the neighbors or surroundings of the pixel being classified. The term *spectral pattern recognition* refers to the family of classification procedures that utilizes this pixel-by-pixel spectral information as the basis for automated land cover classification.

This approach can be extended to make use of many other types of pixel-level data. For example, polarimetric radar imagery could be classified using polarization pattern recognition; multitemporal imagery of any type could be classified using temporal pattern recognition; multi-angle imagery such as that from MISR could be classified based on bidirectional reflectance patterns; and so forth. What all these examples have in common is that each pixel is classified, individually, based on some statistical or deterministic model using the values from multiple data layers (spectral bands, polarization bands, temporal bands, ...) for that one pixel.

Spatial pattern recognition is a very different approach, involving the categorization of image pixels on the basis of their spatial relationship with pixels surrounding them. Spatial classifiers might consider such aspects as image texture, pixel proximity, feature size, shape, directionality, repetition, and context. These types of classifiers attempt to replicate the kind of spatial synthesis done by the human analyst during the visual interpretation process. Accordingly, they tend to be much more complex and computationally intensive than spectral pattern recognition procedures.

These two types of image classifiers may be used in combination in a hybrid mode. For example, *object-based* image analysis (OBIA) involves combined use of both spectral and spatial pattern recognition. It is important to emphasize that there is no single "right" manner in which to approach an image classification problem. The particular approach one might take depends upon the nature of the data being analyzed, the computational resources available, and the intended application of the classified data.

We begin our discussion of image classification with treatment of spectrally oriented procedures for land cover mapping. Historically, spectral approaches have formed the backbone of multispectral classification activities (although with the current wide-scale availability of high resolution data, there is increased use of spatially oriented procedures). First, we describe *supervised classification*. In this type of classification the image analyst "supervises" the pixel categorization process by specifying, to the computer algorithm, numerical descriptors of the various land cover types present in a scene. To do this, representative sample sites of known cover type, called *training areas*, are used to compile a numerical "interpretation key" that describes the spectral attributes for each feature type of interest. Each pixel in the data set is then compared numerically to each category in the interpretation key and labeled with the name of the category it "looks most like." As we see in the next section, there are a number of numerical strategies that can be employed to make this comparison between unknown pixels and training set pixels.

Following our discussion of supervised classification we treat the subject of *unsupervised classification*. Like supervised classifiers, the unsupervised procedures are applied in two separate steps. The fundamental difference between these techniques is that supervised classification involves a training step followed by a classification step. In the unsupervised approach the image data are first classified by aggregating them into the natural spectral groupings, or *clusters*, present in the scene. Then the image analyst determines the land cover identity of these spectral groups by comparing the classified image data to ground reference data. Unsupervised procedures are discussed in Section 7.11.

Building upon the preceding methods, we then discuss *hybrid classification* procedures. Such techniques involve aspects of both supervised and unsupervised classification and are aimed at improving the accuracy or efficiency (or both) of the classification process. Hybrid classification is the subject of Section 7.12. We also look at other specialized topics, such as the classification of mixed pixels (Section 7.13), object-based classification using both spatial and spectral information (Section 7.15), and the use of neural networks for classification (Section 7.16), before concluding this extensive portion of the chapter with a discussion of methods for assessing and reporting the accuracy of image classifications (Section 7.17).

Note that hyperspectral images present their own issues for classification, and as a result some specialized procedures have been developed for classification of these data. We defer coverage of such methods to Section 7.21, along with other analytical procedures for use with hyperspectral data.

7.8 SUPERVISED CLASSIFICATION

We use a hypothetical example to facilitate our discussion of supervised classification. In this example, let us assume that we are dealing with the analysis of five-channel airborne multispectral sensor data. (The identical procedures would apply to Landsat, SPOT, WorldView-2, or virtually any other source of

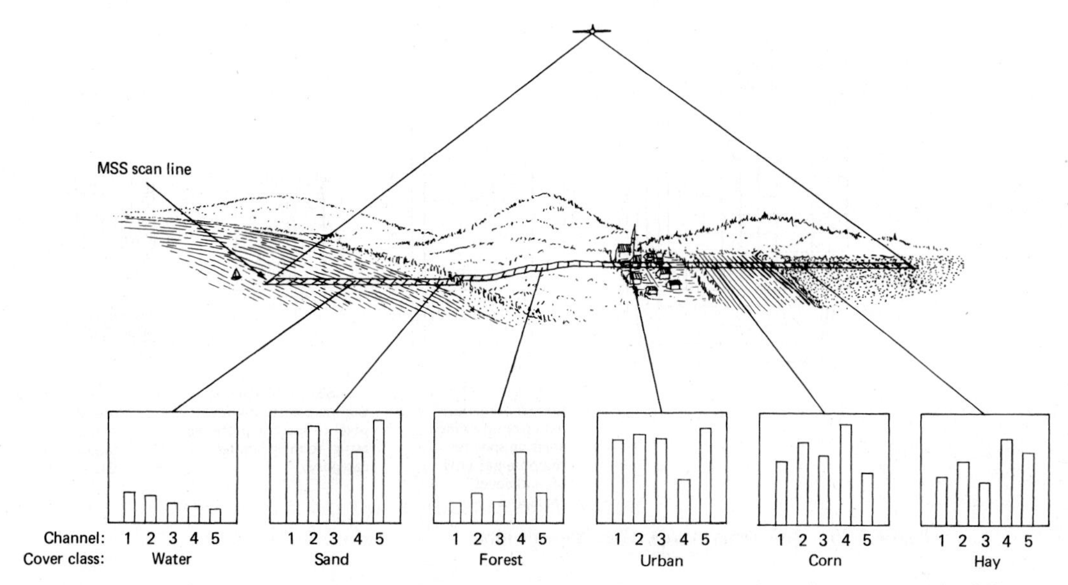

Figure 7.34 Selected multispectral sensor measurements made along one scan line. Sensor covers the following spectral bands: 1, blue; 2, green; 3, red; 4, near infrared; 5, thermal infrared.

multispectral data.) Figure 7.34 shows the location of a single line of data collected for our hypothetical example over a landscape composed of several cover types. For each of the pixels shown along this line, the sensor has measured scene radiance in terms of DNs recorded in each of the five spectral bands of sensing: blue, green, red, near infrared, and thermal infrared. Below the scan line, typical DNs measured over six different land cover types are shown. The vertical bars indicate the relative gray values in each spectral band. These five outputs represent a coarse description of the spectral response patterns of the various terrain features along the scan line. If these spectral patterns are sufficiently distinct for each feature type, they may form the basis for image classification.

Figure 7.35 summarizes the three basic steps involved in a typical supervised classification procedure. In the *training stage* (1), the analyst identifies representative training areas and develops a numerical description of the spectral attributes of each land cover type of interest in the scene. Next, in the *classification stage* (2), each pixel in the image data set is categorized into the land cover class it most closely resembles. If the pixel is insufficiently similar to any training data set, it is usually labeled “unknown.” After all pixels in the input image have been categorized, the results are presented in the *output stage* (3). Being digital in character, the results may be used in a number of different ways. Three typical forms of output products are thematic maps, tables of statistics for the various land cover classes, and digital data files amenable to inclusion in a GIS. In this latter case, the classification “output” becomes a GIS “input.”

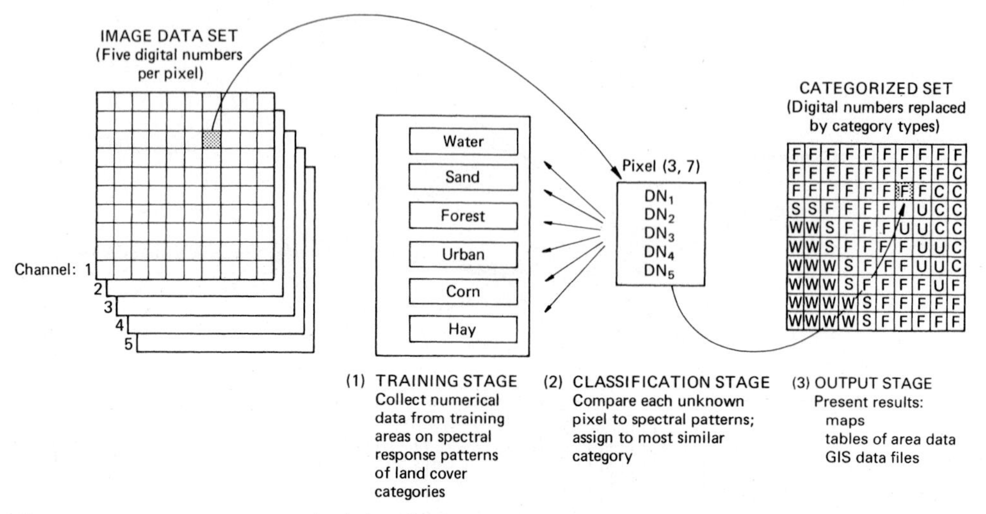

Figure 7.35 Basic steps in supervised classification.

We begin with a discussion of the *classification* stage because it is the heart of the supervised classification process—during this stage a computer-based evaluation of the spectral patterns is made using predefined decision rules to determine the identity of each pixel. Another reason for treating the classification stage first is because familiarity with this step aids in understanding the requirements that must be met in the training stage.

7.9 THE CLASSIFICATION STAGE

Numerous mathematical approaches to spectral pattern recognition have been developed. Our discussion only scratches the surface of this topic.

We illustrate the various classification approaches with a two-channel (bands 3 and 4) subset of our hypothetical five-channel multispectral sensor data set. Rarely are just two channels employed in an analysis, yet this limitation simplifies the graphic portrayal of the various techniques. When implemented numerically, these procedures may be applied to any number of channels of data.

Let us assume that we take a sample of pixel observations from our two-channel digital image data set. The two-dimensional digital values, or *measurement vectors*, attributed to each pixel may be expressed graphically by plotting them on a *scatter diagram* (or *scatter plot*), as shown in Figure 7.36. In this diagram, the band 3 DNs have been plotted on the y axis and the band 4 DNs on the x axis. These two DNs locate each pixel value in the two-dimensional "measurement space" of the graph. Thus, if the band 4 DN for a pixel is 10 and the band

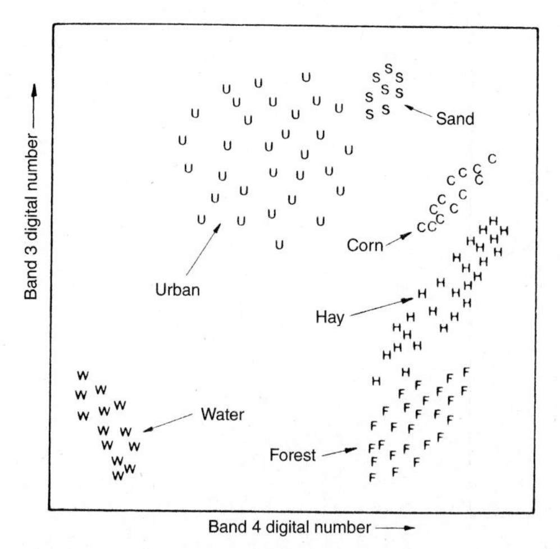

Figure 7.36 Pixel observations from selected training sites plotted on scatter diagram.

3 DN for the same pixel is 68, the measurement vector for this pixel is represented by a point plotted at coordinate (10, 68) in the measurement space.[1]

Let us also assume that the pixel observations shown in Figure 7.36 are from areas of known cover type (that is, from selected training sites). Each pixel value has been plotted on the scatter diagram with a letter indicating the category to which it is known to belong. Note that the pixels within each class do not have a single, repeated spectral value. Rather, they illustrate the natural centralizing tendency—yet variability—of the spectral properties found within each cover class. These "clouds of points" represent multidimensional descriptions of the spectral response patterns of each category of cover type to be interpreted. The following classification strategies use these "training set" descriptions of the category spectral response patterns as interpretation keys by which pixels of unidentified cover type are categorized into their appropriate classes.

Minimum-Distance-to-Means Classifier

Figure 7.37 illustrates one of the simpler classification strategies that may be used. First, the mean, or average, spectral value in each band for each category is determined. These values comprise the *mean vector* for each category. The category means are indicated by + symbols in Figure 7.37. By considering the two-channel pixel values as positional coordinates (as they are portrayed in the scatter

[1] Pattern recognition literature frequently refers to individual bands of data as *features* and scatterplots of data as *feature space plots.*

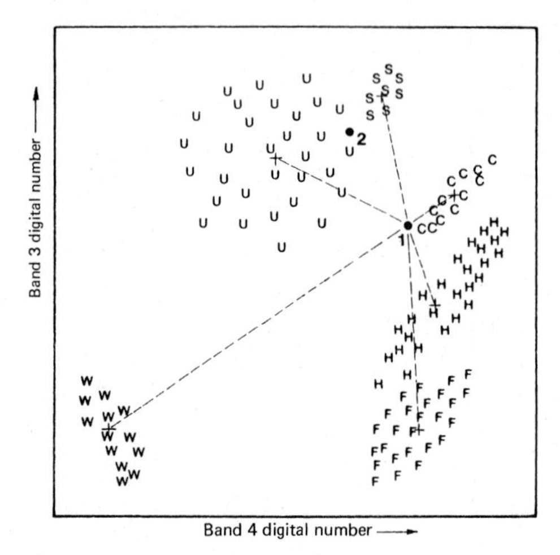

Figure 7.37 Minimum distance to means classification strategy.

diagram), a pixel of unknown identity may be classified by computing the *distance* between the value of the unknown pixel and each of the category means. In Figure 7.37, an unknown pixel value has been plotted at point 1. The distance between this pixel value and each category mean value is illustrated by the dashed lines. After computing the distances, the unknown pixel is assigned to the "closest" class, in this case "corn." If the pixel is farther than an analyst-defined distance from any category mean, it would be classified as "unknown."

The minimum-distance-to-means strategy is mathematically simple and computationally efficient, but it has certain limitations. Most importantly, *it is insensitive to different degrees of variance in the spectral response data*. In Figure 7.37, the pixel value plotted at point 2 would be assigned by the distance-to-means classifier to the "sand" category, in spite of the fact that the greater variability in the "urban" category suggests that "urban" would be a more appropriate class assignment. Because of such problems, this classifier is not widely used in applications where spectral classes are close to one another in the measurement space and have high variance.

Parallelepiped Classifier

We can introduce sensitivity to category variance by considering the *range* of values in each category training set. This range may be defined by the highest and lowest digital number values in each band and appears as a rectangular area in our two-channel scatter diagram, as shown in Figure 7.38. An unknown pixel is classified according to the category range, or *decision region*, in which it lies or as "unknown" if it lies outside all regions. The multidimensional analogs of these rectangular areas are called *parallelepipeds*, and this classification strategy is referred to by that tongue-twisting name. The parallelepiped classifier is also very fast and efficient computationally.

The sensitivity of the parallelepiped classifier to category variance is exemplified by the smaller decision region defined for the highly repeatable "sand" category than for the more variable "urban" class. Because of this, pixel 2 would be appropriately classified as "urban." However, difficulties are encountered when category ranges overlap. Unknown pixel observations that occur in the overlap areas will be classified as "not sure" or be arbitrarily placed in one of the two overlapping classes. Overlap is caused largely because category distributions exhibiting *correlation* or high *covariance* are poorly described by the rectangular decision regions. Covariance is the tendency of spectral values to vary similarly in two bands, resulting in elongated, slanted clouds of observations on the scatter diagram. In our example, the "corn" and "hay" categories have positive covariance (they slant upward to the right), meaning that high values in band 3 are generally associated with high values in band 4, and low values in band 3 are associated with low values in band 4. The water category in our example exhibits *negative covariance* (its distribution slants down to the right), meaning that high values in band 3 are associated with low values in band 4. The "urban" class shows a lack of covariance, resulting in a nearly circular distribution on the scatter diagram.

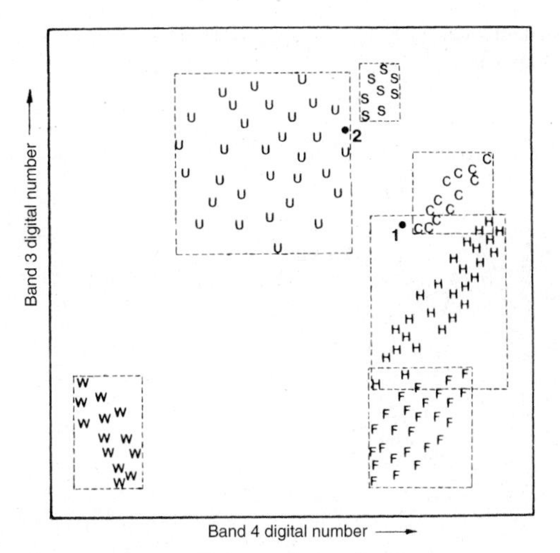

Figure 7.38 Parallelepiped classification strategy.

In the presence of covariance, the rectangular decision regions fit the category training data very poorly, resulting in confusion for a parallelepiped classifier. For example, the insensitivity to covariance would cause pixel 1 to be classified as "hay" instead of "corn."

Unfortunately, spectral response patterns are frequently highly correlated, and high covariance is often the rule rather than the exception.

Gaussian Maximum Likelihood Classifier

The maximum likelihood classifier quantitatively evaluates both the variance and covariance of the category spectral response patterns when classifying an unknown pixel. To do this, an assumption is made that the distribution of the cloud of points forming the category training data is Gaussian (normally distributed). This *assumption of normality* is generally reasonable for common spectral response distributions. Under this assumption, the distribution of a category response pattern can be completely described by the *mean vector* and the *covariance matrix*. Given these parameters, we may compute the statistical probability of a given pixel value being a member of a particular land cover class. Figure 7.39 shows the probability values plotted in a three-dimensional graph. The vertical axis is associated with the probability of a pixel value being a member of one of the classes. The resulting bell-shaped surfaces are called *probability density functions*, and there is one such function for each spectral category.

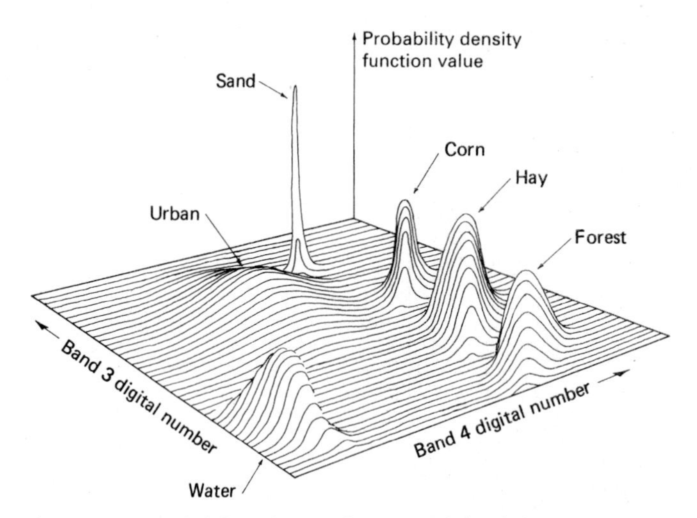

Figure 7.39 Probability density functions defined by a maximum likelihood classifier.

The probability density functions are used to classify an unidentified pixel by computing the probability of the pixel value belonging to each category. That is, the computer would calculate the probability of the pixel value occurring in the distribution of class "corn," then the likelihood of its occurring in class "sand," and so on. After evaluating the probability in each category, the pixel would be assigned to the most likely class (highest probability value) or be labeled "unknown" if the probability values are all below a threshold set by the analyst.

In essence, the maximum likelihood classifier delineates ellipsoidal "equiprobability contours" in the scatter diagram. These decision regions are shown in Figure 7.40. The shape of the equiprobability contours expresses the sensitivity of the likelihood classifier to covariance. For example, because of this sensitivity, it can be seen that pixel 1 would be appropriately assigned to the "corn" category.

An extension of the maximum likelihood approach is the *Bayesian classifier*. This technique applies two weighting factors to the probability estimate. First, the analyst determines the "a priori probability," or the anticipated likelihood of occurrence for each class in the given scene. For example, when classifying a pixel, the probability of the rarely occurring "sand" category might be weighted lightly, and the more likely "urban" class weighted heavily. Second, a weight associated with the "cost" of misclassification is applied to each class. Together, these factors act to minimize the "cost" of misclassifications, resulting in a theoretically optimum classification. In practice, most maximum likelihood classification is performed assuming equal probability of occurrence and cost of misclassification for all classes. If suitable data exist for these factors, the Bayesian implementation of the classifier is preferable.

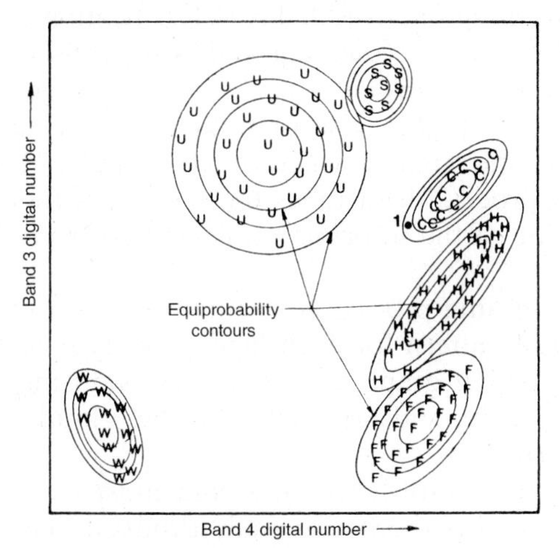

Figure 7.40 Equiprobability contours defined by a maximum likelihood classifier.

The principal drawback of maximum likelihood classification is the large number of computations required to classify each pixel. This is particularly true when either a large number of spectral channels are involved or a large number of spectral classes must be differentiated. In such cases, the maximum likelihood classifier is much slower computationally than the previous techniques. However, with the rapid increase in computational power over the past several decades, this computational complexity is no longer a major consideration for most applications. If an increase in speed is required, one approach is to reduce the dimensionality of the data set used to perform the classification (thereby reducing the complexity of the required computations). As discussed in Section 7.6, principal or canonical component transformations of the original data may be used for this purpose.

Decision tree, *stratified*, or *layered* classifiers have also been utilized to simplify classification computations and maintain classification accuracy. These classifiers are applied in a series of steps, with certain classes being separated during each step in the simplest manner possible. For example, water might first be separated from all other classes based on a simple threshold set in a near-infrared band. Certain other classes may require only two or three bands for categorization and a parallelepiped classifier may be adequate. The use of more bands or the maximum likelihood classifier would then only be required for those land cover categories where residual ambiguity exists between overlapping classes in the measurement space.

7.10 THE TRAINING STAGE

Whereas the actual classification of multispectral image data is a highly automated process, assembling the training data needed for classification is anything but automatic. In many ways, the training effort required in supervised classification is both an art and a science. It requires close interaction between the image analyst and the image data. It also requires substantial reference data and a thorough knowledge of the geographic area to which the data apply. Most importantly, the quality of the training process determines the success of the classification stage and, therefore, the value of the information generated from the entire classification effort.

The overall objective of the training process is to assemble a set of statistics that describe the spectral response pattern for each land cover type to be classified in an image. Relative to our earlier graphical example, it is during the training stage that the location, size, shape, and orientation of the "clouds of points" for each land cover class are determined.

To yield acceptable classification results, training data must be both representative and complete. This means that the image analyst must develop training statistics for all *spectral* classes constituting each *information* class to be discriminated by the classifier. For example, in a final classification output, one might wish to delineate an information class called "water." If the image under analysis contains only one water body and if it has uniform spectral response characteristics over its entire area, then only one training area would be needed to represent the water class. If, however, the same water body contained distinct areas of very clear water and very turbid water, a minimum of two spectral classes would be required to adequately train on this feature. If multiple water bodies occurred in the image, training statistics would be required for each of the other spectral classes that might be present in the water-covered areas. Accordingly, the single information class "water" might be represented by four or five spectral classes. In turn, the four or five spectral classes would eventually be used to classify all the water bodies occurring in the image.

By now it should be clear that the training process can become quite involved. For example, an information class such as "agriculture" might contain several crop types and each crop type might be represented by several spectral classes. These spectral classes could stem from different planting dates, soil moisture conditions, crop management practices, seed varieties, topographic settings, atmospheric conditions, or combinations of these factors. *The point that must be emphasized is that all spectral classes constituting each information class must be adequately represented in the training set statistics used to classify an image*. Depending upon the nature of the information classes sought and the complexity of the geographic area under analysis, it is not uncommon to acquire data from 100 or more training areas to adequately represent the spectral variability in an image.

Figure 7.41 shows the boundaries of several training site polygons that have been delineated in this manner. Note that these polygons have been carefully located to avoid pixels located along the edges between land cover types and to avoid any areas that are visually "rough" in the imagery. The row and column coordinates of the vertices for these polygons are used as the basis for extracting

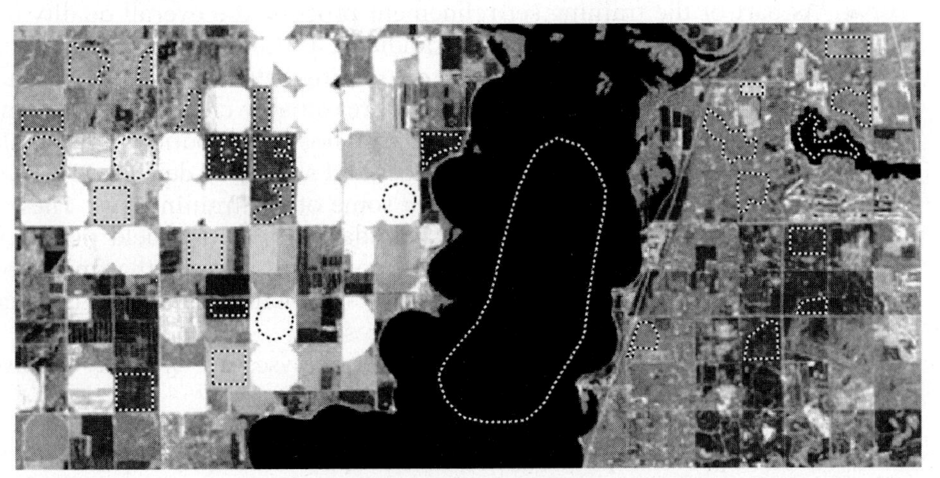

Figure 7.41 Training area polygons delineated on a computer monitor. (Author-prepared figure.)

(from the image file) the digital numbers for the pixels located within each training area boundary. These pixel values then form the sample used to develop the statistical description of each training area (mean vector and covariance matrix in the case of the maximum likelihood classifier).

An alternative to manually delineating training area polygons is the use of a *seed pixel* approach to training. In this case, the display cursor is placed within a prospective training area and a single "seed" pixel is chosen that is thought to be representative of the surrounding area. Then, according to various statistically based criteria, pixels with similar spectral characteristics that are contiguous to the seed pixel are highlighted on the display and become the training samples for that training area.

Irrespective of how training areas are delineated, when using any statistically based classifier (such as the maximum likelihood method), the theoretical lower limit of the number of pixels that must be contained in a training set is $n+1$, where n is the number of spectral bands. In our two-band example, *theoretically* only three observations would be required. Obviously, the use of fewer than three observations would make it impossible to appropriately evaluate the variance and covariance of the spectral response values. In practice, a minimum of $10n$ to $100n$ pixels is used since the estimates of the mean vectors and covariance matrices improve as the number of pixels in the training sets increases. Within reason, the more pixels that can be used in training, the better the statistical representation of each spectral class.

When delineating training set pixels, it is important to analyze several training sites throughout the scene. For example, it would be better to define the training pattern for a given class by analyzing 20 locations containing 40 pixels of a given type than one location containing 800 pixels. Dispersion of the sites throughout the scene increases the chance that the training data will be representative of all the variations in the cover types present in the scene.

As part of the training set refinement process, the overall quality of the data contained in each of the original candidate training areas is assessed, and the spectral separability between pairs of training sets is studied. The analyst confirms that all data sets are unimodal and reasonably close to a Gaussian distribution. Training areas that are bimodal or whose distributions are highly skewed may include more than one spectral class and should be deleted or split. Likewise, extraneous pixels may be deleted from some of the training sets. These might be edge pixels along agricultural field boundaries or within-field pixels containing bare soil rather than the crop trained upon. Training sets that might be merged (or deleted) are identified, and the need to obtain additional training sets for poorly represented spectral classes is addressed.

One or more of the following types of analyses are typically involved in the training set refinement process:

1. **Graphical representation of the spectral response patterns.** The distributions of training area response patterns can be graphically displayed

in many formats. Figure 7.42*a* shows a hypothetical histogram for one of the "hay" category training sites in our five-channel data set. (A similar display would be available for all training areas.) Histogram output is particularly important when a maximum likelihood classifier is used, because it provides a visual check on the normality of the spectral response distributions. Note in the case of the hay category that the data appear to be normally distributed in all bands except band 2, where the distribution is shown to be bimodal. This indicates that the training site data set chosen by the analyst to represent "hay" is in fact composed of two subclasses with slightly different spectral characteristics. These subclasses may represent two different varieties of hay or different illumination conditions, and so on. In any case, the classification accuracy will generally be improved if each of the subclasses is treated as a separate category.

Histograms illustrate the distribution of individual categories very well; yet they do not facilitate comparisons between different category types. To evaluate the spectral separation between categories, it is convenient to use some form of *coincident spectral plot*, as shown in Figure 7.42*b*. This plot illustrates, in each spectral band, the mean spectral response of each category (with a letter) and the variance of the distribution (± 2 standard deviations shown by gray bars). Such plots indicate the overlap between category response patterns. For example, Figure 7.42*b* indicates that the hay and corn response patterns overlap in all spectral bands. The plot also shows which combination of bands might be best for discrimination because of relative reversals of spectral response (such as bands 3 and 5 for hay/corn separation).

The fact that the spectral plots for hay and corn overlap in all spectral bands indicates that the categories could not be accurately classified on any *single* multispectral scanner band. However, this does not preclude successful classification when two or more bands are analyzed (such as bands 3 and 4 illustrated in the last section). Because of this, two-dimensional scatter diagrams (as shown in Figures 7.36 to 7.38 and 7.40) provide better representations of the spectral response pattern distributions.

The utility of scatter diagrams (or scatter plots) is further illustrated in Figures 7.43 to 7.45. Shown in Figure 7.43 are SPOT multispectral HRV images depicting a portion of Madison, Wisconsin. The band 1 (green), band 2 (red), and band 3 (near-IR) images are shown in (*a*), (*b*), and (*c*), respectively. Figure 7.44 shows the histograms for bands 1 and 2 as well as the associated scatter diagram for these two bands. Note that the data in these two bands are highly correlated and a very compact and near-linear "cloud of points" is shown in the scatter diagram.

Figure 7.45 shows the histograms and the scatter diagram for bands 2 and 3. In contrast to Figure 7.44, the scatter diagram in Figure 7.45

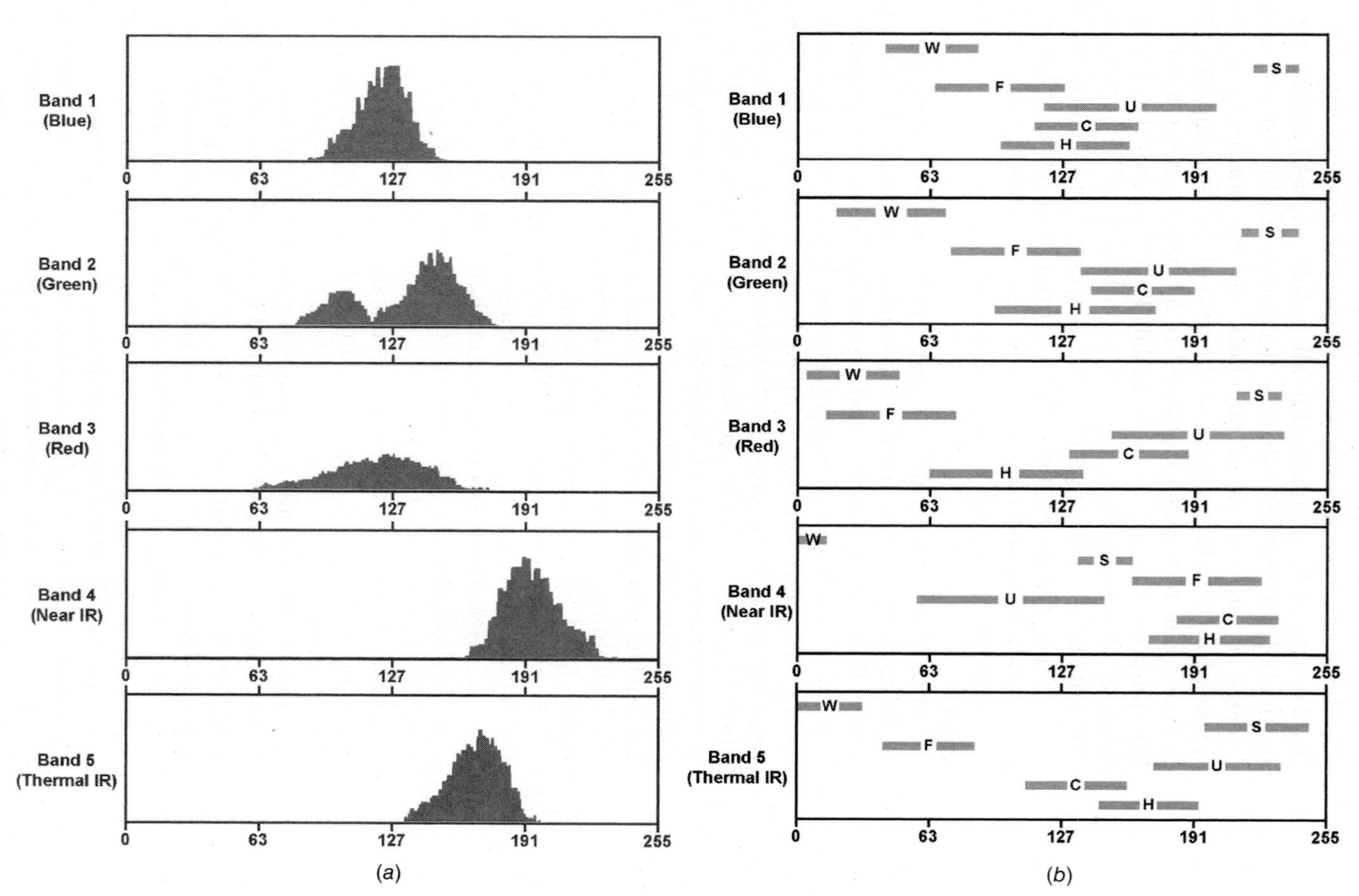

Figure 7.42 Visualization of training plot data. (*a*) Sample histograms for data points included in the training areas for cover type "hay." (*b*) Coincident spectral plots for training data obtained in five bands for six cover types.

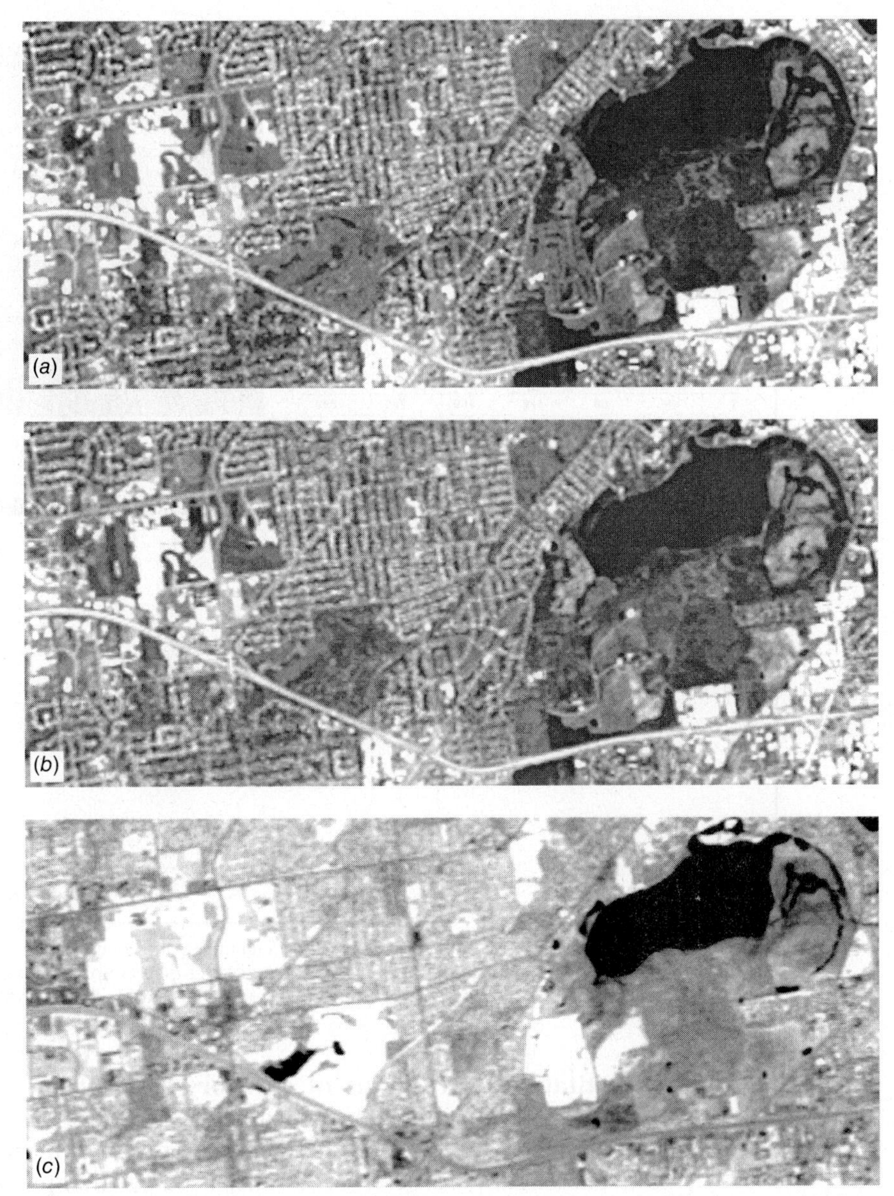

Figure 7.43 SPOT HRV multispectral images of Madison, WI: (*a*) band 1 (green); (*b*) band 2 (red); (*c*) band 3 (near IR). (Author-prepared figure.)

shows that bands 2 and 3 are much less correlated than bands 1 and 2. Whereas various land cover types might overlap one another in bands 1 and 2, they would be much more separable in bands 2 and 3. In fact, these two bands alone may be adequate to perform a generalized land cover classification of this scene.

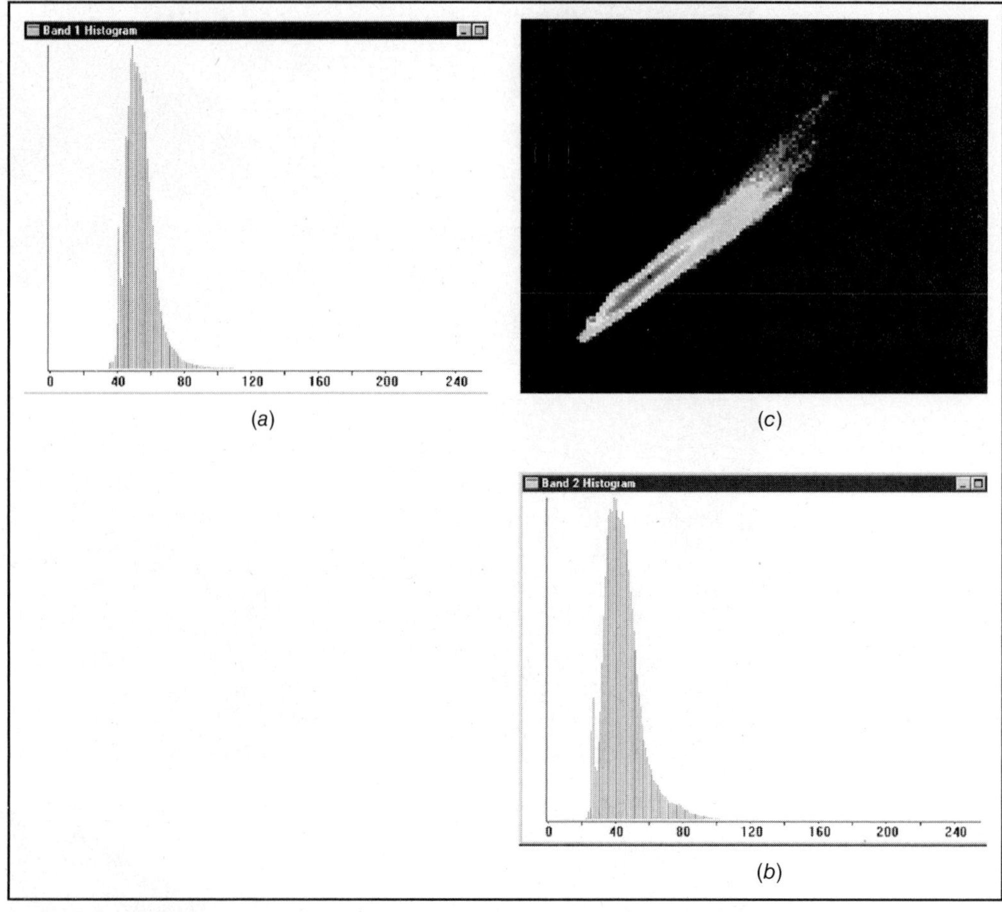

Figure 7.44 Histograms and two-dimensional scatter diagram for the images shown in Figures 7.43*a* and *b*: (*a*) band 1 (green) histogram; (*b*) band 2 (red) histogram; (*c*) scatter diagram plotting band 1 (vertical axis) versus band 2 (horizontal axis). Note the high correlation between these two visible bands. (Author-prepared figure.)

2. **Quantitative expressions of category separation.** A measure of the statistical separation between category response patterns can be computed for all pairs of classes and can be presented in the form of a matrix. One statistical parameter commonly used for this purpose is *transformed divergence*, a covariance-weighted distance between category means. In general, the larger the transformed divergence, the greater the "statistical distance" between training patterns and the higher the probability of correct classification of classes. A portion of a sample matrix of divergence values is shown in Table 7.1. In this example, the maximum possible divergence value is 2000, and values less than 1500 indicate spectrally similar classes. Accordingly, the data in Table 7.1 suggest spectral overlap

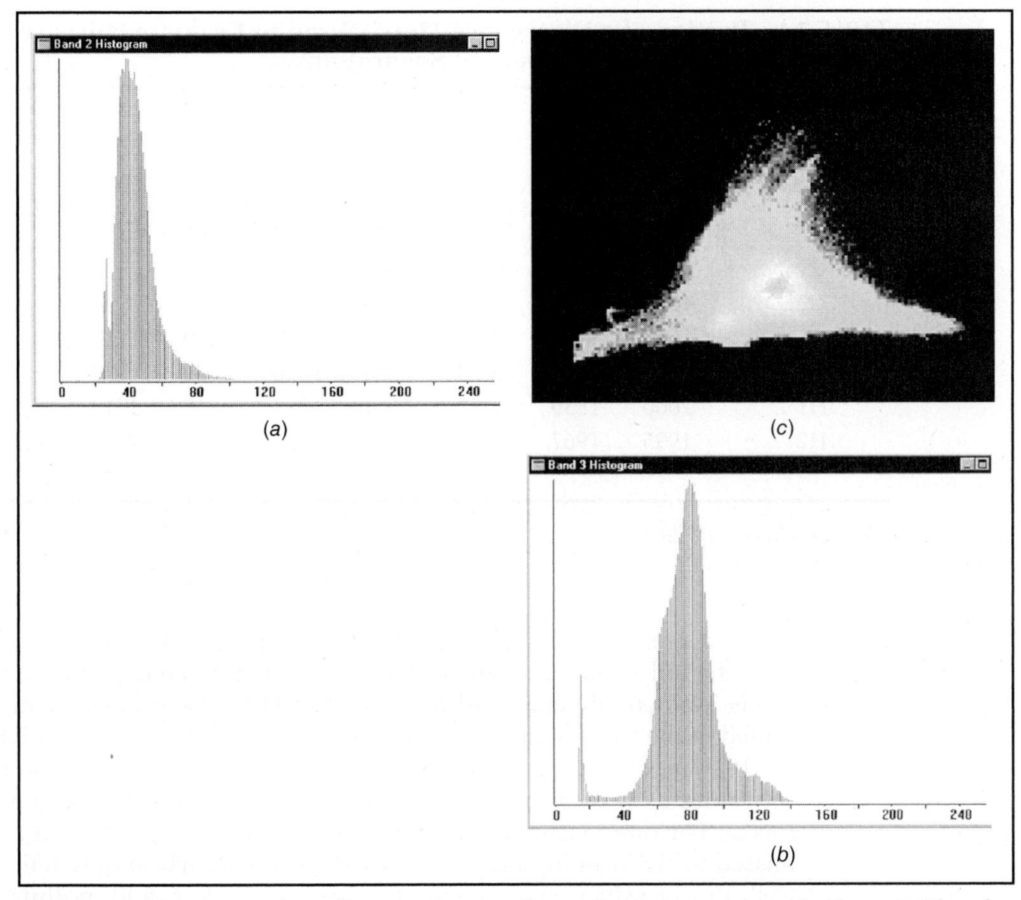

(a) (c) (b)

Figure 7.45 Histograms and two-dimensional scatter diagram for the images shown in Figures 7.43*b* and *c*: (*a*) band 2 (red) histogram; (*b*) band 3 (near-IR) histogram; (*c*) scatter diagram plotting band 2 (vertical axis) versus band 3 (horizontal axis). Note the relative lack of correlation between these visible and near-IR bands. (Author-prepared figure.)

between several pairs of spectral classes. Note that W1, W2, and W3 are all relatively spectrally similar. However, note that this similarity is all among spectral classes from the same information class ("water"). Furthermore, all the "water" classes appear to be spectrally distinct from the spectral classes of the other information classes. More problematic is a situation typified by the divergence between the H1 and C3 spectral classes (860). Here, a "hay" spectral class severely overlaps a "corn" class.

Another statistical distance measure of the separability of two spectral classes is the *Jeffries–Matusita* (*JM*) *distance*. It is similar to transformed divergence in its interpretation but has a maximum value of 1414.

3. **Self-classification of training set data.** Another evaluation of spectral separability is provided by classifying the training set pixels. In such an

TABLE 7.1 Portion of a Divergence Matrix Used to Evaluate Pairwise Training Class Spectral Separability

Spectral Class[a]	W1	W2	W3	C1	C2	C3	C4	H1	H2...
W1	0								
W2	1185	0							
W3	1410	680	0						
C1	1997	2000	1910	0					
C2	1953	1890	1874	860	0				
C3	1980	1953	1930	1340	1353	0			
C4	1992	1997	2000	1700	1810	1749	0		
H1	2000	1839	1911	1410	1123	860	1712	0	
H2	1995	1967	1935	1563	1602	1197	1621	721	0
⋮	⋮								

[a]W, water; C, corn; H, hay.

effort, a preliminary classification of only the training set pixels (rather than the full scene) is made to determine what percentage of the training pixels are actually classified as expected. These percentages can be presented in the form of an *error matrix* (to be described in Section 7.17).

It is important to avoid considering an error matrix based on training set values as a measure of *overall* classification accuracy throughout an image. For one reason, certain land cover classes might be inadvertently missed in the training process. Also, the error matrix simply tells us how well the classifier can classify the *training areas* and nothing more. Because the training areas are usually good, homogeneous examples of each cover type, they can be expected to be classified more accurately than less pure examples that may be found elsewhere in the scene. Overall accuracy can be evaluated only by considering *test areas* that are different from and considerably more extensive than the training areas. This evaluation is generally performed after the entire classification process is complete (as discussed in Section 7.17).

4. **Interactive preliminary classification.** Most modern image processing systems incorporate some provision for interactively displaying how applicable training data are to the full scene to be classified. Often, this involves performing a preliminary classification with a computationally efficient algorithm (e.g., parallelepiped) to provide a visual approximation of the areas that would be classified with the statistics from a given training area. Such areas are typically highlighted in color on the display of the original raw image.

This is illustrated in Plate 30, which shows a partially completed classification of a subset of the data included in Figures 7.43 and 7.45 (bands 2 and 3). Shown in (*a*) are selected training areas delineated on a color infrared composite of bands 1, 2, and 3 depicted as blue, green, and red, respectively. Part (*b*) shows the histograms and scatter plot for bands 2 and 3. Shown in (*c*) are the parallelepipeds associated with the initial training areas an image analyst has chosen to represent four information classes: water, trees, grass, and impervious surfaces. Part (*d*) shows how the statistics from these initial training areas would classify various portions of the original scene.

5. **Representative subscene classification.** Often, an image analyst will perform a classification of a representative subset of the full scene to eventually be classified. The results of this preliminary classification can then be used interactively on an overlay to the original raw image. Selected classes are then viewed individually or in logical groups to determine how they relate to the original image.

In general, the training set refinement process should not be rushed. It is normally an iterative procedure in which the analyst revises the statistical descriptions of the category types until they are sufficiently spectrally separable. That is, the original set of "candidate" training area statistics is revised through merger, deletion, and addition to form the "final" set of statistics used in classification.

Training set refinement for the inexperienced data analyst is often a difficult task. Typically, an analyst has little difficulty in developing the statistics for the distinct "nonoverlapping" spectral classes present in a scene. If there are problems, they typically stem from spectral classes on the borders between information classes—"transition" or "overlapping" classes. In such cases, the impact of alternative deletion and pooling of training classes can be tested by trial and error. In this process the sample size, spectral variances, normality, and identity of the training sets should be rechecked. Problem classes that occur only rarely in the image may be eliminated from the training data so that they are not confused with classes that occur extensively. That is, the analyst may accept misclassification of a class that occurs rarely in the scene in order to preserve the classification accuracy of a spectrally similar class that appears over extensive areas. Furthermore, a classification might initially be developed assuming a particular set of detailed information classes will be maintained. After studying the actual classification results, the image analyst might be faced with aggregating certain of the detailed classes into more general ones (for example, "birch" and "aspen" may have to be merged into a "deciduous" class or "corn" and "hay" into "agriculture").

When multiple images need to be classified—either for a wider area or to study changes in land cover over time—the traditional approach is to extract spectral training data from each image individually. This is necessary because each image will typically have variations in atmospheric conditions, sun angle, and other factors that are sufficient to render the spectral "signatures" extracted from one image unrepresentative of the same classes in the other image(s). If there is a need to classify multiple images with the actual spectral training data from a single image, one approach is to use relative radiometric normalization (Figure 7.5; Section 7.2) to adjust the radiometric characteristics of all the images to match a single "base" image prior to classification. Obviously, the success of this approach is dependent on the accuracy with which the secondary images can be matched to the base image. In addition, this process works best for images of non-vegetated areas, or for images acquired at the same point in the phenological cycle.

One final note to be made here is that training set refinement is essential to improving the accuracy of a classification. However, if certain cover types occurring in an image have inherently similar spectral response patterns, no amount of retraining and refinement will make them spectrally separable! Alternative methods, such as incorporating additional imagery from other sensors, using data resident in a GIS, or performing a visual interpretation, must be used to discriminate these cover types. Multitemporal or spatial pattern recognition procedures may also be applicable in such cases. Increasingly, land cover classification involves some merger of remotely sensed data with ancillary information resident in a GIS.

7.11 UNSUPERVISED CLASSIFICATION

As previously discussed, unsupervised classifiers do *not* utilize training data as the basis for classification. Rather, this family of classifiers involves algorithms that examine the unknown pixels in an image and aggregate them into a number of classes based on the natural groupings or clusters present in the image values. The basic premise is that values within a given cover type should be close together in the measurement space, whereas data in different classes should be comparatively well separated.

The classes that result from unsupervised classification are *spectral classes*. Because they are based solely on the natural groupings in the image values, the identity of the spectral classes will not be initially known. The analyst must compare the classified data with some form of reference data (such as larger scale imagery or maps) to determine the identity and informational value of the spectral classes. Thus, in the *supervised* approach we define useful information categories and then examine their spectral separability; in the *unsupervised* approach we determine spectrally separable classes and then define their informational utility.

We illustrate the unsupervised approach by again considering a two-channel data set. Natural spectral groupings in the data can be visually identified by plotting a scatter diagram. For example, in Figure 7.46 we have plotted pixel values acquired over a forested area. Three groupings are apparent in the scatter diagram. After comparing the classified image data with ground reference data, we might find that one cluster corresponds to deciduous trees, one to conifers, and

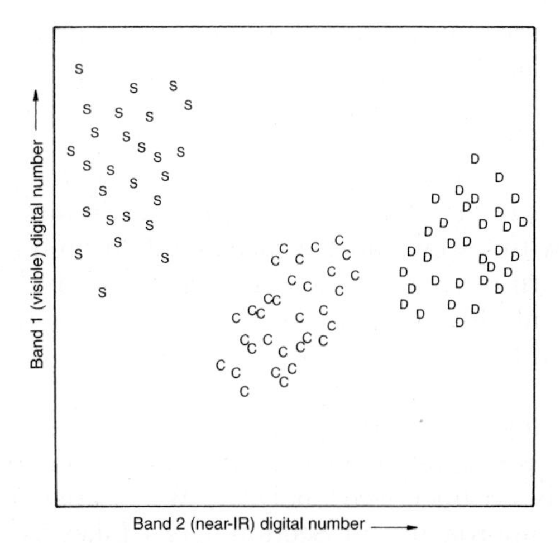

Figure 7.46 Spectral classes in two-channel image data.

one to stressed trees of both types (indicated by D, C, and S in Figure 7.46). In a supervised approach, we may not have considered training for the "stressed" class. This highlights one of the primary advantages of unsupervised classification: The *classifier* identifies the distinct spectral classes present in the image data. Many of these classes might not be initially apparent to the analyst applying a supervised classifier. Likewise, the spectral classes in a scene may be so numerous that it would be difficult to train on all of them. In the unsupervised approach they are found automatically.

There are numerous *clustering* algorithms that can be used to determine the natural spectral groupings present in a data set. One common form of clustering, called the *K-means* approach, accepts from the analyst the number of clusters to be located in the data. The algorithm then arbitrarily "seeds," or locates, that number of cluster centers in the multidimensional measurement space. Each pixel in the image is then assigned to the cluster whose arbitrary mean vector is closest. After all pixels have been classified in this manner, revised mean vectors for each of the clusters are computed. The revised means are then used as the basis to reclassify the image data. The procedure continues until there is no significant change in the location of class mean vectors between successive iterations of the algorithm. Once this point is reached, the analyst determines the land cover identity of each spectral class.

A widely used variant on the K-means method for unsupervised clustering is an algorithm called *Iterative Self-Organizing Data Analysis Techniques A*, or *ISODATA* (Ball and Hall, 1965).[2] This algorithm permits the number of clusters to change from one iteration to the next, by merging, splitting, and deleting clusters.

[2] In the words of its originators, "the *A* was added to make *ISODATA* pronounceable."

The general process follows that described above for K-means. However, in each iteration, following the assignment of pixels to the clusters, the statistics describing each cluster are evaluated. If the distance between the mean points of two clusters is less than some predefined minimum distance, the two clusters are merged together. On the other hand, if a single cluster has a standard deviation (in any one dimension) that is greater than a predefined maximum value, the cluster is split in two. Clusters with fewer than the specified minimum number of pixels are deleted. Finally, as with other variants of K-means, all pixels are then reclassified into the revised set of clusters, and the process repeats, until either there is no significant change in the cluster statistics or some maximum number of iterations is reached.

Data from supervised training areas are sometimes used to augment the results of the above clustering procedure when certain land cover classes are poorly represented in the purely unsupervised analysis. (We discuss other such hybrid supervised/unsupervised approaches in Section 7.12.) Likewise, in some unsupervised classifiers the order in which different feature types are encountered can result in poor representation of some classes. For example, the analyst-specified maximum number of classes may be reached in an image long before the moving window passes throughout the scene.

Often a multistage approach is used with unsupervised classification to improve the representation of certain classes that are imperfectly differentiated in the initial classification. In this approach, two or more clusterings are used to narrow the focus on a particular class of interest. The general sequence is shown in Figure 7.47:

1. **Initial unsupervised classification.** In the first classification, one spectral class is over-broad; it includes pixels that should belong in other spectral classes.
2. **Masking of problem class.** A new image is created in which only pixels from the "problem" class are retained; all others are set to values of no-data.
3. **Second-stage classification of problem class.** A second unsupervised classification is performed, using only pixels from the "problem" class.
4. **Recode output from second-stage classification.** The spectral subclasses from step 3 are reassigned to existing classes from the initial classification in step 1, or to new classes.
5. **Merger of classification results.** The results from the recoded output of the second-stage classification are inserted back into the output image from the initial classification.

The result of this procedure is a modified classification that is identical to the original, except that one over-broad spectral class has been split into two or more other classes. If necessary, more than one "problem" class can be split this way, either simultaneously or in series. Note that this approach is conceptually similar

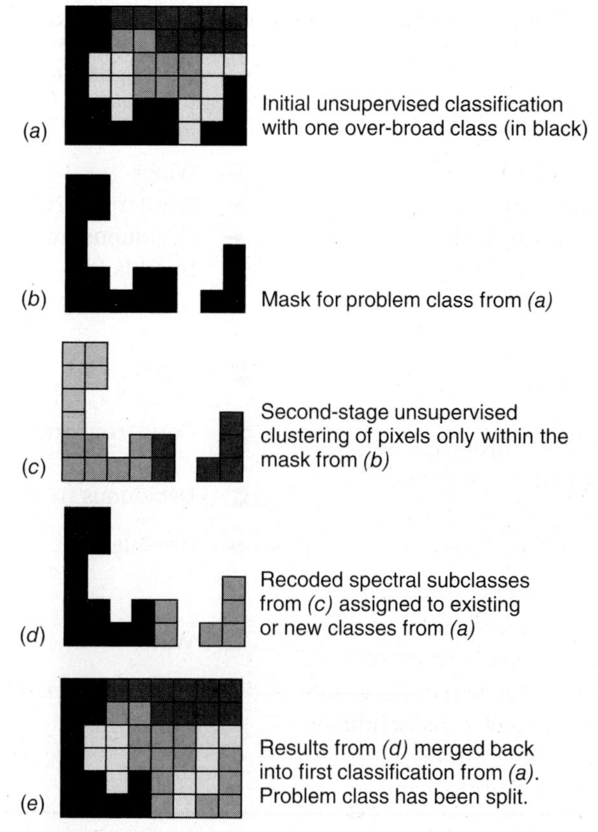

Figure 7.47 The multistage unsupervised classification process.

to the supervised decision-tree (stratified or layered) classifier discussed at the end of Section 7.9, but using unsupervised clustering at each stage rather than a supervised procedure.

Before ending our discussion of unsupervised classification, we reiterate that the result of such efforts is simply the identification of spectrally distinct classes in image data. The analyst must still use reference data to associate the spectral classes with the cover types of interest. This process, like the training set refinement step in supervised classification, can be quite involved.

Table 7.2 illustrates several possible outcomes of associating spectral classes with information classes for data from a scene covering a forested area. The ideal result would be outcome 1, in which each spectral class is found to be associated uniquely with a feature type of interest to the analyst. This outcome will occur only when the features in the scene have highly distinctive spectral characteristics.

A more likely result is presented in outcome 2. Here, several spectral classes are attributable to each information category desired by the analyst. These "subclasses" may be of little informational utility (sunlit versus shaded conifers) or

TABLE 7.2 Spectral Classes Resulting from Clustering a Forested Scene

Spectral Class	Identity of Spectral Class	Corresponding Desired Information Category
Possible Outcome 1		
1	Water →	Water
2	Coniferous trees →	Coniferous trees
3	Deciduous trees →	Deciduous trees
4	Brushland →	Brushland
Possible Outcome 2		
1	Turbid water →	Water
2	Clear water →	
3	Sunlit conifers →	Coniferous trees
4	Shaded hillside conifers →	
5	Upland deciduous →	Deciduous trees
6	Lowland deciduous →	
7	Brushland →	Brushland
Possible Outcome 3		
1	Turbid water →	Water
2	Clear water →	
3	Coniferous trees →	Coniferous trees
4	Mixed coniferous/deciduous →	(Coniferous trees, Deciduous trees)
5	Deciduous trees →	Deciduous trees
6	Deciduous/brushland →	(Deciduous trees, Brushland)
		Brushland

they may provide useful distinctions (turbid versus clear water and upland versus lowland deciduous). In either case, the spectral classes may be aggregated after classification into the smaller set of categories desired by the analyst.

Outcome 3 represents a more troublesome result in which the analyst finds that several spectral classes relate to more than one information category. For example, spectral class 4 was found to correspond to coniferous trees in some locations and deciduous trees in others. Likewise, class 6 included both deciduous trees and brushland vegetation. This means that these information categories are spectrally similar and cannot be differentiated in the given data set.

7.12 HYBRID CLASSIFICATION

Various forms of hybrid supervised/unsupervised classification have been developed to either streamline or improve the accuracy of purely supervised or unsupervised procedures. For example, *unsupervised training areas* might be delineated in an image in order to aid the analyst in identifying the numerous spectral classes that need to be defined in order to adequately represent the land cover information classes to be differentiated in a supervised classification. Unsupervised training areas are image subareas chosen intentionally to be quite different from supervised training areas.

Whereas supervised training areas are located in regions of homogeneous cover type, the unsupervised training areas are chosen to contain numerous cover types at various locations throughout the scene. This ensures that all spectral classes in the scene are represented somewhere in the various subareas. These areas are then clustered independently and the spectral classes from the various areas are analyzed to determine their identity. They are subjected to a pooled statistical analysis to determine their spectral separability and normality. As appropriate, similar clusters representing similar land cover types are combined. Training statistics are developed for the combined classes and used to classify the entire scene (e.g., by a minimum distance or maximum likelihood algorithm).

Hybrid supervised/unsupervised classifiers are particularly valuable in analyses where there is complex variability in the spectral response patterns for individual cover types present. These conditions are quite common in such applications as vegetation mapping. Under these conditions, spectral variability within cover types normally comes about both from variation within cover types (species) and from different site conditions (e.g., soils, slope, aspect, crown closure). *Guided clustering* is a hybrid approach that has been shown to be quite effective in such circumstances (Bauer et al., 1994; Lillesand et al., 1998; Reese et al., 2002; Chipman et al., 2011).

In guided clustering, the analyst delineates numerous "supervised-like" training sets for each cover type to be classified in a scene. Unlike the training sets used in traditional supervised methods, these areas need not be perfectly homogeneous. The data from all the training sites for a given information class are then used in an unsupervised clustering routine to generate several (as many as 20 or more) spectral signatures. These signatures are examined by the analyst; some may be discarded or merged and the remainder are considered to represent spectral subclasses of the desired information class. Signatures are also compared among the different information classes. Once a sufficient number of such spectral subclasses have been acquired for all information classes, a maximum likelihood classification is performed with the full set of refined spectral subclasses. The spectral subclasses are then aggregated back into the original information classes.

Guided clustering may be summarized in the following steps:

1. Delineate training areas for information class X.
2. Cluster all class X training area pixels at one time into spectral subclasses $X_1, \ldots, X_n$ using an automated clustering algorithm.
3. Examine class X signatures and merge or delete signatures as appropriate. A progression of clustering scenarios (e.g., from 3 to 20 cluster

classes) should be investigated, with the final number of clusters and merger and deletion decisions based on such factors as (a) display of a given class on the raw image, (b) multidimensional histogram analysis for each cluster, and (c) multivariate distance measures (e.g., transformed divergence or JM distance).

4. Repeat steps 1 to 3 for all additional information classes.
5. Examine all class signatures and merge or delete signatures as appropriate.
6. Perform maximum likelihood classification on the entire image with the full set of spectral subclasses.
7. Aggregate spectral subclasses back to the original information classes.

Among the advantages of this approach is its ability to help the analyst identify the various spectral subclasses representing an information class "automatically" through clustering. At the same time, the process of labeling the spectral clusters is straightforward because these are developed for one information class at a time. Also, spurious clusters due to such factors as including multiple-cover-type conditions in a single training area can be readily identified (e.g., openings containing understory vegetation in an otherwise closed forest canopy, bare soil in a portion of a crop-covered agricultural field). The method also helps identify situations where mixed pixels might inadvertently be included near the edges of training areas.

7.13 CLASSIFICATION OF MIXED PIXELS

As we have previously discussed (Sections 1.5 and 4.2), mixed pixels result when a sensor's instantaneous field of view (IFOV) includes more than one land cover type or feature on the ground. This can occur for several reasons. First, it is common for two or more different feature types to be widely intermingled across the landscape (e.g., every ground resolution cell within an agricultural field may include both corn plants and bare soil). Second, even when a landscape is made up of more or less homogeneous areas, the ground resolution cells (or image pixels) falling along the boundaries between these homogeneous patches will contain some mixture of the two classes. In either case, these mixed pixels present a difficult problem for image classification, because their spectral characteristics are not representative of any single land cover type. *Spectral mixture analysis* and *fuzzy classification* are two procedures designed to deal with the classification of mixed pixels. They represent means by which "subpixel classification" is accomplished.

Subpixel classification is also frequently referred to as "soft" classification, in that more than one possible land cover class can be associated with each pixel, and information on the proportion of each class might be determined. Figure 7.48

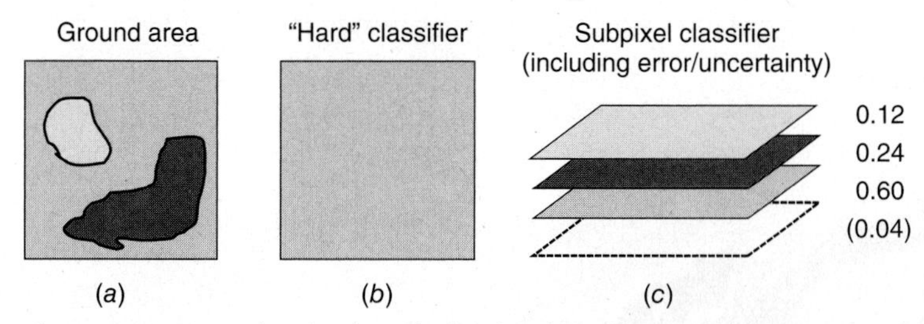

Figure 7.48 Principle of subpixel classification. (*a*) Ground area contained within a single pixel of the image, including patches of three different land cover types. (*b*) Output from a traditional "hard" classifier, representing only a single class. (*c*) Output from a subpixel classifier, including estimated fraction of each class plus error.

contrasts traditional "hard" classification with this subpixel approach. When an image pixel covers a ground area with multiple land cover classes at the subpixel scale, the traditional classifier assigns a single class. As shown in (*b*), this could be the numerically dominant class within the pixel, but it could also be a class that represents a smaller fraction of the area but has a particularly high reflectance. Alternatively, and perhaps worse still, the classifier could interpret the mixture of these three classes as being most similar to some other class that is not present at all in the pixel! (One common real-world example of this problem occurs at the boundaries of water bodies. Because water has very low near-IR reflectance while deciduous forest can have very high reflectance in the near IR, image pixels spanning the border between the two may have an intermediate near IR reflectance. Based on this intermediate value, the "hard" classifier often will conclude that this pixel is some other class entirely, such as evergreen forest or impervious surfaces.)

As shown in Figure 7.48*c*, a subpixel classifier would produce multiple output layers—one for each class, plus (in some cases) an additional estimate of error or uncertainty. For a given pixel, the value in each layer represents the fraction of the pixel that consists of that layer's class. In principle, these fractions should add up to 1.0 (or 100%); depending on the software implementation and on methodological preferences, the analyst may choose to force this "unit sum" constraint, or to allow the sum of all class fractions to depart from 1.0. In the latter case, this "error" may represent the presence of additional land cover types not represented in the set of classes used by the classifier or imperfections in the statistical representation of the existing set of classes.

Figure 7.49 illustrates two class layers (soil and vegetation) resulting from a subpixel classification. In both cases, the brighter the pixel value, the higher the fractional area of that class at the subpixel level. Areas that appear dark in both (*a*) and (*b*) represent landscape units that have low fractions of both soil and vegetation. (Many of those pixels are in water bodies, a third class not shown here.)

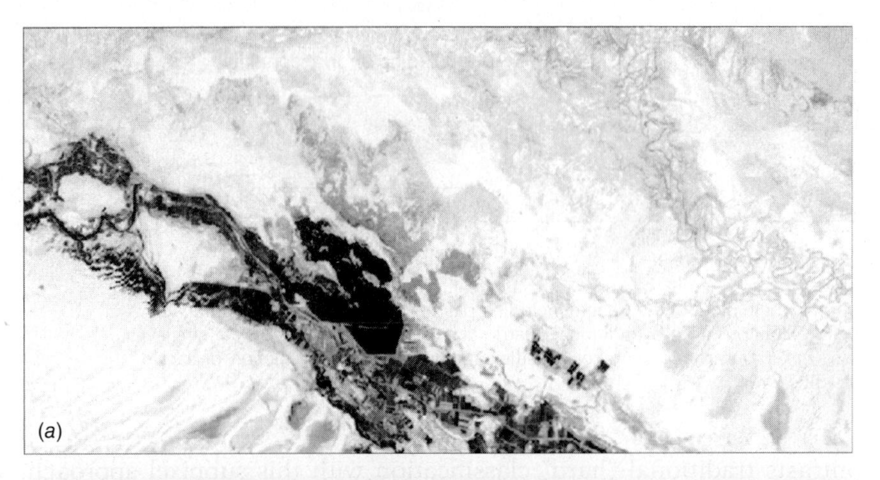

Figure 7.49 Examples of subpixel classification. (*a*) Soil fraction. (*b*) Vegetation fraction. Lighter tones represent higher fractional coverage at the subpixel level. (Author-prepared figure.)

Spectral Mixture Analysis

Spectral mixture analysis involves a range of techniques wherein mixed spectral signatures are compared to a set of "pure" reference spectra (measured in the laboratory, in the field, or from the image itself). The basic assumption is that the spectral variation in an image is caused by mixtures of a limited number of surface materials. The result is an estimate of the approximate proportions of the ground area of each pixel that are occupied by each of the reference classes. For example, Figure 7.50 shows spectra for two "pure" feature types in an EO-1 Hyperion hyperspectral image of Kilauea Volcano, Hawaii. One of the pure spectra is for a bare, unvegetated lava flow (less than 30 years old), and the other is for a nearby tropical rainforest canopy. The third spectrum, located between the other two in Figure 7.50, is for a mixed pixel located on the boundary between a

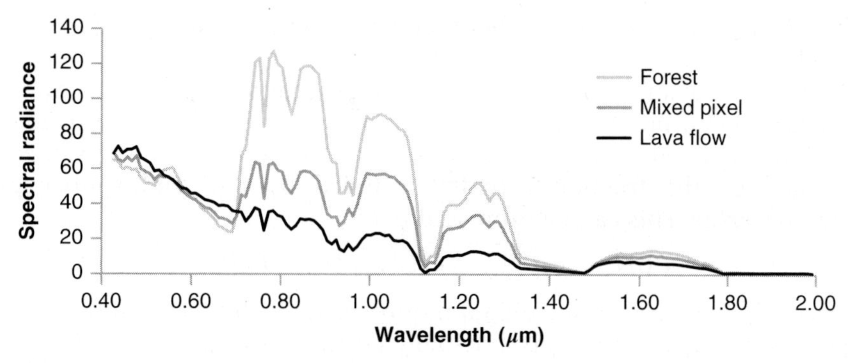

Figure 7.50 Linear mixture of spectra from an EO-1 Hyperion image of tropical rainforest and lava flows in Hawaii.

lava flow and undisturbed forest. From a numerical analysis using the assumption of linear mixing (see below) this mixed pixel's area is estimated to consist of approximately 55% forest and 45% lava flow.

Spectral mixture analysis differs in several ways from other image processing methods for land cover classification. Conceptually, it is a deterministic method rather than a statistical method, because it is based on a physical model of the mixture of discrete spectral response patterns. It provides useful information at the subpixel level, because multiple land cover types can be detected within a single pixel. Many land cover types tend to occur as heterogeneous mixtures even when viewed at very fine spatial scales, thus this method provides a more realistic representation of the true nature of the surface than would be provided by the assignment of a single dominant class to every pixel.

Many applications of spectral mixture analysis make use of linear mixture models, in which the observed spectral response from an area on the ground is assumed to be a *linear* mixture of the individual spectral signatures of the various land cover types present within the area. These pure reference spectral signatures are referred to as *endmembers*, because they represent the cases where 100% of the sensor's field of view is occupied by a single cover type. In this model, the weight for any given endmember signature is the proportion of the area occupied by the class corresponding to the endmember. The input to a linear mixture model consists of a single observed spectral signature for each pixel in an image. The model's output then consists of "abundance" or "fraction" images for each endmember, showing the fraction of each pixel occupied by each endmember.

Linear mixture analysis involves the simultaneous satisfaction of two basic conditions for each pixel in an image. First, the sum of the fractional proportions of all potential endmembers included in a pixel must equal 1. Expressed mathematically,

$$\sum_{i=1}^{N} F_1 = F_1 + F_2 + \cdots + F_N = 1 \tag{7.10}$$

where $F_1, F_2, \ldots, F_N$ represent the fraction of each of N possible endmembers contained in a pixel.

The second condition that must be met is that for a given spectral band λ the observed digital number DN_λ for each pixel represents the sum of the DNs that would be obtained from a pixel that is completely covered by a given endmember weighted by the fraction actually occupied by that endmember plus some unknown error. This can be expressed by

$$DN_\lambda = F_1 DN_{\lambda,1} + F_2 DN_{\lambda,2} + \cdots + F_N DN_{\lambda,N} + E_\lambda \tag{7.11}$$

where DN_λ is the composite digital number actually observed in band λ; $F_1, \ldots, F_N$ equal the fractions of the pixel actually occupied by each of the N endmembers; $DN_{\lambda,1}, \ldots, DN_{\lambda,N}$ equal the digital numbers that would be observed if a pixel were completely covered by the corresponding endmember; and E_λ is the error term.

With multispectral data, there would be one version of Eq. 7.11 for each spectral band. So, for B spectral bands, there would be B equations, plus Eq. 7.10. This means that there are $B+1$ equations available to solve for the various endmember fractions $(F_1, \ldots, F_N)$. If the number of endmember fractions (unknowns) is equal to the number of spectral bands plus 1, the set of equations can be solved simultaneously to produce an exact solution without any error term. If the number of bands $B+1$ is greater than the number of endmembers N, the magnitude of the error term along with the fractional cover for each endmember can be estimated (using the principles of least squares regression). On the other hand, if the number of endmember classes present in a scene exceeds $B+1$, the set of equations will not yield a unique solution.

For example, a spectral mixture analysis of a four-band SPOT-6 multispectral image could be used to find estimates of the fractional proportions of five different endmember classes (with no estimate of the amount of error), or of four, three, or two endmember classes (in which case an estimate of the error would also be produced). Without additional information, this image alone could not be used in linear spectral mixture analysis to derive fractional cover estimates for more than five endmember classes.

Figure 7.51 shows an example of the output from a linear spectral mixture analysis project in which Landsat TM imagery was used to determine the fractional cover of trees, shrubs, and herbaceous plants in the Steese National Conservation Area of central Alaska. Figure 7.51*a* shows a single band (TM band 4, near IR), while 7.51*b* through 7.51*d* show the resulting output for each of the endmember classes. Note that these output images are scaled such that higher fractional cover values appear brighter while lower fractional cover values appear darker.

One drawback of linear mixture models is that they do not account for certain factors such as multiple reflections, which can result in complex nonlinearities in the spectral mixing process. That is, the observed signal from a pixel may include a mixture of spectral signatures from various endmembers, but it may also

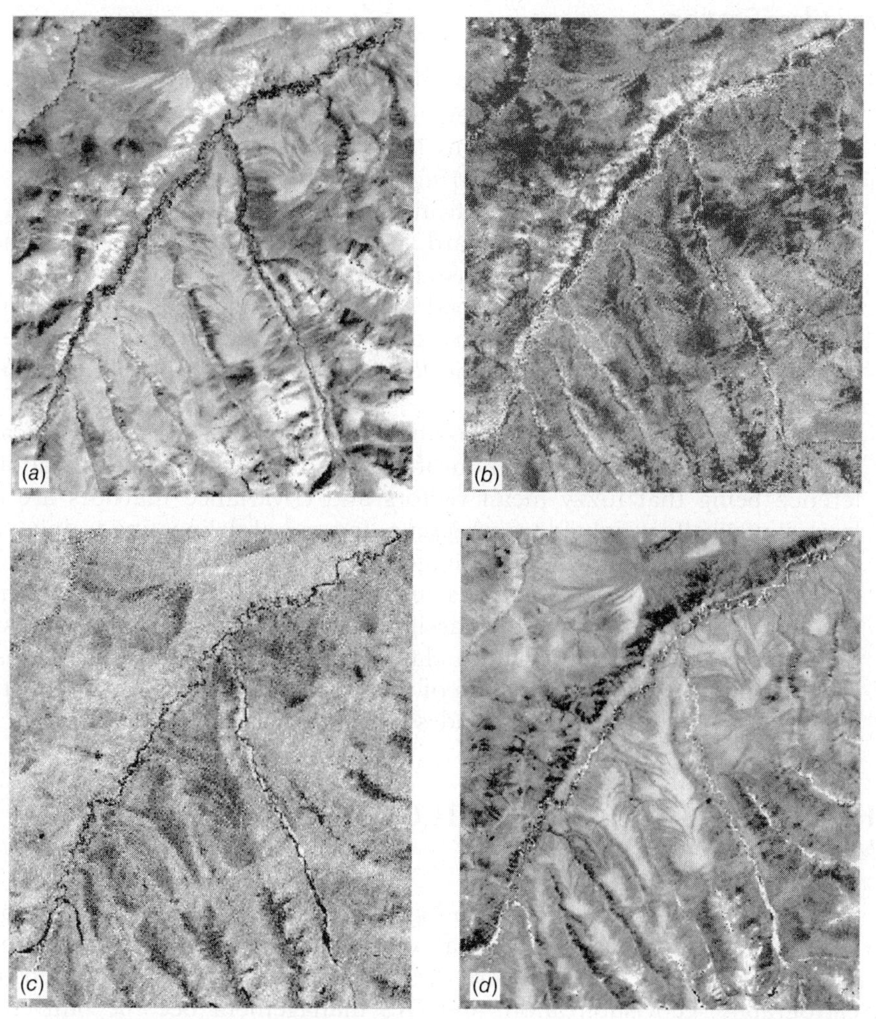

Figure 7.51 Linear spectral mixture analysis of a Landsat TM image including the Steese National Conservation Area of central Alaska: (*a*) band 4 (near IR) of original image; fractional cover images for trees (*b*), shrubs (*c*), and herbaceous plants (*d*). Brighter pixels represent higher fractional cover. (Courtesy Bureau of Land Management–Alaska and Ducks Unlimited, Inc.)

include additional radiance reflected multiple times between scene components such as leaves and the soil surface. In this situation, a more sophisticated *nonlinear* spectral mixture model may be required (Somers et al., 2009). Artificial neural networks (Section 7.16) may be particularly well suited for this task, because they do not require that the input data have a Gaussian distribution and they do not assume that spectra mix linearly (Plaza et al., 2011).

Fuzzy Classification

Fuzzy classification attempts to handle the mixed-pixel problem by employing the fuzzy set concept, in which a given entity (a pixel) may have partial membership in more than one category (Jensen, 2005; Schowengerdt, 2006). One approach to fuzzy classification is *fuzzy clustering*. This procedure is conceptually similar to the K-means unsupervised classification approach described earlier. The difference is that instead of having "hard" boundaries between classes in the spectral measurement space, fuzzy regions are established. So instead of each unknown measurement vector being assigned solely to a single class, irrespective of how close that measurement may be to a partition in the measurement space, *membership grade* values are assigned that describe how close a pixel measurement is to the means of all classes.

Another approach to fuzzy classification is *fuzzy supervised* classification. This approach is similar to application of maximum likelihood classification; the difference being that fuzzy mean vectors and covariance matrices are developed from statistically weighted training data. Instead of delineating training areas that are purely homogeneous, a combination of pure and mixed training sites may be used. Known mixtures of various feature types define the fuzzy training class weights. A classified pixel is then assigned a membership grade with respect to its membership in each information class. For example, a vegetation classification might include a pixel with grades of 0.68 for class "forest," 0.29 for "street," and 0.03 for "grass." (Note that the grades for all potential classes must total 1.)

7.14 THE OUTPUT STAGE AND POSTCLASSIFICATION SMOOTHING

The utility of any image classification is ultimately dependent on the production of output maps, tables, and geospatial data that effectively convey the interpreted information to its end user. Here the boundaries between remote sensing, digital cartography, geovisualization, and GIS management become blurred. Plate 2*a* shows the output from a land cover classification that is being used as input to a model of soil erosion potential within a GIS. For internal use within the GIS, nothing more is needed from the classification output beyond its inherent raster grid of pixel values. On the other hand, as shown in Plate 2*a*, for visualization purposes choices about scale, color selection, shading, and other cartographic design topics become important. In this example, the colors associated with each land cover class are modified with a light hillshading algorithm to convey the shape of local topography, to aid the viewer in their interpretation of the landscape.

One issue that often arises during the output stage is the need for smoothing of classification images to remove isolated misclassified pixels. Classified data often manifest a salt-and-pepper appearance due to the inherent spectral variability encountered by a classifier when applied on a pixel-by-pixel basis (Figure 7.52*a*). For example, in an agricultural area, several pixels scattered throughout a cornfield may be classified as soybeans, or vice versa. In such situations it is often desirable to "smooth" the classified output to show only the dominant (presumably correct) classification. Initially, one might consider the application of the previously described low-pass spatial filters for this purpose. The problem with this approach is that the output from an image classification is an array of pixel locations containing numbers serving the function of *labels*, not *quantities*. That is, a pixel containing land cover 1 may be coded with a 1; a pixel containing land cover 2 may be coded with a 2; and so on. A moving low-pass filter will not properly smooth such data because, for example, the averaging of class 3 and class 5 to arrive at class 4 makes no sense. In short, postclassification smoothing algorithms must operate on the basis of logical operations, rather than simple arithmetic computations.

One means of classification smoothing involves the application of a *majority filter*. In such operations a moving window is passed through the classified data set and the majority class within the window is determined. If the center pixel in the window is not the majority class, its identity is changed to the majority class. If there is no majority class in the window, the identity of the center pixel is not changed. As the window progresses through the data set, the original class codes are continually used, not the labels as modified from the previous window positions. (Figure 7.52*b* was prepared in this manner, applying a 3×3-pixel majority filter to the data shown in Figure 7.52*a*. Figure 7.52*c* was prepared by applying a 5×5-pixel filter.)

Majority filters can also incorporate some form of class and/or spatial weighting function. Data may also be smoothed more than once. Certain algorithms can preserve the boundaries between land cover regions and also involve a

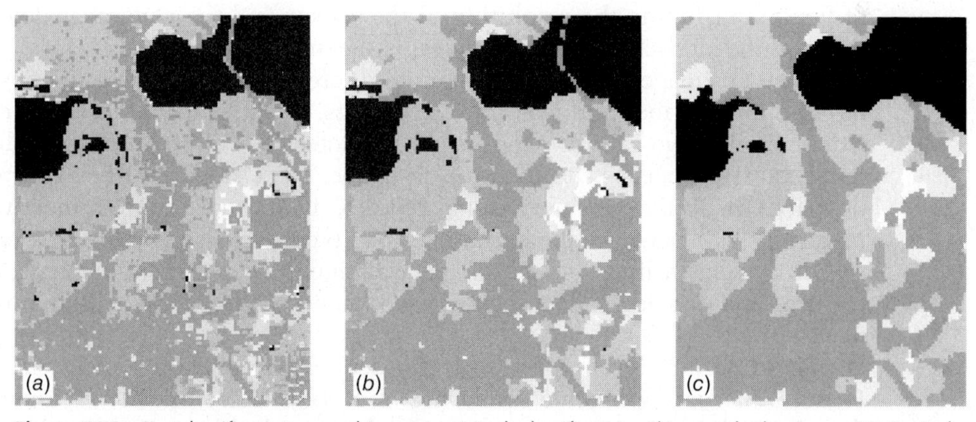

Figure 7.52 Postclassification smoothing: (*a*) original classification; (*b*) smoothed using a 3×3 pixel-majority filter; (*c*) smoothed using a 5×5-pixel majority filter. (Author-prepared figure.)

user-specified minimum area of any given land cover type that will be maintained in the smoothed output.

Another way of obtaining smoother classifications is to integrate the types of logical operations described above directly into the classification process. Object-based classification, the subject of the following section, affords this capability (and much more).

7.15 OBJECT-BASED CLASSIFICATION

All of the classification algorithms discussed to this point have been based solely on the analysis of the spectral characteristics of individual pixels. That is, these *per pixel* procedures generally use spectrally based decision logic that is applied to each pixel in an image individually and in isolation. In contrast, object-based classifiers use both spectral and spatial patterns for image classification (Blaschke, 2010). This is a two-step process involving (1) segmentation of the imagery into discrete objects, followed by (2) classification of those objects. The basic assumption is that the image being classified is made up of relatively homogeneous "patches" that are larger in size than individual pixels. This approach is similar to human visual interpretation of digital images, which works at multiple scales simultaneously and uses color, shape, size, texture, pattern, and context information to group pixels into meaningful objects.

The scale of objects is one of the key variables influencing the image segmentation step in this process. For example, in the case of a forested landscape, at a fine scale the objects being classified might represent individual tree crowns. Segmentation at an intermediate scale would produce objects that correspond to stands of trees of similar species and sizes, while at a still coarser scale, large areas of the forest would be aggregated into a single object. Clearly, the actual scale parameter used for an object-based classification will depend on a number of factors, including the resolution of the sensor and the general scale of the features on the landscape that the analyst is seeking to identify.

Once an image has been segmented, there are many characteristics that can be used to describe (and classify) the objects. Broadly speaking, these characteristics fall into two groups. One set of characteristics is intrinsic to each object—its spectral properties, its texture, its shape, for example. Other characteristics describe the relationships among objects, including their connectivity, their proximity to objects of the same or other types, and so forth. For example, an object that has a linear shape, a spectral signature similar to asphalt, a smooth texture, and topological connectivity to other road objects would probably be itself a road.

Figure 7.53 illustrates object-based segmentation and classification. Figure 7.53*a* is a grayscale rendering of a color-IR composite of Landsat-5 TM data from bands 2, 3, and 4 (originally depicted in blue, green, and red, respectively). This image was acquired over western Vilas County, located in north-central Wisconsin. The major land cover types present in this area include water, deciduous forest, evergreen forest, cranberry bog, wetlands, and clear-cut/barren areas. Figures 7.53*b–d* are a series of segmentations that were produced from the TM data holding most parameters constant (e.g., spectral band weighting, shape factor, smoothness factor), but varying the scale factor.

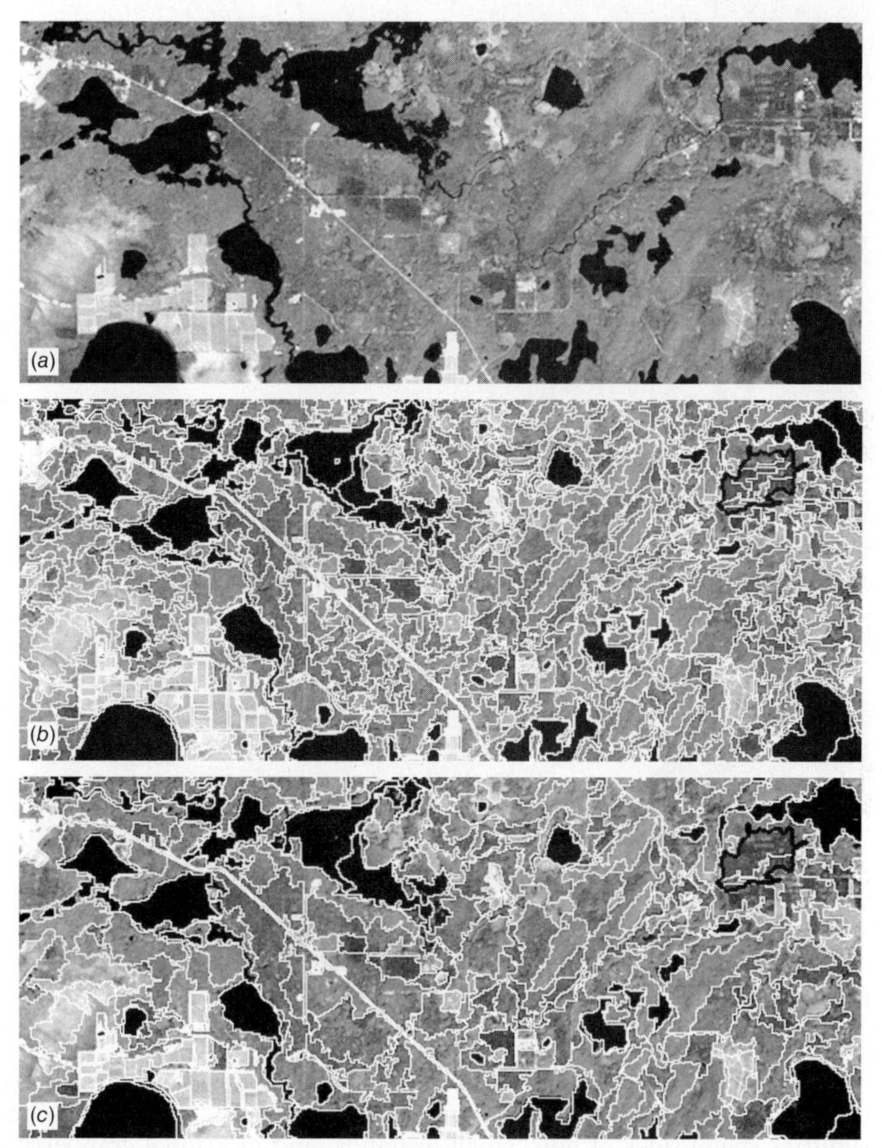

Figure 7.53 Object-oriented image segmentation and classification. (*a*) Landsat-5 TM image obtained over Vilas County, WI. (*b*) Image segmentation using a fine scale factor. (*c*) Image segmentation using an intermediate scale factor. (*d*) Image segmentation using a coarse scale factor. (*e*) Land cover classification using image segmented with a fine scale factor. (*f*) Traditional unsupervised "per pixel" land cover classification of nonsegmented image data. (Author-prepared figure.)

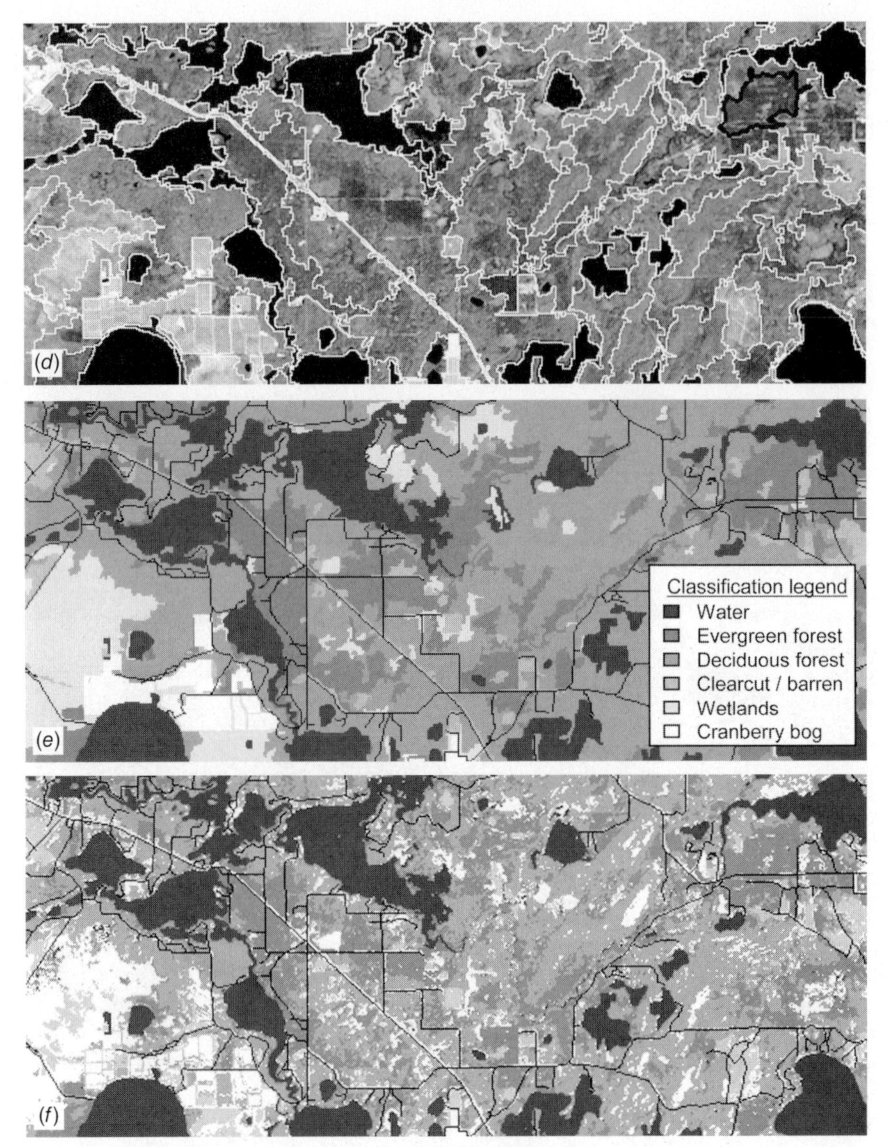

Figure 7.53 *(Continued)*

At a fine scale factor, shown in (*b*), the landscape is divided into small objects corresponding to features such as groups of trees of uniform species composition, sunlit or shady sides of hills, small wetlands, and individual cranberry bog cultivation beds. These objects appear as white polygons superimposed on a backdrop of the original (unclassified) Landsat imagery. The large polygon outlined in black in the upper right portion of the image represents a tree plantation with stands of conifers (red pine and jack pine) of varying ages and sizes. Note that the polygon has been subdivided into multiple objects corresponding to individual, homogeneous stands.

At an intermediate scale factor, shown in (*c*), the objects are larger, and at a coarse scale factor, shown in (*d*), the entire plantation outlined in black is now represented by a single object. Note that the objects at different scales "nest" in a hierarchical fashion, so that all the objects at a fine scale fit perfectly within the boundaries of aggregated objects at a coarser scale.

Once the boundaries of the objects have been established, they can then be classified based on the various characteristics described above in a wide variety of ways. The classification shown in Figure 7.53*e* was produced by first segmenting the image using a fine scale factor and then aggregating the resulting objects into the six land cover information classes listed above. (The roads shown in this image were "burned in" from a transportation GIS layer.) A traditional (non-segmented) unsupervised classification of the image is presented in Figure 7.53*f*. Compared to these per-pixel classifier results, the object-based approach produces a classification that is much smoother in appearance, with relatively homogeneous regions. Note also how the segmentation aids in the discrimination of linear features, such as the road trending from the upper left side to the middle of the lower edge of the study area.

Object-based analysis can also be used to facilitate land cover change in that the approach is capable of preserving the "parent–child" relationships among objects. For example, a large agricultural field containing only one crop type at an early time period might be classified as a single object. If multiple crop types are present in the field at a later date, the parent field will be split into numerous child objects. Spectral data could then be used to classify the identity of the new child objects.

7.16 NEURAL NETWORK CLASSIFICATION

Another increasingly common method for image classification makes use of *artificial neural networks*. Initially inspired by the biological model of neurons in the brain, the use of artificial neural networks was pioneered within computer science's subfields of artificial intelligence, machine learning, and related areas. In later years, implementation of artificial neural networks in computer science focused less on replicating actual biological structures and processes, and instead developed its own statistical and computational logic. Nonetheless, the terminology (including the use of the term "neurons" within these networks) persists, as a legacy of the method's biologically inspired origins.

Neural networks are "self-training" in that they adaptively construct linkages between a given pattern of input data and particular outputs. Neural networks can be used to perform traditional image classification or for more complex operations such as spectral mixture analysis. For image classification, neural networks do not require that the training data have a Gaussian statistical distribution, a requirement that is held by maximum likelihood algorithms. This allows neural networks to be used with a much wider range of types of input data than could be used in a traditional maximum likelihood classification process. (The use of ancillary data from other sources in combination with remotely sensed imagery for classification purposes is discussed in Section 7.20.) In addition, once they have been fully trained, neural networks can perform image classification relatively rapidly, although the training process itself can be quite time consuming. In the following discussion we will focus on back-propagation neural networks, the type most widely used in remote sensing applications, although other types of neural networks have been described.

A neural network consists of a set of three or more layers, each made up of multiple nodes sometimes referred to as neurons. The network's layers include an input layer, an output layer, and one or more hidden layers. The nodes in the input layer represent variables used as input to the neural network. Typically, these might include spectral bands from a remotely sensed image, textural features or other intermediate products derived from such images, or ancillary data describing the region to be analyzed. The nodes in the output layer represent the range of possible output categories to be produced by the network. If the network is being used for image classification, there will be one output node for each class in the classification system.

Between the input and output layers are one or more hidden layers. These consist of multiple nodes, each linked to many nodes in the preceding layer and to many nodes in the following layer. These linkages between nodes are represented by weights, which guide the flow of information through the network. The number of hidden layers used in a neural network is arbitrary. Generally speaking, an increase in the number of hidden layers permits the network to be used for more complex problems but reduces the network's ability to generalize and increases the time required for training. Figure 7.54 shows an example of a neural network that is used to classify land cover based on a combination of spectral, textural, and topographic information. There are seven nodes in the input layer, as follows: nodes 1 to 4 correspond to the four spectral bands of a multispectral image, node 5 corresponds to a textural feature that is calculated from a radar image, and nodes 6 and 7 correspond to terrain slope and aspect, calculated from a digital elevation model. After the input layer, there are two hidden layers, each with nine nodes. Finally, the output layer consists of six nodes, each corresponding to a land cover class (water, sand, forest, urban, corn, and hay). When given any combination of input data, the network will produce the output class that is most likely to result from that set of inputs, based on the network's analysis of previously supplied training data.

Applying a neural network to image classification makes use of an iterative training procedure in which the network is provided with matching sets of input and output data. Each set of input data represents an example of a pattern to be learned, and each corresponding set of output data represents the desired output

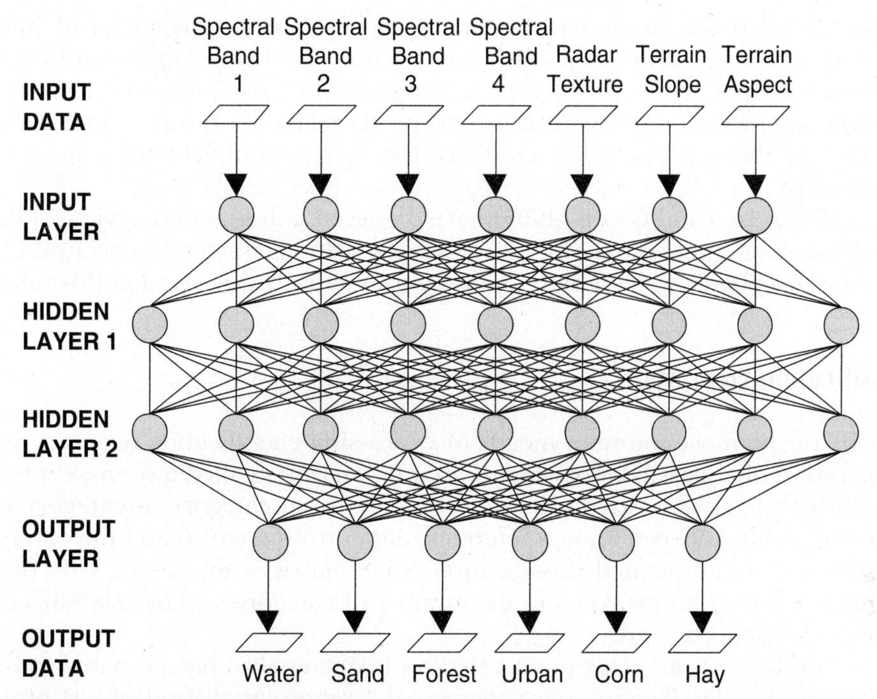

Figure 7.54 Example of an artificial neural network with one input layer, two hidden layers, and one output layer.

that should be produced in response to the input. During the training process, the network autonomously modifies the weights on the linkages between each pair of nodes in such a way as to reduce the discrepancy between the desired output and the actual output.

It should be noted that a back-propagation neural network is not guaranteed to find the ideal solution to any particular problem. During the training process the network may develop in such a way that it becomes caught in a "local minimum" in the output error field, rather than reaching the absolute minimum error. Alternatively, the network may begin to oscillate between two slightly different states, each of which results in approximately equal error. A variety of strategies have been proposed to help push neural networks out of these pitfalls and enable them to continue development toward the absolute minimum error.

7.17 CLASSIFICATION ACCURACY ASSESSMENT

The process of producing a land cover classification from remotely sensed data can be lengthy. At the end of the process, when the classification algorithm has done its work, any problematic classes have been split or refined, and the analyst has recoded the larger set of spectral classes into a reduced set of information classes, it is tempting to publish the resulting land cover map and then move on to other work. However, before moving on to other endeavors, it is essential to document the *categorical accuracy* of the classification results. This need is embodied in the expression "A classification is not complete until its accuracy is assessed."

Congalton and Green (2009) have prepared a thorough overview of the principles and practices currently in use for assessing classification accuracy. Many of the concepts we present here in brief are more fully described in this reference.

Classification Error Matrix

One of the most common means of expressing classification accuracy is the preparation of a classification *error matrix* (sometimes called a *confusion matrix* or a *contingency table*). Error matrices compare, on a category-by-category basis, the relationship between known reference data (ground truth) and the corresponding results of an automated classification. Such matrices are square, with the number of rows and columns equal to the number of categories whose classification accuracy is being assessed.

Table 7.3 is an error matrix that an image analyst has prepared to determine how well a classification has categorized a representative set of test pixels whose "true" classes were determined from *ground truth* data. This ground truth provides information about the actual land cover types present on the ground at the test pixel locations. (Sampling procedures for collecting these ground truth data will be discussed later in this section.) The matrix in Table 7.3 stems from classifying the test pixels and listing the known ("true") cover types from ground truth (columns) versus the mapped cover type from the classifier (rows).

Many characteristics about classification performance are expressed by an error matrix. For example, one can study the various classification errors of omission (exclusion) and commission (inclusion). Note in Table 7.3 that the test pixels that are classified into the proper land cover categories are located along the major diagonal of the error matrix (running from upper left to lower right). All nondiagonal elements of the matrix represent errors of omission or commission. Omission errors correspond to nondiagonal column elements (e.g., seven pixels that should have been classified as "water" were omitted from that category—three were misclassified as "forest," two as "urban," and one each as "corn" and "hay"). Commission errors are represented by nondiagonal row elements (e.g., 92 "urban" pixels plus one "corn" pixel were improperly included in the "sand" category).

Several other descriptive measures can be obtained from the error matrix. For example, the *overall accuracy* is computed by dividing the total number of correctly classified pixels (i.e., the sum of the elements along the major diagonal) by the total number of reference pixels. Likewise, the accuracies of individual

TABLE 7.3 Error Matrix Resulting from Classifying Test Pixels

	Reference Data[a]						
	W	S	F	U	C	H	Row Total
Classification data							
W	226	0	0	12	0	1	239
S	0	216	0	92	1	0	309
F	3	0	360	228	3	5	599
U	2	108	2	397	8	4	521
C	1	4	48	132	190	78	453
H	1	0	19	84	36	219	359
Column total	233	328	429	945	238	307	2480

Producer's Accuracy	**User's Accuracy**
W = 226/233 = 97%	W = 226/239 = 94%
S = 216/328 = 66%	S = 216/309 = 70%
F = 360/429 = 84%	F = 360/599 = 60%
U = 397/945 = 42%	U = 397/521 = 76%
C = 190/238 = 80%	C = 190/453 = 42%
H = 219/307 = 71%	H = 219/359 = 61%

Overall accuracy = (226 + 216 + 360 + 397 + 190 + 219)/2480 = 65%

[a] W, water; S, sand; F, forest; U, urban; C, corn; H, hay.

categories can be calculated by dividing the number of correctly classified pixels in each category by either the total number of pixels in the corresponding row or column.

What are often termed *producer's accuracies* result from dividing the number of correctly classified pixels in each category (on the major diagonal) by the number of test set pixels used for that category (the column total). This figure indicates how well test set pixels of the given cover type are classified.

User's accuracies are computed by dividing the number of correctly classified pixels in each category by the total number of pixels that were classified in that category (the row total). This figure is a measure of commission error and indicates the probability that a pixel classified into a given category actually represents that category on the ground.

Note that the error matrix in Table 7.3 indicates an overall accuracy of 65%. However, producer's accuracies range from just 42% ("urban") to 97% ("water"), and user's accuracies vary from, again, 42% ("corn") to 94% ("water").

From a statistical perspective, it is essential to consider the independence of the test data used for assessing the final accuracy of the classification. Generally speaking, these data should not have been used previously in any stage of the classification process. In some cases, analysts will compute an error matrix based on the same data used for training the classifier (in a supervised classification process) or based on the same data used as guidance for the analyst when assigning spectral clusters to information classes in an unsupervised classification. *It should be remembered that such procedures only indicate how well the statistics extracted from these areas can be used to categorize the same areas!* If the results are good, it means nothing more than that the training areas are homogeneous, the training classes are spectrally separable, and the classification strategy being employed works well in the training areas. Examination of such an error matrix (created using the training data, rather than independent test data) can aid in the training set refinement process, but it indicates little about how the classifier performs elsewhere. One should expect training area accuracies to be overly optimistic, especially if they are derived from limited data sets. (Nevertheless, training area accuracies are sometimes used in the literature as an indication of overall accuracy. They should not be!)

Evaluating Classification Error Matrices

A number of features are readily apparent from inspection of the error matrix included in Table 7.3. First, we can begin to appreciate the need for considering overall, producer's, and user's accuracies simultaneously. In this example, the overall accuracy of the classification is 65%. However, if the primary purpose of the classification is to map the locations of the "forest" category, we might note that the producer's accuracy of this class is quite good (84%). This would potentially lead one to the conclusion that although the overall accuracy of the classification was poor (65%), it is adequate for the purpose of mapping the forest class. The problem with this conclusion is the fact that the user's accuracy for this class is only 60%. That is, even though 84% of the forested areas have been correctly identified as "forest," only 60% of the areas identified as "forest" within the classification are truly of that category. Accordingly, although the producer of the classification can reasonably claim that 84% of the time an area that was forested was identified as such, a user of this classification would find that only 60% of the time will an area visited on the ground that the classification says is "forest" actually be "forest." In fact, the only highly reliable category associated with this classification from both a producer's and a user's perspective is "water."

A further point to be made about interpreting classification accuracies is the fact that even a completely random assignment of pixels to classes will produce some correct values in the error matrix. In fact, if the number of classes is small, such a random assignment could result in a surprisingly good apparent classification result—a two-category classification could be expected to be 50% correct solely due to random chance. The $\hat{k}$ (*"kappa"* or *"KHAT"*) statistic is a measure of the difference between the actual agreement between reference data and an automated classifier and the chance agreement between the reference data and a random classifier. Conceptually, $\hat{k}$ can be defined as

$$\hat{k} = \frac{\text{observed accuracy} - \text{chance agreement}}{1 - \text{chance agreement}} \tag{7.12}$$

This statistic serves as an indicator of the extent to which the percentage correct values of an error matrix are due to "true" agreement versus "chance" agreement. As true agreement (observed) approaches 1 and chance agreement approaches 0, $\hat{k}$ approaches 1. This is the ideal case. In reality, $\hat{k}$ usually ranges between 0 and 1. For example, a $\hat{k}$ value of 0.67 can be thought of as an indication that an observed classification is 67% better than one resulting from chance. A $\hat{k}$ of 0 suggests that a given classification is no better than a random assignment of pixels. In cases where chance agreement is large enough, $\hat{k}$ can take on negative values—an indication of very poor classification performance.

The $\hat{k}$ statistic is computed as

$$\hat{k} = \frac{N\sum_{i=1}^{r} x_{ii} - \sum_{i=1}^{r}(x_{i+} \cdot x_{+i})}{N^2 - \sum_{i=1}^{r}(x_{i+} \cdot x_{+i})} \tag{7.13}$$

where

r = number of rows in the error matrix
x_{ii} = number of observations in row i and column i (on the major diagonal)
x_{i+} = total of observations in row i (shown as marginal total to right of the matrix)
x_{+i} = total of observations in column i (shown as marginal total at bottom of the matrix)
N = total number of observations included in matrix

To illustrate the computation of kappa for the error matrix included in Table 7.3,

$$\sum_{i=1}^{r} x_{ii} = 226 + 216 + 360 + 397 + 190 + 219 = 1608$$

$$\sum_{i=1}^{r}(x_{i+} \cdot x_{+i}) = (239 \cdot 233) + (309 \cdot 328) + (599 \cdot 429) + (521 \cdot 945) + (453 \cdot 238) + (359 \cdot 307) = 1{,}124{,}382$$

$$\hat{k} = \frac{2480(1608) - 1{,}124{,}382}{(2480)^2 - 1{,}124{,}382} = 0.57$$

Note that the $\hat{k}$ value (0.57) obtained in the above example is somewhat lower than the overall accuracy (0.65) computed earlier. Differences in these two measures are to be expected in that each incorporates different forms of information from the error matrix. The overall accuracy only includes the data along the major diagonal and excludes the errors of omission and commission. On the other hand, $\hat{k}$ incorporates the nondiagonal elements of the error matrix as a product of the row and column marginal. Accordingly, it is not possible to give definitive advice as to when each measure should be used in any given application. Normally, it is desirable to compute and analyze both of these values.

One of the principal advantages of computing $\hat{k}$ is the ability to use this value as a basis for determining the statistical significance of any given matrix or the differences among matrices. For example, one might wish to compare the error matrices resulting from different dates of images, different classification techniques, or different individuals performing the classification. Such tests are based on computing an estimate of the variance of $\hat{k}$, then using a Z test to determine if an individual matrix is significantly different from a random result and if $\hat{k}$ values from two separate matrices are significantly different from each other. This is a somewhat complicated process because the calculation for the variance of $\hat{k}$ depends on the sampling design used to collect the test data. The interested reader is encouraged to consult the literature for details; for example, Stehman (1996) provides the formula for the variance of $\hat{k}$ under stratified random sampling.

Sampling Considerations

Test areas are areas of representative, uniform land cover that are different from and (usually) more numerous than training areas. They are often located during the training stage of supervised classification by intentionally collecting ground truth for more sites than are actually needed for training data. A subset of these may then be withheld for the postclassification accuracy assessment. The accuracies obtained in these areas represent at least a first approximation to classification performance throughout the scene. However, being homogeneous, test areas might not provide a valid indication of classification accuracy at the individual pixel level of land cover variability. In addition, training data are often collected using an "ad-hoc" sampling design that is neither systematic nor random, instead emphasizing the opportunistic acquisition of ground truth at sites that are readily accessible on the ground, readily interpretable in higher-resolution imagery, or otherwise conducive to use in the training process. Such a sampling scheme may be inadequate for the statistical validity that is desired in the accuracy assessment process.

One way that would appear to ensure adequate accuracy assessment at the pixel level of specificity would be to compare the land cover classification at every pixel in an image with a reference source. While such "wall-to-wall" comparisons may have value in research situations, assembling reference land cover information for an entire project area is expensive and defeats the whole purpose of performing a remote-sensing-based classification in the first place.

Random sampling of pixels circumvents the above problems, but it is plagued with its own set of limitations. First, collection of reference data for a large sample of randomly distributed points is often very difficult and costly. For example, travel distance and access to random sites might be prohibitive. Second, the validity of random sampling depends on the ability to precisely register the reference data to the image data. This is often difficult to do. One way to overcome this problem is to sample only pixels whose identity is not influenced by potential registration errors (for example, points at least several pixels away from field boundaries), but any such "interventions" will again affect the interpretation of the resulting accuracy statistics.

Another consideration is making certain that the randomly selected test pixels or areas are geographically representative of the data set under analysis. Simple random sampling tends to undersample small but potentially important areas. *Stratified random sampling*, where each land cover category may be considered a stratum, is frequently used in such cases. Clearly, the sampling approach appropriate for an agricultural inventory would differ from that of a wetlands mapping activity. Each sample design must account for the area being studied and the cover type being classified.

One common means of accomplishing random sampling is to overlay classified output data with a grid. Test cells within the grid are then selected randomly and groups of pixels within the test cells are evaluated. The cover types present are determined through ground verification (or other reference data) and compared to the classification data.

Other sampling designs have also been described, such as those combining both random and systematic sampling. Such a technique may use systematically sampled areas to collect some accuracy assessment data early in a project (perhaps as part of the training area selection process) and random sampling within strata after the classification is complete.

Consideration must also be given to the *sample unit* employed in accuracy assessment. Depending upon the application, the appropriate sample unit might be individual pixels, clusters of pixels, or polygons. Polygon sampling is the most common approach in current use. One frequent mistake is to use test areas consisting of multiple pixels and treat each pixel as an independent entry in the error matrix. This is inappropriate, due to the extremely high spatial autocorrelation (local similarity) among adjacent pixels in most remotely sensed data sets.

Sample size must also weigh heavily in the development and interpretation of classification accuracy figures. As a broad guideline, it has been suggested that a minimum of 50 samples of each land cover category be included in the error matrix. Classifications covering large areas (over a million acres) or with more than 12 land cover categories require more thorough sampling, typically 75 to 100 samples per category (Congalton and Green, 2009, p. 75). Similarly, the number of samples for each category might be adjusted based on the relative importance of that category for a particular application (i.e., more samples taken in more important categories). Also, sampling might be allocated with respect to the variability within each category (i.e., more samples taken in more variable categories such as wetlands and fewer in less variable categories such as open water). As noted earlier, however, all decisions about sampling design must include consideration of the statistical validity and tractability of the accuracy assessment process.

Final Thoughts on Accuracy Assessment

There are three other facets of classification accuracy assessment that we wish to emphasize before leaving the subject. The first relates to the fact that the quality of any accuracy estimate is only as good as the information used to establish the "true" land cover types present in the test sites. To the extent possible, some estimate of the errors present in the reference data should be incorporated into the accuracy assessment process. It is not uncommon to have the accuracy of the reference data influenced by such factors as spatial misregistration, image interpretation errors, data entry errors, and changes in land cover between the date of the classified image and the date of the reference data. The second point to be made is that the accuracy assessment procedure must be designed to reflect the intended use of the classification. For example, a single pixel misclassified as "wetland" in the midst of a "corn" field might be of little significance in the development of a regional land use plan. However, this same error might be unacceptable if the classification forms the basis for land taxation or for enforcement of wetland preservation legislation. Finally, it should be noted that remotely sensed data are normally just a small subset of many possible forms of data resident in a GIS. The propagation of errors through the multiple layers of information in a GIS is beyond the scope of this work; interested readers may consult the literature on GIS and spatial analysis.

7.18 CHANGE DETECTION

One of the most powerful advantages of remote sensing images is their ability to capture and preserve a record of conditions at different points in time, to enable the identification and characterization of changes over time. This process is referred to as *change detection* and is among the most common uses of digital image analysis. The types of changes that might be of interest can range from nearly instantaneous (movement of a vehicle or animal) to long-term phenomena such as urban fringe development or desertification. Ideally, change detection procedures should involve data acquired by the same (or similar) sensor and be recorded using the same spatial resolution, viewing geometry, spectral bands, radiometric resolution, and time of day. For detecting changes at time scales greater than one year, *anniversary dates* are preferred to minimize sun angle and seasonal differences. Accurate spatial registration of the various dates of imagery (ideally to within ¼ to ½ pixel) is also a requirement for effective change detection. Clearly, when misregistration is greater than one pixel, numerous errors will result when comparing the images.

The reliability of the change detection process may also be strongly influenced by various environmental factors that might change between image dates. In addition to atmospheric effects, such factors as water level, tidal stage, wind, or soil moisture condition might also be important. Even with the use of anniversary dates of imagery, such influences as different planting dates and season-to-season changes in plant phenology must be considered.

One way of discriminating changes over time is to employ *postclassification comparison*. In this approach, two dates of imagery are independently classified and registered, and pixels whose class changed between dates are identified. In addition, statistics (and change maps) can be compiled to express the specific nature of the changes between the dates of imagery, for example, by highlighting only those areas that changed from *Class A* to *Class B*. Obviously, the accuracy of such procedures depends upon the accuracy of each of the independent classifications used in the analysis. The errors present in each of the initial classifications are compounded in the change detection process.

Figure 7.55 shows a representative example of postclassification change detection in a rapidly changing landscape. The original classifications for two dates appear in (*a*) and (*b*). In Figure 7.55*c*, any pixel that changed its class is highlighted—in this dynamic landscape, that includes nearly 50% of the area. Figure 7.55*d* focuses on one particular type of change, from shrubland to agriculture (approximately 15% of the area experienced this change).

Another approach to change detection using spectral pattern recognition is simply the *classification of multitemporal data sets*. The spectral bands for two (or more) dates of imagery are "stacked" together to make a new multitemporal data set; for example, two dates of four-band QuickBird multispectral imagery could be stacked together in a new eight-band data set. Supervised or unsupervised classification is then used to classify the combined image into temporal-categorical "classes"; some of these classes will be stable (i.e., the same class on both dates), while other classes will represent individual types of change. The success of such efforts depends upon the extent to which "change classes" are significantly different spectrally from the "nonchange" classes. Also, the dimensionality and complexity of the classification can be quite great, and if all bands from each date are used, there may be substantial redundancy in their information content.

Principal components analysis is sometimes used to analyze multidate image composites for change detection purposes. In this approach, two (or more) images are registered to form a new multiband image containing various bands from each date. Several of the uncorrelated principal components computed from the combined data set can often be related to areas of change. One disadvantage to this process is that it is often difficult to interpret and identify the specific nature of the changes involved. Plate 31 illustrates the application of multidate principal components analysis to the process of assessing tornado

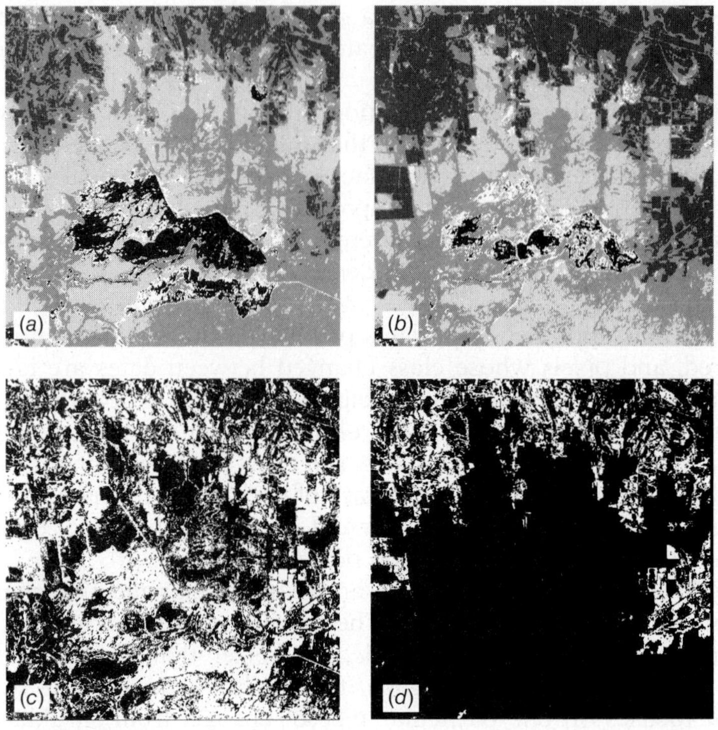

Figure 7.55 Postclassification change detection: (*a*) classification on date 1; (*b*) classification on date 2; (*c*) all change pixels highlighted in white; (*d*) only pixels that changed from class "shrubland" to class "agriculture." (Author-prepared figure.)

damage from "before" and "after" images of the tornado's path of destruction. The "before" image shown in (*a*), is a Landsat-7 ETM+ composite of bands, 1, 2, and 5 shown as blue, green, and red, respectively. The "after" image, shown in (*b*), was acquired 32 days later than the image shown in (*a*), on the day immediately following the tornado. While the damage path from the tornado is fairly discernable in (*b*), it is most distinct in the principal component image shown in (*c*). This image depicts the second principal component image computed from the six-band composite formed by registering bands, 1, 2, and 5 from both the "before" and "after" images.

Temporal image differencing is yet another common approach to change detection. In the image differencing procedure, DNs from one date are simply subtracted from those of the other. The difference in areas of no change will be very small (approaching zero), and areas of change will manifest larger negative or positive values. If 8-bit images are used, the possible range of values for the difference image is –255 to +255, so normally a constant (e.g., 255) is added to each difference image value for display purposes. Figure 7.56*c* illustrates the

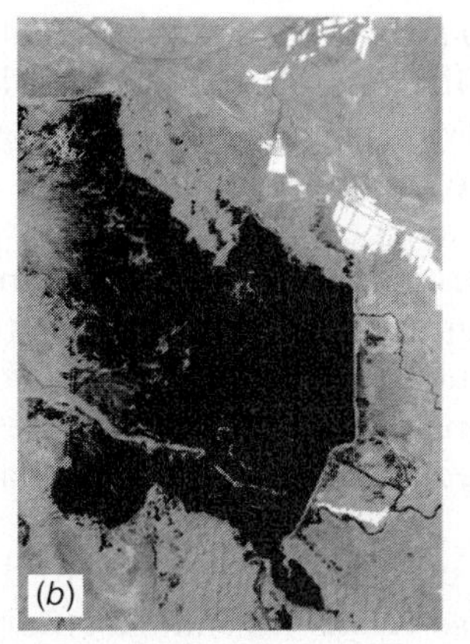

Figure 7.56 Temporal image differencing: (*a*) near-IR band from date 1; (*b*) near-IR band from date 2; (*c*) difference of (*b*) minus (*a*). (Author-prepared figure.)

result of temporal image differencing, with the two original (near-IR) images shown in (*a*) and (*b*). In this rendition, areas with no change have a mid-gray tone; areas whose near-IR reflectance increased are shown in lighter tones; and areas where the near-IR reflectance decreased are in darker tones. (Positive and negative changes are scaled symmetrically around the middle of the range.)

Temporal image ratioing involves computing the ratio of the data from two dates of imaging. Ratios for areas of no change tend toward 1 and areas of change will have higher or lower ratio values. Again, the ratioed data are normally scaled for display purposes (Section 7.6). One of the advantages to the ratioing technique is that it tends to normalize the data for changes in such extraneous factors as sun angle and shadows.

Whether image differencing or ratioing is employed, the analyst must find a meaningful "change–no change threshold" within the data. This can be done by compiling a histogram for the differenced or ratioed image data and noting that the change areas will reside within the tails of the distribution. A variance from the mean can then be chosen and tested empirically to determine if it represents a reasonable threshold. The threshold can also be varied interactively in most image analysis systems so the analyst can obtain immediate visual feedback on the suitability of a given threshold.

In lieu of using raw DNs to prepare temporal difference or ratio images, it is often desirable to correct for illumination and atmospheric effects and to transform the image data into physically meaningful quantities such as radiances or reflectances (Section 7.2). Also, the images may be prepared using spatial filtering or transformations such as principal components or vegetation components. Likewise, linear regression procedures may be used to compare the two dates of imagery. In this approach a linear regression model is applied to predict data values for date 2 based on those of date 1. Again, the analyst must set a threshold for detecting meaningful change in land cover between the dates of imaging.

Change vector analysis is a change detection procedure that is a conceptual extension of image differencing. Figure 7.57 illustrates the basis for this approach in two dimensions. Two spectral variables (e.g., data from two bands, two vegetation components) are plotted at dates 1 and 2 for a given pixel. The vector connecting these two data sets describes both the magnitude and direction of spectral change between dates. A threshold on the magnitude can be established as the basis for determining areas of change, and the direction of the spectral change vector often relates to the type of change. For example, Figure 7.57*b* illustrates the differing directions of the spectral change vector for vegetated areas that have been recently cleared versus those that have experienced regrowth between images.

One of the more efficient approaches to delineating change in multidate imagery is use of a *change-versus-no-change binary mask to guide multidate classification*. This method begins with a traditional classification of one image as a reference (time 1). Then, one of the spectral bands from this date is registered to the same band in a second date (time 2). This two-band data set is then analyzed using one of the earlier described algebraic operations (e.g., image differencing or ratioing). A threshold is then set to separate areas that have changed between dates from those that have not. This forms the basis for creating a binary mask of change-versus-no-change areas. This mask is then applied to the multiband image acquired at time 2 and only the areas of change are then classified for time 2. A traditional postclassification comparison is then performed in the areas known to have changed between dates.

This discussion has only provided a brief overview of the diversity of methods that have been developed for detection of changes between pairs of remotely

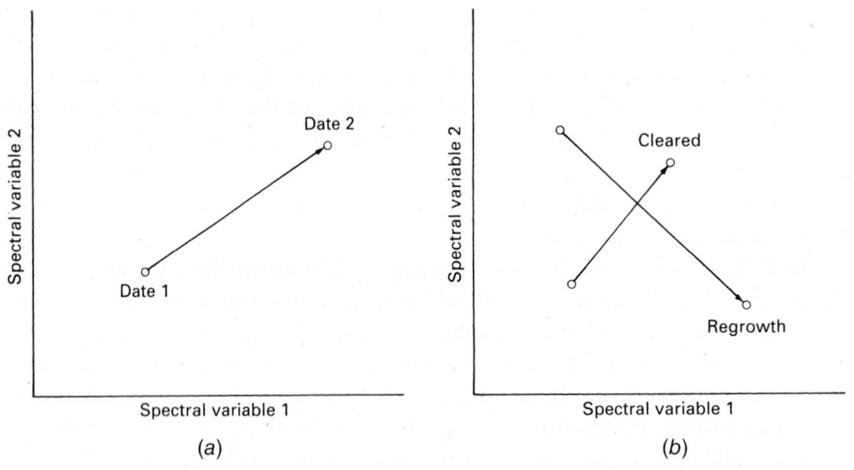

Figure 7.57 Spectral change vector analysis: (*a*) spectral change vector observed for a single land cover type; (*b*) length and direction of spectral change vectors for hypothetical "cleared" and "regrowth" areas.

sensed images. The automation of change detection, and of methods to interpret the significance of changes, is an area of great research interest. Given the ever-increasing constellation of airborne and spaceborne remote sensing platforms (with the role of UAVs expanding particularly rapidly), the volume of data being collected is immense, and automated methods to detect and characterize changes are correspondingly more important than ever.

7.19 IMAGE TIME SERIES ANALYSIS

The change detection methods discussed in the previous section are most commonly applied to small numbers of images from distinct times. The availability of long-term, systematically collected global data from sensors such as AVHRR, MODIS, and others provides the opportunity for obtaining much denser and richer temporal information. For any given location on the Earth, images from these "global monitoring systems" can be used to produce a time series of imagery and derived products (e.g., vegetation indices) at daily to weekly to monthly intervals, over a period of many years. These image time series are the basis for many studies of seasonal to decadal variations in agricultural, ecological, hydrological, and other regional-scale systems.

Figures 7.58 and 7.59 illustrate the concepts involved in a dense time series of imagery. The image in Figure 7.58 is a single date of NDVI (vegetation index) data derived from MODIS imagery in June 2011 for a site in China's far western Xinjiang region. Although it contains the usual two-dimensional information about the distribution of NDVI values on the landscape, this image also can be seen as just one link in a long chain of 544 images extending forward and backward in time. (Additional NDVI images of this same area from Landsat imagery at finer scale were shown earlier in Figure 7.22.)

For any given pixel or group of pixels in this series of 544 dates of imagery, a "temporal signature" can be extracted, showing the seasonal to decadal evolution of NDVI within the pixel(s). Figure 7.59 shows four such series, for the points labeled *a* through *d* on the image in Figure 7.58. Each of these points represents a unique landscape feature with its own cyclical patterns and long-term trends in the vegetation index data. Location *a* represents existing farm fields planted with cotton (a crop that requires heavy irrigation in this arid region). The seasonal cycle in NDVI is distinct, and obviously derives from the recurring annual cycle of planting, growth, harvest, and senescence in the cotton fields. Location *b* began the decade as undeveloped, sparsely vegetated land with scattered desert shrubs growing on sandy soil. In 2006–2007, new farm fields were constructed in this area, and from 2008 onward, this site shows a pattern that is clearly similar to the existing agricultural fields at *a*. The area around *c* also was undeveloped land at the beginning of the decade, with shrubs (tamarisk) and scattered patches of *Populus euphratica* along current and former stream channels. In mid-2000, due to abnormally high flow in the main rivers in this area, there was extensive

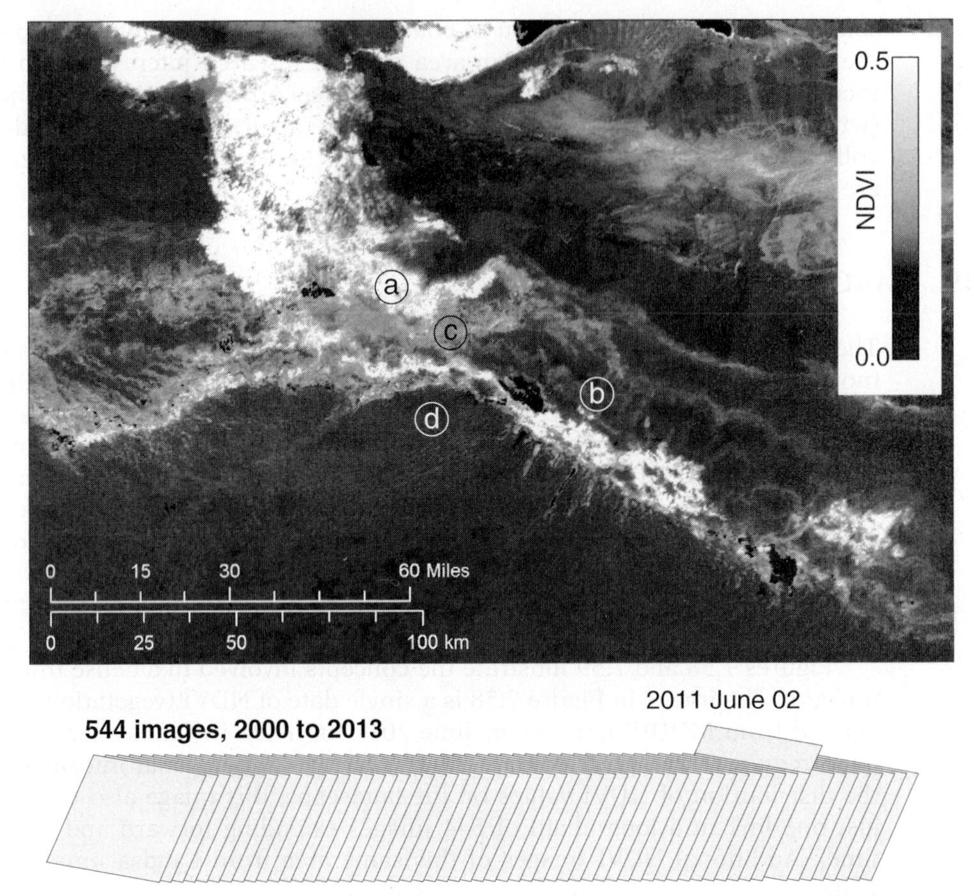

Figure 7.58 One typical date of MODIS NDVI imagery extracted from a time series of 544 images for the Tarim River area in Xinjiang, China, from February 2000 through February 2013. Labeled points (*a–d*) are sources for time-series graphs shown in Figure 7.59. (Author-prepared figure.)

flooding around site *c*. This addition of water produced a sudden increase in vegetation greenness over the next few years, manifested in the time series as an increase in the annual average NDVI, annual peak NDVI, and seasonal amplitude of NDVI. However, with no further recurrence of wide-scale flooding during the subsequent decade, the NDVI signal gradually declined back toward its previous state. For comparison, the temporal signature at *d*—basically, a flat line—is representative of unvegetated areas (in this case, large sand dunes in the Taklimakan Desert south of the river).

Plate 18 shows another analysis of an NDVI time series, in this case for the annual cycle of green-up and senescence across the United States. The analysis of these series of images representing spatial-temporal variation in vegetation indices (or other remotely sensed indices) is sometimes referred to as *landscape*

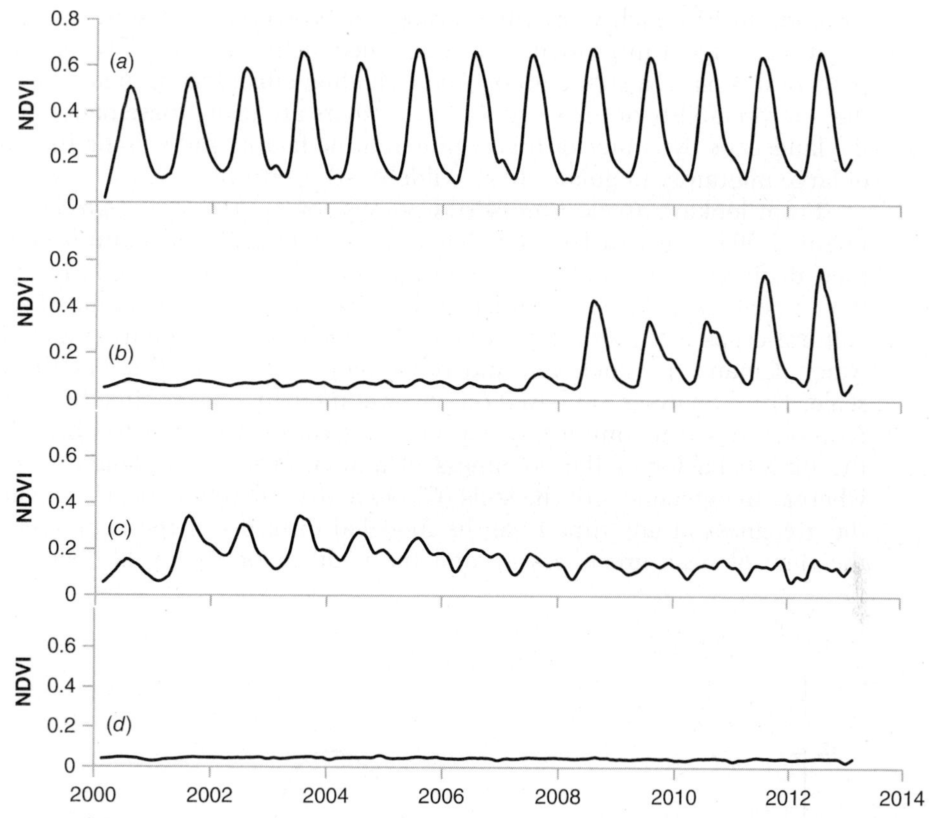

Figure 7.59 Time-series plots of MODIS NDVI for points shown in Figure 7.58: (*a*) established cotton farms; (*b*) new cotton farms; (*c*) tamarisk shrubland following an episode of flooding; (*d*) sand dunes.

phenology. It is increasingly being used as the basis for research on regional-scale linkages between climate and ecosystems (White et al., 2009), as well as for tasks as diverse as agricultural yield forecasting, hydrologic monitoring, and studies of changes in snow and ice cover.

Plate 32 shows several representations of a 12-year (2000 to 2011) time series of MODIS enhanced vegetation index (EVI; see Section 7.6) data for the region around Tarangire National Park in northern Tanzania. In (*a*), the long-term mean EVI value is shown for each pixel, with greener hues representing more dense vegetation. This region consists of savannas at lower elevations, with forests growing at middle elevations on isolated volcanic peaks such as Mt Kilimanjaro (upper right). Using the full time-series of images, a linear decadal trend was calculated for every pixel (*b*). The greenest areas in Plate 32*b* represent increasing trends of over 0.1 EVI units per decade, while the reddest areas represent correspondingly large negative trends. Finally, part (*c*) shows the amplitude of the normal seasonal cycle in each pixel, with blue areas having very little peak-to-trough variation in EVI each year, while orange or brown areas manifest a much stronger annual signal in growth and senescence at the landscape scale. (In all three parts of this plate, a subtle topographic shading effect has been added, to enhance the interpretability of the images.) Researchers are using these data to understand the landscape dynamics of this region and the factors influencing the movements of large migratory ungulates (e.g., wildebeest) over time.

From looking at the time series graphs (or *multitemporal profiles*) shown in Figure 7.59*a–c*, it can be seen that both agricultural and natural systems have their distinctive seasonal patterns of greenness, often with regularly timed cycles that repeat each year. Scientists use these profiles for classification (of agricultural crops and natural ecosystems) as well as for estimating factors such as water demand, productivity, and other biophysical quantities at the landscape scale. These analyses are based on physical modeling of the time behavior of each crop or ecological community's spectral response pattern. It has been found that the time behavior of the greenness of annual crops is sigmoidal (Figure 7.60), whereas the greenness of the soils (G_0) in a given region is nearly constant. Thus, the greenness at any time t can be modeled in terms of the peak greenness G_m, the time of peak greenness t_p, and the width σ of the profile between its two

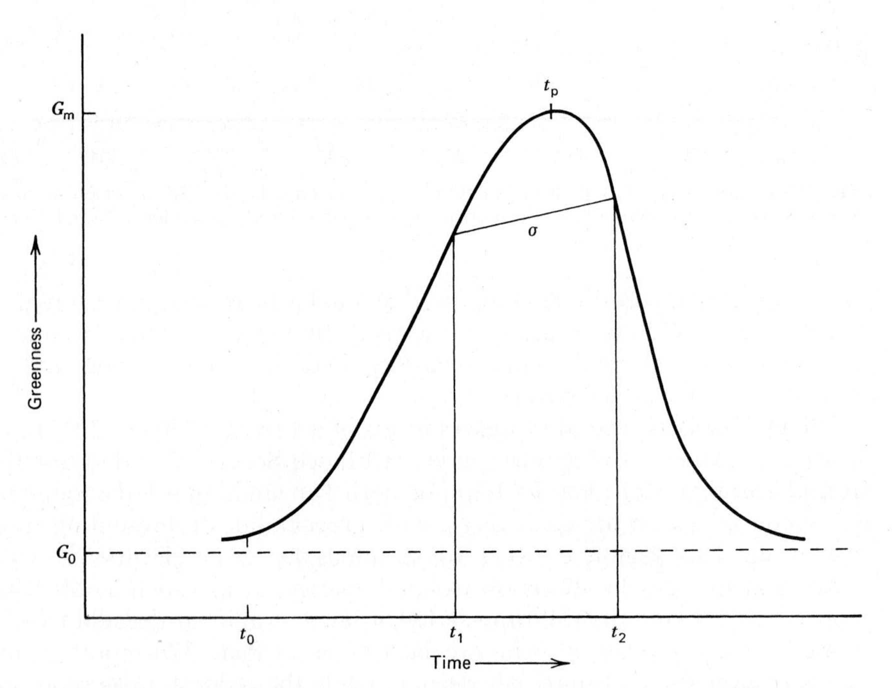

Figure 7.60 Temporal profile model for greenness. Key parameters include spectral emergence date (t_0), time (t_p) of peak greenness (G_m), and width of the profile (σ). (Adapted from Bauer, 1985, after Badhwar, 1985.)

inflection points. (The inflection points, t_1 and t_2, are related to the rates of change in greenness early in the growing season and at the onset of senescence.) The features G_m, t_p, and σ account for more than 95% of the information in the original data and can therefore be used for classification and modeling, instead of the original spectral response patterns. These three features are important because they not only reduce the dimensionality of the original data but also provide variables directly relatable to agrophysical parameters.

7.20 DATA FUSION AND GIS INTEGRATION

Many applications of digital image processing are enhanced through the merger or *fusion* of multiple data sets covering the same geographical area. These data sets can be of a virtually unlimited variety of forms. We have already discussed, or alluded to, several applications of data fusion in remote sensing. For example, one frequently applied form of data merger is the combining of multiresolution data acquired by the same sensor, such as the "pan-sharpening" approach shown in Plate 16 that merges IKONOS 1-m panchromatic and 4-m multispectral data. In another example, in Figure 7.54, an artificial neural network was used to combine multispectral imagery with radar-derived texture and topographic slope and aspect (from a DEM) for image classification.

Often, remotely sensed data and their derived products are integrated with other spatial data in a GIS. In Plate 2, we illustrated the merger of automated land cover classification data with soil erodibility and slope information in a GIS environment in order to assist in the process of soil erosion potential mapping. Over time, the boundaries between digital image processing and GIS operations have become blurred, and fully integrated spatial analysis systems have become the norm. In the following discussion, we begin with the simplest case—merging multiple data sets from the same type of sensor—then proceed to discuss multisensor image fusion, and finally the integration of remotely sensed data with other types of data.

Multitemporal Data Fusion

In Sections 7.18 and 7.19, we examined the use of multitemporal data for detecting changes on the landscape and for studying seasonal to interannual variability in landscape properties such as vegetation cover. In other cases, multitemporal images are "fused" together to create a product useful for visual interpretation. For example, agricultural crop interpretation is often facilitated through merger of images taken early and late in the growing season. In early season images from mid-latitude temperate agricultural regions, bare soils often appear that later will be planted in such crops as corn or soybeans. At the same time, early season images might show perennial alfalfa or winter wheat in an advanced state of maturity. In the late season images, substantial changes in the appearance of the crops present in the scene are typical. Merging various combinations of bands from the two dates to create color composites can aid the interpreter in discriminating the various crop types present.

Plate 33 illustrates two examples of multitemporal NDVI data fusion for visualization purposes. Shown in (*a*) is the use of this technique to aid in mapping invasive plant species, in this case reed canary grass (*Phalaris arundinacea* L.). This part of the plate consists of a multitemporal color composite of NDVI values derived from Landsat-7 ETM+ images of southern Wisconsin on March 7 (blue), April 24 (green), and October 15 (red). Reed canary grass, which invades native wetland communities, tends to have a relatively high NDVI in the fall (October) compared to the native species and hence appears bright red to pink in the multitemporal composite. It can also be seen that features such as certain agricultural crops also manifest such tones. To eliminate the interpretation of such areas as "false positives" (identifying the areas as reed canary grass when they are not of this cover type), a GIS-derived wetland boundary layer (shown in yellow) has been overlain on the image. In this manner, the image analyst can readily focus solely upon those pink to red areas known to be included in wetlands.

Plate 33*b* represents a multitemporal color composite that depicts the three northernmost lakes shown in Plate 33*a*. In this case, a slightly different set of dates and color assignments has been used to produce the color composite of NDVI values. These include April 24 (blue), October 31 (green), and October 15 (red). It so happened that the timing of the October 31 image corresponded with the occurrence of algal blooms in two of the three lakes shown. These blooms appear as the bright green features within the lakes.

As one would suspect, automated land cover classification is often enhanced through the use of multidate data sets. In fact, in many applications the use of multitemporal data is required to obtain satisfactory cover type discrimination. The extent to which use of multitemporal data improves classification accuracy and/or categorized detail is clearly a function of the particular cover types involved and both the number and timing of the various dates of imagery used.

Various strategies can be employed to combine multitemporal data in automatic land cover classification. One approach is to simply register all spectral bands from all dates of imaging into one master data set for classification. For example, bands 1–7 of a Landsat-8 OLI image from one date might be combined with the same seven bands for an image acquired on another date, resulting in a 14-band data set to be used in the classification. Alternatively, principal components analysis (Section 7.6) can be used to reduce the dimensionality of the combined data set prior to classification. For example, the first three principal components for each image could be computed separately and then merged to create a final six-band data set for classification. The six-band image can be stored, manipulated, and classified with much greater efficiency than the original 14-band image.

Multisensor Image Fusion

In many cases, visualization and analysis can be enhanced by combining imagery from different types of sensors. For example, moderate resolution multispectral imagery (in the form of a color composite of spectral bands) might be fused with radar imagery. The objective of this fusion would be to combine the color content of the spectral data (which often relates to molecular-level absorptance and thus the chemistry of materials) with the texturally rich spatial content of the radar data (which is largely affected by surface roughness, larger-scale structures, and dielectric properties of the surface). The fusion might be accomplished in any number of ways, for example using an IHS transformation (Section 7.6) where the multispectral data are transformed from RGB-space to IHS-space and a (possibly enhanced) version of the radar imagery is substituted for the "intensity" band before the transformation is reversed. This example demonstrates how multisensor image merging often results in a composite image product that offers greater interpretability than an image from any one sensor alone.

Other such examples could be considered, using combinations of images from sensors with differing spatial and spectral resolutions, or operating in different regions of the spectrum (visible/near-infrared, thermal, microwave), or with different viewing geometries. Most commonly, these fused composite products are used primarily for visual image interpretation. In other cases, merging data from different types of sensors may lead to increased accuracy in the image classification process or may contribute to other quantitative analysis methodologies.

Merging of Image Data with Ancillary Data

Probably one of the most important forms of data merger employed in digital image processing is the registration of image data with "nonimage," or ancillary, data sets. This latter type of data set can vary, ranging from soil type, to elevation data, to assessed property valuation. The only requirement is that the ancillary data be amenable to accurate geocoding so that they can be registered with the image data to a common geographic base. Usually, although not always, the merger is made in a GIS environment (Jensen and Jensen, 2013).

Digital elevation models (DEMs) have been combined with image data for a number of different purposes (Section 1.5). In many cases, shaded relief, elevation contours, or other representations of topography will be superimposed on (or fused with) image data to improve the viewer's ability to interpret landscape features. Plates 2 and 32 both combine image-derived data with shaded relief from a DEM.

Another application of the fusion of DEMs and image data is for the production of *synthetic stereoscopic images*. With a co-registered DEM and remotely sensed image, photogrammetric software can be used to create a pair of synthetic images that can be viewed in stereo with an analyst-specified degree of vertical exaggeration. The elevation at each pixel position is used to offset the pixel according to its relative elevation. When this synthetic image pair is viewed stereoscopically, a three-dimensional effect is perceived. Such images are particularly valuable in applications where landform analysis is central to the interpretation process.

A related technique is the production of perspective-view images, such as Figure 7.61, in which a panchromatic digital orthophotograph is merged with a DEM for Meteor Crater (also known as Barringer Crater), Arizona. In Figure 7.61*a*, this synthetic 3D scene is viewed from a point above the terrain and northwest of the impact crater. The depth to which the crater was excavated below ground level by the impact event can be seen in 7.61*b*, in which the synthetic scene is viewed from the side.

Merging topographic information and image data is often useful in image classification. For example, topographic information is often important in

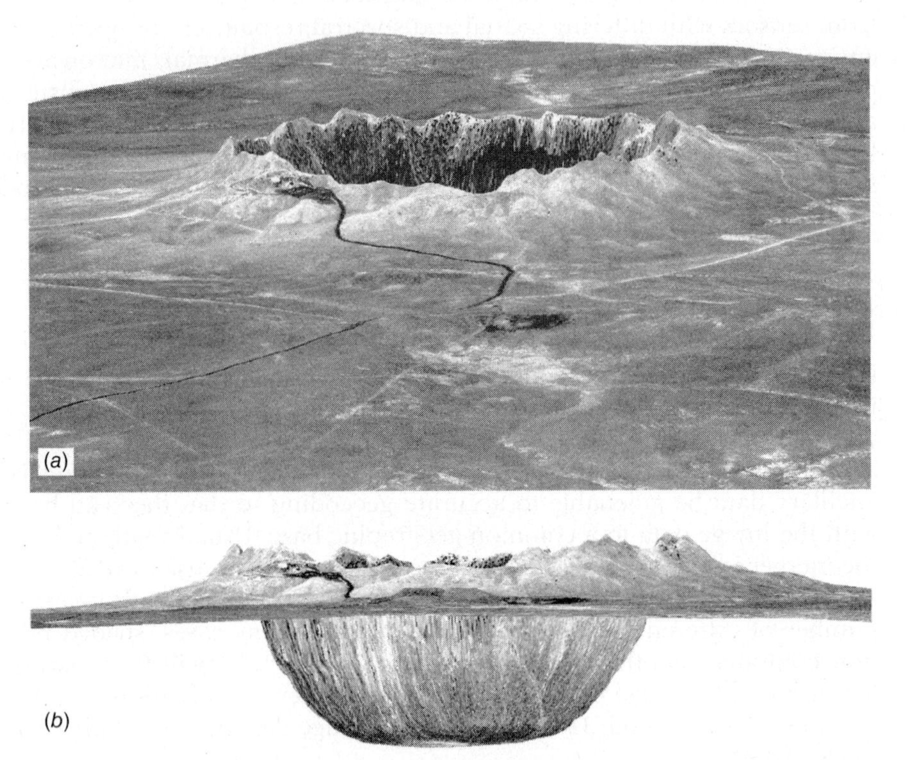

Figure 7.61 Perspective 3D views of Meteor Crater (Barringer Crater), Arizona. (*a*) Viewed from above the terrain, looking southeast. (*b*) Viewed from the side, showing the depth of the crater floor below ground level. (Author-prepared figure.)

forest-type mapping in mountainous regions. In such situations, species that have very similar spectral characteristics might occupy quite different elevation ranges, slopes, or aspects. Thus, the topographic information—in the form of a digital raster layer of slope, aspect, or elevation—might serve as another "channel" of data in the classification directly or as a postclassification basis upon which to discriminate between only the spectrally similar classes in an image. In either case, the key to improving the classification is being able to define and model the various associations between the cover types present in a scene and their habitats.

Incorporating GIS Data in Automated Land Cover Classification

Obviously, topographic information is not the only type of ancillary data that might be resident in a GIS and useful as an aid in image classification. For example, data as varied as soil types, census statistics, ownership boundaries, and zoning districts have been used extensively in the classification process. The basic premise of any such operation is that the accuracy and/or the categorical detail of a classification based on image *and* ancillary data will be an improvement over a classification based on either data source alone. Ancillary data are often used to perform *geographic stratification* of an image prior to classification. As with the use of topographic data, the aim of this process is to subdivide an image into a series of relatively homogeneous geographic areas (strata) that are then classified separately. The basis of stratification need not be a single variable (e.g., upland versus wetland, urban versus rural) but can also be such factors as landscape units or ecoregions that combine several interrelated variables (e.g., local climate, soil type, vegetation, landform).

There are an unlimited number of data sources and ways of combining them in the classification process. Similarly, the ancillary data can be used either prior to, during, or after the image classification process (or even some combination of these choices might be employed in a given application). The particular sources of data used and how and when they are employed are normally determined through the formulation of *multisource image classification decision rules* developed by the image analyst. These rules are most often formulated on a case-by-case basis through careful consideration of the form, quality, and logical interrelationship among the data sources available. For example, the "roads" in a land cover classification might be extracted from current digital line graph (DLG) data rather than from an image source. Similarly, a particular spectral class might be labeled "alfalfa" or "grass" in different locations of a classification depending upon whether it occurs within an area zoned as agricultural or residential.

The concepts involved in this type of decision-rule-based classification can be illustrated in a case study. Consider a situation where several sources of imagery and vector GIS data are being used to analyze land cover in a midwestern U.S.

agricultural region, with a particular focus on agricultural crops and wildlife habitat:

1. A preliminary supervised classification of Landsat imagery acquired in early May (spring).
2. A preliminary supervised classification of Landsat imagery acquired in late June (summer).
3. A preliminary supervised classification of both dates combined using a principal components analysis.
4. A wetlands GIS layer obtained from the state Department of Natural Resources (DNR).
5. A roads GIS layer ("DLG") obtained from the U.S. Geological Survey.

Used alone, none of the above data sources could provide the classification accuracy or detail needed for the purpose of monitoring the wildlife habitat characteristics of the study area. However, when all of the data are integrated in a GIS, the data analyst is able to develop a series of postclassification decision rules utilizing the various data sources in combination. Simply put, these decision rules are based on the premise that certain cover types are better classified in one classification than the others. In such cases, the optimal classification for that category is used for assigning that cover type to the composite classification. For example, the water class has nearly 100% accuracy in the May classification. Therefore, all pixels having a classification of water in the May scene are assigned to that category in the composite classification.

Other categories in the above example are assigned to the composite classification using other decision rules. For example, early attempts to automatically discriminate roads in any of the Landsat classifications suggested that this class would be very poorly represented on the basis of the (30-meter resolution) satellite data. Accordingly, the road class was dropped from the training process and the roads are simply included as a DLG overlay to the composite classification or converted to raster format and "burned into" the classification itself.

The wetland GIS layer is used in yet another way, namely to aid in the discrimination between deciduous upland and deciduous wetland vegetation. None of the Landsat classifications can adequately distinguish between these two classes on their own. Accordingly, any pixel categorized as deciduous in the May imagery is assigned either to the upland or wetland class in the composite classification based on whether that pixel is outside or within a wetland according to the wetland GIS layer. A similar procedure is used to discriminate between the grazed upland and grazed wetland classes in the principal components classification.

Several of the land cover categories in this example are discriminated using rules that involve comparison among the various classifications for a given pixel in each of the three preliminary classifications. For example, hay is classified well in the May scene and the principal components image but with some errors of commission into the grazed upland, cool season grass, and old field categories. Accordingly, a pixel is classified as hay if it is hay in the May or principal components classification but at the same time is not classified as grazed upland or cool season grass in the principal components classification or as old field in the June classification. Similarly, pixels are assigned to the oats class in the composite classification if they are classified as oats, corn, peas, or beans in the May scene but oats in the June scene.

Table 7.4 lists a representative sample of the various decision rules used in the decision-tree classification discussed in this example. They are presented to illustrate the basic manner in which GIS data and spatial analysis techniques can be combined with digital image processing to improve the accuracy and categorical detail of land cover classifications. The integration of remote sensing, GIS,

TABLE 7.4 Basis for Sample Decision Rules Used for a Decision-Tree Classification

Sample Class in Composite Classification	Classification			GIS Data	
	May	June	PC	Wetlands	Roads
Water	Yes				
Roads					Yes
Deciduous upland	Yes			Outside	
Deciduous wetland	Yes			Inside	
Grazed upland			Yes	Outside	
Grazed wetland			Yes	Inside	
Hay	Yes	Old field—no	Yes Grazed upland—no Cool season grass—no		
Oats	Oats—yes Corn—yes Peas—yes Beans—yes	Oats—yes			
Peas	Oats—yes Corn—yes Peas—yes Beans—yes		Peas—yes		
Beans	Oats—yes Corn—yes Peas—yes Beans—yes	Beans—yes			
Reed canary grass			Yes		
Warm season grass	Yes				
Cool season grass			Yes		

and "expert system" techniques for such purposes is an active area of current research. Indeed, these combined technologies are resulting in the development of increasingly "intelligent" information systems.

7.21 HYPERSPECTRAL IMAGE ANALYSIS

The hyperspectral sensors discussed in Chapters 4 and 5 differ from other optical sensors in that they typically produce contiguous, high resolution radiance spectra rather than discrete measurements of average radiance over isolated, wide spectral bands. As a result, these sensors can potentially provide vast amounts of information about the physical and chemical composition of the surface under observation as well as insight into the characteristics of the atmosphere between the sensor and the surface. While most multispectral sensors merely *discriminate* among various earth surface features, hyperspectral sensors afford the opportunity to *identify* and *determine many characteristics about* such features. However, these sensors have their disadvantages as well, including an increase in the volume of data to be processed, relatively poor signal-to-noise ratios, and increased susceptibility to atmospheric interference if such effects are not corrected for. As a result, the image processing techniques that are used to analyze hyperspectral imagery differ somewhat from those discussed previously for multispectral sensors. In general, hyperspectral image analysis requires more attention to issues of atmospheric correction and relies more heavily on physical and biophysical models rather than on purely statistical techniques such as maximum likelihood classification.

Atmospheric Correction of Hyperspectral Images

Atmospheric constituents such as gases and aerosols have two types of effects on the radiance observed by a hyperspectral sensor. The atmosphere absorbs (or attenuates) light at particular wavelengths, thus decreasing the radiance that can be measured. At the same time, the atmosphere also scatters light into the sensor's field of view, thus adding an extraneous source of radiance that is unrelated to the properties of the surface. The magnitude of absorption and scattering will vary from place to place (even within a single image) and time to time depending on the concentrations and particle sizes of the various atmospheric constituents. The end result is that the "raw" radiance values observed by a hyperspectral sensor cannot be directly compared to either laboratory spectra or remotely sensed hyperspectral imagery acquired at other times or places. Before such comparisons can be performed, an atmospheric correction process must be used to compensate for the transient effects of atmospheric absorption and scattering (Gao et al., 2009).

One significant advantage of hyperspectral sensors is that the contiguous, high resolution spectra that they produce contain a substantial amount of information about atmospheric characteristics at the time of image acquisition. In some cases, atmospheric models can be used with the image data themselves to compute quantities such as the total atmospheric column water vapor content and other atmospheric correction parameters. Alternatively, ground measurements of atmospheric transmittance or optical depth, obtained by instruments such as sunphotometers, may also be incorporated into the atmospheric correction models. Such methods are beyond the scope of this discussion.

Hyperspectral Image Analysis Techniques

Many of the techniques used for the analysis of remotely sensed hyperspectral imagery are derived from the field of spectroscopy, wherein the molecular composition of a particular material is related to the distinctive patterns in which the material absorbs and reflects light at individual wavelengths. Once a hyperspectral image has been corrected for the effects of atmospheric absorption and scattering, the reflectance "signature" of each pixel can be compared to previously acquired spectra for known material types. Many "libraries" of spectral reference data have been collected in the laboratory and in the field, representing primarily minerals, soils, and vegetation types. (See Section 4.13.)

Numerous approaches have been taken to compare remotely sensed hyperspectral image data to known reference spectra. At the simplest level, individual wavelength-specific absorption features can be identified in the image and compared to similar features in the reference spectra. This can be done by selecting one spectral band that occurs in the "trough," or low point, of an absorption feature and two "bracket" bands located on either side of the absorption feature. The default spectral reflectance that would be observed if the absorption feature were not present is calculated by interpolating between the two bracket bands, and the value measured in the trough band is then subtracted from this default value (or divided by this default value) to obtain an estimate of the strength of the absorption that is occurring. This process can be applied to both the image spectra and the reference spectra in the same fashion.

With the increase in computational power that has occurred in recent years, the above method has evolved to the direct comparison of entire spectral signatures, rather than individual absorption features within a signature. One such approach is referred to as *spectral angle mapping* (*SAM*) (Kruse et al., 1993). This technique is based on the idea that an observed reflectance spectrum can be considered as a vector in a multidimensional space, where the number of dimensions equals the number of spectral bands. If the overall illumination increases or decreases (perhaps due to the presence of a mix of sunlight and shadows), the length of this vector will increase or decrease, but its angular orientation will remain constant. This is shown in Figure 7.62*a*, where a two-band "spectrum" for a particular material will lie somewhere along a line passing through the origin of a two-dimensional space. Under low illumination conditions, the length of the

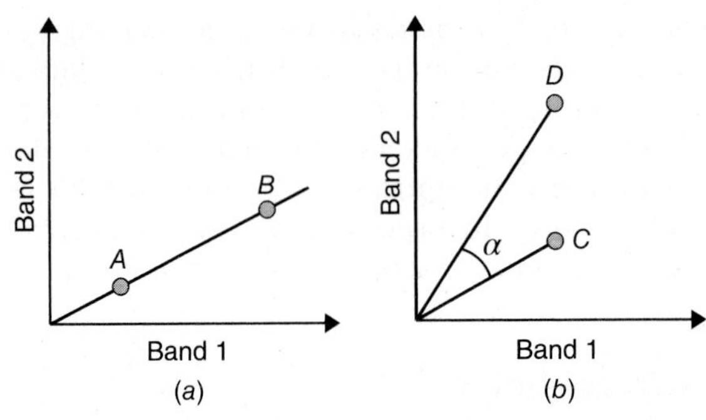

Figure 7.62 Spectral angle mapping concept. (*a*) For a given feature type, the vector corresponding to its spectrum will lie along a line passing through the origin, with the magnitude of the vector being smaller (*A*) or larger (*B*) under lower or higher illumination, respectively. (*b*) When comparing the vector for an unknown feature type (*C*) to a known material with laboratory-measured spectral vector (*D*), the two features match if the angle α is smaller than a specified tolerance value. (After Kruse et al., 1993.)

vector will be short, and the point will be located closer to the origin of the multidimensional space (point *A* in Figure 7.62*a*). If the illumination is increased, the length of the vector will increase, and the point will move farther from the origin (point *B*).

To compare two spectra, such as an image pixel spectrum and a library reference spectrum, the multidimensional vectors are defined for each spectrum and the angle between the two vectors is calculated. If this angle is smaller than a given tolerance level, the spectra are considered to match, even if one spectrum is much brighter than the other (farther from the origin) overall. For example, Figure 7.62*b* shows vectors for a pair of two-band spectra. If angle α is acceptably small, then the spectrum corresponding to vector *C* (derived from an unknown pixel in the image) will be considered to match the spectrum of vector *D* (derived from a laboratory measurement).

It should be noted that the examples shown in Figures 7.62*a* and *b* are for a hypothetical sensor with only two spectral bands. Needless to say, for a real sensor with 200 or more spectral bands, the dimensionality of the space in which the vectors are located becomes too large for human visualization.

These and other techniques for comparing image and reference spectra can be further developed by the incorporation of prior knowledge about the likely characteristics of materials on the ground. This leads to the area of expert systems, whereby the existing knowledge of expert scientists in a particular field, such as geology, is incorporated as a set of constraints for the results of the spectral analysis process. The development of these systems requires an extensive effort on the part of both the "expert" and the programmer. Despite this difficulty, such techniques have been used successfully, particularly in mineralogical applications (Kruse et al., 1993).

As discussed earlier in Section 7.13, spectral mixture analysis provides a powerful, physically based method for estimating the proportions of multiple surface components within a single pixel. As might be expected, this analytical technique is particularly well suited to hyperspectral imagery. Because the maximum number of endmembers that can be included in a spectral mixture analysis is directly proportional to the number of spectral bands +1, the vastly increased dimensionality of a hyperspectral sensor effectively removes the sensor-related limit on the number of endmembers available. The fact that the number of channels far exceeds the likely number of endmembers for most applications readily permits the exclusion from the analysis of any bands with low signal-to-noise ratios or with significant atmospheric absorption effects. The endmember spectra may be determined by examining image pixels for known areas on the landscape or by adopting reference spectra from laboratory or field measurements. Other approaches (analogous to unsupervised clustering of multispectral imagery) seek to identify "natural" endmembers within the n-dimensional cloud of hyperspectral data. In any case, the end result of the process is an estimate of the proportion of each pixel in the image that is represented by each endmember. Spectral mixture analysis is fast becoming one of the most widely used methods for extracting biophysical information from remotely sensed hyperspectral images.

Derivative analysis, another method derived from the field of spectroscopy, has also been adopted for use with hyperspectral imagery. This approach makes use of the fact that the derivative of a function tends to emphasize changes irrespective of the mean level. The advantage of spectral derivatives is their ability to locate and characterize subtle spectral details. A common disadvantage of this method is its extreme sensitivity to noise. For this reason, the derivative computation is typically coupled with spectral smoothing, the most common method being the optimized Savitsky–Golay procedure (Savitsky and Golay, 1964). When the smoothing and the derivative are well matched to a spectral feature, the method is very effective in isolating and characterizing spectral details independent of the magnitude of the signal. Hence, radiance data that are uncorrected for atmospheric effects may be analyzed in this manner.

Numerous other physically based methods have been developed for analysis of hyperspectral imagery and their comparison to reference spectra obtained from spectral libraries, field measurements, or targets internal to the imagery (e.g., the USGS Spectroscopy Laboratory's "Tetracorder" algorithm; Clark et al., 2003). When working with hyperspectral imagery, a typical sequence of activities that might be performed would be as follows (after Clark and Swayze, 1995):

Step 1. For a particular pixel or region in an image, identify the spectral features that are present, such as chlorophyll absorption and clay mineral absorption features.

Step 2. Using an unmixing algorithm, determine the proportion of the area covered by plants.

Step 3. Using the plant spectrum, determine the water content of the leaves, remove this component from the plant spectrum, and calculate the lignin–nitrogen ratio.

Step 4. Derive a soil spectrum by removing the plant spectrum from the original spectrum and search for various minerals present.

For other applications, such as water quality monitoring, the steps in this list would be replaced with similar analyses that are relevant to the application of interest. It should be emphasized that the successful use of software systems such as Tetracorder require the user to be familiar with the principles of spectroscopy as well as with the potential material types and environmental conditions that are likely to be found in the area under investigation.

It is also possible to apply traditional image classification methods such as maximum likelihood classification to hyperspectral data. However, there are significant disadvantages to this approach. First, the increased number of spectral bands results in a vast increase in the computational load for the statistical classification algorithms. In addition, maximum likelihood classification requires that every training set must include at least one more pixel than there are bands in the sensor, and more often it is desirable to have the number of pixels per training set be between 10 times and 100 times as large as the number of sensor bands in order to reliably derive class-specific covariance matrices. Obviously, this means that the shift from a seven-band Landsat image to a 224-band AVIRIS hyperspectral image will result in a major increase in the number of pixels required for each training set.

One classification approach that has been widely used with hyperspectral imagery is based on artificial neural networks (Section 7.16). In fact, pulling together several themes from this chapter, researchers have used hyperspectral imagery in a neural network for subpixel classification (Licciardi and Del Frate, 2011) and also for object-based classification (Zhang and Xie, 2012). However, like the traditional maximum-likelihood and unsupervised classification algorithms, neural networks are a fundamentally empirical (statistical) approach to image analysis, as opposed to the physically based methods that have been discussed previously. Thus, in one sense they represent a potential loss of the valuable information that is provided in the continuous reflectance spectrum about the physical and chemical characteristics of the features under analysis.

7.22 BIOPHYSICAL MODELING

Digital remote sensing data have been used extensively in the realm of quantitative biophysical modeling. The intent of such operations is to relate quantitatively the data recorded by a remote sensing system to biophysical features and phenomena measured on the earth's surface. For example, remotely sensed data might be used in applications as varied as crop yield estimation, defoliation measurement, biomass prediction, water depth determination, and pollution concentration estimation.

Three basic approaches can be employed to relate digital remote sensing data to biophysical variables. In *physical modeling*, the data analyst attempts to account mathematically for all known parameters affecting the radiometric characteristics of the remote sensing data (e.g., earth–sun distance, solar elevation, atmospheric effects, sensor gain and offset, viewing geometry). Alternatively, *empirical modeling* can be employed. In this approach, the quantitative relationship between the remote sensing data and ground-based data is calibrated by interrelating known points of coincident observation of the two (e.g., field-based measurement of forest defoliation conditions precisely at the time a satellite image is acquired). Statistical regression procedures are often used in this process. Combinations of physical and empirical techniques are also possible.

Earlier (Section 7.6) we discussed the process of empirically relating the greenness component of a tasseled cap transformation of Landsat TM data to the percent impervious area in urban or developed areas. Likewise, Figure 6.58 shows a map of Arctic sea ice developed from a model using passive microwave imagery. Similar sea ice maps are operationally produced on a daily basis by several organizations worldwide.

Plate 34 illustrates the application of biophysical modeling with MODIS imagery to map chlorophyll concentrations in lakes along the US/Canada border region. Shown in (*a*) is a small portion of a MODIS image acquired over an area in northern Minnesota and adjacent portions of Canada. A MODIS band 2 image (near IR) was used to create a binary mask to separate water from land. (A vector GIS hydrography layer could also have been used for this purpose.) In (*a*), such a mask has been applied to portray water areas as a true-color composite (MODIS bands 1, 4, and 3 shown as red, green, and blue, respectively) with non-water areas in grayscale. Red dots on this image represent the locations where volunteers participating in the research project took water samples for chlorophyll (and other limnological data). This extensive volunteer effort was led and coordinated by the North American Lake Management Society (NALMS) as part of a larger program of promoting research on the role of remote sensing in lake management.

Note the relatively large variation in hue and brightness in the water pixels in Plate 34*a*. It has been shown that in the case of most Upper Midwest U.S. lakes, the ratio of the radiance measured in MODIS band 3 (blue) to that measured in band 1 (red) can be used to predict several optical properties of lakes, in this case including chlorophyll concentration. This is shown in Figure 7.63, which is a plot of the natural logarithm of chlorophyll concentration (from the volunteers' field sample measurements) versus the band 3/band 1 ratio measured over the corresponding area in each field-sampled lake. A regression model based on this relationship is then applied to all water pixels in the image. The result is shown in Plate 34*b*, where pixel-by-pixel colors correspond to the various chlorophyll concentrations predicted by the model shown in Figure 7.63.

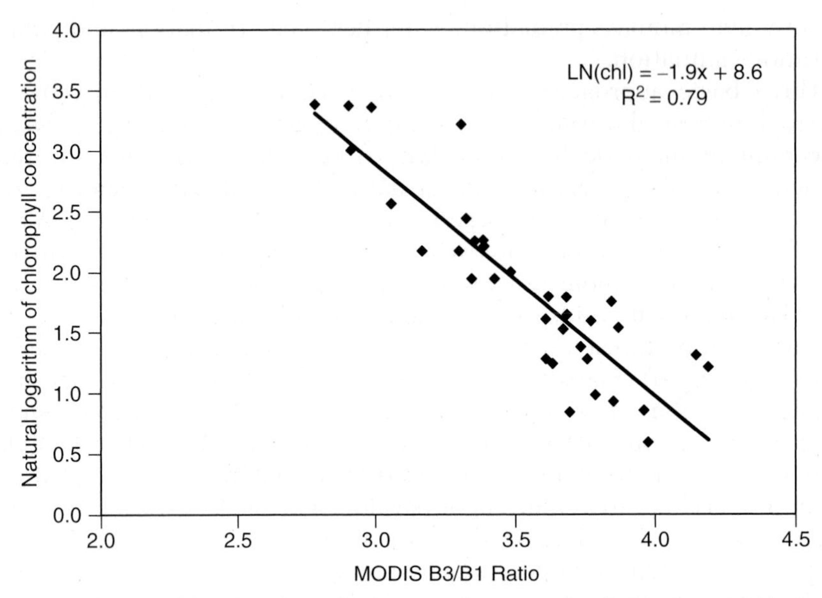

Figure 7.63 Modeling the relationship between the MODIS band 3/band 1 ratio (blue/red bands) and the natural logarithm of chlorophyll concentration in lakes.

Increasingly, remotely sensed data are being used in concert with GIS techniques in the context of *environmental modeling*. The objective of environmental modeling is to simulate the processes functioning within environmental systems and to predict and understand their behavior under altered conditions. Over the past several decades, improved understanding of environmental systems, in conjunction with increased remote sensing and GIS capabilities, has permitted highly accurate and detailed spatial representations of the behavior of environmental systems (e.g., hydrologic systems, terrestrial ecosystems). Remotely sensed data normally form only one source of input to environmental models. GIS procedures permit the bringing together of models of numerous different processes, often using data from different sources, with different structures, formats, and levels of accuracy.

Environmental modeling using remote sensing and GIS is an active area of continuing research and development. Progress to date suggests that these combined technologies will play an increasingly important role in effective resource management, environmental risk assessment, and predicting and analyzing the impacts of global environmental change.

Scale Effects

One of the most important considerations in the use of remotely sensed data in environmental modeling is that of spatial scale. The definition of the term *scale* can be problematic when specialists from varying disciplines are involved in modeling activities. First, there is a distinction to be made between temporal scale and spatial scale. The former refers to the frequency with which a measurement or observation is made. The latter has various, often confusing, meanings to individuals from different backgrounds. For example, a remote sensing specialist uses the term *scale* to refer to the relationship between the size of a feature on a map or image to the corresponding dimensions on the ground. In contrast, an ecologist typically uses the term *spatial scale* to infer two characteristics of data collection: *grain*, the finest spatial resolution within which data are collected, and *extent*, the size of the study area. In essence, these terms are synonymous with what we have defined in this book as spatial resolution and area of coverage, respectively.

Another potential source of confusion in terminology associated with spatial scale is the varying use of the adjectives "small" and "large." Again, to the remote sensing specialist, small and large scale refer to the relative relationship between map and image dimensions and their counterparts on the earth's surface. Under this convention, small scale infers relatively coarser spatial portrayal of features than does the term large scale. The ecologist (or other scientist) will usually use these terms with the reverse meaning. That is, a small scale study in this context would relate to an analysis performed in relatively great spatial detail (usually over a small area). Similarly, a large scale analysis would encompass a large area (usually with less spatial detail).

Again, in this book *grain* and *spatial resolution* are used synonymously. What is important to recognize is that the appropriate resolution to be used in any particular biophysical modeling effort is both a function of the spatial structure of the environment under investigation and the kind of information desired. For example, the spatial resolution required in the study of an urban area is usually much different than that needed to study an agricultural area or the open ocean.

Figure 7.64 illustrates the effects of changing the spatial resolution (grain) with which an area in northeastern Wisconsin is recorded in the near infrared. This area contains numerous lakes of various sizes, and Figure 7.64 shows a series of binary masks produced to portray the lakes (dark areas) versus the land (light areas) in the scene. Part (*a*) was produced using data from the Landsat TM (band 4) acquired at a resolution of 30 m. The remaining parts of this figure were generated by simulating successively coarser grain sizes through the application of a pixel aggregation algorithm. This "majority rules" algorithm involved clumping progressively larger numbers of the 30-m pixels shown in (*a*) into progressively larger synthesized ground resolution cells. The grain sizes so developed were 60, 120, 240, 480, and 960 m, shown in (*b*) to (*f*), respectively.

After the above masks were produced, three basic landscape parameters were extracted from each: (1) the percentage of the scene covered by water, (2) the number of lakes included, and (3) the mean surface area of the lakes. Figure 7.65 includes graphs of the results of this data extraction process. Note how sensitive

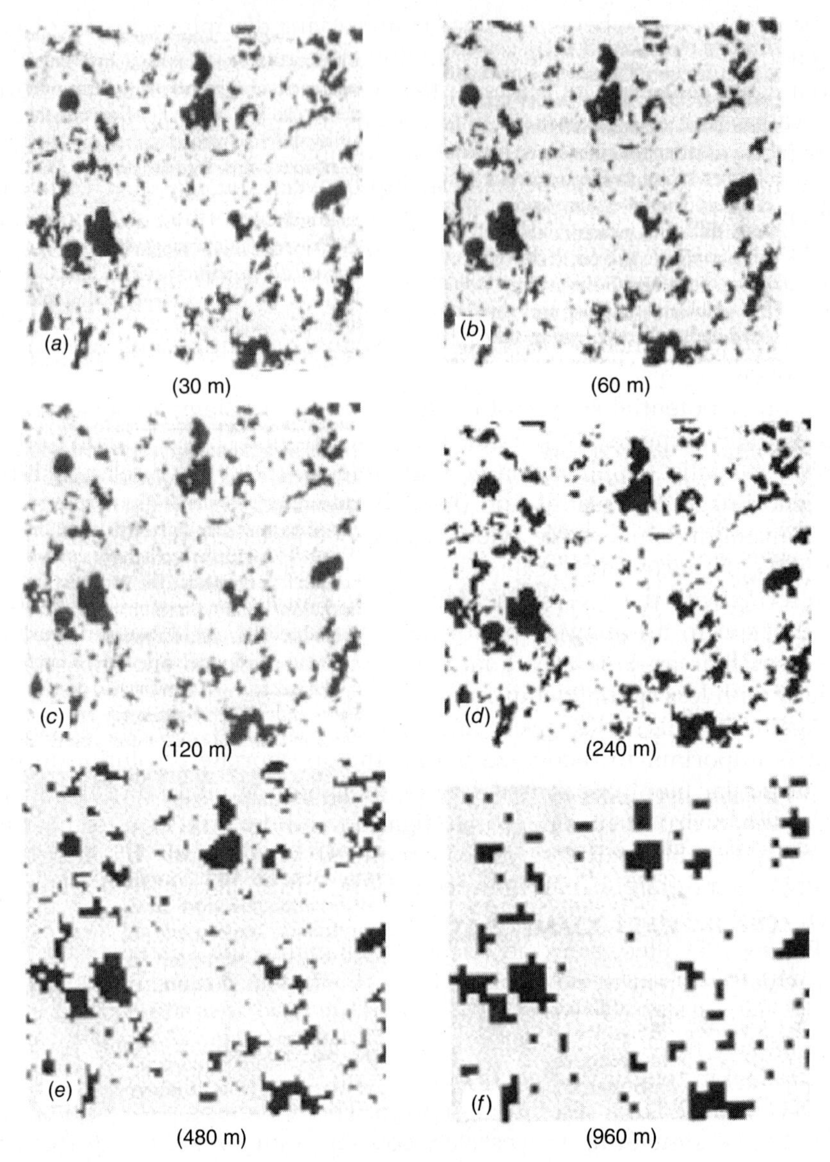

Figure 7.64 Land versus water binary masks resulting from analyzing successive aggregations of 30-m-resolution Landsat TM (band 4) data. (Adapted from Benson and Mackenzie, 1995.)

all of the landscape parameters are to changes in grain. As the ground resolution cell size increases from 30 to 960 m, the percentage of the scene masked into the water class at first increases slightly and then continually decreases. At the same time, the number of lakes decreases in a nonlinear fashion and the mean lake surface area increases nearly linearly.

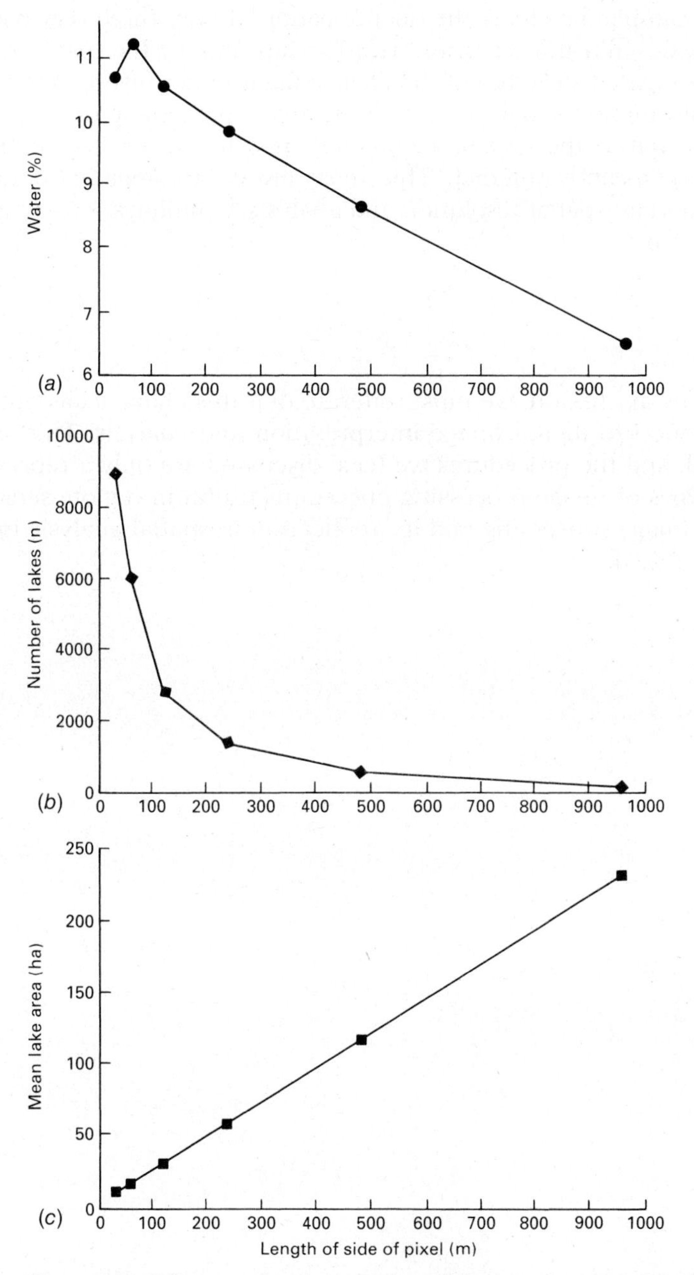

Figure 7.65 Influence of grain (spatial resolution) on measurement of selected landscape parameters for the scene included in Figure 7.64. Aggregation of 30-m-resolution Landsat TM pixels was used to simulate progressively larger ground resolution cell sizes. (*a*) Percentage of water. (*b*) Number of lakes. (*c*) Mean lake area. (Adapted from Benson and Mackenzie, 1995.)

The above example, involving the quantification of only three very basic landscape characteristics, illustrates the interrelationships among the spatial resolution of the sensor, the spatial structure of the environment under investigation, and the nature of the information sought in any given image processing operation. These three factors, as well as the specific techniques used to extract information from digital imagery, constantly interact. This must always be kept in mind as one selects the appropriate spatial resolution and analysis techniques for any given setting and application.

7.23 CONCLUSION

In concluding this discussion, we must reiterate that this chapter has only been a general introduction to digital image interpretation and analysis. This subject is extremely broad, and the procedures we have discussed are only a representative sample of the types of image processing operations useful in remote sensing. The scope of digital image processing and its application in spatial analysis in general are virtually unlimited.

8 APPLICATIONS OF REMOTE SENSING

8.1 INTRODUCTION

Since 1858, when the first aerial photographs were taken by French balloonist Nadar, the technology of acquiring and processing remotely sensed imagery has progressed rapidly. There would have been little incentive for this technological development, however, had remote sensing remained purely an intellectual curiosity. The *application* of remotely sensed imagery in many areas of scientific, economic, and social activity has provided the motivating force that led to the development of the diverse range of sensors we have today, from high-resolution digital cameras (Chapter 2) to multispectral, hyperspectral, and thermal imaging systems (Chapter 4) and radar and lidar systems (Chapter 6). The fields of human activity within which remote sensing might be used are virtually limitless. In this book, we focus on *civilian* applications of remote sensing, although many advances in remote sensing have been driven by military research.

Whether they involve an urban planner using lidar data to construct 3D models of cities, an ecologist using high-resolution satellite imagery to map the extent of invasive plants in a wetland, or a mining company using hyperspectral imagery to prospect for economically valuable minerals, all these applications follow the same general process: Remotely sensed *data* are turned into useful *information*, through a combination of visual image interpretation and/or quantitative analysis. In this process, the importance of *subject area expertise* cannot be stressed too highly as a necessary prerequisite for accurate and insightful interpretation of features in the imagery.

The availability of remotely sensed data for civilian use became increasingly widespread over the course of the past century. In the United States, routine use of aerial photography for applications such as agricultural land management, timber surveys, and mineral exploration expanded rapidly from the 1930s onward. One important step forward occurred in 1937, when the U.S. Department of Agriculture's Agricultural Stabilization and Conservation Service (USDA-ASCS) began photographing selected counties of the United States on a repetitive basis. After World War II, the routine collection of aerial photography became common in many regions, particularly for mapping and managing land resources.

Images from space with various levels of detail have been available since the 1960s. The earliest of these images were of low resolution and were often at oblique angles. Since the advent of the Landsat satellite program in the 1970s and the SPOT satellite program in the 1980s (see Chapter 5), near-vertical multispectral and panchromatic images with resolutions useful for earth resources mapping have been available. Satellite images having resolutions of 1 m or better have become widely available since the launch of the first commercially developed satellite system, IKONOS, in 1999. Such images are now used in a broad range of scientific, commercial, and governmental applications.

Given the scope and breadth of this field, we cannot hope to cover all possible applications of remote sensing, let alone all the numerous ways that image interpretation and analysis are used within those areas of application. We will, however, cover in varying degrees of detail all the major categories of civilian applications of remote sensing technology. In the remainder of this chapter, Sections 8.2 through 8.15, we discuss applications in land use/land cover mapping, geologic and soil mapping, agriculture, forestry, rangeland, water resources, snow and ice mapping, urban and regional planning, wetland mapping, wildlife ecology, archaeology, environmental assessment, disaster assessment, and landform identification and evaluation.

The methods used in these diverse applications include both visual image interpretation (discussed in Chapter 1, with examples throughout the book) and computer-assisted quantitative analysis (Chapter 7). The data to which these methods are applied could come from any of the types of sensors discussed in Chapters 2 through 6. Thus, in a sense, this concluding chapter on applications of remote sensing integrates all the rest of the book, pulling together many different strands that have hitherto been covered individually.

8.2 LAND USE/LAND COVER MAPPING

Timely and accurate information on *land use and land cover* is important for many planning and management activities and is considered an essential element for modeling and understanding the earth as a system. While the terms "land cover" and "land use" are sometimes employed interchangeably, they refer to fundamentally different concepts. The term *land cover* relates to the type of feature present on the surface of the earth. Cornfields, lakes, maple trees, and concrete highways are all examples of land cover types. The term *land use* relates to the human activity or economic function associated with a specific piece of land. As an example, a tract of land on the fringe of an urban area may be used for single-family housing. Depending on the level of mapping detail, its *land use* could be described as urban use, residential use, or single-family residential use. The same tract of land would have a *land cover* consisting of roofs, pavement, grass, and trees. For a study of the socioeconomic aspects of land use planning (school requirements, municipal services, tax income, etc.), it would be important to know that the use of this land is for single-family dwellings. For a hydrologic study of rainfall-runoff characteristics, it would be important to know the spatial area and distribution of rooftops, pavement, grass, and trees in this tract. Thus, a knowledge of both land use and land cover can be important for land planning and land management activities.

While land cover information can be directly interpreted from appropriate remote sensing images, information about human activity on the land (land use) cannot always be inferred directly from the imagery alone. As an example, extensive recreational activities covering large tracts of land are not particularly amenable to interpretation from aerial photographs or satellite images. For instance, hunting is a common and pervasive recreational use occurring on land that would be classified as some type of forest, range, wetland, or agricultural land during either a ground survey or image interpretation. Thus, when land *use* is being mapped, additional information sources are sometimes needed to supplement the remotely sensed image data. Supplemental information is particularly necessary for determining the use of such lands as parks, game refuges, or water conservation districts that may have land uses coincident with administrative boundaries not usually identifiable on remote sensor images. Often, ancillary GIS data about land ownership and management will be incorporated alongside the remotely sensed imagery when land use is being mapped.

Remotely sensed data have been used for land use and land cover mapping at global to local scales, based on methods ranging from visual image interpretation to the spectral and object-based image classification algorithms discussed in Chapter 7. It is important to emphasize that the choice of a classification system (or list of land use/land cover categories to be mapped) is dependent on the scale, image characteristics, and analytical methods available for use. The U.S. Geological Survey (USGS) devised a land use and land cover classification system for use with remote sensor data (Anderson et al., 1976), consisting of a series of nested, hierarchical levels of increasing categorical detail, such that categories from one or more levels can be selected that should be appropriate for any scale of analysis, from global to local. The basic concepts and structure of this system are still valid today. A number of more recent land use/land cover mapping efforts follow these basic concepts and, although their mapping units may be more detailed or more specialized, and they may use more recent remote sensing systems as data sources, they still follow the basic structure originally set forth by the USGS. In the remainder of this section, we first explain the USGS land use and land cover classification system, then describe some ongoing land use/land cover mapping efforts in the United States and elsewhere.

Ideally, land use and land cover information should be presented on separate maps and not intermixed. From a practical standpoint, however, it is often most efficient to mix the two systems when remote sensing data form the principal data source for such mapping activities, and the USGS classification system does include elements of both land use and land cover. Recognizing that some information cannot be derived from remote sensing data in isolation, the USGS system emphasizes categories that can be reasonably interpreted from aerial or space imagery.

The USGS land use and land cover classification system was designed according to the following criteria: (1) The minimum level of interpretation accuracy using remotely sensed data should be at least 85 percent, (2) the accuracy of interpretation for the several categories should be about equal, (3) repeatable results should be obtainable from one interpreter to another and from one time of sensing to another, (4) the classification system should be applicable over extensive areas, (5) the categorization should permit land use to be inferred from the land cover types, (6) the classification system should be suitable for use with remote sensor data obtained at different times of the year, (7) categories should be divisible into more detailed subcategories that can be obtained from large-scale imagery or ground surveys, (8) aggregation of categories must be possible, (9) comparison with future land use and land cover data should be possible, and (10) multiple uses of land should be recognized when possible.

It is important to note that these criteria were developed prior to the widespread use of satellite imagery and computer-assisted classification techniques. While most of the 10 criteria have withstood the test of time, experience has shown that the first two criteria regarding overall and per class consistency and accuracy are not always attainable when mapping land use and land cover over large, complex geographic areas. In particular, when using computer-assisted classification methods, it is frequently not possible to map consistently at a single level of the USGS hierarchy. This is typically due to the occasionally ambiguous relationship between land cover and spectral response and the implications of land use on land cover.

The basic USGS land use and land cover classification system for use with remote sensor data is shown in Table 8.1. The system is designed to use four levels

TABLE 8.1 USGS Land Use/Land Cover Classification System for Use with Remote Sensor Data

Level I	Level II
1 Urban or built-up land	11 Residential
	12 Commercial and service
	13 Industrial
	14 Transportation, communications, and utilities
	15 Industrial and commercial complexes
	16 Mixed urban or built-up land
	17 Other urban or built-up land
2 Agricultural land	21 Cropland and pasture
	22 Orchards, groves, vineyards, nurseries, and ornamental horticultural areas
	23 Confined feeding operations
	24 Other agricultural land
3 Rangeland	31 Herbaceous rangeland
	32 Shrub and brush rangeland
	33 Mixed rangeland
4 Forest land	41 Deciduous forest land
	42 Evergreen forest land
	43 Mixed forest land
5 Water	51 Streams and canals
	52 Lakes
	53 Reservoirs
	54 Bays and estuaries
6 Wetland	61 Forested wetland
	62 Nonforested wetland
7 Barren land	71 Dry salt flats
	72 Beaches
	73 Sandy areas other than beaches
	74 Bare exposed rock
	75 Strip mines, quarries, and gravel pits
	76 Transitional areas
	77 Mixed barren land
8 Tundra	81 Shrub and brush tundra
	82 Herbaceous tundra
	83 Bare ground tundra
	84 Wet tundra
	85 Mixed tundra
9 Perennial snow or ice	91 Perennial snowfields
	92 Glaciers

of information, two of which are detailed in Table 8.1. As noted earlier, this multilevel system has been devised because different degrees of detail are appropriate in different contexts, depending on the scale, the sensor, and the methods used.

The USGS classification system also provides for the inclusion of more detailed land use/land cover categories in Levels III and IV. Levels I and II, with classifications specified by the USGS (Table 8.1), are principally of interest to users who desire information on a nationwide, interstate, or statewide basis. Levels III and IV can be utilized to provide information at a resolution appropriate for regional (multicounty), county, or local planning and management activities. Again, as shown in Table 8.1, Level I and II categories are specified by the USGS. It is intended that Levels III and IV be designed by the local users of the USGS system, keeping in mind that the categories in each level must aggregate into the categories in the next higher level. Figure 8.1 illustrates a sample aggregation of classifications for Levels III, II, and I.

For mapping at Levels III and IV, substantial amounts of supplemental information, in addition to that obtained from the imagery, may need to be acquired. At these levels, a resolution of 1 to 5 m (Level III) or finer (Level IV) is appropriate. Both aerial photographs and high-resolution satellite data can be used as data sources at this level.

The USGS definitions for Level I classes are set forth in the following paragraphs. This system is intended to account for 100% of the earth's land surface (including inland water bodies). Each Level II subcategory is explained in Anderson et al. (1976) but is not detailed here.

Urban or *built-up land* is composed of areas of intensive use with much of the land covered by structures. Included in this category are cities; towns; villages; strip developments along highways; transportation, power, and communication facilities; and areas such as those occupied by mills, shopping centers, industrial and commercial complexes, and institutions that may, in some instances, be isolated from urban areas. This category takes precedence over others when the criteria for more than one category are met. For example, residential areas that have sufficient tree cover to meet *forest land* criteria should be placed in the urban or built-up land category.

Agricultural land may be broadly defined as land used primarily for production of food and fiber. The category includes the following uses: cropland and pasture, orchards, groves and vineyards, nurseries and ornamental horticultural areas, and confined feeding operations. Where farming activities are limited by soil wetness, exact boundaries may be difficult to locate and *agricultural land* may grade into *wetland*. When wetlands are drained for agricultural purposes, they are included in the *agricultural land* category. When such drainage enterprises fall into disuse and if wetland vegetation is reestablished, the land reverts to the *wetland* category.

Rangeland historically has been defined as land where the potential natural vegetation is predominantly grasses, grasslike plants, forbs, or shrubs and where natural grazing was an important influence in its presettlement state.

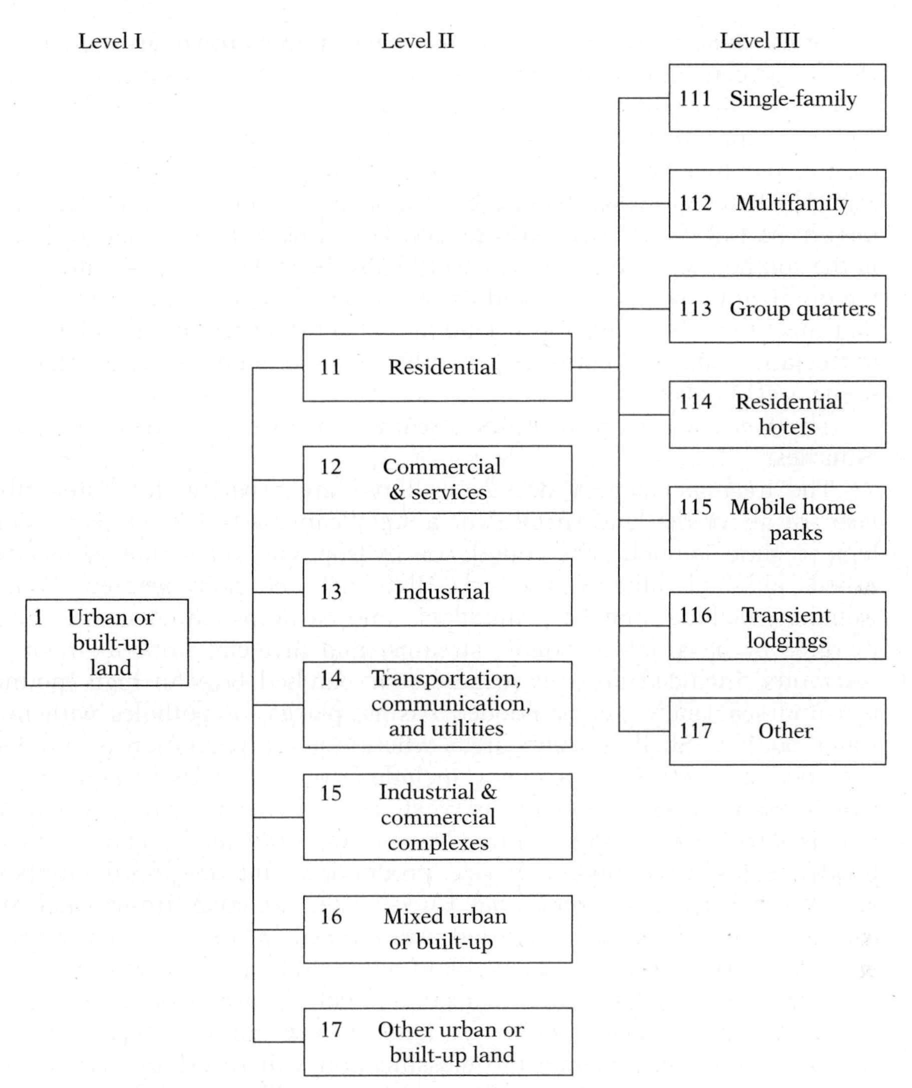

Figure 8.1 An example aggregation of land use/land cover types.

Under this traditional definition, most of the rangelands in the United States are in the western range, the area to the west of an irregular north–south line that cuts through the Dakotas, Nebraska, Kansas, Oklahoma, and Texas. Rangelands also are found in additional regions, such as the Flint Hills (eastern Kansas), the southeastern states, and Alaska. The historical connotation of rangeland is expanded in the USGS classification to include those areas in the eastern states called brushlands.

Forest land represents areas that have a tree-crown areal density (crown closure percentage) of 10% or more, are stocked with trees capable of producing timber or other wood products, and exert an influence on the climate or water regime. Lands from which trees have been removed to less than 10% crown closure but that have not been developed for other uses are also included. For example, lands on which there are rotation cycles of clearcutting and blockplanting are part of the forest land category. Forest land that is extensively grazed, as in the southeastern United States, would also be included in this category because the dominant cover is forest and the dominant activities are forest related. Areas that meet the criteria for forest land and also urban and built-up land are placed in the latter category. Forested areas that have wetland characteristics are placed in the *wetland* class.

The *water* category includes streams, canals, lakes, reservoirs, bays, and estuaries.

The *wetland* category designates those areas where the water table is at, near, or above the land surface for a significant part of most years. The hydrologic regime is such that aquatic or hydrophytic vegetation is usually established, although alluvial and tidal flats may be nonvegetated. Examples of wetlands include marshes, mudflats, and swamps situated on the shallow margins of bays, lakes, ponds, streams, and artificial impoundments such as reservoirs. Included are wet meadows or perched bogs in high mountain valleys and seasonally wet or flooded basins, playas, or potholes with no surface water outflow. Shallow water areas where aquatic vegetation is submerged are classified as *water* and are not included in the *wetland* category. Areas in which soil wetness or flooding is so short-lived that no typical wetland vegetation is developed belong in other categories. Cultivated wetlands such as the flooded fields associated with rice production and developed cranberry bogs are classified as *agricultural land*. Uncultivated wetlands from which wild rice, cattails, and so forth are harvested are retained in the *wetland* category, as are wetlands grazed by livestock. Wetland areas drained for any purpose belong to the other land use/land cover categories such as urban or built-up land, agricultural land, rangeland, or forest land. If the drainage is discontinued and wetland conditions resume, the classification will revert to *wetland*. Wetlands managed for wildlife purposes are properly classified as *wetland*.

Barren land is land of limited ability to support life and in which less than one-third of the area has vegetation or other cover. This category includes such areas as dry salt flats, beaches, bare exposed rock, strip mines, quarries, and gravel pits. Wet, nonvegetated barren lands are included in the wetland category. Agricultural land temporarily without vegetative cover because of cropping season or tillage practices is considered *agricultural land*. Areas of intensively managed forest land that have clear-cut blocks evident are classified as *forest land*.

Tundra is the term applied to the treeless regions beyond the geographic limit of the boreal forest and above the altitudinal limit of trees in high mountain ranges. In North America, tundra occurs primarily in Alaska and northern Canada and in isolated areas of the high mountain ranges.

Perennial snow or *ice* areas occur because of a combination of environmental factors that cause these features to survive the summer melting season. In so doing, they persist as relatively permanent features on the landscape.

As noted above, some parcels of land could be placed into more than one category, and specific definitions are necessary to explain the classification priorities. This comes about because the USGS land use/land cover classification system contains a mixture of land activity, land cover, and land condition attributes.

Several land use/land cover mapping efforts that have been undertaken in the United States and elsewhere use the USGS land use/land cover classification system, or variations thereof.

The Multi-Resolution Land Characteristics (MRLC) Consortium is a group of federal agencies working together to acquire Landsat imagery for the conterminous United States and to develop a land cover dataset called the National Land Cover Dataset (with "Dataset" later replaced by "Database"; abbreviated NLCD in either case). Over the past two decades, a series of comprehensive land cover maps for the nation have been produced, using imagery circa 1992, 2001, 2006, and 2011. The MRLC consortium is specifically designed to meet the needs of federal agencies for nationally consistent satellite remote sensing and land cover data. However, the consortium also provides imagery and land cover data as public domain information, all of which can be accessed through the MRLC website. Federal agencies included in the MRLC are the USGS, Environmental Protection Agency (EPA), Forest Service (USFS), National Oceanic and Atmospheric Administration (NOAA), National Aeronautics and Space Administration (NASA), National Park Service (NPS), National Agricultural Statistics Service (NASS), Bureau of Land Management (BLM), Fish and Wildlife Service (USFWS), and Army Corps of Engineers (USACE).

Details on the NLCD data from 1992–2011 are available from the MRLC website; in Homer et al. (2012), a summary of the program; and in papers on the individual yearly datasets (Vogelmann et al., 2001; Homer et al., 2004; Xian et al., 2009). In addition to the land cover data, other related products (such as maps of percent impervious surface area and percent tree canopy cover) have been produced as well. All of these datasets can be downloaded through the USGS *National Map* website.

The USGS *National Gap Analysis Program* (GAP) is a state-, regional-, and national-level program established in 1987 to provide map data and other information about natural vegetation, distribution of native vertebrate species, and land ownership. Gap analysis is a scientific method for identifying the degree to which native animal species and natural plant communities are represented in our present-day network of conservation lands. Those species and communities not adequately represented constitute "gaps" in conservation lands and efforts.

Data products include land cover maps, species distribution maps, land stewardship maps, and state project reports.

Similar large-area land cover/land use mapping programs have been implemented elsewhere in the world. Several examples follow.

The *CORINE* (Coordination of Information on the Environment) land cover initiative is led by the European Environment Agency. Its principal aim is to compile information on the state of the environment. Features of the environment under study include the geographical distribution and state of natural areas, the geographical distribution and abundance of wild fauna and flora, the quality and abundance of water resources, land cover structure and the state of the soil, the quantities of toxic substances discharged into environments, and information on natural hazards. Land cover data have been produced across Europe for the reference years 1990, 2000, 2006, and 2012, with the first two cycles employing Landsat imagery and the latter two a mixture of imagery from SPOT, IRS, and RapidEye satellites (Chapter 5).

The *Africover* project of the United Nations Food and Agriculture Organization (FAO) is developing country-by-country land cover datasets derived primarily from visual interpretation of digitally enhanced Landsat imagery. These datasets, referred to as the *Multipurpose Africover Database for Environmental Resources*, are intended to provide reliable and geo-referenced information on natural resources and land management at subnational, national, and regional levels.

Also developed by the FAO is a series of regional to continental scale datasets of *Land Use Systems of the World* (LUS). These are distributed at a spatial resolution of 5 arcminutes, and are intended to provide broad-scale information related to agriculture, arid land management, and food security.

On a global scale, NASA's Land Processes Distributed Active Archive Center (LP-DAAC) distributes a set of standard land cover products derived from imagery from the MODIS instrument on Terra and Aqua (Chapter 5). These are produced on an annual basis from 2000 to the present, at spatial resolutions of 500 m and 5600 m. In an innovative development, each pixel is assigned class codes in five different land cover classification systems. The primary class is based on a set of 17 categories defined by the International Geosphere Biosphere Programme (IGBP); the second classification system was developed at the University of Maryland; and the remaining three classification schemes are based on Leaf Area Index (LAI), Net Primary Productivity (NPP), and Plant Functional Type (PFT) information. The classification process itself is performed using a supervised decision-tree method.

8.3 GEOLOGIC AND SOIL MAPPING

The earth has a highly complex and variable surface whose topographic relief and material composition reflect the bedrock and unconsolidated materials that underlie each part of the surface as well as the agents of change that have acted on them. Each type of rock, each fracture or other effect of internal movement, and each erosional and depositional feature bear the imprint of the processes that produced them. Persons who seek to describe and explain earth materials and structures must understand geomorphological principles and be able to recognize the surface expressions of the various materials and structures. Remote sensing applications in geologic and soil mapping focus on identifying these materials and structures and evaluating their conditions and characteristics. Geologic and soil mapping usually requires a considerable amount of field exploration, but the mapping process can be greatly facilitated through the use of visual image interpretation and digital image analysis. In this section, we provide an overview of remote sensing applications in geologic and soil mapping. Section 8.15 provides a more detailed discussion of landform identification and evaluation, with a focus on visual image interpretation of stereoscopic aerial photographs.

Geologic Mapping

The first aerial photographs taken from an airplane for geologic mapping purposes were used to construct a mosaic covering Benghazi, Libya, in 1913. From the 1940s onward, the use of aerial (and later, satellite) imagery in geologic mapping has become widespread. The earliest and perhaps most common use of imagery in geologic mapping has typically been as a base map for manual interpretation and compilation of geologic units. Topographic products derived from remotely sensed imagery, such as digital elevation models and shaded relief maps, are also critical to many geologic mapping efforts.

Geologic mapping involves the identification of landforms, rock types, and rock structures (folds, faults, fractures) and the portrayal of geologic units and structures on a map or other display in their correct spatial relationship with one another. Mineral resource exploration is an important type of geologic mapping activity. Because most of the surface and near-surface mineral deposits in accessible regions of the earth have been found, current emphasis is on the location of deposits far below the earth's surface or in inaccessible regions. Geophysical methods that provide deep penetration into the earth are generally needed to locate potential deposits, and drill holes are required to confirm their existence. However, much information about potential areas for mineral exploration can be provided by interpretation of surface features on aerial photographs and satellite images.

One important role of remotely sensed imagery in this field is as a logistical aid to field-based geologic mapping projects, especially in remote regions. Aerial and satellite imagery provides an efficient, comparatively low-cost means of targeting key areas for the much more expensive ground-based field surveys. Images are used to locate areas where rocks are exposed at the surface and are thus accessible to the geologist for study and to trace key geologic units across the landscape. Images also allow the geologist to make important distinctions between landforms, relate them to the geologic processes that formed them, and thus interpret the geologic history of the area.

Multistage image interpretation is typically utilized in geologic studies. The interpreter may begin by making interpretations of satellite images at scales of 1:250,000 to 1:1,000,000, then examining high-altitude stereoscopic aerial photographs at scales from 1:58,000 to 1:130,000. For detailed mapping, stereoscopic aerial photographs at scales as large as 1:20,000 may be utilized.

Small-scale mapping often involves the mapping of *lineaments*, regional linear features that are caused by the linear alignment of morphological features, such as streams, escarpments, and mountain ranges, and tonal features that in many areas are the surface expressions of fractures or fault zones. Increased moisture retention within the fractured material of a fracture or fault zone is often manifested as distinctive vegetation or small impounded bodies known as "sag ponds." Major lineaments can range from a few to hundreds of kilometers in length. Note the clear expression of the linear Garlock and San Andreas faults in Figure 8.2*a*, a small-scale satellite image that covers an area 127 by 165 km in size. The mapping of lineaments is also important in mineral resource studies because many ore deposits are located along fracture zones.

Several factors influence the detection of lineaments and other topographic features of geologic significance. One of the most important is the angular relationship between the feature and the illumination source. In general, features that trend parallel to the illumination source are not detected as readily as those that are oriented perpendicularly. Moderately low illumination angles are preferred for the detection of subtle topographic linear features. In optical imagery, the illumination angle is inherently determined by the position of the sun. One alternative is to use imaging radar (Chapter 8), whose side-looking configuration can be used to accentuate these subtle topographic features. Another alternative is to construct a digital elevation model (DEM) from photogrammetry, lidar data, or radar interferometry, and then produce shaded-relief maps in which the angle of simulated illumination is chosen to optimize the interpretability of the topography.

Many geologic features can be identified in such shaded-relief images. Figure 8.3 shows surficial geology in coastal Maine (New England, USA) in an image derived from high-resolution airborne lidar data. In this example, the simulated illumination is from the northwest, a condition that would never actually occur with solar illumination in this northern location. Researchers with the Maine Geological Survey have identified a wide variety of glacial and postglacial landforms in this area, many of which are difficult to detect in any other source of imagery (or even on the ground) due to the dense tree cover. In the area at B (and elsewhere) prominent linear ridges are closely spaced swarms of narrow moraines, believed to represent annual deposition where the retreating ice margin temporarily stabilized during the winter months (Thompson, 2011). These moraines were deposited in shallow water, with marine clays occurring in the gaps between the narrow ridges. Smaller moraines are also visible at A, cut by a series of scarps representing former marine shorelines that have been uplifted by

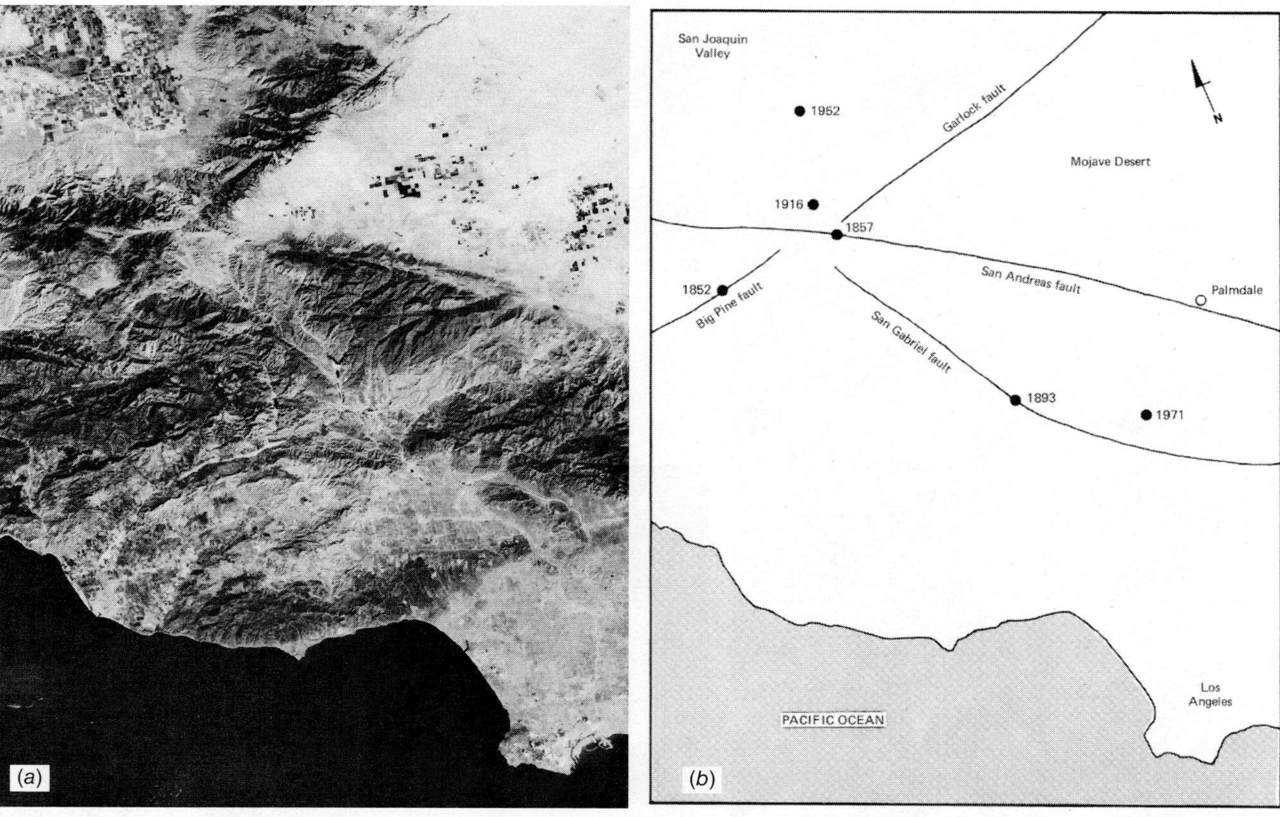

Figure 8.2 Extensive geologic features visible on satellite imagery. (*a*) Landsat MSS image, Los Angeles, CA, and vicinity. Scale 1:1,500,000. (Courtesy USGS.) (*b*) Map showing major geologic faults and major earthquake sites. (Adapted from Williams and Carter, 1976.)

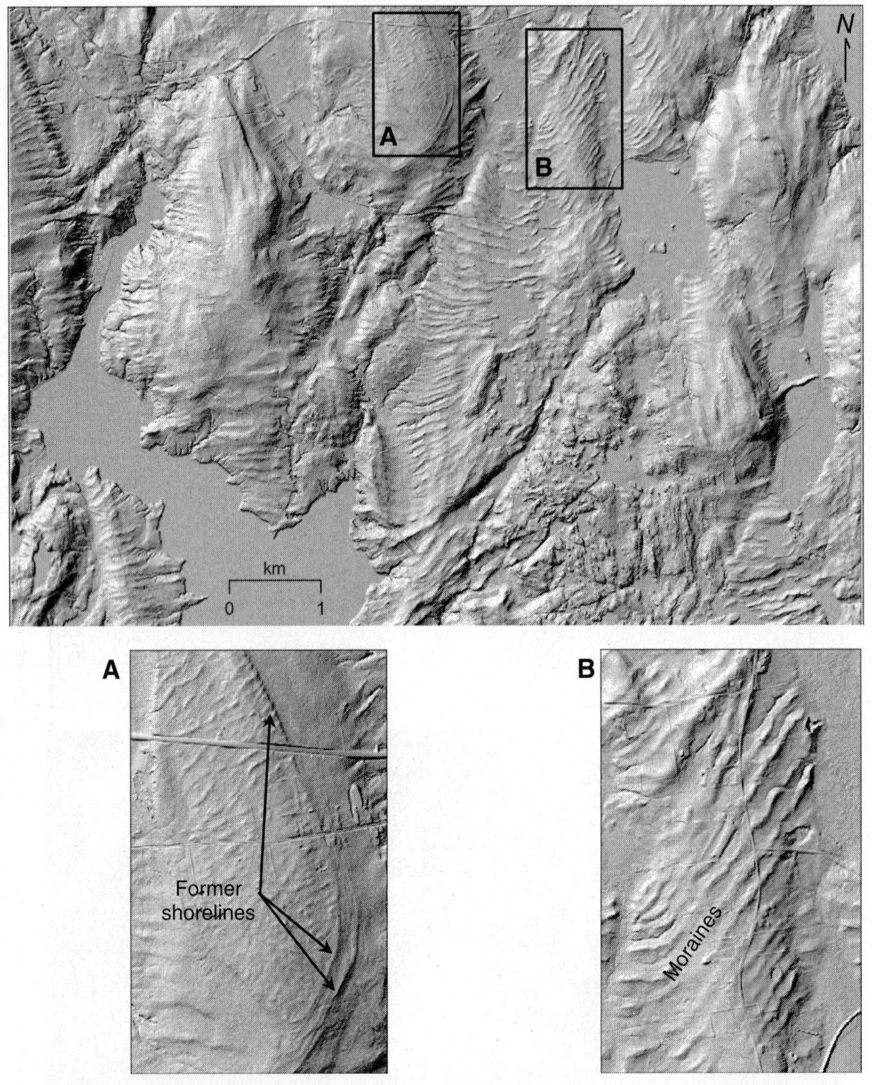

Figure 8.3 Interpreting surficial geology in high-resolution lidar data, Waldoboro, ME. Scale 1:60,000 in main image, 1:20,000 in insets. (Lidar data courtesy Maine State Office of GIS; adapted from Thompson, 2011.)

postglacial isostatic rebound. Numerous other surficial and bedrock geology features can be mapped in this high-quality lidar dataset.

Many geologists believe that reflection in spectral bands around 1.6 and 2.2 μm is particularly important for mineral exploration and lithologic mapping. These bands cannot be photographed, but they can be sensed with various multispectral and hyperspectral sensors (see Chapters 4 and 5, including Plate 11).

Also, the examination of multiple narrow bands in the thermal infrared spectral region shows great promise in discriminating rock and mineral types; multiband thermal imagery is available from ASTER (Plate 22) and some airborne sensors. Extensive research has gone into the development of spectral libraries and spectral mapping methods for automated classification of minerals (where rocks are exposed on the surface), particularly with hyperspectral sensors, and these techniques have been widely employed in mineral exploration.

Although monoscopic viewing is often suitable for lineament mapping, *lithologic mapping*, the mapping of rock units, is greatly enhanced by the use of stereoscopic viewing. As outlined in Section 8.15, the process of rock unit identification and mapping involves the stereoscopic examination of images to determine the topographic form (including drainage pattern and texture), image tone, and natural vegetative cover of the area under study. In unvegetated areas, many lithologic units are distinguishable on the basis of their topographic form and spectral properties. In vegetated areas, identification is much more difficult because the rock surface is obscured, and some of the more subtle aspects of changes in vegetative cover must be considered.

Because some 70% of the earth's land surface is covered with vegetation, a geobotanical approach to geologic unit discrimination is important. The basis of *geobotany* is the relationship between a plant's nutrient requirements and two interrelated factors—the availability of nutrients in the soil and the physical properties of the soil, including the availability of soil moisture. The distribution of vegetation often is used as an indirect indicator of the composition of the underlying soil and rock materials. A geobotanical approach to geologic mapping using remotely sensed images suggests a cooperative effort among geologists, soil scientists, and field-oriented botanists, each of whom should be familiar with remote sensing. An especially important aspect of this approach is the identification of vegetation anomalies related to mineralized areas. Geobotanical anomalies may be expressed in a number of ways: (1) anomalous distribution of species and/or plant communities, (2) stunted or enhanced growth and/or anomalously sparse or dense ground cover, (3) alteration of leaf pigment and/or physiological processes that produce leaf color changes, and (4) anomalous changes in the phenologic cycle, such as early foliage change or senescence in the fall, alteration of flowering periods, and/or late leaf flush in the spring. Such vegetation anomalies are best identified by analyzing images acquired several times during the year, with emphasis placed on the growing period, from leaf flush in the spring to fall senescence. Using this approach, "normal" vegetation conditions can be established, and anomalous conditions can be more readily identified.

Soil Mapping

Detailed soil surveys form a primary source of resource information about an area. Hence, they are used heavily in such activities as comprehensive land use planning. Understanding soil suitability for various land use activities is essential to preventing environmental deterioration associated with misuse of land. In short, if planning is to be an effective tool for guiding land use, it must be premised on a thorough inventory of the natural resource base; soil data are an essential facet of such inventories.

Detailed soil surveys are the product of an intensive study of soil resources by trained scientists. The delineation of soil units has traditionally utilized airphoto or high-resolution satellite image interpretation coupled with extensive field work. Soil scientists traverse the landscape on foot, identify soils, and delineate soil boundaries. This process involves the field examination of numerous soil profiles (cross sections) and the identification and classification of soil units. The soil scientist's experience and training are relied on to evaluate the relationship of soils to vegetation, geologic parent material, landform, and landscape position. Airphoto interpretation has been utilized since the early 1930s to facilitate the soil mapping process. Typically, panchromatic aerial photographs at scales ranging from 1:15,840 to 1:40,000 have been used as mapping bases; with the advent of high-resolution panchromatic and multispectral satellite images, new possibilities are now available to aid in the interpretation process.

Agricultural soil survey maps have been prepared for portions of the United States by the USDA since about the year 1900. Most of the soil surveys published since 1957 contain soil maps printed on a photomosaic base at a scale of 1:24,000, 1:20,000, or 1:15,840. Beginning in the mid-1980s, soil survey map information for many counties has been made available both as line maps and as digital files that can be incorporated into geographic information systems. The original purpose of these surveys was to provide technical assistance to farmers and ranchers for cropland and grazing operations. Soil surveys published since 1957 contain information about the suitability of each mapped soil unit for a variety of uses. They contain information for such purposes as estimating yields of common agricultural crops; evaluating rangeland suitability; determining woodland productivity; assessing wildlife habitat conditions; judging suitability for various recreational uses; and determining suitability for various developmental uses, such as highways, local streets and roads, building foundations, and septic tank absorption fields.

The USDA Natural Resources Conservation Service (formerly the Soil Conservation Service) provides soil survey maps in digital form for many areas of the United States. Since 1994 it has provided nationwide detailed soil information by means of the *National Soil Information System*, an online soil attribute database system.

A portion of a 1:15,840 scale USDA soil map printed on a photomosaic base is shown as Figure 8.4. Table 8.2 shows a sampling of the kind of soil information and interpretations contained in USDA soil survey reports. This map and table show that the nature of soil conditions and, therefore, the appropriateness of land areas for various uses can vary greatly over short distances. As with soil map data,

Figure 8.4 Portion of a USDA–ASCS soil map, Dane County, WI. Original scale 1:15,840 (4 in. = 1 mile). Printed here at a scale of 1:20,000. (Courtesy USDA-ASCS.)

much of the interpretive soil information (such as shown in Table 8.2) is available online in digital formats.

As described in Section 1.4, the reflection of sunlight from bare (unvegetated) soil surfaces depends on many interrelated factors, including soil moisture content, soil texture, surface roughness, the presence of iron oxide, and the organic matter content. A unit of bare soil may manifest significantly different image tones on different days, depending especially on its moisture content. Also, as the area of vegetated surfaces (e.g., leaves) increases during the growing season, the reflectance from the scene is more the result of vegetative characteristics than the soil type.

Plate 36 illustrates the dramatically changing appearance of one agricultural field during one growing season, as well as the presence of fine-scale variability in soil conditions. Except for a small area at the upper right, the entire 15-hectare field is mapped as one soil type by the USDA (map unit BbB, as shown in Figure 8.4 and described in Table 8.2). The soil parent materials in this field consist of glacial meltwater deposits of stratified sand and gravel overlain by 45 to 150 cm of loess (wind-deposited silt). Maximum relief is about 2 m and slope

TABLE 8.2 Soil Information and Interpretation for Five Soils Shown in Figure 8.4

					Predicted Degree of Limitations for Use As		
Map Unit (Figure 8.4)	Soil Name	Soil Description	Depth to Groundwater Table (cm)	Predicted Corn Yield (kg/ha)	Septic Tank Absorption Fields	Dwellings with Basements	Sites for Golf Course Fairways
BbB	Batavia silt loam, gravelly substratum, 2–6% slope	100–200 cm silt over stratified sand and gravel	>150	8700	Moderate	Slight	Slight
Ho	Houghton muck, 0–2% slope	Muck at least 150 cm deep	0–30	8100 (when drained)	Very severe	Very severe	Severe
KrE2	Kidder soils, 20–35% slope	About 60 cm silt over sandy loam glacial till	>150	Not suited	Severe	Severe	Severe
MdB	McHenry silty loam, 2–6% slope	25–40 cm silt over sandy loam glacial till	>150	7000	Slight	Slight	Slight
Wa	Wacousta silty clay loam, 0–2% slope	Silty clay loam and silt loam glacial lakebed materials	0–30	7000	Very severe	Very severe	Severe

Source: From U.S. Department of Agriculture.

ranges from 0 to 6 percent. This field was planted to corn (*Zea mays* L.) in May and harvested in November.

Plates 36*a*, *b*, and *c* illustrate the change in surface moisture patterns visible on the cultivated soil over a span of 48 hr in early summer. During this period, the corn plants were only about 10 cm tall, and consequently most of the field surface was bare soil. The area received about 2.5 cm of rain on June 29. On June 30, when the photo in Plate 36*a* was exposed, the moist soil had a nearly uniform surface tone. By July 2 (36*c*), distinct patterns of dry soil surface (light image tone) could be differentiated from areas of wet soil surface (darker image tones). The dry areas have relatively high infiltration capacity and are slight mounds of 1 to 2 m relief with gentle slopes. Lower areas remain wet longer because they have relatively low infiltration capacity and receive additional runoff from the higher areas.

Plates 36*d*, *e*, and *f* illustrate changes in the appearance of the corn crop during the growing season. By August 11 (36*d*), the corn had grown to a height of 2 m and completely covered the soil surface, giving the field had a very uniform appearance. However, by September 17 (36*e*), distinct tonal patterns were again evident. Very little rain fell on this field during July, August, and early September, and growth of the corn during this period was dependent on moisture stored in the soil. In the dry areas, shown in light tan-yellow, the leaves and stalks of the corn were drying out and turning brown. In the wetter areas of pink and red photo colors, the corn plants were still green and continuing to grow. Note the striking similarity of the pattern of wet and dry soils in (36*c*) versus the "green" and brown areas of corn in (36*e*). The pattern seen in the September photograph (36*e*) persists

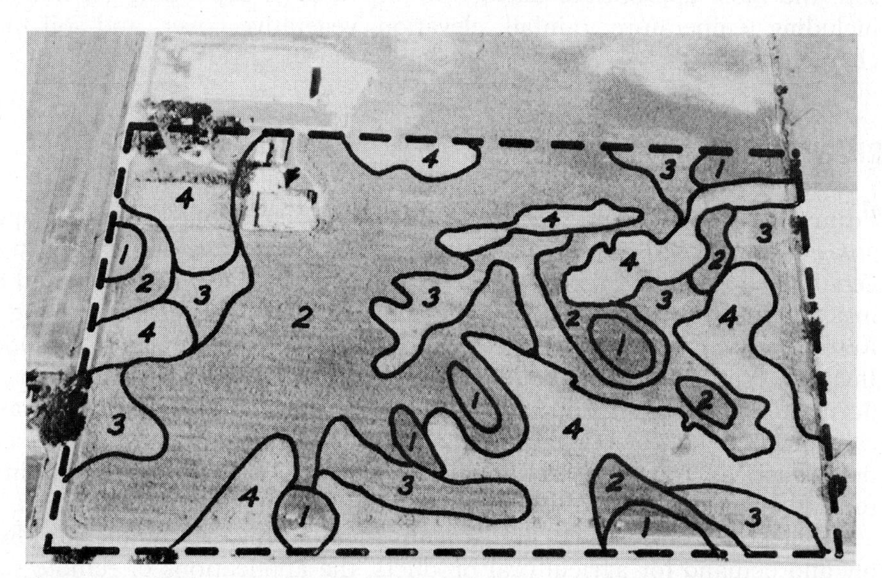

Figure 8.5 Oblique aerial photograph, September 17, with overlay showing four levels of soil moisture (see also Plate 36e), Dane County, WI. Scale approximately 1:4000 at photo center. (Author-prepared figure.)

TABLE 8.3 Selected Characteristics of the Four Soil Units Shown in Figure 8.5

Characteristic	Unit 1	Unit 2	Unit 3	Unit 4
Thickness of silt over sand and gravel	At least 150 cm	105–135 cm	90–120 cm	45–105 cm
Soil drainage class (see Section 8.16)	Somewhat poorly drained	Moderately well drained	Moderately well to well drained	Well drained
Average corn yield (kg/ha)	Not sampled	9100	8250	5850

in the October photograph (36*f*); however, there are larger areas of dry corn in October.

Based on these photographs, a soil scientist was able to divide the soil moisture conditions in this field into four classes, as shown in Figure 8.5. Field inspection of selected sites in each of the four units produced the information in Table 8.3. Note that the corn yield is more than 50% greater in unit 2 than in unit 4.

This sequence of photographs taken during one growing season illustrates that certain times of the year are better suited to image acquisition for soil mapping (and crop management) purposes than others. In any given region and season, the most appropriate dates will vary widely, depending on many factors, including temperature, rainfall, elevation, vegetative cover, and soil infiltration characteristics.

8.4 AGRICULTURAL APPLICATIONS

From the 1930s to the present, agriculture has been among the largest application areas of remote sensing. Much of the systematic coverage of high-resolution imagery for the United States has been driven by agricultural inventory and monitoring programs, from the first large-area airphoto collections by the Natural Resources Conservation Service (NRCS) in 1937, to the current National Aerial Imagery Program (NAIP) collecting high-resolution digital imagery nationwide during the growing season. Worldwide, agricultural crop forecasting has historically been the largest single use of Landsat imagery, while new satellite constellations such as RapidEye are explicitly oriented toward providing timely and reliable imagery at high resolution for agricultural management.

When one considers the components involved in studying the worldwide supply and demand for agricultural products, the applications of remote sensing in general are indeed many and varied. Among the most important uses of imagery in this area are crop-type classification, assessment of crop conditions and early detection of problems, and geospatial support for "precision farming" and other forms of crop management.

Crop-type classification (and area inventory) is based on the premise that specific crop types can be identified by their spectral response patterns and image texture. Successful identification of crops requires knowledge of the developmental stages of each crop in the area to be inventoried. This information is typically summarized in the form of a *crop calendar* that lists the expected developmental status and appearance of each crop in an area throughout the year. Because of changes in crop characteristics during the growing season, it is often desirable to use images acquired on several dates during the growing cycle for crop identification. Often, crops that appear very similar on one date will look quite different on another date, and several dates of image acquisition may be necessary to obtain unique spectral response patterns from each crop type. The use of color-infrared photography, or multispectral imagery with spectral bands in the near- and mid-infrared, generally allows more detailed and accurate crop classification than panchromatic or true-color imagery.

Assessment of crop conditions can in some cases be performed visually, but increasingly involves the use of multispectral or hyperspectral imagery as an input to bio-optical models that use particular combinations of spectral bands to generate estimated values for parameters such as leaf area index (LAI), fraction of absorbed photosynthetically active radiation (fPAR), leaf water content, and other variables. These models require atmospherically corrected imagery to ensure that spectral signatures in the imagery are representative of spectral reflectance at the surface. Figure 8.6 shows an example of a map of LAI from Terra MODIS imagery, for the U.S. northern Great Plains region, one of the great agricultural breadbaskets of the world. The MODIS data for this image were acquired over a 16-day period in early September, were atmospherically corrected to yield surface reflectance, and were then used as input to a bio-optical model that produces estimated values of LAI and fPAR at 1 km resolution. Similar products are produced on a routine basis from MODIS imagery worldwide.

Figure 8.7 shows another example of the use of remote sensing in assessing agricultural crop conditions, for a site in northern France. In this case, high-resolution WorldView-2 imagery (with eight spectral bands) was used to produce maps of nine vegetation and soil condition variables, including those related to chlorophyll and other pigment concentrations, leaf water content, leaf mass and leaf area index, and others. This type of information could be used to forecast crop yield, to detect potential problems such as water stress, disease, or nutrient deficiencies, or as input to a system for precision crop management.

Precision farming or "precision crop management (PCM)," has been defined as an information- and technology-based agricultural management system to identify, analyze, and manage site-soil spatial and temporal variability within fields for optimum profitability, sustainability, and protection of the environment (Robert, 1997). An essential component of precision farming, or PCM, is the use of *variable rate technology* (*VRT*). This refers to the application of various crop production inputs

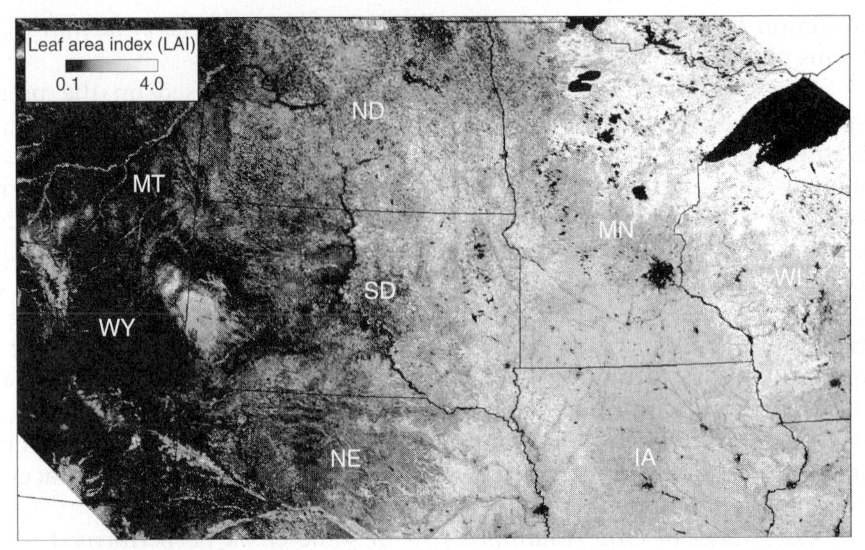

Figure 8.6 Leaf area index (LAI) for the U.S. northern Great Plains region, early September. From 16-day composite of Terra MODIS imagery, early September. Scale 1:12,000,000. (Author-prepared figure.)

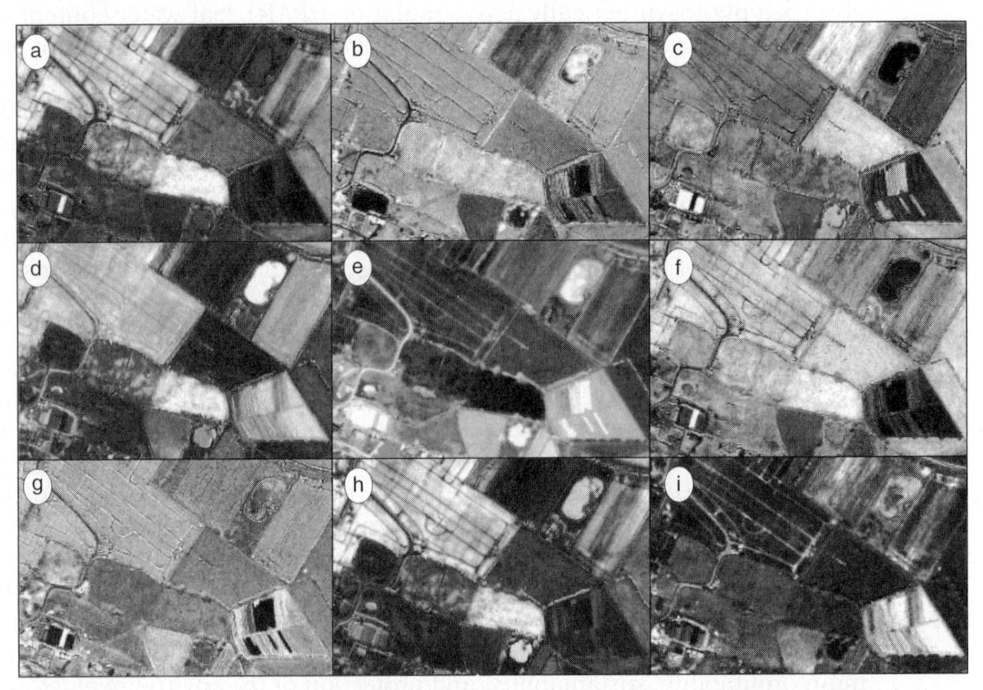

Figure 8.7 Agricultural crop parameters from eight-band WorldView-2 imagery, Lefaux, France. (*a*) Chlorophyll content. (*b*) Carotenoid content. (*c*) Brown pigment content. (*d*) Equivalent water thickness. (*e*) Leaf mass. (*f*) Structure coefficient. (*g*) Average leaf angle. (*h*) Leaf area index. (*i*) Soil dryness. (Images courtesy Christoph Borel-Donohue and DigitalGlobe.)

on a location-specific basis. Variable rate equipment has been developed for a range of materials, including herbicide, fertilizer, water, insecticide, and seeds. The within-field position of such equipment is typically determined through the use of global positioning system technology (GPS; Section 1.7). In turn, the GPS guidance system is often linked to a GIS, which provides the "intelligence" necessary to determine the rate at which a given material should be applied at a given location within a field. The data used to determine application rates for various materials usually come from a range of sources (e.g., remote sensing images, detailed soil maps, historical yield records, equipment-mounted sensors). The goal of PCM is to target application rates to the location-specific conditions occurring within a field and thereby maximize profitability and minimize energy waste and surface and ground water pollution.

Whereas PCM is a relatively recent development, visual image interpretation has a long history of contributing to agricultural crop management. For example, large-scale images have proven useful for documenting deleterious conditions due to crop disease, insect damage, plant stress from other causes, and disaster damage. The most successful applications have utilized large-scale color infrared images taken on specific dates, often two or more dates during the growing season. In addition to "stress detection," such images can provide many other forms of information important to crop management. Some of the plant diseases that have been detected using visual image interpretation are southern corn leaf blight, bacterial blight of field beans, potato wilt, sugar beet leaf spot, stem rust of wheat and oats, late blight fungus of potatoes, fusarium wilt and downy mildew in tomato and watermelon, powdery mildew on lettuce and cucumber, powdery mildew in barley, root rotting fungus in cotton, vineyard *Armillaria mellea* soil fungus, pecan root rot, and coconut wilt. Some types of insect damage that have been detected are aphid infestation in corn fields, phylloxera root feeding damage to vineyards, red mite damage to peach tree foliage, and plant damage due to fire ants, harvester ants, leaf cutting ants, army worms, and grasshoppers. Other types of plant damage that have been detected include those from moisture stress, iron deficiency, nitrogen deficiency, excessive soil salinity, wind and water erosion, rodent activity, road salts, air pollution, and cultivator damage.

Image interpretation for crop condition assessment is a much more difficult task than image interpretation for crop type and area inventory. Ground reference data are essential, and in most studies to date, comparisons have been made between healthy and stressed vegetation growing in adjacent fields or plots. Under these conditions, interpreters might discriminate between finer differences in spectral response than would be possible in a noncomparative analysis—that is, the level of success would be lower if they did not know a stress existed in an area. It would also be more difficult to differentiate among the effects of disease, insect damage, nutrient deficiencies, or drought from variations caused by plant variety, plant maturity, planting rate, or background soil color differences. Because many stress effects are most apparent during dry spells, images should not be acquired too soon after rainy weather. Several days of dry weather may be required in order for the stress effects to become observable.

8.5 FORESTRY APPLICATIONS

Forestry is concerned with the management of forests for wood, forage, water, wildlife, recreation, and other values. Because the principal raw product from forests is wood, forestry is especially concerned with timber management, maintenance and improvement of existing forest stands, and fire control. Forests of one type or another cover nearly a third of the world's land area. They are distributed unevenly and their resource value varies widely.

The earliest applications of remote sensing in forest management focused on visual image interpretation, and this continues to play a central role in many forestry activities, particularly in forest inventory, stand management, and forest operations planning. In recent years, however, technological advances have led to a greatly increased role for digital image analysis and computer-based extraction of information from remotely sensed data. Following the historical evolution of this field, we will begin by discussing the role of visual image interpretation in forestry applications and then proceed with computer-based methods.

The visual image interpretation process for *tree species identification* is generally more complex than for agricultural crop identification. A given area of forest land is often occupied by a complex mixture of many tree species, as contrasted with agricultural land where large, relatively uniform fields typically are encountered. Also, foresters may be interested in the species composition of the "forest understory," which is often blocked from view on aerial and satellite images by the crowns of the large trees.

Tree species can be identified on aerial and satellite images through the process of elimination. The first step is to eliminate those species whose presence in an area is impossible or improbable because of location, physiography, or climate. The second step is to establish which groups of species do occur in the area, based on knowledge of the common species associations and their requirements. The final stage is the identification of individual tree species using basic image interpretation principles.

The image characteristics of shape, size, pattern, shadow, tone, and texture, as described in Chapter 1, are used by interpreters in tree species identification. Individual tree species have their own characteristic crown *shape* and *size*. As illustrated in Figure 8.8, some species have rounded crowns, some have cone-shaped crowns, and some have star-shaped crowns. Variations of these basic crown shapes also occur. In dense stands, the arrangement of tree crowns produces a *pattern* that is distinct for many species. When trees are isolated, *shadows* often provide a profile image of trees that is useful in species identification. Toward the edges of aerial images, relief displacement can afford somewhat of a profile view of trees. *Image tone* depends on many factors, and it is not generally possible to correlate absolute tonal values with individual tree species. Relative tones on a single image, or a group of images, may be of great value in delineating adjacent stands of different species. Variations in crown *texture* are important in species

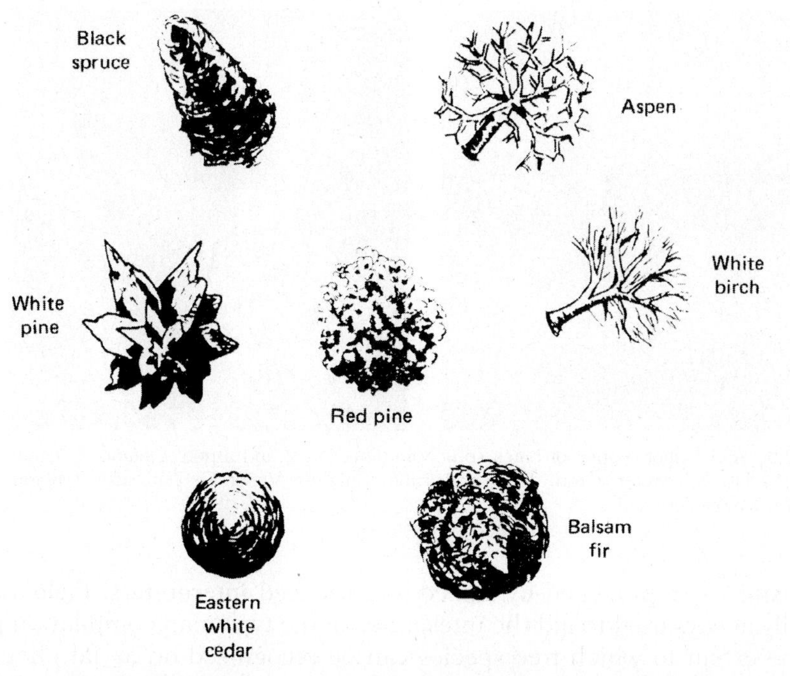

Figure 8.8 Aerial views of tree crowns. Note that most of these trees are shown with radial displacement. (From Sayn-Wittgenstein, 1961. Copyright © 1961, American Society of Photogrammetry. Reproduced with permission.)

identification. Some species have a tufted appearance, others appear smooth, and still others look billowy. Image texture is very scale dependent, so textural characteristics of a given species will vary when looking at single crowns or a continuous forest canopy.

Figure 8.9 illustrates how the above-described image characteristics can be used to identify tree species. A pure stand of black spruce (outlined area) surrounded by aspen is shown in Figure 8.9. Black spruce are coniferous trees with very slender crowns and pointed tops. In pure stands, the canopy is regular in pattern and the tree height is even or changes gradually with the quality of the site. The crown texture of dense black spruce stands is carpetlike in appearance. In contrast, aspen are deciduous trees with rounded crowns that are more widely spaced and more variable in size and density than the spruce trees. The striking difference in image texture between black spruce and aspen is apparent in Figure 8.9.

The process of tree species identification using visual image interpretation is not as simple as might be implied by the straightforward examples shown in these figures. Naturally, the process is easiest to accomplish when dealing with pure, even-aged stands. Under other conditions, species identification can be as much of an art as a science. Identification of tree species has, however, been very

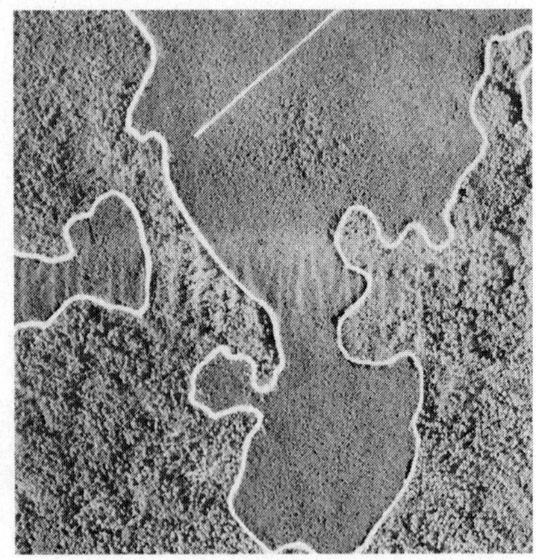

Figure 8.9 Aerial photographs of black spruce (outlined area) and aspen, Ontario, Canada. Scale 1:15,840. Stereopair. (From Zsilinszky, 1966. Courtesy Victor G. Zsilinszky, Ontario Centre for Remote Sensing.)

successful when practiced by skilled, experienced interpreters. Field visitation is virtually always used to aid the interpreter in the type map compilation process.

The extent to which tree species can be recognized on aerial photographs is largely determined by the scale and quality of the images, as well as the variety and arrangement of species on the image. The characteristics of tree form, such as crown shape and branching habit, are heavily used for identification on large-scale images. The interpretability of these characteristics becomes progressively less as the scale is decreased. Eventually, the characteristics of individual trees become so indistinct that they are replaced by overall stand characteristics in terms of image tone, texture, and shadow pattern. On images at extremely large scales (such as 1:600), most species can be recognized almost entirely by their morphological characteristics. At this scale, twig structure, leaf arrangement, and crown shape are important clues to species recognition. At scales of 1:2400 to 1:3000, small and medium branches are still visible and individual crowns can be clearly distinguished. At 1:8000, individual trees can still be separated, except when growing in dense stands, but it is not always possible to describe crown shape. At 1:15,840 (Figure 8.9), crown shape can still be determined from tree shadows for large trees growing in the open. At scales smaller than 1:20,000, individual trees generally cannot be recognized when growing in stands, and stand tone and texture become the important identifying criteria (Sayn-Wittgenstein, 1961).

It is difficult to develop visual image interpretation keys for tree species identification because individual stands vary considerably in appearance depending on age, site conditions, geographic location, geomorphic setting, and other factors. However, a number of elimination keys have been developed for use with aerial photographs that have proven to be valuable interpretive tools when utilized by experienced image interpreters. Table 8.4 is an example of this type of key.

Phenological correlations are useful in tree species identification. Changes in the appearance of trees in the different seasons of the year sometimes enable discrimination of species that are indistinguishable on single dates. The most obvious example is the separation of deciduous and evergreen trees that is easily made on images acquired when the deciduous foliage has fallen. This distinction can also be discerned on spring images acquired shortly after the flushing of leaves or on fall images acquired after the trees have turned color. For example, in the summer, panchromatic and color photographs show little difference in

TABLE 8.4 Airphoto Interpretation Key for the Identification of Hardwoods in Summer

1. Crowns compact, dense, large	
2. Crowns very symmetrical and very smooth, oblong or oval; trees form small portion of stand	Basswood
2. Crowns irregularly rounded (sometimes symmetrical), billowy, or tufted	
3. Surface of crown not smooth, but billowy	Oak
3. Crowns rounded, sometimes symmetrical, smooth surfaced	Sugar maple,[a] beech[a]
3. Crowns irregularly rounded or tufted	Yellow birch[a]
1. Crowns small or, if large, open or multiple	
6. Crowns small or, if large, open and irregular, revealing light-colored trunk	
7. Trunk chalk white, often forked; trees tend to grow in clumps	White birch
7. Trunk light, but not white, undivided trunk reaching high into crown, generally not in clumps	Aspen
6. Crown medium sized or large; trunk dark	
8. Crown tufted or narrow and pointed	
9. Trunk often divided, crown tufted	Red maple
9. Undivided trunk, crown narrow	Balsam poplar
8. Crowns flat topped or rounded	
10. Crowns medium sized, rounded; undivided trunk; branches ascending	Ash
10. Crowns large, wide; trunk divided into big spreading branches	
11. Top of crown appears pitted	Elm
11. Top of crown closed	Silver maple

[a]A local tone-key showing levels 4 and 5 is usually necessary to distinguish these species.

Source: From Sayn-Wittgenstein, 1961.

tone between deciduous and evergreen trees (Figure 1.8*a*). Differences in tones are generally quite striking, however, on summer color infrared and black and white infrared photographs (Figure 1.8*b*).

In spring images, differences in the time at which species leaf out can provide valuable clues for species recognition. For example, trembling aspen and white birch consistently are among the first trees to leaf out, while the oaks, ashes, and large-tooth aspen are among the last. These two groups could be distinguished on images acquired shortly after trembling aspen and white birch have leafed out. Tone differences between hardwoods, which are small during the summer, become definite during the fall, when some species turn yellow and others red or brown.

Visual image interpretation has been used extensively for "timber cruising." The primary objective of such operations is to determine the volume of timber that might be harvested from an individual tree or (more commonly) a stand of trees. To be successful, image-based timber cruising requires a highly skilled interpreter working with both aerial or satellite and ground data. Image measurements on individual trees or stands are statistically related to ground measurements of tree volume in selected plots. The results are then extrapolated to large areas. The image measurements most often used are (1) tree height or stand height, (2) tree-crown diameter, (3) density of stocking, and (4) stand area.

The height of an individual tree, or the mean height of a stand of trees, is normally determined by measuring relief displacement or image parallax. The task of measuring tree-crown diameters is no different from obtaining other distance measurements on images. Ground distances are obtained from image distances via the scale relationship. The process is expedited by the use of special-purpose overlays similar to dot grids. Overlays are also used to measure the density of stocking in an area in terms of the crown closure or percentage of the ground area covered by tree crowns. Alternatively, some measure of the number of individual crowns per unit area may be made.

Once data on individual trees or stands are extracted from images, they are statistically related (using multiple regression) with ground data on timber volume to prepare *volume tables*. The volume of *individual* trees is normally determined as a function of species, crown diameter, and height. This method of timber volume estimation is practical only on large-scale images and is normally used to measure the volume of scattered trees in open areas. More frequently, *stand volumes* are of interest. Stand volume tables are normally based on combinations of species, height, crown diameter, and crown closure.

In some ways, the development and use of digital image processing in forestry mimics the historical development of visual image interpretation methods. Remotely sensed multispectral (or hyperspectral) imagery can be classified into broad species groups using either spectral or object-based classification methods (Chapter 7). Additional information—for example, using carefully timed multitemporal imagery to take advantage of phenological changes in forest canopy spectral response patterns—may allow more reliable discrimination of individual tree species (e.g., Wolter et al., 1995). One particular example of the use of image classification in forest management is mapping the distribution of invasive or nuisance species, such as tamarisk in the U.S. Southwest. Merging optical imagery with other types of remotely sensed data, such as radar images or structural information from airborne lidar, can also improve the accuracy and/or specificity of forest type classification projects.

Lidar, in particular, is playing an increasingly important role in applications from forest inventory to forest operations management. Plate 1 and Figure 1.21 illustrate the separation of lidar data into models of the forest floor and canopy, allowing tree height to be measured quickly and accurately over large areas. Further analysis of the three-dimensional lidar point cloud, and of the full waveform of returned lidar pulses, can provide much more information about the structure of individual trees and stands. Lidar data are discussed in more detail in Chapter 6, and are illustrated in a forest applications context in Figures 6.61, 6.62, and 6.64. Needless to say, the production of topographic data, whether from photogrammetry, radar interferometry, or lidar, contributes to a host of forest management logistics, including road construction, harvest operations planning, wildfire risk modeling and fire response, watershed management and erosion control, and recreation.

Remote sensing can play a particularly important role in planning for and responding to wildfires on forest lands. Remotely sensed data and derived products on tree cover, species, and volume/biomass estimates, can be used as inputs to spatially explicit models of fire fuel loadings. Once a wildfire has become active, near-real-time imagery can be critical to the efforts to contain and control the fire. Plate 9 (discussed in Chapter 4) illustrates the use of the Ikhana UAV for monitoring active wildfires in the western United States, but many sources of aerial and satellite imagery have been used for this purpose, and on a global scale, sensors such as MODIS are used to compile a fully automated, real-time detection system for "hotspots" associated with fires.

Many of the applications of remote sensing in agriculture, discussed in the previous section, are also relevant to forest resources. For example, multispectral or hyperspectral imagery can be used to detect stress in tree foliage associated with insect or pathogen outbreaks, nutrient imbalances, herbivory and other wildlife impacts, environmental degradation (due to ozone, acid deposition, smog, and other factors), and meteorological conditions (moisture stress, drought, and storm events). Examples of tree disease damage due to bacteria, fungus, virus, and other agents that have been detected using visual image interpretation are ash dieback, beech bark disease, Douglas fir root rot, Dutch elm disease, maple dieback, oak wilt, and white pine blister rust. Some types of insect damage that have been detected are those caused by the balsam wooly aphid, black-headed budworm, Black Hills bark beetle, Douglas fir beetle, gypsy moth larva, pine butterfly, mountain pine beetle, southern pine beetle, spruce budworm, western hemlock looper, western pine beetle, and white pine weevil. In particular, the continental-scale epidemic of mountain pine beetles that has damaged tens of millions of hectares from Mexico to western Canada over the past 20 years has been the subject of numerous remote sensing-based studies, and aerial and satellite imagery are playing a critical role in tracking the evolution of this ongoing event.

8.6 RANGELAND APPLICATIONS

Rangeland has historically been defined as land where the potential natural vegetation is predominantly grasses, grasslike plants, forbs, or shrubs and where animal grazing was an important influence in its presettlement state. Rangelands not only provide forage for domestic and wild animals, they also represent areas potentially supporting land uses as varied as intensive agriculture, recreation, and housing.

Rangeland management utilizes rangeland science and practical experience for the purpose of the protection, improvement, and continued welfare of the basic rangeland resources, including soils, vegetation, endangered plants and animals, wilderness, water, and historical sites.

Rangeland management places emphasis on the following: (1) determining the suitability of vegetation for multiple uses, (2) designing and implementing vegetation improvements, (3) understanding the social and economic effects of alternative land uses, (4) controlling range pests and undesirable vegetation, (5) determining multiple-use carrying capacities, (6) reducing or eliminating soil erosion and protecting soil stability, (7) reclaiming soil and vegetation on disturbed areas, (8) designing and controlling livestock grazing systems, (9) coordinating rangeland management activities with other resource managers, (10) protecting and maintaining environmental quality, (11) mediating land use conflicts, and (12) furnishing information to policymakers (Heady and Child, 1994).

Given the expanse and remoteness of rangelands and the diversity and intensity of pressures upon them, remotely sensed imagery has been shown to be a valuable range management tool. A physical-measurement-oriented list of range management activities that have some potential for being accomplished by image interpretation techniques includes: (1) inventory and classification of rangeland vegetation; (2) determination of carrying capacity of rangeland plant communities; (3) determination of the productivity of rangeland plant communities; (4) condition classification and trend monitoring; (5) determination of forage and browse utilization; (6) determination of range readiness for grazing; (7) kind, class, and breed of livestock using a range area; (8) measurement of watershed values, including measurements of erosion; (9) making wildlife censuses and evaluations of rangelands for wildlife habitat; (10) detection and monitoring the spread of invasive or nuisance species; (11) evaluating the recreational use of rangelands; (12) judging and measuring the improvement potential of various range sites; (13) implementing intensive grazing management systems; and (14) planning for and responding to rangeland wildfires (list modified from Tueller, 1996).

Because much of rangeland management involves characterizing and monitoring the presence and condition of vegetation, including grasses, shrubs, and trees, many of the applications of remote sensing in agriculture and forestry also apply to rangeland management, although there are important differences. Where vegetation cover is relatively sparse and soil reflectance contributes significantly to the integrated spectral response pattern of image pixels, algorithms designed to characterize vegetation cover and condition become dependent on modeling or compensating for the soil component.

Many rangelands are characterized by broad spatial scales, low topographic relief, and sparse or no human presence. Given these conditions, the use of UAVs in monitoring rangelands may be particularly efficient (Rango et al., 2009). Plate 35 (from Laliberte et al., 2011) illustrates the use of high spatial resolution imagery from a UAV for species-level vegetation classification on a rangeland in semi-arid New Mexico, USA. The sensor was a lightweight multispectral instrument with six spectral bands in the visible and near infrared; at a flying height of 210 m above ground level, it had a spatial resolution of 14 cm. At this fine spatial resolution, traditional per-pixel spectral classification methods are less effective, so an object-based classifier was used (Laliberte, 2011), yielding an 87% overall accuracy for eight shrub classes and four nonvegetation classes. As the range and autonomy of UAVs improve and the legal and administrative guidelines under which they may be used become clearer, it is expected that the use of this type of remote sensing platform will become increasingly common in rangelands settings (among many others).

8.7 WATER RESOURCE APPLICATIONS

Water is the essential ingredient for almost every human activity. At the outset of the 21st century, scarce water resources were already in high demand for agriculture, domestic consumption, hydropower, sanitation, industrial manufacturing, transportation, recreational activities, and ecosystem services. Yet new needs for water continue to evolve—particularly in mining and petroleum resource development, where vast quantities of water are now being employed in everything from extraction and processing of unconventional oil from tar sands to hydraulic fracturing ("fracking") in which high-pressure fluids are injected into wells to expand networks of fractures in the bedrock, allowing oil and natural gas to be collected from formations where they would otherwise be unobtainable. At the same time, water can also be a hazard—whether from flooding, pollution, or the spread of waterborne disease. The uses of remote sensing in managing water resources are numerous and important and are illustrated in many examples throughout the previous seven chapters of this book. Here, we first review the basic principles of remote sensing of water, before moving on to discuss methods for studying water from a hydrologic perspective (water quantity and distribution) and from a water-quality perspective (characterizing conditions within water bodies). An overall summary of remote sensing methods in lake management applications is provided by Chipman et al. (2009).

In general, most of the sunlight that enters a clear water body is absorbed within about 2 m of the surface. As discussed in Chapter 1, the degree of absorption is highly dependent on wavelength. Near-infrared wavelengths are absorbed in only a few tenths of a meter of water, resulting in very dark image tones of even shallow water bodies on near-infrared images. Absorption in the visible portion of the spectrum varies quite dramatically with the characteristics of the water body under study. From the standpoint of detecting objects within or at the bottom of a column of clear water, the best light penetration is achieved at relatively short wavelengths (e.g., 0.48 to 0.60 μm). Penetration of up to about 20 m in clear, calm ocean water has been reported in this wavelength band (Jupp et al., 1984; Smith and Jensen, 1998). Although shorter wavelengths at the blue end of the spectrum penetrate well, they are extensively scattered and an "underwater haze" results. Nonetheless, spectral bands below 0.48 μm (sometimes labeled as "coastal" bands) are increasingly being added to sensors, as on Landsat-8 OLI and WorldView-2. Red wavelengths penetrate only a few meters in water, but may be particularly important in many water quality applications as an indicator of the presence of suspended sediment or algal pigments such as chlorophyll.

Remote sensing systems outside the visible/near infrared range are also widely used in water resources monitoring. Thermal sensors are particularly well suited for this application area, since the high and nearly uniform thermal emissivity of water makes it possible to reliably calculate the actual kinetic temperature of the water surface (Chapter 4). In the microwave spectrum (Chapter 6), radar signals are affected by the roughness of the water surface and by the presence of moisture in the upper layers of unvegetated soils, while passive microwave radiometers are also sensitive to soil moisture.

Water Quantity and Distribution

Remotely sensed imagery—both optical and radar—is readily used to map the location and spatial extent of water bodies, including lakes, ponds, rivers, and inundated wetlands. Along with this two-dimensional information, the third dimension (water level) can be obtained using radar altimeters or lidar. Over time, monitoring either surface area or water level can indicate the expansion or contraction of water bodies due to flooding or drought, and combining the two methods can provide quantitative measurements of changes in water volume.

The use of aerial and satellite images for *flood assessment* is illustrated in Figures 8.10 to 8.12. Such images can help determine the extent of flooding and the need for disaster relief efforts, when appropriate, and can be utilized by insurance agencies to assist in assessing the monetary value of property loss.

Figure 8.10 illustrates the use of satellite images for monitoring regional-scale flooding. Here we see a grayscale composite of the near- and mid-infrared bands

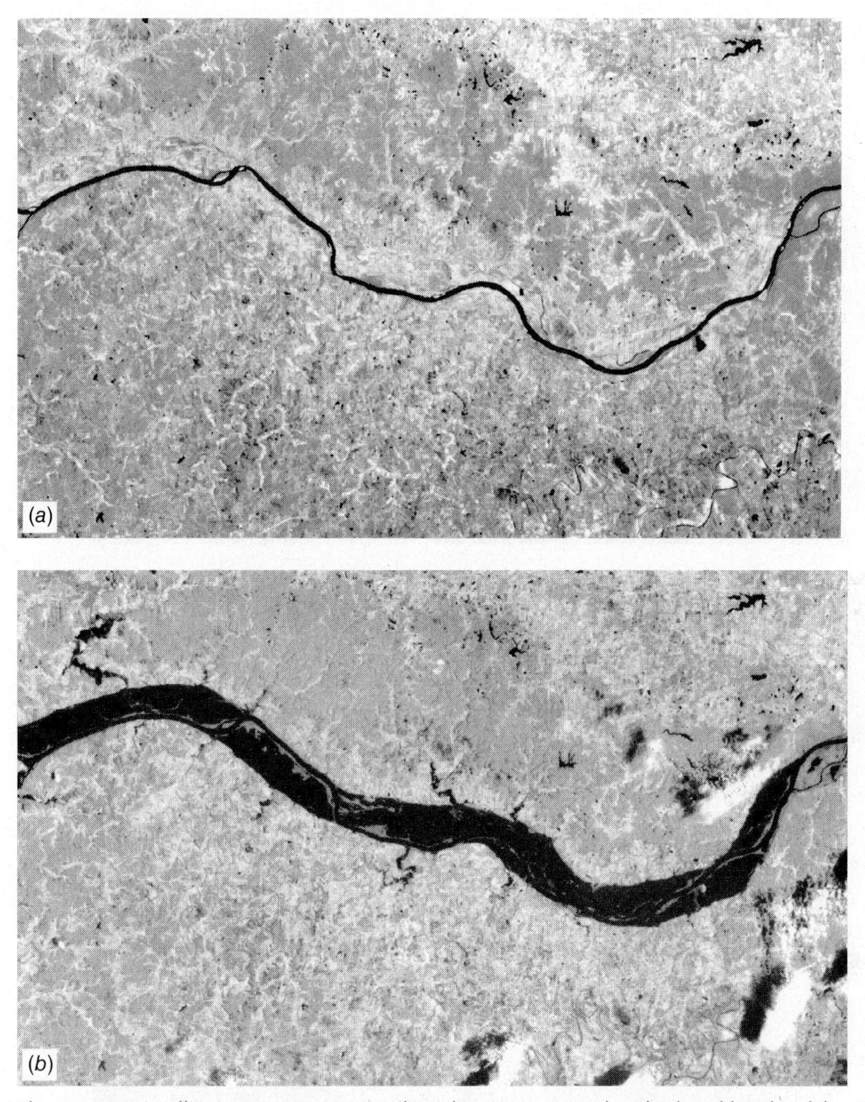

Figure 8.10 Satellite images (composite of Landsat TM near- and mid-infrared bands) of the Missouri River near St. Louis. Scale 1:640,000. (*a*) Normal flow conditions, July 1988. (*b*) Flooding of a severity expected only once every 100 years, July 1993. (Courtesy USGS.)

of Landsat TM (Chapter 5), bands in which water appears very dark as contrasted with the surrounding vegetation. In (*a*) we see the Missouri River under low flow conditions, with a river width of about 500 m. In (*b*) we see a major flood in progress, with a river width of about 3000 m. This historic 1993 flood, with a severity expected only once every 100 years, had a devastating impact on many communities along the Missouri and Mississippi Rivers.

Figure 8.11 Black-and-white copies of color infrared aerial photographs showing flooding and its aftereffects, Pecatonica River near Gratiot, WI: (*a*) Oblique color infrared photograph, June 30. (*b*) Oblique color infrared photograph, July 22. The flying height was 1100 m. (Author-prepared figure.)

The pair of photographs in Figure 8.11 illustrate river flooding and its aftereffects along the Pecatonica River as it meanders through cropland in southern Wisconsin. Figure 8.11*a* was taken on the day after a storm in which more than 150 mm of rain fell in a 2.5-hr period on the Pecatonica River watershed, which encompasses roughly 1800 km^2 upstream from this site, causing the river to rise

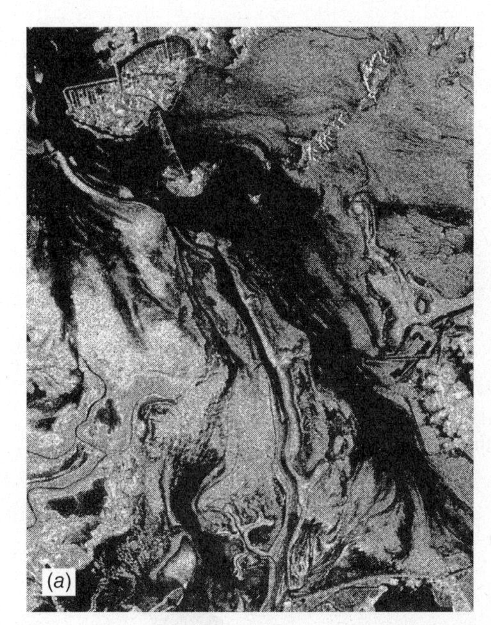

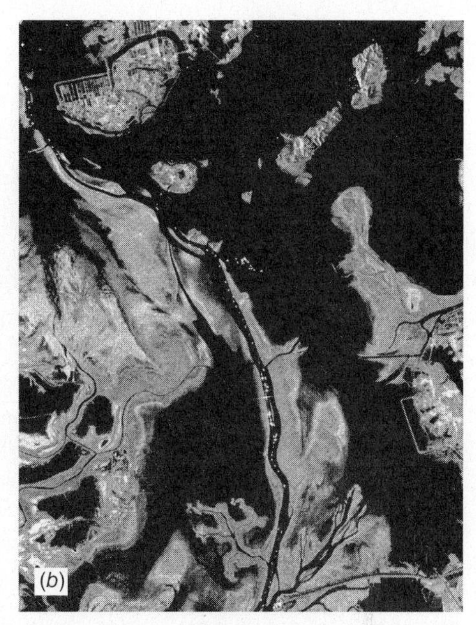

Figure 8.12 Synthetic aperture radar images of inundation in Poyang Lake, China: (*a*) Envisat ASAR image, April 2008. (*b*) Sentinel-1A image, May 2014. (Courtesy ESA.)

rapidly and spill out over its banks. The floodwater is about 3 m deep in the area at the center of this photograph. Figure 8.25*b* shows the same area three weeks after flooding. The soils in the flooded area are moderately well drained to poorly drained silt loam alluvial soils that are high in fertility and in moisture-supplying capacity. The darkest soil tones in 8.25*b* correspond to the poorly drained areas that are still quite wet three weeks after flooding, while light tones in the formerly flooded area show widespread crop damage.

Figure 8.12 shows the use of synthetic aperture radar (SAR; Chapter 6) images for monitoring inundation in China's Poyang Lake. The "lake" consists of a vast expanse of wetlands that are intermittently flooded, with extreme intra- and interannual variation in the extent of flooding. This region provides essential habitat for migratory bird species, including Siberian cranes. Figure 8.12*a* was acquired via the ASAR radar instrument on Envisat in April 2008, while 8.12*b* was acquired by Sentinel-1A in May 2014. Because flooding often occurs during periods of heavy cloud cover, SAR images with their all-weather capability can be a particularly useful data source for monitoring flooding on a routine basis.

Figures 8.13 through 8.15 show a variety of remotely sensed data being used to monitor a chain of large lakes in southern Egypt, formed by the overflow of water from the Nile River through a spillway constructed to prevent overtopping of the Aswan Dam. First spotted in 1998 by astronauts aboard the Space Shuttle *Discovery*, the Toshka Lakes fluctuated in size and shape in their early years, then began a steady decline as upstream water levels dropped. While the smaller lakes

Figure 8.13 MODIS image of newly formed Toshka Lakes in southern Egypt. Approximate scale 1:4,000,000. September 14, 2002. (Author-prepared figure.)

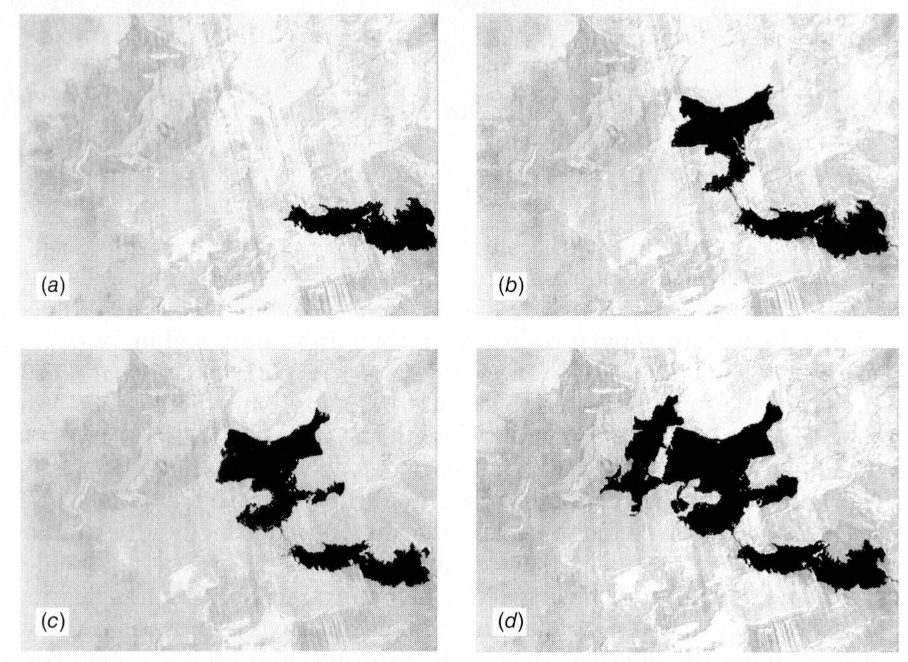

Figure 8.14 Landsat-7 ETM+ images of newly formed Toshka Lakes in southern Egypt. Approximate scale 1:2,200,000. (*a*) November 9, 1999, (*b*) January 12, 2000, (*c*) August 23, 2000, and (*d*) December 16, 2001. (Author-prepared figure.)

have temporarily dried up, future high water levels on the Nile could cause the lakes to reappear. Figure 8.13 shows the location of these lakes relative to Lake Nasser, the lake formed by the Aswan High Dam on the Nile River. Figure 8.14 illustrates how these lakes grew in size and number during the two-year period beginning November 1999. The use of ICESat GLAS spaceborne lidar data for

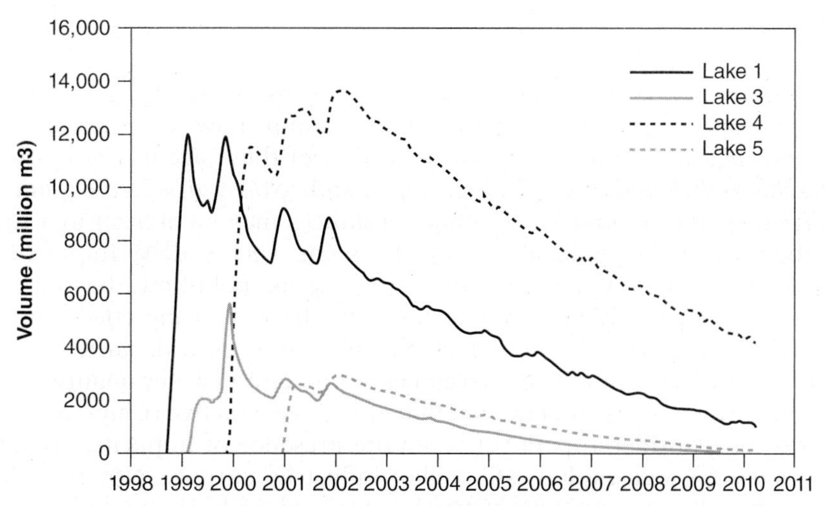

Figure 8.15 Evolution of lake volume over time for four of the Toshka Lakes, Egypt. Volume is calculated based on lake surface area from Terra and Aqua MODIS and lake water level from a combination of ICESat GLAS laser altimetry, basin topography from a DEM, and numerical modeling.

measuring water levels in lakes was discussed in Section 6.25 and illustrated in Figure 6.67, for the westernmost Toshka Lake shown in Figure 8.14*d*. Finally, Figure 8.15 shows the result of a numerical modeling effort to monitor changes in the volume of water in these lakes over time, based entirely on remotely sensed data. Lake surface area was measured at an eight-day interval over the entire decade using MODIS (with AVHRR for 1998–1999). Lake water level was measured intermittently with the ICESat GLAS lidar (as per Figure 6.67), with a DEM (from radar interferometry) and modeling used to derive water levels for dates without ICESat data. The lake surface area and lake water level data were then combined to calculate volume.

Although this discussion has focused on remote sensing of surface water quantity and spatial distribution, *groundwater monitoring* is important for both water supply and pollution control analysis. The identification of topographic and vegetation indicators of groundwater and the determination of the location of *groundwater discharge areas* (springs and seeps) can assist in the location of potential well sites. Also, it is important to be able to identify *groundwater recharge zones* in order to protect these areas from activities that would pollute the groundwater supply. Plate 25*c* illustrated the use of interferometric radar for monitoring land subsidence due to groundwater pumping in Las Vegas, Nevada. Over broader areas—regional to continental scales—changes in groundwater can be inferred from models of the earth's gravity field, derived from satellites such as GRACE. While these satellite geodesy missions are outside the scope of this book, they do provide a unique view of hydrologic changes at very broad spatial scales.

Water Quality

All naturally occurring water contains other substances. When these substances affect the optical properties of the water column, they are referred to as *color-producing agents*. Among the most important of these agents are *suspended sediment, chlorophyll and other algal pigments*, and *colored dissolved organic matter* or *dissolved organic carbon*. Many other substances may be present in a water body, and their origins, significance, and effects can vary widely. Impurities may be naturally occurring or the result of anthropogenic pollution; they may be hazardous to life or entirely benign; and they may have a strong effect on the optical properties of the water, or no detectable effect at all. A wide variety of visual and automated methods have been developed for assessing water quality using remote sensing, for those aspects of quality that are optically active (Chipman et al., 2009).

Water is considered polluted when the presence of impurities is sufficient to limit its use for a given domestic and/or industrial purpose. Not all pollutants are the result of human activity. Natural sources of pollution include minerals leached from soil and decaying vegetation. When dealing with water pollution, it is appropriate to consider two types of sources: point and nonpoint. *Point sources* are highly localized, such as industrial outfalls. *Nonpoint sources*, such as fertilizer and sediment runoff from agricultural fields, have large and dispersed source areas.

Each of the following categories of materials, when present in excessive amounts, can result in water pollution: (1) organic wastes contributed by domestic sewage and industrial wastes of plant and animal origin that remove oxygen from the water through decomposition; (2) infectious agents contributed by domestic sewage and by certain kinds of industrial wastes that may transmit disease; (3) plant nutrients that promote nuisance growths of aquatic plant life such as algae and water weeds; (4) synthetic-organic chemicals such as detergents and pesticides resulting from chemical technology that are toxic to aquatic life and potentially toxic to humans; (5) inorganic chemical and mineral substances resulting from mining, manufacturing processes, oil plant operations, and agricultural practices that interfere with natural stream purification, destroy fish and aquatic life, cause excessive hardness of water supplies, produce corrosive effects, and in general add to the cost of water treatment; (6) sediments that fill streams, channels, harbors, and reservoirs, cause abrasion of hydroelectric power and pumping equipment, affect the fish and shellfish population by blanketing fish nests, spawn, and food supplies, and increase the cost of water treatment; (7) radioactive pollution resulting from the mining and processing of radioactive ores, the use of refined radioactive materials, and fallout following nuclear testing or nuclear accidents; and (8) temperature increases that result from the use of water for cooling purposes by steam electric power plants and industries and from impoundment of water in reservoirs, have harmful effects on fish and aquatic life, and reduce the capacity of the receiving water to assimilate wastes.

Heavy sediment loads in rivers are often clearly depicted on aerial and space images. Figure 8.16 shows the silt-laden waters of the Po River discharging into the Adriatic Sea. The seawater has a low reflectance of sunlight, similar to that for "Water" shown in Figure 1.9. The spectral response pattern of the suspended solids resembles that of "Soil" shown in Figure 1.9. Because the spectral response pattern of the suspended materials is distinct from that of the natural seawater, these two materials can be readily distinguished on the image.

One effect of these sediment plumes is that they often carry nutrients (from agricultural lands, urban areas, and other terrestrial sources) that can alter the *trophic state* (nutrient state) of a lake or other water body. A lake choked with aquatic weeds or a lake with extreme-nuisance algal blooms is called a *eutrophic* (nutrient-rich) lake. A lake with very clear water is called an *oligotrophic* (low nutrient, high oxygen) lake. The general process by which lakes age is referred to as *eutrophication*. Eutrophication is a natural process that plays out over geologic time. However, when influenced by human activity, the process is greatly accelerated and may result in "polluted" water conditions. Such processes are termed *cultural eutrophication* and are intimately related to land use/land cover.

Figure 8.16 Space photograph (taken with the Large Format Camera) showing the silt-laden Po River discharging its sediments into the Adriatic Sea. Scale 1:610,000. (Courtesy NASA and ITEK Optical Systems.)

What constitutes an unacceptable degree of eutrophication is a function of who is making the judgment. Most recreational users of water bodies prefer clear water free of excessive *macrophytes* (large aquatic plants) and *algae*. Swimmers, boaters, and water skiers prefer lakes relatively free of submersed macrophytes, while persons fishing for bass and similar fish generally prefer some macrophytes. Large concentrations of *blue-green algae* (or *cyanobacteria*) have an unpleasant odor that is offensive to most people, especially during "blooms," or periods following active algal growth. These cyanobacterial blooms can also release a wide variety of neurotoxins and other potentially harmful chemicals, occasionally leading to injuries or fatalities among animals and humans. When cyanobacterial blooms occur in water bodies used as drinking water sources, water treatment becomes more challenging and expensive. In contrast, *green algae* tend to be less bothersome, unless present in large quantities.

The use of visual image interpretation coupled with selective field observations is an effective technique for mapping aquatic macrophytes. Macrophyte community mapping can be aided by the use of image interpretation keys. More detailed information regarding total plant biomass or plant density can often be achieved by utilizing quantitative techniques (Chapter 7).

Concentrations of free-floating algae are a good indicator of a lake's trophic status. Excessive concentrations of blue-green algae are especially prevalent under eutrophic conditions. Seasonally, blooms of blue-green algae occur during warm water conditions in late summer, whereas diatoms are more common in the cold water of spring and fall. Green algae are typically present at any point in the seasonal cycle of lakes. Because the different broad classes of algae have somewhat different spectral response patterns, they can be distinguished by aerial and space imaging. However, the wavelengths corresponding to peak reflectance of blue-green and green algae are often close together, and the most positive results can be obtained using multispectral or hyperspectral scanners (Chapter 4) with at least several narrow bands in the 0.45- to 0.60-μm wavelength range. Algae blooms floating on or very near the water surface also reflect highly in the near infrared, as illustrated in Figure 8.17. This figure shows a Landsat image of a massive cyanobacterial bloom occurring on Lake Winnebago, a wide and shallow lake in east-central Wisconsin. At left, a river enters the lake at the city of Oshkosh. The plume of water at the river mouth contrasts sharply with the highly textured, bright appearance of the floating cyanobacterial bloom on the lake itself. The texture within the bloom comes from wind acting on the floating algae, while long, narrow linear features are the wakes from boats. Another example of satellite imagery of an algal bloom is shown in Plate 33*b*.

Materials that form films on the water surface, such as oil films, can also be detected through the use of remotely sensed images. Oil enters the world's water bodies from a variety of sources, including natural seeps, municipal and industrial waste discharges, urban runoff, and refinery and shipping losses and accidents. Thick *oil slicks* have a distinct brown or black color. Thinner *oil sheens* and *oil rainbows* have a characteristic silvery sheen or iridescent color banding but do

Figure 8.17 Grayscale version of a true-color Landsat image composite, showing a large cyanobacterial bloom, Lake Winnebago, WI. (Author-prepared figure.)

not have a distinct brown or black color. The principal reflectance differences between water bodies and oil films in the visible spectrum occur between 0.30 and 0.45 μm. Therefore, the best results are obtained when the imaging systems are sensitive to these wavelengths, which is not the case for many multispectral systems. Oil slicks can also be detected using imaging radar because of the dampening effect of oil slicks on waves (Chapter 6). As a result, radar images have been used as part of the petroleum exploration process, looking for small natural seeps of oil in offshore regions that may be indicative of larger deposits nearby.

8.8 SNOW AND ICE APPLICATIONS

Much of the earth's surface is covered for at least part of the year by snow, ice, or frozen ground, particularly at high latitudes and/or high elevations. This domain in which water occurs in its frozen state is referred to as the *cryosphere*. As those who live in such regions can testify, snow and ice come in an astonishing diversity of forms—soft or hard, smooth or rough, bright white or very dark. We think of snow and ice as cold, but they can also provide insulation. Ice can feel as hard as rock, but under pressure in the depths of a glacier it can flow and move, reshaping the land itself. In this section, we will briefly review the ways snow and ice interact with electromagnetic radiation, then consider some of the many ways that remote sensing is used to study the distribution and characteristics of these materials.

The spectral properties of snow are discussed briefly in Chapter 1 and are illustrated in Figures 5.16 and 7.23 and Plates 13 and 38. Overall, snow is usually highly reflective in the visible spectrum, with stronger absorption in portions of the mid-infrared. However, its spectral response pattern can vary depending on density, grain size, liquid water content, and the presence of impurities such as dust, soot, and algae. In general, as snow cover ages, its overall reflectance (often referred to by cryosphere scientists as *albedo*) decreases. In part, this is due to structural changes in the snow crystals themselves and in part due to the accumulation and concentration of the abovementioned impurities. The spectral reflectance of ice can be highly variable. Ice's appearance depends on the roughness and age of its surface, its thickness (and for thin ice, the reflectance properties of the surface beneath it), and the presence or absence of air bubbles, along with many other factors. Snow and ice tend to act as specular reflectors, particularly when the surface is smooth; rough surfaces may scatter light more diffusely. Both have relatively high emissivity in the thermal infrared, typically 0.85 to 0.90 for dry snow, and 0.97 to 0.98 for rough ice. If anything, the response of snow and ice to radar signals is even more complicated than in the visible/infrared spectrum. This is partly due to the aforementioned diversity of forms and structures that can occur in snow and ice, and partly due to the fact that their dielectric constant can vary radically depending on the density and presence or absence of liquid water. The dielectric constant of dry snow is typically somewhere between the values for ice (3.2) and air (1.0), depending on density, while that for wet snow will be strongly influenced by water's high dielectric constant (80).

Remote sensing is widely used for monitoring seasonal snow cover at mid- to high latitudes. Such data are essential for water resources management (where downstream water supplies are fed by higher-elevation snowpack), flood prediction, and modeling the regional energy balance. Remote sensing of snow extent is typically done over large areas using relatively coarse resolution optical sensors (AVHRR, MODIS, VIIRS, and others) but site-specific studies may use higher resolution sensors. Errors in satellite-derived snow maps are generally largest in forested areas, while their reliability increases in more open terrain. Passive microwave systems are also used for continental- to global-scale snow maps.

Land ice, particularly in montane glaciers and larger ice sheets, is also the subject of remote sensing studies. Historical aerial photographs (or historical Corona satellite photographs), coupled with modern high-resolution satellite images, are often used to measure long-term changes in the terminus positions of glaciers as a component of glacier mass balance studies (the rate at which the terminus of a glacier advances or retreats is affected by the glacier's positive or negative mass balance, which in turn indicates whether it is gaining or losing ice). The velocity of ice flow within a glacier can be measured using feature-tracking algorithms with multitemporal optical or radar images, or using differential radar interferometry (Chapter 6), and variability in the rate of flow can be quantified (Joughin et al., 2010). Changes in the elevation (and thickness) of a glacier or ice sheet can be measured with lidar, radar altimetry, or radar interferometry (Rignot et al., 2008; Siegfried et al., 2011). *Borehole optical stratigraphy* uses digital video cameras lowered into holes drilled in the top of a glacier or ice sheet to measure the optical properties of ice and to quantify the rate at which firn (partially consolidated snow) is compacted into ice within the glacier or ice sheet (Hawley et al., 2011). Finally, identification of glacial landforms such as moraines, eskers, outwash plains, and other features in aerial or satellite imagery can help geoscientists infer the history and past climates of postglacial landscapes, as discussed in Section 8.3 and shown in Figure 8.3.

Figure 8.18 shows a simulated perspective view of high-elevation tropical glaciers on Mt Stanley (elev. 5109 m) in the Rwenzori Mountains of Central Africa. In this example, a WorldView-1 panchromatic image (50 cm resolution) has been draped over a DEM with a simulated viewpoint looking northeast. In addition to the glaciers on Mt Stanley itself, other small glaciers can be seen on Mt Speke in the background. Mt Stanley lies along the border between Uganda (upper right portion of Figure 8.18) and the Democratic Republic of Congo (lower left). The Rwenzori (sometimes referred to as the "Mountains of the Moon") formerly included over 40 glaciers, many of which have disappeared over the past century. Scientists are using this imagery, along with extensive field surveys, to map glacial landforms indicating when and where the Rwenzori glaciers have advanced and retreated since the late-Pleistocene Last Glacial Maximum, in an effort to expand the relatively sparse paleoclimate record in this region of Central Africa (Kelly et al., 2014).

Figure 8.18 Simulated perspective view of glaciers on Mt Stanley, Rwenzori Mountains. (Original WorldView-1 imagery courtesy DigitalGlobe.)

In terms of spatial extent, much of the world's ice occurs not on land but as sea ice on the Arctic Ocean and its neighboring bodies of water, and in the Southern Ocean surrounding Antarctica. Sea ice plays a prominent role in the planetary climate, acting both to reflect solar radiation away from the Earth in summer and to insulate the ocean below the ice from the colder atmosphere in winter. It provides habitat for Arctic wildlife such as seals, walrus, and polar bears, and imposes serious constraints on a host of economic activities at high latitudes, from shipping (via the Northwest Passage in the Canadian Arctic, or the Northern Sea Route along Russia's Arctic coast) to offshore oil exploration. As discussed in Chapter 6, sea ice is routinely monitored using passive microwave remote sensing (Figure 6.58), with optical sensors and active microwave (radar) systems employed as well.

8.9 URBAN AND REGIONAL PLANNING APPLICATIONS

Urban and regional planners require nearly continuous acquisition of data to formulate governmental policies and programs. The role of planning agencies is becoming increasingly more complex and is extending to a wider range of activities in the social, economic, and environmental domains. Consequently, there is an increased need for these agencies to have timely, accurate, and cost-effective sources of spatial information. Much of the information used by planning agencies is described elsewhere in this chapter—for example, land use/land cover data, discussed at length in Section 8.2, and maps of geology, soils, and landforms, covered in Sections 8.3 and 8.15. Here we discuss several other ways in which remotely sensed imagery can provide useful information for urban and regional planning.

In economically developed countries, population statistics are widely available from national censuses, but in much of the Global South these statistics are often sparse, obsolete, or inaccurate. *Population estimates* can be indirectly obtained through visual image interpretation, or through correlation between population and the intensity of upwelling radiance measured by optical sensors at night ("nighttime lights"). For visual estimation of population at the neighborhood scale, the traditional practice involves use of medium- to large-scale aerial photographs or high-resolution satellite imagery to estimate the number of dwelling units of each housing type in an area (single-family, two-family, multiple-family) and then multiply the number of dwelling units by the average family size per dwelling unit for each housing type.

Nighttime lights images from the Defense Meteorological Satellite Program (DMSP) Operational Linescan System (OLS), and more recently the Suomi NPP VIIRS instrument (Chapter 5) have been used to examine the earth's urban development from space. These instruments are sensitive enough to detect low levels of visible and near-IR energy at night. Among the light sources that show up in these nocturnal images are cities, towns, industrial sites, gas flares, fishing fleets, and fires. Figure 5.30 shows examples of nighttime lights image composites from VIIRS. Care must be taken when attempting to compare data from different parts of the world. The brightest areas of the earth are the most urbanized, but not necessarily the most heavily populated. Analysts have developed statistical methods to compensate for these regional disparities in the relationships among population, economic development, and nighttime light intensity, and nighttime lights data provide a valuable data source for tracking the growth of the earth's urbanization over time.

Thermal imagery has been used to assess the intensity of urban heat islands, areas in which dense coverage of pavement, concrete, and other nonvegetated surfaces leads to the buildup of higher temperatures than in surrounding less urbanized areas. Urban heat islands may be associated with reduced air quality and other health hazards and often afflict less affluent neighborhoods with proportionately less access to parks and other forms of cooling greenspace.

Visual image interpretation can also assist in *housing quality studies*, *traffic and parking studies*, and *site selection*. Characteristics of a neighborhood's housing can be inferred based on observable features in high-resolution imagery, including both the structures themselves and their surroundings). Parked vehicle numbers and the spatial distribution of parking density can be determined from high-resolution imagery, while moving vehicles in traffic can be counted and their average speeds estimated using sequential images (or video). Visual interpretation of all of these can be hindered by shadows from nearby buildings, and in particularly dense areas the street grid may be partially obscured by relief displacement of large buildings.

The site selection process involves identifying optimal locations for particular buildings, utilities, transportation routes, parks, and other infrastructure. This process typically employs a wide range of spatial data on social, economic, and environmental factors. Many of these data will be interpreted from remotely sensed sources. The task of analyzing the data is greatly facilitated by the use of a GIS.

Urban change detection mapping and analysis can be facilitated through the interpretation of multidate aerial and satellite images, such as the images shown in Figure 8.19, which illustrate the changes in an urban fringe area over a period of 67 years. The 1937 photograph (*a*) shows the area to be entirely agricultural land. The 1955 photograph (*b*) shows that a "beltline" highway has been constructed across the top of the area and that a gravel pit has begun operation in a glacial outwash plain at lower left. The 1968 photograph (*c*) shows that commercial development has begun at upper left and that extensive single-family housing development has begun at lower right. A school has been constructed at lower center and the gravel pit continues operation. The 2004 image (*d*) shows that the commercial and single-family development has continued. Multiple-family housing units have been constructed at left. The gravel pit site is now a city park that was a sanitary landfill site for a number of years between the dates of images (*c*) and (*d*).

Figure 8.19 Multidate images illustrating urban change, southwest Madison, WI (scale 1:35,000): (*a*) 1937, (*b*) 1955, (*c*) 1968, (*d*) 2008. [(*a–c*) Courtesy USDA–ASCS photos. (*d*) Courtesy DigitalGlobe.]

8.10 WETLAND MAPPING

The value of the world's wetland systems has gained increased recognition. Wetlands contribute to a healthy environment in many ways. They act to retain water during dry periods, thus keeping the water table high and relatively stable. During periods of flooding, they act to reduce flood levels and to trap suspended solids and attached nutrients. Thus, streams flowing into lakes by way of wetland areas will transport fewer suspended solids and nutrients to the lakes than if they flow directly into the lakes. The loss of such wetland systems through urbanization or other processes typically causes lake water quality to worsen. In addition, wetlands provide habitat for wildlife and a stopping place and refuge for migratory waterfowl. As with any natural habitat, wetlands are important in supporting species diversity and have a complex and important food web. Scientific values of wetlands include a record of biological and botanical events of the past, a place to study biological relationships, and a place for teaching. It is especially easy to obtain a feel for the biological world by studying a wetland. Other human uses include low intensity recreation and esthetic enjoyment.

The increased interest in wetlands has led to a new emphasis on inventory and monitoring of these ecosystems. To perform a wetlands inventory, a classification system must be devised that will provide the information necessary to the inventory users. The system should be based primarily on enduring wetland characteristics so that the inventory does not become outdated too quickly, but the classification should also accommodate user information requirements for ephemeral wetland characteristics. In addition, the inventory system must provide a detailed description of specifically what is considered to be a wetland. If the wetland definition used for various "wetland maps" is not clearly stated, then it is not possible to tell if apparent wetland changes noted between maps of different ages result from actual wetland changes or are due to differences in concepts of what is considered a wetland.

At the federal level in the United States, four principal agencies are involved with wetland identification and delineation: (1) the Environmental Protection Agency, (2) the Army Corps of Engineers, (3) the Natural Resources Conservation Service, and (4) the Fish and Wildlife Service. In 1989, these four agencies produced a *Federal Manual for Identifying and Delineating Jurisdictional Wetlands* (Federal Interagency Committee for Wetland Delineation, 1989), which provides a common basis for identifying and delineating wetlands. There is general agreement on the three basic elements for identifying wetlands: (1) hydrophytic vegetation, (2) hydric soils, and (3) wetland hydrology. *Hydrophytic vegetation* is defined as macrophytic plant life growing in water, soil, or substrate that is at least periodically deficient in oxygen as a result of excessive water content. *Hydric soils* are defined as soils that are saturated, flooded, or ponded long enough during the growing season to develop anaerobic (lacking free oxygen) conditions in the upper part. In general, hydric soils are flooded, ponded, or saturated for one week or more during the period when soil temperatures are above biologic zero (5°C) and usually support hydrophytic vegetation. *Wetland hydrology* refers to conditions of permanent or periodic inundation, or soil saturation to the surface, at least seasonally, hydrologic conditions that are the driving forces behind wetland formation. Numerous factors influence the wetness of an area, including precipitation, stratigraphy, topography, soil permeability, and plant cover. All wetlands typically have at least a seasonal abundance of water that may come from direct precipitation, overbank flooding, surface water runoff resulting from precipitation or snow melt, groundwater discharge, or tidal flooding.

Imagery including both visible and infrared bands—whether color infrared photography, or multispectral imagery—has been the preferred source for wetlands image interpretation. It provides interpreters with a high level of contrast in image tone and color between wetland and nonwetland environments, and moist soil spectral reflectance patterns contrast more distinctively with less moist soils in infrared imagery. Multispectral sensors with bands in the mid-infrared may also be useful for identifying wetlands soils and vegetation.

An example of wetland mapping is shown in Figures 8.20 and 8.21. Figure 8.20 is a 5.0× enlargement of a color infrared airphoto that was used for wetland vegetation mapping at an original scale of 1:60,000. The wetland vegetation map (Figure 8.21) shows the vegetation in this scene grouped into nine classes. The smallest units mapped at the original scale of 1:60,000 are a few distinctive stands of reed canary grass and cattails about 1/3 ha in size. Most of the units mapped are much larger.

Another example of wetland mapping can be seen in Plate 33*a*, which illustrates multitemporal data merging as an aid in mapping invasive plant species, in this case reed canary grass.

At the federal level, the U.S. Fish and Wildlife Service is responsible for a *National Wetlands Inventory* (NWI) that provides current geospatially referenced

Figure 8.20 Black-and-white copy of a color infrared aerial photograph of Sheboygan Marsh, WI. Scale 1:13,000 (enlarged 4.6 times from 1:60,000). Grid ticks appearing in image are from a reseau grid included in camera focal plane. (Courtesy NASA.)

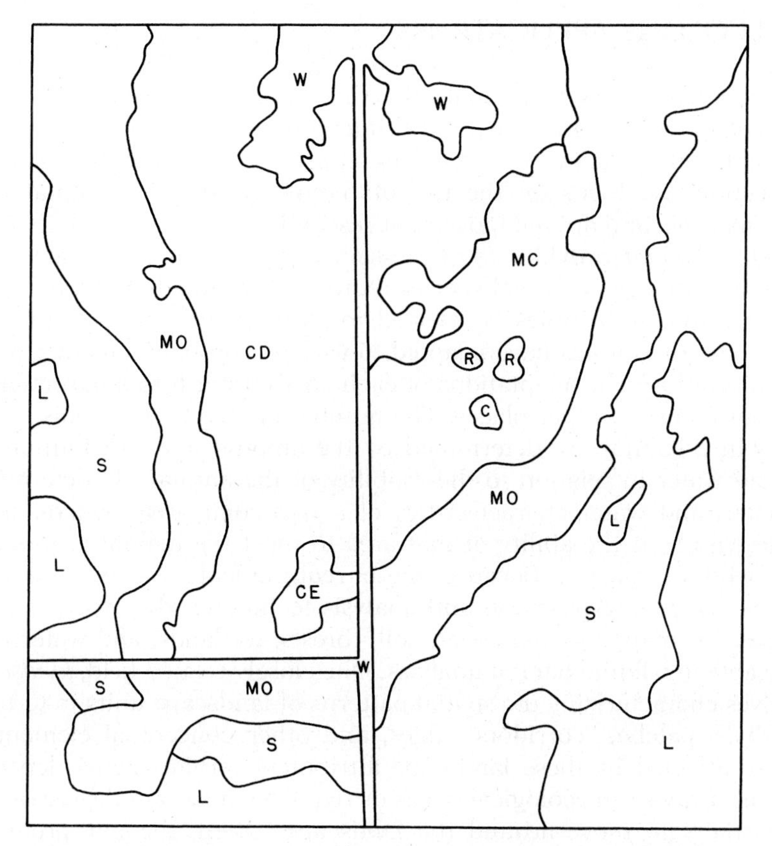

Figure 8.21 Vegetation classes in Sheboygan Marsh (scale 1:13,000): W = open water, D = deep water emergents, E = shallow water emergents, C = cattail (solid stand), O = sedges and grasses, R = reed canary grass (solid stand), M = mixed wetland vegetation, S = shrubs, L = lowland conifer forest.

information on the status, extent, characteristics, and functions of wetland, riparian, deepwater, and related aquatic habitats in priority areas to promote the understanding and conservation of these resources. The NWI strategy is focusing on three program goals: (1) strategic updating of maps in areas of the United States experiencing substantial development pressure and providing these products to the public over the Internet; (2) analyzing changes and trends to wetlands and other aquatic habitats at ecosystem, regional, or local scales; and (3) analyzing and disseminating resource information to improve identification of threats and risks to important wetland and aquatic habitats in order to promote sound decision making. Information on map and digital coverage is available on the Internet.

8.11 WILDLIFE ECOLOGY APPLICATIONS

The term *wildlife* refers to animals that live in a wild, undomesticated state. *Wildlife ecology* is concerned with the interactions between wildlife and their environment. Related activities are *wildlife conservation* and *wildlife management*. In this section, we focus on the use of remote sensing for habitat mapping, wildlife censusing, and animal movement research.

A *wildlife habitat* provides the necessary combination of climate, substrate, and vegetation that each animal species requires. Within a habitat, the functional area that an animal occupies is referred to as its *niche*. Throughout evolution, various species of animals have adapted to various combinations of physical factors and vegetation. The adaptations of each species suit it to a particular habitat and rule out its use of other places. The number and type of animals that can be supported in a habitat are determined by the amount and distribution of food, shelter, and water in relation to the mobility of the animal. By determining the food, shelter, and water characteristics of a particular area, general inferences can be drawn about the ability of that area to meet the habitat requirements of different wildlife species. Because these requirements involve many natural factors, the image interpretation and analysis techniques described elsewhere in this chapter for mapping land cover, soil, forests, wetlands, and water resources are applicable to wildlife habitat analysis. One closely related field, *landscape ecology*, involves characterizing the spatial patterns of landscape units in terms of features such as patches, corridors, edges, and other conceptual elements. Many species are affected by these landscape attributes, perhaps being dependent on "core" areas away from ecological edges or requiring a certain degree of topological connectivity to move around the landscape. Often, the interpreted habitat characteristics are incorporated in GIS-based modeling of the relationship between the habitat and the number and behavior of various species.

Wildlife censusing can be accomplished by ground surveys, aerial visual observations, or aerial imaging. Ground surveys rely on statistical sampling techniques and are often tedious, time consuming, and difficult to extrapolate. Many of the wildlife areas to be sampled are often nearly inaccessible on the ground. Aerial visual observations involve attempting to count the number of individuals of a species while flying over a survey area. Although this can be a low-cost and relatively rapid type of survey, there are many problems involved. Aerial visual observations require quick decisions on the part of the observer regarding numbers, species composition, and percentages of various age and sex classes. Aggregations of mammals or birds may be too large for accurate counting in the brief time period available. In addition, low-flying aircraft almost invariably disturb wildlife, with much of the population taking cover before being counted.

In many cases, vertical aerial imaging has been the best method of accurately censusing wildlife populations. If the mammals or birds are not disturbed by the aircraft, the images will permit very accurate counts to be undertaken. In addition, normal patterns of spatial distributions of individuals within groups will be apparent. Aerial images provide a permanent record that can be examined any number of times. Prolonged study of the images may reveal information that could not have been otherwise understood.

A variety of mammals and birds have been successfully censused using vertical aerial imaging, including moose, elephants, camels, whales, elk, sheep, deer, antelope, sea lions, caribou, beavers, seals, geese, ducks, flamingos, gulls, oyster catchers, and penguins. Vertical aerial imaging obviously cannot be used to census all wildlife populations. The size of individual animals must be large enough for reliable identification given the resolution of the imagery. Likewise, only those species that frequent relatively open areas during daylight hours can be counted in optical imagery. (Thermal sensing can be used to detect large animals in open areas at nighttime.)

Figure 8.22 shows a large group of beluga whales (small white whales) that have congregated in an arctic estuarine environment principally for the purpose of calving. At the image scales shown here, it is possible to determine the number and characteristics of individual whales and to measure their lengths. On the full 230 $\times$ 230-mm aerial photograph from which Figure 8.22 was rephotographed, a total of about 1600 individual whales were counted. At the original scale of 1:2000, the average adult length was measured as 4 m and the average calf length was measured as 2 m. Numerous adults with calves can be seen, especially in the enlargement (Figure 8.22*b*). "Bachelor groups" of eight and six males can be seen at the lower left and lower right of Figure 8.22*b*.

Determination of the movement of individuals or groups of animals over time is often important in wildlife ecology. Remote sensing can be used to directly contribute to this process through the use of newly developed image matching techniques for repeated identification of individual animals. Photographs taken of animals at different locations and different times are digitally compared to identify pairs of images that appear to include the same individual. These methods work best for animals with distinctive individual patterns of highly contrasting markings, such as giraffe, wildebeest, and others. Figure 8.23 illustrates two photographs of a single female wildebeest, photographed in two different areas of northern Tanzania a year apart (this region is shown in Plate 32). Wildebeest are notable for their unique patterns of dark and light stripes on the neck and shoulders. The patterns of these markings, highlighted in the boxes in Figure 8.23, can be used in an image matching program to search a catalog of previously photographed individuals and highlight any matches. In this case, the specific animal was determined to have been photographed in the Simanjiro Plains outside Tarangire National Park in June 2006, and again in Tanzania's Northern Plains in June 2007.

For a more complete record of the movement behavior of animals, wildlife ecologists often make use of GPS tracking. In some cases the data are stored in a device carried along with the GPS unit for later retrieval, while in other cases they are transmitted via satellite for real time tracking. In either case, the GPS tracks can then be superimposed on remotely sensed imagery to assess how the animal

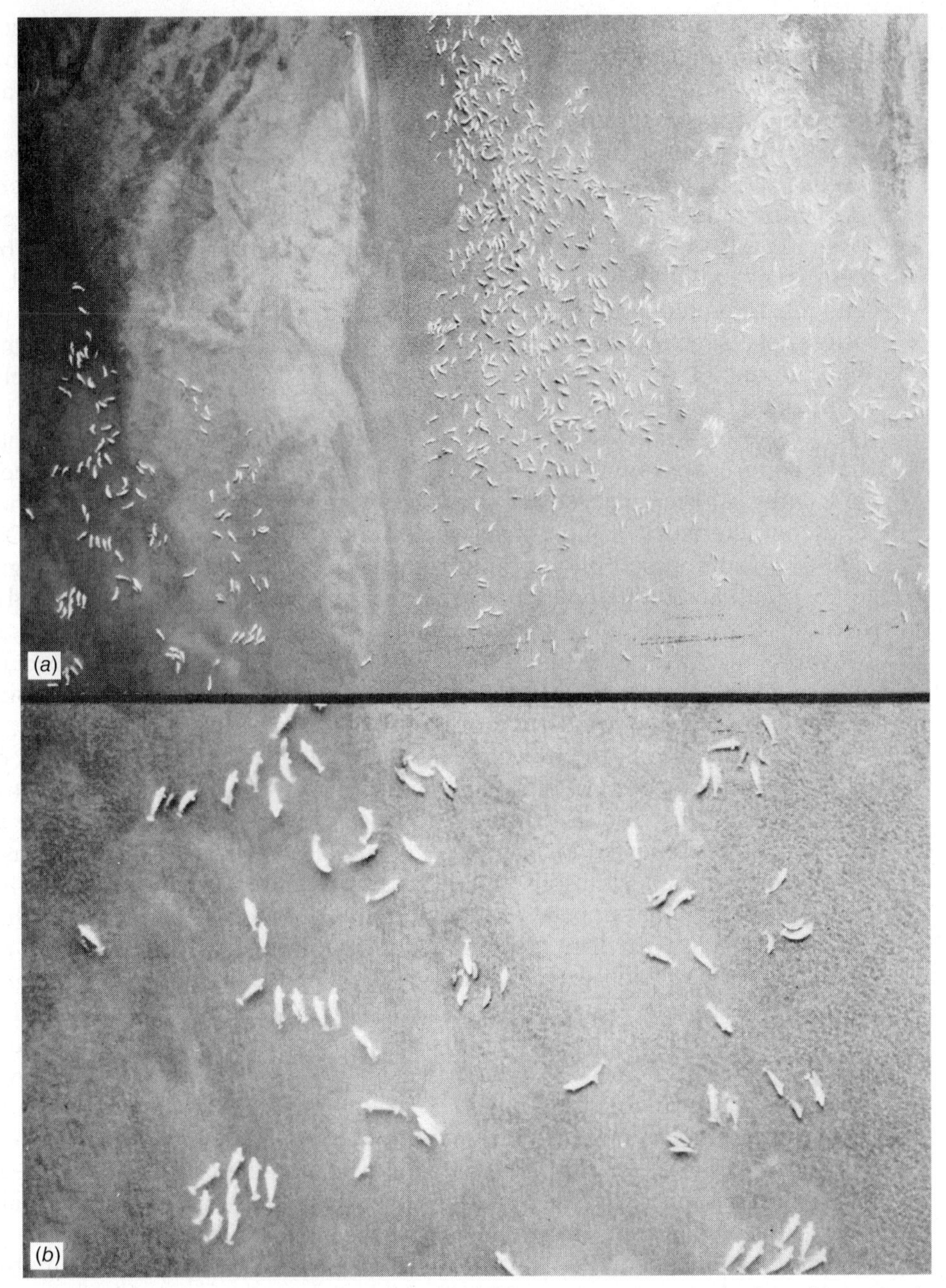

Figure 8.22 Large group of beluga whales, Cunningham Inlet, Somerset Island, northern Canada (black-and-white copy of photograph taken with Kodak Water Penetration Color Film, SO-224): (*a*) 1:2400, (*b*) 1:800. (*b*) is a three times enlargement of the lower left portion of (*a*). (Courtesy J.D. Heyland, Metcalfe, Ontario/Canadian Forest Service Publications.)

Figure 8.23 Automated matching of repeat photographs of a wildebeest in Tanzania. Photo (*a*) was taken in the Simanjiro Plains in June 2006; photo (*b*) was taken in the Northern Plains in June 2007. (Courtesy Thomas Morrison.)

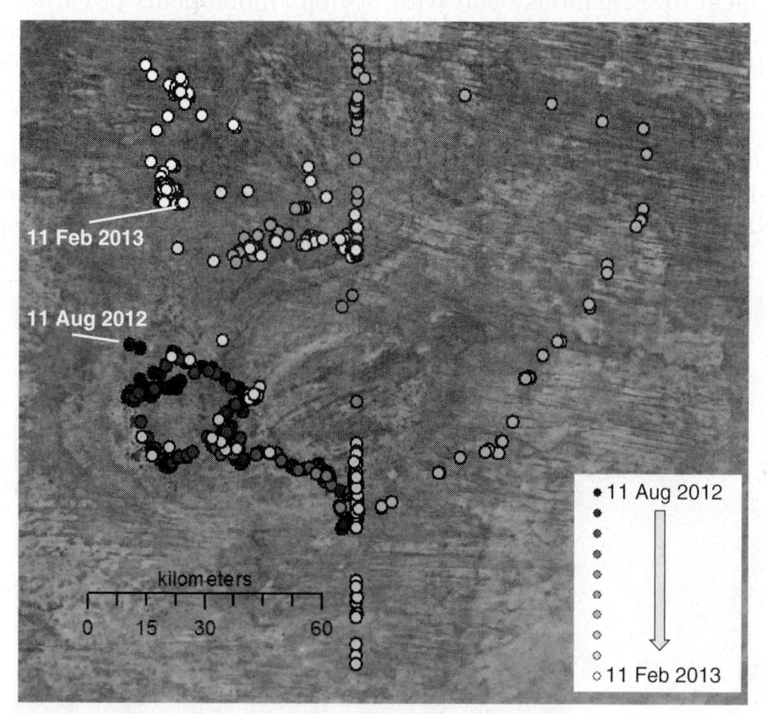

Figure 8.24 Satellite tracking of GPS-collared giraffe, Namibia and Botswana. (Author-prepared figure.)

moves through the landscape, what types of land cover it seeks out or avoids, and other aspects of its movement behavior. Figure 8.24 shows a map of the movement of one GPS-collared giraffe in southern Africa, as it moved across the landscape over a six-month period. Each dot represents one transmitted location. The prominent north–south line of dots in the center of the image represents the border between the nations of Namibia (left) and Botswana (right). The collared giraffe followed this fenceline for a long distance before briefly crossing into Botswana, making a large loop to the east (moving rapidly, as shown by the larger gaps between points), then turning south and west, and recrossing back into Namibia. Over the course of the entire six months, examination of satellite imagery (background of Figure 8.24) and the GPS track allows wildlife ecologists to formulate and test hypotheses about the land cover, plant communities, soil conditions, and other landscape features that individual animals make use of.

8.12 ARCHAEOLOGICAL APPLICATIONS

Archaeology is concerned with the scientific study of historic or prehistoric peoples by excavation and analysis of the remains of their existence. The earliest archaeological investigations dealt with obvious monuments of earlier societies. The existence of these sites was often known from historical accounts. Visual image interpretation has proven particularly useful in locating sites whose existence has been lost to history. Both surface and subsurface features of interest to archaeologists have been detected using visual image interpretation.

Surface features include visible ruins, mounds, rock piles, and various other surface markings ranging from small pictographs to large patterns such as the ancient Nazca Lines in Peru. These lines (of which a part is shown in Figure 8.25)

Figure 8.25 Vertical photomosaic showing Nazca Lines, Peru. (From Kosok, 1965. Courtesy Long Island University Press.)

are estimated to have been made between 1300 and 2200 years ago and cover an area of about 500 km^2. Many geometric shapes have been found, as well as narrow straight lines that extend for as long as 8 km. They were made by clearing away literally millions of rocks to expose the lighter toned ground beneath. The cleared rocks were piled around the outer boundaries of the "lines." Although the markings may have been observed from the ground previously, they were first brought to the world's attention after being seen from the air during the 1920s. There is no consensus among experts on the reason for their construction, although many hypotheses have been suggested.

Subsurface archaeological features include buried ruins of buildings, ditches, canals, and roads. When such features are covered by agricultural fields or native vegetation, they may be revealed on aerial or satellite images by tonal anomalies resulting from subtle differences in soil moisture or crop growth. On occasion, such features have been revealed by ephemeral differences in frost patterns.

The sites of more than a thousand Roman villas have been discovered in northern France through the use of 35-mm aerial photography. The buildings were destroyed in the third century C.E., but their foundation materials remain in the soil. In Figure 8.26, we see an outline of a villa foundation due to differences

Figure 8.26 Oblique 35-mm airphoto of a cereal crop field in northern France. Differences in crop vigor reveal the foundation of a Roman villa. (Photograph by R. Agache. Reproduced by permission of the Council for British Archaeology.)

in crop vigor. The area shown in this figure has recently been converted from pasture to cropland. In the early years following such conversion, farmers applied little or no fertilizer to the fields. The cereal crops over the foundation materials are light toned, owing to both the lack of fertilizer and a period of drought prior to the date of photography. The crops are darker toned over the remainder of the field. The main building (in the foreground) was 95 × 60 m.

One of the biggest advances in archaeological remote sensing has been the use of imaging radar and/or airborne lidar for high-resolution mapping of sites, including those hidden beneath a forest canopy. Figure 8.27 shows a shaded-relief image of the Maya urban site of Caracol, in what is now southern Belize. This site, extending over 200 km^2, contains a vast number of buildings believed to serve residential, administrative, and religious purposes, as well as agricultural terraces, causeways, and "chultuns" or underground storage units (Chase et al., 2012). This heavily developed and spatially integrated landscape suggests that ancient Caracol supported a large population under conditions that could be described as "low-density urbanism" (Fletcher 2009; Chase et al., 2012). The shaded-relief image was derived from airborne lidar data, allowing the first spatially continuous

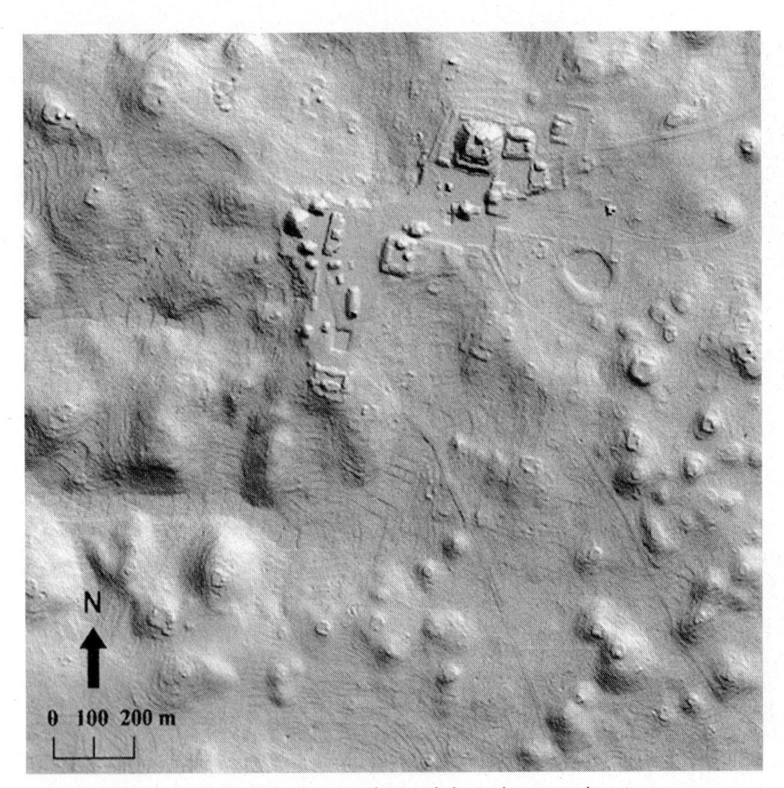

Figure 8.27 Shaded-relief image from lidar dataset showing numerous archaeological sites at Caracol in southern Belize. (Copyright © A.F. Chase and the National Academy of Sciences; used with permission.)

view to be compiled of this vast agro-metropolis long hidden beneath the trees. As the use of lidar (and interferometric radar) becomes more widespread in archaeology, researchers will gain the ability not merely to find and identify previously unknown or "lost" sites, but to better understand the spatial connections and linkages among what previously might have been considered isolated or only vaguely related locales.

8.13 ENVIRONMENTAL ASSESSMENT AND PROTECTION

Many human activities produce potentially adverse environmental effects. Examples include the construction and operation of highways, railroads, pipelines, airports, industrial sites, power plants, and transmission lines; subdivision and commercial developments; sanitary landfill and hazardous waste disposal operations; and timber harvesting and strip mining operations.

Dating as far back as 1969, the U.S. *National Environmental Policy Act* (NEPA) established as national policy the creation and maintenance of conditions that encourage harmony between people and their environment and minimize environmental degradation. This act requires that *environmental impact statements* be prepared for any federal action having significant impact on the environment. The key items to be evaluated in an environmental impact statement are (1) the environmental impact of the proposed action, (2) any adverse environmental effects that cannot be avoided should the action be implemented, (3) alternatives to the proposed action, (4) the relationship between local short-term uses of the environment and the maintenance and enhancement of long-term productivity, and (5) any irreversible and irretrievable commitments of resources that would be involved in the proposed action should it be implemented. Since the passage of NEPA, many other federal and state laws have been passed with environmental assessment as a primary component.

Environmental assessments involve, at a minimum, comprehensive inventory of physiographic, geologic, soil, cultural, vegetative, wildlife, watershed, and airshed conditions. Such assessments will typically draw on expertise of persons from many areas such as civil engineering, forestry, landscape architecture, land planning, geography, geology, archaeology, environmental economics, rural sociology, ecology, seismology, soils engineering, pedology, botany, biology, zoology, hydrology, water chemistry, aquatic biology, environmental engineering, meteorology, air chemistry, and air pollution engineering. Many of the remote sensing and image interpretation techniques set forth in this book can be utilized to assist in the conduct of such assessments. Overall, the applications of remote sensing in environmental monitoring and assessment are virtually limitless, ranging from environmental impact assessment to emergency response planning, landfill monitoring, permitting and enforcement, and natural disaster mitigation, to name but a few.

Effectively, real-time imaging is used in such applications as responding to the spillage or accidental release of hazardous materials. Such images are used to determine the extent and location of visible spillage and release, vegetation damage, and threats to natural drainage and human welfare. On the other hand, historical images are often used to conduct intensive site analyses of waste sites, augmenting these with current images when necessary. These analyses may include characterizing changes in surface drainage conditions through time; identifying the location of landfills, waste treatment ponds, and lagoons and their subsequent burial and abandonment; and detecting and identifying drums containing waste materials, or checking for indications of septic tank failure. In the aquatic realm, tracking sediment plumes, harmful algae blooms, and other manifestations of degraded water quality are covered in Section 8.7. Monitoring the spread of oil spills on water was likewise discussed in Section 8.7 and also in Chapter 6 (using imaging radar). In particular, the response to the 2010 Deepwater Horizon oil spill in the Gulf of Mexico involved widespread use of remote sensing, in a variety of different forms, for tracking the movement of the oil spill and assessing its impact on coastal systems.

Figure 8.28 shows a plume of chlorine gas resulting from a train derailment. When such incidents occur, there is immediate need to assess downwind susceptibility of human exposure and the other potential impacts of the event. When available, remote sensing imagery and numerous forms of GIS data are used in conjunction with ancillary information (such as wind speed and direction) to develop emergency evacuation and response plans in the general vicinity of the

Figure 8.28 Low-altitude, oblique aerial photograph of chlorine spill resulting from train derailment near Alberton, MT. (Courtesy RMP Systems.)

incident. In the specific case of Figure 8.28, the GIS and remote sensing initiated for immediate response was also very useful and important for the longer-term monitoring of the spill site.

Figure 8.29 shows a massive smoke plume rising from a fire at a tire recycling facility in Watertown, Wisconsin, that involved an estimated one million tires.

Figure 8.29 Massive smoke plume rising from a fire at a tire recycling facility in Watertown, WI. (*a*) Oblique airphoto looking to the south. (Courtesy Mike DeVries/The Capital Times.) (*b*) Landsat Thematic Mapper image (north is to the top). Scale 1:97,000. (*a*) and (*b*) are images from the same day, but are not simultaneous. (Author-prepared figure.)

The smoke plume extended 150 km to the southeast, stretching across Milwaukee and over central Lake Michigan, and was visible in both Landsat and MODIS images (see Section 7.6 and Figure 7.27).

8.14 NATURAL DISASTER ASSESSMENT

Many forms of natural and human-induced disasters have caused loss of life, property damage, and damage to natural features. A variety of remote sensing systems can be used to detect, monitor, and respond to natural disasters, as well as assess disaster vulnerability. Here, we discuss wildfires, severe storms, floods, volcanic eruptions, dust and smoke, earthquakes, tsunamis, shoreline erosion, and landslides. NASA, NOAA, and the USGS maintain websites devoted to natural hazards. The USGS Natural Hazards website, for example, has information about earthquakes, hurricanes, tsunamis, floods, landslides, volcanoes, and wildfires.

Throughout this section, there are numerous references to various satellites that are described in Chapters 5 and 6, including DMC, IKONOS, Landsat, MODIS, QuickBird, Radarsat, SeaWiFS, and SPOT. For further information on these satellites, see the relevant sections of those chapters.

Internationally, the *Disaster Monitoring Constellation* (DMC), a consortium of nations (Algeria, Nigeria, Turkey, the United Kingdom, and China) operates a coordinated constellation of innovative, small, low-cost satellites that provide daily imaging capability. These microsatellites produce images of large areas (up to 600 km swath width) with a ground sampled distance (see Chapter 5) of 22 to 32 m in three spectral bands (green, red, and near-IR), and several include higher-resolution panchromatic imagery. As of late 2014, five DMC satellites were operational, four had been retired, and three more were scheduled for launch in the near future. Since its inception, the DMC has contributed to efforts in response to the Indian Ocean tsunami (2004), China's Wenchuan earthquake (2008), and the Tōhoku earthquake and tsunami (2011). The DMC has provided imagery for a variety of civil and commercial uses. For example, the United Nations has used such imagery to select and map camp sites occupied by internally displaced people in Darfur, Sudan. Similarly, such imagery has been used for tracking locust breeding grounds and estimating population sizes in Algeria.

Wildfires

Wildfires are a serious and growing hazard over much of the world. They pose a great threat to life and property, especially when they move into populated areas. Wildfires are a natural process, and their suppression is now recognized to have created greater fire hazards than in the past. Wildfire suppression has also disrupted natural plant succession and wildlife habitat in many areas. Figure 8.30 is a MODIS image of wildfires in the Yakutia region of northeastern Siberia

Figure 8.30 MODIS image of wildfires in Siberia, July 10, 2012. Scale 1:3,150,000. (Courtesy NASA.)

acquired on July 10, 2012. At that time, more than 25,000 ha were burning, and by the end of the summer Russia had recorded one of its worst fire seasons in known history. Smoke plumes from the fire reached across the Pacific and led to degraded air quality in western North America.

Severe Storms

Severe storms can take many forms, including tornados and cyclones. Tornadoes are rotating columns of air, usually with a funnel-shaped vortex several hundred meters in diameter, whirling destructively at speeds up to about 500 km/hr. Tornadoes occur most often in association with thunderstorms during the spring and summer in the midlatitudes of both the Northern and Southern Hemispheres. Cyclones are atmospheric systems characterized by the rapid, inward circulation of air masses about a low-pressure center, accompanied by stormy, often destructive, weather. Cyclones circulate counterclockwise in the Northern Hemisphere, and clockwise in the Southern Hemisphere. Hurricanes are severe tropical cyclones originating in the equatorial regions of the Atlantic Ocean or Caribbean Sea. Typhoons are tropical cyclones occurring in the western Pacific or Indian Oceans.

Tornado intensity is commonly estimated by analyzing damage to structures and then correlating it with the wind speeds required to produce such destruction. Tornado intensity is most often determined using the *Fujita Scale*, or *F-Scale*

(Table 8.5). Although very few (about 2%) of all tornadoes reach F4 and F5 intensities, they account for about 65% of all deaths.

Plate 37 is a large-scale digital camera image showing a portion of the aftermath of an F4 tornado that struck Haysville, Kansas. This tornado was responsible for six deaths, 150 injuries, and over $140 million in property damage.

Plate 31 contains "before" and "after" Landsat-7 satellite images of the damage caused by an F3 tornado that struck Burnett County, Wisconsin. This storm resulted in three deaths, eight serious injuries, complete destruction of 180 homes and businesses, and damage to 200 others. Figure 8.31 is a large-scale aerial photograph showing the "blowdown" of trees by this tornado. Tornado damage is also shown in Figure 6.33, a radar image of a forested area in northern Wisconsin.

Hurricane intensity is most often determined using the Saffir–Simpson Hurricane Scale, which rates hurricane intensity on a scale from 1 to 5 and is used to give an estimate of potential property damage and flooding along the coast from a hurricane landfall. The strongest hurricanes are Category 5 hurricanes, with winds greater than 249 km/hr. Category 5 hurricanes typically have a storm surge 5.5 m above normal sea level (this value varies widely with ocean bottom topography) and result in complete roof failure on many residences and some complete building failures. Many shrubs, trees, and signs are blown down. Severe and extensive window and door damage can occur, as can complete destruction of

TABLE 8.5 Fujita Scale (F-Scale) of Tornado Intensity

F-Scale Value	Wind Speed (km/hr)	Typical damage	Description
F0	<117	Light	Some damage to chimneys; branches broken off trees; shallow-rooted trees pushed over; sign boards damaged
F1	117–180	Moderate	Peels surface off roofs; mobile homes pushed off foundations or overturned; moving autos blown off roads
F2	181–251	Considerable	Roofs torn off frame houses; mobile homes demolished; boxcars overturned; large trees snapped or uprooted; light-object missiles generated; cars lifted off ground
F3	252–330	Severe	Roofs and some walls torn off well-constructed houses; trains overturned; most trees in forest uprooted; heavy cars lifted off the ground and thrown
F4	331–416	Devastating	Well-constructed houses leveled; structures with weak foundations blown away some distance; cars thrown and large missiles generated
F5	>416	Incredible	Strong frame houses leveled off foundations and swept away; automobile-sized missiles fly through the air in excess of 100 meters; trees debarked

Source: NOAA.

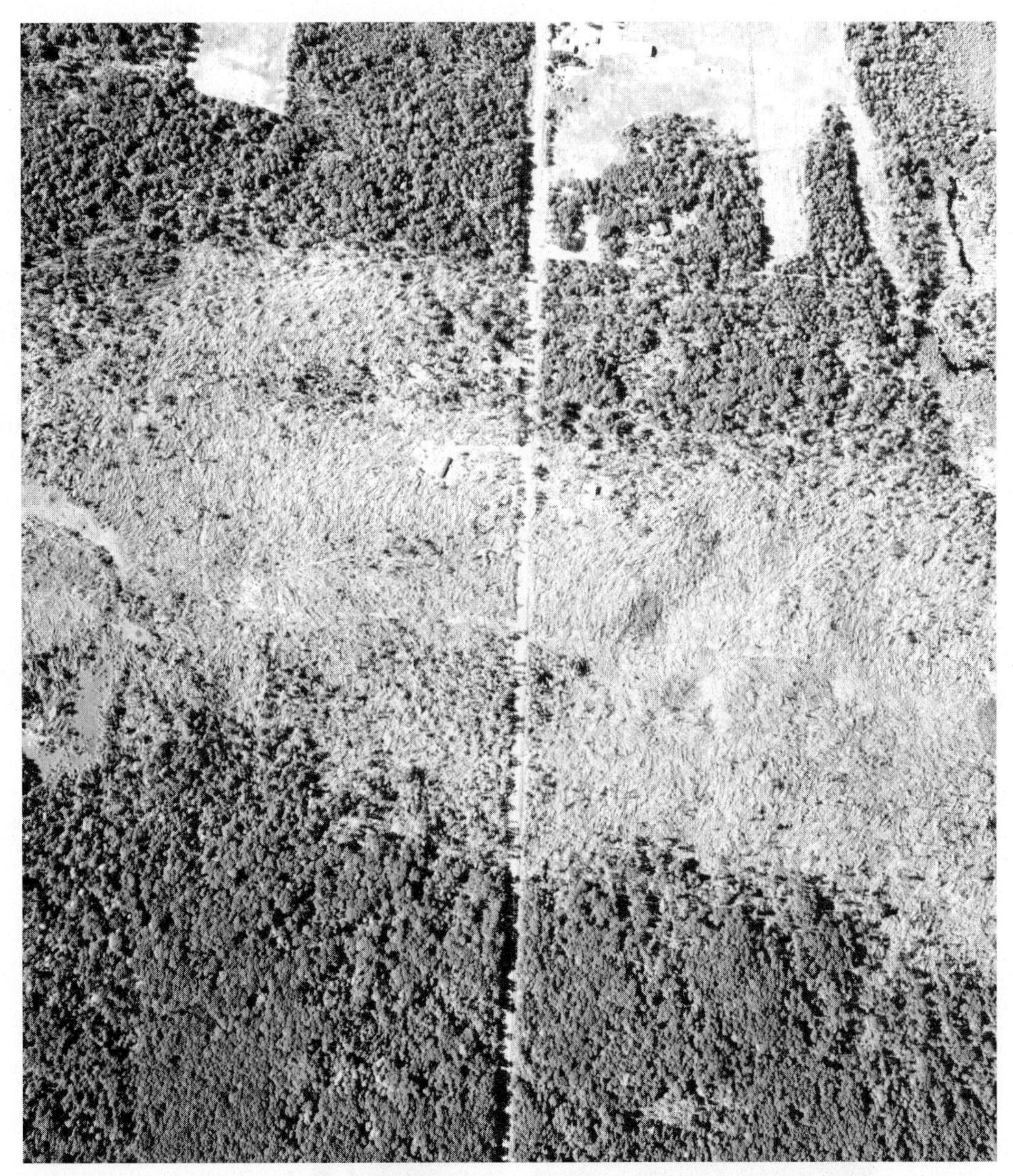

Figure 8.31 Aerial photograph showing the destruction of hundreds of trees by an F3 tornado that struck Burnett County, WI, on June 18, 2001. Scale 1:8100. (Courtesy Burnett County Land Information Office, University of Wisconsin-Madison Environmental Remote Sensing Center, and NASA Regional Earth Science Applications Center Program.)

mobile homes. Also, there is typically major damage to the lower floors of all structures located less than 5 m above sea level and within 500 m of the shoreline. Before the 1940s, many hurricanes went undetected. At present every hurricane is detected and tracked by satellite imaging. Note that the term *hurricane* is used specifically for large tropical storms in the Atlantic and Northeast Pacific; equivalent storms in the Northwest Pacific are called *typhoons,* and those in the South Pacific and Indian Oceans are called *cyclones,* although the latter term is also used generically.

Figure 8.32 shows Typhoon Haiyan as it approached the Philippines on November 7, 2013. Haiyan (known as Typhoon Yolanda in the Philippines) wreaked havoc across the western Pacific and Southeast Asia and is unofficially regarded as the most powerful typhoon on record. Shortly before this image was acquired, Haiyan's sustained wind speeds were measured at 280 km/hr. Over 6000 people were killed following the storm's landfall in the Philippines, and further damage occurred in other countries in the region.

Figure 8.32 MODIS image of Typhoon Haiyan as it approached the Philippines on November 7, 2013. (Courtesy NASA.)

Floods

Over-the-bank river flooding delivers valuable topsoil and nutrients to farmland and brings life to otherwise infertile regions of the world, such as the Nile River Valley. On the other hand, flash floods and abnormally large-scale flood events are responsible for more deaths than tornadoes or hurricanes and cause great amounts of property damage.

Various examples of flooding can be found elsewhere in this book, including in Section 8.7 and Chapter 6.

Volcanic Eruptions

Volcanic eruptions are one of earth's most dramatic and violent agents of change. Eruptions often force people living near volcanoes to abandon their land and homes, sometimes forever. Volcanic activity in the last 300 years has killed more than 250,000 people, destroyed entire cities and forests, and severely disrupted local economies. Volcanoes can present a major hazard to those who live near them, for a variety of reasons: (1) Pyroclastic eruptions can smother large areas of the landscape with hot ash, dust, and smoke within a span of minutes to hours; (2) red hot rocks spewed from the mouth of a volcano can ignite fires in nearby forests and towns, while rivers of molten lava can consume almost anything in their paths as they reshape the landscape; (3) heavy rains or rapidly melting summit snowpacks can trigger lahars, sluices of mud that can flow for miles, overrunning roads and villages; and (4) large plumes of ash and gas ejected high into the atmosphere can influence climate, sometimes on a global scale (based on USGS and NASA Natural Hazards websites).

Plate 26 shows an extensive volcanic terrain in central Africa. Numerous lava flows can be seen on the slopes of Nyamuragiro volcano, which dominates the lower portion of the image. To the upper right of Nyamuragiro volcano is Nyiragongo volcano, which erupted in 2002 with loss of life and great property damage in and around the city of Goma, located on the shore of Lake Kivu, which can be seen at the right edge of Plate 26.

Plate 38 is a Landsat-8 OLI image showing the eruption near Bárðarbunga, Iceland on September 6, 2014. Plate 38*a* shows an enlargement of the area around the main fissure, located between Bárðarbunga itself (a large stratovolcano under the Vatnajökull ice cap, to the south of the fissure) and the Askja caldera to the north. The bright red-orange areas along the fissure are flowing lava on the surface and molten lava within the fissure. This image is a composite of Landsat OLI bands 3 (sensitive to green wavelengths of light), 5 (near IR), and 7 (mid-IR), displayed in blue, green, and red, respectively. Because the molten lava emits very little energy in green wavelengths, and a great deal of energy in the mid-IR, it appears in Plate 38 with a red-orange color. (Note that Plate 5*b* showed flowing lava as photographed with color infrared film.) A smoke plume can be seen blowing downwind (east) from the fissure. At the lower left corner of Plate 38*a*, the terminus of one of the outlet glaciers from Vatnajökull is visible, looking relatively dark due to the absence of fresh snow at its low elevation.

Plate 38*b* shows a broader context for this eruption, from the same Landsat OLI image as in *(a)*. The large, bright, cyan-colored feature at the center of Plate 38*b* is the Vatnajökull ice cap, the largest such feature in Europe. Most of the ice cap is covered with fresh snow, but lower elevations around the edges appear darker due to bare (or dirty) ice. The semitransparent mottled white pattern at left is an area of cloud cover. The landscape in the northern part of Plate 38*b*, including the area around the active fissure, consists of essentially uninhabited, barren lava desert. Closer to the coast, greener hues indicate the presence of more vegetation, with its characteristic elevated reflectance in the near infrared.

As discussed in Chapter 6, differential radar interferometry can be used to monitor broad-scale changes in topography, including those associated with the movement of magma beneath a volcano (Plate 25*b*).

Dust and Smoke

Aerosols are small particles suspended in the air. Some occur naturally, originating from volcanoes, dust storms, and forest and grassland fires. Human activities, such as the burning of fossil fuels, prescribed fires, and the alteration of natural land surface cover (e.g., slash-and-burn activities), also generate aerosols. Many human-produced aerosols are small enough to be inhaled, so they can present a serious health hazard around industrial centers or even hundreds of miles downwind. Additionally, thick dust or smoke plumes severely limit visibility and can make it hazardous to travel by air or road. Smoke plumes are visible in Plate 38 and Figures 7.27 and 8.29.

Dust plumes have been observed in many arid regions around the globe. They can be extensive and travel great distances. For example, using satellite images, dust plumes originating near the west coast of Africa have been observed reaching the east coast of South America.

Earthquakes

Earthquakes occur in many parts of the world and can cause considerable property damage and loss of life. Most naturally occurring earthquakes are related to the tectonic nature of the earth. The earth's lithosphere is a patchwork of plates in slow but constant motion caused by heat in the earth's mantle and core. Plate boundaries glide past each other, creating frictional stress. When the frictional stress exceeds a critical value, a sudden failure occurs, resulting in an

TABLE 8.6 Richter Scale of Earthquake Magnitudes

Classification	Richter Magnitude	Earthquake Effects
Micro	Less than 3.0	Generally not felt, but may be recorded
Minor	3.0–3.9	Often felt, but rarely cause damage
Light	4.0–4.9	Noticeable shaking of indoor items, rattling noises Significant damage unlikely
Moderate	5.0–5.9	Some damage to poorly constructed buildings At most, slight damage to well-constructed buildings
Strong	6.0–6.9	Can be destructive in areas up to about 100 km across in populated areas
Major	7.0–7.9	Can cause serious damage over larger areas
Great	8.0–8.9	Can cause serious damage over areas several hundred km across
Rare Great	9.0 or higher	Devastating in areas several thousand km across

earthquake. The subduction of one plate under another can also build up stresses that are relieved as earthquakes. The severity of earthquakes can be described by magnitude or intensity. Magnitude measures the energy released at the source of the earthquake and is most often measured using the logarithmic *Richter Scale*. Table 8.6 describes the effect of earthquakes of various magnitudes on the Richter Scale. Intensity is most often measured using the *Mercalli Intensity Scale*, which measures the strength of shaking produced by the earthquake at a certain location and is determined from effects on people, human structures, and the natural environment.

Figure 8.2 illustrated extensive geologic features visible on satellite imagery that can be correlated with major geologic faults and showed major earthquake sites. Mapping faults from satellite imagery is an important application of remote sensing in seismology. Shaded-relief maps may also be useful for this purpose, but note that the angle of illumination has a large impact on the interpretability of faults (see Section 8.3).

Radar imagery, specifically interferometric radar, can help assess the pattern of ground movement during an earthquake. Plate 25*a* showed a radar interferogram depicting the seismic effects of the 2011 Tōhoku earthquake.

Tsunamis

A *tsunami* is a series of waves that occurs when a body of water, such as an ocean, is rapidly displaced on a massive scale. Earthquakes, mass movements above or below water (often occurring at tectonic plate boundaries), volcanic eruptions and other underwater explosions, and large meteorite impacts are among the possible causes of tsunamis. The effects of a large tsunami can be devastating. Large tsunamis contain immense energy, propagate at high speeds, and can travel great transoceanic distances with little overall energy loss. Most of the damage that results from tsunamis is caused by the huge mass of water behind the initial wave front, as the height of the ocean keeps rising fast and floods powerfully into the coastal area.

Plate 39 shows before- and after-tsunami QuickBird images of an area near Gleebruk, Indonesia, that resulted from the 2004 *Indian Ocean Earthquake*, known by the scientific community as the Sumatra-Andaman Earthquake. This earthquake originated in the Indian Ocean and devastated the shores of Indonesia, Sri Lanka, India, and Thailand, with waves up to 30 m high. Entire villages were destroyed, and it is estimated that nearly 300,000 people lost their lives. The earthquake was caused by an estimated 1200-km-long fault line that slipped about 15 m along the subduction zone where the India Plate dives under the Burma Plate. The resulting earthquake was one of the largest ever recorded on earth, with an estimated magnitude of 9.3 on the Richter Scale. The tsunami of December 26, 2004, caused by the 2004 Indian Ocean Earthquake, struck hardest along the northwest coast of Sumatra. The coast was drowned in waves up to 15 m high, and the water was channeled inland through low-lying areas such as stream floodplains. As seen in Plate 39, the force of the tsunami scoured the area; buildings, trees, and beaches were swept away by the force of the wave, and a bridge was washed away as well.

Similar images were acquired in northeastern Japan, following the devastating tsunami associated with the aforementioned 2011 Tōhoku earthquake. Perhaps the single most widely covered impact of this tsunami was the extensive damage to the Fukushima nuclear power plant.

Shoreline Erosion

Driven by rising sea and lake levels, large storms, flooding, and powerful ocean waves, erosion wears away the beaches and bluffs along the world's shorelines. Bluff erosion rates vary widely, depending on geologic setting, waves, and weather. The erosion rate for a bluff can be regular over the years, or it can change from near zero for decades to several meters in a matter of seconds. Remote sensing studies of shoreline erosion have used a variety of platforms, from cameras in microlight aircraft to satellite data. Historical data using aerial photographs dating back to the 1930s, as well as more recent satellite data, can be used to document shoreline erosion over time. Lidar data (Chapter 6) are also a particularly effective resource for mapping shoreline erosion.

Landslides

Landslides are mass movements of soil or rock down slopes and a major natural hazard because they are widespread. Globally, landslides cause an estimated 1000 deaths per year and great property damage. They commonly occur in conjunction with other major natural disasters, such as earthquakes, floods, and volcanic eruptions. Landslides can also be caused by excessive precipitation or human activities, such as deforestation or developments that disturb natural slope stability. They do considerable damage to infrastructure, especially highways, railways, waterways, and pipelines.

Historically, aerial photographs have been used extensively to characterize landslides and to produce landslide inventory maps, particularly because of their stereoviewing capability and high spatial resolution. High-resolution satellite images are now being used as well. Radar interferometry techniques (Chapter 6) have also been used in landslide studies in mountainous areas (Singhroy et al., 1998).

One particularly devastating landslide occurred on March 22, 2014, in northwest Washington State, about 6 km east of the town of Oso. The Oso landslide was among the deadliest single landslides in U.S. history, with more than 40 deaths confirmed and nearly 50 homes destroyed. The landslide occurred in an area known for its past landslide activity, but the slide was larger and had greater destructive force than previously experienced landslides.

Plate 40 shows lidar images and natural color photographs before and after the landslide. Analysis of the lidar images showed that the area covered by the landslide was about 120 ha and that about 4,000,000 cubic meters of material moved downslope.

These images were analyzed under the direction of the Washington State Department of Transportation (WSDOT). A number of deliverable items were produced, including 3D CADD modeling, DEM/DTM creation (Chapter 1), and 30-foot (9 m) GRID Debris Depth measurements and calculations. Accurate imagery and 3D mapping data and analysis results were prepared for a variety of organizations, including the Washington State Department of Natural Resources, the Army Corps of Engineers, FEMA, and various WSDOT offices.

As noted above, the area around the Oso landslide is known to have experienced many previous slides. Figure 8.33*a* shows a shaded-relief image of the area around the Oso slide, derived from airborne lidar. In Figure 8.33*b*, the extents of many previous landslides have been delineated, extending over the past 14,000 years (Haugerud, 2014). Clearly, this landscape is highly prone to episodic large landslides, a fact that should be taken into account by regional planners (Section 8.9) as well as those living near the slopes of these hills.

For additional information on landslide hazards, see the USGS Natural Hazards website at www.usgs.gov.

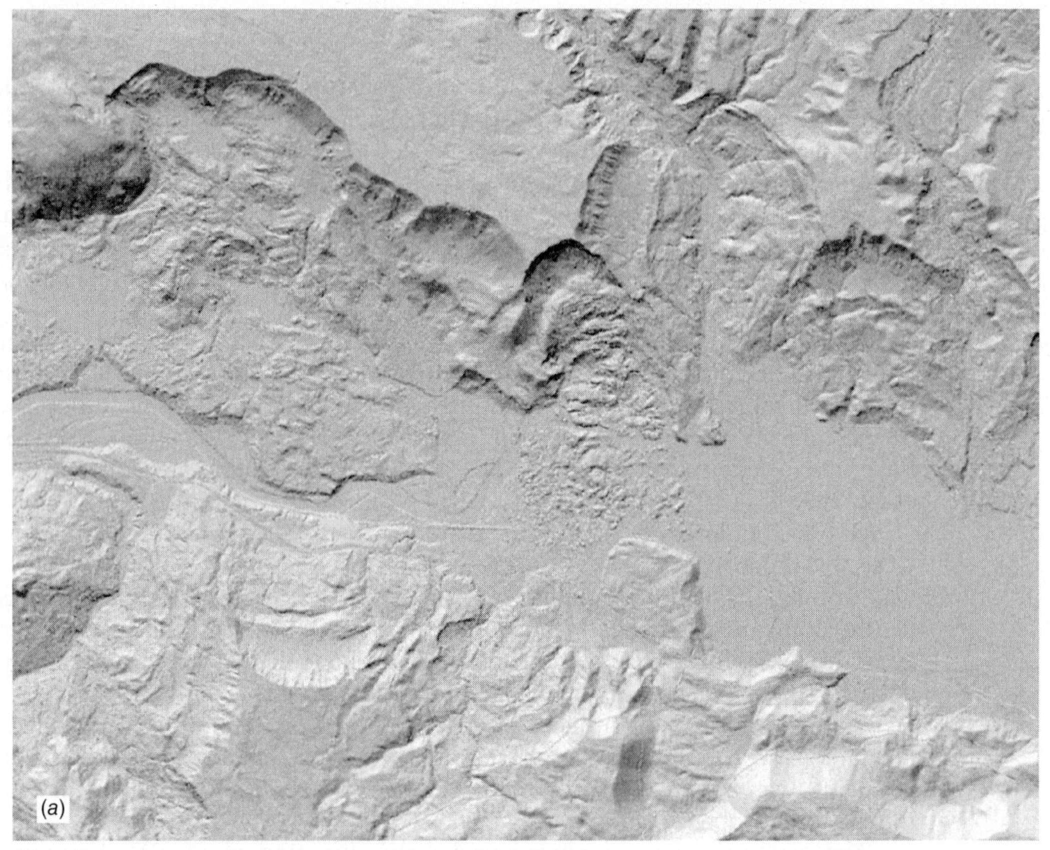

Figure 8.33 Recent and historical landslides near Oso, WA, USA. (*a*) Shaded relief from airborne lidar data. (*b*) With interpretation of historical landslides (labels A through D indicate relative age, from youngest to oldest). (Courtesy USGS, preliminary interpretation by R.A. Haugerud.)

8.15 PRINCIPLES OF LANDFORM IDENTIFICATION AND EVALUATION

Various terrain characteristics are important to soil scientists, geologists, geographers, civil engineers, urban and regional planners, landscape architects, real estate developers, and others who wish to evaluate the suitability of the terrain for various land uses. Because terrain conditions strongly influence the capability of the land to support various species of vegetation, an understanding of image interpretation for terrain evaluation is also important for botanists, conservation biologists, foresters, wildlife ecologists, and others concerned with vegetation mapping and evaluation.

The principal terrain characteristics that can be estimated by means of visual image interpretation are bedrock type, landform, soil texture, site drainage conditions, susceptibility to flooding, and depth of unconsolidated materials over

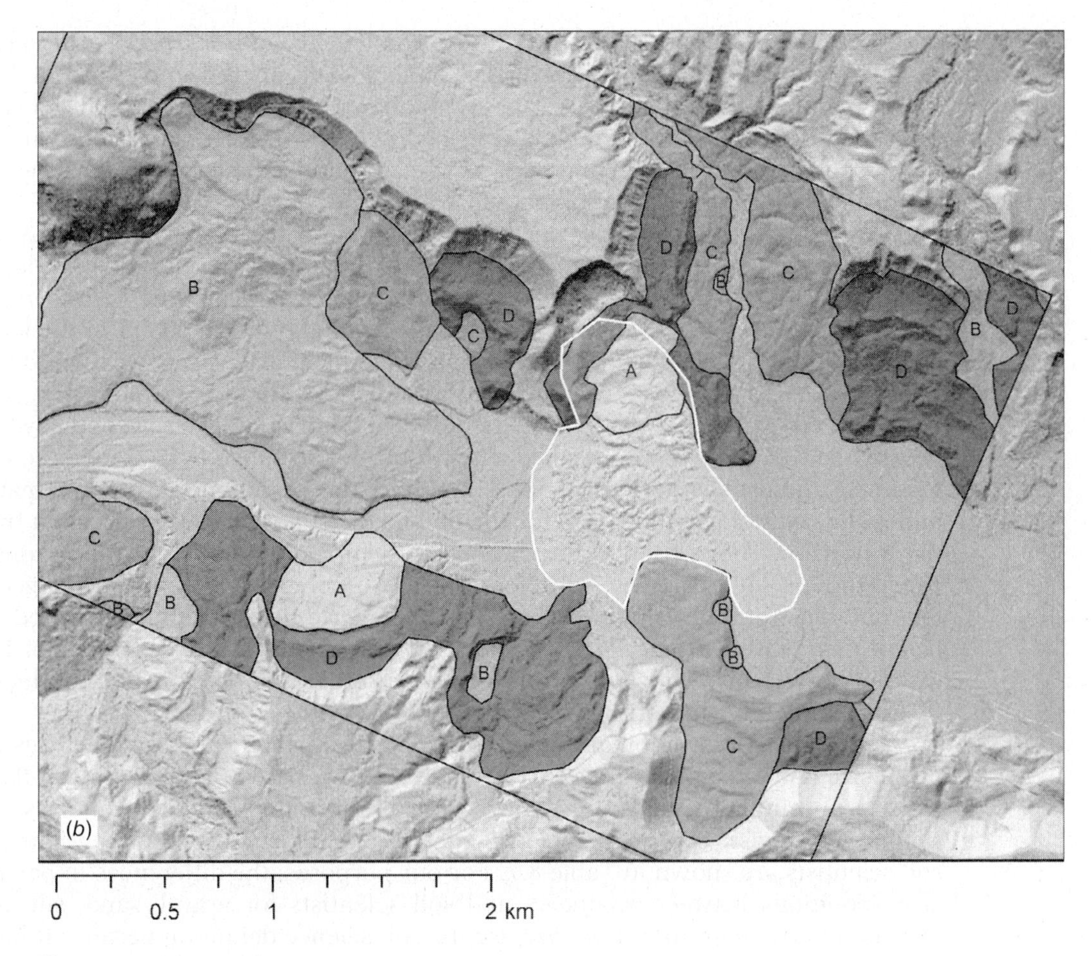

Figure 8.33 *(Continued)*

bedrock. In addition, the slope of the land surface can be estimated by stereo-image viewing and measured by photogrammetric methods.

Space limits the image interpretation process described to the assessment of terrain characteristics that are visible on medium-scale stereoscopic aerial photographs. Similar principles apply to nonphotographic and spaceborne sources.

Soil Characteristics

The term *soil* has specific scientific connotations to different groups involved with soil surveying and mapping. Most engineers consider all unconsolidated earth material lying above bedrock to be "soil." Agricultural soil scientists regard soil as a material that develops from a geologic parent material through the natural process of weathering and contains a certain amount of organic material and other constituents that support plant life. For example, a 10-m-thick deposit of glacial till over bedrock might be extensively weathered and altered to a depth of 1 m. The remaining 9 m would be relatively unaltered. An engineer would consider this a soil deposit 10 m thick lying over bedrock. A soil scientist would consider this a soil layer 1 m thick lying over glacial till parent material. We use the soil science (pedological) concept of soil in this chapter.

Through the processes of weathering, including the effects of climate and plant and animal activity, unconsolidated earth materials develop distinct layers that soil scientists call *soil horizons*. The top layer is designated the *A horizon* and called the *surface soil*, or *topsoil*. It can range from about 0 to 60 cm in thickness and is typically 15 to 30 cm. The A horizon is the most extensively weathered horizon. It contains the most organic matter of any horizon and has had some of its fine-textured particles washed down into lower horizons. The second layer is designated the *B horizon* and called the *subsoil*. It can range from 0 to 250 cm in thickness and usually is 45 to 60 cm. The B horizon contains some organic matter and is the layer of accumulation for the fine-textured particles washed down from the A horizon. The portion of the soil profile occupied by the A and B horizons is called the *soil* (or *solum*) by soil scientists. The *C horizon* is the underlying geologic material from which the A and B horizons have developed and is called the *parent material* (or *initial material*). The concept of soil profile development into distinct horizons is vitally important for agricultural soils mapping and productivity estimation as well as for many developmental uses of the landscape.

Soils consist of combinations of solid particles, water, and air. Particles are given size names, such as gravel, sand, silt, and clay, based on particle size. Particle size terminology is not standardized for all disciplines and several classification systems exist. Typical particle size definitions for engineers and agricultural soil scientists are shown in Table 8.7. For our purposes, the differences in particle size definitions between engineers and soil scientists for gravel, sand, silt, and clay are relatively unimportant. We use the soil science definition because it has a convenient system for naming combinations of particle sizes.

We consider materials containing more than 50% silt and clay to be *fine textured* and materials containing more than 50% sand and gravel to be *coarse textured*.

Soils have characteristic drainage conditions that depend on surface runoff, soil permeability, and internal soil drainage. We use the USDA soil drainage classification

TABLE 8.7 Soil Particle Size Designations

	Soil Particle Size (mm)	
Soil Particle Size Name	Engineering Definition	Agricultural Soil Science Definition
Gravel	2.0–76.2	2.0–76.2
Sand	0.074–2.0	0.05–2.0
Silt	0.005–0.074	0.002–0.05
Clay	Below 0.005	Below 0.002

system (U.S. Department of Agriculture, Soil Survey Staff, 1997) for soils in their natural condition, with the seven soil drainage classes described as follows.

1. *Very poorly drained.* Natural removal of water from the soil is so slow that the water table remains at or near the surface most of the time. Soils of this drainage class usually occupy level or depressed sites and are frequently ponded.
2. *Poorly drained.* Natural removal of water from the soil is so slow that it remains wet for a large part of the time. The water table is commonly at or near the ground surface during a considerable part of the year.
3. *Somewhat poorly drained.* Natural removal of water from the soil is slow enough to keep it wet for significant periods, but not all the time.
4. *Moderately well drained.* Natural removal of water from the soil is somewhat slow so that the soil is wet for a small but significant part of the time.
5. *Well drained.* Natural removal of water from the soil is at a moderate rate without notable impedance.
6. *Somewhat excessively drained.* Natural removal of water from the soil is rapid. Many soils of this drainage class are sandy and very porous.
7. *Excessively drained.* Natural removal of water from the soil is very rapid. Excessively drained soils may be on steep slopes, very porous, or both.

Land Use Suitability Evaluation

Terrain information can be used to evaluate the suitability of land areas for a variety of land uses. As an example, we consider the land suitability for developmental purposes, principally urban and suburban land uses, but many other types of "suitability" could be studied.

The topographic characteristics of an area are one of the most important determinants of the suitability of an area for residential development. Slopes in the 2 to 6% range are steep enough to provide for good surface drainage and interesting siting and yet flat enough so that few significant site development problems will be encountered provided the soil is well drained. Some drainage problems may be encountered in the 0 to 2% range, but these can be overcome unless there is a large expanse of absolutely flat land with insufficient internal drainage. Slopes over 12% present problems in street development and lot design and also pose serious problems when septic tanks are used for domestic sewage disposal. For industrial park and commercial sites, slopes of not more than 5% are preferred.

The soil texture and drainage conditions also affect land use suitability. Well-drained, coarse-textured soils present few limitations to development. Poorly drained or very poorly drained, fine-textured soils can present severe limitations. Shallow groundwater tables and poor soil drainage conditions cause problems in septic tank installation and operation, in basement and foundation excavation, and in keeping basements water free after construction. In general, depths to the water table of at least 2 m are preferred. Depths of 1 to 2 m may be satisfactory where public sewage disposal is provided and buildings are constructed without basements.

Shallow depths to bedrock cause problems in septic tank installation and maintenance, in utility line construction, in basement and foundation excavation, and in street location and construction, especially when present in combination with steep slopes. Depths to bedrock over 2 m are preferred. Excavation costs are increased where basements and public sewage disposal facilities are to be constructed. A depth to bedrock of less than 1 m presents serious limitations to development and is an unsatisfactory condition in almost all cases of land development.

Slope stability problems occur with certain soil-slope conditions. Although we will not discuss techniques for slope stability analysis using image interpretation, it should be mentioned that numerous areas of incipient landslide failure have been detected by image interpretation.

Despite the emphasis here on land development, it must be recognized that many land areas are worthy of preservation in their natural state because of outstanding topographic or geologic characteristics or because rare or endangered plant or animal species occupy those areas. The potential alteration of the hydrology of an area must also be kept in mind. In addition, the maintenance of prime agricultural land for agricultural rather than developmental use must be an important consideration in all land use planning decisions. Similar concerns also apply to the preservation of forested areas and wetland systems.

Elements of Image Interpretation for Landform Identification and Evaluation

Image interpretation for landform identification and evaluation is based on a systematic observation and evaluation of key elements that are studied stereoscopically. These are topography, drainage pattern and texture, erosion, image tone, and vegetation and land use.

Topography

Each landform and bedrock type described here has its own characteristic topographic form, including a typical size and shape. In fact, there is often a distinct topographic change at the boundary between two different landforms.

With vertical photographs having a normal 60% overlap, most individuals see the terrain exaggerated in height about four times. Consequently, slopes appear steeper than they actually are. The specific amount of vertical exaggeration observed in any given stereopair is a function of the geometric conditions under which the photographs are viewed and taken.

Drainage Pattern and Texture

The drainage pattern and texture seen on aerial and space images are indicators of landform and bedrock type and also suggest soil characteristics and site drainage conditions.

Six of the most common drainage patterns are illustrated in Figure 8.34. The *dendritic drainage pattern* is a well-integrated pattern formed by a main stream with its tributaries branching and rebranching freely in all directions; it occurs on relatively homogeneous materials such as horizontally bedded sedimentary

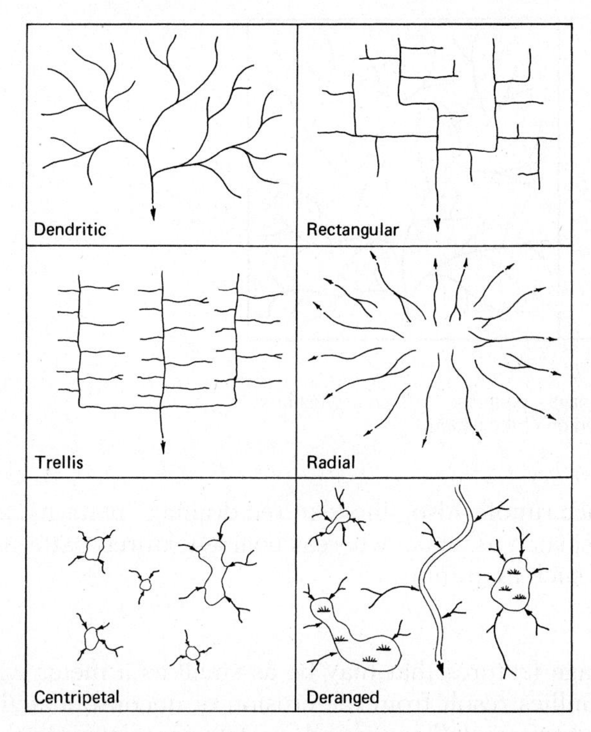

Figure 8.34 Six basic drainage patterns.

rock and granite. The *rectangular drainage pattern* is basically a dendritic pattern modified by structural bedrock control such that the tributaries meet at right angles; it is typical of flat-lying massive sandstone formations with a well-developed joint system. The *trellis drainage pattern* consists of streams having one dominant direction, with subsidiary directions of drainage at right angles; this occurs in areas of folded sedimentary rocks. The *radial drainage pattern* is formed by streams that radiate outward from a central area as is typical of volcanoes and domes. The *centripetal drainage pattern* is the reverse of the radial drainage pattern (drainage is directed toward a central point) and occurs in areas of limestone sinkholes, glacial kettle holes, volcanic craters, and other depressions. The *deranged drainage pattern* is a disordered pattern of aimlessly directed short streams, ponds, and wetland areas typical of ablation glacial till areas.

The previously described drainage patterns are all "erosional" drainage patterns resulting from the erosion of the land surface; they should not be confused with "depositional" drainage features that are remnants of the mode of origin of landforms such as alluvial fans and glacial outwash plains.

Coupled with drainage pattern is drainage texture. Figure 8.35 shows *coarse-textured* and *fine-textured* drainage patterns. Coarse-textured patterns develop where the soils and rocks have good internal drainage with little surface runoff. Fine-textured patterns develop where the soils and rocks have poor internal

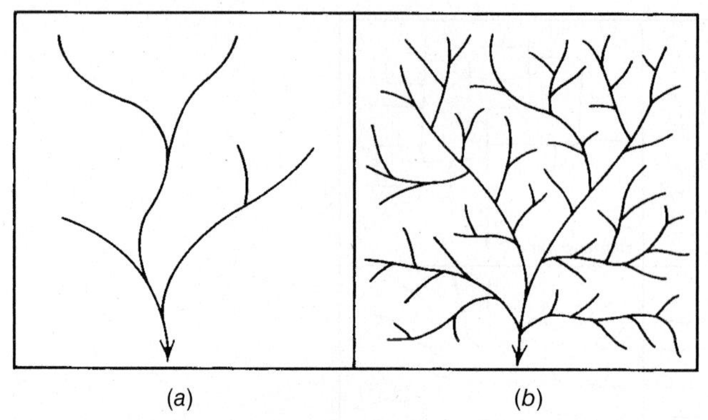

Figure 8.35 Illustrative drainage patterns: (*a*) coarse-textured dendritic pattern; (*b*) fine-textured dendritic pattern.

drainage and high surface runoff. Also, fine-textured drainage patterns develop on soft, easily eroded rocks, such as shale, whereas coarse-textured patterns develop on hard, massive rocks, such as granite.

Erosion

Gullies are small drainage features that may be as small as a meter wide and a hundred meters long. Gullies result from the erosion of unconsolidated material by runoff and develop where rainfall cannot adequately percolate into the ground but instead collects and flows across the surface in small rivulets. These initial rivulets enlarge and take on a particular shape characteristic of the material in which they are formed. As illustrated in Figures 8.36 and 8.37, short gullies with V-shaped cross sections tend to develop in sand and gravel; gullies with U-shaped cross sections tend to develop in silty soils; and long gullies with gently rounded cross sections tend to develop in silty clay and clay soils.

Image Tone

The term *image tone* refers to the "brightness" at any point on an aerial or space image. The absolute value of the image tone depends not only on terrain

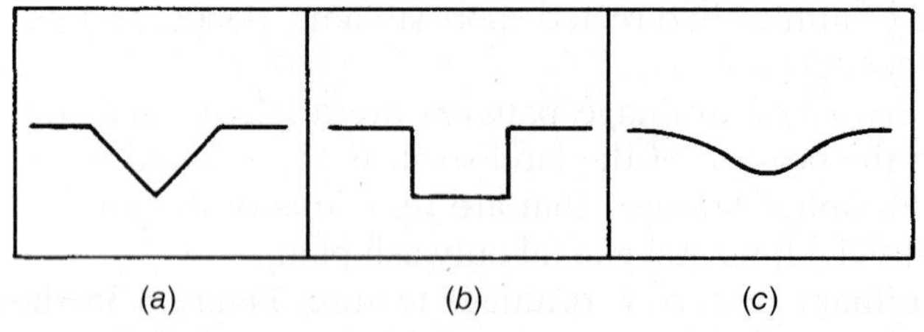

Figure 8.36 Illustrative gully cross sections: (*a*) sand and gravel, (*b*) silt, (*c*) silty clay or clay.

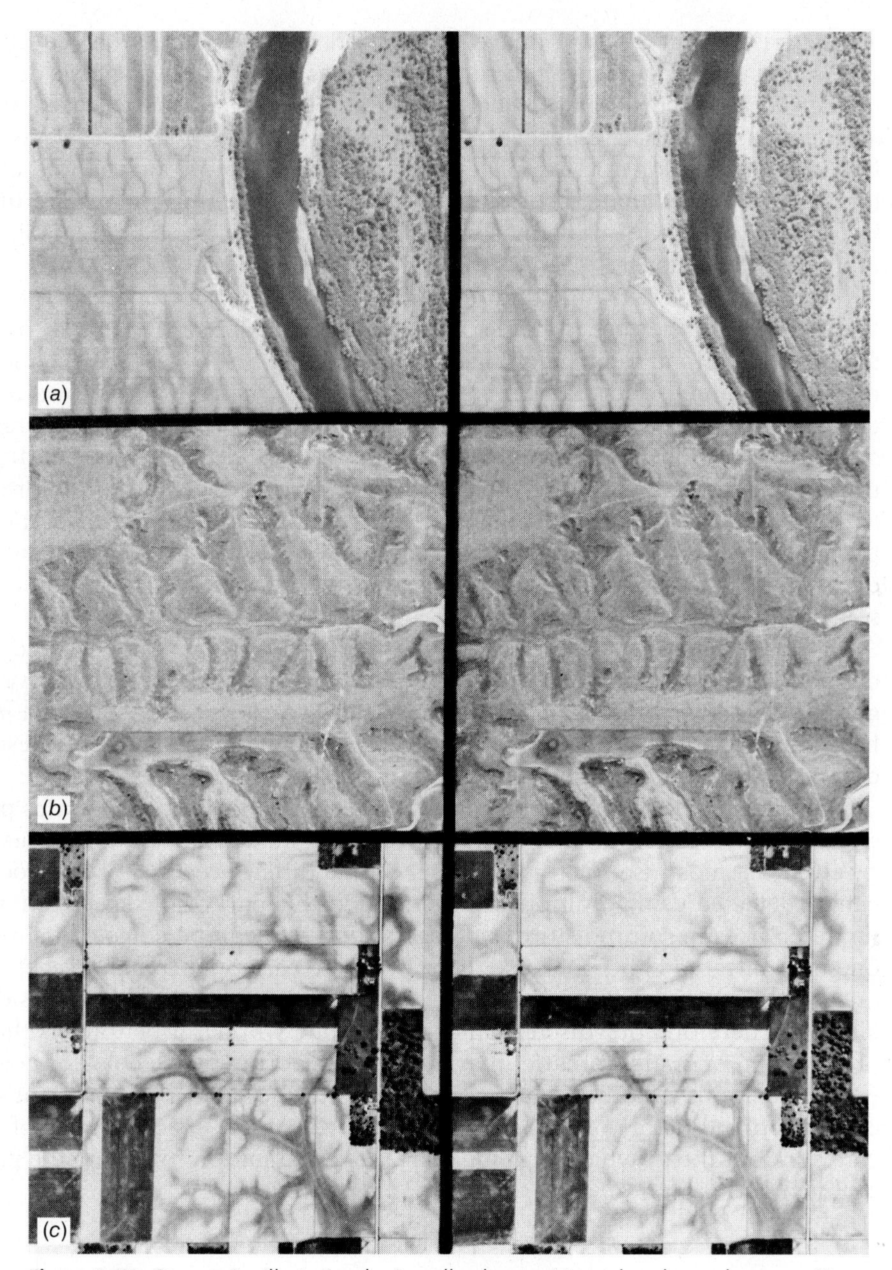

Figure 8.37 Stereopairs illustrating basic gully shapes: (*a*) sand and gravel terrace, Dunn County, WI; (*b*) loess (wind-deposited silt), Buffalo County, NE; (*c*) silty clay loam glacial till, Madison County, IN. Scale 1:20,000. (Courtesy USDA–ASCS panchromatic photos.)

characteristics but also on image acquisition factors such as choice of spectral bands and filters, exposure, and image processing. Image tone also depends on meteorological and climatological factors such as atmospheric haze, sun angle, and cloud shadows. Because of the effect of these non-terrain-related factors, image interpretation for terrain evaluation must rely on an analysis of *relative* tone values, rather than absolute tone values. Relative tone values are important because they often form distinct image patterns that may be of great significance in image interpretation.

The effect of terrain conditions on relative image tone can be seen in Figure 8.37*c*. In the case of bare soils (nonvegetated soils), the lighter toned areas tend to have a topographically higher position, a coarser soil texture, a lower soil moisture content, and a lower organic content. Figure 8.37*c* shows a striking tonal pattern often seen on fine-textured glacial till soils. The tonal differences are caused by differences in sunlight reflection due principally to the varying moisture content of the soil. The lighter toned areas are somewhat poorly drained silt loam soils on rises to 1 m above the surrounding darker toned areas of very poorly drained silty clay loam soils. The degree of contrast between lighter and darker toned bare soils varies depending on the overall moisture conditions of the soil, as illustrated in Plate 36.

The sharpness of the boundary between lighter and darker toned areas is often related to the soil texture. Coarser textured soils will generally have sharper gradations between light and dark tones while finer textured soils will generally have more gradual gradations. These variations in tonal gradients result from differences in capillary action occurring in soils of different textures.

Our discussion of image interpretation for terrain evaluation relates primarily to the use of panchromatic imagery because this image type has historically received the most use for this purpose. Subtle differences in soil and rock colors can be detected using multiple bands in the visible part of the spectrum, and subtle differences in soil moisture and vegetation vigor can be detected using at least one near-infrared band. Because there is a wide variety of soil and vegetation colors possible on color and color infrared images, it is not possible to consider them all here. Therefore, our discussion of image tone will describe tone as the shades of gray seen on panchromatic images. Persons working with color or color infrared photographs (or other sensors such as multispectral or hyperspectral scanners or side-looking radar) of specific geographic regions at specific times of the year can work out their own criteria for image tone evaluation following the principles outlined in this section.

Vegetation and Land Use

Differences in natural or cultivated vegetation often indicate differences in terrain conditions. For example, orchards and vineyards are generally located on well-drained soils, whereas truck farming activities often take place on highly organic soils such as muck and peat deposits. In many cases, however, vegetation and land use obscure differences in terrain conditions and the interpreter must be careful to draw inferences only from meaningful differences in vegetation and land use.

The Image Interpretation Process

Through an analysis of the elements of image interpretation (topography, drainage pattern and texture, erosion, image tone, vegetation, and land use), the image interpreter can identify different terrain conditions and can determine the boundaries between them. Initially, image interpreters will need to consider carefully each of the above elements individually and in combination in order to estimate terrain conditions. After some experience, these elements are often applied subconsciously as the interpreter develops the facility to recognize certain recurring image patterns almost instantaneously. In complex areas, the interpreter should not make snap decisions about terrain conditions but should carefully consider the topography, drainage pattern and texture, erosion, image tone, vegetation, and land use characteristics exhibited on the aerial and space images.

In the remainder of this section, we examine several of the principal bedrock types common on the earth's surface. For each of these, we consider geologic origin and formation, soil and/or bedrock characteristics, implications for land use planning, and image identification using the elements of image interpretation for terrain evaluation. Our illustrations are limited to occurrences in the United States. We emphasize the recognition of clear-cut examples of various bedrock types. In nature, there are many variations to each type. Interpreters working in specific localities can use the principles set forth here to develop their own image interpretation keys.

In cases where distinctions in image appearance must be made for different climatic situations, we will speak of "humid" and "arid" climates. We will consider *humid climates* to occur in areas that receive 50 cm or more rainfall per year and *arid climates* to occur in areas that receive less than 50 cm/year rainfall. In the United States, farming without irrigation is generally feasible in areas with a rainfall of about 50 cm/year or more. Areas receiving less than 50 cm/year rainfall typically require irrigation for farming.

Even the most searching and capable image analysis can benefit from field verification because the image interpretation process is seldom expected to stand alone. The image interpreter should consult existing topographic, geologic, and soil maps and should conduct a selective field check. The principal benefits of image interpretation for terrain evaluation should be a savings in time, money, and effort. The use of image interpretation techniques can allow for terrain mapping during periods of unsuitable weather for field mapping and can provide for more efficient field operations.

In order to illustrate the process of image interpretation for landform identification and evaluation, we will consider the terrain characteristics and image identification of several common bedrock types. Specifically, we treat the analysis of selected sedimentary and igneous rocks. The first three editions of this book treated the subject of landform identification and evaluation in greater detail by including discussions of aeolian landforms, glacial landforms, fluvial landforms, and organic soils (the first and second editions contain the most detailed coverage).

Sedimentary Rocks

The principal sedimentary rock types to be considered are sandstone, shale, and limestone. Sedimentary rocks are by far the most common rock type exposed at the earth's surface and extend over approximately 75% of the earth's land surface (igneous rocks extend over approximately 20% and metamorphic rocks over about 5%).

Sedimentary rocks are formed by the consolidation of layers of sediments that have settled out of water or air. Sediments are converted into coherent rock masses by lithification, a process that involves cementation and compaction by the weight of overlying sediments.

Clastic sedimentary rocks are rocks containing discrete particles derived from the erosion, transportation, and deposition of preexisting rocks and soils. The nature of the constituent particles and the way in which they are bound together determine the texture, permeability, and strength of the rocks. Clastic sedimentary rocks containing primarily sand-sized particles are called *sandstone*, those containing primarily silt-sized particles are called *siltstone*, and those containing primarily clay-sized particles are called *shale*.

Limestone has a high calcium carbonate content and is formed from chemical or biochemical action. The distinction between the two methods of formation is as follows. Chemically formed limestone results from the precipitation of calcium carbonate from water. Biochemically formed limestone results from chemical processes acting on shells, shell fragments, and plant materials.

The principal sedimentary rock characteristics that affect the appearance of the terrain on aerial and space images are *bedding*, *jointing*, and *resistance to erosion*.

Sedimentary rocks are typically stratified or layered as the result of variations in the depositional process. The individual strata or layers are called *beds*. The top and bottom of each bed have more or less distinct surfaces, called bedding planes, that delineate the termination of one bed and the beginning of another with somewhat different characteristics. Individual beds may range in thickness from a few millimeters to many meters. Beds in their initial condition usually are nearly horizontal but may be tilted to any angle by subsequent movements of the earth's crust.

Joints are cracks through solid bodies of rock with little or no movement parallel to joint surfaces. Joints in sedimentary rocks are primarily perpendicular to bedding planes and form plane surfaces that may intersect other joint planes. Several systematic joints constitute a joint set, and when two or more sets are recognized in an area, the overall pattern is called a joint system. Because joints are planes of weakness in rocks, they often form surfaces clearly visible on aerial and space images, especially in the case of sandstone. Streams often follow joint lines and may zigzag from one joint line to another.

The *resistance to erosion* of sedimentary rocks depends on rock strength, permeability, and solubility. Rock strength depends principally on the strength of the bonding agent holding the individual sediment particles together and on the thickness of the beds. Thick beds of sandstone cemented by quartz are very strong and may be used as building materials. Thin beds of shale are often so weak that they can be shattered by hand into flakes and plates. Rock permeability refers to the ability of the rock mass to transmit water and depends on the size of the pore spaces between sediment particles and on the continuity of their connections. Sandstone is generally a very permeable rock. Shale is usually quite impermeable and water moves principally along joint planes rather than in sediment void spaces. Limestones high in calcium carbonate are soluble in water and may dissolve under the action of rainfall and ground water movement.

Here, we describe the characteristics of sandstone, shale, and limestone with a horizontally bedded attitude. The first and second editions of this book include a discussion of interbedded sedimentary rocks (both horizontally bedded and tilted).

Sandstone

Sandstone deposits commonly occur in beds a few meters thick interbedded with shale and/or limestone. Here we are concerned primarily with sandstone formations about 10 m or more in thickness.

Sandstone *bedding* is often prominent on images, especially when the sandstone beds occur over softer, more easily eroded formations such as shale. *Jointing* is prominent, with a joint system consisting of two or three dominant directions. The *resistance to erosion* varies, depending on the strength of the cementing agent. Sandstone cemented with iron compounds and silica is typically very strong, whereas sandstone cemented with carbonates is generally quite weak. Because sandstone is very permeable, most rainfall percolates downward through the rock rather than becoming erosion-producing surface runoff. Sandstone cemented with carbonates may weaken as percolating water dissolves the cementing agent.

In arid areas, there is seldom a residual soil cover over sandstone because any weathered sand particles are removed by wind erosion. In humid areas, the depth of residual soil cover depends on the strength of the cementing agent but is commonly less than 1 m and seldom more than 2 m.

Areas of massive sandstone beds with a residual soil cover are commonly undeveloped because of a combination of their typically rugged topography and shallow depths to bedrock. Buried sandstone strata are often an excellent source of groundwater for both individual homeowners and municipalities. Well-cemented sandstone rock is often used as building stone for residential construction.

Image Identification of Horizontally Bedded Sandstone

Topography: Bold, massive, relatively flat-topped hills with nearly vertical or very steep hillsides. *Drainage:* Coarse-textured, joint-controlled, modified dendritic pattern; often a rectangular pattern caused by perpendicular directions of joint sets. *Erosion:* Few gullies; V-shaped if present in residual soil. *Photographic image tone:* Generally light toned due to light rock color and excellent internal drainage of both residual soil and sandstone rock. Reddish sandstone in arid areas may have a somewhat dark tone on panchromatic imagery. A dense tree cover over sandstone in humid areas generally appears dark, but in this case the interpreter is looking at the tree canopy rather than the soil or rock surface. *Vegetation and land use:* Sparse vegetation in arid areas. Often forested in humid areas because residual soil is too well drained to support crops. In a humid climate, flat-topped sandstone ridges with a loess cover are often farmed. *Other:* Sandstone is sometimes mistakenly identified as granite.

Figure 8.38 shows horizontally bedded sandstone in an arid climate interbedded with a few thin shale beds. The bedding can best be seen by inspecting the valley walls where the deeply incised stream cuts across the terrain. The direction of the major joint set is nearly vertical on the page; a secondary direction is perpendicular to the major joint set. These joint sets only partially control the direction of flow of the major stream but strongly influence the direction of secondary drainage.

Shale

Deposits of shale are common throughout the world as both thick deposits and thin deposits interbedded with sandstone and limestone. Shale *bedding* is very extensive with beds typically 1 to 20 cm in thickness. Bedding is not always visible on aerial and space images. However, if beds with a distinct difference in color or resistance to erosion are present or if shale is interbedded with sandstone or limestone, bedding may be seen. The effect of *jointing* is not always strong enough to alter the surface drainage system into a significantly joint-controlled pattern. The *resistance to erosion* is low, compared with other sedimentary rock types. Because shale is relatively impermeable, most rainfall runs off the ground surface, causing extensive erosion.

The depth of residual soil cover is generally less than 1 m and seldom more than 2 m. The residual soil is high in silt and clay, with textures typically silty loam, silty clay loam, silty clay, and clay. Internal soil drainage is typically moderately well drained or poorer, depending on soil texture and on soil and rock structure.

Image Identification of Horizontally Bedded Shale

Topography: In an arid climate, minutely dissected terrain with steep stream/gully side slopes resulting from rapid surface runoff associated with short duration heavy rainfall. In a humid climate, gently to moderately sloping, softly rounded hills. *Drainage:* A dendritic pattern with gently curving streams; fine textured in arid climates and medium to fine textured in humid climates. *Erosion:* Gullies in residual soil have gently rounded cross sections. *Photographic image tone:* Varies widely, generally dark toned compared with sandstone and limestone. Differences

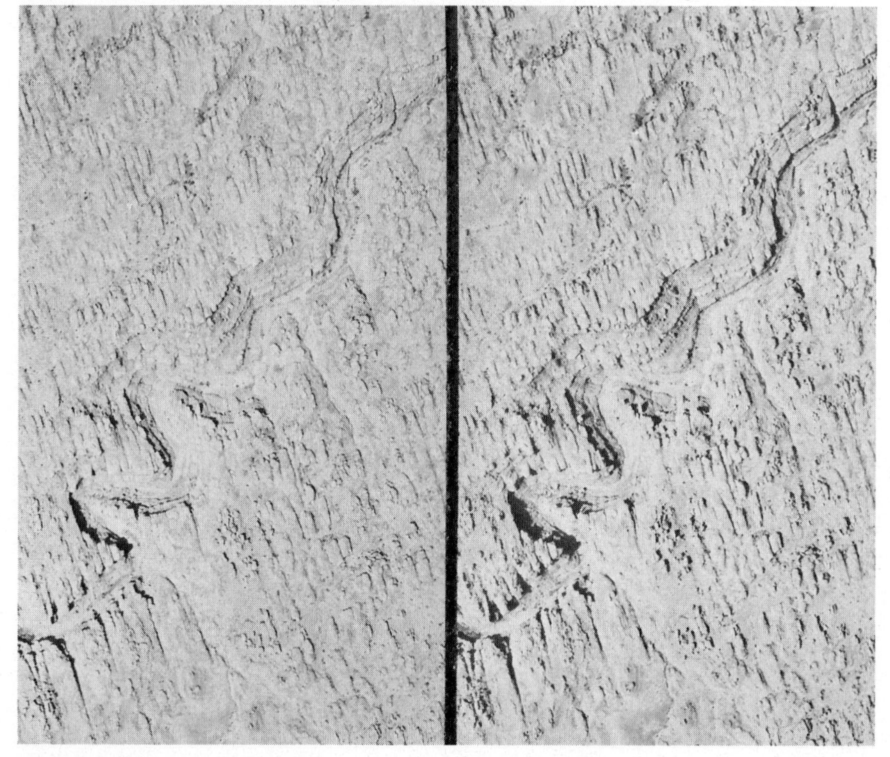

Figure 8.38 Horizontally bedded sandstone in an arid climate, southern Utah. Scale 1:20,000. (Courtesy USGS panchromatic photos.)

in image tone may outline bedding. *Vegetation and land use:* Arid areas usually barren, except for desert vegetation. Humid areas intensively cultivated or heavily forested. *Other:* Shale is sometimes mistakenly identified as loess.

Figure 8.39 shows horizontally bedded shale in an arid climate. A comparison with Figure 8.38 illustrates the contrast in bedding, jointing, and resistance to erosion between shale and sandstone.

Limestone

Limestone consists mainly of calcium carbonate, which is soluble in water. Limestone that contains a significant amount of calcium carbonate and magnesium carbonate (or calcium magnesium carbonate) is called dolomitic limestone, or dolomite, and is less soluble in water. Limestone occurs throughout the world. For example, an area of very soluble limestone occurs in the United States in a region spanning portions of Indiana, Kentucky, and Tennessee.

Limestone *bedding* is generally not prominent on images unless the limestone is interbedded with sandstone or shale. *Jointing* is strong and determines the location of many of the pathways for subsurface drainage. However, jointing is generally not prominent on images of limestone in a humid climate. The *resistance to*

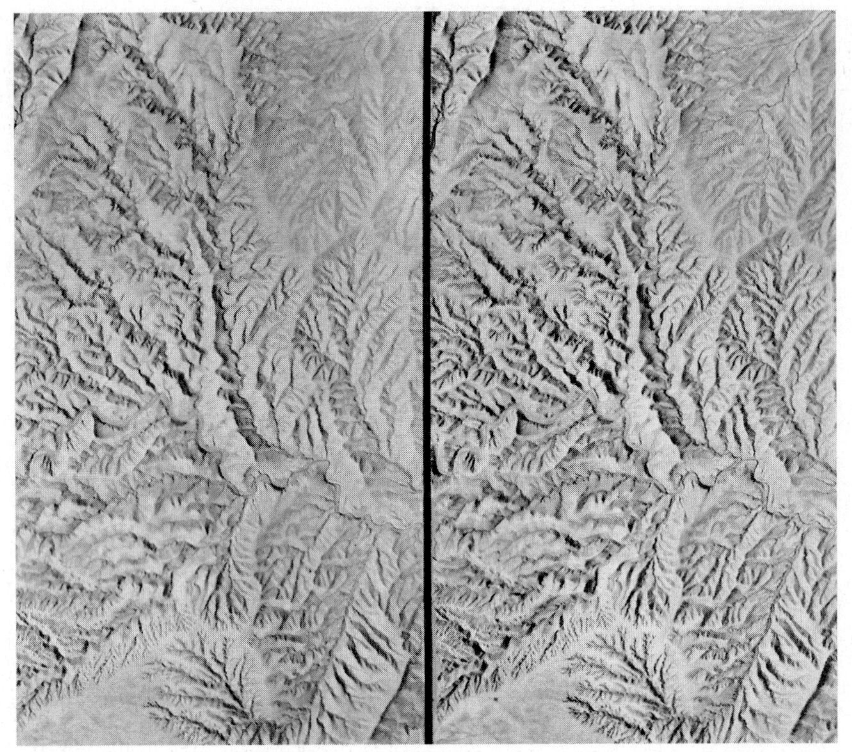

Figure 8.39 Horizontally bedded shale in an arid climate, Utah. Scale 1:26,700. (Courtesy USGS panchromatic photos.)

erosion varies, depending on the solubility and jointing of the rock. Because calcium carbonate is soluble in water, many limestone areas have been severely eroded by rainfall and groundwater action.

The ground surface in areas of soluble limestone in humid climates is typically dotted with literally thousands of roughly circular depressions called *sinkholes*. They form when surface runoff drains vertically through the rock along joint planes and the intersections of joint planes, gradually enlarging the underground drainage ways by solution and causing the ground surface to collapse and form sinkholes.

There is generally only a shallow residual soil cover over limestone in arid areas where limestone often caps ridges and plateaus. In humid areas, the depth of residual soil cover is extremely variable and depends on the amount of solution weathering. Generally, residual soil depth ranges from 2 to 4 m for soluble limestone (which typically occurs as valleys or plains) and is somewhat less for dolomite (which may cap ridges and plateaus).

Image Identification of Horizontally Bedded Limestone

This discussion refers to soluble limestone in humid climates. *Topography:* A gently rolling surface broken by numerous roughly circular sinkholes that are typically 3 to 15 m in depth and 5 to 50 m in diameter. *Drainage:* Centripetal drainage into individual sinkholes. Very few surface streams. Surface streams from adjacent landforms or rock types may disappear underground via sinkholes when streams reach the limestone. *Erosion:* Gullies with gently rounded cross sections develop in the fine-textured residual soil. *Photographic image tone:* Mottled tone due to extensive sinkhole development. *Vegetation and land use:* Typically farmed, except for sinkhole bottoms that are often wet or contain standing water a portion of the year. *Other:* Limestone with extensive sinkhole development might be mistakenly identified as ablation till. Dolomitic limestone is more difficult to identify than soluble limestone. It is generally well drained and has subtle sinkholes.

Figure 8.40 shows horizontally bedded soluble limestone in a humid climate. Note the extensive sinkhole development (up to 40 sinkholes per square kilometer

Figure 8.40 Horizontally bedded soluble limestone in a humid climate, Harrison County, IN. Scale 1:20,000. (Courtesy USDA-ASCS panchromatic photos.)

are present) and the complete lack of surface streams. The residual soils here are well-drained silty clay loam and silty clay 1.5 to 3 m deep over limestone bedrock.

Igneous Rocks

Igneous rocks are formed by the cooling and consequent solidification of magma, a molten mass of rock material. Igneous rocks are divided into two groups: intrusive and extrusive. *Intrusive igneous rocks* are formed when magma does not reach the earth's surface but solidifies in cavities or cracks it has made by pushing the surrounding rock apart or by melting or dissolving it. *Extrusive igneous rocks* are formed when magma reaches the ground surface.

Intrusive igneous rocks commonly occur in large masses in which the molten magma has cooled very slowly and solidified into large crystals. The crystal grains interlock closely to produce a dense, strong rock that is free of cavities. Erosion of overlying materials exposes intrusive igneous rocks.

Extrusive igneous rocks occur as various volcanic forms, including various types of lava flows, cones, and ash deposits. These rocks have cooled more rapidly than intrusive rocks and consequently have smaller crystals.

Intrusive Igneous Rocks

Intrusive igneous rocks range from granite, a light-colored, coarse-grained rock consisting principally of quartz and feldspar, to gabbro, a dark-colored, coarse-grained rock consisting principally of ferromagnesian minerals and feldspar. There are many intrusive igneous rocks intermediate between granite and gabbro in composition, such as granodiorite and diorite. We consider only the broad class of intrusive igneous rocks called *granitic rocks*, a term used to describe any coarse-grained, light-colored, intrusive igneous rock.

Granitic rocks occur as massive, *unbedded* formations such as the Sierra Nevada Mountains and the Black Hills of South Dakota. They are often strongly fractured into a series of irregularly oriented *joints* as a result of cooling from a molten state and/or pressure relief as overburden is eroded. Granitic rocks have a high *resistance to erosion*. As they weather, they tend to break or peel in concentric sheets through a process called exfoliation.

Image Identification of Granite Rocks

Topography: Massive, rounded, unbedded, domelike hills with variable summit elevations and steep side slopes. Often strongly jointed with an irregular and sometimes gently curving pattern. Joints may form topographic depressions in which soil and vegetation accumulate and along which water tends to flow. *Drainage and erosion:* Coarse-textured dendritic pattern with a tendency for streams to curve around the bases of domelike hills. Secondary drainage channels form along joints. Few gullies, except in areas of deeper residual soil. *Photographic image tone:* Light toned due to light rock color. Darker toned in depressions that form along joints. *Vegetation and land use:* Sparse vegetation in an arid climate.

Often forested with some bare rock outcrops in a humid climate. Vegetation may be concentrated in depressions that form along some joints. *Other:* Granitic rocks are sometimes mistakenly identified as horizontally bedded sandstone. The principal difference in image identification of granitic rocks versus sandstone can be summarized as follows. (1) *Evidences of bedding:* Granitic rocks are unbedded; sandstone is bedded. (2) *Topography:* Granitic outcrops have variable summit elevations, sandstone caprocks form plateaus; granitic rocks have rounded cliffs, sandstone has vertical cliffs; granitic microfeatures are rounded, sandstone microfeatures are blocky. (3) *Joint pattern:* Granitic rocks have an irregular joint pattern with some distinct linear depressions; sandstone has a joint system consisting of two or three principal directions.

Figure 8.41 shows granitic rocks in an arid climate with very little soil or vegetative cover. Note the massive, unbedded formation with rounded cliffs. Note also that a number of joints are enlarged and form depressions with some soil and vegetative cover.

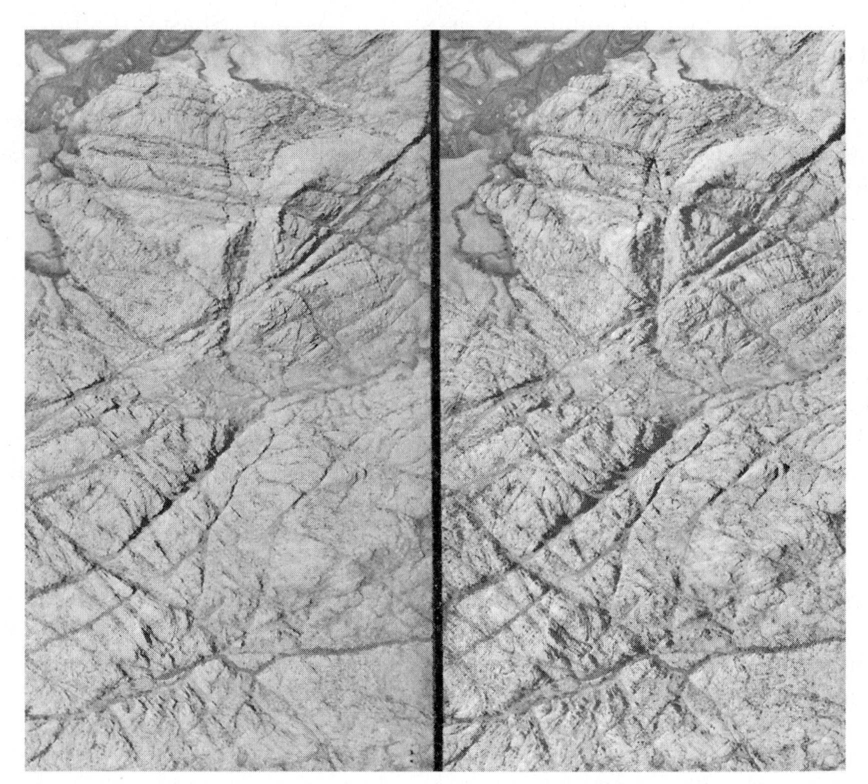

Figure 8.41 Granitic rock in an arid climate, Wyoming. Scale 1:37,300. (Courtesy USGS panchromatic photos.)

Extrusive Igneous Rocks

Extrusive igneous rocks consist principally of lava flows and pyroclastic materials. Lava flows are the rock bodies formed from the solidification of molten rock that issued from volcanic cones or fissures with little or no explosive activity. In contrast, pyroclastic materials, such as cinders and ash, were ejected from volcanic vents.

The form of lava flows depends principally on the viscosity of the flowing lava. The viscosity of lava increases with the proportion of silica (SiO_2) and alumina (Al_2O_3) in the lava. The least viscous (most fluid) lavas are the basaltic lavas, which contain about 65% silica and alumina. Andesitic lavas are intermediate in viscosity and contain about 75% silica and alumina. Rhyolitic lavas are very viscous and contain about 85% silica and alumina. Several basic volcanic forms are recognized.

Strato volcanoes (also called composite volcanoes) are steep-sided, cone-shaped volcanoes composed of alternating layers of lava and pyroclastic materials. The lava is typically andesitic or rhyolitic and side slopes can be 30° or more. Many strato volcanoes are graceful cones of striking beauty and grandeur. Each of the following mountains is a strato volcano: Shasta (California), Hood (Oregon), Ranier (Washington), St. Helens (Washington), Fuji (Japan), Vesuvius (Italy), and Kilimanjaro (Tanzania).

Shield volcanoes (also called Hawaiian-type volcanoes) are broad, gently sloping volcanic cones of flat domical shape built chiefly of overlapping basaltic lava flows. Side slopes generally range from about 4° to 10°. The Hawaiian volcanoes Haleakala, Mauna Kea, Mauna Loa, and Kilauea are shield volcanoes.

Flood basalt (also called plateau basalt) consists of large-scale eruptions of very fluid basalt that build broad, nearly level plains, some of which are at high elevation. Extensive flood basalt flows form the Columbia River and Snake River plains of the northwest United States.

Image Identification of Lava Flows

Topography: A series of tonguelike flows that may overlap and interbed, often with associated cinder and spatter cones. Viscous lavas (andesite and rhyolite) form thick flows with prominent, steep edges. Fluid lavas (basalt) form thin flows, seldom exceeding 15 m in thickness. *Drainage and erosion:* Lava is well drained internally and there is seldom a well-developed drainage pattern. *Photographic image tone and vegetation:* The color of unweathered, unvegetated lava is dark toned in the case of basalt, medium toned for andesite, and light toned for rhyolite. In general, recent unvegetated flows are darker toned than weathered, vegetated flows. *Land use:* Recent flows are seldom farmed or developed.

Here, we illustrate only one example of a lava flow issuing from a volcano. Figure 8.42 shows a viscous lava flow that emanated from Mt. Shasta, CA, a strato volcano. This flow is 60 m thick and has a 30° slope on its front face.

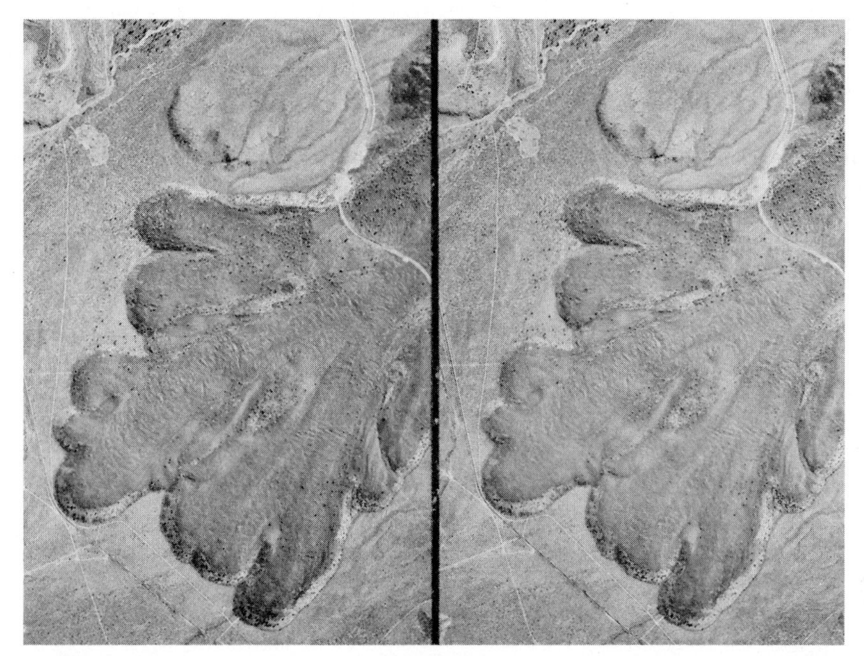

Figure 8.42 Viscous lava flow in an arid climate, Siskiyou County, CA. Scale 1:33,000. (Courtesy USDA-ASCS panchromatic photos.)

Metamorphic Rocks

Common metamorphic rocks are quartzite, slate, marble, gneiss, and schist. They are formed from preexisting sedimentary or igneous rocks due principally to the action of heat and pressure. Occasionally, chemical action or shearing stresses are also involved.

Most metamorphic rocks have a distinct banding that can be seen via field observations and that sets them apart from sedimentary and igneous rocks.

Metamorphic rocks can be found throughout the world. However, because their extent is limited, the identification of metamorphic rocks is not covered here. In addition, the image identification of metamorphic rocks is more difficult than for sedimentary and igneous rocks, and interpretive techniques for metamorphic rocks are not well established.

8.16 CONCLUSION

In this chapter we have summarized only the major current applications of remote sensing. Given the push of technological change and the pull of the increased demand for geospatial data across the local to global continuum, the scope of remote sensing applications is rapidly increasing and virtually open-ended. Our goal as authors in writing this book has been to provide you, the reader, with an understanding of the basic principles underpinning this important and dynamic field. We hope you will welcome the personal challenge and opportunity to continue to broaden the applications of this technology in science, government, and the private sector alike.

WORKS CITED

Aber, J.S., et al., *Small-Format Aerial Photography: Principles, Techniques, and Geoscience Applications*, Oxford, UK: Elsevier, 2010.

Agfa-Gevaert Group, *Specialty Products, Aerial Photography-Online Publications*, available at http://www.agfa.com, 2014.

American Society of Photogrammetry (ASP), *Manual of Remote Sensing*, 2nd ed., Falls Church, VA: ASP, 1983.

American Society for Photogrammetry and Remote Sensing (ASPRS), *Manual of Photographic Interpretation*, 2nd ed., Bethesda, MD: ASPRS, 1997.

American Society for Photogrammetry and Remote Sensing (ASPRS), *Corona between the Sun and the Earth: The First NRO Reconnaissance Eye in Space*, Bethesda, MD: ASPRS, 1997.

American Society for Photogrammetry and Remote Sensing (ASPRS), *Principles and Applications of Imaging Radar, Manual of Remote Sensing*, 3rd ed., vol. 2, New York: Wiley, 1998.

American Society for Photogrammetry and Remote Sensing (ASPRS), *Manual of Photogrammetry*, 5th ed., Bethesda, MD: ASPRS, 2004.

American Society for Photogrammetry and Remote Sensing (ASPRS), *Remote Sensing of the Marine Environment, Manual of Remote Sensing*, 3rd ed., vol. 6, Bethesda, MD: ASPRS, 2006.

American Society for Photogrammetry and Remote Sensing (ASPRS), *Manual of Airborne Topographic Lidar*, Bethesda, MD: ASPRS, 2012.

Andersen, H.-E., R.J. McGaughey, and S.E. Reutebuch, "Estimating Forest Canopy Fuel Parameters using LIDAR Data," *Remote Sensing of Environment*, vol. 94, no. 4, 2005, pp. 441–449.

Andersen, H.-E., S.E. Reutebuch, and R.J. McGaughey, "Active Remote Sensing," in G. Shao and K.M. Reynolds, Eds., *Computer Applications in Sustainable Forest Management*, Berlin: Springer-Verlag, 2006.

Andersen, H.-E., S.E. Reutebuch, and G.F. Schreuder, "Bayesian Object Recognition for the Analysis of Complex Forest Scenes in Airborne Laser Scanner Data," *ISPRS Commission III Symposium*, September 9–13, 2002, Graz, Austria, Part 3A, pp. 35–41.

Anderson, J.R., et al., "A Land Use and Land Cover Classification System for Use with Remote Sensor Data," *Geological Survey Professional Paper 964*, U.S. Government Printing Office, Washington, DC, 1976.

Avery, T.E., and G.L. Berlin, *Fundamentals of Remote Sensing and Airphoto Interpretation*, New York: Macmillan, 1992.

Badhwar, G.D., "Classification of Corn and Soybeans Using Multitemporal Thematic Mapper Data," *Remote Sensing of Environment*, vol. 16, 1985, pp. 175–181.

Baker, S., "San Francisco in Ruins: The 1906 Aerial Photographs of George R. Lawrence," *Landscape*, vol. 30, no. 2, 1989, pp. 9–14.

Ball, G.H., and D.J. Hall, *ISODATA: A Novel Method of Data Analysis and Pattern Recognition*, Technical report, Stanford Research Institute, Menlo Park, CA, 1965.

Bauer, M.E., "Spectral Inputs to Crop Identification and Condition Assessment," *Proceedings of the IEEE*, vol. 73, no. 6, 1985, pp. 1071–1085.

Bauer, M.E., and B. Wilson, "Satellite Tabulation of Impervious Surface Areas," *Lakeline*, Spring, 2005, pp. 17–20.

Bauer, M.E., et al., "Field Spectroscopy of Agricultural Crops," *IEEE Transactions on Geoscience and Remote Sensing*, vol. GE-24, no. 1, 1986, pp. 65–75.

Bauer, M.E., et al., "Satellite Inventory of Minnesota Forest Resources," *Photogrammetric Engineering and Remote Sensing*, vol. 60, no. 3, 1994, pp. 287–298.

Benson, B.J., and M.D. Mackenzie, "Effects of Sensor Spatial Resolution on Landscape Structure Parameters," *Landscape Ecology*, vol. 10, no. 2, 1995, pp. 113–120.

Blaschke, T., "Object Based Image Analysis for Remote Sensing," *ISPRS Journal of Photogrammetry and Remote Sensing*, vol. 65, no. 1, 2010, pp 2–16.

Boerner, W.-M., et al., "Polarimetry in Radar Remote Sensing: Basic and Applied Concepts," *Principles and Applications of Imaging Radar, Manual of Remote Sensing*, 3rd ed., vol. 2, New York: Wiley, 1998, pp. 271–357.

Bolger, D.T., et al., "A Computer-Assisted System for Photographic Mark–Recapture Analysis," *Methods in Ecology and Evolution*, vol. 3, no. 5, 2012, pp. 813–822.

Bowker, D.E., et al., *Spectral Reflectances of Natural Targets for Use in Remote Sensing Studies*, Washington, DC: National Aeronautics and Space Administration, 1985.

Campbell, J.B., *Introduction to Remote Sensing*, 3rd ed., New York: Guilford Press, 2002.

Campbell, J.B., and R. Wynne, *Introduction to Remote Sensing*, 5th ed., New York: Guilford Press, 2011.

Caselles, V., et al., "Thermal Band Selection for the PRISM Instrument 3," *Journal of Geophysical Research*, vol. 103, 1998.

Chase, A.F., et al., "Geospatial Revolution and Remote Sensing LiDAR in Mesoamerican Archaeology," *Proceedings of the National Academy of Sciences*, vol. 109, no. 32, 2012, pp. 12916–12921.

Chipman, J.W., and T.M. Lillesand, "Satellite-Based Assessment of the Dynamics of New Lakes in Southern Egypt," *International Journal of Remote Sensing*, vol. 28, no. 19, 2007, pp. 4365–4379.

Chipman, J.W., L.G. Olmanson, and A.A. Gitelson, *Remote Sensing Methods for Lake Management*, Madison, WI: North American Lake Management Society, 2009.

Chipman, J.W., T. Morrison, and D. Bolger, "Land Cover Variability across Spatial and Temporal Scales: Implications for Wild Ungulate Populations in Tanzania," Annual conference, American Society for Photogrammetry and Remote Sensing (ASPRS), 4 May 2011, Milwaukee, WI.

Clark, R.N., and G.A. Swayze, "Mapping Minerals, Amorphous Materials, Environmental Materials, Vegetation, Water, Ice, and Snow, and Other Materials: The USGS Tricorder Algorithm," *Summaries of the Fifth Annual JPL Airborne Earth Science Workshop, JPL Publication 95-1,* Jet Propulsion Laboratory, Pasadena, CA, 1995, pp. 39–40.

Clark, R.N., et al., "Imaging Spectroscopy: Earth and Planetary Remote Sensing with the USGS Tetracorder and Expert Systems," *Journal of Geophysical Research: Planets*, vol. 108, no. E12, 2003.

Clark, M.L., D.B. Clark, and D.A. Roberts, "Small-Footprint Lidar Estimation of Sub-canopy Elevation and Tree Height in a Tropical Rain Forest Landscape," *Remote Sensing of Environment*, vol. 91, no. 1, 2004, pp. 68–89.

Comer, R.P., et al., "Talking Digital," *Photogrammetric Engineering and Remote Sensing*, vol. 64, no. 12, 1998, pp. 1139–1142.

Congalton, R., and K. Green, *Assessing the Accuracy of Remotely Sensed Data: Principles and Practices*, 2nd ed., Boca Raton, FL: CRC/Taylor & Francis, 2009.

Curlander, J.C., and R.N. McDonough, *Synthetic Aperture Radar: Systems and Signal Processing*, New York: Wiley, 1991.

Dash, J., and P.J. Curran, "The MERIS Terrestrial Chlorophyll Index," *International Journal of Remote Sensing*, vol. 25, no. 23, 2004, pp. 5403–5413.

Doyle, F.J., "The Large Format Camera on Shuttle Mission 41-G," *Photogrammetric Engineering and Remote Sensing*, vol. 51, no. 2, 1985, pp. 200–203.

Dozier, J., and T.H. Painter, "Multispectral and Hyperspectral Remote Sensing of Alpine Snow Properties," *Annual Review of Earth and Planetary Sciences*, vol. 32, 2004, pp. 465–494.

Eastman Kodak Company, *Literature and Publications: Kodak Aerial Products*, 2014. Available at http://www.kodak.com.

Eastman Kodak Company, *Kodak Photographic Filters Handbook*, Rochester, NY: Eastman Kodak Company, 1990.

Eastman Kodak Company, *Kodak Data for Aerial Photography*, 6th ed., Rochester, NY: Eastman Kodak Company, 1992.

Eidenshink, J.C., and J.L. Faundeen, "The 1-km AVHRR Global Land Data Set: First Stages in Implementation," *International Journal of Remote Sensing*, vol. 15, no. 17, 1994, pp. 3443–3462.

Elachi, C., *Introduction to the Physics and Techniques of Remote Sensing*, Hoboken, NJ: Wiley, 1987.

Elachi, C., *Spaceborne Radar Remote Sensing: Applications and Techniques*, New York: IEEE Press, 1987.

Federal Interagency Committee for Wetland Delineation, *Federal Manual for Identifying and Delineating Jurisdictional Wetlands*, U.S. Army Corps of Engineers, U.S. Environmental Protection Agency, U.S. Fish and Wildlife Service, and USDA Soil Conservation Service, Washington, DC: Cooperative Technical Publication, 1989.

Fletcher, R., "Low-Density, Agrarian-Based Urbanism: A Comparative View," *Insights*, vol. 2, no. 4, 2009, pp. 1–19.

Fu, L., and B. Holt, *Seasat Views Oceans and Sea Ice with Synthetic-Aperture Radar*, JPL Publ. 81–120, Pasadena, CA: NASA Jet Propulsion Laboratory, 1982.

Gao, B.-C., et al., "Atmospheric Correction Algorithms for Hyperspectral Remote Sensing Data of Land and Ocean," *Remote Sensing of Environment*, vol. 113, 2009, pp. S17–S24.

Gitelson, A.A., et al., "Remote Estimation of Phytoplankton Density in Productive Waters," *Archives of Hydrobiology Special Issues in Advanced Limnology*, vol. 55, February 2000, pp. 121–136.

Goetz, A.H., et al., "Imaging Spectrometry for Earth Remote Sensing," *Science*, vol. 228, no. 4704, June 7, 1985, pp. 1147–1153.

Graham, R., and A. Koh, *Digital Aerial Survey: Theory and Practice*, Latheronwheel, Scotland: Whittles Publishing, 2002.

Hall, D.K., et al., "Assessment of Snow-Cover Mapping Accuracy in a Variety of Vegetation-Cover Densities in Central Alaska," *Remote Sensing of Environment*, vol. 66, no. 2, 1998, pp. 129–137.

Haugerud, R.A., "Preliminary Interpretation of Pre-2014 Landslide Deposits in the Vicinity of Oso, Washington," *U.S. Geological Survey Open-File Report 2014–1065*, 2014, doi:10.3133/ofr20141065.

Hawley, R.L., and E.D. Waddington, "In situ Measurements of Firn Compaction Profiles Using Borehole Optical Stratigraphy," *Journal of Glaciology*, vol. 57, no. 202, 2011, pp. 289–294.

Heady, H.F., and R.D. Child, *Introductory Geographic Information Systems*, Boulder, CO: Westview Press, 1994.

Helfert, M., and K. Lulla, Ed., Special Issue on "Human Directed Observation of the Earth from Space," *Geocarto International Journal*, vol. 4, no. 1, 1989.

Hoffer, R.M., P.W. Mueller, and D.F. Lozano-Garcia, "Multiple Incidence Angle Shuttle Imaging Radar Data for Discriminating Forest Cover Types,"

Technical Papers of the American Society for Photogrammetry and Remote Sensing, ACSM–ASPRS Fall Technical Meeting, September 1985, pp. 476–485.

Homer, C., et al., "Development of a 2001 National Land-Cover Database for the United States," *Photogrammetric Engineering and Remote Sensing*, vol. 70, no. 7, 2004, pp. 829–840.

Homer, C.H., J.A. Fry, and C.A. Barnes, The National Land Cover Database, U.S. Geological Survey Fact Sheet no. 2012-3020, 2012.

Homer, C., et al., "Completion of the 2001 National Land Cover Database for the Conterminous United States," *Photogrammetric Engineering and Remote Sensing*, vol. 73, no. 4, 2007, p. 337.

Hudson, R.D., Jr., *Infrared System Engineering*, Wiley, New York, 1969.

Huete, A.R., "A Soil-Adjusted Vegetation Index (SAVI)," *Remote Sensing of Environment*, vol. 25, no. 3, 1988, pp. 295–309.

Irons, J.R., J.L. Dwyer, and J.A. Barsi, "The Next Landsat Satellite: The Landsat Data Continuity Mission," *Remote Sensing of Environment*, vol. 122, 2012.

Jensen, J.R., *Introductory Digital Image Processing: A Remote Sensing Perspective*, 3rd ed., Upper Saddle River, NJ: Prentice Hall, 2005.

Jensen, J.R., and R.R. Jensen, *Introductory Geographic Information Systems*, Glenview, IL: Pearson Education, Inc., 2013.

Jet Propulsion Laboratory, *ASTER Spectral Library*, Pasadena: California Institute of Technology, 1999.

Jezek, K.C., "Observing the Antarctic Ice Sheet Using the RADARSAT-1 Synthetic Aperture Radar," *Polar Geography*, vol. 27, no. 3, 2003.

Jiménez-Muñoz, J.-C., and J.A. Sobrino, "Feasibility of Retrieving Land-Surface Temperature from ASTER TIR Bands Using Two-Channel Algorithms: A Case Study of Agricultural Areas," *IEEE Geoscience and Remote Sensing Letters*, vol. 5, no. 4, 2008.

Joughin, I., et al., "Greenland Flow Variability from Ice-Sheet-Wide Velocity Mapping," *Journal of Glaciology*, vol. 56, no. 197, 2010, pp. 415–430.

Jupp, D.L.B., and A.H. Strahler, "A Hotspot Model for Leaf Canopies," *Remote Sensing of Environment*, vol. 38, 1991, pp. 193–210.

Jupp, D.L.B., et al., "The Application and Potential of Remote Sensing to Planning and Managing the Great Barrier Reef of Australia," *Proceedings: Eighteenth International Symposium on Remote Sensing of Environment*, Paris, France, 1984, pp. 121–137.

Kaufmann, et al., *Science Plan for the Environmental Mapping and Analysis Program (EnMap)*, Deutsches GeoForschungsZentrum GFZ, 65 pp., Scientific Technical Report, Potsdam, 2012.

Kelly, M.A., et al., "Expanded Glaciers During a Dry and Cold Last Glacial Maximum in Equatorial East Africa," *Geology*, vol. 42, no. 6, 2014, pp. 519–522.

Kessler, D.J., and B.G. Cour-Palais, "Collision Frequency of Artificial Satellites: The Creation of a Debris Belt," *Journal of Geophysical Research*, vol. 83, no A6, 1978.

Kim, Y., and J.J. van Zyl, "A Time-Series Approach to Estimate Soil Moisture Using Polarimetric Radar Data," *IEEE Transactions on Geoscience and Remote Sensing*, vol. 47, no. 8, 2009, pp. 2519–2527.

Kosok, P., *Life, Land and Water in Ancient Peru*, New York: Long Island University Press, 1965.

Kruse, F.A., "Mapping Surface Mineralogy Using Imaging Spectrometry," *Geomorphology*, vol. 137, no. 1, January 15, 2012, pp. 41–56.

Kruse, F.A., A.B. Lefkoff, and J.B. Dietz, "Expert System-Based Mineral Mapping in Northern Death Valley, California/Nevada Using the Airborne Visible/Infrared Imaging Spectrometer (AVIRIS)," *Remote Sensing of Environment*, vol. 44, 1993, pp. 309–336.

Kruse, F., et al., "The Spectral Image Processing System (SIPS)—Interactive Visualization and Analysis of Imaging Spectrometer Data," *Remote Sensing of Environment*, vol. 44, 1993, pp. 145–163.

Laliberte, A., et al., "Acquisition, Orthorectification, and Object-based Classification of Unmanned Aerial Vehicle (UAV) Imagery for Rangeland Monitoring," *Photogrammetric Engineering and Remote Sensing*, vol. 76, no. 6, 2010, pp. 661–672.

Laliberte, A., et al., "Multispectral Remote Sensing from Unmanned Aircraft: Image Processing Workflows and Applications for Rangeland Environments," *Remote Sensing*, vol. 3, no. 11, 2011, pp. 2529–2551.

Lauer, D.T., S.A. Morain, and V.V. Salomonson, "The Landsat Program: Its Origins, Evolution, and Impacts," *Photogrammetric Engineering and Remote Sensing*, vol. 63, no. 7, pp. 831– 838.

Leberl, F.W., *Radargrammetric Image Processing*, Norwood, MA: Artech House Inc., 1990.

Leberl, F.W., "Radargrammetry," *Principles and Applications of Imaging Radar, Manual of Remote Sensing*, 3rd ed., vol. 2, New York: Wiley, 1998, pp. 183–269.

Lewis, A.J., Ed., "Geoscience Applications of Imaging Radar Systems," *Remote Sensing of the Electromagnetic Spectrum*, vol. 3, no. 3, 1976.

Lewis, A.J., F.M. Henderson, and D.W. Holcomb, "Radar Fundamentals: The Geoscience Perspective," *Principles and Applications of Imaging Radar, Manual of Remote Sensing*, 3rd ed., vol. 2, New York: Wiley, 1998, pp. 131–181.

Liang, S., *Quantitative Remote Sensing of Land Surfaces*, Hoboken, NJ: Wiley, 2004.

Licciardi, G.A., and F. Del Frate, "Pixel Unmixing in Hyperspectral Data by Means of Neural Networks," *IEEE Transactions on Geoscience and Remote Sensing*, vol. 49, no. 11, 2011, pp. 4163–4172.

Lillesand, T.M., et al., *Upper Midwest Gap Analysis Program Image Processing Protocol*, USGS Environmental Management Technical Center, Onalaska, WI, 1998.

Mallet, C., and F. Bretar, "Full-Waveform Topographic Lidar: State-of-the-Art," *ISPRS Journal of Photogrammetry and Remote Sensing*, vol. 64, no. 1, 2009, pp. 1–16.

Maune, D.F., Ed., *Digital Elevation Model Technologies and Applications: The DEM Users Manual*, 2nd ed., Bethesda, MD: American Society for Photogrammetry and Remote Sensing, 2007.

Measures, R.M., *Laser Remote Sensing*, New York: Wiley, 1984.

Meigs, A.D., et al., "Ultraspectral Imaging: A New Contribution to Global Virtual Presence," *Aerospace and Electronic Systems*, vol. 23, no. 10, Oct. 2008, pp. 11–17.

Mikhail, E.M., J.S. Bethel, and J.C. McGlone, *Introduction to Modern Photogrammetry*, New York: Wiley, 2001.

Morrison, T.A., et al., "Estimating Survival in Photographic Capture–Recapture Studies: Overcoming Misidentification Error," *Methods in Ecology and Evolution*, vol. 2, no. 5, 2011, pp. 454–463.

NASA, *The Gateway to Astronaut Photography of Earth*, available at http://earth.jsc.nasa.gov/sseop/clickmap, 2006.

National Research Council (NRC) Committee on Implementation of a Sustained Land Imaging Program, *Landsat and Beyond: Sustaining and Enhancing the Nation's Land Imaging Program*, National Academy Press, available at http://www.nap.edu, 2013.

National Research Council (NRC) Committee on the Assessment of NASA's Orbital Debris Programs, *Limiting Future Collision Risk to Spacecraft: An Assessment on NASA's Meteoroid and Orbital Debris Programs*, National Academy Press, available at http://www.nap.edu, 2011.

NOAA, "Visual Interpretation of TM Band Combinations Being Studied," *Landsat Data Users Notes*, no. 30, March 1984.

NOAA GVI, *Global Vegetation Index Products*, available at http://www.osdpd.noaa.gov/PSB/IMAGES/gvi.html, 2006.

Office of Science and Technology Policy (OSTP), *National Plan for Civil Earth Observations*, available at http://www.whitehouse.gov, July, 2014.

Olson, Charles E., Jr., "Elements of Photographic Interpretation Common to Several Sensors," *Photogrammetric Engineering*, vol. 26, no. 4, 1960, pp. 651–656.

Photogrammetric Engineering and Remote Sensing (PERS), Special Issue on Landsat, *Photogrammetric Engineering and Remote Sensing*, vol. 72, no. 10, 2006.

Plaza, A., et al., "Recent Developments in Endmember Extraction and Spectral Unmixing," In S. Prasad, et al., Eds., *Optical Remote Sensing*, Berlin: Springer, 2009, pp. 235–267.

Raney, R.K., "Radar Fundamentals: Technical Perspective," *Principles and Applications of Imaging Radar, Manual of Remote Sensing*, 3rd ed., vol. 2, New York: Wiley, 1998, pp. 9–130.

Rango, A., et al., "Unmanned Aerial Vehicle-Based Remote Sensing for Rangeland Assessment, Monitoring, and Management," *Journal of Applied Remote Sensing*, vol. 3, no. 1, 2009, paper 033542.

Reese, H.M., et al., "Statewide Land Cover Derived from Multiseasonal Landsat TM Data: A Retrospective of the WISCLAND Project," *Remote Sensing of Environment*, vol. 82, nos. 2–3, 2002, pp. 224–237.

Rignot, E., et al., "Recent Antarctic Ice Mass Loss from Radar Interferometry and Regional Climate Modelling," *Nature Geoscience*, vol. 1, no. 2, 2008, pp. 106–110.

Robert, P.C., "*Remote Sensing*: A Potentially Powerful Technique for Precision Agriculture," *Proceedings: Land Satellite Information in the Next Decade II: Sources and Applications,* Bethesda, MD: American Society for Photogrammetry and Remote Sensing, 1997, pp. 19–25 (CD-ROM).

Robinson, A.H., et al., *Elements of Cartography*, 6th ed., New York: Wiley, 1995.

Rutzinger, M., et al., "Object-Based Point Cloud Analysis of Full-Waveform Airborne Laser Scanning Data for Urban Vegetation Classification," *Sensors*, vol. 8, no. 8, 2008, pp. 4505–4528, doi:10.3390/s8084505.

Sabins, F.F., Jr., *Remote Sensing—Principles and Interpretation*, 3rd ed., New York: W.H. Freeman, 1997.

Salisbury, J.W., and D.M. D'Aria, "Emissivity of Terrestrial Materials in the 8–14 mm Atmospheric Window," *Remote Sensing of Environment*, vol. 42, 1992, pp. 83–106.

Savitsky, A., and M.J.E. Golay, "Smoothing and Differentiation of Data by Simplified Least Squares Procedures," *Analytical Chemistry*, vol. 36, no. 8, 1964, pp. 1627–1639.

Sayn-Wittgenstein, L., "Recognition of Tree Species on Air Photographs by Crown Characteristics," *Photogrammetric Engineering*, vol. 27, no. 5, 1961, pp. 792–809.

Schott, J.R., *Remote Sensing: The Image Chain Approach*, 2nd ed., New York: Oxford University Press, 2007.

Schowengerdt, R.A., *Remote Sensing Models and Methods for Image Processing*, 2nd ed., New York: Academic Press, 1997.

Schowengerdt, R.A., *Remote Sensing Models and Methods for Image Processing*, 3rd ed., New York: Academic Press, 2006.

Siegfried, M.R., R.L. Hawley, and J.F. Burkhart, "High-Resolution Ground-Based GPS Measurements Show Inter-Campaign Bias in ICESat Elevation Data," *IEEE Transactions on Geoscience and Remote Sensing*, vol. 49, no. 10, 2011.

Singhroy, V., K. Mattar, and A.L. Gray, "Landslide Characterization in Canada Using Interferometric SAR and Combined SAR and TM Images," *Advances in Space Research*, 1998, vol. 21, no. 3, pp. 465–476.

Smith, F.G.F., and J.R. Jensen, "The Multispectral Mapping of Seagrass: Application of Band Transformations for Minimization of Water Attenuation Using Landsat TM," Technical Papers, ASPRS-RTI 1998 Annual Conference, Bethesda, MD: American Society for Photogrammetry and Remote Sensing, 1998, pp. 592–603.

Somers, B., et al., "Nonlinear Hyperspectral Mixture Analysis for Tree Cover Estimates in Orchards," *Remote Sensing of Environment*, vol. 113, no. 6, 2009, pp. 1183–1193.

Spreen, G., L. Kaleschke, and G. Heygster, "Sea Ice Remote Sensing Using AMSR-E 89 GHz Channels," *Journal of Geophysical Research*, vol. 113, 2008, C02S03, doi:10.1029/2005JC003384.

Stehman, S.V., "Estimating Standard Errors of Accuracy Assessment Statistics Under Cluster Sampling," *Remote Sensing of Environment*, vol. 60, no. 3, 1997, pp. 258–269.

Thompson, W., "Lidar Imagery Reveals Maine's Land Surface in Unprecedented Detail," *Maine Geological Survey report*, Department of Agriculture, Conservation & Forestry, December 2011.

Tueller, P.T., "Rangeland Management," *The Remote Sensing Core Curriculum, Applications in Remote Sensing*, vol. 4, on the Internet at http://www.asprs.org, 1996.

U.S. Department of Agriculture, *Soil Survey of Dane County, Wisconsin*, Washington, DC: U.S. Government Printing Office, 1977.

U.S. Department of Agriculture, Soil Survey Staff, Natural Resources Conservation Service, *National Soil Survey Handbook*, Title 430-VI, Washington, DC: U.S. Government Printing Office, available at http://www.soils.usda.gov, 1997.

Vane, G., "High Spectral Resolution Remote Sensing of the Earth," *Sensors*, 1985, no. 2, pp. 11–20.

Viña, A., G.M. Henebry, and A.A. Gitelson, "Satellite Monitoring of Vegetation Dynamics: Sensitivity Enhancement by the Wide Dynamic Range Vegetation Index," *Geophysical Research Letters*, vol. 31, 2004, L04503, doi:10. 1029/2003GL019034.

Vogelmann, J.E., et al., "Completion of the 1990s National Land Cover Data Set for the Conterminous United States from Landsat Thematic Mapper Data and Ancillary Data Sources," *Photogrammetric Engineering and Remote Sensing*, vol. 67, no. 6, 2001, pp. 650–661.

Warner, W.S., et al., *Small Format Aerial Photography*, Bethesda, MD: American Society for Photogrammetry and Remote Sensing, 1996.

White, M.A., et al., "Intercomparison, Interpretation, and Assessment of Spring Phenology in North America Estimated from Remote Sensing for 1982 to 2006," *Global Change Biology*, 2009, doi:10.1111/j.1356-2486.2009.01910.x.

Williams, R.S., and W.D. Carter, Eds., *ERTS-1, A New Window on Our Planet, USGS Professional Paper 929*, Washington, DC, 1976.

Wilson, D.R., Ed., *Aerial Reconnaissance for Archaeology*, Research Report No. 12, London: The Council for British Archaeology, 1975.

Wolf, P.R., B.A. Dewitt, and B.E. Wilkinson, *Elements of Photogrammetry with Applications in GIS*, 4th ed., New York: McGraw-Hill, 2013.

Wolter, P.T., et al., "Improved Forest Classification in the Northern Lake States Using Multi-Temporal Landsat Imagery," *Photogrammetric Engineering and Remote Sensing*, vol. 61, no. 9, 1995, pp. 1129–1144.

Wulder, M.A., and J.G. Masek, "Landsat Legacy" Special Issue, *Remote Sensing of Environment*, vol. 122, 2012.

Wright, C.W., and J. Brock, "EAARL: A Lidar for Mapping Shallow Coral Reefs and Other Coastal Environments," *Proceedings of the Seventh International Conference on Remote Sensing for Marine and Coastal Environments, Miami, May* 20–22, 2002.

Wynne, R.H., et al., "Satellite Monitoring of Lake Ice Breakup on the Laurentian Shield (1980–1994)," *Photogrammetric Engineering and Remote Sensing*, vol. 64, no. 6, 1998, pp. 607–617.

Xian, G., C. Homer, and J. Fry, "Updating the 2001 National Land Cover Database Land Cover Classification to 2006 by using Landsat Imagery Change Detection Methods," *Remote Sensing of Environment*, vol. 113, no. 6, 2009, pp. 1133–1147.

Zhang, C., and Z. Xie, "Combining Object-Based Texture Measures with a Neural Network for Vegetation Mapping in the Everglades from Hyperspectral Imagery," *Remote Sensing of Environment*, vol. 124, 2012, pp. 310–320.

Zsilinszky, V.G., *Photographic Interpretation of Tree Species in Ontario*, Ottawa: Ontario Department of Lands and Forests, 1966.

INDEX

Pages listed with prefixes A., B., or C. are found in online Appendices A, B, or C.

SI UNITS FREQUENTLY USED IN REMOTE SENSING

TABLE 1 Fundamental Units

Quantity	SI Unit	cgs Unit	fps Unit
Length (L)	meter (m)	centimeter (cm)	foot (ft)
Mass (M)	kilogram (kg)	gram (g)	slug
Time (T)	second (sec)	second (sec)	second (sec)
Force (MLT^{-2})	newton (N)	dyne	pound (lb)
Energy, work (ML^2T^{-2})	joule (J)	erg	foot-pound (ft-lb)
Power (ML^2T^{-3})	watt (W)	erg $\cdot$ sec^{-1}	horsepower (hp)

TABLE 2 Unit Prefix Notation

Multiplier	Prefix
10^{12}	tera (T)
10^{9}	giga (G)
10^{6}	mega (M)
10^{3}	kilo (k)
10^{-2}	centi (c)
10^{-3}	milli (m)
10^{-6}	micro (μ)
10^{-9}	nano (n)
10^{-12}	pico (p)

TABLE 3 Common Units of Wavelength (λ)

Unit	Equivalent
Centimeter (cm)	10^{-2} m
Millimeter (mm)	10^{-3} m
Micron (μ)	10^{-6} m
Micrometer (μm)	10^{-6} m
Nanometer (nm)	10^{-9} m
Ångstrom (Å)	10^{-10} m

TABLE 4 Helpful Conversion Factors

To Convert from	To	Multiply by
acre	hectare (ha)	4.046873×10^{-1}
acre	square meter (m^2)	4.046873×10^{3}
degree (angle)	radian (rad)	1.745329×10^{-2}
degree Fahrenheit	Celsius (°C)	°C = 5/9 × (°F −32)
degree Celsius	Kelvin (K)	K = °C + 273.15
erg	joule (J)	1×10^{-7}
foot (U.S. survey)	meter (m)	3.048006×10^{-1}
hectare	square meter (m^2)	1×10^{4}
inch	meter (m)	2.54×10^{-2}
kilometer per hour	meter per second (m/sec)	2.777778×10^{-1}
mile (U.S. survey)	kilometer (km)	1.609347
mile per hour	meter per second (m/sec)	4.470409×10^{-1}
nautical mile (international)	kilometer (km)	1.852
pound	kilogram (kg)	4.535924×10^{-1}
square foot	square meter (m^2)	9.290341×10^{-2}
square inch	square meter (m^2)	6.4516×10^{-4}
square mile	square meter (m^2)	2.589998×10^{6}
wavelength (in μm)	frequency (Hz)	$\nu = (2.9979 \times 10^{14})/\lambda$